DIFFERENTIAL EQUATIONS FOR ENGINEERS

This book presents a systematic and comprehensive introduction to ordinary differential equations for engineering students and practitioners. Mathematical concepts and various techniques are presented in a clear, logical, and concise manner. Various visual features are used to highlight focus areas. Complete illustrative diagrams are used to facilitate mathematical modeling of application problems. Readers are motivated by a focus on the relevance of differential equations through their applications in various engineering disciplines. Studies of various types of differential equations are determined by engineering applications. Theory and techniques for solving differential equations are then applied to solve practical engineering problems. Detailed step-by-step analysis is presented to model the engineering problems using differential equations from physical principles and to solve the differential equations using the easiest possible method. Such a detailed, step-by-step approach, especially when applied to practical engineering problems, helps the readers to develop problem-solving skills.

This book is suitable for use not only as a textbook on ordinary differential equations for undergraduate students in an engineering program but also as a guide to self-study. It can also be used as a reference after students have completed learning the subject.

Wei-Chau Xie is a Professor in the Department of Civil and Environment Engineering and the Department of Applied Mathematics at the University of Waterloo. He is the author of *Dynamic Stability of Structures* and has published numerous journal articles on dynamic stability, structural dynamics and random vibration, nonlinear dynamics and stochastic mechanics, reliability and safety analysis of engineering systems, and seismic analysis and design of engineering structures. He has been teaching differential equations to engineering students for almost twenty years. He received the Teaching Excellence Award in 2001 in recognition of his exemplary record of outstanding teaching, concern for students, and commitment to the development and enrichment of engineering education at Waterloo. He is the recipient of the Distinguished Teacher Award in 2007, which is the highest formal recognition given by the University of Waterloo for a superior record of continued excellence in teaching.

DIFFERENTIAL EQUATIONS FOR ENGINEERS

Differential Equations for Engineers

Wei-Chau Xie
University of Waterloo

CAMBRIDGE
UNIVERSITY PRESS

CAMBRIDGE UNIVERSITY PRESS
Cambridge, New York, Melbourne, Madrid, Cape Town,
Singapore, São Paulo, Delhi, Mexico City

Cambridge University Press
32 Avenue of the Americas, New York NY 10013-2473, USA

Published in the United States of America by Cambridge University Press, New York

www.cambridge.org
Information on this title: www.cambridge.org/9781107632950

First published 2010
First paperback edition 2013

A catalogue record for this publication is available from the British Library

Library of Congress Cataloguing in Publication Data

Xie, Wei-Chau, 1964–
Differential equations for engineers / Wei-Chau Xie.
p. cm.
Includes bibliographical references and index.
ISBN 978-0-521-19424-2
1. Differential equations. 2. Engineering mathematics. I. Title.
TA347.D45X54 2010
620.001′515352–dc22 2010001101

ISBN 978-0-521-19424-2 Hardback
ISBN 978-1-107-63295-0 Paperback

TO

My Family

Contents

Preface xiii

1 Introduction 1

1.1 Motivating Examples 1
1.2 General Concepts and Definitions 6

2 First-Order and Simple Higher-Order Differential Equations . 16

2.1 The Method of Separation of Variables 16
2.2 Method of Transformation of Variables 20
2.2.1 Homogeneous Equations 20
2.2.2 Special Transformations 25
2.3 Exact Differential Equations and Integrating Factors 31
2.3.1 Exact Differential Equations 32
2.3.2 Integrating Factors 39
2.3.3 Method of Inspection 45
2.3.4 Integrating Factors by Groups 48
2.4 Linear First-Order Equations 55
2.4.1 Linear First-Order Equations 55
2.4.2 Bernoulli Differential Equations 58
2.5 Equations Solvable for the Independent or Dependent Variable 61
2.6 Simple Higher-Order Differential Equations 68
2.6.1 Equations Immediately Integrable 68
2.6.2 The Dependent Variable Absent 70
2.6.3 The Independent Variable Absent 72
2.7 Summary 74
Problems 78

3 Applications of First-Order and Simple Higher-Order Equations 87

3.1 Heating and Cooling 87
3.2 Motion of a Particle in a Resisting Medium 91
3.3 Hanging Cables 97

3.3.1 The Suspension Bridge 97
3.3.2 Cable under Self-Weight 102
3.4 Electric Circuits 108
3.5 Natural Purification in a Stream 114
3.6 Various Application Problems 120
Problems 130

4 **Linear Differential Equations** 140
4.1 General Linear Ordinary Differential Equations 140
4.2 Complementary Solutions 143
4.2.1 Characteristic Equation Having Real Distinct Roots 143
4.2.2 Characteristic Equation Having Complex Roots 147
4.2.3 Characteristic Equation Having Repeated Roots 151
4.3 Particular Solutions 153
4.3.1 Method of Undetermined Coefficients 153
4.3.2 Method of Operators 162
4.3.3 Method of Variation of Parameters 173
4.4 Euler Differential Equations 178
4.5 Summary 180
Problems 183

5 **Applications of Linear Differential Equations** 188
5.1 Vibration of a Single Degree-of-Freedom System 188
5.1.1 Formulation—Equation of Motion 188
5.1.2 Response of a Single Degree-of-Freedom System 193
5.1.2.1 Free Vibration—Complementary Solution 193
5.1.2.2 Forced Vibration—Particular Solution 200
5.2 Electric Circuits 209
5.3 Vibration of a Vehicle Passing a Speed Bump 213
5.4 Beam-Columns 218
5.5 Various Application Problems 223
Problems 232

6 **The Laplace Transform and Its Applications** 244
6.1 The Laplace Transform 244

6.2 The Heaviside Step Function 249
6.3 Impulse Functions and the Dirac Delta Function 254
6.4 The Inverse Laplace Transform 257
6.5 Solving Differential Equations Using the Laplace Transform 263
6.6 Applications of the Laplace Transform 268
6.6.1 Response of a Single Degree-of-Freedom System 268
6.6.2 Other Applications 275
6.6.3 Beams on Elastic Foundation 283
6.7 Summary 289
Problems 291

7 Systems of Linear Differential Equations 300
7.1 Introduction 300
7.2 The Method of Operator 304
7.2.1 Complementary Solutions 304
7.2.2 Particular Solutions 307
7.3 The Method of Laplace Transform 318
7.4 The Matrix Method 325
7.4.1 Complementary Solutions 326
7.4.2 Particular Solutions 334
7.4.3 Response of Multiple Degrees-of-Freedom Systems 344
7.5 Summary 347
7.5.1 The Method of Operator 347
7.5.2 The Method of Laplace Transform 348
7.5.3 The Matrix Method 349
Problems 351

8 Applications of Systems of Linear Differential Equations . . . 357
8.1 Mathematical Modeling of Mechanical Vibrations 357
8.2 Vibration Absorbers or Tuned Mass Dampers 366
8.3 An Electric Circuit 372
8.4 Vibration of a Two-Story Shear Building 377
8.4.1 Free Vibration—Complementary Solutions 378
8.4.2 Forced Vibration—General Solutions 380
Problems 384

9 **Series Solutions of Differential Equations** 390

9.1 Review of Power Series 391

9.2 Series Solution about an Ordinary Point 394

9.3 Series Solution about a Regular Singular Point 403

9.3.1 Bessel's Equation and Its Applications 408

9.3.1.1 Solutions of Bessel's Equation 408

9.3.2 Applications of Bessel's Equation 418

9.4 Summary 424

Problems 426

10 **Numerical Solutions of Differential Equations** 431

10.1 Numerical Solutions of First-Order Initial Value Problems 431

10.1.1 The Euler Method or Constant Slope Method 432

10.1.2 Error Analysis 434

10.1.3 The Backward Euler Method 436

10.1.4 Improved Euler Method—Average Slope Method 437

10.1.5 The Runge-Kutta Methods 440

10.2 Numerical Solutions of Systems of Differential Equations 445

10.3 Stiff Differential Equations 449

10.4 Summary 452

Problems 454

11 **Partial Differential Equations** 457

11.1 Simple Partial Differential Equations 457

11.2 Method of Separation of Variables 458

11.3 Application—Flexural Motion of Beams 465

11.3.1 Formulation—Equation of Motion 465

11.3.2 Free Vibration 466

11.3.3 Forced Vibration 471

11.4 Application—Heat Conduction 473

11.4.1 Formulation—Heat Equation 473

11.4.2 Two-Dimensional Steady-State Heat Conduction 476

11.4.3 One-Dimensional Transient Heat Conduction 480

11.4.4 One-Dimensional Transient Heat Conduction on a Semi-Infinite Interval 483

11.4.5 Three-Dimensional Steady-State Heat Conduction 488
11.5 Summary 492
Problems 493

12 Solving Ordinary Differential Equations Using *Maple* 498
12.1 Closed-Form Solutions of Differential Equations 499
12.1.1 Simple Ordinary Differential Equations 499
12.1.2 Linear Ordinary Differential Equations 506
12.1.3 The Laplace Transform 507
12.1.4 Systems of Ordinary Differential Equations 509
12.2 Series Solutions of Differential Equations 512
12.3 Numerical Solutions of Differential Equations 517
Problems 526

Appendix A Tables of Mathematical Formulas 531
A.1 Table of Trigonometric Identities 531
A.2 Table of Derivatives 533
A.3 Table of Integrals 534
A.4 Table of Laplace Transforms 537
A.5 Table of Inverse Laplace Transforms 539
Index 542

Preface

Background

Differential equations have wide applications in various engineering and science disciplines. In general, modeling of the variation of a physical quantity, such as temperature, pressure, displacement, velocity, stress, strain, current, voltage, or concentration of a pollutant, with the change of time or location, or both would result in differential equations. Similarly, studying the variation of some physical quantities on other physical quantities would also lead to differential equations. In fact, many engineering subjects, such as mechanical vibration or structural dynamics, heat transfer, or theory of electric circuits, are founded on the theory of differential equations. It is practically important for engineers to be able to model physical problems using mathematical equations, and then solve these equations so that the behavior of the systems concerned can be studied.

I have been teaching differential equations to engineering students for the past two decades. Most, if not all, of the textbooks are written by mathematicians with little engineering background. Based on my experience and feedback from students, the following lists some of the gaps frequently seen in current textbooks:

- A major focus is put on explaining mathematical concepts

  For engineers, the purpose of learning the theory of differential equations is to be able to solve practical problems where differential equations are used. For engineering students, it is more important to know the applications and techniques for solving application problems than to delve into the nuances of mathematical concepts and theorems. Knowing the appropriate applications can motivate them to study the mathematical concepts and techniques. However, it is much more challenging to model an application problem using physical principles and then solve the resulting differential equations than it is to merely carry out mathematical exercises.

- Insufficient emphasis is placed on the step-by-step problem solving techniques

  Engineering students do not usually have the same mathematical background and interest as students who major in mathematics. Mathematicians are more interested if: (1) there are solutions to a differential equation or a system of differential equations; (2) the solutions are unique under a certain set of conditions; and (3) the differential equations can be solved. On the other hand, engineers are more interested in mathematical modeling of a practical problem and actually solving the equations to find the solutions using the easiest possible method. Hence, a detailed step-by-step approach, especially applied to practical engineering problems, helps students to develop problem solving skills.

- Presentations are usually formula-driven with little variation in visual design

  It is very difficult to attract students to read boring formulas without variation of presentation. Readers often miss the points of importance.

Objectives

This book addresses the needs of engineering students and aims to achieve the following objectives:

- To motivate students on the relevance of differential equations in engineering through their applications in various engineering disciplines. Studies of various types of differential equations are motivated by engineering applications; theory and techniques for solving differential equations are then applied to solve practical engineering problems.
- To have a balance between theory and applications. This book could be used as a reference after students have completed learning the subject. As a reference, it has to be reasonably comprehensive and complete. Detailed step-by-step analysis is presented to model the engineering problems using differential equations and to solve the differential equations.
- To present the mathematical concepts and various techniques in a clear, logical and concise manner. Various visual features, such as side-notes (preceded by the ✍ symbol), different fonts and shades, are used to highlight focus areas. Complete illustrative diagrams are used to facilitate mathematical modeling of application problems. This book is not only suitable as a textbook for classroom use but also is easy for self-study. As a textbook, it has to be easy to understand. For self-study, the presentation is detailed with all necessary steps and useful formulas given as side-notes.

Scope

This book is primarily for engineering students and practitioners as the main audience. It is suitable as a textbook on ordinary differential equations for undergraduate students in an engineering program. Such a course is usually offered in the second year after students have taken calculus and linear algebra in the first year. Although it is assumed that students have a working knowledge of calculus and linear algebra, some important concepts and results are reviewed when they are first used so as to refresh their memory.

Chapter 1 first presents some motivating examples, which will be studied in detail later in the book, to illustrate how differential equations arise in engineering applications. Some basic general concepts of differential equations are then introduced.

In Chapter 2, various techniques for solving first-order and simple higher-order ordinary differential equations are presented. These methods are then applied in Chapter 3 to study various application problems involving first-order and simple higher-order differential equations.

Chapter 4 studies linear ordinary differential equations. Complementary solutions are obtained through the characteristic equations and characteristic numbers. Particular solutions are obtained using the method of undetermined coefficients, the operator method, and the method of variation of parameters. Applications involving linear ordinary differential equations are presented in Chapter 5.

Solutions of linear ordinary differential equations using the Laplace transform are studied in Chapter 6, emphasizing functions involving Heaviside step function and Dirac delta function.

Chapter 7 studies solutions of systems of linear ordinary differential equations. The method of operator, the method of Laplace transform, and the matrix method are introduced. Applications involving systems of linear ordinary differential equations are considered in Chapter 8.

In Chapter 9, solutions of ordinary differential equations in series about an ordinary point and a regular singular point are presented. Applications of Bessel's equation in engineering are considered.

Some classical methods, including forward and backward Euler method, improved Euler method, and Runge-Kutta methods, are presented in Chapter 10 for numerical solutions of ordinary differential equations.

In Chapter 11, the method of separation of variables is applied to solve partial differential equations. When the method is applicable, it converts a partial differential equation into a set of ordinary differential equations. Flexural vibration of beams and heat conduction are studied as examples of application.

Solutions of ordinary differential equations using *Maple* are presented in Chapter 12. Symbolic computation software, such as *Maple*, is very efficient in solving problems involving ordinary differential equations. However, it cannot replace learning and thinking, especially mathematical modeling. It is important to develop analytical skills and proficiency through "hand" calculations, as has been done in previous chapters. This will also help the development of insight into the problems and appreciation of the solution process. For this reason, solutions of ordinary differential equations using *Maple* is presented in the last chapter of the book instead of a scattering throughout the book.

The book covers a wide range of materials on ordinary differential equations and their engineering applications. There are more than enough materials for a one-term (semester) undergraduate course. Instructors can select the materials according to the curriculum. Drafts of this book were used as the textbook in a one-term undergraduate course at the University of Waterloo.

Acknowledgments

First and foremost, my sincere appreciation goes to my students. It is the students who give me a stage where I can cultivate my talent and passion for teaching. It is for the students that this book is written, as my small contribution to their success in academic and professional careers. My undergraduate students who have used the draft of this book as a textbook have made many encouraging comments and constructive suggestions.

I am very grateful to many people who have reviewed and commented on the book, including Professor Hong-Jian Lai of West Virginia University, Professors S.T. Ariaratnam, Xin-Zhi Liu, Stanislav Potapenko, and Edward Vrscay of the University of Waterloo.

My graduate students Mohamad Alwan, Qinghua Huang, Jun Liu, Shunhao Ni, and Richard Wiebe have carefully read the book and made many helpful and critical suggestions.

My sincere appreciation goes to Mr. Peter Gordon, Senior Editor, Engineering, Cambridge University Press, for his encouragement, trust, and hard work to publish this book.

Special thanks are due to Mr. John Bennett, my mentor, teacher, and friend, for his advice and guidance. He has also painstakingly proofread and copyedited this book.

Without the unfailing love and support of my mother, who has always believed in me, this work would not have been possible. In addition, the care, love, patience, and understanding of my wife Cong-Rong and lovely daughters Victoria and Tiffany have been of inestimable encouragement and help. I love them very much and appreciate all that they have contributed to my work.

I appreciate hearing your comments through email (xie@uwaterloo.ca) or regular correspondence.

Wei-Chau Xie
Waterloo, Ontario, Canada

C H A P T E R 1

Introduction

1.1 Motivating Examples

Differential equations have wide applications in various engineering and science disciplines. In general, modeling variations of a physical quantity, such as temperature, pressure, displacement, velocity, stress, strain, or concentration of a pollutant, with the change of time t or location, such as the coordinates (x, y, z), or both would require differential equations. Similarly, studying the variation of a physical quantity on other physical quantities would lead to differential equations. For example, the change of strain on stress for some viscoelastic materials follows a differential equation.

It is important for engineers to be able to model physical problems using mathematical equations, and then solve these equations so that the behavior of the systems concerned can be studied.

In this section, a few examples are presented to illustrate how practical problems are modeled mathematically and how differential equations arise in them.

Motivating Example 1

First consider the projectile of a mass m launched with initial velocity v_0 at angle θ_0 at time $t = 0$, as shown.

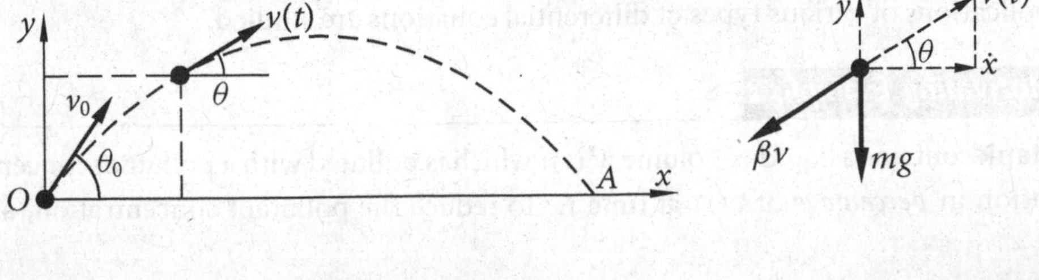

The atmosphere exerts a resistance force on the mass, which is proportional to the instantaneous velocity of the mass, i.e., $R = \beta v$, where β is a constant, and is opposite to the direction of the velocity of the mass. Set up the Cartesian coordinate system as shown by placing the origin at the point from where the mass m is launched.

At time t, the mass is at location $\big(x(t), y(t)\big)$. The instantaneous velocity of the mass in the x- and y-directions are $\dot{x}(t)$ and $\dot{y}(t)$, respectively. Hence the velocity of the mass is $v(t) = \sqrt{\dot{x}^2(t) + \dot{y}^2(t)}$ at the angle $\theta(t) = \tan^{-1}\big[\dot{y}(t)/\dot{x}(t)\big]$.

The mass is subjected to two forces: the vertical downward gravity mg and the resistance force $R(t) = \beta v(t)$.

The equations of motion of the mass can be established using Newton's Second Law: $F = \sum ma$. The x-component of the resistance force is $-R(t)\cos\theta(t)$. In the y-direction, the component of the resistance force is $-R(t)\sin\theta(t)$. Hence, applying Newton's Second Law yields

$$\begin{aligned}
&x\text{-direction:} \quad ma_x = \textstyle\sum F_x \quad \Longrightarrow \quad m\ddot{x}(t) = -R(t)\cos\theta(t),\\
&y\text{-direction:} \quad ma_y = \textstyle\sum F_y \quad \Longrightarrow \quad m\ddot{y}(t) = -mg - R(t)\sin\theta(t).
\end{aligned}$$

Since

$$\theta(t) = \tan^{-1}\frac{\dot{y}(t)}{\dot{x}(t)} \quad \Longrightarrow \quad \cos\theta = \frac{\dot{x}(t)}{\sqrt{\dot{x}^2(t) + \dot{y}^2(t)}}, \qquad \sin\theta = \frac{\dot{y}(t)}{\sqrt{\dot{x}^2(t) + \dot{y}^2(t)}},$$

the equations of motion become

$$\begin{aligned}
m\ddot{x}(t) &= -\beta v(t) \cdot \frac{\dot{x}(t)}{\sqrt{\dot{x}^2(t) + \dot{y}^2(t)}} \quad &&\Longrightarrow \quad m\ddot{x}(t) + \beta\dot{x}(t) = 0,\\
m\ddot{y}(t) &= -mg - \beta v(t) \cdot \frac{\dot{y}(t)}{\sqrt{\dot{x}^2(t) + \dot{y}^2(t)}} \quad &&\Longrightarrow \quad m\ddot{y}(t) + \beta\dot{y}(t) = -mg,
\end{aligned}$$

in which the initial conditions are at time $t = 0$: $x(0) = 0$, $y(0) = 0$, $\dot{x}(0) = v_0\cos\theta_0$, $\dot{y}(0) = v_0\sin\theta_0$. The equations of motion are two equations involving the first- and second-order derivatives $\dot{x}(t)$, $\dot{y}(t)$, $\ddot{x}(t)$, and $\ddot{y}(t)$. These equations are called, as will be defined later, a system of two second-order ordinary differential equations.

Because of the complexity of the problems, in the following examples, the problems are described and the governing equations are presented without detailed derivation. These problems will be investigated in details in later chapters when applications of various types of differential equations are studied.

Motivating Example 2

A tank contains a liquid of volume $V(t)$, which is polluted with a pollutant concentration in *percentage* of $c(t)$ at time t. To reduce the pollutant concentration, an inflow of rate Q_{in} is injected to the tank. Unfortunately, the inflow is also polluted but to a lesser degree with a pollutant concentration c_{in}. It is assumed that the inflow is perfectly mixed with the liquid in the tank instantaneously. An outflow of rate Q_{out} is removed from the tank as shown. Suppose that, at time $t=0$, the volume of the liquid is V_0 with a pollutant concentration of c_0.

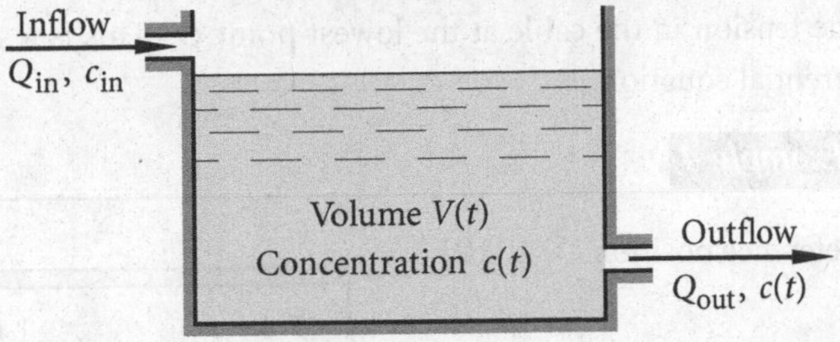

The equation governing the pollutant concentration $c(t)$ is given by

$$[V_0+(Q_{\text{in}}-Q_{\text{out}})t]\frac{dc(t)}{dt}+Q_{\text{in}}c(t)=Q_{\text{in}}c_{\text{in}},$$

with initial condition $c(0)=c_0$. This is a first-order ordinary differential equation.

Motivating Example 3

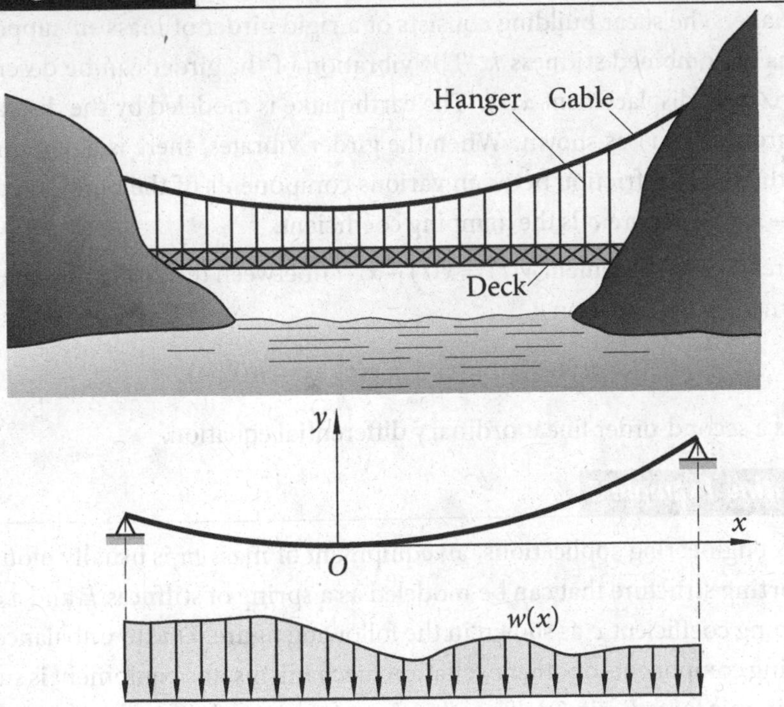

Consider the suspension bridge as shown, which consists of the main cable, the hangers, and the deck. The self-weight of the deck and the loads applied on the deck are transferred to the cable through the hangers.

Set up the Cartesian coordinate system by placing the origin O at the lowest point of the cable. The cable can be modeled as subjected to a distributed load $w(x)$. The equation governing the shape of the cable is given by

$$\frac{d^2y}{dx^2} = \frac{w(x)}{H},$$

where H is the tension in the cable at the lowest point O. This is a second-order ordinary differential equation.

Motivating Example 4

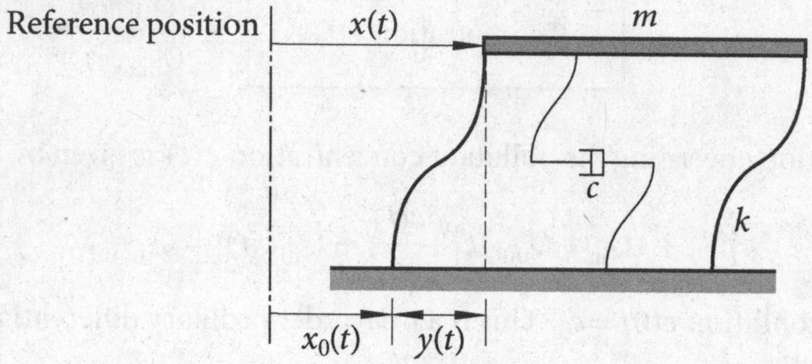

Consider the vibration of a single-story shear building under the excitation of earthquake. The shear building consists of a rigid girder of mass m supported by columns of combined stiffness k. The vibration of the girder can be described by the horizontal displacement $x(t)$. The earthquake is modeled by the displacement of the ground $x_0(t)$ as shown. When the girder vibrates, there is a damping force due to the internal friction between various components of the building, given by $c\left[\dot{x}(t)-\dot{x}_0(t)\right]$, where c is the damping coefficient.

The relative displacement $y(t)=x(t)-x_0(t)$ between the girder and the ground is governed by the equation

$$m\ddot{y}(t) + c\dot{y}(t) + ky(t) = -m\ddot{x}_0(t),$$

which is a second-order linear ordinary differential equation.

Motivating Example 5

In many engineering applications, an equipment of mass m is usually mounted on a supporting structure that can be modeled as a spring of stiffness k and a damper of damping coefficient c as shown in the following figure. Due to unbalanced mass in rotating components or other excitation mechanisms, the equipment is subjected to a harmonic force $F_0 \sin \Omega t$. The vibration of the mass is described by the vertical displacement $x(t)$. When the excitation frequency Ω is close to $\omega_0=\sqrt{k/m}$, which is the natural circular frequency of the equipment and its support, vibration of large amplitudes occurs.

In order to reduce the vibration of the equipment, a vibration absorber is mounted on the equipment. The vibration absorber can be modeled as a mass m_a, a spring of stiffness k_a, and a damper of damping coefficient c_a. The vibration of the absorber is described by the vertical displacement $x_a(t)$.

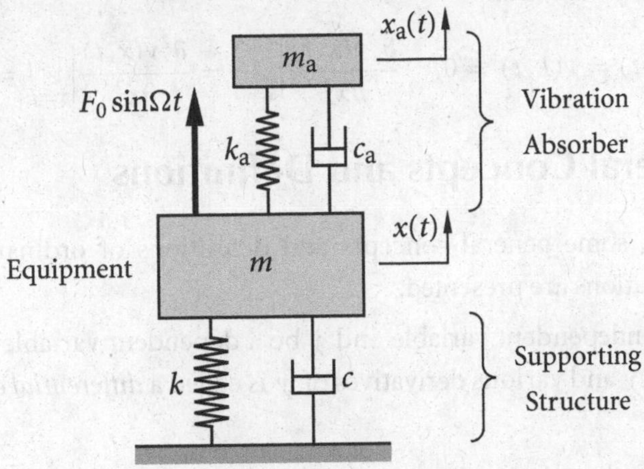

The equations of motion governing the vibration of the equipment and the absorber are given by

$$m\ddot{x} + (c+c_a)\dot{x} + (k+k_a)x - c_a\dot{x}_a - k_a x_a = F_0 \sin \Omega t,$$
$$m_a\ddot{x}_a + c_a\dot{x}_a + k_a x_a - c_a\dot{x} - k_a x = 0,$$

which comprises a system of two coupled second-order linear ordinary differential equations.

Motivating Example 6

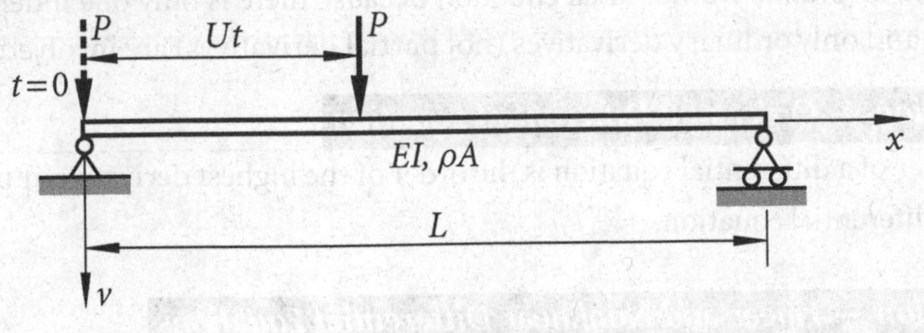

A bridge may be modeled as a simply supported beam of length L, mass density per unit length ρA, and flexural rigidity EI as shown. A vehicle of weight P crosses the bridge at a constant speed U. Suppose at time $t=0$, the vehicle is at the left end of the bridge and the bridge is at rest. The deflection of the bridge is $v(x,t)$, which is a function of both location x and time t. The equation governing $v(x,t)$ is the partial differential equation

$$\rho A \frac{\partial^2 v(x,t)}{\partial t^2} + EI \frac{\partial^4 v(x,t)}{\partial x^4} = P\delta(x-Ut),$$

where $\delta(x-a)$ is the Dirac delta function. The equation of motion satisfies the initial conditions

$$v(x,0)=0, \qquad \left.\frac{\partial v(x,t)}{\partial t}\right|_{t=0}=0,$$

and the boundary conditions

$$v(0,t)=v(L,t)=0, \qquad \left.\frac{\partial^2 v(x,t)}{\partial x^2}\right|_{x=0}=\left.\frac{\partial^2 v(x,t)}{\partial x^2}\right|_{x=L}=0.$$

1.2 General Concepts and Definitions

In this section, some general concepts and definitions of ordinary and partial differential equations are presented.

Let x be an independent variable and y be a dependent variable. An equation that involves x, y and various derivatives of y is called a *differential equation* (DE). For example,

$$\frac{\mathrm{d}y}{\mathrm{d}x}=2y+\sin x, \qquad \left(\frac{\mathrm{d}y}{\mathrm{d}x}\right)^3+\mathrm{e}^x+2=\frac{\mathrm{d}^2y}{\mathrm{d}x^2}$$

are differential equations.

Definition — Ordinary Differential Equation

In general, an equation of the form

$$F\Big(x,\ y,\ \frac{\mathrm{d}y}{\mathrm{d}x},\ \ldots,\ \frac{\mathrm{d}^n y}{\mathrm{d}x^n}\Big)=0$$

is an *Ordinary Differential Equation* (ODE).

It is called an *ordinary* differential equation because there is only one independent variable and only ordinary derivatives (not partial derivatives) are involved.

Definition — Order of a Differential Equation

The *order* of a differential equation is the order of the highest derivative appearing in the differential equation.

Definition — Linear and Nonlinear Differential Equations

If y and its various derivatives y', y'', ... appear linearly in the equation, it is a *linear* differential equation; otherwise, it is *nonlinear*.

For example,

$\dfrac{\mathrm{d}^2y}{\mathrm{d}x^2}+\omega^2y=\sin x, \quad \omega=\text{constant},$ ✍ Second-order, linear

$\left(\dfrac{\mathrm{d}y}{\mathrm{d}x}\right)^2+4y=\cos x,$ ✍ First-order, nonlinear because of the term $\left(\dfrac{\mathrm{d}y}{\mathrm{d}x}\right)^2$

$$x^3\frac{d^3y}{dx^3}+5x\frac{dy}{dx}+6y=e^x,$$ ✍ Third-order, linear

$$\frac{d^2y}{dx^2}+y\frac{dy}{dx}+2y=x.$$ ✍ Second-order, nonlinear because of the term $y\dfrac{dy}{dx}$

Sometimes, the roles of independent and dependent variables can be exchanged to render a differential equation linear. For example,

$$\frac{d^2x}{dy^2}-x\sqrt{y}=5$$

is a second-order linear equation with y being regarded as the independent variable and x the dependent variable.

In some applications, the roles of independent and dependent variables are obvious. For example, in a differential equation governing the variation of temperature T with time t, the time variable t is the independent variable and the temperature T is the dependent variable; time t cannot be the dependent variable. In other applications, the roles of independent and dependent variables are interchangeable. For example, in a differential equation governing the relationship between temperature T and pressure p, the temperature T can be considered as the independent variable and the pressure p the dependent variable, or vice versa.

Definition — Linear Ordinary Differential Equations

The general form of an nth-order linear ordinary differential equation is

$$a_n(x)\frac{d^ny}{dx^n}+a_{n-1}(x)\frac{d^{n-1}y}{dx^{n-1}}+\cdots+a_1(x)\frac{dy}{dx}+a_0(x)y=f(x).$$

If $a_0(x), a_1(x), \ldots, a_n(x)$ are constants, the ordinary differential equation is said to have *constant coefficients*; otherwise it is said to have *variable coefficients*.

For example,

$$\frac{d^2y}{dx^2}+0.1\frac{dy}{dx}+4y=10\cos 2x,$$ ✍ Second-order linear, constant coefficients

$$x^2\frac{d^2y}{dx^2}+x\frac{dy}{dx}+(x^2-\nu^2)y=0, \quad x>0, \quad \nu\geqslant 0 \text{ is a constant.}$$

✍ Second-order linear, variable coefficients (Bessel's equation)

Definition — Homogeneous and Nonhomogeneous Differential Equations

A differential equation is said to be *homogeneous* if it has zero as a solution; otherwise, it is *nonhomogeneous*.

For example,

$$\frac{d^2y}{dx^2} + 0.1\frac{dy}{dx} + 4y = 0, \quad \text{✍ Homogeneous}$$

$$\frac{d^2y}{dx^2} + 0.1\frac{dy}{dx} + 4y = 2\sin 2x + 5\cos 3x. \quad \text{✍ Nonhomogeneous}$$

Note that a homogeneous differential equation may have distinctively different meanings in different situations (see Section 2.2).

Partial Differential Equations

Definition — Partial Differential Equations

If the dependent variable u is a function of more than one independent variable, say $x_1, x_2, \ldots, x_m$, an equation involving the variables $x_1, x_2, \ldots, x_m$, u and various partial derivatives of u with respect to $x_1, x_2, \ldots, x_m$ is called a *Partial Differential Equation* (PDE).

For example,

$$\frac{\partial^2 u}{\partial x^2} = \frac{1}{\alpha}\frac{\partial u}{\partial t}, \quad \alpha = \text{constant}, \quad \text{✍ Heat equation in one-dimension}$$

$$\frac{\partial^2 u}{\partial x^2} + \frac{\partial^2 u}{\partial y^2} = f(x, y), \quad \text{✍ Poisson's equation in two-dimensions; Laplace's equation if } f(x, y) = 0$$

$$\frac{\partial^4 u}{\partial x^4} + 2\frac{\partial^4 u}{\partial x^2 \partial y^2} + \frac{\partial^4 u}{\partial y^4} = 0, \quad \text{✍ Biharmonic equation in two-dimensions}$$

$$\frac{\partial^2 u}{\partial x^2} + \frac{\partial^2 u}{\partial y^2} + \frac{\partial^2 u}{\partial z^2} = \frac{1}{\alpha}\frac{\partial u}{\partial t}, \quad \text{✍ Heat equation in three-dimensions}$$

$$\frac{\partial^2 u}{\partial x^2} + \frac{\partial^2 u}{\partial y^2} + \frac{\partial^2 u}{\partial z^2} = 0. \quad \text{✍ Laplace's equation in three-dimensions}$$

General and Particular Solutions

Definition — Solution of a Differential Equation

For an nth-order ordinary differential equation $F(x, y, y', \ldots, y^{(n)}) = 0$, a function $y = y(x)$, which is n times differentiable and satisfies the differential equation in some interval $a < x < b$ when substituted into the equation, is called a *solution* of the differential equation over the interval $a < x < b$.

Consider the first-order differential equation

$$\frac{dy}{dx} = 3.$$

Integrating with respect to x yields the general solution

$$y = 3x + C, \qquad C = \text{constant}.$$

The general solution of the differential equation, which includes all possible solutions, is a family of straight lines with slope equal to 3. On the other hand, $y = 3x$ is a particular solution passing through the origin, with the constant C being 0.

Consider the differential equation

$$\frac{\mathrm{d}^3 y}{\mathrm{d}x^3} = 48x.$$

Integrating both sides of the equation with respect to x gives

$$\frac{\mathrm{d}^2 y}{\mathrm{d}x^2} = 24x^2 + C_1.$$

Integrating with respect to x again yields

$$\frac{\mathrm{d}y}{\mathrm{d}x} = 8x^3 + C_1 x + C_2.$$

Integrating with respect to x once more results in the general solution

$$y = 2x^4 + \tfrac{1}{2} C_1 x^2 + C_2 x + C_3,$$

where C_1, C_2, C_3 are arbitrary constants. When the constants C_1, C_2, C_3 take specific values, one obtains particular solutions. For example,

$$y = 2x^4 + 3x^2 + 1, \qquad C_1 = 6,\ C_2 = 0,\ C_3 = 1,$$

$$y = 2x^4 + x^2 + 3x + 5, \qquad C_1 = 2,\ C_2 = 3,\ C_3 = 5,$$

are two particular solutions.

Remarks: In general, an nth-order ordinary differential equation will contain n arbitrary constants in its general solution. Hence, for an nth-order ordinary differential equation, n conditions are required to determine the n constants to yield a particular solution.

In applications, there are usually two types of conditions that can be used to determine the constants.

Illustrative Example

Consider the motion of an object dropped vertically at time $t = 0$ from $x = 0$ as shown in the following figure. Suppose that there is no resistance from the medium.

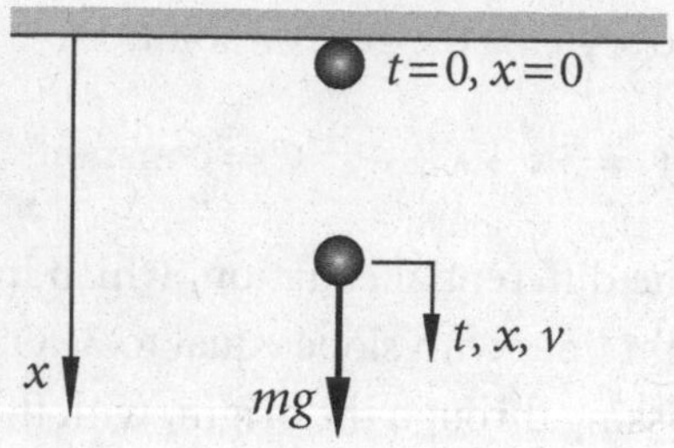

The equation of motion is given by

$$\frac{d^2x}{dt^2} = g,$$

and the general solution is, by integrating both sides of the equation with respect to t twice,

$$x(t) = C_0 + C_1 t + \frac{1}{2} g t^2.$$

The following are two possible ways of specifying the conditions.

Initial Value Problem

If the object is dropped with initial velocity v_0, the conditions required are

$$\text{at time } t = 0: \quad x(0) = 0, \quad \dot{x}(0) = \left.\frac{dx}{dt}\right|_{t=0} = v_0.$$

The constants C_0 and C_1 can be determined from these two conditions and the solution of the differential equation is

$$x(t) = v_0 t + \frac{1}{2} g t^2.$$

In this case, the differential equation is required to satisfy conditions specified at one value of t, i.e., $t = 0$.

Definition — Initial Value Problem

If a differential equation is required to satisfy conditions on the dependent variable and its derivatives specified at *one value* of the independent variable, these conditions are called *initial conditions* and the problem is called an *initial value problem.*

Boundary Value Problem

If the object is required to reach $x = L$ at time $t = T$, $L \geqslant \frac{1}{2} g T^2$, the conditions can be specified as

$$\text{at time } t = 0: \; x(0) = 0; \qquad \text{at time } t = T: \; x(T) = L.$$

The solution of the differential equation is

$$x(t) = \left(\frac{L}{T} - \frac{1}{2}gT\right)t + \frac{1}{2}gt^2.$$

In this case, the differential equation is required to satisfy conditions specified at two values of t, i.e., $t=0$ and $t=T$.

Definition — Boundary Value Problem

If a differential equation is required to satisfy conditions on the dependent variable and possibly its derivatives specified at *two or more values* of the independent variable, these conditions are called *boundary conditions* and the problem is called a *boundary value problem.*

Existence and Uniqueness of Solutions

Note that y' is the slope of curve $y=y(x)$ on the x-y plane. Hence, solving differential equation $y'=f(x,y)$ means finding curves whose slope at any given point (x, y) is equal to $f(x, y)$. Solving the initial value problem $y'=f(x,y)$ with $y(x_0)=y_0$ means finding curves passing through point (x_0, y_0) whose slope at any given point (x, y) is equal to $f(x, y)$.

This can be better visualized using direction fields. At a given point (x, y) in region R, one can draw a short straight line whose slope is $f(x, y)$. A direction field as shown in Figure 1.1 is then obtained if this is done for a large number of points.

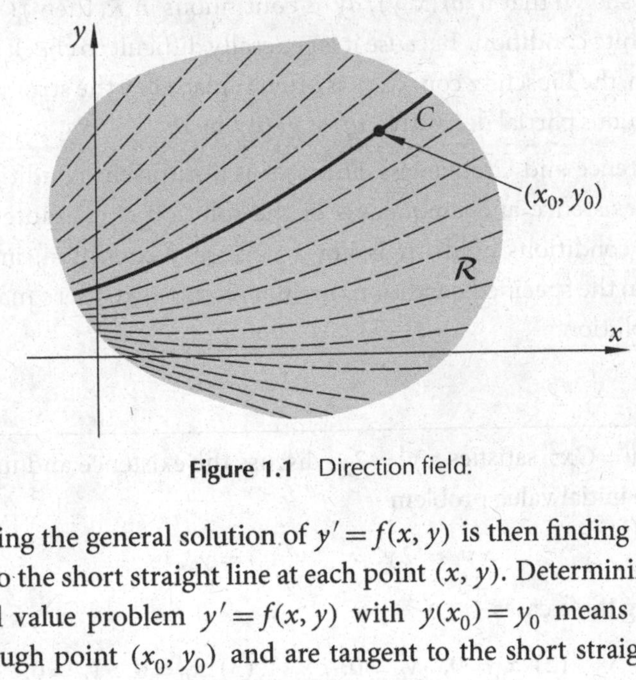

Figure 1.1 Direction field.

Determining the general solution of $y'=f(x,y)$ is then finding the curves that are tangent to the short straight line at each point (x, y). Determining the solution of the initial value problem $y'=f(x,y)$ with $y(x_0)=y_0$ means finding curves passing through point (x_0, y_0) and are tangent to the short straight line at each point (x, y).

Theorem — Existence and Uniqueness

Consider the initial value problem

$$y' = f(x, y), \quad y(x_0) = y_0,$$

where $f(x, y)$ is a continuous function in the rectangular region

$$R: \quad |x - x_0| \leqslant a, \quad |y - y_0| \leqslant b, \quad a > 0, \quad b > 0.$$

Suppose $f(x, y)$ also satisfies the *Lipschitz condition* with respect to y in R, i.e., there exists a constant $L > 0$ such that, for every (x, y_1) and (x, y_2) in R,

$$|f(x, y_1) - f(x, y_2)| \leqslant L|y_1 - y_2|.$$

Then *there exists a unique solution* $y = \varphi(x)$, continuous on $|x - x_0| \leqslant h$ and satisfying the initial condition $\varphi(x_0) = y_0$, where

$$h = \min\left(a, \frac{b}{M}\right), \qquad M = \max|f(x, y)| \text{ in } R.$$

The graphical interpretation of the Existence and Uniqueness Theorem is that, in region R in which the specified conditions hold, passing through any given point (x_0, y_0) *there exists one and only one* curve C such that the slope of curve C at any point (x, y) in R is equal to $f(x, y)$.

Remarks:

- It can be shown that if $\partial f(x,y)/\partial y$ is continuous in R, then $f(x, y)$ satisfies the Lipschitz condition. Because it is generally difficult to check the Lipschitz condition, the Lipschitz condition is often replaced by the stronger condition of continuous partial derivative $\partial f(x,y)/\partial y$ in R.
- The Existence and Uniqueness Theorem is a sufficient condition, meaning that the existence and uniqueness of the solution is guaranteed when the specified conditions hold. It is not a necessary condition, implying that, even when the specified conditions are not all satisfied, there may still exist a unique solution.

Example 1.1

Knowing that $y = Cx^2$ satisfies $xy' = 2y$, discuss the existence and uniqueness of solutions of the initial value problem

$$xy' = 2y, \qquad y(x_0) = y_0,$$

for the following three cases

(1) $x_0 \neq 0$; (2) $x_0 = 0$, $y_0 = 0$; (3) $x_0 = 0$, $y_0 \neq 0$.

Since $$f(x,y)=\frac{2y}{x}, \qquad \frac{\partial f(x,y)}{\partial y}=\frac{2}{x},$$
the conditions of the Existence and Uniqueness Theorem are not satisfied in a region including points with $x=0$.

With the help of the direction field as shown in the following figure, the solution of the initial value problem can be easily obtained.

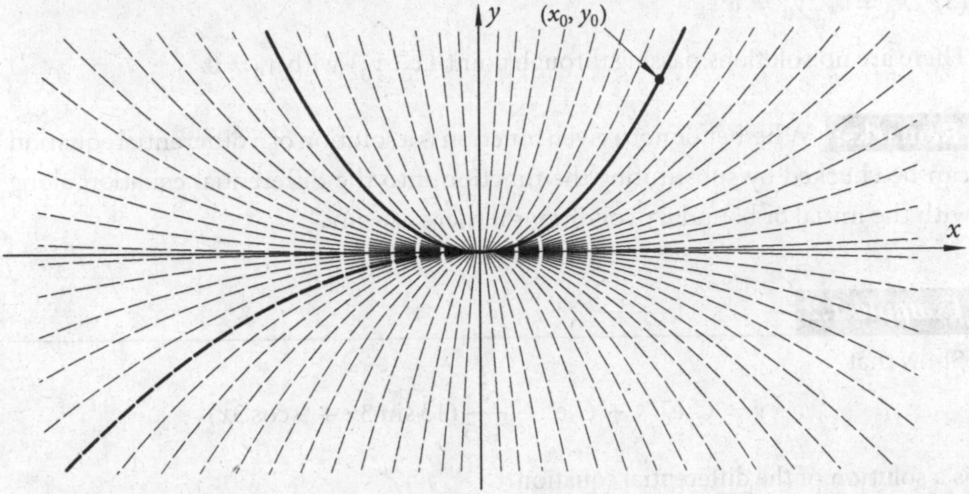

(1) $x_0 \neq 0$

(a) If $x_0>0$, then, in the region R with $x>0$, there exists a unique solution to the initial value problem
$$y=\frac{y_0}{x_0^2}x^2, \quad x>0.$$

(b) If $x_0<0$, then, in the region R with $x<0$, there exists a unique solution to the initial value problem
$$y=\frac{y_0}{x_0^2}x^2, \quad x<0.$$

(c) If $x_0>0$, then, in the region R including $x=0$, the solution to the initial value problem is not unique
$$y=\begin{cases}\dfrac{y_0}{x_0^2}x^2, & x\geqslant 0,\\ ax^2, & x<0, \quad a \text{ is a constant.}\end{cases}$$

(d) If $x_0<0$, then, in the region R including $x=0$, the solution to the initial value problem is not unique
$$y=\begin{cases}ax^2, & x>0, \quad a \text{ is a constant,}\\ \dfrac{y_0}{x_0^2}x^2, & x\leqslant 0.\end{cases}$$

(2) $x_0 = 0,\ y_0 = 0$

Passing through (0, 0), there are infinitely many solutions

$$y = \begin{cases} ax^2, & x \geqslant 0, \\ bx^2, & x < 0, \end{cases} \qquad a,\ b \text{ are constants.}$$

(3) $x_0 = 0,\ y_0 \neq 0$

There are no solutions passing through point (x_0, y_0) with $y_0 \neq 0$.

Remarks: Whether or not a given function is a solution of a differential equation can be checked by substituting the function into the differential equation along with the initial or boundary conditions if there are any.

Example 1.2

Show that

$$y = C_1 e^{-4x} + C_2 e^x - \frac{3}{125}(13 \sin 3x + 9 \cos 3x)$$

is a solution of the differential equation

$$y'' + 3y' - 4y = 6 \sin 3x.$$

Differentiating y successively twice yields

$$y' = -4C_1 e^{-4x} + C_2 e^x - \frac{3}{125}(39 \cos 3x - 27 \sin 3x),$$

$$y'' = 16C_1 e^{-4x} + C_2 e^x - \frac{3}{125}(-117 \sin 3x - 81 \cos 3x).$$

Substituting into the differential equation gives

$$\begin{aligned} y'' + 3y' - 4y = {} & 16C_1 e^{-4x} + C_2 e^x - \frac{3}{125}(-117 \sin 3x - 81 \cos 3x) \\ & - 12C_1 e^{-4x} + 3C_2 e^x - \frac{3}{125}(-81 \sin 3x + 117 \cos 3x) \\ & - 4C_1 e^{-4x} - 4C_2 e^x - \frac{3}{125}(-52 \sin 3x - 36 \cos 3x) \\ = {} & 6 \sin 3x. \end{aligned}$$

Hence

$$y = C_1 e^{-4x} + C_2 e^x - \frac{3}{125}(13 \sin 3x + 9 \cos 3x)$$

is a solution of the differential equation.

Example 1.3

Show that

$$u(x,t) = 2\sin\frac{3\pi x}{L}\exp\left(-\frac{\pi^2}{L^2}t\right)$$

is a solution of the partial differential equation

$$\frac{1}{9}\frac{\partial^2 u}{\partial x^2} = \frac{\partial u}{\partial t},$$

with the initial condition

$$u(x,0) = 2\sin\frac{3\pi x}{L}, \qquad \text{for } 0 \leqslant x \leqslant L;$$

and the boundary conditions

$$u(0,t) = 0, \quad u(L,t) = 0, \qquad \text{for } t > 0.$$

Evaluate the partial derivatives

$$\frac{\partial u}{\partial t} = 2\sin\frac{3\pi x}{L}\cdot\left(-\frac{\pi^2}{L^2}\right)\exp\left(-\frac{\pi^2}{L^2}t\right),$$

$$\frac{\partial u}{\partial x} = 2\cdot\left(\frac{3\pi}{L}\right)\cos\frac{3\pi x}{L}\cdot\exp\left(-\frac{\pi^2}{L^2}t\right),$$

$$\frac{\partial^2 u}{\partial x^2} = 2\cdot(-)\left(\frac{3\pi}{L}\right)^2\sin\frac{3\pi x}{L}\cdot\exp\left(-\frac{\pi^2}{L^2}t\right).$$

Substitute into the differential equation

$$\left.\begin{aligned} \text{L.H.S.} &= \frac{1}{9}\frac{\partial^2 u}{\partial x^2} = -\frac{2\pi^2}{L^2}\sin\frac{3\pi x}{L}\exp\left(-\frac{\pi^2}{L^2}t\right) \\ \text{R.H.S.} &= \frac{\partial u}{\partial t} = -\frac{2\pi^2}{L^2}\sin\frac{3\pi x}{L}\exp\left(-\frac{\pi^2}{L^2}t\right) \end{aligned}\right\} \Longrightarrow \text{L.H.S.} = \text{R.H.S.}$$

Check the initial and boundary conditions

$$u(x,0) = 2\sin\frac{3\pi x}{L}\exp\left(-\frac{\pi^2}{L^2}\cdot 0\right) = 2\sin\frac{3\pi x}{L}, \quad \text{satisfied,}$$

$$u(0,t) = 2\sin\frac{3\pi\cdot 0}{L}\exp\left(-\frac{\pi^2}{L^2}t\right) = 0, \quad \text{satisfied,}$$

$$u(L,t) = 2\sin\frac{3\pi\cdot L}{L}\exp\left(-\frac{\pi^2}{L^2}t\right) = 0, \quad \text{satisfied.}$$

Hence

$$u(x,t) = 2\sin\frac{3\pi x}{L}\exp\left(-\frac{\pi^2}{L^2}t\right)$$

is a solution of the partial differential equation with the given initial and boundary conditions.

CHAPTER 2

First-Order and Simple Higher-Order Differential Equations

There are various techniques for solving first-order and simple higher-order ordinary differential equations. The key in the application of the specific technique hinges on the identification of the type of a given equation. The objectives of this chapter are to introduce various types of first-order and simple higher-order differential equations and the corresponding techniques for solving these differential equations.

In this chapter, it is assumed that x is the independent variable and y is the dependent variable. Solutions in the explicit form $y=\eta(x)$ or in the implicit form $u(x, y)=0$ are sought.

2.1 The Method of Separation of Variables

Consider a first-order ordinary differential equation of the form

$$\frac{\mathrm{d}y}{\mathrm{d}x} = F(x, y).$$

Suppose that the right-hand side $F(x, y)$, which is a function of x and y, can be written as a product of a function of x and a function of y, i.e.,

$$F(x, y) = f(x)\cdot\phi(y).$$

For example, the functions

$$\mathrm{e}^{x+y^2} = \mathrm{e}^x\cdot\mathrm{e}^{y^2}, \qquad xy+x+2y+2 = (x+2)\cdot(y+1)$$

can be separated into a product of a function of x and a function of y, but the following functions cannot be separated

$$\ln(x+2y), \qquad \sin(x^2+y), \qquad xy^2+x^2.$$

This type of differential equation is called *variable separable* or *separable* differential equations. The equations can be solved by the method of separation of variables. Rewrite the equation as

$$\frac{dy}{dx} = f(x)\cdot\phi(y).$$

Case 1. If $\phi(y)\neq 0$, moving terms involving variable y to the left-hand side and terms of variable x to the right-hand side yields

$$\underbrace{g(y)\,dy}_{\text{function of } y \text{ only}} = \underbrace{f(x)\,dx}_{\text{function of } x \text{ only}}, \qquad g(y) = \frac{1}{\phi(y)}.$$

Integrating both sides of the equation results in the general solution

$$\int g(y)\,dy = \int f(x)\,dx + C,$$

where C is an arbitrary constant.

Remarks: When dividing a differential equation by a function, it is important to ensure that the function is not zero. Otherwise, solutions may be lost in the process. Hence, the case when the function is zero should be considered separately to determine if it yields extra solutions.

Case 2. If $\phi(y)=0$, solve for the roots of this equation. Let $y=y_0$ be one of the solutions of equation $\phi(y)=0$. Then $y=y_0$ is a solution of the differential equation. Note that sometimes the solution $y=y_0$ may already be included in the general solution obtained from Case 1.

Remarks: It should be emphasized that, only when one side of the equation contains only variable x and the other side of the equation contains only variable y, the equation can be integrated to obtain the general solution.

Example 2.1

Solve $\quad \dfrac{dy}{dx} + \dfrac{1}{y}\,e^{y^2+3x} = 0, \quad y\neq 0.$

Separating the variables yields

$$-ye^{-y^2}\,dy = e^{3x}\,dx.$$

Integrating both sides to obtain the general solution leads to

$$-\int ye^{-y^2}\,dy = \int e^{3x}\,dx + C,$$

$$\frac{1}{2}\int e^{-y^2}\,d(-y^2) = \frac{1}{3}\int e^{3x}\,d(3x) + C, \qquad d(-y^2) = -2y\,dy$$

$$\therefore \quad \frac{1}{2}e^{-y^2} = \frac{1}{3}e^{3x} + C. \qquad \int e^x\,dx = e^x \quad \text{General solution}$$

Example 2.2

Solve $\tan y\,dx - \cot x\,dy = 0, \quad \cos y \neq 0, \quad \sin x \neq 0.$

The equation can be written as

$$\frac{\sin y}{\cos y}\,dx = \frac{\cos x}{\sin x}\,dy.$$

To separate the variables, multiply the differential equation by $\dfrac{\sin x \cos y}{\cos x \sin y}$; it is required that $\sin y$ not be zero.

Case 1. If $\sin y \neq 0$, separating the variables yields

$$\frac{\sin x}{\cos x}\,dx = \frac{\cos y}{\sin y}\,dy.$$

Integrating both sides results in the general solution

$$\int \frac{\sin x}{\cos x}\,dx = \int \frac{\cos y}{\sin y}\,dy + C,$$

$$-\int \frac{1}{\cos x}\,d(\cos x) = \int \frac{1}{\sin y}\,d(\sin y) + C, \qquad \begin{aligned} d(\sin x) &= \cos x\,dx \\ d(\cos x) &= -\sin x\,dx \end{aligned}$$

$$\therefore \quad -\ln|\cos x| = \ln|\sin y| + C. \qquad \int \frac{1}{x}\,dx = \ln|x|$$

The result can be simplified as follows

$$\ln|\cos x \cdot \sin y| = -C, \qquad \ln a + \ln b = \ln(a\cdot b)$$

$$|\cos x \cdot \sin y| = e^{-C} \quad \Longrightarrow \quad \cos x \cdot \sin y = A. \qquad \pm e^{-C} \Rightarrow A$$

Since C is an arbitrary constant, A is an arbitrary constant, which can in turn be renamed as C. The general solution becomes

$$\cos x \cdot \sin y = C. \qquad \text{General solution}$$

Case 2. If $\sin y = 0$, one has $y = k\pi$, $k = 0, \pm 1, \pm 2, \ldots$. It is obvious that $y = k\pi$ or $\sin y = 0$ is a solution of the differential equation. However, $\sin y = 0$ is already included in the general solution $\cos x \cdot \sin y = C$, with $C = 0$, obtained in Case 1.

Example 2.3

Solve $\quad xy^3\,\mathrm{d}x + (y+1)\,\mathrm{e}^{-x}\,\mathrm{d}y = 0.$

Case 1. If $y \neq 0$, separating the variables leads to

$$x\,\mathrm{e}^x\,\mathrm{d}x = -\frac{y+1}{y^3}\,\mathrm{d}y.$$

Integrating both sides results in the general solution

$$\int x\,\mathrm{e}^x\,\mathrm{d}x = -\int \frac{y+1}{y^3}\,\mathrm{d}y + C.$$

The integrals are evaluated as

$$\int x\,\mathrm{e}^x\,\mathrm{d}x = \int x\,\mathrm{d}(\mathrm{e}^x) \qquad ✍\ \mathrm{d}(\mathrm{e}^x) = \mathrm{e}^x\mathrm{d}x, \quad \text{Integration by parts}$$

$$= x\,\mathrm{e}^x - \int \mathrm{e}^x\,\mathrm{d}x = x\,\mathrm{e}^x - \mathrm{e}^x,$$

$$-\int \frac{y+1}{y^3}\,\mathrm{d}y = -\int (y^{-2} + y^{-3})\,\mathrm{d}y$$

$$= -\left(\frac{y^{-1}}{-1} + \frac{y^{-2}}{-2}\right) = \frac{1}{y} + \frac{1}{2y^2}. \qquad ✍\ \int x^n\,\mathrm{d}x = \frac{x^{n+1}}{n+1}$$

The general solution becomes

$$(x-1)\,\mathrm{e}^x = \frac{1}{y} + \frac{1}{2y^2} + C.$$

Case 2. It is easy to verify that $y = 0$ is a solution of the differential. This solution cannot be obtained from the general solution for any value of the constant C.

Definition — Singular Solution

Any solutions of a differential equation that cannot be obtained from the general solution for any values of the arbitrary constants are called *singular solutions*.

Hence, combining Cases 1 and 2, the solutions of the differential equation are

$$(x-1)\,\mathrm{e}^x = -\frac{1}{y} - \frac{1}{2y^2} + C, \qquad ✍\ \text{General solution}$$

$$y = 0. \qquad ✍\ \text{Singular solution}$$

Remarks: A variable separable equation is very easy to identify, and it is easy to express the general solution in terms of integrals. However, the actual evaluation of the integrals may sometimes be quite challenging.

2.2 Method of Transformation of Variables

2.2.1 Homogeneous Equations

Equations of the type

$$\frac{dy}{dx} = f\left(\frac{y}{x}\right) \tag{1}$$

are called *homogeneous differential equations*. For example,

$$g(x, y) = \frac{x^2+3y^2}{x^2-xy+y^2} = \frac{1+3\left(\frac{y}{x}\right)^2}{1-\left(\frac{y}{x}\right)+\left(\frac{y}{x}\right)^2} = f\left(\frac{y}{x}\right),$$

$$g(x, y) = \ln x - \ln y = \ln\left(\frac{x}{y}\right) = -\ln\left(\frac{y}{x}\right) = f\left(\frac{y}{x}\right).$$

Remarks: A homogeneous differential equation has several distinct meanings:

- A first-order ordinary differential equation of the form $\frac{dy}{dx}=f\left(\frac{y}{x}\right)$ is of the type of homogeneous equation.
- A homogeneous differential equation, defined in Chapter 1, means that the differential equation has zero as a solution.

A homogeneous equation can be converted to a variable separable equation using a transformation of variables. Let $v=\frac{y}{x}$ be the new dependent variable, while x is still the independent variable. Hence

$$y = xv \implies \frac{dy}{dx} = v + x\frac{dv}{dx}.$$

Substituting into differential equation (1) leads to

$$v + x\frac{dv}{dx} = f(v) \implies x\frac{dv}{dx} = f(v) - v.$$

Case 1. $f(v)-v=0$. One has $y=v_0x$, where v_0 is the solution of $f(v_0)-v_0=0$.

Case 2. $f(v)-v\neq 0$. Separating the variables leads to

$$\frac{dv}{f(v)-v} = \frac{dx}{x}.$$

The transformed differential equation is variable separable. Integrating both sides gives the general solution

$$\int \frac{dv}{f(v)-v} = \int \frac{dx}{x} + C.$$

Example 2.4

Solve $\quad \dfrac{dy}{dx} + \dfrac{x}{y} + 2 = 0, \quad y \neq 0, \quad y(0) = 1.$

The differential equation is homogeneous. Letting $v = \dfrac{y}{x}$,

$$y = xv \implies \frac{dy}{dx} = v + x\frac{dv}{dx},$$

the differential equation becomes

$$v + x\frac{dv}{dx} + \frac{1}{v} + 2 = 0 \implies x\frac{dv}{dx} = -\Big(v + \frac{1}{v} + 2\Big) \implies x\frac{dv}{dx} = -\frac{(v+1)^2}{v}.$$

Case 1. $v = -1 \implies y = -x$. But it does not satisfy the condition $y(0) = 1$.

Case 2. $v \neq -1$, separating the variables yields

$$\frac{v}{(v+1)^2}\,dv = -\frac{1}{x}\,dx.$$

Integrating both sides gives

$$\int \frac{v}{(v+1)^2}\,dv = -\int \frac{1}{x}\,dx + C.$$

Since

$$\int \frac{v}{(v+1)^2}\,dv = \int \frac{(v+1)-1}{(v+1)^2}\,dv = \int \Big\{\frac{1}{v+1} - \frac{1}{(v+1)^2}\Big\}\,dv$$

$$= \int \frac{1}{v+1}\,d(v+1) - \int \frac{1}{(v+1)^2}\,d(v+1)$$

$$= \ln|v+1| + \frac{1}{v+1}, \qquad ✍ \int \frac{1}{x}\,dx = \ln|x|, \quad \int \frac{1}{x^2}\,dx = -\frac{1}{x}$$

one obtains

$$\ln|v+1| + \frac{1}{v+1} = -\ln|x| + C.$$

Converting back to the original variables x and y results in the general solution

$$\ln\left|\frac{y}{x} + 1\right| + \ln|x| + \frac{1}{\frac{y}{x} + 1} = C, \qquad ✍ \ln a + \ln b = \ln(a \cdot b)$$

$$\therefore \quad \ln|y+x| + \frac{x}{y+x} = C. \qquad ✍ \text{ General solution}$$

The constant C is determined using the initial condition $y(0) = 1$

$$\ln|1+0| + \frac{0}{1+0} = C \implies C = 0.$$

The particular solution satisfying $y(0)=1$ is

$$\ln|y+x| + \frac{x}{y+x} = 0. \quad \text{Particular solution}$$

Example 2.5

Solve $\quad x(\ln x - \ln y)\,dy - y\,dx = 0, \quad x>0, \quad y>0.$

Dividing both sides of the equation by x gives

$$\left(\ln\frac{x}{y}\right)dy - \frac{y}{x}\,dx = 0 \implies \frac{dy}{dx} = \frac{\dfrac{y}{x}}{-\ln\dfrac{y}{x}},$$

which is homogeneous. Putting $v=\dfrac{y}{x}$, $v>0$,

$$y = xv \implies \frac{dy}{dx} = v + x\frac{dv}{dx},$$

the equation becomes

$$v + x\frac{dv}{dx} = \frac{v}{-\ln v} \implies x\frac{dv}{dx} = -\frac{v}{\ln v} - v = -v\frac{1+\ln v}{\ln v}.$$

Case 1. For $\ln v \neq -1$, the equation can be separated as

$$\frac{\ln v}{v(1+\ln v)}\,dv = -\frac{1}{x}\,dx.$$

Integrating both sides yields

$$\int \frac{\ln v}{v(1+\ln v)}\,dv = -\int \frac{1}{x}\,dx + C.$$

Since $d(\ln v) = \dfrac{1}{v}\,dv$, one has

$$\int \frac{\ln v}{v(1+\ln v)}\,dv = \int \frac{(1 + \ln v) - 1}{1+\ln v}\,d(\ln v)$$
$$= \int \left(1 - \frac{1}{1+\ln v}\right)d(\ln v) = \ln v - \ln|1+\ln v|.$$

Hence,

$$\ln\left|\frac{v}{1+\ln v}\right| = -\ln x + C.$$

Replacing v by the original variables results in the general solution

$$\ln\left|\frac{y/x}{1+\ln(y/x)}\right| + \ln x = C,$$

$$\ln\left|\frac{y}{1+\ln y-\ln x}\right| = \ln C,$$ ✍ Since C is an arbitrary constant, it is rewritten as $\ln C$.

$$\therefore\quad \frac{y}{1+\ln y-\ln x} = C.$$

Case 2. If $\ln v=-1$, one has $v=\mathrm{e}^{-1}$. Hence, $v=y/x=\mathrm{e}^{-1}$ or $\mathrm{e}y=x$ is a solution. This solution cannot be obtained from the general solution for any value of the constant C and is therefore a singular solution.

Combining Cases 1 and 2, the solutions of the differential equations are

$$\frac{y}{1+\ln y-\ln x} = C,$$ ✍ General solution

$$\mathrm{e}\,y = x.$$ ✍ Singular solution

Example 2.6

Solve $(y+x)\,\mathrm{d}y + (x-y)\,\mathrm{d}x = 0.$

Since $y=-x$ is not a solution, the equation can be written as

$$\frac{\mathrm{d}y}{\mathrm{d}x} = -\frac{x-y}{y+x} = -\frac{1-\dfrac{y}{x}}{\dfrac{y}{x}+1},$$ ✍ Dividing both the numerator and denominator by x, $x\neq 0$

which is a homogeneous equation. Letting $v=\dfrac{y}{x}$,

$$y = xv \implies \frac{\mathrm{d}y}{\mathrm{d}x} = v + x\frac{\mathrm{d}v}{\mathrm{d}x},$$

the equation becomes

$$v + x\frac{\mathrm{d}v}{\mathrm{d}x} = -\frac{1-v}{v+1},$$

$$x\frac{\mathrm{d}v}{\mathrm{d}x} = -\frac{1-v}{v+1} - v = -\frac{1-v+v(v+1)}{v+1} = -\frac{v^2+1}{v+1}.$$

Since $v^2+1\neq 0$, separating the variables gives

$$\frac{v+1}{v^2+1}\,\mathrm{d}v = -\frac{1}{x}\,\mathrm{d}x.$$

Integrating both sides yields

$$\int \frac{v}{v^2+1}\,\mathrm{d}v + \int \frac{1}{v^2+1}\,\mathrm{d}v = -\int \frac{1}{x}\,\mathrm{d}x + C,$$

$$\frac{1}{2}\int \frac{1}{v^2+1}\,\mathrm{d}(v^2+1) + \tan^{-1}v = -\ln|x| + C,$$ ✍ $\displaystyle\int \frac{1}{x^2+1}\,\mathrm{d}x = \tan^{-1}x$

$$\frac{1}{2}\ln\left|v^2+1\right| + \tan^{-1}v = -\ln|x| + C. \qquad ✍ \int \frac{1}{x}\,\mathrm{d}x = \ln|x|$$

Replacing v by the original variables x and y results in the general solution

$$\ln\left|\left(\frac{y}{x}\right)^2+1\right| + 2\ln|x| + 2\tan^{-1}\frac{y}{x} = 2C, \qquad ✍ \begin{array}{l} a\ln b = \ln b^a \\ \ln a + \ln b = \ln(a\cdot b) \end{array}$$

$$\therefore \quad \ln(y^2+x^2) + 2\tan^{-1}\frac{y}{x} = C. \qquad ✍ \text{ 2C is renamed as C.}$$

Example 2.7

Solve $\dfrac{\mathrm{d}y}{\mathrm{d}x} = \dfrac{2x-y+1}{x-2y+1}, \quad x-2y+1 \neq 0.$

The equation is not homogeneous because of the constant terms in both the numerator and denominator. $2x-y+1=0$ and $x-2y+1=0$ are the equations of straight lines. The coordinates of the point of intersection are the solution of the equations

$$\left.\begin{array}{r} 2x-y+1=0 \\ x-2y+1=0 \end{array}\right\} \implies x = -\tfrac{1}{3}, \quad y = \tfrac{1}{3}.$$

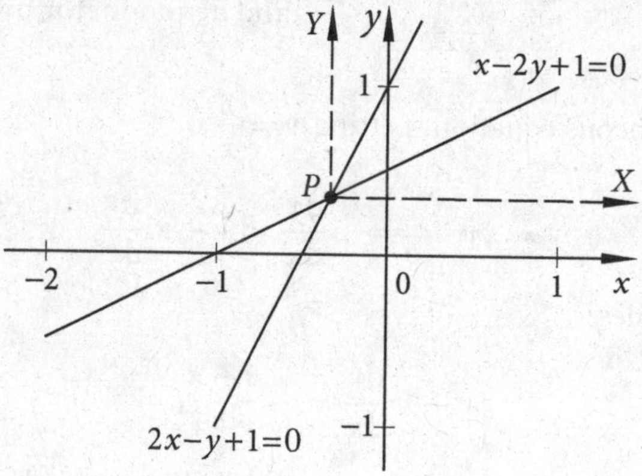

Hence, the point of intersection is $P(h,k)$, $(h,k) = \left(-\frac{1}{3}, \frac{1}{3}\right)$. Shift to new axes (X, Y) through point $P(h,k)$. Then

$$x = X+h = X - \tfrac{1}{3}, \quad y = Y+k = Y+\tfrac{1}{3}, \quad \text{and} \quad \frac{\mathrm{d}y}{\mathrm{d}x} = \frac{\mathrm{d}Y}{\mathrm{d}X}.$$

In the new variables X and Y, the constant terms are removed

$$\frac{\mathrm{d}Y}{\mathrm{d}X} = \frac{2(X-\frac{1}{3}) - (Y+\frac{1}{3}) + 1}{(X-\frac{1}{3}) - 2(Y+\frac{1}{3}) + 1} = \frac{2X-Y}{X-2Y} = \frac{2-\dfrac{Y}{X}}{1-2\dfrac{Y}{X}},$$

✍ Dividing both the numerator and denominator by X, $X \neq 0$

which is now a homogeneous equation. Putting $v=\dfrac{Y}{X}$,

$$Y = Xv \implies \frac{dY}{dX} = v + X\frac{dv}{dX},$$

and the equation becomes

$$v + X\frac{dv}{dX} = \frac{2-v}{1-2v},$$

$$X\frac{dv}{dX} = \frac{2-v}{1-2v} - v = \frac{2-v-(v-2v^2)}{1-2v} = 2\frac{v^2-v+1}{1-2v}.$$

Since $v^2-v+1 \neq 0$, separating the variables yields

$$\frac{1-2v}{v^2-v+1}\,dv = \frac{2}{X}\,dX.$$

Integrating both sides leads to

$$\int \frac{1-2v}{v^2-v+1}\,dv = 2\int \frac{1}{X}\,dX + C.$$

Since $d(v^2-v+1) = (2v-1)\,dv$, one obtains

$$-\int \frac{1}{v^2-v+1}\,d(v^2-v+1) = 2\int \frac{1}{X}\,dX + C,$$

$$-\ln|v^2-v+1| = 2\ln|X| + C.$$ ✍ $\int \frac{1}{x}\,dx = \ln|x|$

Replacing v by X and Y gives

$$\ln|X^2| + \ln\left|\left(\frac{Y}{X}\right)^2 - \left(\frac{Y}{X}\right) + 1\right| = \ln|C|,$$ ✍ $-C$ is written as $\ln|C|$.

$$\ln|Y^2 - XY + X^2| = \ln|C| \implies Y^2 - XY + X^2 = C.$$

In terms of the original variables x and y, the general solution becomes

$$\left(y-\tfrac{1}{3}\right)^2 - \left(x+\tfrac{1}{3}\right)\left(y-\tfrac{1}{3}\right) + \left(x+\tfrac{1}{3}\right)^2 = C.$$

2.2.2 *Special Transformations*

Example 2.8

Solve $\dfrac{dy}{dx} = \dfrac{x-y+5}{2x-2y-2}, \quad x-y-1 \neq 0.$

Unlike the previous example, the two lines $x-y+5=0$ and $2x-2y-2=0$ are parallel so that there is no finite point of intersection.

Because both the numerator and denominator have the term $x-y$, take a new dependent variable $v=x-y$,

$$v=x-y \implies \frac{dv}{dx}=1-\frac{dy}{dx} \implies \frac{dy}{dx}=1-\frac{dv}{dx}.$$

The differential equation becomes

$$1-\frac{dv}{dx}=\frac{v+5}{2v-2},$$

$$\frac{dv}{dx}=1-\frac{v+5}{2v-2}=\frac{2v-2-v-5}{2v-2}=\frac{v-7}{2(v-1)}.$$

Case 1. $v\neq 7$, separating the variables gives

$$\frac{v-1}{v-7}\,dv=\frac{1}{2}\,dx.$$ ✍ Variable separable

Integrating both sides yields

$$\int\frac{(v-7)+6}{v-7}\,dv=\int\frac{1}{2}\,dx+C \implies \int\Big(1+\frac{6}{v-7}\Big)dv=\frac{1}{2}x+C,$$

$$v+6\ln|v-7|=\tfrac{1}{2}x+C.$$

Replacing v by the original variables gives the general solution

$$x-y+6\ln|x-y-7|=\tfrac{1}{2}x+C,$$

$$\therefore\quad \tfrac{1}{2}x-y+6\ln|x-y-7|=C.$$

Case 2. $v=7 \implies x-y=7$. It can be easily verified that $x-y=7$ is a solution of the differential equation. This solution cannot be obtained from the general solution for any value of the constant C and is therefore a singular solution.

Combining Cases 1 and 2, the solutions of the differential equation are

$$\tfrac{1}{2}x-y+6\ln|x-y-7|=C,$$ ✍ General solution

$$x-y=7.$$ ✍ Singular solution

Summary

Consider a differential equation of the form

$$\frac{dy}{dx}=\frac{ax+by+c}{\alpha x+\beta y+\gamma},\qquad ab\neq 0,\ \alpha\beta\neq 0.$$

Case 1. $\dfrac{a}{\alpha}\neq\dfrac{b}{\beta}$

The two straight lines $ax+by+c=0$ and $\alpha x+\beta y+\gamma=0$ intersect at point $P(h,k)$, where (h,k) is the solution of

$$ah+bk+c=0, \qquad \alpha h+\beta k+\gamma=0.$$

Letting

$$x=X+h, \qquad y=Y+k,$$

the differential equation becomes

$$\frac{dY}{dX}=\frac{aX+bY}{\alpha X+\beta Y}. \qquad \text{✍ Homogeneous}$$

Case 2. $\dfrac{a}{\alpha}=\dfrac{b}{\beta}=\dfrac{1}{r} \implies \alpha=ra, \ \beta=rb$

Letting

$$v=ax+by \implies \frac{dv}{dx}=a+b\frac{dy}{dx} \implies \frac{dy}{dx}=\frac{1}{b}\Big(\frac{dv}{dx}-a\Big),$$

the differential equation becomes

$$\frac{1}{b}\Big(\frac{dv}{dx}-a\Big)=\frac{v+c}{rv+\gamma} \implies \frac{dv}{dx}=\frac{b(v+c)}{rv+\gamma}+a. \qquad \text{✍ Variable separable}$$

Example 2.9

Solve $\dfrac{dy}{dx}=(x+y)^2.$

This equation is neither variable separable nor homogeneous. The "special" term in the equation is $x+y$. Hence, letting $v=x+y$ be the new dependent variable

$$\frac{dv}{dx}=1+\frac{dy}{dx} \implies \frac{dy}{dx}=\frac{dv}{dx}-1,$$

the equation becomes

$$\frac{dv}{dx}-1=v^2 \implies \frac{dv}{dx}=v^2+1.$$

Since $v^2+1\neq 0$, separating the variables gives

$$\frac{1}{v^2+1}dv=dx. \qquad \text{✍ Variable separable}$$

Integrating both sides leads to

$$\int\frac{1}{v^2+1}dv=\int dx+C \implies \tan^{-1}v=x+C.$$

Replacing v by the original variable results in the general solution

$$\tan^{-1}(x+y) = x + C, \quad \text{or} \quad x + y = \tan(x+C).$$

Remarks: There are no systematic procedures to follow in applying the method of special transformations. It is important to carefully inspect the differential equation to uncover the "special" term and then determine the transformation accordingly.

Example 2.10

Solve $\quad \dfrac{dy}{dx} = \dfrac{y^6-2x^2}{2xy^5+x^2y^2}, \quad x \neq 0, \ y \neq 0, \ 2y^3+x \neq 0.$

The differential equation can be rewritten as

$$\frac{dy}{dx} = \frac{y^6-2x^2}{xy^2(2y^3+x)} = \frac{\left(\dfrac{y^3}{x}\right)^2 - 2}{y^2\left[2\left(\dfrac{y^3}{x}\right)+1\right]}.$$

Dividing both the numerator and denominator by x^2, $x \neq 0$

The "special" term in the equation is $\dfrac{y^3}{x}$. Hence, letting

$$v = \frac{y^3}{x} \implies y^3 = xv \implies 3y^2\frac{dy}{dx} = v + x\frac{dv}{dx},$$

the differential equation becomes

$$\frac{1}{3}\left(v + x\frac{dv}{dx}\right) = \frac{v^2-2}{2v+1},$$

$$x\frac{dv}{dx} = \frac{3(v^2-2)}{2v+1} - v = \frac{3v^2-6-2v^2-v}{2v+1} = \frac{v^2-v-6}{2v+1}.$$

Case 1. $v^2-v-6=0$, i.e., $(v-3)(v+2)=0 \implies v=-2$ or $v=3$, which gives

$$y^3 = -2x \quad \text{or} \quad y^3 = 3x.$$

Case 2. $v^2-v-6 \neq 0$, separating the variables gives

$$\frac{2v+1}{v^2-v-6}\,dv = \frac{1}{x}\,dx.$$

Variable separable

Integrating both sides yields

$$\int \frac{2v+1}{v^2-v-6}\,dv = \int \frac{1}{x}\,dx + C.$$

The first integral can be evaluated using partial fractions (see pages 259–261 for a brief review on partial fractions)

$$\frac{2v+1}{v^2-v-6}=\frac{2v+1}{(v-3)(v+2)}=\frac{A}{v-3}+\frac{B}{v+2}.$$

Using the cover-up method, the coefficients A and B can be easily determined

$$A=\left.\frac{2v+1}{v+2}\right|_{v=3}=\frac{7}{5},\qquad B=\left.\frac{2v+1}{v-3}\right|_{v=-2}=\frac{3}{5},$$

$$\int\frac{2v+1}{v^2-v-6}\,dv=\frac{1}{5}\int\left(\frac{7}{v-3}+\frac{3}{v+2}\right)dv=\tfrac{7}{5}\ln|v-3|+\tfrac{3}{5}\ln|v+2|.$$

Hence

$$\tfrac{7}{5}\ln|v-3|+\tfrac{3}{5}\ln|v+2|=\ln|x|+C,$$

$$7\ln|v-3|+3\ln|v+2|=5\ln|x|+\ln|C|,\qquad ✍\ 5C\Rightarrow\ln|C|$$

$$\ln\left|(v-3)^7(v+2)^3\right|=\ln\left|Cx^5\right|,\qquad ✍\ \ln a+\ln b=\ln(a\cdot b)$$

$$\therefore\quad (v-3)^7(v+2)^3=Cx^5.$$

Replacing v by the original variables results in the general solution

$$\left(\frac{y^3}{x}-3\right)^7\left(\frac{y^3}{x}+2\right)^3=Cx^5\implies (y^3-3x)^7(y^3+2x)^3=Cx^{15}.$$

Note that the solutions $y^3=-2x$ and $y^3=3x$ obtained in Case 1 are contained in the general solution $(y^3-3x)^7(y^3+2x)^3=Cx^{15}$ obtained in Case 2, with $C=0$. Hence, the solution of the differential equation is

$$(y^3-3x)^7(y^3+2x)^3=Cx^{15}.\qquad ✍\ \text{General solution}$$

Example 2.11

1. Show that equations of the form
$$\frac{x}{y}\frac{dy}{dx}=f(xy),\quad y\neq 0,$$
can be converted to variable separable by the transformation $xy=v$.
2. Using the result obtained above, solve
$$\frac{x}{y}\frac{dy}{dx}=\frac{2+x^2y^2}{2-x^2y^2}.$$

---

1. Letting $xy=v$,

$$y+x\frac{dy}{dx}=\frac{dv}{dx}\implies \frac{x}{y}\frac{dy}{dx}=\frac{1}{y}\frac{dv}{dx}-1=\frac{x}{v}\frac{dv}{dx}-1,$$

the differential equation becomes

$$\frac{x}{v}\frac{dv}{dx} - 1 = f(v) \implies \frac{x}{v}\frac{dv}{dx} = f(v) + 1.$$

Case 1. $f(v)+1=0$. If v_0 is a root of $f(v_0)+1=0$, then a solution is $xy=v_0$.
Case 2. $f(v)+1\neq 0$. Separating the variables gives

$$\frac{1}{v\left[f(v)+1\right]}\,dv = \frac{1}{x}\,dx. \qquad \text{✍ Variable separable}$$

Hence, the transformation $xy=v$ converts the original differential equation to variable separable.

2. In this case,

$$f(v) = \frac{2+v^2}{2-v^2} \implies \frac{x}{v}\frac{dv}{dx} = \frac{4}{2-v^2},$$

and separating the variables gives

$$\frac{2-v^2}{2v}\,dv = \frac{2}{x}\,dx.$$

Integrating both sides leads to

$$\int\left(\frac{1}{v} - \frac{v}{2}\right)dv = 2\int\frac{1}{x}\,dx + C \implies \ln|v| - \tfrac{1}{4}v^2 = 2\ln|x| + C.$$

Replacing v by the original variables gives

$$\ln\left|\frac{y}{x}\right| - \tfrac{1}{4}(xy)^2 = C.$$

Example 2.12

Solve $\quad \dfrac{dy}{dx} = \dfrac{\sqrt{x+y}-\sqrt{x-y}}{\sqrt{x+y}+\sqrt{x-y}}, \quad x>0,\ x\geqslant |y|.$

The differential equation is a homogeneous equation. However, it can be solved more easily using a special transformation. The equation can be written as

$$\frac{dy}{dx} = \frac{(\sqrt{x+y}-\sqrt{x-y})^2}{(\sqrt{x+y}+\sqrt{x-y})(\sqrt{x+y}-\sqrt{x-y})} = \frac{(x+y)-2\sqrt{x^2-y^2}+(x-y)}{(x+y)-(x-y)}$$

$$= \frac{x-\sqrt{x^2-y^2}}{y}.$$

The "special" term is x^2-y^2. In order to remove the square root, let $x^2-y^2=v^2$

$$x^2-y^2=v^2 \implies 2x-2y\frac{dy}{dx} = 2v\frac{dv}{dx} \implies y\frac{dy}{dx} = x - v\frac{dv}{dx}.$$

The differential equation becomes

$$x - v\frac{\mathrm{d}v}{\mathrm{d}x} = x - v \implies v\left(\frac{\mathrm{d}v}{\mathrm{d}x} - 1\right) = 0.$$

Case 1. $v \neq 0 \implies \dfrac{\mathrm{d}v}{\mathrm{d}x} - 1 = 0 \implies \mathrm{d}v = \mathrm{d}x$. Integrating both sides yields

$$v = x + C \implies v^2 = (x+C)^2.$$

Replacing v by the original variables results in the general solution

$$x^2 - y^2 = (x+C)^2 \implies y^2 + 2Cx + C^2 = 0.$$

Case 2. $v = 0 \implies x^2 - y^2 = 0 \implies y = \pm x$. This solution cannot be obtained from the general solution for any value of the constant C and is therefore a singular solution.

Combining Cases 1 and 2, the solutions of the differential equation are

$$y^2 + 2Cx + C^2 = 0, \quad \text{✍ General solution}$$

$$y = \pm x. \quad \text{✍ Singular solution}$$

2.3 Exact Differential Equations and Integrating Factors

Consider differential equations of the form

$$M(x, y)\,\mathrm{d}x + N(x, y)\,\mathrm{d}y = 0, \qquad \text{or} \qquad \frac{\mathrm{d}y}{\mathrm{d}x} = -\frac{M(x, y)}{N(x, y)}, \quad N(x, y) \neq 0, \qquad (1)$$

where $\dfrac{\partial M}{\partial y}$ and $\dfrac{\partial N}{\partial x}$ are continuous. Suppose the solution of equation (1) is $u(x, y) = C$, $C = \text{constant}$. Taking the differential yields

$$\mathrm{d}u = \frac{\partial u}{\partial x}\mathrm{d}x + \frac{\partial u}{\partial y}\mathrm{d}y = \mathrm{d}C = 0 \implies \frac{\partial u}{\partial x}\mathrm{d}x + \frac{\partial u}{\partial y}\mathrm{d}y = 0. \qquad (2)$$

Equation (2) should be the same as equation (1) if $u(x, y) = C$ is the solution of (1), except for a common factor $\mu(x, y)$, i.e., the coefficients of $\mathrm{d}x$ and $\mathrm{d}y$ in equations (1) and (2) are proportional

$$\frac{\dfrac{\partial u}{\partial x}}{M(x, y)} = \frac{\dfrac{\partial u}{\partial y}}{N(x, y)} = \mu(x, y) \implies \frac{\partial u}{\partial x} = \mu M, \qquad \frac{\partial u}{\partial y} = \mu N.$$

Substituting into equation (2) gives

$$\mu M\,dx + \mu N\,dy = 0. \tag{2'}$$

Since the left-hand side is an exact differential of some function $u(x, y)$,

$$\therefore \quad du(x, y) = 0 \implies u(x, y) = C.$$

Hence, if one could find a function $\mu(x, y)$, called an *integrating factor* (IF) multiplying it to equation (1) yields an *exact differential equation* (2′), which means that the left-hand side is the *exact differential* of some function. The resulting differential equation can then be easily solved.

Motivating Example

Solve $\quad (y + 2xy^2)\,dx + (2x + 3x^2y)\,dy = 0.$

It happens that an integrating factor is y. Multiplying the differential equation by y results in

$$(y^2 + 2xy^3)\,dx + (2xy + 3x^2y^2)\,dy = 0.$$

The left-hand side is the exact differential of $u(x, y) = xy^2 + x^2y^3$. Hence,

$$d(xy^2 + x^2y^3) = 0 \implies xy^2 + x^2y^3 = C. \qquad \text{✍ General solution}$$

2.3.1 Exact Differential Equations

If the differential equation

$$M(x, y)\,dx + N(x, y)\,dy = 0 \tag{1}$$

is exact, then there exists a function $u(x, y)$ such that

$$du = M(x, y)\,dx + N(x, y)\,dy. \tag{2}$$

But, by definition of differential,

$$du = \frac{\partial u}{\partial x}\,dx + \frac{\partial u}{\partial y}\,dy. \tag{3}$$

Comparing equations (2) and (3) leads to

$$M = \frac{\partial u}{\partial x}, \quad N = \frac{\partial u}{\partial y} \implies \frac{\partial M}{\partial y} = \frac{\partial^2 u}{\partial y \partial x}, \quad \frac{\partial N}{\partial x} = \frac{\partial^2 u}{\partial x \partial y}.$$

If $\dfrac{\partial M}{\partial y}$ and $\dfrac{\partial N}{\partial x}$ are continuous, one has

$$\frac{\partial^2 u}{\partial y \partial x} = \frac{\partial^2 u}{\partial x \partial y}. \tag{4}$$

Hence, a necessary condition for exactness is, from equations (4),

$$\frac{\partial M}{\partial y} = \frac{\partial N}{\partial x}.$$

It can be shown that this condition is also sufficient.

Exact Differential Equations

Consider the differential equation

$$M(x, y)\,\mathrm{d}x + N(x, y)\,\mathrm{d}y = 0.$$

If

$$\frac{\partial M}{\partial y} = \frac{\partial N}{\partial x}, \qquad \text{Exactness condition}$$

then the differential equation is *exact*, meaning that the left-hand side is the *exact differential* of some function.

Example 2.13

Solve $\quad (6xy^2 + 4x^3y)\,\mathrm{d}x + (6x^2y + x^4 + \mathrm{e}^y)\,\mathrm{d}y = 0.$

The differential equation is of the form

$$M(x, y)\,\mathrm{d}x + N(x, y)\,\mathrm{d}y = 0,$$

where

$$M(x, y) = 6xy^2 + 4x^3y, \qquad N(x, y) = 6x^2y + x^4 + \mathrm{e}^y.$$

Test for exactness:

$$\frac{\partial M}{\partial y} = 12xy + 4x^3, \qquad \frac{\partial N}{\partial x} = 12xy + 4x^3,$$

$$\therefore \quad \frac{\partial M}{\partial y} = \frac{\partial N}{\partial x} \quad \Longrightarrow \quad \text{The differential equation is exact.}$$

Two methods are introduced in the following to find the general solution.

Method 1: Since the differential equation is exact, there exists a function $u(x, y)$ such that

$$\mathrm{d}u = \frac{\partial u}{\partial x}\mathrm{d}x + \frac{\partial u}{\partial y}\mathrm{d}y = (6xy^2 + 4x^3y)\,\mathrm{d}x + (6x^2y + x^4 + \mathrm{e}^y)\,\mathrm{d}y,$$

i.e.,

$$\frac{\partial u}{\partial x} = 6xy^2 + 4x^3y, \tag{1}$$

$$\frac{\partial u}{\partial y} = 6x^2y + x^4 + \mathrm{e}^y. \tag{2}$$

To determine $u(x, y)$, integrate equation (1) with respect to x

$$u(x, y) = \int (6xy^2 + 4x^3y)\,dx + f(y)$$

✍ When integrating w.r.t. x, y is treated as constant or fixed.

$$= 3x^2y^2 + x^4y + f(y). \tag{3}$$

Differentiating equation (3) with respect to y and comparing with equation (2) yield

$$\frac{\partial u}{\partial y} = 6x^2y + x^4 + \frac{df(y)}{dy}$$

✍ When differentiating w.r.t. y, x is treated as constant or fixed.

$$= 6x^2y + x^4 + e^y;$$

✍ Equation (2)

hence,

$$\frac{df(y)}{dy} = e^y \implies f(y) = e^y.$$

Substituting into equation (3) leads to

$$u(x, y) = 3x^2y^2 + x^4y + e^y.$$

The general solution is then given by

$$u(x, y) = C \implies 3x^2y^2 + x^4y + e^y = C.$$

Method 2: The Method of Grouping Terms

The essence of Method 1 is to determine function $u(x, y)$ by

- integrating the coefficient of dx with respect to x,
- differentiating the result with respect to y and comparing with the coefficient of dy.

This procedure can be recast to result in the *method of grouping terms*, which is noticeably more succinct and is the preferred method. The method is illustrated step-by-step as follows:

1. Pick up a term, for example $6xy^2\,dx$.
   - Since the term has dx, integrate the coefficient $6xy^2$ with respect to x to yield $3x^2y^2$.
   - Differentiate the result with respect to y to yield the coefficient of dy term, i.e., $6x^2y$.
   - The two terms $6xy^2\,dx + 6x^2y\,dy$ are grouped together.

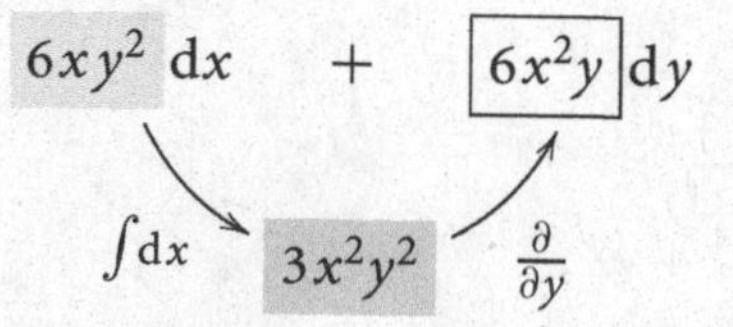

✍ $\int dx$ stands for integrating w.r.t. x,

✍ $\frac{\partial}{\partial y}$ denotes differentiating w.r.t. y.

2. Pick up one of the remaining terms, for example $4x^3y\,dx$.

  - Similarly, since the term has dx, integrate the coefficient $4x^3y$ with respect to x to yield x^4y.
  - Differentiate the result with respect to y to yield the coefficient of dy term, i.e., x^4.
  - The two terms $4x^3y\,dx + x^4\,dy$ are grouped together.

$$4x^3y\ dx\ \ +\ \ \boxed{x^4}\,dy$$
$$\underset{\int dx}{\searrow}\ \underset{x^4y}{}\ \underset{\frac{\partial}{\partial y}}{\nearrow}$$

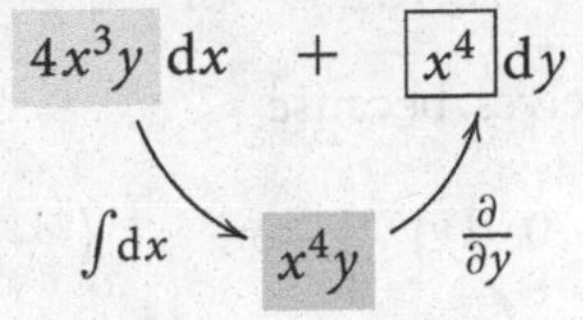

3. Pick up one of the remaining terms. Since there is only one term left, $e^y\,dy$ is picked.

  - Since the term has dy, integrate the coefficient e^y with respect to y to yield e^y.
  - Differentiate the result with respect to x to yield the coefficient of dx term, i.e., 0.
  - The term $e^y\,dy$ is in a group by itself.

$$e^y\ dy\ \ +\ \ \boxed{0}\cdot dx$$
$$\underset{\int dy}{\searrow}\ \underset{e^y}{}\ \underset{\frac{\partial}{\partial x}}{\nearrow}$$

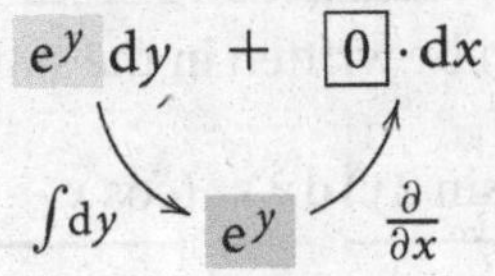

4. All the terms on the left-hand side of the equation have now been grouped.

5. Steps 1 to 3 can be combined to give a single expression as follows

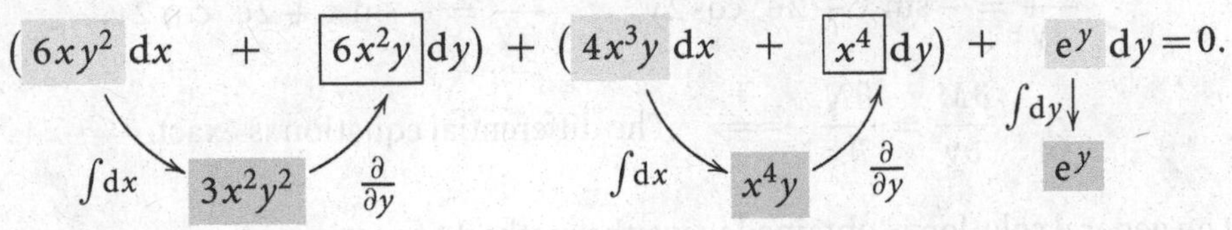

$$\big(\,6xy^2\ dx\ \ +\ \ \boxed{6x^2y}\,dy\big)\ +\ \big(\,4x^3y\ dx\ \ +\ \ \boxed{x^4}\,dy\big)\ +\ \ e^y\ dy = 0.$$
$$\underset{\int dx}{\searrow}\ \underset{3x^2y^2}{}\ \underset{\frac{\partial}{\partial y}}{\nearrow}\qquad \underset{\int dx}{\searrow}\ \underset{x^4y}{}\ \underset{\frac{\partial}{\partial y}}{\nearrow}\qquad \underset{\int dy\downarrow}{}\ \underset{e^y}{}$$

6. Hence
$$d(3x^2y^2 + x^4y + e^y) = 0,$$

which gives the general solution
$$3x^2y^2 + x^4y + e^y = C.$$

Remarks:

  - The method of grouping terms is easier to apply. The sum of the functions in the second row is the required function $u(x, y)$, and the general solution can be readily obtained as $u(x, y) = C$. Hence, the method of grouping terms is the preferred method.

- If a differential equation is exact, then all the terms on the left-hand side of the equation will be grouped. If there are terms left that cannot be grouped, there must be mistakes made in the calculation.
- Terms of the form

$$f(x)\,\mathrm{d}x \qquad \text{or} \qquad g(y)\,\mathrm{d}y$$

are in groups by themselves, because

$$\big(f(x)\,\mathrm{d}x \quad + \quad 0\cdot\mathrm{d}y\big) \qquad \text{or} \qquad \big(g(y)\,\mathrm{d}y \quad + \quad 0\cdot\mathrm{d}x\big).$$

$$\int\mathrm{d}x \searrow \int f(x)\,\mathrm{d}x \nearrow \frac{\partial}{\partial y} \qquad\qquad \int\mathrm{d}y \searrow \int g(y)\,\mathrm{d}y \nearrow \frac{\partial}{\partial x}$$

Example 2.14

Solve $\dfrac{\mathrm{d}y}{\mathrm{d}x} = \dfrac{y\sin x - \mathrm{e}^x \sin 2y}{\cos x + 2\mathrm{e}^x \cos 2y}$.

The differential equation can be written in the standard form $M\,\mathrm{d}x + N\,\mathrm{d}y = 0$:

$$\underbrace{(-y\sin x + \mathrm{e}^x \sin 2y)}_{M(x,\,y)}\mathrm{d}x + \underbrace{(\cos x + 2\mathrm{e}^x \cos 2y)}_{N(x,\,y)}\mathrm{d}y = 0.$$

Test for exactness:

$$\frac{\partial M}{\partial y} = -\sin x + 2\mathrm{e}^x \cos 2y, \qquad \frac{\partial N}{\partial x} = -\sin x + 2\mathrm{e}^x \cos 2y,$$

$$\therefore \quad \frac{\partial M}{\partial y} = \frac{\partial N}{\partial x} \quad \Longrightarrow \quad \text{The differential equation is exact.}$$

The general solution is obtained using the method of grouping terms:

$$\big(-y\sin x\,\mathrm{d}x \quad + \quad \boxed{\cos x}\,\mathrm{d}y\big) + \big(\mathrm{e}^x \sin 2y\,\mathrm{d}x \quad + \quad \boxed{2\mathrm{e}^x \cos 2y}\,\mathrm{d}y\big) = 0.$$

$$\int\mathrm{d}x \searrow y\cos x \nearrow \frac{\partial}{\partial y} \qquad\qquad \int\mathrm{d}x \searrow \mathrm{e}^x \sin 2y \nearrow \frac{\partial}{\partial y}$$

Hence, by summing up the terms in the second row one obtains the function $u(x, y)$, and the general solution is given by

$$y\cos x + \mathrm{e}^x \sin 2y = C.$$

Example 2.15

Solve $\quad 2x(3x + y - y\mathrm{e}^{-x^2})\,\mathrm{d}x + (x^2 + 3y^2 + \mathrm{e}^{-x^2})\,\mathrm{d}y = 0.$

The differential equation is of the standard form $M\,\mathrm{d}x + N\,\mathrm{d}y = 0$, where

$$M(x, y) = 6x^2 + 2xy - 2xy\mathrm{e}^{-x^2}, \qquad N(x, y) = x^2 + 3y^2 + \mathrm{e}^{-x^2}.$$

Test for exactness:

$$\frac{\partial M}{\partial y} = 2x - 2x\mathrm{e}^{-x^2}, \qquad \frac{\partial N}{\partial x} = 2x - 2x\mathrm{e}^{-x^2},$$

$$\therefore \quad \frac{\partial M}{\partial y} = \frac{\partial N}{\partial x} \quad \Longrightarrow \quad \text{The differential equation is exact.}$$

The general solution is determined using the method of grouping terms:

$$(2xy\,\mathrm{d}x + x^2\,\mathrm{d}y) + (\mathrm{e}^{-x^2}\,\mathrm{d}y + -2xy\mathrm{e}^{-x^2}\,\mathrm{d}x)$$

$\int \mathrm{d}x$: x^2y, $\frac{\partial}{\partial y}$; $\int \mathrm{d}y$: $y\mathrm{e}^{-x^2}$, $\frac{\partial}{\partial x}$

$$+ \quad 6x^2\,\mathrm{d}x + 3y^2\,\mathrm{d}y = 0,$$

$\int \mathrm{d}x \downarrow$ $2x^3$; $\int \mathrm{d}y \downarrow$ y^3

which gives

$$x^2y + y\mathrm{e}^{-x^2} + 2x^3 + y^3 = C. \qquad \text{General solution}$$

Remarks: In the second group of terms above, it is easier to pick up the term $\mathrm{e}^{-x^2}\,\mathrm{d}y$ first, integrate its coefficient with respect to y, and then differentiate the result with respect to x to find the matching term.

Example 2.16

Solve $\quad \left(\frac{1}{y}\sin\frac{x}{y} - \frac{y}{x^2}\cos\frac{y}{x} + 1\right)\mathrm{d}x + \left(\frac{1}{x}\cos\frac{y}{x} - \frac{x}{y^2}\sin\frac{x}{y} + \frac{1}{y^2}\right)\mathrm{d}y = 0.$

The differential equation implies that $x \neq 0$ and $y \neq 0$. The equation is of the standard form $M\,\mathrm{d}x + N\,\mathrm{d}y = 0$, where

$$M(x, y) = \frac{1}{y}\sin\frac{x}{y} - \frac{y}{x^2}\cos\frac{y}{x} + 1, \quad N(x, y) = \frac{1}{x}\cos\frac{y}{x} - \frac{x}{y^2}\sin\frac{x}{y} + \frac{1}{y^2}.$$

Test for exactness:

$$\frac{\partial M}{\partial y} = -\frac{1}{y^2}\sin\frac{x}{y} + \frac{1}{y}\cos\frac{x}{y}\cdot\left(-\frac{x}{y^2}\right) - \frac{1}{x^2}\cos\frac{y}{x} - \frac{y}{x^2}\left(-\sin\frac{y}{x}\cdot\frac{1}{x}\right)$$

$$= -\frac{1}{y^2}\sin\frac{x}{y} - \frac{x}{y^3}\cos\frac{x}{y} - \frac{1}{x^2}\cos\frac{y}{x} + \frac{y}{x^3}\sin\frac{y}{x},$$

$$\frac{\partial N}{\partial x} = -\frac{1}{x^2}\cos\frac{y}{x} + \frac{1}{x}\left[-\sin\frac{y}{x}\cdot\left(-\frac{y}{x^2}\right)\right] - \frac{1}{y^2}\sin\frac{x}{y} - \frac{x}{y^2}\left(\cos\frac{x}{y}\cdot\frac{1}{y}\right)$$

$$= -\frac{1}{x^2}\cos\frac{y}{x} + \frac{y}{x^3}\sin\frac{y}{x} - \frac{1}{y^2}\sin\frac{x}{y} - \frac{x}{y^3}\cos\frac{x}{y},$$

$$\therefore \quad \frac{\partial M}{\partial y} = \frac{\partial N}{\partial x} \implies \text{The differential equation is exact.}$$

The general solution is determined using the method of grouping terms

$$\left(\frac{1}{y}\sin\frac{x}{y}\,dx \quad + \quad \boxed{-\frac{x}{y^2}\sin\frac{x}{y}}\,dy\right) + \quad 1\,dx + \quad \frac{1}{y^2}\,dy$$

$$\int dx \searrow \quad -\cos\frac{x}{y} \quad \nearrow \frac{\partial}{\partial y} \qquad \int dx\downarrow \qquad \int dy\downarrow$$

$$x \qquad -\frac{1}{y}$$

$$+ \left(\frac{1}{x}\cos\frac{y}{x}\,dy \quad + \quad \boxed{-\frac{y}{x^2}\cos\frac{y}{x}}\,dx\right) = 0,$$

$$\int dy \searrow \quad \sin\frac{y}{x} \quad \nearrow \frac{\partial}{\partial x}$$

which gives

$$-\cos\frac{x}{y} + x - \frac{1}{y} + \sin\frac{y}{x} = C. \qquad \text{✍ General solution}$$

Note that in the fourth group of terms above, the term $\left(\frac{1}{x}\cos\frac{y}{x}\,dy\right)$ is picked first, because it is easier to integrate the coefficient $\left(\frac{1}{x}\cos\frac{y}{x}\right)$ with respect to y.

Remarks: When applying the method of grouping terms, whether to pick a term $f(x, y)\,dx$ or $g(x, y)\,dy$ first depends on whether it is easier to integrate

$$\int f(x, y)\,dx \quad \text{or} \quad \int g(x, y)\,dy.$$

2.3.2 *Integrating Factors*

Consider the differential equation

$$M(x, y)\,\mathrm{d}x + N(x, y)\,\mathrm{d}y = 0. \tag{1}$$

- If $\dfrac{\partial M}{\partial y} = \dfrac{\partial N}{\partial x}$, the differential equation is exact.
- If $\dfrac{\partial M}{\partial y} \neq \dfrac{\partial N}{\partial x}$, the differential equation can be rendered exact by multiplying by a function $\mu(x, y)$, known as an *integrating factor* (IF), i.e.,

$$\mu(x, y)M(x, y)\,\mathrm{d}x + \mu(x, y)N(x, y)\,\mathrm{d}y = 0, \tag{2}$$

is exact.

To find an integrating factor $\mu(x, y)$, apply the exactness condition on equation (2)

$$\frac{\partial(\mu M)}{\partial y} = \frac{\partial(\mu N)}{\partial x},$$

i.e.,

$$M\frac{\partial \mu}{\partial y} + \mu\frac{\partial M}{\partial y} = N\frac{\partial \mu}{\partial x} + \mu\frac{\partial N}{\partial x} \implies \mu\Big(\frac{\partial M}{\partial y} - \frac{\partial N}{\partial x}\Big) = N\frac{\partial \mu}{\partial x} - M\frac{\partial \mu}{\partial y}. \tag{3}$$

This is a partial differential equation for the unknown function $\mu(x, y)$, which is usually more difficult to solve than the original ordinary differential equation (1). However, for some special cases, equation (3) can be solved for an integrating factor.

Special Cases:

If μ is a function of x only, i.e., $\mu = \mu(x)$, then

$$\frac{\partial \mu}{\partial x} = \frac{\mathrm{d}\mu}{\mathrm{d}x}, \qquad \frac{\partial \mu}{\partial y} = 0,$$

and equation (3) becomes

$$N\frac{\mathrm{d}\mu}{\mathrm{d}x} = \mu\Big(\frac{\partial M}{\partial y} - \frac{\partial N}{\partial x}\Big) \implies \frac{1}{\mu}\frac{\mathrm{d}\mu}{\mathrm{d}x} = \frac{1}{N}\Big(\frac{\partial M}{\partial y} - \frac{\partial N}{\partial x}\Big). \tag{4}$$

Since $\mu(x)$ is a function of x only, the left-hand side is a function of x only. Hence, if an integrating factor of the form $\mu = \mu(x)$ is to exist, the right-hand side must also be a function of x only. Equation (4) is variable separable, which can be solved easily by integration

$$\ln \mu = \int \frac{1}{N}\Big(\frac{\partial M}{\partial y} - \frac{\partial N}{\partial x}\Big)\mathrm{d}x \implies \mu(x) = \exp\Big[\int \frac{1}{N}\Big(\frac{\partial M}{\partial y} - \frac{\partial N}{\partial x}\Big)\mathrm{d}x\Big]. \tag{5}$$

Note that, since only *one* integrating factor is sought, there is no need to include a constant of integration C.

Interchanging M and N, and x and y in equation (5), one obtains an integrating factor for another special case

$$\mu(y) = \exp\left[\int \underbrace{\frac{1}{M}\left(\frac{\partial N}{\partial x} - \frac{\partial M}{\partial y}\right)}_{\text{function of } y \text{ only}} \mathrm{d}y\right]. \tag{6}$$

Integrating Factors

Consider the differential equation

$$M(x, y)\,\mathrm{d}x + N(x, y)\,\mathrm{d}y = 0.$$

- If $\dfrac{1}{N}\left(\dfrac{\partial M}{\partial y} - \dfrac{\partial N}{\partial x}\right)$ is a function of x only,

$$\mu(x) = \exp\left[\int \frac{1}{N}\left(\frac{\partial M}{\partial y} - \frac{\partial N}{\partial x}\right)\mathrm{d}x\right].$$

- If $\dfrac{1}{M}\left(\dfrac{\partial N}{\partial x} - \dfrac{\partial M}{\partial y}\right)$ is a function of y only,

$$\mu(y) = \exp\left[\int \frac{1}{M}\left(\frac{\partial N}{\partial x} - \frac{\partial M}{\partial y}\right)\mathrm{d}y\right].$$

Example 2.17

Solve $\quad 3(x^2 + y^2)\,\mathrm{d}x + x(x^2 + 3y^2 + 6y)\,\mathrm{d}y = 0.$

The differential equation is of the standard form $M\,\mathrm{d}x + N\,\mathrm{d}y = 0$, where

$$M(x, y) = 3(x^2 + y^2), \quad N(x, y) = x^3 + 3xy^2 + 6xy.$$

Test for exactness:

$$\frac{\partial M}{\partial y} = 6y, \qquad \frac{\partial N}{\partial x} = 3x^2 + 3y^2 + 6y,$$

$$\therefore \quad \frac{\partial M}{\partial y} \neq \frac{\partial N}{\partial x} \quad \Longrightarrow \quad \text{The differential equation is } \textit{not} \text{ exact.}$$

Since

$$\frac{1}{M}\left(\frac{\partial N}{\partial x} - \frac{\partial M}{\partial y}\right) = \frac{1}{3(x^2 + y^2)}\left[(3x^2 + 3y^2 + 6y) - 6y\right]$$

$$= 1, \qquad \text{✍ A function of } y \text{ only}$$

$$\therefore \quad \mu(y) = \exp\left[\int \frac{1}{M}\Big(\frac{\partial N}{\partial x} - \frac{\partial M}{\partial y}\Big)\mathrm{d}y\right] = \exp\Big[\int 1\cdot \mathrm{d}y\Big] = \mathrm{e}^y.$$

Multiplying the differential equation by the integrating factor $\mu(y) = \mathrm{e}^y$ yields

$$(3x^2\mathrm{e}^y + 3y^2\mathrm{e}^y)\,\mathrm{d}x + (x^3\mathrm{e}^y + 3xy^2\mathrm{e}^y + 6xy\mathrm{e}^y)\,\mathrm{d}y = 0.$$

The general solution is determined using the method of grouping terms

$$\big(3x^2\mathrm{e}^y\,\mathrm{d}x + \boxed{x^3\mathrm{e}^y}\,\mathrm{d}y\big) + \big[3y^2\mathrm{e}^y\,\mathrm{d}x + \boxed{(6xy\mathrm{e}^y + 3xy^2\mathrm{e}^y)}\,\mathrm{d}y\big] = 0,$$

$$\int \mathrm{d}x \searrow \; x^3\mathrm{e}^y \; \nearrow \frac{\partial}{\partial y} \qquad\qquad \int \mathrm{d}x \searrow \; 3xy^2\mathrm{e}^y \; \nearrow \frac{\partial}{\partial y}$$

which gives

$$x^3\mathrm{e}^y + 3xy^2\mathrm{e}^y = C \implies x\mathrm{e}^y(x^2 + 3y^2) = C. \quad \text{✍ General solution}$$

Example 2.18

Solve $\quad y(2x - y + 2)\,\mathrm{d}x + 2(x - y)\,\mathrm{d}y = 0.$

The differential equation is of the standard form $M\,\mathrm{d}x + N\,\mathrm{d}y = 0$, where

$$M(x, y) = 2xy - y^2 + 2y, \quad N(x, y) = 2(x - y).$$

Test for exactness:

$$\frac{\partial M}{\partial y} = 2x - 2y + 2, \quad \frac{\partial N}{\partial x} = 2 \implies \frac{\partial M}{\partial y} \neq \frac{\partial N}{\partial x} \implies \text{The DE is } \textit{not} \text{ exact.}$$

Since

$$\frac{1}{N}\Big(\frac{\partial M}{\partial y} - \frac{\partial N}{\partial x}\Big) = \frac{1}{2(x - y)}\big[(2x - 2y + 2) - 2\big] = 1, \quad \text{✍ A function of } x \text{ only}$$

$$\therefore \quad \mu(x) = \exp\left[\int \frac{1}{N}\Big(\frac{\partial M}{\partial y} - \frac{\partial N}{\partial x}\Big)\mathrm{d}x\right] = \exp\Big[\int 1\cdot \mathrm{d}x\Big] = \mathrm{e}^x.$$

Multiplying the differential equation by the integrating factor $\mu(x) = \mathrm{e}^x$ yields

$$(2xy\mathrm{e}^x - y^2\mathrm{e}^x + 2y\mathrm{e}^x)\,\mathrm{d}x + (2x\mathrm{e}^x - 2y\mathrm{e}^x)\,\mathrm{d}y = 0.$$

The general solution is determined using the method of grouping terms

$$\big[2x\mathrm{e}^x\,\mathrm{d}y + \boxed{(2y\mathrm{e}^x + 2xy\mathrm{e}^x)}\,\mathrm{d}x\big] + \big(-y^2\mathrm{e}^x\,\mathrm{d}x + \boxed{-2y\mathrm{e}^x}\,\mathrm{d}x\big] = 0,$$

$$\int \mathrm{d}y \searrow \; 2xy\mathrm{e}^x \; \nearrow \frac{\partial}{\partial x} \qquad\qquad \int \mathrm{d}x \searrow \; -y^2\mathrm{e}^x \; \nearrow \frac{\partial}{\partial y}$$

which gives

$$2xy\mathrm{e}^x - y^2\mathrm{e}^x = C \implies y\mathrm{e}^x(2x - y) = C. \quad \text{✍ General solution}$$

Example 2.19

Solve $\quad y\,(\cos^3 x + y\sin x)\,\mathrm{d}x + \cos x\,(\sin x\cos x + 2y)\,\mathrm{d}y = 0.$

The differential equation is of the standard form $M\,\mathrm{d}x + N\,\mathrm{d}y = 0$, where

$$M(x,y) = y\cos^3 x + y^2\sin x, \quad N(x,y) = \sin x\cos^2 x + 2y\cos x.$$

Test for exactness:

$$\frac{\partial M}{\partial y} = \cos^3 x + 2y\sin x, \qquad \frac{\partial N}{\partial x} = \cos^3 x - 2\sin^2 x\cos x - 2y\sin x,$$

$$\therefore \quad \frac{\partial M}{\partial y} \neq \frac{\partial N}{\partial x} \implies \text{The differential equation is } \textit{not} \text{ exact.}$$

Since

$$\frac{1}{N}\left(\frac{\partial M}{\partial y} - \frac{\partial N}{\partial x}\right) = \frac{(\cos^3 x + 2y\sin x) - (\cos^3 x - 2\sin^2 x\cos x - 2y\sin x)}{\sin x\cos^2 x + 2y\cos x}$$

$$= \frac{2\sin x\,(2y + \sin x\cos x)}{\cos x\,(2y + \sin x\cos x)} = \frac{2\sin x}{\cos x}, \quad \text{✍ A function of } x \text{ only}$$

$$\therefore \quad \mu(x) = \exp\left[\int \frac{1}{N}\left(\frac{\partial M}{\partial y} - \frac{\partial N}{\partial x}\right)\mathrm{d}x\right] = \exp\left[\int \frac{2\sin x}{\cos x}\mathrm{d}x\right]$$

$$= \exp\left[-2\int \frac{1}{\cos x}\mathrm{d}(\cos x)\right] = \exp\left[-2\ln|\cos x|\right] = \frac{1}{\cos^2 x}.$$

Multiplying the differential equation by the integrating factor $\mu(x) = \dfrac{1}{\cos^2 x}$ yields

$$\left(y\cos x + \frac{y^2\sin x}{\cos^2 x}\right)\mathrm{d}x + \left(\sin x + \frac{2y}{\cos x}\right)\mathrm{d}y = 0.$$

The general solution is determined using the method of grouping terms

$$\left(y\cos x\,\mathrm{d}x + \boxed{\sin x}\,\mathrm{d}y\right) + \left(\frac{2y}{\cos x}\,\mathrm{d}y + \boxed{\frac{y^2\sin x}{\cos^2 x}}\,\mathrm{d}x\right) = 0,$$

$$y\cos x \xrightarrow{\int \mathrm{d}x} y\sin x \xrightarrow{\frac{\partial}{\partial y}} \sin x, \qquad \frac{2y}{\cos x} \xrightarrow{\int \mathrm{d}y} \frac{y^2}{\cos x} \xrightarrow{\frac{\partial}{\partial x}} \frac{y^2\sin x}{\cos^2 x}$$

which gives

$$y\sin x + \frac{y^2}{\cos x} = C. \qquad \text{✍ General solution}$$

Example 2.20

1. Show that if the equation $M(x, y)\,dx + N(x, y)\,dy = 0$ is such that

$$\frac{x^2}{xM + yN}\left(\frac{\partial N}{\partial x} - \frac{\partial M}{\partial y}\right) = F\left(\frac{y}{x}\right)$$

then an integrating factor is given by

$$\mu(x, y) = \exp\left\{\int F(u)\,du\right\}, \qquad u = \frac{y}{x}.$$

2. Using the result of Part 1, solve the differential equation

$$(2x - y + 2xy - y^2)\,dx + (x + x^2 + xy)\,dy = 0.$$

1. If

$$\mu = \exp\left\{\int F(u)\,du\right\}, \quad u = \frac{y}{x},$$

is an integrating factor, then $\mu M\,dx + \mu N\,dy = 0$ is an exact differential equation, and the exactness condition must be satisfied

$$\frac{\partial(\mu M)}{\partial y} - \frac{\partial(\mu N)}{\partial x} = \left(\frac{\partial \mu}{\partial y}M + \mu\frac{\partial M}{\partial y}\right) - \left(\frac{\partial \mu}{\partial x}N + \mu\frac{\partial N}{\partial x}\right)$$

$$= \exp\left\{\int F(u)\,du\right\}F(u)\frac{1}{x}\cdot M - \exp\left\{\int F(u)\,du\right\}F(u)\left(-\frac{y}{x^2}\right)\cdot N$$

$$+ \mu\left(\frac{\partial M}{\partial y} - \frac{\partial N}{\partial x}\right)$$

$$= \mu F(u)\left(\frac{M}{x} + \frac{yN}{x^2}\right) - \mu\left(\frac{\partial N}{\partial x} - \frac{\partial M}{\partial y}\right)$$

$$= \mu\left[\frac{F(u)}{x^2}(xM + yN) - \left(\frac{\partial N}{\partial x} - \frac{\partial M}{\partial y}\right)\right] = 0.$$

Since $\mu \neq 0$, one has

$$\frac{F(u)}{x^2}(xM + yN) - \left(\frac{\partial N}{\partial x} - \frac{\partial M}{\partial y}\right) = 0,$$

i.e.,

$$\frac{x^2}{xM + yN}\left(\frac{\partial N}{\partial x} - \frac{\partial M}{\partial y}\right) = F(u), \quad u = \frac{y}{x}.$$

2. The differential equation is of the standard form $M\,dx + N\,dy = 0$, where

$$M(x, y) = 2x + 2xy - y - y^2, \quad N(x, y) = x^2 + x + xy.$$

It can be easily evaluated

$$\frac{\partial N}{\partial x} - \frac{\partial M}{\partial y} = (2x + 1 + y) - (2x - 1 - 2y) = 2 + 3y,$$

$$\begin{aligned} xM + yN &= (2x^2 + 2x^2y - xy - xy^2) + (x^2y + xy + xy^2) \\ &= 2x^2 + 3x^2y = x^2(2 + 3y), \end{aligned}$$

$$\frac{x^2}{xM + yN}\left(\frac{\partial N}{\partial x} - \frac{\partial M}{\partial y}\right) = \frac{x^2}{x^2(2 + 3y)}(2 + 3y) = 1 = F(u),$$

$$\mu(x, y) = \exp\left\{\int F(u)\,\mathrm{d}u\right\} = \exp\left\{\int 1\cdot\mathrm{d}u\right\} = \exp(u) = \exp\left(\frac{y}{x}\right).$$

Multiplying the differential equation by the integrating factor yields

$$(2x - y + 2xy - y^2)\exp\left(\frac{y}{x}\right)\mathrm{d}x + (x + x^2 + xy)\exp\left(\frac{y}{x}\right)\mathrm{d}y = 0.$$

Note the following integrals

$$\int x\exp\left(\frac{y}{x}\right)\mathrm{d}y = x^2\exp\left(\frac{y}{x}\right),$$

$$\int (x^2 + xy)\exp\left(\frac{y}{x}\right)\mathrm{d}y = x^3\exp\left(\frac{y}{x}\right) + x^2\int y\,\mathrm{d}\left[\exp\left(\frac{y}{x}\right)\right]$$

$$= x^3\exp\left(\frac{y}{x}\right) + x^2\left[y\exp\left(\frac{y}{x}\right) - \int \exp\left(\frac{y}{x}\right)\mathrm{d}y\right]$$ ✍ Integration by parts

$$= x^3\exp\left(\frac{y}{x}\right) + x^2\left[y\exp\left(\frac{y}{x}\right) - x\exp\left(\frac{y}{x}\right)\right] = x^2y\exp\left(\frac{y}{x}\right).$$

The general solution is determined using the method of grouping terms

$$\left[x\exp\left(\frac{y}{x}\right)\mathrm{d}y + (2x - y)\exp\left(\frac{y}{x}\right)\mathrm{d}x\right]$$

$\int \mathrm{d}y$ → $x^2\exp\left(\frac{y}{x}\right)$ → $\frac{\partial}{\partial x}$

$$+\left[(x^2 + xy)\exp\left(\frac{y}{x}\right)\mathrm{d}y + (2xy - y^2)\exp\left(\frac{y}{x}\right)\mathrm{d}x\right] = 0,$$

$\int \mathrm{d}y$ → $x^2y\exp\left(\frac{y}{x}\right)$ → $\frac{\partial}{\partial x}$

which gives

$$x^2(1 + y)\exp\left(\frac{y}{x}\right) = C.$$ ✍ General solution

2.3.3 *Method of Inspection*

Useful Formulas

1. $$\mathrm{d}(xy) = y\,\mathrm{d}x + x\,\mathrm{d}y,$$

2. $$\mathrm{d}\Big(\frac{y}{x}\Big) = \frac{-y\,\mathrm{d}x + x\,\mathrm{d}y}{x^2}, \qquad \mathrm{d}\Big(\frac{x}{y}\Big) = \frac{y\,\mathrm{d}x - x\,\mathrm{d}y}{y^2},$$

3. $$\mathrm{d}\Big(\tan^{-1}\frac{y}{x}\Big) = \frac{-y\,\mathrm{d}x + x\,\mathrm{d}y}{x^2+y^2}, \qquad \mathrm{d}\Big(\tan^{-1}\frac{x}{y}\Big) = \frac{y\,\mathrm{d}x - x\,\mathrm{d}y}{x^2+y^2},$$

4. $$\mathrm{d}\Big[\frac{1}{2}\ln(x^2+y^2)\Big] = \frac{x\,\mathrm{d}x + y\,\mathrm{d}y}{x^2+y^2},$$

5. $$\mathrm{d}\big(\sqrt{x^2 \pm y^2}\big) = \frac{x\,\mathrm{d}x \pm y\,\mathrm{d}y}{\sqrt{x^2 \pm y^2}}.$$

By rearrangement of terms, multiplication or division of suitable functions, an integrating factor may be determined using these formulas.

Example 2.21

Solve $(3x^4 + y)\,\mathrm{d}x + (2x^2y - x)\,\mathrm{d}y = 0.$

The differential equation can be rearranged as

$$x^2(3x^2\,\mathrm{d}x + 2y\,\mathrm{d}y) + (y\,\mathrm{d}x - x\,\mathrm{d}y) = 0.$$

Dividing the equation by x^2 yields

$$3x^2\,\mathrm{d}x + 2y\,\mathrm{d}y - \frac{-y\,\mathrm{d}x + x\,\mathrm{d}y}{x^2} = 0.$$

Hence

$$\mathrm{d}(x^3) + \mathrm{d}(y^2) - \mathrm{d}\Big(\frac{y}{x}\Big) = 0,$$

and the general solution is given by

$$x^3 + y^2 - \frac{y}{x} = C.$$

Remarks: Using the normal procedure as introduced in Section 2.3.2, it is easy to determine that $\mu(x) = 1/x^2$ is an integrating factor.

Example 2.22

Solve $(x + x^2y + y^3)\,dx + (y + x^3 + xy^2 - x^2y^2 - y^4)\,dy = 0.$

The differential equation can be rearranged as

$$\left[x + y(x^2+y^2)\right]dx + \left[y + x(x^2+y^2) - y^2(x^2+y^2)\right]dy = 0.$$

Dividing the equation by (x^2+y^2) yields

$$\left(\frac{x}{x^2+y^2} + y\right)dx + \left(\frac{y}{x^2+y^2} + x - y^2\right)dy = 0.$$

Rearranging the equation leads to

$$\frac{x\,dx + y\,dy}{x^2+y^2} + (y\,dx + x\,dy) - y^2\,dy = 0.$$

Hence

$$d\left[\tfrac{1}{2}\ln(x^2+y^2)\right] + d(xy) - \tfrac{1}{3}d(y^3) = 0,$$

and the general solution is given by

$$\tfrac{1}{2}\ln(x^2+y^2) + xy - \tfrac{1}{3}y^3 = C.$$

Example 2.23

Solve $(2x\sqrt{x} + x^2 + y^2)\,dx + 2y\sqrt{x}\,dy = 0.$

Rearrange the differential equation as

$$2\sqrt{x}\,(x\,dx + y\,dy) + (x^2+y^2)\,dx = 0.$$

Dividing the equation by $\sqrt{x}\,(x^2+y^2)$ leads to

$$2\,\frac{x\,dx + y\,dy}{x^2+y^2} + \frac{1}{\sqrt{x}}\,dx = 0,$$

which gives

$$2\,d\left[\tfrac{1}{2}\ln(x^2+y^2)\right] + d(2\sqrt{x}) = 0.$$

Hence, the general solution is

$$\ln(x^2+y^2) + 2\sqrt{x} = C.$$

Example 2.24

Solve $y^2\,dx + (xy + y^2 - 1)\,dy = 0.$

The differential equation can be rearranged as

$$y(y\,dx + x\,dy) + (y^2 - 1)\,dy = 0.$$

It is easy to see that $y=0$ is a solution of the differential equation. For $y \neq 0$, dividing the equation by y leads to

$$y\,dx + x\,dy + \left(y - \frac{1}{y}\right)dy = 0,$$

which gives

$$d(xy) + d(\tfrac{1}{2}y^2) - d(\ln|y|) = 0.$$

Hence, the solutions are given by

$$xy + \tfrac{1}{2}y^2 - \ln|y| = C, \quad y=0.$$

Remarks: Using the normal procedure as introduced in Section 2.3.2, it is easy to determine that $\mu(y) = 1/y$ is an integrating factor.

Example 2.25

Solve $\quad xy\,dx - (x^2 + x^2y + y^3)\,dy = 0.$

Rearrange the differential equation as

$$x(y\,dx - x\,dy) - y(x^2 + y^2)\,dy = 0.$$

Divide the equation by $x^2 + y^2$

$$x\,\frac{y\,dx - x\,dy}{x^2+y^2} - y\,dy = 0 \implies x\,d\left(\tan^{-1}\frac{x}{y}\right) - y\,dy = 0.$$

It is easy to see that $y=0$ is a solution of the differential equation. For $y \neq 0$, dividing the equation by y yields

$$\frac{x}{y}\,d\left(\tan^{-1}\frac{x}{y}\right) - dy = 0.$$

Since

$$\int u\,d(\tan^{-1}u) = \int u \cdot \frac{1}{1+u^2}\,du = \frac{1}{2}\int \frac{1}{1+u^2}\,d(1+u^2) = \frac{1}{2}\ln(1+u^2),$$

one has

$$u\,d(\tan^{-1}u) = \tfrac{1}{2}\,d[\ln(1+u^2)] \implies \frac{x}{y}\,d\left(\tan^{-1}\frac{x}{y}\right) = \tfrac{1}{2}\,d\left[\ln\left(1+\frac{x^2}{y^2}\right)\right].$$

Hence

$$\tfrac{1}{2}\,d\left[\ln\left(1+\frac{x^2}{y^2}\right)\right] - dy = 0 \implies \ln\left(1+\frac{x^2}{y^2}\right) - 2y = C.$$ ✍ General solution

2.3.4 Integrating Factors by Groups

Theorem

If $\mu(x, y)$ is an integrating factor of the differential equation

$$M(x, y)\,dx + N(x, y)\,dy = 0, \tag{1}$$

which implies that

$$\mu(x, y)M(x, y)\,dx + \mu(x, y)N(x, y)\,dy = 0$$

is an exact differential equation, i.e.,

$$\mu(x, y)M(x, y)\,dx + \mu(x, y)N(x, y)\,dy = dv(x, y),$$

then $\mu(x, y) \cdot g\big[v(x, y)\big]$ is also an integrating factor of equation (1), where $g(\cdot)$ is any differentiable nonzero function.

Using this theorem, the following method of integrating factors by groups can be derived.

Method of Integrating Factors by Groups

Suppose equation (1) can be separated into two groups

$$\underbrace{\big[M_1(x, y)\,dx + N_1(x, y)\,dy\big]}_{\text{First group}} + \underbrace{\big[M_2(x, y)\,dx + N_2(x, y)\,dy\big]}_{\text{Second group}} = 0. \tag{2}$$

If the first and second groups have integrating factors $\mu_1(x, y)$ and $\mu_2(x, y)$, respectively, such that

$$\mu_1 M_1\,dx + \mu_1 N_1\,dy = dv_1,$$

$$\mu_2 M_2\,dx + \mu_2 N_2\,dy = dv_2,$$

then from the theorem, for any differentiable functions g_1 and g_2,

- $\mu_1(x, y) \cdot g_1\big[v_1(x, y)\big]$ is the integrating factor of the first group, and
- $\mu_2(x, y) \cdot g_2\big[v_2(x, y)\big]$ is the integrating factor of the second group.

If one can choose g_1 and g_2 suitably such that

$$\mu_1(x, y) \cdot g_1\big[v_1(x, y)\big] = \mu_2(x, y) \cdot g_2\big[v_2(x, y)\big] \implies \mu(x, y),$$

then $\mu(x, y)$ is an integrating factor of equation (1).

Example 2.26

Solve $(4xy + 3y^4)\,dx + (2x^2 + 5xy^3)\,dy = 0$.

The differential equation is of the standard form $M\,dx + N\,dy = 0$, where

$$M(x, y) = 4xy + 3y^4, \quad N(x, y) = 2x^2 + 5xy^3.$$

Test for exactness:

$$\frac{\partial M}{\partial y} = 4x + 12y^3, \qquad \frac{\partial N}{\partial x} = 4x + 15y^3.$$

It can be seen that there does not exist an integrating factor that is a function of x only or y only.

Separate the differential equation into two groups as

$$\underbrace{(4xy\,dx + 2x^2\,dy)}_{\text{First group}} + \underbrace{(3y^4\,dx + 5xy^3\,dy)}_{\text{Second group}} = 0.$$

❧ For the first group: $M_1 = 4xy$, $N_1 = 2x^2$,

$$\frac{\partial M_1}{\partial y} = 4x, \qquad \frac{\partial N_1}{\partial x} = 4x;$$

hence, the first group is exact or an integrating factor is $\mu_1 = 1$.

It is easy to find that

$$\boxed{4xy\,dx} \quad + \quad \boxed{2x^2}\,dy \implies 2d(x^2 y) \implies v_1(x, y) = x^2 y.$$

$$\int dx \qquad 2x^2 y \qquad \frac{\partial}{\partial y}$$

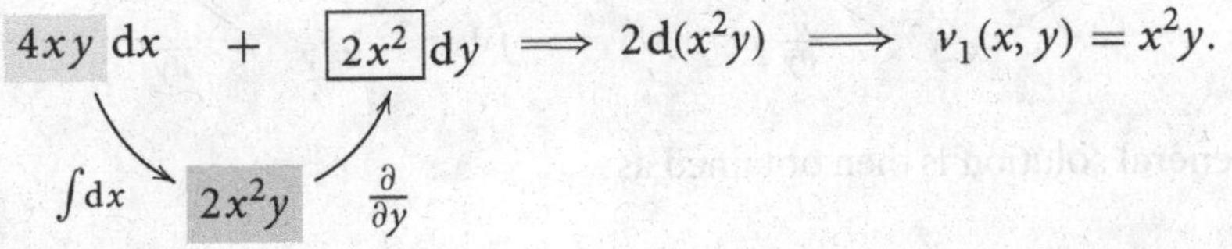

❧ For the second group: $M_2 = 3y^4$, $N_2 = 5xy^3$,

$$\frac{\partial M_2}{\partial y} = 12y^3, \qquad \frac{\partial N_2}{\partial x} = 5y^3,$$

$$\frac{1}{N_2}\left(\frac{\partial M_2}{\partial y} - \frac{\partial N_2}{\partial x}\right) = \frac{1}{5xy^3}(12y^3 - 5y^3) = \frac{7}{5x}, \quad \text{✍ A function of } x \text{ only}$$

$$\mu_2(x) = \exp\left[\int \frac{1}{N_2}\left(\frac{\partial M_2}{\partial y} - \frac{\partial N_2}{\partial x}\right)dx\right] = \exp\left(\frac{7}{5}\int \frac{1}{x}\,dx\right)$$

$$= \exp\left(\frac{7}{5}\ln|x|\right) = x^{\frac{7}{5}}.$$

Multiplying $\mu_2(x)$ to the second group yields

$$3x^{\frac{7}{5}}y^4\,\mathrm{d}x \quad + \quad \boxed{5x^{\frac{12}{5}}y^3}\,\mathrm{d}y \implies \tfrac{5}{4}\mathrm{d}\big(x^{\frac{12}{5}}y^4\big) \implies v_2(x,y) = x^{\frac{12}{5}}y^4.$$

$$\int \mathrm{d}x \searrow \quad \frac{5}{4}x^{\frac{12}{5}}y^4 \quad \nearrow \frac{\partial}{\partial y}$$

To find an integrating factor for the original differential equation, one needs to find functions g_1 and g_2 such that

$$\mu_1 \cdot g_1(v_1) = \mu_2 \cdot g_2(v_2) \implies 1 \cdot g_1(x^2y) = x^{\frac{7}{5}} \cdot g_2(x^{\frac{12}{5}}y^4).$$

Letting $g_1(v_1) = v_1^{\alpha}$ and $g_2(v_2) = v_2^{\beta}$ leads to

$$(x^2y)^{\alpha} = x^{\frac{7}{5}}\,(x^{\frac{12}{5}}y^4)^{\beta} \implies x^{2\alpha}y^{\alpha} = x^{\frac{12\beta+7}{5}}y^{4\beta},$$

$$\therefore \quad \left.\begin{aligned} 2\alpha &= \frac{12\beta+7}{5} \\ \alpha &= 4\beta \end{aligned}\right\} \implies \left\{\begin{aligned} \alpha &= 1, \\ \beta &= \tfrac{1}{4}. \end{aligned}\right.$$

Hence, an integrating factor is given by

$$\mu(x,y) = \mu_1 \cdot g_1(v_1) = 1 \cdot (x^2y)^1 = x^2y.$$

Multiplying $\mu = x^2y$ to the original differential equation leads to

$$\big(4x^3y^2\,\mathrm{d}x \quad + \quad \boxed{2x^4y}\,\mathrm{d}y\big) + \big(3x^2y^5\,\mathrm{d}x \quad + \quad \boxed{5x^3y^4}\,\mathrm{d}y\big) = 0.$$

$$\int \mathrm{d}x \searrow \quad x^4y^2 \quad \nearrow \frac{\partial}{\partial y} \qquad\qquad \int \mathrm{d}x \searrow \quad x^3y^5 \quad \nearrow \frac{\partial}{\partial y}$$

The general solution is then obtained as

$$x^4y^2 + x^3y^5 = C.$$

Example 2.27

Solve $\quad (5xy - 3y^3)\,\mathrm{d}x + (3x^2 - 7xy^2)\,\mathrm{d}y = 0.$

The differential equation is of the standard form $M\,\mathrm{d}x + N\,\mathrm{d}y = 0$, where

$$M(x,y) = 5xy - 3y^3, \quad N(x,y) = 3x^2 - 7xy^2.$$

Test for exactness:

$$\frac{\partial M}{\partial y} = 5x - 9y^2, \qquad \frac{\partial N}{\partial x} = 6x - 7y^2,$$

$$\frac{\partial M}{\partial y} - \frac{\partial N}{\partial x} = (5x - 9y^2) - (6x - 7y^2) = -x - 2y^2.$$

It can be seen that there does not exist an integrating factor that is a function of x only or y only.

Separate the differential equation into two groups as

$$\underbrace{(5xy\,\mathrm{d}x + 3x^2\,\mathrm{d}y)}_{\text{First group}} + \underbrace{(-3y^3\,\mathrm{d}x - 7xy^2\,\mathrm{d}y)}_{\text{Second group}} = 0.$$

- For the first group: $M_1 = 5xy,\ N_1 = 3x^2,$

$$\frac{\partial M_1}{\partial y} = 5x, \qquad \frac{\partial N_1}{\partial x} = 6x,$$

$$\frac{1}{N_1}\Big(\frac{\partial M_1}{\partial y} - \frac{\partial N_1}{\partial x}\Big) = \frac{1}{3x^2}(5x - 6x) = -\frac{1}{3x}, \quad \text{✍ A function of } x \text{ only}$$

$$\mu_1(x) = \exp\Big[\int \frac{1}{N_1}\Big(\frac{\partial M_1}{\partial y} - \frac{\partial N_1}{\partial x}\Big)\mathrm{d}x\Big] = \exp\Big(-\frac{1}{3}\int \frac{1}{x}\,\mathrm{d}x\Big)$$

$$= \exp\big(-\tfrac{1}{3}\ln|x|\big) = x^{-\frac{1}{3}}.$$

Multiplying $\mu_1(x)$ to the first group yields

$$\boxed{5x^{\frac{2}{3}}y}\,\mathrm{d}x + \boxed{3x^{\frac{5}{3}}}\,\mathrm{d}y \implies 3\,\mathrm{d}(x^{\frac{5}{3}}y) \implies v_1(x,y) = x^{\frac{5}{3}}y.$$

$\int \mathrm{d}x$ $3x^{\frac{5}{3}}y$ $\dfrac{\partial}{\partial y}$

- For the second group: $M_2 = -3y^3,\ N_2 = -7xy^2,$

$$\frac{\partial M_2}{\partial y} = -9y^2, \qquad \frac{\partial N_2}{\partial x} = -7y^2,$$

$$\frac{1}{N_2}\Big(\frac{\partial M_2}{\partial y} - \frac{\partial N_2}{\partial x}\Big) = \frac{1}{-7xy^2}(-9y^2 + 7y^2) = \frac{2}{7x}, \quad \text{✍ A function of } x \text{ only}$$

$$\mu_2(x) = \exp\Big[\int \frac{1}{N_2}\Big(\frac{\partial M_2}{\partial y} - \frac{\partial N_2}{\partial x}\Big)\mathrm{d}x\Big] = \exp\Big(\frac{2}{7}\int \frac{1}{x}\,\mathrm{d}x\Big)$$

$$= \exp\big(\tfrac{2}{7}\ln|x|\big) = x^{\frac{2}{7}}.$$

Multiplying $\mu_2(x)$ to the second group yields

$$-3x^{\frac{2}{7}}y^3\,\mathrm{d}x \quad + \quad \boxed{-7x^{\frac{9}{7}}y^2}\,\mathrm{d}y \Longrightarrow -\tfrac{7}{3}\,\mathrm{d}\big(x^{\frac{9}{7}}y^3\big) \Longrightarrow v_2(x,y)=x^{\frac{9}{7}}y^3.$$

$$\int \mathrm{d}x \searrow \quad -\tfrac{7}{3}\,x^{\frac{9}{7}}y^3 \quad \nearrow \frac{\partial}{\partial y}$$

To find an integrating factor for the original differential equation, one needs to find functions g_1 and g_2 such that

$$\mu_1\cdot g_1(v_1)=\mu_2\cdot g_2(v_2) \quad\Longrightarrow\quad x^{-\frac{1}{3}}\cdot g_1\big(x^{\frac{5}{3}}y\big)=x^{\frac{2}{7}}\cdot g_2\big(x^{\frac{9}{7}}y^3\big).$$

Letting $g_1(v_1)=v_1^{\alpha}$ and $g_2(v_2)=v_2^{\beta}$ leads to

$$x^{-\frac{1}{3}}\,\big(x^{\frac{5}{3}}y\big)^{\alpha}=x^{\frac{2}{7}}\,\big(x^{\frac{9}{7}}y^3\big)^{\beta} \quad\Longrightarrow\quad x^{\frac{5\alpha-1}{3}}\,y^{\alpha}=x^{\frac{9\beta+2}{7}}\,y^{3\beta},$$

$$\therefore\quad \left.\begin{aligned}\frac{5\alpha-1}{3}&=\frac{9\beta+2}{7}\\ \alpha&=3\beta\end{aligned}\right\} \quad\Longrightarrow\quad \left\{\begin{aligned}\alpha&=\tfrac{1}{2},\\ \beta&=\tfrac{1}{6}.\end{aligned}\right.$$

Hence, an integrating factor is given by

$$\mu(x,y)=\mu_1\cdot g_1(v_1)=x^{-\frac{1}{3}}\cdot\big(x^{\frac{5}{3}}y\big)^{\frac{1}{2}}=x^{\frac{1}{2}}y^{\frac{1}{2}}.$$

Multiplying $\mu=x^{\frac{1}{2}}y^{\frac{1}{2}}$ to the original differential equation leads to

$$\Big(5x^{\frac{3}{2}}y^{\frac{3}{2}}\,\mathrm{d}x \quad + \quad \boxed{3x^{\frac{5}{2}}y^{\frac{1}{2}}}\,\mathrm{d}y\Big) + \Big(-3x^{\frac{1}{2}}y^{\frac{7}{2}}\,\mathrm{d}x \quad + \quad \boxed{-7x^{\frac{3}{2}}y^{\frac{5}{2}}}\,\mathrm{d}y\Big)$$

$$\int \mathrm{d}x \searrow \quad 2x^{\frac{5}{2}}y^{\frac{3}{2}} \quad \nearrow \frac{\partial}{\partial y} \qquad\qquad \int \mathrm{d}x \searrow \quad -2x^{\frac{3}{2}}y^{\frac{7}{2}} \quad \nearrow \frac{\partial}{\partial y}$$

$$=0.$$

The general solution is then obtained as

$$2x^{\frac{5}{2}}y^{\frac{3}{2}}-2x^{\frac{3}{2}}y^{\frac{7}{2}}=C \quad\Longrightarrow\quad x^{\frac{3}{2}}y^{\frac{3}{2}}\,(x-y^2)=C.$$

Remarks: In the following two examples, the techniques used in the method of inspection can be combined with the method of integrating factors by groups to result in an efficient way of finding an integrating factor.

Example 2.28

Solve $\quad (y\,\mathrm{d}x-x\,\mathrm{d}y)+\sqrt{xy}\,\mathrm{d}y=0, \quad x>0, \quad y>0.$

Dividing the equation by x^2 leads to

$$\frac{y\,dx - x\,dy}{x^2} + x^{-\frac{3}{2}}y^{\frac{1}{2}}\,dy = 0, \tag{1}$$

$$\therefore \quad \underbrace{d\left(\frac{y}{x}\right)}_{\text{First group}} + \underbrace{x^{-\frac{3}{2}}y^{\frac{1}{2}}\,dy}_{\text{Second group}} = 0. \tag{1'}$$

Obviously, the first group has an integrating factor $\mu_1 = 1$ and the corresponding $v_1 = y/x$. The second group has an integrating factor $\mu_2 = x^{\frac{3}{2}}y^{-\frac{1}{2}}$ and the corresponding $v_2 = y$.

To find an integrating factor for the original differential equation, one needs to find functions g_1 and g_2 such that

$$\mu_1 \cdot g_1(v_1) = \mu_2 \cdot g_2(v_2) \implies 1 \cdot g_1\left(\frac{y}{x}\right) = x^{\frac{3}{2}}y^{-\frac{1}{2}} \cdot g_2(y).$$

Letting $g_1(v_1) = v_1^{\alpha}$ and $g_2(v_2) = v_2^{\beta}$ leads to

$$1 \cdot \left(\frac{y}{x}\right)^{\alpha} = x^{\frac{3}{2}}y^{-\frac{1}{2}} \cdot (y)^{\beta} \implies x^{-\alpha}y^{\alpha} = x^{\frac{3}{2}}y^{\beta-\frac{1}{2}},$$

$$\therefore \quad \left.\begin{aligned} -\alpha &= \frac{3}{2} \\ \alpha &= \beta - \frac{1}{2} \end{aligned}\right\} \implies \left\{\begin{aligned} \alpha &= \frac{3}{2}, \\ \beta &= -1. \end{aligned}\right.$$

Hence, an integrating factor is given by

$$\mu(x, y) = \mu_1 \cdot g_1(v_1) = x^{\frac{3}{2}}y^{-\frac{3}{2}}.$$

Multiplying $\mu = x^{\frac{3}{2}}y^{-\frac{3}{2}}$ to equation (1) leads to

$$\left(x^{-\frac{1}{2}}y^{-\frac{1}{2}}\,dx + -x^{\frac{1}{2}}y^{-\frac{3}{2}}\,dy\right) + y^{-1}\,dy = 0.$$

$$x^{-\frac{1}{2}}y^{-\frac{1}{2}} \xrightarrow{\int dx} 2x^{\frac{1}{2}}y^{-\frac{1}{2}} \xrightarrow{\frac{\partial}{\partial y}} -x^{\frac{1}{2}}y^{-\frac{3}{2}}, \qquad y^{-1} \xrightarrow{\int dy} \ln y$$

The general solution is then obtained as

$$2\sqrt{\frac{x}{y}} + \ln y = C.$$

Example 2.29

Solve $\quad (xye^{\frac{x}{y}} + y^4)\,dx - x^2 e^{\frac{x}{y}}\,dy = 0, \quad y \neq 0.$

Rearrange the equation as

$$x\mathrm{e}^{\frac{x}{y}}(y\,\mathrm{d}x - x\,\mathrm{d}y) + y^4\,\mathrm{d}x = 0.$$

Dividing the equation by xy^2 leads to

$$\mathrm{e}^{\frac{x}{y}}\,\frac{y\,\mathrm{d}x - x\,\mathrm{d}y}{y^2} + \frac{y^2}{x}\,\mathrm{d}x = 0 \implies \mathrm{e}^{\frac{x}{y}}\,\mathrm{d}\Big(\frac{x}{y}\Big) + \frac{y^2}{x}\,\mathrm{d}x = 0,$$

$$\therefore \quad \underbrace{\mathrm{d}(\mathrm{e}^{\frac{x}{y}})}_{\text{First group}} + \underbrace{\frac{y^2}{x}\,\mathrm{d}x}_{\text{Second group}} = 0. \tag{1}$$

Obviously, the first group has an integrating factor $\mu_1 = 1$ and the corresponding $v_1 = \mathrm{e}^{\frac{x}{y}}$. The second group has an integrating factor $\mu_2 = \dfrac{x}{y^2}$ and the corresponding $v_2 = x$.

To find an integrating factor for the original differential equation, one needs to find functions g_1 and g_2 such that

$$\mu_1 \cdot g_1(v_1) = \mu_2 \cdot g_2(v_2) \implies 1 \cdot g_1\big(\mathrm{e}^{\frac{x}{y}}\big) = \frac{x}{y^2} \cdot g_2(x).$$

To remove the exponential function on the left-hand side, one must take g_1 as a logarithmic function. Because of $\dfrac{1}{y^2}$ on the right-hand side, letting $g_1(v_1) = \big[\ln|v_1|\big]^2$ and $g_2(v_2) = v_2$ leads to

$$1 \cdot \Big(\frac{x}{y}\Big)^2 = \frac{x}{y^2} \cdot x \implies \mu = \frac{x^2}{y^2}.$$

Multiplying $\mu = \dfrac{x^2}{y^2}$ to equation (1) leads to

$$\Big(\frac{x}{y}\Big)^2 \mathrm{d}\big(\mathrm{e}^{\frac{x}{y}}\big) + x\,\mathrm{d}x = 0.$$

Since

$$\begin{aligned}
\int \Big(\frac{x}{y}\Big)^2 \mathrm{d}\big(\mathrm{e}^{\frac{x}{y}}\big) &= \int z^2\,\mathrm{d}(\mathrm{e}^z), \qquad z = \frac{x}{y} \\
&= z^2\mathrm{e}^z - \int \mathrm{e}^z \cdot 2z\,\mathrm{d}z \qquad \text{✍ Integrating by parts} \\
&= z^2\mathrm{e}^z - 2\int z\,\mathrm{d}(\mathrm{e}^z) = z^2\mathrm{e}^z - 2\Big(z\mathrm{e}^z - \int \mathrm{e}^z\,\mathrm{d}z\Big) \\
&= \mathrm{e}^z(z^2 - 2z + 2) = \mathrm{e}^{\frac{x}{y}}\Big[\Big(\frac{x}{y}\Big)^2 - 2\Big(\frac{x}{y}\Big) + 2\Big],
\end{aligned}$$

one has

$$\mathrm{d}\left\{\mathrm{e}^{\frac{x}{y}}\left[\left(\frac{x}{y}\right)^2-2\left(\frac{x}{y}\right)+2\right]\right\}+\mathrm{d}\left(\frac{x^2}{2}\right)=0.$$

The general solution is then obtained as

$$\mathrm{e}^{\frac{x}{y}}\left[\left(\frac{x}{y}\right)^2-2\left(\frac{x}{y}\right)+2\right]+\frac{x^2}{2}=C.$$

2.4 Linear First-Order Equations

2.4.1 Linear First-Order Equations

Linear first-order equations occur in many engineering applications and are of the form

$$\frac{\mathrm{d}y}{\mathrm{d}x}+P(x)\cdot y=Q(x). \tag{1}$$

The equation is first-order because the highest order of derivative present is first $\frac{\mathrm{d}y}{\mathrm{d}x}$, and it is linear because y and $\frac{\mathrm{d}y}{\mathrm{d}x}$ appear linearly.

Sometimes, the roles of x and y may be exchanged to result in a linear first-order equation of the form

$$\frac{\mathrm{d}x}{\mathrm{d}y}+P(y)\cdot x=Q(y),$$

in which x is treated as a function of y, and x and $\frac{\mathrm{d}x}{\mathrm{d}y}$ appear linearly.

Equation (1) can be written in the form $M\mathrm{d}x+N\mathrm{d}y=0$:

$$\big[P(x)y-Q(x)\big]\mathrm{d}x+\mathrm{d}y=0, \tag{2}$$

in which $M(x,y)=P(x)y-Q(x)$, $N(x,y)=1$. Test for exactness:

$$\frac{\partial M}{\partial y}=P(x),\qquad \frac{\partial N}{\partial x}=0 \implies \text{Differential equation (2) is } \textit{not} \text{ exact.}$$

However, since

$$\frac{1}{N}\left(\frac{\partial M}{\partial y}-\frac{\partial N}{\partial x}\right)=P(x)$$

is a function of x only, there exists an integrating factor that is a function of x only given by

$$\mu(x)=\exp\left[\int\frac{1}{N}\left(\frac{\partial M}{\partial y}-\frac{\partial N}{\partial x}\right)\mathrm{d}x\right]=\mathrm{e}^{\int P(x)\,\mathrm{d}x}.$$

Multiplying equation (2) by the integrating factor $\mu(x)$ yields

$$\big[P(x)y-Q(x)\big]\mathrm{e}^{\int P(x)\,\mathrm{d}x}\,\mathrm{d}x+\mathrm{e}^{\int P(x)\,\mathrm{d}x}\,\mathrm{d}y=0.$$

The general solution can be determined using the method of grouping terms

$$\Big[e^{\int P(x)\,dx}\, dy + \boxed{P(x)\, y\, e^{\int P(x)\,dx}}\, dx \Big] + -Q(x)\, e^{\int P(x)\,dx}\, dx = 0,$$

$$\int dy \;\rightarrow\; y\, e^{\int P(x)\,dx} \;\xrightarrow{\;\frac{\partial}{\partial x}\;}\; P(x)\, y\, e^{\int P(x)\,dx}, \qquad \int dx \;\downarrow\; -\int Q(x)\, e^{\int P(x)\,dx}\, dx$$

which results in

$$y e^{\int P(x)\,dx} - \int Q(x)\, e^{\int P(x)\,dx}\, dx = C.$$ ✍ General solution

The above results can be summarized as follows.

Linear First-Order Equations

1. $\dfrac{dy}{dx} + P(x)\cdot y = Q(x) \implies y = e^{-\int P(x)\,dx}\Big[\displaystyle\int Q(x)\, e^{\int P(x)\,dx}\, dx + C\Big].$

2. $\dfrac{dx}{dy} + P(y)\cdot x = Q(y) \implies x = e^{-\int P(y)\,dy}\Big[\displaystyle\int Q(y)\, e^{\int P(y)\,dy}\, dy + C\Big].$

Example 2.30

Solve $\quad y' = 1 + 3y\tan x.$

The differential equation can be written as

$$y' - 3\tan x\cdot y = 1,$$ ✍ Linear first-order

which is of the form $y' + P(x)\cdot y = Q(x)$, $P(x) = -3\tan x$, $Q(x) = 1$. The following quantities can be evaluated

$$\int P(x)\,dx = -3\int \tan x\,dx = 3\ln|\cos x|,$$

$$e^{\int P(x)\,dx} = e^{3\ln|\cos x|} = \cos^3 x, \qquad e^{-\int P(x)\,dx} = \frac{1}{\cos^3 x},$$

$$\int Q(x)\, e^{\int P(x)\,dx}\, dx = \int 1\cdot\cos^3 x\,dx = \int \cos^2 x\cdot\cos x\,dx$$

$$= \int (1-\sin^2 x)\, d(\sin x) = \sin x - \tfrac{1}{3}\sin^3 x.$$

The general solution of the differential equation is

$$y = e^{-\int P(x)\,dx}\Big[\int Q(x)\, e^{\int P(x)\,dx}\, dx + C\Big] = \frac{1}{\cos^3 x}\big(\sin x - \tfrac{1}{3}\sin^3 x + C\big).$$

Example 2.31

Solve $\quad y\,dx-(e^y+2xy-2x)\,dy=0.$

It is easy to see that $y=0$ is a solution of the differential equation. For $y\neq 0$, the differential equation can be written as

$$\frac{dx}{dy}+\frac{2(1-y)}{y}\cdot x=\frac{e^y}{y},$$ ✍ Linear first-order

which is of the form

$$\frac{dx}{dy}+P(y)\cdot x=Q(y),\qquad P(y)=\frac{2(1-y)}{y},\quad Q(y)=\frac{e^y}{y}.$$

The following quantities can be evaluated

$$\int P(y)\,dy=2\int\Big(\frac{1}{y}-1\Big)dy=2\big(\ln|y|-y\big)=\ln\big|y^2e^{-2y}\big|,$$

$$e^{\int P(y)\,dy}=e^{\ln|y^2e^{-2y}|}=y^2e^{-2y},\qquad e^{-\int P(y)\,dy}=\frac{e^{2y}}{y^2},$$

$$\int Q(y)\,e^{\int P(y)\,dy}dy=\int\frac{e^y}{y}\cdot y^2e^{-2y}\,dy=\int ye^{-y}\,dy=-\int y\,d(e^{-y})$$

$$=-\Big(ye^{-y}-\int e^{-y}\,dy\Big)$$ ✍ Integration by parts

$$=-(ye^{-y}+e^{-y})=-e^{-y}(y+1).$$

The general solution of the differential equation is

$$x=e^{-\int P(y)\,dy}\Big[\int Q(y)\,e^{\int P(y)\,dy}dy+C\Big]=\frac{e^{2y}}{y^2}\big[-e^{-y}(y+1)+C\big],$$

$$\therefore\quad xy^2=-e^y(y+1)+Ce^{2y}.$$ ✍ General solution

Note that the solution $y=0$ is included in the general solution with $C=1$.

Example 2.32

Solve $\quad \dfrac{dy}{dx}=\dfrac{1}{x\cos y+\sin 2y},\quad \cos y\neq 0.$

The differential equation can be written as

$$\frac{dx}{dy}-\cos y\cdot x=\sin 2y,$$ ✍ Linear first-order

which is of the form

$$\frac{dx}{dy}+P(y)\cdot x=Q(y),\qquad P(y)=-\cos y,\quad Q(y)=\sin 2y.$$

The following quantities can be evaluated

$$\int P(y)\,\mathrm{d}y = -\int \cos y\,\mathrm{d}y = -\sin y, \quad \mathrm{e}^{\int P(y)\,\mathrm{d}y} = \mathrm{e}^{-\sin y}, \quad \mathrm{e}^{-\int P(y)\,\mathrm{d}y} = \mathrm{e}^{\sin y},$$

$$\int Q(y)\,\mathrm{e}^{\int P(y)\,\mathrm{d}y}\mathrm{d}y = \int \sin 2y\,\mathrm{e}^{-\sin y}\,\mathrm{d}y = 2\int \sin y \cos y \cdot \mathrm{e}^{-\sin y}\,\mathrm{d}y$$

$$= -2\int \sin y\,\mathrm{d}(\mathrm{e}^{-\sin y}) \qquad \text{Integration by parts}$$

$$= -2\Big[\sin y \cdot \mathrm{e}^{-\sin y} - \int \mathrm{e}^{-\sin y}\,\mathrm{d}(\sin y)\Big] = -2\mathrm{e}^{-\sin y}(\sin y+1).$$

The general solution of the differential equation is

$$x = \mathrm{e}^{-\int P(y)\,\mathrm{d}y}\Big[\int Q(y)\,\mathrm{e}^{\int P(y)\,\mathrm{d}y}\mathrm{d}y + C\Big] = \mathrm{e}^{\sin y}\big[-2\mathrm{e}^{-\sin y}(\sin y+1) + C\big],$$

i.e.,

$$x = -2(\sin y+1) + C\,\mathrm{e}^{\sin y}. \qquad \text{General solution}$$

2.4.2 Bernoulli Differential Equations

A differential equation of the form

$$\frac{\mathrm{d}y}{\mathrm{d}x} + P(x)\cdot y = Q(x)\cdot y^n, \quad n = \text{Constant}, \tag{1}$$

is called a Bernoulli differential equation. Except when $n=0$ or 1, the equation is nonlinear.

- When $n=0$, the equation reduces to a linear first-order equation

$$\frac{\mathrm{d}y}{\mathrm{d}x} + P(x)\cdot y = Q(x).$$

- When $n=1$, the equation can be written as

$$\frac{\mathrm{d}y}{\mathrm{d}x} = \big[Q(x) - P(x)\big]\,y. \qquad \text{Variable separable}$$

- In the following, the case when $n \neq 0, 1$ is considered.

For $n<0$, $y\neq 0$. When $n>0$, $y=0$ is a solution of the differential equation. For $y\neq 0$, dividing both sides of equation (1) by y^n yields

$$y^{-n}\frac{\mathrm{d}y}{\mathrm{d}x} + P(x)\cdot y^{1-n} = Q(x).$$

Letting $u=y^{1-n}$, when it is defined, one has

$$\frac{\mathrm{d}u}{\mathrm{d}x} = (1-n)\,y^{-n}\frac{\mathrm{d}y}{\mathrm{d}x} \implies y^{-n}\frac{\mathrm{d}y}{\mathrm{d}x} = \frac{1}{1-n}\frac{\mathrm{d}u}{\mathrm{d}x},$$

and the equation becomes

$$\frac{1}{1-n}\frac{du}{dx} + P(x)\,u = Q(x),$$

or

$$\frac{du}{dx} + \underbrace{(1-n)P(x)}_{\bar{P}(x)} \cdot u = \underbrace{(1-n)Q(x)}_{\bar{Q}(x)}, \qquad n \neq 1. \tag{2}$$

Hence, a Bernoulli differential equation is transformed to a linear first-order equation (2) in the new variable u.

Remarks: It is important to consider exchanging the roles of x and y so that a differential equation can be cast into a linear first-order equation or a Bernoulli differential equation.

Bernoulli Differential Equations

1. $\dfrac{dy}{dx} + P(x)\cdot y = Q(x)\cdot y^n$

$$u = y^{1-n} \implies \frac{du}{dx} + \underbrace{(1-n)P(x)}_{\bar{P}(x)} \cdot u = \underbrace{(1-n)Q(x)}_{\bar{Q}(x)}, \quad \text{✍ Linear first-order}$$

2. $\dfrac{dx}{dy} + P(y)\cdot x = Q(y)\cdot x^n$

$$u = x^{1-n} \implies \frac{du}{dy} + \underbrace{(1-n)P(y)}_{\bar{P}(y)} \cdot u = \underbrace{(1-n)Q(y)}_{\bar{Q}(y)}, \quad \text{✍ Linear first-order}$$

Example 2.33

Solve $\quad 2xyy' = y^2 - 2x^3, \quad y(1) = 2.$

The differential equation can be written as

$$y' - \frac{1}{2x}\cdot y = -x^2 \cdot \frac{1}{y}, \qquad \text{✍ Bernoulli DE with } n = -1$$

Multiplying both sides of the equation by y yields

$$yy' - \frac{1}{2x}\cdot y^2 = -x^2.$$

Letting $u=y^2 \implies \dfrac{du}{dx}=2y\dfrac{dy}{dx}$, the equation becomes

$$\frac{1}{2}\frac{du}{dx}-\frac{1}{2x}\cdot u=-x^2 \implies \frac{du}{dx}-\frac{1}{x}\cdot u=-2x^2.$$

The equation is linear first-order of the form

$$\frac{du}{dx}+\bar{P}(x)\cdot u=\bar{Q}(x), \qquad \bar{P}(x)=-\frac{1}{x}, \qquad \bar{Q}(x)=-2x^2.$$

The following quantities can be evaluated

$$\int \bar{P}(x)\,dx=\int -\frac{1}{x}\,dx=-\ln|x|, \quad e^{\int \bar{P}(x)\,dx}=e^{-\ln|x|}=\frac{1}{x}, \quad e^{-\int \bar{P}(x)\,dx}=x,$$

$$\int \bar{Q}(x)\,e^{\int \bar{P}(x)\,dx}\,dx=\int -2x^2\cdot\frac{1}{x}\,dx=-x^2.$$

Hence

$$u=e^{-\int \bar{P}(x)\,dx}\left[\int \bar{Q}(x)\,e^{\int \bar{P}(x)\,dx}\,dx+C\right]=x(-x^2+C),$$

i.e.,

$$y^2=x(-x^2+C). \qquad \text{✍ General solution}$$

The constant C can be determined from the initial condition $y(1)=2$

$$2^2=1\cdot(-1^2+C) \implies C=5.$$

The particular solution is

$$y^2=x(5-x^2). \qquad \text{✍ Particular solution}$$

Example 2.34

Solve $\quad 3y\,dx-x\,(3x^3y\ln|y|+1)\,dy=0, \quad y\neq 0.$

The differential equation can be written as

$$\frac{dx}{dy}-\frac{1}{3y}\cdot x=\ln|y|\cdot x^4. \qquad \text{✍ Bernoulli DE with } n=4$$

Dividing both sides of the equation by x^4 yields

$$\frac{1}{x^4}\frac{dx}{dy}-\frac{1}{3y}\cdot\frac{1}{x^3}=\ln|y|.$$

Letting $u=\dfrac{1}{x^3} \implies \dfrac{du}{dy}=-\dfrac{3}{x^4}\dfrac{dx}{dy}$, the equation becomes

$$-\frac{1}{3}\frac{du}{dy}-\frac{1}{3y}\cdot u=\ln|y| \implies \frac{du}{dy}+\frac{1}{y}\cdot u=-3\ln|y|.$$

The equation is linear first-order of the form

$$\frac{du}{dy} + \bar{P}(y)\cdot u = \bar{Q}(y), \qquad \bar{P}(y) = \frac{1}{y}, \quad \bar{Q}(y) = -3\ln|y|.$$

The following quantities can be evaluated

$$\int \bar{P}(y)\,dy = \int \frac{1}{y}\,dy = \ln|y|, \quad e^{\int \bar{P}(y)\,dy} = e^{\ln|y|} = y, \quad e^{-\int \bar{P}(y)\,dy} = \frac{1}{y},$$

$$\int \bar{Q}(y)\,e^{\int \bar{P}(y)\,dy}\,dy = \int -3\ln|y|\cdot y\,dy = -\tfrac{3}{2}\int \ln|y|\,d(y^2) \qquad \text{Integration by parts}$$

$$= -\tfrac{3}{2}\left(y^2\ln|y| - \int y^2\cdot\frac{1}{y}\,dy\right) = -\tfrac{3}{2}\left(y^2\ln|y| - \tfrac{1}{2}y^2\right).$$

Hence

$$u = e^{-\int \bar{P}(y)\,dy}\left[\int \bar{Q}(y)\,e^{\int \bar{P}(y)\,dy}\,dy + C\right] = \frac{1}{y}\left[-\tfrac{3}{4}y^2\left(2\ln|y| - 1\right) + C\right].$$

Replacing u by the original variable results in the general solution

$$\frac{4}{x^3} = -3y\left(2\ln|y| - 1\right) + \frac{C}{y}.$$

2.5 Equations Solvable for the Independent or Dependent Variable

A general first-order differential equation is of the form

$$F(x, y, y') = 0, \tag{1}$$

or

$$F(x, y, p) = 0, \quad y' = p. \tag{1'}$$

In the following, cases when variable x or y can be solved are considered.

Case 1. Equation Solvable for Variable y

Suppose, from equation (1′), variable y can be expressed explicitly as a function of x and p to yield

$$y = f(x, p). \tag{2}$$

Differentiating equation (2) with respect to x gives

$$\frac{dy}{dx} = \frac{\partial f}{\partial x} + \frac{\partial f}{\partial p}\frac{dp}{dx} \quad\Longrightarrow\quad p = f_x + f_p\frac{dp}{dx}, \tag{3}$$

which is a differential equation between x and p. If equation (3) can be solved to obtain the general solution

$$\phi(x, p, C) = 0, \tag{4}$$

then the general solution of equation (1) can be obtained as follows.

- Eliminate variable p between equations (2) and (4) to obtain the solution in terms of x and y.
- If it is difficult to eliminate p between equations (2) and (4), then equations (2) and (4) can be treated as parametric equations with p being the parameter.

For example, consider the parametric equations

$$x = a + r\cos\theta, \quad y = b + r\sin\theta,$$

where a, b, and r are constants, and θ is the parameter. Rewrite the equations as

$$\frac{x-a}{r} = \cos\theta, \quad \frac{y-b}{r} = \sin\theta.$$

Using the trigonometric identity $\cos^2\theta + \sin^2\theta = 1$, the parameter θ can be eliminated to yield

$$\cos^2\theta + \sin^2\theta = \left(\frac{x-a}{r}\right)^2 + \left(\frac{y-b}{r}\right)^2 = 1 \implies (x-a)^2 + (y-b)^2 = r^2,$$

which is the equation of a circle with center at (a, b) and radius r.

Case 2. Equation Solvable for Variable x

Suppose, from equation (1′), variable x can be expressed explicitly as a function of y and p to yield

$$x = g(y, p). \tag{5}$$

Differentiating equation (5) with respect to y gives

$$\frac{\mathrm{d}x}{\mathrm{d}y} = \frac{\partial g}{\partial y} + \frac{\partial g}{\partial p}\frac{\mathrm{d}p}{\mathrm{d}y} \implies \frac{1}{p} = g_y + g_p\frac{\mathrm{d}p}{\mathrm{d}y}, \tag{6}$$

which is a differential equation between y and p. If equation (6) can be solved to obtain the general solution

$$\psi(y, p, C) = 0, \tag{7}$$

then the general solution of equation (1) can be obtained as follows.

- Eliminate variable p between equations (5) and (7) to obtain the solution in terms of x and y.
- If it is difficult to eliminate p between equations (5) and (7), then equations (5) and (7) can be treated as parametric equations with p being the parameter.

Example 2.35

Solve $x = \dfrac{dy}{dx} + \left(\dfrac{dy}{dx}\right)^4.$

Letting $p = \dfrac{dy}{dx}$, the equation can be written as

$$x = p + p^4 = f(p),$$

which is the case of equation solvable for x. Differentiating with respect to y yields

$$\frac{1}{p} = \frac{df}{dp}\frac{dp}{dy} = (1+4p^3)\frac{dp}{dy}, \qquad \frac{dx}{dy} = \frac{1}{p},$$

which can be written as

$$dy = (p + 4p^4)\, dp. \qquad \text{Variable separable}$$

Integrating both sides leads to

$$y = \tfrac{1}{2}p^2 + \tfrac{4}{5}p^5 + C.$$

Hence, the general solution is given by the parametric equations

$$\begin{cases} x = p + p^4, \\ y = \tfrac{1}{2}p^2 + \tfrac{4}{5}p^5 + C, \end{cases}$$

where p is a parameter.

Example 2.36

Solve $x\left(\dfrac{dy}{dx}\right)^2 - 2y\dfrac{dy}{dx} - x = 0.$

Solution 1. Letting $p = \dfrac{dy}{dx}$, one has

$$xp^2 - 2yp - x = 0. \tag{1}$$

Since $p=0$ is not a solution, one has $p \neq 0$. Variable y can be expressed explicitly in terms of x and p to yield

$$y = \frac{1}{2}x\left(p - \frac{1}{p}\right) = f(x, p). \tag{2}$$

Differentiating equation (2) with respect to x gives

$$p = \frac{\partial f}{\partial x} + \frac{\partial f}{\partial p}\frac{dp}{dx} = \frac{1}{2}\left(p - \frac{1}{p}\right) + \frac{1}{2}x\left(1 + \frac{1}{p^2}\right)\frac{dp}{dx},$$

which can be simplified as

$$\frac{p^2+1}{p}\left(\frac{x}{p}\frac{\mathrm{d}p}{\mathrm{d}x}-1\right)=0.$$

Since $(p^2+1)/p\neq 0$, one has

$$\frac{x}{p}\frac{\mathrm{d}p}{\mathrm{d}x}-1=0 \implies \frac{1}{p}\mathrm{d}p=\frac{1}{x}\mathrm{d}x, \qquad \text{✍ Variable separable}$$

which can be solved to give

$$\int\frac{1}{p}\mathrm{d}p=\int\frac{1}{x}\mathrm{d}x+C \implies \ln|p|=\ln|x|+\ln|C|,$$

$$\therefore\quad \ln|p|=\ln|Cx| \implies p=Cx.$$

Substituting into equation (2) results in the general solution

$$y=\frac{1}{2}x\left(Cx-\frac{1}{Cx}\right) \implies 2y=Cx^2-\frac{1}{C}.$$

Solution 2. Since $p=\pm 1$ is not a solution, one has $p\neq\pm 1$. From equation (1), variable x can also be expressed explicitly in terms of y and p to yield

$$x=\frac{2yp}{p^2-1}=g(y,p). \tag{2$'$}$$

Differentiating equation (2′) with respect to y yields

$$\frac{1}{p}=\frac{\partial g}{\partial y}+\frac{\partial g}{\partial p}\frac{\mathrm{d}p}{\mathrm{d}y}=\frac{2p}{p^2-1}-2y\frac{p^2+1}{(p^2-1)^2}\frac{\mathrm{d}p}{\mathrm{d}y},$$

which can be simplified as

$$-\frac{p^2+1}{p(p^2-1)}=-2y\frac{p^2+1}{(p^2-1)^2}\frac{\mathrm{d}p}{\mathrm{d}y} \implies \frac{2p}{p^2-1}\mathrm{d}p=\frac{1}{y}\mathrm{d}y.$$

The equation is variable separable and can be solved by integrating both sides

$$\int\frac{1}{p^2-1}\mathrm{d}(p^2-1)=\int\frac{1}{y}\mathrm{d}y+D,$$

$$\ln|p^2-1|=\ln|2Cy|, \qquad \text{✍ Set } D=\ln|2C| \text{ for the purpose of comparing the results of two methods.}$$

$$\therefore\quad p^2-1=2Cy. \tag{3}$$

Parameter p can be eliminated between equations (2′) and (3). Substituting equation (3) into (2′) yields

$$x=\frac{2yp}{2Cy} \implies p=Cx.$$

Substituting into equation (3) results in the general solution

$$C^2x^2-1=2Cy.$$

Example 2.37

Solve $$y = x\left\{\frac{dy}{dx} + \sqrt{1+\left(\frac{dy}{dx}\right)^2}\right\}.$$

Letting $p = \dfrac{dy}{dx}$, the differential equation becomes

$$y = x\left(p + \sqrt{1+p^2}\right) = f(x, p). \tag{1}$$

Differentiating equation (1) with respect to x yields

$$p = \frac{\partial f}{\partial x} + \frac{\partial f}{\partial p}\frac{dp}{dx} = p + \sqrt{1+p^2} + x\left(1 + \frac{p}{\sqrt{1+p^2}}\right)\frac{dp}{dx},$$

which can be simplified as

$$-\sqrt{1+p^2} = x\left(1 + \frac{p}{\sqrt{1+p^2}}\right)\frac{dp}{dx},$$

$$\therefore \quad -\frac{1}{x}dx = \left(\frac{1}{\sqrt{1+p^2}} + \frac{p}{1+p^2}\right)dp. \quad \text{✍ Variable separable}$$

Integrating both sides leads to

$$-\int \frac{1}{x}dx = \int \left(\frac{1}{\sqrt{1+p^2}} + \frac{p}{1+p^2}\right)dp + C,$$

$$-\ln|x| = \ln\left|p + \sqrt{1+p^2}\right| + \tfrac{1}{2}\ln\left|1+p^2\right| + \ln|D|, \quad \text{✍ } C \Rightarrow \ln|D|$$

$$\therefore \quad \frac{1}{x} = D\left(p + \sqrt{1+p^2}\right)\sqrt{1+p^2}. \tag{2}$$

The parameter p can be eliminated between equations (1) and (2). From equation (1), one has

$$p + \sqrt{1+p^2} = \frac{y}{x}. \tag{3}$$

Substituting into equation (2) yields

$$\frac{1}{x} = D\sqrt{1+p^2}\cdot\frac{y}{x} \implies \sqrt{1+p^2} = \frac{C}{y} \quad \text{or} \quad p^2 = \left(\frac{C}{y}\right)^2 - 1. \quad \text{✍ } \frac{1}{D} \Rightarrow C$$

Substituting into equation (3) leads to

$$p = \frac{y}{x} - \sqrt{1+p^2} \implies p^2 = \left(\frac{y}{x} - \sqrt{1+p^2}\right)^2$$

$$\therefore \quad \left(\frac{C}{y}\right)^2 - 1 = \left(\frac{y}{x} - \frac{C}{y}\right)^2,$$

which can be further simplified as

$$\left(\frac{C}{y}\right)^2-\left(\frac{y}{x}-\frac{C}{y}\right)^2=1,$$

$$\left[\frac{C}{y}+\left(\frac{y}{x}-\frac{C}{y}\right)\right]\left[\frac{C}{y}-\left(\frac{y}{x}-\frac{C}{y}\right)\right]=1, \qquad ✍\ a^2-b^2=(a+b)(a-b)$$

$$\therefore\quad \frac{y}{x}\left(\frac{2C}{y}-\frac{y}{x}\right)=1 \implies \frac{x}{y}+\frac{y}{x}=\frac{C}{y}. \qquad ✍\ \text{General solution, } 2C\Rightarrow C$$

Example 2.38

Solve $\quad y^2\left(\dfrac{dy}{dx}\right)^2+3x\left(\dfrac{dy}{dx}\right)-y=0.$

Letting $p=\dfrac{dy}{dx}$, the differential equation can be written as

$$y^2p^2+3xp-y=0.$$

If $p=0$, one must have $y=0$, which is a solution of the differential equation.

For $p\neq 0$, solving for x gives

$$x=-\frac{1}{3}\left(y^2p-\frac{y}{p}\right)=g(y,p). \tag{1}$$

Differentiating equation (1) with respect to y yields

$$\frac{1}{p}=\frac{\partial g}{\partial y}+\frac{\partial g}{\partial p}\frac{dp}{dy}=-\frac{1}{3}\left(2yp-\frac{1}{p}\right)-\frac{1}{3}\left(y^2+\frac{y}{p^2}\right)\frac{dp}{dy}.$$

Multiplying both sides by $-3p$ and arranging yield

$$(yp^2+1)\left(2+\frac{y}{p}\frac{dp}{dy}\right)=0.$$

Case 1. $\quad 2+\dfrac{y}{p}\dfrac{dp}{dy}=0$

The equation can be written as

$$\frac{1}{p}dp=-\frac{2}{y}dy. \qquad ✍\ \text{Variable separable}$$

Integrating both sides yields

$$\int\frac{1}{p}dp=-2\int\frac{1}{y}dy+C \implies \ln|p|=-2\ln|y|+\ln|C| \implies p=\frac{C}{y^2}.$$

Substituting into equation (1) results in the general solution

$$x = -\frac{1}{3\left(\frac{C}{y^2}\right)}\left[y^2\left(\frac{C}{y^2}\right)^2 - y\right] \implies 3Cx - y^3 + C^2 = 0. \quad \text{General solution}$$

Case 2. $yp^2 + 1 = 0 \implies p^2 = -\frac{1}{y}$

Substituting into equation (1) results in

$$x^2 = \frac{1}{9p^2}(y^2p^2 - y)^2 \implies x^2 = \frac{1}{9\left(-\frac{1}{y}\right)}\left[y^2\left(-\frac{1}{y}\right) - y\right]^2,$$

$$\therefore \quad 9x^2 + 4y^3 = 0. \quad \text{Singular solution}$$

This solution is not obtainable from the general solution for any value of C and is therefore a singular solution.

The Clairaut Equation

A first-order differential equation of the form

$$y = xy' + f(y'),$$

or

$$y = xp + f(p), \qquad p = \frac{dy}{dx}.$$

is called the Clairaut equation. The equation is of the type of equation solvable for variable y and can be solved using the approach introduced in this section.

Example 2.39

Solve $\quad y = xy' + \frac{a^2}{y'}.$

Letting $y' = p$, the differential equation becomes

$$y = xp + \frac{a^2}{p} = f(x, p). \tag{1}$$

Differentiating with respect to x yields

$$p = \frac{\partial f}{\partial x} + \frac{\partial f}{\partial p}\frac{dp}{dx} = p + \left(x - \frac{a^2}{p^2}\right)\frac{dp}{dx} \implies \left(x - \frac{a^2}{p^2}\right)\frac{dp}{dx} = 0.$$

Case 1. $\frac{dp}{dx} = 0 \implies p = C.$

Substituting into equation (1) results in the general solution

$$y = Cx + \frac{a^2}{C}, \qquad \text{✍ General solution} \tag{2}$$

which is a family of straight lines with slope C and y-intercept a^2/C.

Case 2. $x - \dfrac{a^2}{p^2} = 0 \implies p^2 = \dfrac{a^2}{x}$.

Substituting into equation (1) gives

$$py = xp^2 + a^2 \implies p^2y^2 = (xp^2 + a^2)^2,$$

$$\frac{a^2}{x} \cdot y^2 = \left(x \cdot \frac{a^2}{x} + a^2\right)^2 \implies y^2 = 4a^2x. \qquad \text{✍ Singular solution}$$

This solution is a parabola, which cannot be obtained from the general solution (2) for any value of C and is therefore a singular solution.

2.6 Simple Higher-Order Differential Equations

2.6.1 *Equations Immediately Integrable*

An nth-order differential equation of the form

$$\frac{d^n y}{dx^n} = f(x)$$

can be solved easily by integrating n times the function $f(x)$ with respect to x. The general solution of an nth-order differential equation contains n constants of integration.

Example 2.40

Solve $\quad x^2 y''' = 120x^3 + 8x^2 e^{2x} + 1, \quad x \neq 0.$

Dividing both sides of the equation by x^2 yields

$$y''' = 120x + 8e^{2x} + \frac{1}{x^2}.$$

Integrating the equation with respect to x once gives

$$y'' = 60x^2 + 4e^{2x} - \frac{1}{x} + C_1.$$

Integrating the equation with respect to x again leads to

$$y' = 20x^3 + 2e^{2x} - \ln|x| + C_1 x + C_2. \tag{1}$$

Since

$$\int \ln|x|\,dx = x\ln|x| - \int x\cdot\frac{1}{x}\,dx = x\big(\ln|x|-1\big),$$ ✍ Integrating by parts

integrating equation (1) with respect to x results in the general solution

$$y = 5x^4 + e^{2x} - x\big(\ln|x|-1\big) + \tfrac{1}{2}C_1x^2 + C_2x + C_3,$$

$$\therefore \quad y = C_0 + C_1x + C_2x^2 + 5x^4 + e^{2x} - x\ln|x|.$$ ✍ General solution

Example 2.41

Solve $\quad e^{-x}y'' - \sin x = 2, \quad y(0) = y'(0) = 1.$

The equation can be written as

$$y'' = e^x\sin x + 2e^x. \tag{1}$$

It can be easily determined

$$\int e^x\sin x\,dx = \int \sin x\,d(e^x) = e^x\sin x - \int e^x\cos x\,dx$$ ✍ Integrating by parts

$$= e^x\sin x - \int \cos x\,d(e^x) = e^x\sin x - \Big(e^x\cos x + \int e^x\sin x\,dx\Big),$$

$$\Longrightarrow \int e^x\sin x\,dx = \tfrac{1}{2}e^x(\sin x - \cos x),$$

$$\int e^x\cos x\,dx = \int \cos x\,d(e^x) = e^x\cos x + \int e^x\sin x\,dx$$

$$= e^x\cos x + \tfrac{1}{2}e^x(\sin x - \cos x) = \tfrac{1}{2}e^x(\sin x + \cos x).$$

Integrating equation (1) with respect to x yields

$$y' = \tfrac{1}{2}e^x(\sin x - \cos x) + 2e^x + C_1.$$

The constant C_1 can be determined from the initial condition $y'(0) = 1$:

$$1 = \tfrac{1}{2}e^0(\sin 0 - \cos 0) + 2e^0 + C_1 \quad\Longrightarrow\quad C_1 = -\tfrac{1}{2}.$$

Hence

$$y' = \tfrac{1}{2}e^x(\sin x - \cos x) + 2e^x - \tfrac{1}{2}.$$

Integrating with respect to x again gives

$$y = \tfrac{1}{2}\Big[\tfrac{1}{2}e^x(\sin x - \cos x) - \tfrac{1}{2}e^x(\sin x + \cos x)\Big] + 2e^x - \tfrac{1}{2}x + C_2$$

$$= -\tfrac{1}{2}e^x\cos x + 2e^x - \tfrac{1}{2}x + C_2.$$

The constant C_2 is determined from the initial condition $y(0)=1$:

$$1 = -\tfrac{1}{2}\mathrm{e}^0 \cos 0 + 2\mathrm{e}^0 - \tfrac{1}{2}\cdot 0 + C_2 \implies C_2 = -\tfrac{1}{2}.$$

$$\therefore \quad y = -\tfrac{1}{2}\mathrm{e}^x \cos x + 2\mathrm{e}^x - \tfrac{1}{2}x - \tfrac{1}{2}. \quad \text{✍ Particular solution}$$

2.6.2 *The Dependent Variable Absent*

In general, an nth-order differential equation is of the form

$$f\big(x;\, y, y', y'', \dots, y^{(n)}\big) = 0.$$

If the dependent variable y is absent, or more generally, $y, y', y'', \dots, y^{(k-1)}$ are absent, the differential equation is of the form

$$f\big(x;\, y^{(k)},\, y^{(k+1)}, \dots,\, y^{(n)}\big) = 0.$$

Let $u = y^{(k)}$ be the new dependent variable, the differential equation becomes

$$f\big(x;\, u,\, u',\, u'', \dots,\, u^{(n-k)}\big) = 0,$$

which is an $(n-k)$th-order equation. Hence the order of the differential equation is reduced from n to $(n-k)$.

Remarks: It should be emphasized that for the differential equation to be considered as $y^{(k-1)}$ absent, all derivatives with order lower than $(k-1)$ must also be absent.

For example, equation

$$y'''' + y' = g(x)$$

is of the type y absent, whereas equation

$$y'''' + y'' = g(x)$$

is of the type y' absent.

Example 2.42

Solve $\quad 2y'' = (y')^3 \sin 2x, \quad y(0) = y'(0) = 1.$

The equation is of the type y absent. Let $y'=u$, $y''=u'$. Since $y'=u=0$ does not satisfy the initial condition $y'(0)=1$, one has $u \neq 0$. The equation becomes

$$2\frac{\mathrm{d}u}{\mathrm{d}x} = u^3 \sin 2x \implies \frac{2}{u^3}\,\mathrm{d}u = \sin 2x\,\mathrm{d}x. \qquad \text{✍ Variable separable}$$

Integrating both sides yields

$$\int \frac{2}{u^3}\,\mathrm{d}u = \int \sin 2x\,\mathrm{d}x + C_1 \implies -\frac{1}{u^2} = -\frac{\cos 2x}{2} + C_1.$$

The constant C_1 can be determined from the initial condition $u(0)=y'(0)=1$:

$$-1=-\tfrac{1}{2}+C_1 \implies C_1=-\tfrac{1}{2}.$$

Hence

$$-\frac{1}{u^2}=-\frac{1+\cos 2x}{2}=-\cos^2 x \implies u=\pm\frac{1}{\cos x}.$$

Since $u(0)=1$, only the positive sign is taken, which leads to

$$u=\frac{1}{\cos x} \implies \frac{dy}{dx}=\sec x.$$ ✍ Immediately integrable

Integrating both sides results in

$$y=\int \sec x\,dx+C_2=\ln\big|\sec x+\tan x\big|+C_2.$$

The constant C_2 can be determined using the initial condition $y(0)=1$:

$$1=\ln|1+0|+C_2 \implies C_2=1,$$

$$\therefore\quad y=\ln\big|\sec x+\tan x\big|+1.$$ ✍ Particular solution

Example 2.43

Solve $xy''-(y')^3-y'=0.$

The equation is of the type y absent. Denoting $y'=u$, $y''=u'$, one has

$$xu'-u^3-u=0 \implies u'-\frac{1}{x}\cdot u=\frac{1}{x}\cdot u^3.$$ ✍ Bernoulli DE

Case 1. $u=0 \implies y=C$ is a solution of the differential equation.
Case 2. $u\neq 0$. Dividing both sides by u^3 leads to

$$\frac{1}{u^3}u'-\frac{1}{x}\cdot\frac{1}{u^2}=\frac{1}{x}.$$

Letting $v=\dfrac{1}{u^2}$, $\dfrac{dv}{dx}=-\dfrac{2}{u^3}\dfrac{du}{dx} \implies \dfrac{1}{u^3}\dfrac{du}{dx}=-\dfrac{1}{2}\dfrac{dv}{dx}$, the equation becomes

$$-\frac{1}{2}\frac{dv}{dx}-\frac{1}{x}\cdot v=\frac{1}{x} \implies \frac{dv}{dx}+\frac{2}{x}\cdot v=-\frac{2}{x},$$

which is linear first-order with

$$P(x)=\frac{2}{x},\qquad Q(x)=-\frac{2}{x}.$$

The following quantities can be easily determined

$$\int P(x)\,dx=\int\frac{2}{x}\,dx=2\ln|x|,\quad e^{\int P(x)\,dx}=e^{2\ln|x|}=x^2,\quad e^{-\int P(x)\,dx}=\frac{1}{x^2},$$

$$\int Q(x)\,\mathrm{e}^{\int P(x)\,\mathrm{d}x}\mathrm{d}x = \int -\frac{2}{x}\cdot x^2\,\mathrm{d}x = -x^2.$$

Hence, the solution is

$$v = \mathrm{e}^{-\int P(x)\,\mathrm{d}x}\Big[\int Q(x)\,\mathrm{e}^{\int P(x)\,\mathrm{d}x}\mathrm{d}x + C_1\Big] = \frac{1}{x^2}(-x^2 + C_1),$$

$$\frac{1}{u^2} = \frac{C_1 - x^2}{x^2} \implies u = \frac{\mathrm{d}y}{\mathrm{d}x} = \pm\frac{x}{\sqrt{C_1 - x^2}}.$$ ✍ Immediately integrable

Integrating both sides results in the general solution

$$y = \pm\int\frac{x}{\sqrt{C_1 - x^2}}\mathrm{d}x + C_2 = \pm\sqrt{C_1 - x^2} + C_2 \implies y - C_2 = \pm\sqrt{C_1 - x^2}.$$

Squaring both sides gives

$$(y - C_2)^2 = C_1 - x^2 \implies x^2 + (y - C_2)^2 = C_1.$$ ✍ General solution

2.6.3 The Independent Variable Absent

When the independent variable x does not appear explicitly, an nth-order differential equation is of the form

$$f\big(y, y', y'', \ldots, y^{(n)}\big) = 0. \tag{1}$$

Let y be the new independent variable and $u = y'$ the new dependent variable. Using the chain rule, it is easy to show that

$$\frac{\mathrm{d}y}{\mathrm{d}x} = u,$$

$$\frac{\mathrm{d}^2y}{\mathrm{d}x^2} = \frac{\mathrm{d}u}{\mathrm{d}x} = \frac{\mathrm{d}u}{\mathrm{d}y}\frac{\mathrm{d}y}{\mathrm{d}x} = u\frac{\mathrm{d}u}{\mathrm{d}y},$$ ✍ Chain rule

$$\frac{\mathrm{d}^3y}{\mathrm{d}x^3} = \frac{\mathrm{d}}{\mathrm{d}x}\Big(\frac{\mathrm{d}^2y}{\mathrm{d}x^2}\Big) = \frac{\mathrm{d}}{\mathrm{d}y}\Big(u\frac{\mathrm{d}u}{\mathrm{d}y}\Big)\frac{\mathrm{d}y}{\mathrm{d}x} = u\Big(\frac{\mathrm{d}u}{\mathrm{d}y}\Big)^2 + u^2\frac{\mathrm{d}^2u}{\mathrm{d}y^2},$$

$$\cdots\cdots$$

It may be shown that $\dfrac{\mathrm{d}^ky}{\mathrm{d}x^k}$ may be expressed in terms of u, $\dfrac{\mathrm{d}u}{\mathrm{d}y}, \ldots, \dfrac{\mathrm{d}^{k-1}u}{\mathrm{d}y^{k-1}}$, for $k \leqslant n$. Hence, differential equation (1) becomes

$$f\Big[y;\; u,\; u\frac{\mathrm{d}u}{\mathrm{d}y},\; u\Big(\frac{\mathrm{d}u}{\mathrm{d}y}\Big)^2 + u^2\frac{\mathrm{d}^2u}{\mathrm{d}y^2},\; \ldots\Big] = 0,$$

or

$$g\Big(y;\; u,\; \frac{\mathrm{d}u}{\mathrm{d}y},\; \frac{\mathrm{d}^2u}{\mathrm{d}y^2},\; \ldots,\; \frac{\mathrm{d}^{n-1}u}{\mathrm{d}y^{n-1}}\Big) = 0,$$

in which the order of the differential equation is reduced by 1.

Example 2.44

Solve $\quad 3yy'y'' - (y')^3 + 1 = 0.$

The equation is of the type x absent. Let y be the new independent variable and $u = y'$ the new dependent variable, $y'' = u\,\mathrm{d}u/\mathrm{d}y$.

Case 1. $u^3 = 1 \implies u = 1 \implies y = x$, which is a solution of the equation.

Case 2. For $u \neq 1$, the equation becomes

$$3y \cdot u \cdot u\frac{\mathrm{d}u}{\mathrm{d}y} - u^3 + 1 = 0 \implies \frac{3u^2}{u^3-1}\mathrm{d}u = \frac{1}{y}\mathrm{d}y. \quad \text{Variable separable}$$

Integrating both sides yields

$$\int \frac{1}{u^3-1}\mathrm{d}(u^3-1) = \int \frac{1}{y}\mathrm{d}y + C_1 \implies \ln|u^3-1| = \ln|y| + \ln|C_1|,$$

$$u^3 - 1 = C_1 y \implies \frac{\mathrm{d}y}{\mathrm{d}x} = u = (C_1 y + 1)^{\frac{1}{3}}.$$

Since $y = \text{constant}$ is not a solution, one has $(C_1 y + 1) \neq 0$; hence

$$(C_1 y + 1)^{-\frac{1}{3}}\mathrm{d}y = \mathrm{d}x. \quad \text{Variable separable}$$

Integrating both sides results in the general solution

$$\frac{1}{C_1}\int (C_1 y + 1)^{-\frac{1}{3}}\mathrm{d}(C_1 y + 1) = x + C_2 \implies \frac{1}{C_1}\cdot\frac{3}{2}(C_1 y + 1)^{\frac{2}{3}} = x + C_2,$$

$$\therefore \quad 3(C_1 y + 1)^{\frac{2}{3}} - 2C_1 x = C_2. \quad \text{General solution}$$

Example 2.45

Solve $\quad yy'' = (y')^2(1 - y'\sin y - yy'\cos y).$

The equation is of the type x absent. Let y be the new independent variable and $u = y'$ the new dependent variable, $y'' = u\,\mathrm{d}u/\mathrm{d}y$. The equation becomes

$$y \cdot u\frac{\mathrm{d}u}{\mathrm{d}y} = u^2(1 - u\sin y - yu\cos y).$$

Case 1. $\quad u = 0 \implies y' = 0 \implies y = C.$

Case 2. $\quad y\dfrac{\mathrm{d}u}{\mathrm{d}y} = u(1 - u\sin y - yu\cos y).$

The differential equation can be written as

$$\frac{\mathrm{d}u}{\mathrm{d}y} - \frac{1}{y}\cdot u = -\frac{\sin y + y\cos y}{y}\cdot u^2. \quad \text{Bernoulli DE}$$

Dividing both sides by u^2 yields

$$\frac{1}{u^2}\frac{du}{dy} - \frac{1}{y}\cdot\frac{1}{u} = -\frac{\sin y + y\cos y}{y}.$$

Letting $v = \dfrac{1}{u}$, $\dfrac{dv}{dy} = -\dfrac{1}{u^2}\dfrac{du}{dy}$, one obtains

$$\frac{dv}{dy} + \underbrace{\frac{1}{y}}_{P(y)}\cdot v = \underbrace{\frac{\sin y + y\cos y}{y}}_{Q(y)}.$$ ✍ Linear first-order

The following quantities can be easily determined

$$\int P(y)\,dy = \int \frac{1}{y}\,dy = \ln|y|, \quad e^{\int P(y)\,dy} = e^{\ln|y|} = y, \quad e^{-\int P(y)\,dy} = \frac{1}{y},$$

$$\int Q(y)e^{\int P(y)\,dy}dy = \int \frac{\sin y + y\cos y}{y}\cdot y\,dy = \int (\sin y + y\cos y)\,dy$$

$$= -\cos y + \int y\,d(\sin y) = -\cos y + y\sin y - \int \sin y\,dy = y\sin y.$$

Hence, the solution is

$$v = e^{-\int P(y)\,dy}\Big[\int Q(y)e^{\int P(y)\,dy}dy + C_1\Big] = \frac{1}{y}(y\sin y + C_1) = \sin y + \frac{C_1}{y},$$

$$\therefore \quad v = \frac{1}{u} = \frac{dx}{dy} = \sin y + \frac{C_1}{y}.$$ ✍ Immediately integrable

Integrating both sides yields

$$x = \int\Big(\sin y + \frac{C_1}{y}\Big)dy + C_2 \implies x = -\cos y + C_1\ln|y| + C_2.$$

Therefore, the solutions are $y = C$, $x = -\cos y + C_1\ln|y| + C_2$.

2.7 Summary

In this chapter, various types of first-order and simple higher-order ordinary differential equations and the associated methods are introduced. The key in solving such equations depends on identifying the type of the equations. Sometimes, an equation can be classified as more than one type and solved using more than one method.

When one is given a differential equation, the following steps may be followed to identify and solve the equation.

☞ When dividing a differential equation by a function, care must be taken to ensure that the function is not zero. The case when the function is zero should be considered separately to determine if it will give extra solutions.

First-Order Ordinary Differential Equation

1. Method of Separation of Variables

If the differential equation can be written as

$$\frac{dy}{dx} = f(x)g(y) \qquad \text{or} \qquad f_1(x)g_1(y)\,dx + f_2(x)g_2(y)\,dy = 0,$$

then the equation is variable separable. Moving all terms involving variable x to one side and all terms involving variable y to the other side of the equation gives

$$\frac{1}{g(y)}\,dy = f(x)\,dx \qquad \text{or} \qquad \frac{g_2(y)}{g_1(y)}\,dy = -\frac{f_1(x)}{f_2(x)}\,dx.$$

Integrating both sides of the equation yields the general solution

$$\int \frac{1}{g(y)}\,dy = \int f(x)\,dx + C \qquad \text{or} \qquad \int \frac{g_2(y)}{g_1(y)}\,dy = -\int \frac{f_1(x)}{f_2(x)}\,dx + C.$$

This solution is valid for $g(y) \neq 0$ or $g_1(y) \neq 0$. Check if $g(y) = 0$ or $g_1(y) = 0$ is a solution of the differential equation. If it is a solution and is not included in the general solution, then it is a singular solution.

2. Homogeneous Equations

If the differential equation can be written in the form $\dfrac{dy}{dx} = f\Big(\dfrac{y}{x}\Big)$, then it is a homogeneous equation. Applying the transformation

$$\frac{y}{x} = v \implies y = xv \implies \frac{dy}{dx} = v + x\frac{dv}{dx},$$

the differential equation becomes variable separable $\implies x\dfrac{dv}{dx} = f(v) - v.$

3. Linear First-Order Equations and Bernoulli Equations

Rewrite the differential equation in the form

$$\frac{dy}{dx} + P(x)\cdot y = f(x,y) \qquad \text{or} \qquad \frac{dx}{dy} + P(y)\cdot x = f(x,y).$$

Depending on the right-hand side, the equation may be linear first-order

$$\frac{dy}{dx} + P(x)\cdot y = Q(x) \implies y = e^{-\int P(x)\,dx}\Big[\int Q(x)\,e^{\int P(x)\,dx}\,dx + C\Big],$$

$$\frac{dx}{dy} + P(y)\cdot x = Q(y) \implies x = e^{-\int P(y)\,dy}\Big[\int Q(y)\,e^{\int P(y)\,dy}\,dy + C\Big],$$

or Bernoulli equations, for $n \neq 0, 1$, which can be converted to linear first-order equations using a change of variable,

$$\frac{dy}{dx} + P(x) \cdot y = Q(x) \cdot y^n \quad \xrightarrow{u = y^{1-n}} \quad \frac{du}{dx} + \underbrace{(1-n)P(x)}_{\bar{P}(x)} \cdot u = \underbrace{(1-n)Q(x)}_{\bar{Q}(x)},$$

$$\frac{dx}{dy} + P(y) \cdot x = Q(y) \cdot x^n \quad \xrightarrow{u = x^{1-n}} \quad \frac{du}{dy} + \underbrace{(1-n)P(y)}_{\bar{P}(y)} \cdot u = \underbrace{(1-n)Q(y)}_{\bar{Q}(y)}.$$

4. Exact Differential Equations and Integrating Factors

Consider the differential equation $M(x, y)\,dx + N(x, y)\,dy = 0$.

Exact Differential Equations: If $\dfrac{\partial M}{\partial y} = \dfrac{\partial N}{\partial x}$, then the equation is exact.

The method of grouping terms can be applied to find the general solution

$$M\,dx + N\,dy = 0 \quad \xrightarrow{\text{Grouping terms}} \quad du(x, y) = 0 \implies u(x, y) = C.$$

Integrating Factors: If $\dfrac{\partial M}{\partial y} \neq \dfrac{\partial N}{\partial x}$, an integrating factor $\mu(x, y)$ may be determined so that $\mu M\,dx + \mu N\,dy = 0$ is exact.

- If $\dfrac{1}{N}\left(\dfrac{\partial M}{\partial y} - \dfrac{\partial N}{\partial x}\right) = g(x) \implies \mu(x) = \exp\left\{\int g(x)\,dx\right\}$.
- If $\dfrac{1}{M}\left(\dfrac{\partial N}{\partial x} - \dfrac{\partial M}{\partial y}\right) = g(y) \implies \mu(y) = \exp\left\{\int g(y)\,dy\right\}$.
- **Method of Inspection**—By rearranging terms, multiplying or dividing suitable functions, an integrating factor may be found

1. $d(xy) = y\,dx + x\,dy$,
2. $d\left(\dfrac{y}{x}\right) = \dfrac{-y\,dx + x\,dy}{x^2}, \qquad d\left(\dfrac{x}{y}\right) = \dfrac{y\,dx - x\,dy}{y^2}$,
3. $d\left(\tan^{-1}\dfrac{y}{x}\right) = \dfrac{-y\,dx + x\,dy}{x^2 + y^2}, \qquad d\left(\tan^{-1}\dfrac{x}{y}\right) = \dfrac{y\,dx - x\,dy}{x^2 + y^2}$,
4. $d\left[\dfrac{1}{2}\ln(x^2 + y^2)\right] = \dfrac{x\,dx + y\,dy}{x^2 + y^2}$,
5. $d\left(\sqrt{x^2 \pm y^2}\right) = \dfrac{x\,dx \pm y\,dy}{\sqrt{x^2 \pm y^2}}$.

❧ **Integrating Factors by Groups**—Separate the equation into two groups

$$\big[M_1(x,y)\,\mathrm{d}x + N_1(x,y)\,\mathrm{d}y\big] + \big[M_2(x,y)\,\mathrm{d}x + N_2(x,y)\,\mathrm{d}y\big] = 0.$$

Suppose $\mu_1(x,y)$ and $\mu_2(x,y)$ are the integrating factors for the first and second groups, respectively, and

$$\mu_1 M_1\,\mathrm{d}x + \mu_1 N_1\,\mathrm{d}y = \mathrm{d}v_1, \qquad \mu_2 M_2\,\mathrm{d}x + \mu_2 N_2\,\mathrm{d}y = \mathrm{d}v_2.$$

If functions g_1 and g_2 can be found such that

$$\mu_1(x,y)\cdot g_1\big[v_1(x,y)\big] = \mu_2(x,y)\cdot g_2\big[v_2(x,y)\big] = \mu(x,y),$$

then $\mu(x,y)$ is an integrating factor.

5. Equation Solvable for the Independent or Dependent Variable

❧ **Equation Solvable for Variable y:** $\quad y = f(x,p), \quad p = y'$

Differentiate with respect to x

$$p = \frac{\partial f}{\partial x} + \frac{\partial f}{\partial p}\frac{\mathrm{d}p}{\mathrm{d}x} \quad\Longrightarrow\quad \phi(x,p,C) = 0.$$

General solution (parametric): $y = f(x,p)$, $\phi(x,p,C) = 0$, $p =$ parameter.

❧ **Equation Solvable for Variable x:** $\quad x = g(y,p), \quad p = y'$

Differentiate with respect to y

$$\frac{1}{p} = \frac{\partial g}{\partial y} + \frac{\partial g}{\partial p}\frac{\mathrm{d}p}{\mathrm{d}y} \quad\Longrightarrow\quad \psi(y,p,C) = 0.$$

General solution (parametric): $x = g(y,p)$, $\psi(y,p,C) = 0$, $p =$ parameter.

6. Method of Special Transformations

There are no systematic procedures and rules to follow in applying the method of special transformations. The key is to uncover the "special" term in a given equation and determine a transformation accordingly.

Simple Higher-Order Differential Equation

1. Equations Immediately Integrable

$$\frac{\mathrm{d}^n y}{\mathrm{d}x^n} = f(x) \quad \xrightarrow{\text{Integrate } n \text{ times w.r.t. } x} \quad \text{General solution.}$$

2. The Dependent Variable Absent

$$f\big(x;\, y^{(k)},\, y^{(k+1)},\, \dots,\, y^{(n)}\big) = 0 \quad\Longrightarrow\quad y, y', y'', \dots, y^{(k-1)} \text{ absent.}$$

Let $u = y^{(k)} \implies f\big(x;\, u,\, u',\, u'',\, \ldots,\, u^{(n-k)}\big) = 0$. The order of the equation is reduced by k.

3. The Independent Variable Absent

$$f\big(y, y', \ldots, y^{(n)}\big) = 0 \implies x \text{ absent.}$$

Let y be the new independent variable and $u = y'$ the new dependent variable,

$$\frac{dy}{dx} = u, \quad \frac{d^2y}{dx^2} = u\frac{du}{dy}, \quad \frac{d^3y}{dx^3} = u\left(\frac{du}{dy}\right)^2 + u^2\frac{d^2u}{dy^2}, \quad \ldots .$$

The order of the differential equation is reduced by 1.

Problems

Solve the following differential equations, if not stated otherwise.

Variable Separable

2.1 $\cos^2 y\, dx + (1+e^{-x}) \sin y\, dy = 0$ ANS $\ln(e^x+1) = -\dfrac{1}{\cos y} + C;\ \cos y = 0$

2.2 $\dfrac{dy}{dx} = \dfrac{x^3 e^{x^2}}{y \ln y}$ ANS $y^2\big(\ln y - \tfrac{1}{2}\big) = e^{x^2}(x^2-1) + C$

2.3 $x \cos^2 y\, dx + e^x \tan y\, dy = 0$ ANS $e^{-x}(x+1) = \dfrac{1}{2\cos^2 y} + C;\ \cos y = 0$

2.4 $x\,(y^2+1)\, dx + (2y+1)\, e^{-x}\, dy = 0$

ANS $(x-1)\, e^x + \ln(y^2+1) + \tan^{-1} y = C$

2.5 $x y^3\, dx + e^{x^2}\, dy = 0$ ANS $e^{-x^2} + \dfrac{1}{y^2} = C;\ y = 0$

2.6 $x \cos^2 y\, dx + \tan y\, dy = 0$ ANS $x^2 + \tan^2 y = C$

2.7 $x y^3\, dx + (y+1)\, e^{-x}\, dy = 0$ ANS $e^x\,(x-1) - \dfrac{1}{y} - \dfrac{1}{2y^2} = C;\ y = 0$

Homogeneous and Special Transformations

2.8 $\dfrac{dy}{dx} + \dfrac{x}{y} + 2 = 0$ ANS $\ln|x+y| + \dfrac{x}{x+y} = C;\ y = -x$

2.9 $x\, dy - y\, dx = x \cot\left(\dfrac{y}{x}\right) dx$ ANS $\cos\left(\dfrac{y}{x}\right) = \dfrac{C}{x}$

2.10 $\left[x \cos^2\left(\dfrac{y}{x}\right) - y\right] dx + x\, dy = 0$ ANS $\ln|x| + \tan\dfrac{y}{x} = C;\ \cos\dfrac{y}{x} = 0$

2.11 $x\, dy = y(1 + \ln y - \ln x)\, dx$ ANS $y = x e^{Cx}$

2.12 $xy\,dx + (x^2 + y^2)\,dy = 0$ ANS $y^2(2x^2 + y^2) = C$

2.13 $\left[1 + \exp\left(-\frac{y}{x}\right)\right] dy + \left(1 - \frac{y}{x}\right) dx = 0$ ANS $x \exp\left(\frac{y}{x}\right) + y = C$

2.14 $(x^2 - xy + y^2)\,dx - xy\,dy = 0$ ANS $(y - x)e^{y/x} = C$

2.15 $(3 + 2x + 4y)y' = 1 + x + 2y$

ANS $8y - 4x + \ln|4x + 8y + 5| = C;\quad 4x + 8y + 5 = 0$

2.16 $y' = \dfrac{2x + y - 1}{x - y - 2}$ ANS $\sqrt{2}\tan^{-1}\dfrac{y + 1}{\sqrt{2}(x - 1)} = \ln[(y+1)^2 + 2(x-1)^2] + C$

2.17 $(y + 2)\,dx = (2x + y - 4)\,dy$ ANS $(y+2)^2 = C(x + y - 1);\quad y = 1 - x$

2.18 $y' = \sin^2(x - y)$ ANS $x = \tan(x - y) + C;\; x - y = \frac{\pi}{2} \pm k\pi,\; k = 0, 1, 2, \ldots$

2.19 $\dfrac{dy}{dx} = (x+1)^2 + (4y+1)^2 + 8xy + 1$ ANS $\frac{2}{3}(x + 4y + 1) = \tan(6x + C)$

Exact Differential Equations

2.20 $(3x^2 + 6xy^2)\,dx + (6x^2y + 4y^3)\,dy = 0$ ANS $x^3 + 3x^2y^2 + y^4 = C$

2.21 $(2x^3 - xy^2 - 2y + 3)\,dx - (x^2y + 2x)\,dy = 0$

ANS $x^4 - x^2y^2 - 4xy + 6x = C$

2.22 $(xy^2 + x - 2y + 3)\,dx + x^2y\,dy = 2(x + y)\,dy$

ANS $x^2y^2 + x^2 + 6x - 4xy - 2y^2 = C$

2.23 $3y(x^2 - 1)\,dx + (x^3 + 8y - 3x)\,dy = 0$, when $x = 0$, $y = 1$

ANS $x^3y - 3xy + 4y^2 = 4$

2.24 $(x^2 + \ln y)\,dx + \dfrac{x}{y}\,dy = 0$ ANS $\frac{1}{3}x^3 + x\ln y = C$

2.25 $2x(3x + y - ye^{-x^2})\,dx + (x^2 + 3y^2 + e^{-x^2})\,dy = 0$

ANS $2x^3 + x^2y + ye^{-x^2} + y^3 = C$

2.26 $(3 + y + 2y^2\sin^2 x)\,dx + (x + 2xy - y\sin 2x)\,dy = 0$

ANS $3x + xy + xy^2 - \frac{1}{2}y^2\sin 2x = C$

2.27 $(2xy + y^2)\,dx + (x^2 + 2xy + y^2)\,dy = 0$ ANS $x^2y + xy^2 + \frac{1}{3}y^3 = C$

Integrating Factors

2.28 $(x^2 - \sin^2 y)\,dx + x\sin 2y\,dy = 0$ ANS $x + \dfrac{\sin^2 y}{x} = C$

2.29 $y(2x-y+2)\,dx + 2(x-y)\,dy = 0$ ANS $ye^x(2x-y) = C$

2.30 $(4xy+3y^2-x)\,dx + x(x+2y)\,dy = 0$ ANS $4x^4y + 4x^3y^2 - x^4 = C$

2.31 $y\,dx + x(y^2+\ln x)\,dy = 0$ ANS $3y\ln x + y^3 = C$

2.32 $(x^2+2x+y)\,dx + (3x^2y-x)\,dy = 0$ ANS $x+2\ln|x| - \dfrac{y}{x} + \dfrac{3}{2}y^2 = C$

2.33 $y^2\,dx + (xy+y^2-1)\,dy = 0$ ANS $xy + \frac{1}{2}y^2 - \ln|y| = C$

2.34 $3(x^2+y^2)\,dx + x(x^2+3y^2+6y)\,dy = 0$ ANS $xe^y(x^2+3y^2) = C$

2.35 $2y(x+y+2)\,dx + (y^2-x^2-4x-1)\,dy = 0$

ANS $x^2 + 4x + 2xy + y^2 + 1 = Cy$

2.36 $(2+y^2+2x)\,dx + 2y\,dy = 0$ ANS $e^x(2x+y^2) = C$

2.37 $(2xy^2-y)\,dx + (y^2+x+y)\,dy = 0$ ANS $x^2 - \dfrac{x}{y} + y + \ln|y| = C$

2.38 $y(x+y)\,dx + (x+2y-1)\,dy = 0$ ANS $e^x(xy+y^2-y) = C$

2.39 $2x(x^2-\sin y+1)\,dx + (x^2+1)\cos y\,dy = 0$ ANS $\ln(x^2+1) + \dfrac{\sin y}{x^2+1} = C$

2.40 Consider a homogeneous differential equation of the form

$$M(u)\,dx + N(u)\,dy = 0, \qquad u = \frac{y}{x}.$$

If $Mx+Ny=0$, i.e., $M(u)+N(u)u=0$, show that $\dfrac{1}{xM}$ is an integrating factor.

Method of Inspection

2.41 $(x^2+y+y^2)\,dx - x\,dy = 0$ ANS $x - \tan^{-1}\dfrac{y}{x} = C$

2.42 $\big(x-\sqrt{x^2+y^2}\big)\,dx + \big(y-\sqrt{x^2+y^2}\big)\,dy = 0$ ANS $\sqrt{x^2+y^2} - x - y = C$

2.43 $y\sqrt{1+y^2}\,dx + \big(x\sqrt{1+y^2}-y\big)\,dy = 0$ ANS $xy - \sqrt{1+y^2} = C$

2.44 $y^2\,dx - (xy+x^3)\,dy = 0$ ANS $\dfrac{1}{2}\left(\dfrac{y}{x}\right)^2 + y = C$

2.45 $y\,dx - x\,dy - 2x^3\tan\dfrac{y}{x}\,dx = 0$ ANS $\sin\dfrac{y}{x} = Ce^{-x^2}$

2.46 $(2x^2y^2+y)\,dx + (x^3y-x)\,dy = 0$ ANS $x^2y + \ln\left|\dfrac{x}{y}\right| = C$

2.47 $y^2\,dx + \big[xy+\tan(xy)\big]\,dy = 0$ ANS $y\sin(xy) = C$

2.48 $(2x^2y^4-y)\,dx + (4x^3y^3-x)\,dy = 0$ ANS $2xy^2 + \dfrac{1}{xy} = C$

Integrating Factors by Groups

2.49 $(x^2y^3+y)\,dx+(x^3y^2-x)\,dy=0$ ANS $\frac{1}{2}x^2y^2+\ln\left|\frac{x}{y}\right|=C$

2.50 $y(y^2+1)\,dx+x(y^2-x+1)\,dy=0$ ANS $\frac{1}{xy}-\frac{1}{y}-\tan^{-1}y=C$

2.51 $y^2\,dx+(e^x-y)\,dy=0$ ANS $-ye^{-x}+\ln|y|=C$

2.52 $(x^2y^2-2y)\,dx+(x^3y-x)\,dy=0$ ANS $\ln|xy|+\frac{1}{x^2y}=C$

2.53 $(2x^3y+y^3)\,dx-(x^4+2xy^2)\,dy=0$ ANS $4x^{-\frac{1}{3}}y^{\frac{2}{3}}-x^{\frac{8}{3}}y^{-\frac{4}{3}}=C$

Linear First-Order Equations

2.54 $(1+y\cos x)\,dx-\sin x\,dy=0$ ANS $y=-\cos x+C\sin x$

2.55 $(\sin^2 y+x\cot y)y'=1$ ANS $x=\sin y(C-\cos y)$

2.56 $dx-(y-2xy)\,dy=0$ ANS $2x=1+C\exp(-y^2)$

2.57 $dx-(1+2x\tan y)\,dy=0$ ANS $2x\cos^2 y=y+\sin y\cos y+C$

2.58 $\frac{dy}{dx}\left(y^3+\frac{x}{y}\right)=1$ ANS $x=\frac{1}{3}y^4+Cy$

2.59 $dx+(x-y^2)\,dy=0$ ANS $x=y^2-2y+2+Ce^{-y}$

2.60 $y^2\,dx+(xy+y^2-1)\,dy=0$ ANS $y^2+2xy-2\ln|y|=C$

2.61 $y\,dx=(e^y+2xy-2x)\,dy$ ANS $y^2x=Ce^{2y}-(y+1)e^y$

2.62 $(2x+3)y'=y+(2x+3)^{1/2},\ y(-1)=0$ ANS $2y=\sqrt{2x+3}\,\ln|2x+3|$

2.63 $y\,dx+(y^2e^y-x)\,dy=0$ ANS $x=Cy-ye^y$

2.64 $y'=1+3y\tan x$ ANS $y=\frac{1}{\cos^3 x}\left(\sin x-\frac{1}{3}\sin^3 x+C\right)$

2.65 $(1+\cos x)y'=\sin x(\sin x+\sin x\cos x-y)$

ANS $y=(1+\cos x)(x-\sin x+C)$

2.66 $y'=(\sin^2 x-y)\cos x$ ANS $y=\sin^2 x-2\sin x+2+Ce^{-\sin x}$

2.67 $xy'-ny-x^{n+2}e^x=0,\ n=\text{constant}$ ANS $y=x^n\left[e^x(x-1)+C\right]$

2.68 $(1+x)\frac{dy}{dx}-y=x(1+x)^2$ ANS $y=(1+x)\left(\frac{1}{2}x^2+C\right)$

2.69 $(1+y)\,dx+\left[x-y(1+y)^2\right]dy=0$ ANS $x=\frac{1}{1+y}\left(\frac{y^4}{4}+\frac{2y^3}{3}+\frac{y^2}{2}+C\right)$

2.70 Consider the first-order differential equation

$$\frac{dy}{dx} = \alpha(x)\,F(y) + \beta(x)\,G(y)$$

If $\dfrac{G'(y)F(y) - G(y)F'(y)}{F(y)} = a = \text{constant}$, then the transformation $u = \dfrac{G(y)}{F(y)}$ reduces the differential equation to a first-order linear differential equation. Show that the general solution of the differential equation is given by

$$\frac{G(y)}{F(y)} = \exp\Big[a\int\beta(x)dx\Big]\Big\{a\int\alpha(x)\exp\Big[-a\int\beta(x)dx\Big]dx + C\Big\}.$$

2.71 The *Riccati equation* is given by $y' = \alpha(x)y^2 + \beta(x)y + \gamma(x)$.

1. If one solution of this equation, say $y_1(x)$, is known, then the general solution can be found by using the transformation $y = y_1 + \dfrac{1}{u}$, where u is a new dependent variable. Show that u is given by
$$u = e^{-\int P(x)\,dx\,dx}\Big[\int Q(x)\,e^{\int P(x)\,dx\,dx}\,dx + C\Big],$$
where $P(x) = 2\alpha(x)y_1(x) + \beta(x)$ and $Q(x) = -\alpha(x)$.
2. For the differential equation $y' + y^2 = 1 + x^2$, first guess a solution $y_1(x)$ and then use the result of Part 1 to find the general solution $y(x)$.

ANS $y = x + \dfrac{e^{-x^2}}{\displaystyle\int e^{-x^2}dx + C}$

Bernoulli Differential Equations

2.72 $3xy' - 3xy^4\ln x - y = 0$ ANS $\dfrac{1}{y^3} = -\dfrac{3}{4}x(2\ln x - 1) + \dfrac{C}{x};\ y = 0$

2.73 $\dfrac{dy}{dx} = \dfrac{4x^3y^2}{x^4y + 2}$ ANS $x^4 = -\dfrac{1}{y} + Cy;\ y = 0$

2.74 $y(6y^2 - x - 1)\,dx + 2x\,dy = 0$ ANS $\dfrac{1}{y^2} = \dfrac{1}{x}(6 + Ce^{-x});\ y = 0$

2.75 $(1 + x)(y' + y^2) - y = 0$ ANS $\dfrac{1}{y} = \dfrac{1}{1+x}\Big(\dfrac{x^2}{2} + x + C\Big);\ y = 0$

2.76 $xyy' + y^2 - \sin x = 0$ ANS $x^2y^2 = -2x\cos x + 2\sin x + C$

2.77 $(2x^3 - y^4)\,dx + xy^3\,dy = 0$ ANS $y^4 = 8x^3 + Cx^4$

2.78 $y' - y\tan x + y^2\cos x = 0$ ANS $\dfrac{1}{y} = \cos x\,(x + C);\ y = 0$

2.79 $6y^2\,dx - x(2x^3 + y)\,dy = 0$ ANS $(y - 2x^3)^2 = Cyx^6;\ y = 0$

Equation Solvable for the Independent or Dependent Variable

2.80 $xy'^3 - yy'^2 + 1 = 0$ ANS $y = Cx + \dfrac{1}{C^2}$; $4y^3 = 27x^2$

2.81 $y = xy' + y'^3$ ANS $y = Cx + C^3$; $4x^3 + 27y^2 = 0$

2.82 $x(y'^2 - 1) = 2y'$ ANS $x = \dfrac{2p}{p^2-1}$, $y = \dfrac{2}{p^2-1} - \ln|p^2-1| + C$

2.83 $xy'(y'+2) = y$ ANS $y = -x$; $y = \pm 2C\sqrt{x} + C^2$

2.84 $x = y'\sqrt{y'^2+1}$ ANS $x = p\sqrt{p^2+1}$, $3y = \sqrt{p^2+1}(2p^2-1) + C$

2.85 $2y'^2(y - xy') = 1$ ANS $y = Cx + \dfrac{1}{2C^2}$; $8y^3 = 27x^2$

2.86 $y = 2xy' + y^2y'^3$ ANS $y^2 = 2Cx + C^3$; $32x^3 + 27y^4 = 0$; $y = 0$

2.87 $y'^3 + y^2 = xyy'$ ANS $p^3 + y^2 = xyp$, $\dfrac{p^4}{2y^2} - p = C$; $y = 0$

2.88 $2xy' - y = y'\ln(yy')$ ANS $2x = 1 + 2\ln|y|$; $y^2 = 2Cx - C\ln C$

2.89 $y = xy' - x^2y'^3$ ANS $y = 0$; $xp^2 = C\sqrt{|p|} - 1$, $y = xp - x^2p^3$

2.90 $y(y - 2xy')^3 = y'^2$ ANS $27x^2y^2 = 1$; $y^2 = 2C^3x + C^2$

2.91 $y + xy' = 4\sqrt{y'}$ ANS $x = \dfrac{\ln p + C}{\sqrt{p}}$, $y = \sqrt{p}\,(4 - \ln p - C)$; $y = 0$

2.92 $2xy' - y = \ln y'$ ANS $x = \dfrac{1}{p} + \dfrac{C}{p^2}$, $y = 2\left(1 + \dfrac{C}{p}\right) - \ln p$

Simple Higher-Order Differential Equations

2.93 $y'' = 2yy'^3$ ANS $y = C$; $3x + y^3 + C_1y = C_2$

2.94 $yy'' = y'^2 - y'^3$ ANS $y = C$; $C_1\ln|y| + y = x + C_2$

2.95 $xy''' = (1-x)y''$ ANS $y = C_1(x+2)e^{-x} + C_2x + C_3$

2.96 $y'' = e^x y'^2$ ANS $y = \dfrac{1}{C_1}\ln|1 + C_1e^{-x}| + C_2$; $y = C$

2.97 $yy'' + y'^2 = 0$ ANS $y = C$; $\tfrac{1}{2}y^2 = C_1x + C_2$

2.98 $1 + y'^2 = 2yy''$ ANS $4(C_1y - 1) = C_1^2(x + C_2)^2$

2.99 $xy'' = y'(\ln y' - \ln x)$

ANS $y = \dfrac{1}{C_1}e^{C_1x+1}\left(x - \dfrac{1}{C_1}\right) + C_2$; $y = \tfrac{1}{2}ex^2 + C$

2.100 $3yy'y'' - y'^3 + 1 = 0$ ANS $3(C_1y+1)^{2/3} - 2C_1x = C_2;\ \ y = x$

2.101 $y'' - y'^2 - 1 = 0$ ANS $y = -\ln\left|\cos(x+C_1)\right| + C_2$

2.102 $x^3y'' - x^2y' = 3 - x^2$ ANS $y = \dfrac{1}{x} + x + C_1x^2 + C_2$

2.103 $2y'' = y'^3 \sin 2x,\ y(0) = 1,\ y'(0) = 1$ ANS $y = 1 + \ln\left|\sec x + \tan x\right|$

2.104 $x\dfrac{d^2y}{dx^2} = 2 - \dfrac{dy}{dx}$ ANS $y = 2x + C_1\ln|x| + C_2$

2.105 $y'' = 3\sqrt{y},\ \ y(0) = 1,\ y'(0) = 2$ ANS $y = \left(\pm\tfrac{1}{2}x + 1\right)^4$

2.106 $x\dfrac{d^2y}{dx^2} = \dfrac{dy}{dx} + x\sin\left(\dfrac{1}{x}\cdot\dfrac{dy}{dx}\right)$

ANS $y = \left(x^2 + \dfrac{1}{C_1^2}\right)\tan^{-1}C_1x - \dfrac{x}{C_1} + C_2;\ \ y = \dfrac{k\pi}{2}x^2 + C,\ k = 0, \pm1, \pm2, \ldots$

2.107 $yy'' = y'^2(1 - y'\sin y - yy'\cos y)$

ANS $y = C;\ \ x = -\cos y + C_1\ln|y| + C_2$

2.108 $y'' + xy' = x$ ANS $y = x + C_1\displaystyle\int e^{-\frac{1}{2}x^2}\,dx + C_2$

2.109 $xy'' - y'^3 - y' = 0$ ANS $x^2 + (y - C_1)^2 = C_2;\ \ y = C$

2.110 $y(1 - \ln y)y'' + (1 + \ln y)y'^2 = 0$

ANS $y = C;\ \ (C_1x + C_2)(\ln y - 1) + 1 = 0$

Review Problems

2.111 $xy^2(xy' + y) = 1$ ANS $2x^3y^3 - 3x^2 = C$

2.112 $5y + y'^2 = x(x + y')$ ANS $4y = x^2;\ 5y = -5x^2 + 5Cx - C^2$

2.113 $y' = \dfrac{y+2}{x+1} + \tan\dfrac{y-2x}{x+1}$

ANS $\sin\dfrac{y-2x}{x+1} = C(x+1);\ \ \ \dfrac{y+2}{x+1} = n\pi + 2,\ n = 0, \pm1, \pm2, \ldots$

2.114 $y''(e^x + 1) + y' = 0$ ANS $y = C_1(x - e^{-x}) + C_2$

2.115 $xy' = y - xe^{y/x}$ ANS $y = -x\ln\left|\ln|Cx|\right|$

2.116 $(1 + y^2\sin 2x)\,dx - 2y\cos^2 x\,dy = 0$ ANS $x - y^2\cos^2 x = C$

2.117 $(2\sqrt{xy} - y)\,dx - x\,dy = 0,\ x>0,\ y>0$ ANS $\sqrt{xy} - x = C$

2.118 $y'' + y'^2 = 2e^{-y}$ ANS $e^y + C_1 = (x + C_2)^2$

2.119 $y' = e^{xy'/y}$ ANS $y = ex;\quad Cx = \ln|Cy|$

2.120 $(2x^3y^2 - y)\,dx + (2x^2y^3 - x)\,dy = 0$ ANS $x^2 + y^2 + \dfrac{1}{xy} = C$

2.121 $(y-1-xy)\,dx + x\,dy = 0$ ANS $xy + 1 = Ce^x$

2.122 $xy' - y = x\tan\dfrac{y}{x}$ ANS $\sin\dfrac{y}{x} = Cx$

2.123 $y' + \dfrac{y}{x} = e^{xy}$ ANS $-e^{-xy} = \tfrac{1}{2}x^2 + C$

2.124 $yy'' - yy' = (y')^2$ ANS $\ln\big|\ln|y| + C_2\big| = x + C_1;\quad y = C$

2.125 $2y\,dx - x\big[\ln(x^2y) - 1\big]\,dy = 0$ ANS $\ln(x^2y) - Cy = 0$

2.126 $y' = \dfrac{1}{xy + x^3y^3}$ ANS $\dfrac{1}{x^2} = 1 - y^2 + Ce^{-y^2}$

2.127 $y' = 2\left(\dfrac{y+2}{x+y-1}\right)^2$ ANS $y + 2 = C\exp\left[-2\tan^{-1}\dfrac{y+2}{x-3}\right]$

2.128 $(e^x + 3y^2)\,dx + 2xy\,dy = 0$ ANS $e^x(x^2 - 2x + 2) + x^3y^2 = C$

2.129 $(xy + 2x^3y)\,dx + x^2dy = 0$ ANS $xye^{x^2} = C$

2.130 $x(y')^2 - 2yy' + 4x = 0$ ANS $y = \dfrac{C}{2}x^2 + \dfrac{2}{C};\quad y = \pm 2x$

2.131 $y''' = 2(y'' - 1)\cot x$ ANS $y = (C_2+1)\dfrac{x^2}{2} + \dfrac{C_2}{4}\cos 2x + C_1x + C_0$

2.132 $(y + 3x^4y^2)\,dx + (x + 2x^2y^3)\,dy = 0$ ANS $-\dfrac{1}{xy} + x^3 + y^2 = C;\quad y = 0$

2.133 $xy' = y + \sqrt{x^2 - y^2},\ x > 0,\ |y| \leqslant |x|$ ANS $\sin^{-1}\dfrac{y}{x} = \ln|x| + C;\ y = \pm x$

2.134 $2y(xe^{x^2} + y\sin x\cos x)\,dx + (2e^{x^2} + 3y\sin^2 x)\,dy = 0$

ANS $y^2e^{x^2} + y^3\sin^2 x = C$

2.135 $\cos y\,dx + \sin y(x - \sin y\cos y)\,dy = 0$

ANS $x = \cos y\big(\ln|\sec y + \tan y| - \sin y + C\big)$

2.136 $y^3\,dx + (3x^2 - 2xy^2)\,dy = 0$ ANS $y^3 = Ce^{y^2/x}$

2.137 $(y' + 1)\ln\dfrac{y+x}{x+3} = \dfrac{y+x}{x+3}$ ANS $\ln\dfrac{y+x}{x+3} = 1 + \dfrac{C}{x+y}$

2.138 $2x^3yy' + 3x^2y^2 + 7 = 0$ ANS $x^3y^2 + 7x = C$

2.139 $\left(x - y\cos\dfrac{y}{x}\right)dx + x\cos\dfrac{y}{x}\,dy = 0$ ANS $\sin\dfrac{y}{x} = C - \ln|x|$

2.140 $x^2(x\,dy - y\,dx) = (x+y)y\,dx$ ANS $\ln\left|\dfrac{x}{y}+1\right| = \dfrac{1}{x} + C$

2.141 $(y^4+xy)\,dx + (xy^3-x^2)\,dy = 0$ ANS $2xy+\left(\dfrac{x}{y}\right)^2 = C;\ \ y=0$

2.142 $(x^2 + 3\ln y)\,dx - \dfrac{x}{y}\,dy = 0$ ANS $x^2 + \ln y = Cx^3$

2.143 $xy'' = y' + x$ ANS $4y = x^2\left(2\ln|x| + C_1\right) + C_2$

2.144 $y\,dx + (xy - x - y^3)\,dy = 0$ ANS $x = y(y-1) + Cye^{-y};\ \ y = 0$

2.145 $y + 2y^3y' = (x + 4y\ln y)\,y'$ ANS $\dfrac{x}{y} + y^2 - 2(\ln y)^2 = C$

2.146 $y\ln x\,\ln y\,dx + dy = 0$ ANS $x\ln x - x + \ln|\ln y| = C$

2.147 $(2x\sqrt{x} + x^2 + y^2)\,dx + 2y\sqrt{x}\,dy = 0$ ANS $\ln(x^2+y^2) + 2\sqrt{x} = C$

2.148 $\left[2x + y\cos(xy)\right]dx + x\cos(xy)\,dy = 0$ ANS $x^2 + \sin(xy) = C$

2.149 $yy'' - y^2y' - y'^2 = 0$ ANS $y = C;\ \ \dfrac{1}{C_1}\ln\left|\dfrac{y}{y+C_1}\right| = x + C_2$

2.150 $2y' + x = 4\sqrt{y}$ ANS $(2\sqrt{y}-x)\ln\left|C(2\sqrt{y}-x)\right| = x;\ \ 2\sqrt{y} = x$

2.151 $2y'^3 - 3y'^2 + x = y$ ANS $y = x-1;\ \ 4(x+C)^3 = 27(y+C)^2$

2.152 $y' - 6xe^{x-y} - 1 = 0$ ANS $e^{y-x} = 3x^2 + C$

2.153 $(1 + y^2)y'' + y'^3 + y' = 0$

ANS $-C_1y - (1 + C_1^2)\ln|y - C_1| = x + C_2;\ \ y = C$

2.154 $(y\sin x + \cos^2 x)\,dx - \cos x\,dy = 0$ ANS $-y\cos x + \dfrac{x}{2} + \dfrac{\sin 2x}{4} = C$

2.155 $y(6y^2 - x - 1)\,dx + 2x\,dy = 0$ ANS $\dfrac{x}{y^2} = 6 + Ce^{-x};\ \ y = 0$

2.156 $y'(x - \ln y') = 1$ ANS $x = \ln p + p^{-1},\ \ y = p - \ln p + C$

2.157 $(1 + \cos x)y' + \sin x\,(\sin x + \sin x\cos x - y) = 0$

ANS $\frac{1}{2}x - \frac{1}{4}\sin 2x + \frac{1}{3}\sin^3 x + y\cos x + y = C$

2.158 $x\,dx + \sin^2\!\left(\dfrac{y}{x}\right)(y\,dx - x\,dy) = 0$ ANS $\dfrac{y}{2x} - \dfrac{1}{4}\sin\dfrac{2y}{x} = \ln|x| + C$

2.159 $(2xy^4e^y + 2xy^3 + y)\,dx + (x^2y^4e^y - x^2y^2 - 3x)\,dy = 0$

ANS $x^2e^y + \dfrac{x^2}{y} + \dfrac{x}{y^3} = C$

2.160 $(xy^3 - 1)\,dx + x^2y^2\,dy = 0$ ANS $2x^3y^3 - 3x^2 = C$

CHAPTER 3

Applications of First-Order and Simple Higher-Order Equations

In this chapter, a number of examples are studied to illustrate the application of first-order and simple higher-order differential equations in various science and engineering disciplines.

3.1 Heating and Cooling

Problems involving heating and cooling follow Newton's Law of Cooling.

Newton's Law of Cooling

The rate of change in the temperature $T(t)$, dT/dt, of a body in a medium of temperature T_m is proportional to the temperature difference between the body and the medium, i.e.,

$$\frac{dT}{dt} = -k(T - T_m),$$

where $k > 0$ is a constant of proportionality.

Example 3.1 — Body Cooling in Air

A body cools in air of constant temperature $T_m = 20°C$. If the temperature of the body changes from 100°C to 60°C in 20 minutes, determine how much more time it will need for the temperature to fall to 30°C.

Newton's Law of Cooling requires that

$$\frac{dT}{dt} = -k(T - T_m).$$ ✍ Variable separable

The general solution is

$$\int \frac{dT}{T-T_m} = -k\int dt + C \implies \ln|T-T_m| = -kt + \ln C$$

$$\therefore \quad T = T_m + Ce^{-kt}.$$

At $t=0$, $T=100°$C:

$$100 = 20 + Ce^{-k\cdot 0} = 20 + C \implies C = 80.$$

At $t=20$ min, $T=60°$C:

$$60 = 20 + 80e^{-k\cdot 20} \implies k = -\frac{1}{20}\ln\frac{60-20}{80} = 0.03466.$$

Hence

$$T = 20 + 80e^{-0.03466t} \qquad \text{or} \qquad t = -\frac{1}{0.03466}\ln\frac{T-20}{80} \text{ min.}$$

When $T=30°$C:

$$t = -\frac{1}{0.03466}\ln\frac{30-20}{80} = 60 \text{ min.}$$

Hence, it will need another $60-20=40$ minutes for the temperature to fall to 30°C.

Example 3.2 — Heating in a Building

The rate of heat loss from a building is equal to $K_1\big[T_B(t)-T_A(t)\big]$, where $T_B(t)$ and $T_A(t)$ are the temperatures of the building and the atmosphere at time t, respectively, and K_1 is a constant. The rate of heat supplied to the building by the heating system is given by $Q+K_2\big[T_S-T_B(t)\big]$, where T_S is the "set" temperature of the building, and Q and K_2 are constants. The value of Q is such that the building is maintained at the "set" temperature when the atmosphere is at constant temperature T_0. The thermal capacity of the building is c.

1. Set up the differential equation governing the temperature of the building $T_B(t)$.
2. If the atmospheric temperature fluctuates sinusoidally about the mean value T_0 with an amplitude of T_1 (°C) and a period of $2\pi/\omega$ (hour), i.e.,

$$T_A = T_0 + T_1 \sin\omega t,$$

   determine the amplitude of temperature variation of the building due to atmospheric temperature fluctuation.
3. Suppose $T_1=12°$C, $c/K_1=4$ hour, the atmospheric temperature fluctuates with a period of 24 hours, i.e., $\omega=\pi/12$. The temperature of the building is required to remain within 3°C of the set value, i.e., the amplitude of

temperature fluctuation is less than or equal to 3°C. Show that the value of the ratio K_2/K_1 must satisfy

$$\frac{K_2}{K_1} \geqslant 4\sqrt{1-\frac{\pi^2}{144}} - 1.$$

---

1. Consider a time period from t to $t+\Delta t$, the Principle of Conservation of Energy requires

$$(\text{Heat supply in time } \Delta t) - (\text{Heat loss in time } \Delta t) = c\,\Delta T_B,$$

where

$$\begin{aligned}\text{Heat supply in time } \Delta t &= \big\{Q + K_2\big[T_S - T_B(t)\big]\big\}\,\Delta t,\\ \text{Heat loss in time } \Delta t &= K_1\big[T_B(t) - T_A(t)\big]\,\Delta t.\end{aligned}$$

Hence,

$$\big\{Q + K_2\big[T_S - T_B(t)\big]\big\}\,\Delta t - K_1\big[T_B(t) - T_A(t)\big]\,\Delta t = c\,\Delta T_B.$$

Dividing the equation by Δt and taking the limit as $\Delta t \to 0$ lead to

$$Q + K_2\big[T_S - T_B(t)\big] - K_1\big[T_B(t) - T_A(t)\big] = c\,\frac{dT_B}{dt},$$

or

$$\frac{c}{K_1}\,\frac{dT_B(t)}{dt} + \Big(1+\frac{K_2}{K_1}\Big)T_B(t) = \frac{Q}{K_1} + \frac{K_2}{K_1}T_S + T_A(t).$$

2. Since $T_A = T_0 + T_1 \sin\omega t$, the differential equation becomes

$$\frac{c}{K_1}\,\frac{dT_B(t)}{dt} + \Big(1+\frac{K_2}{K_1}\Big)T_B(t) = \Big(\frac{Q}{K_1} + \frac{K_2}{K_1}T_S + T_0\Big) + T_1 \sin\omega t,$$

or

$$\frac{dT_B(t)}{dt} + \underbrace{\frac{K_1}{c}\Big(1+\frac{K_2}{K_1}\Big)}_{k}T_B(t) = \underbrace{\frac{K_1}{c}\Big(\frac{Q}{K_1} + \frac{K_2}{K_1}T_S + T_0\Big)}_{\alpha_0} + \underbrace{\frac{K_1T_1}{c}}_{\alpha_1}\sin\omega t.$$

The differential equation is linear first-order of the form

$$\frac{dT_B(t)}{dt} + P(t)\cdot T_B(t) = Q(t), \quad P(t) = k, \quad Q(t) = \alpha_0 + \alpha_1 \sin\omega t,$$

where

$$k = \frac{K_1}{c}\Big(1+\frac{K_2}{K_1}\Big), \quad \alpha_0 = \frac{K_1}{c}\Big(\frac{Q}{K_1} + \frac{K_2}{K_1}T_S + T_0\Big), \quad \alpha_1 = \frac{K_1T_1}{c}.$$

It is easy to evaluate

$$\int P(t)\,\mathrm{d}t = kt, \qquad \mathrm{e}^{\int P(t)\,\mathrm{d}t} = \mathrm{e}^{kt}, \qquad \mathrm{e}^{-\int P(t)\,\mathrm{d}t} = \mathrm{e}^{-kt},$$

$$\begin{aligned}\int Q(t)\,\mathrm{e}^{\int P(t)\,\mathrm{d}t}\,\mathrm{d}t &= \int (\alpha_0 + \alpha_1 \sin\omega t)\,\mathrm{e}^{kt}\,\mathrm{d}t\\ &= \frac{\alpha_0}{k}\,\mathrm{e}^{kt} + \frac{\alpha_1}{k^2+\omega^2}\,\mathrm{e}^{kt}(k\sin\omega t - \omega\cos\omega t)\\ &= \mathrm{e}^{kt}\left[\frac{\alpha_0}{k} + \frac{\alpha_1}{\sqrt{k^2+\omega^2}}\sin(\omega t-\varphi)\right], \qquad \varphi = \tan^{-1}\frac{\omega}{k}.\end{aligned}$$

✍ See the Remarks on page 195 on finding the amplitude of $A\cos\omega t + B\sin\omega t$.

Hence, the general solution is

$$\begin{aligned}T_B(t) &= \mathrm{e}^{-\int P(t)\,\mathrm{d}t}\left[\int Q(t)\,\mathrm{e}^{\int P(t)\,\mathrm{d}t}\,\mathrm{d}t + C\right]\\ &= \mathrm{e}^{-kt}\left\{\mathrm{e}^{kt}\left[\frac{\alpha_0}{k} + \frac{\alpha_1}{\sqrt{k^2+\omega^2}}\sin(\omega t-\varphi)\right] + C\right\}\\ &= \frac{\alpha_0}{k} + \underbrace{\frac{\alpha_1}{\sqrt{k^2+\omega^2}}\sin(\omega t-\varphi)}_{\text{Variation due to atmospheric temperature fluctuation}} + C\,\mathrm{e}^{-kt}.\end{aligned}$$

The amplitude of temperature variation due to atmospheric temperature fluctuation is

$$a = \frac{\alpha_1}{\sqrt{k^2+\omega^2}} = \frac{\dfrac{K_1 T_1}{c}}{\sqrt{\left[\dfrac{K_1}{c}\left(1+\dfrac{K_2}{K_1}\right)\right]^2 + \omega^2}} = \frac{T_1}{\sqrt{\left(1+\dfrac{K_2}{K_1}\right)^2 + \left(\dfrac{c\omega}{K_1}\right)^2}}.$$

3. That the amplitude of temperature fluctuation is less than or equal to 3°C means $a \leqslant 3$, i.e.,

$$\frac{T_1}{\sqrt{\left(1+\dfrac{K_2}{K_1}\right)^2 + \left(\dfrac{c\omega}{K_1}\right)^2}} \leqslant 3,$$

which gives

$$\left(\frac{T_1}{3}\right)^2 \leqslant \left(1+\frac{K_2}{K_1}\right)^2 + \left(\frac{c\omega}{K_1}\right)^2 \implies \frac{K_2}{K_1} \geqslant \sqrt{\left(\frac{T_1}{3}\right)^2 - \left(\frac{c\omega}{K_1}\right)^2} - 1,$$

$$\therefore \quad \frac{K_2}{K_1} \geqslant \sqrt{\left(\frac{12}{3}\right)^2 - \left(4\cdot\frac{\pi}{12}\right)^2} - 1 = 4\sqrt{1-\frac{\pi^2}{144}} - 1.$$

3.2 Motion of a Particle in a Resisting Medium

Newton's Second Law and D'Alembert's Principle

Newton's Second Law: **The product of the mass of an object and its acceleration is equal to the sum of forces applied on the object,** i.e., $ma = \sum F$.

D'Alembert's Principle: Rewrite Newton's Second Law as $\sum F - ma = 0$. Treat $-ma$ as a force, known as the *inertia force.* **An object is in (dynamic) equilibrium under the action of all the forces applied, including the inertia force.** This is known as *D'Alembert's Principle*, which transforms a problem in dynamics into a problem of static equilibrium.

Impulse-Momentum Principle

For a system of particles, the change in momentum of the system is equal to the total impulse on the system, i.e.,

(Momentum at time t_2) − (Momentum at time t_1) = (Impulse during $t_2 - t_1$).

The momentum of a mass m moving at velocity v is equal to mv. The impulse of a force F during time interval Δt is equal to $F\Delta t$.

Consider the motion of a particle moving in a resisting medium, such as air or water. The medium exerts a resisting force R on the particle. In many applications, the resisting force R is proportional to v^n, where v is the velocity of the particle and $n > 0$, and is opposite to the direction of the velocity. Hence, the resisting force can be expressed as $R = \beta v^n$, where β is a constant. For particles moving in an unbounded viscous medium at low speed, the resisting force is $R = \beta v$, i.e., $n = 1$.

In the following, the case with $R = \beta v$ will be studied for motion in the vertical direction and specific initial conditions.

Case I: Upward Motion

Consider an object being launched vertically at time $t = 0$ from $x = 0$ with initial velocity v_0 as shown in Figure 3.1.

The displacement x, the velocity $v = \dot{x}$, and the acceleration $a = \dot{v} = \ddot{x}$ are taken as positive in the upward direction. The particle is subjected to two forces: the downward gravity mg and the resisting force from the medium $R = \beta v$, which is opposite to the direction of the velocity and hence is downward.

From Newton's Second Law, the equation of motion is

$$\uparrow\ ma = \sum F: \quad m\frac{dv}{dt} = -R - mg, \quad R = \beta v, \quad m = \frac{w}{g},$$

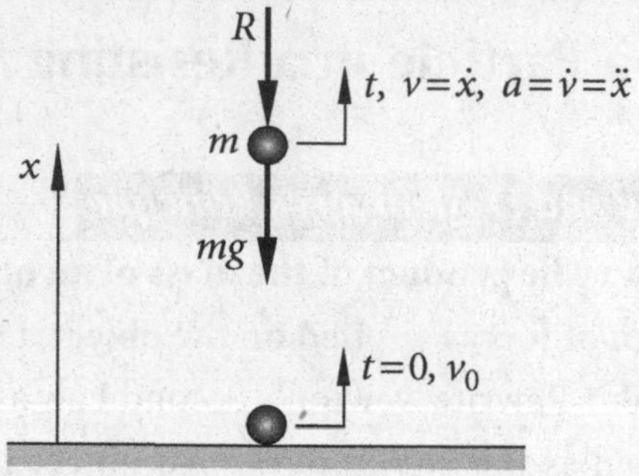

Figure 3.1 Upward motion of a particle in a resisting medium.

$$\therefore \quad \frac{\mathrm{d}v}{\mathrm{d}t} = -g(\alpha v + 1), \quad \alpha = \frac{\beta}{w} > 0, \qquad \text{✍ Variable separable}$$

$$\int \frac{\mathrm{d}v}{\alpha v + 1} = -\int g\,\mathrm{d}t + C \implies \frac{1}{\alpha}\ln(\alpha v + 1) = -gt + C. \tag{1}$$

Constant C is determined from the initial condition $t=0$, $v=v_0$:

$$\frac{1}{\alpha}\ln(\alpha v_0 + 1) = 0 + C.$$

Substituting into equation (1) yields

$$\frac{1}{\alpha}\ln(\alpha v + 1) = -gt + \frac{1}{\alpha}\ln(\alpha v_0 + 1),$$

$$\frac{1}{\alpha}\ln\left(\frac{\alpha v + 1}{\alpha v_0 + 1}\right) = -gt \implies \frac{\alpha v + 1}{\alpha v_0 + 1} = \mathrm{e}^{-\alpha g t}.$$

Solving for v leads to

$$\alpha v = \mathrm{e}^{-\alpha g t}(\alpha v_0 + 1) - 1. \tag{2}$$

When the object reaches the maximum height at time $t = t_{\max}$, $v=0$, and

$$\alpha \cdot 0 = \mathrm{e}^{-\alpha g t_{\max}}(\alpha v_0 + 1) - 1 \implies t_{\max} = \frac{1}{\alpha g}\ln(\alpha v_0 + 1).$$

To determine the displacement $x(t)$, note that $v = \dfrac{\mathrm{d}x}{\mathrm{d}t}$ and use equation (2)

$$\frac{\mathrm{d}x}{\mathrm{d}t} = \frac{1}{\alpha}(\alpha v_0 + 1)\,\mathrm{e}^{-\alpha g t} - \frac{1}{\alpha}. \qquad \text{✍ Immediately integrable}$$

Integrating with respect to x gives

$$x = \frac{1}{\alpha}(\alpha v_0 + 1)\int \mathrm{e}^{-\alpha g t}\mathrm{d}t - \frac{t}{\alpha} + D = -\frac{1}{\alpha^2 g}(\alpha v_0 + 1)\,\mathrm{e}^{-\alpha g t} - \frac{t}{\alpha} + D.$$

Constant D is determined from the initial condition $t=0$, $x=0$:

$$0 = -\frac{1}{\alpha^2 g}(\alpha v_0 + 1)\cdot \mathrm{e}^0 - 0 + D \implies D = \frac{1}{\alpha^2 g}(\alpha v_0 + 1).$$

Hence

$$x = -\frac{1}{\alpha^2 g}(\alpha v_0 + 1)(e^{-\alpha g t} - 1) - \frac{t}{\alpha}.$$

At time $t = t_{max}$, the object reaches the maximum height given by

$$x = x_{max} = x(t)\Big|_{t=t_{max}} = -\frac{1}{\alpha^2 g}(\alpha v_0 + 1)\left[e^{-\ln(\alpha v_0+1)} - 1\right] - \frac{1}{\alpha}\cdot\frac{1}{\alpha g}\ln(\alpha v_0 + 1)$$

$$= \frac{1}{\alpha^2 g}\left[\alpha v_0 - \ln(\alpha v_0 + 1)\right].$$

Case II: Downward Motion

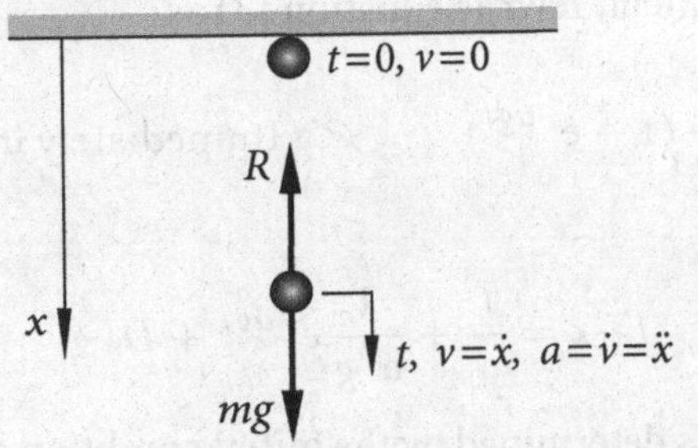

Figure 3.2 Downward motion of a particle in a resisting medium.

Consider an object being released and dropped at time $t = 0$ from $x = 0$ with $v = 0$ as shown in Figure 3.2. In this case, it is more convenient to take x, v, and a as positive in the downward direction. Newton's Second Law requires

$$\downarrow ma = \sum F: \qquad m\frac{dv}{dt} = mg - R, \quad R = \beta v, \quad m = \frac{w}{g},$$

$$\therefore \quad \frac{dv}{dt} = g - \alpha g v, \quad \alpha = \frac{\beta}{w} > 0.$$ ✍ Variable separable

The equation can be solved easily as

$$\int \frac{dv}{1 - \alpha v} = \int g\,dt + C \implies -\frac{1}{\alpha}\ln\left|1 - \alpha v\right| = gt + C,$$

where the constant C is determined from the initial condition $t = 0$, $v = 0$:

$$-\frac{1}{\alpha}\ln 1 = 0 + C \implies C = 0,$$

$$\therefore \quad -\frac{1}{\alpha}\ln\left|1 - \alpha v\right| = gt \implies v = \frac{1}{\alpha}(1 - e^{-\alpha g t}). \qquad (3)$$

When time t approaches infinity, the velocity approaches a constant, the so-called terminal velocity,

$$v = v_{terminal} = \lim_{t\to\infty} v = \frac{1}{\alpha}.$$

The change of velocity with time is shown in Figure 3.3.

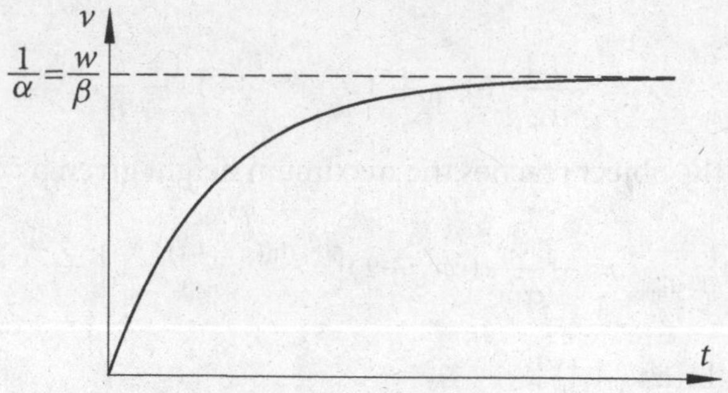

Figure 3.3 Velocity of a particle moving downward in a resisting medium.

To find the displacement, rewrite equation (3) as

$$\frac{dx}{dt} = \frac{1}{\alpha}(1 - e^{-\alpha g t}). \qquad \text{✍ Immediately integrable}$$

Integrating yields

$$x = \frac{t}{\alpha} + \frac{1}{\alpha^2 g}\, e^{-\alpha g t} + D,$$

where the constant D is determined by the initial condition $t=0$, $x=0$:

$$0 = 0 + \frac{1}{\alpha^2 g}\cdot e^0 + D \Longrightarrow D = -\frac{1}{\alpha^2 g}.$$

Hence the displacement is given by

$$x = \frac{t}{\alpha} + \frac{1}{\alpha^2 g}(e^{-\alpha g t} - 1).$$

Example 3.3 — Bullet through a Plate

A bullet is fired perpendicularly into a plate at an initial speed of $v_0 = 100$ m/sec. When the bullet exits the plate, its speed is $v_1 = 80$ m/sec. It is known that the thickness of the plate is $b = 0.1$ m and the resistant force of the plate on the bullet is proportional to the square of the speed of the bullet, i.e., $R = \beta v^2$. Determine the time T that the bullet takes to pass through the plate.

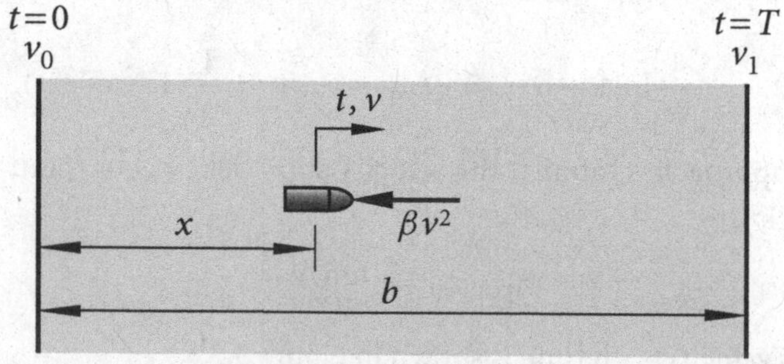

Applying Newton's Second Law to the bullet as shown yields

$$\to\ ma=\sum F:\quad m\frac{dv}{dt}=-\beta v^2. \qquad \text{✍ Variable separable}$$

The general solution is given by

$$\int -\frac{dv}{v^2}=\int \frac{\beta}{m}\,dt+C \implies \frac{1}{v}=kt+C,\quad k=\frac{\beta}{m},$$

where the constant C is determined from the initial condition $t=0$, $v=v_0$:

$$\frac{1}{v_0}=k\cdot 0+C \implies C=\frac{1}{v_0}.$$

Hence

$$\frac{1}{v}=kt+\frac{1}{v_0} \implies v=\frac{dx}{dt}=\frac{1}{kt+\dfrac{1}{v_0}}. \qquad \text{✍ Immediately integrable} \quad (1)$$

Integrating with respect to t leads to

$$x=\frac{1}{k}\ln\Big(kt+\frac{1}{v_0}\Big)+D,$$

where the constant D is determined from the initial condition $t=0$, $x=0$:

$$0=\frac{1}{k}\ln\Big(k\cdot 0+\frac{1}{v_0}\Big)+D \implies D=-\frac{1}{k}\ln\frac{1}{v_0}.$$

Hence

$$x=\frac{1}{k}\ln\Big(kt+\frac{1}{v_0}\Big)-\frac{1}{k}\ln\frac{1}{v_0}. \qquad (2)$$

From equation (1),

$$t=T,\ v=v_1:\qquad \frac{1}{v_1}=kT+\frac{1}{v_0}. \qquad (3)$$

From equation (2), $t=T$, $x=b$:

$$b=\frac{1}{k}\ln\Big(kT+\frac{1}{v_0}\Big)-\frac{1}{k}\ln\frac{1}{v_0}=\frac{1}{k}\ln\frac{1}{v_1}-\frac{1}{k}\ln\frac{1}{v_0}=\frac{1}{k}\ln\frac{v_0}{v_1} \implies k=\frac{1}{b}\ln\frac{v_0}{v_1}.$$

Using equation (3),

$$T=\frac{1}{k}\Big(\frac{1}{v_1}-\frac{1}{v_0}\Big)=b\,\frac{\dfrac{1}{v_1}-\dfrac{1}{v_0}}{\ln\dfrac{v_0}{v_1}}=0.1\times\frac{\dfrac{1}{80}-\dfrac{1}{100}}{\ln\dfrac{100}{80}}=0.000819\ \text{sec}.$$

Example 3.4 — Object Falling in Air

An object of mass m falls against air resistance which is proportional to the speed (i.e., $R=\beta v$) and under gravity g.

1. If v_0 and v_E are the initial and final (terminal) speeds, and v is the speed at time t, show that
$$\frac{v-v_E}{v_0-v_E}=\mathrm{e}^{-kt}, \qquad k=\frac{\beta}{m}.$$
2. The speed of the object is found to be 30, 40, 45 m/sec at times $t=1$, 2, and 3 sec, respectively, after starting. Find v_E and v_0.
3. At what time will the speed of the object be 49 m/sec?

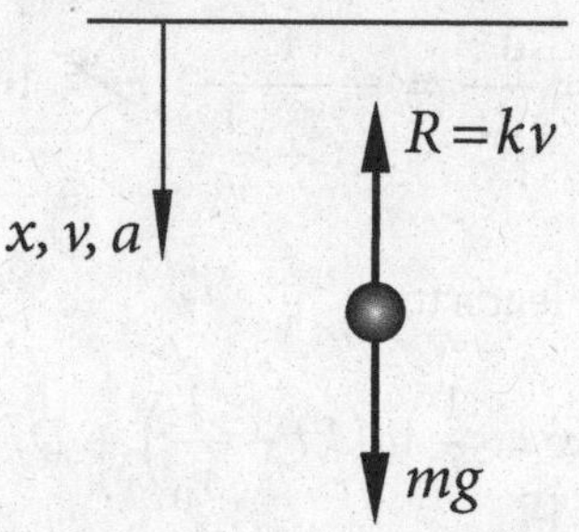

1. The object is subjected to two forces as shown: the downward gravity mg and the upward air resistance βv. Newton's Second Law requires

$$\downarrow\ ma=\sum F: \quad m\frac{\mathrm{d}v}{\mathrm{d}t}=mg-\beta v \implies \frac{\mathrm{d}v}{\mathrm{d}t}=g-kv, \quad k=\frac{\beta}{m}.$$

Noting that $g-kv>0$, the equation is variable separable and the solution is

$$\int\frac{\mathrm{d}v}{g-kv}=\int\mathrm{d}t+C \implies -\frac{1}{k}\ln(g-kv)=t+C \implies v=\frac{g}{k}-C\mathrm{e}^{-kt}.$$

Constant C is determined from the initial condition $t=0$, $v=v_0$:

$$v_0=\frac{g}{k}-C\mathrm{e}^0 \implies C=\frac{g}{k}-v_0.$$

When $t\to\infty$, $v=v_E \Rightarrow v_E=\dfrac{g}{k} \Rightarrow C=v_E-v_0$. Hence the velocity is given by

$$v=v_E-(v_E-v_0)\mathrm{e}^{-kt} \implies \frac{v-v_E}{v_0-v_E}=\mathrm{e}^{-kt}. \tag{1}$$

2. From equation (1),

$$t=1, \quad v=30: \qquad \frac{30-v_E}{v_0-v_E}=\mathrm{e}^{-k}, \tag{2}$$

$$t=2, \quad v=40: \qquad \frac{40-v_E}{v_0-v_E}=e^{-2k}, \tag{3}$$

$$t=3, \quad v=45: \qquad \frac{45-v_E}{v_0-v_E}=e^{-3k}. \tag{4}$$

Since $e^{-k} \cdot e^{-3k} = e^{-4k} = (e^{-2k})^2$,

$$\text{Eq}(2) \times \text{Eq}(4) = \text{Eq}(3)^2: \quad \left(\frac{30-v_E}{v_0-v_E}\right)\left(\frac{45-v_E}{v_0-v_E}\right)=\left(\frac{40-v_E}{v_0-v_E}\right)^2,$$

$$(30-v_E)(45-v_E)=(40-v_E)^2 \implies 1350-75v_E+v_E^2=1600-80v_E+v_E^2,$$

$$\therefore \quad 5v_E=250 \implies v_E=50\text{ m/sec}.$$

$$\frac{\text{Eq}(2)^2}{\text{Eq}(3)}: \quad \frac{\left(\dfrac{30-v_E}{v_0-v_E}\right)^2}{\dfrac{40-v_E}{v_0-v_E}}=\frac{e^{-2k}}{e^{-2k}}=1 \implies \frac{(30-v_E)^2}{(v_0-v_E)(40-v_E)}=1,$$

$$\therefore \quad v_0=\frac{(30-v_E)^2}{40-v_E}+v_E=\frac{(30-50)^2}{40-50}+50=10\text{ m/sec}.$$

3. From equation (2),

$$k=-\ln\left|\frac{30-v_E}{v_0-v_E}\right|=-\ln\left|\frac{30-50}{10-50}\right|=-\ln\frac{1}{2}=\ln 2,$$

$$e^{-kt}=\frac{v-v_E}{v_0-v_E}=\frac{49-50}{10-50}=\frac{1}{40} \implies t=-\frac{1}{k}\ln\frac{1}{40}=\frac{\ln 40}{\ln 2}=5.32\text{ sec}.$$

3.3 Hanging Cables

3.3.1 The Suspension Bridge

A typical suspension bridge consists of cables, piers (towers), anchors, hangers (suspenders), and deck (stiffening girder) as shown in Figure 3.4. Normally the self-weights of the cables are negligible compared with the load they carry. The load on the cables is from the load on the deck, which includes the self-weight of the deck and traffic load, and is transmitted by the hangers.

Consider a cable supported at two supports A and B as shown in Figure 3.5(a). The load on the cable is modeled as a distributed load $w(x)$. Set up the Cartesian coordinate system by placing the origin at the lowest point of the cable.

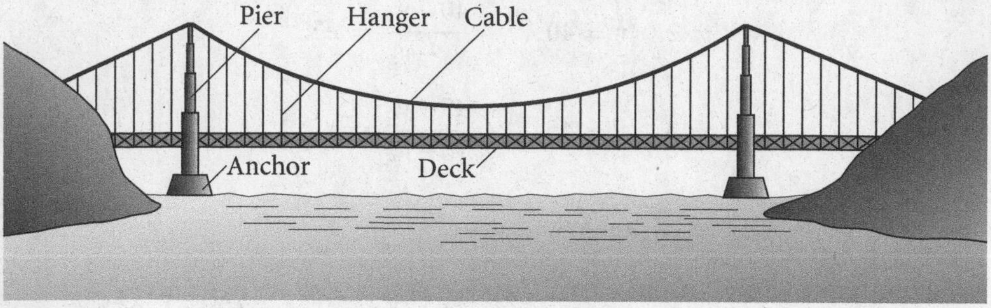

Figure 3.4 A suspension bridge.

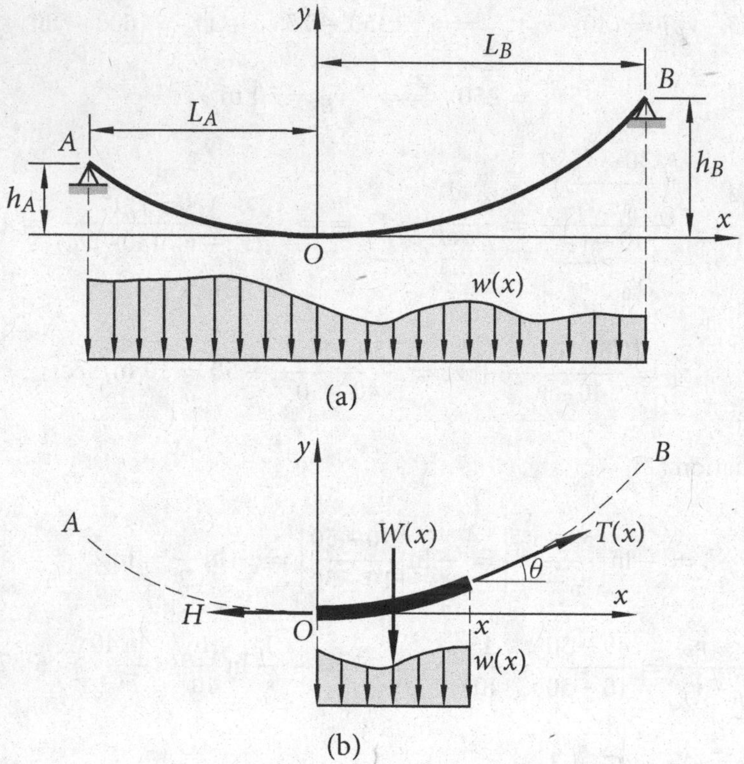

Figure 3.5 A cable under distributed load.

To establish the governing differential equation, consider the equilibrium of a segment of cable between 0 and x as shown in Figure 3.5(b). The cable is subjected to three forces:

- the horizontal tension force H at the left end,
- the tension force $T(x)$ tangent to the cable at the right end, and
- the portion of the distributed load $w(x)$ between 0 and x. It can be replaced by its resultant $W(x)$ applied at the centroid of the area enclosed by the load intensity curve $w(x)$ (the shaded area).

The equilibrium of the segment requires that

$$\rightarrow \sum F_x = 0: \quad T(x)\cos\theta - H = 0, \tag{1}$$

$$\uparrow \sum F_y = 0: \quad T(x)\sin\theta - W(x) = 0, \quad W(x) = \int_0^x w(x)\,\mathrm{d}x. \tag{2}$$

Eliminating $T(x)$ from these two equations yields

$$\frac{T(x)\sin\theta}{T(x)\cos\theta} = \frac{W(x)}{H} \quad \Longrightarrow \quad \tan\theta = \frac{W(x)}{H}.$$

From geometry, one has

$$\tan\theta = \frac{\mathrm{d}y}{\mathrm{d}x} \quad \Longrightarrow \quad \frac{\mathrm{d}y}{\mathrm{d}x} = \frac{W(x)}{H}.$$

Differentiating with respect to x leads to

$$\frac{\mathrm{d}^2y}{\mathrm{d}x^2} = \frac{1}{H}\frac{\mathrm{d}W(x)}{\mathrm{d}x} = \frac{w(x)}{H}, \qquad \frac{\mathrm{d}W(x)}{\mathrm{d}x} = w(x).$$

Suppose that the load is uniformly distributed, i.e., $w(x) = w$. The differential equation becomes

$$\frac{\mathrm{d}^2y}{\mathrm{d}x^2} = \frac{w}{H}. \tag{3}$$

✍ Second-order DE
Immediately integrable

Since the origin is taken at the lowest point, one has $x=0$, $y=0$, $\dfrac{\mathrm{d}y}{\mathrm{d}x}=0$.

Integrating equation (3) once yields

$$\frac{\mathrm{d}y}{\mathrm{d}x} = \frac{w}{H}x + C,$$

where the constant C is determined from the initial condition $x=0$, $\dfrac{\mathrm{d}y}{\mathrm{d}x}=0$:

$$0 = 0 + C \quad \Longrightarrow \quad C = 0.$$

Integrating again leads to

$$y = \frac{w}{2H}x^2 + D,$$

where the constant D is determined from the initial condition $x=0$, $y=0$:

$$0 = 0 + D \quad \Longrightarrow \quad D = 0.$$

Hence the shape of the cable is a parabola given by

$$y = \frac{w}{2H}x^2. \tag{4}$$

The sags h_A and h_B can be determined from equation (4)

$$\text{when } x=-L_A,\ y=h_A: \qquad h_A=\frac{w}{2H}L_A^2 \implies H=\frac{wL_A^2}{2h_A}, \tag{5a}$$

$$\text{when } x=L_B,\ y=h_B: \qquad h_B=\frac{w}{2H}L_B^2 \implies H=\frac{wL_B^2}{2h_B}. \tag{5b}$$

From equations (5), one obtains the relationship among L_A, L_B, h_A, and h_B:

$$H=\frac{wL_A^2}{2h_A}=\frac{wL_B^2}{2h_B} \implies \frac{L_A^2}{h_A}=\frac{L_B^2}{h_B}.$$

To determine the tension at any point, use equations (1) and (2)

$$T(x)\cos\theta=H, \qquad T(x)\sin\theta=W(x).$$

Squaring both sides of these two equations and adding them lead to

$$T^2\cos^2\theta+T^2\sin^2\theta=H^2+W^2(x) \implies T^2=H^2+W^2(x).$$

Since the load is uniformly distributed, $w(x)=w$,

$$W(x)=\int_0^x w(x)\,\mathrm{d}x=w\int_0^x \mathrm{d}x=wx.$$

Therefore, the tension at any point is given by

$$T=\sqrt{H^2+W^2(x)}=\sqrt{H^2+w^2x^2}. \tag{6}$$

The tension T is maximum when $|x|$ is maximum. Hence the tension is maximum at the higher support.

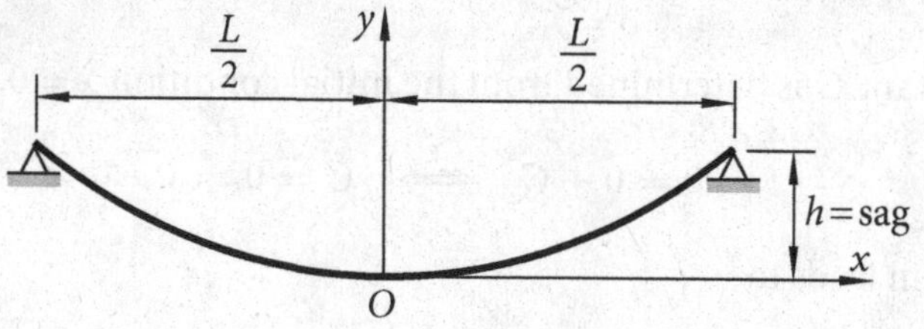

Figure 3.6 A suspension bridge cable with supports at equal height.

For a suspension bridge cable with supports at equal height $h_A=h_B=h$, one has $L_A=L_B=L/2$, where L is the span length, as shown in Figure 3.6. Then equations (5) give the relationship between the sag h and the horizontal tension H at the lowest point:

$$h=\frac{wL^2}{8H} \quad \text{or} \quad H=\frac{wL^2}{8h}. \tag{5$'$}$$

Example 3.5 — Cable of a Suspension Bridge

Consider the main cable of a suspension bridge carrying a uniformly distributed load of intensity w. The two supports of the cable are at the same height. The span of the cable is L, the sag is h, and the axial rigidity is EA.

1. Derive a formula for the elongation δ of the cable.
2. One of the main cables of the central span of the Golden Gate Bridge has the following properties: $L=1,280$ m, $h=143$ m, $w=200$ kN/m, $E=200$ GPa. The cable consists of 27,572 parallel wires of diameter 5 mm. Determine the elongation of this cable.

---

1. Consider a small segment of cable of length $\mathrm{d}s$ as shown. It is subjected to the axial tension forces: T at the left end and $T+\mathrm{d}T$ at the right end.

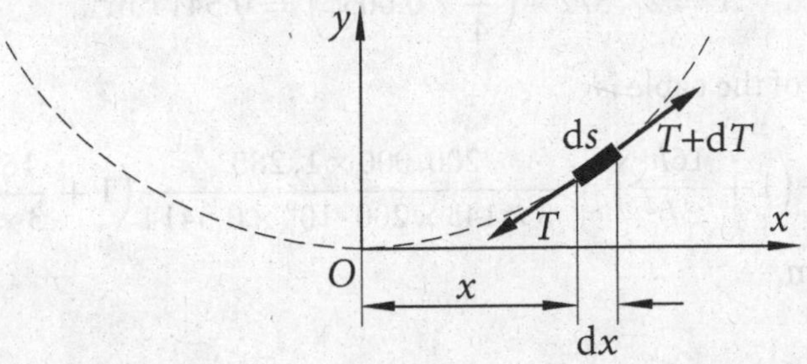

Using the formula of elongation of an axially loaded member in Mechanics of Solids

$$\delta = \frac{TL}{EA},$$

where T is the axial force, L is the length of the member, E is the Young's modulus, and A is the cross-sectional area, the elongation of the cable segment $\mathrm{d}s$ is

$$\mathrm{d}\delta = \frac{T\,\mathrm{d}s}{EA}.$$

Since the length of the cable segment is

$$\mathrm{d}s = \sqrt{1+\left(\frac{\mathrm{d}y}{\mathrm{d}x}\right)^2} = \sqrt{1+\left(\frac{w}{H}x\right)^2}\,\mathrm{d}x, \quad \text{✍ Using equation (4), } y=\frac{w}{2H}x^2$$

and the tension is given by equation (6)

$$T = \sqrt{H^2+w^2x^2},$$

one has

$$\mathrm{d}\delta = \frac{\sqrt{H^2+w^2x^2}}{EA}\sqrt{1+\left(\frac{w}{H}x\right)^2}\,\mathrm{d}x = \frac{H}{EA}\left[1+\left(\frac{w}{H}x\right)^2\right]\mathrm{d}x.$$

Integrating over the length of the span yields the elongation of the cable

$$\delta = 2\int_0^{L/2} \frac{H}{EA}\left[1+\left(\frac{w}{H}x\right)^2\right]\mathrm{d}x$$
$$= \frac{2H}{EA}\left[x+\frac{w^2}{3H^2}x^3\right]_0^{L/2} = \frac{2H}{EA}\left(\frac{L}{2}+\frac{w^2}{3H^2}\cdot\frac{L^3}{8}\right).$$

Using equation (5′) to express the horizontal tension in terms of the sag h, one obtains

$$\delta = \frac{2}{EA}\left(\frac{wL^2}{8h}\right)\left[\frac{L}{2}+\frac{w^2}{3\left(\frac{wL^2}{8h}\right)^2}\cdot\frac{L^3}{8}\right] = \frac{wL^3}{8hEA}\left(1+\frac{16h^2}{3L^2}\right).$$

2. The cross-sectional area of the cable is

$$A = 27{,}572\times\left(\frac{\pi}{4}\times 0.005^2\right) = 0.5414\ \mathrm{m}^2.$$

The elongation of the cable is

$$\delta = \frac{wL^3}{8hEA}\left(1+\frac{16h^2}{3L^2}\right) = \frac{200{,}000\times 1{,}280^3}{8\times 143\times 200\cdot 10^9\times 0.5414}\left(1+\frac{16\times 143^2}{3\times 1{,}280^2}\right)$$
$$= 3.61\ \mathrm{m}.$$

3.3.2 Cable under Self-Weight

There are many applications, such as cloth lines and power transmission cables as shown in Figure 3.7, in which cables are suspended between two supports under own weights.

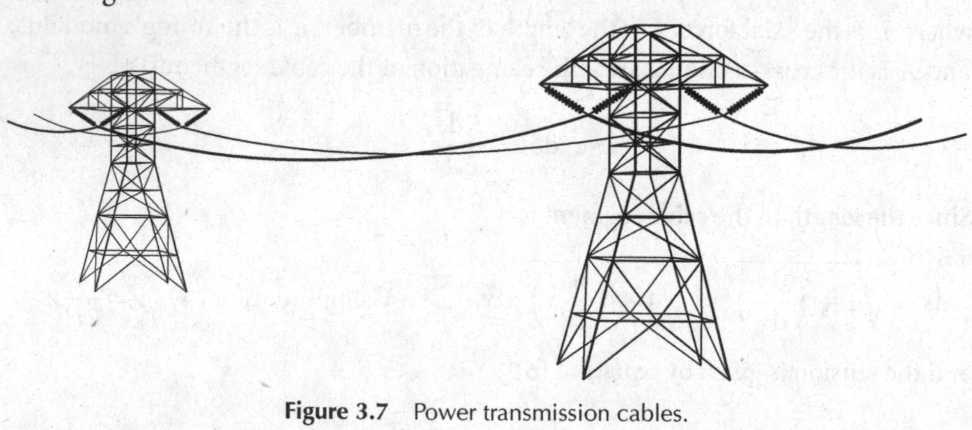

Figure 3.7 Power transmission cables.

Consider a cable suspended between two supports as shown in Figure 3.8(a). The cable is hung under its own weight. Set up the Cartesian coordinate system by placing the origin under the lowest point of the cable. The length of the cable s is measured from the lowest point.

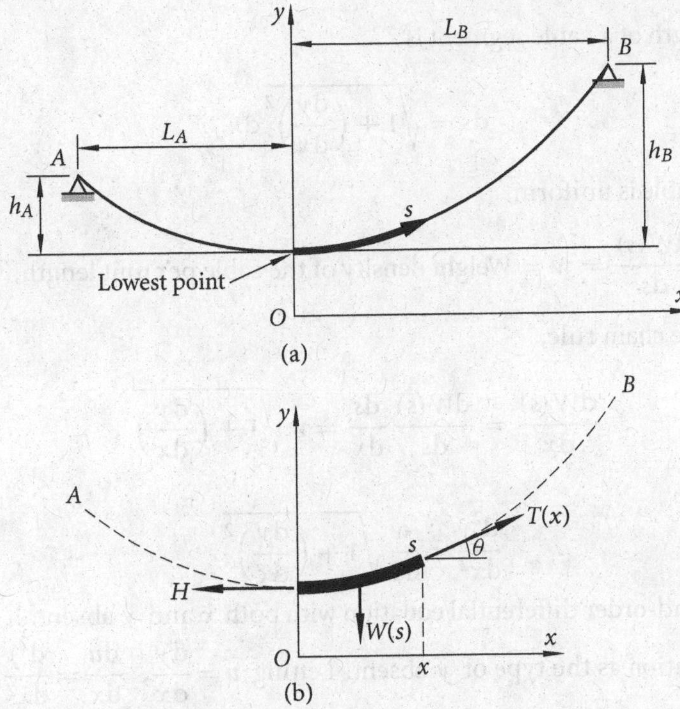

Figure 3.8 A cable under self-weight.

To establish the governing differential equation, consider the equilibrium of a segment of the cable of length s as shown in Figure 3.8(b). The cable segment is subjected to three forces: the self-weight $W(s)$, the horizontal tension force H at the left end, and the tension force T tangent to the cable at the right end. The equilibrium of this cable segment requires

$$\rightarrow \sum F_x = 0: \quad T(x)\cos\theta - H = 0, \tag{1}$$

$$\uparrow \sum F_y = 0: \quad T(x)\sin\theta - W(s) = 0. \tag{2}$$

Dividing equation (2) by equation (1) yields

$$\frac{T(x)\sin\theta}{T(x)\cos\theta} = \frac{W(s)}{H} \implies \tan\theta = \frac{W(s)}{H}.$$

Since $\tan\theta = \dfrac{dy}{dx}$, one obtains

$$\frac{dy}{dx} = \frac{W(s)}{H},$$

or, after differentiating the equation with respect to x,

$$\frac{d^2y}{dx^2} = \frac{1}{H}\frac{dW(s)}{dx}. \tag{3}$$

Since the length of a cable segment is

$$\mathrm{d}s = \sqrt{1+\left(\frac{\mathrm{d}y}{\mathrm{d}x}\right)^2}\,\mathrm{d}x,$$

then, if the cable is uniform,

$$\frac{\mathrm{d}W(s)}{\mathrm{d}s} = w = \text{Weight density of the cable per unit length},$$

and, using the chain rule,

$$\frac{\mathrm{d}W(s)}{\mathrm{d}x} = \frac{\mathrm{d}W(s)}{\mathrm{d}s}\,\frac{\mathrm{d}s}{\mathrm{d}x} = w\sqrt{1+\left(\frac{\mathrm{d}y}{\mathrm{d}x}\right)^2},$$

which leads to

$$\frac{\mathrm{d}^2y}{\mathrm{d}x^2} = \frac{w}{H}\sqrt{1+\left(\frac{\mathrm{d}y}{\mathrm{d}x}\right)^2}. \tag{4}$$

This is a second-order differential equation with both x and y absent. It is easier to solve the equation as the type of y absent. Letting $u = \dfrac{\mathrm{d}y}{\mathrm{d}x}$, $\dfrac{\mathrm{d}u}{\mathrm{d}x} = \dfrac{\mathrm{d}^2y}{\mathrm{d}x^2}$, equation (4) becomes

$$\frac{\mathrm{d}u}{\mathrm{d}x} = \frac{w}{H}\sqrt{1+u^2}, \qquad \text{✍ Variable separable}$$

$$\int \frac{\mathrm{d}u}{\sqrt{1+u^2}} = \frac{w}{H}\int \mathrm{d}x + C \implies \sinh^{-1}u = \frac{w}{H}x + C, \quad ✍ \int \frac{\mathrm{d}x}{\sqrt{a^2+x^2}} = \sinh^{-1}\frac{x}{a}$$

$$\therefore \quad u = \sinh\left(\frac{w}{H}x + C\right).$$

✍ Some properties of the hyperbolic functions are summarized on page 145.

Constant C is determined from the initial condition $x=0$, $\dfrac{\mathrm{d}y}{\mathrm{d}x}=0$:

$$0 = \sinh\left(\frac{w}{H}\cdot 0 + C\right) \implies 0 = \sinh C \implies C = 0.$$

Hence

$$u = \frac{\mathrm{d}y}{\mathrm{d}x} = \sinh\left(\frac{w}{H}x\right). \qquad \text{✍ First-order DE Immediately integrable}$$

Integrating leads to

$$y = \frac{H}{w}\cosh\left(\frac{w}{H}x\right) + D. \qquad ✍ \int \sinh ax\,\mathrm{d}x = \frac{1}{a}\cosh ax$$

When $x=0$,

$$y = \frac{H}{w}\cosh 0 + D = \frac{H}{w} + D.$$

To simplify the expression of y, choose the origin such that $y=\dfrac{H}{w}$ when $x=0$, which results in $D=0$.

Hence, the shape of the cable as shown in Figure 3.9, which is called *catenary*, is

$$y=\frac{H}{w}\cosh\left(\frac{w}{H}x\right). \tag{5}$$

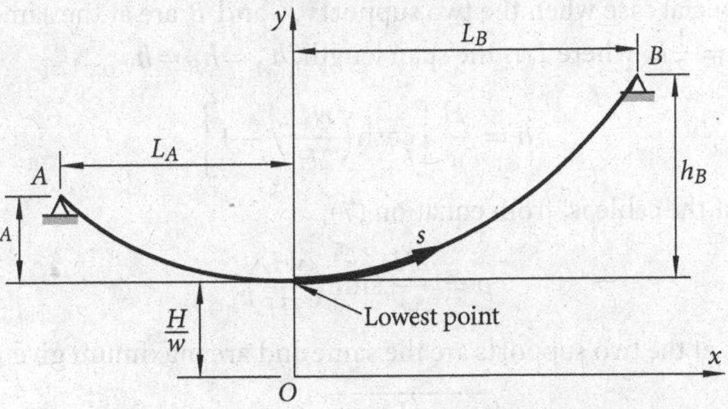

Figure 3.9 Shape of a cable under self-weight.

The sag of the cable can be easily determined in terms of h_A and h_B

$$\text{at } x=-L_A,\ y=\frac{H}{w}+h_A:\quad \frac{H}{w}+h_A=\frac{H}{w}\cosh\left[\frac{w(-L_A)}{H}\right],$$

$$\text{at } x=L_B,\ y=\frac{H}{w}+h_B:\quad \frac{H}{w}+h_B=\frac{H}{w}\cosh\left(\frac{wL_B}{H}\right),$$

$$\therefore\quad h_A=\frac{H}{w}\left[\cosh\left(\frac{wL_A}{H}\right)-1\right],\qquad h_B=\frac{H}{w}\left[\cosh\left(\frac{wL_B}{H}\right)-1\right]. \tag{6}$$

The length of the cable is given by

$$p=\text{perimeter}=\int_{-L_A}^{L_B}\frac{\mathrm{d}s}{\mathrm{d}x}\mathrm{d}x=\int_{-L_A}^{L_B}\sqrt{1+\left(\frac{\mathrm{d}y}{\mathrm{d}x}\right)^2}\,\mathrm{d}x$$

$$=\int_{-L_A}^{L_B}\sqrt{1+\sinh^2\left(\frac{w}{H}x\right)}\,\mathrm{d}x=\int_{-L_A}^{L_B}\cosh\left(\frac{w}{H}x\right)\mathrm{d}x \quad ✍\ 1+\sinh^2x=\cosh^2x$$

$$=\frac{H}{w}\sinh\left(\frac{w}{H}x\right)\Bigg|_{-L_A}^{L_B} \qquad ✍\ \int\cosh ax\,\mathrm{d}x=\frac{1}{a}\sinh ax$$

$$=\frac{H}{w}\left[\sinh\left(\frac{wL_B}{H}\right)+\sinh\left(\frac{wL_A}{H}\right)\right]. \tag{7}$$

To determine the tension at any point, use equations (1) and (2)

$$\text{Eq(1)}^2+\text{Eq(2)}^2:\ T^2\cos^2\theta+T^2\sin^2\theta=H^2+W^2(s)\implies T=\sqrt{H^2+W^2(s)},$$

where s is the length of the cable between the point of interest and the lowest point, $W(s)=ws$ is the weight of this segment of cable. Hence

$$T=\sqrt{H^2+w^2s^2}. \tag{8}$$

The tension T is maximum when s is maximum. Hence the tension is maximum at the higher support.

For the special case when the two supports A and B are at the same height, one has $L_A=L_B=\frac{1}{2}L$, where L is the span length, $h_A=h_B=h$.

$$h=\frac{H}{w}\left[\cosh\left(\frac{wL}{2H}\right)-1\right]. \tag{6$'$}$$

The length of the cable is, from equation (7),

$$p=\frac{2H}{w}\sinh\left(\frac{wL}{2H}\right). \tag{7$'$}$$

The tensions at the two supports are the same and are maximum given by

$$T_{\max}=\sqrt{H^2+\tfrac{1}{4}w^2p^2}, \qquad s_{\max}=\tfrac{1}{2}p. \tag{8$'$}$$

Example 3.6 — Hanging Cable

A cable of weight density of 50 N/m is suspended at two supports of equal height. The supports are 10 m apart and the sag is 2 m. Determine the following:

(1) the horizontal tension at the lowest point;

(2) the tension at the support;

(3) the length of the cable.

The following parameters are known: $w=50$ N/m, $L=10$ m, $h=2$ m.

From equation (6′),

$$h=\frac{H}{w}\left[\cosh\left(\frac{wL}{2H}\right)-1\right] \implies 2=\frac{H}{50}\left[\cosh\left(\frac{250}{H}\right)-1\right].$$

This is a transcendental equation and a numerical method is required to determine its root. For example, use `fsolve` in *Maple* (see Example 12.19 for details on using `fsolve` for root-finding). It is found that $H=327.93$ N.

Using equation (7′), the length of the cable is given by

$$p=\frac{2H}{w}\sinh\left(\frac{wL}{2H}\right)=\frac{2\times 327.93}{50}\sinh\left(\frac{50\times 10}{2\times 327.93}\right)=11.00 \text{ m}.$$

The tension at the support is, from equation (8′),

$$T_{\max}=\sqrt{H^2+\tfrac{1}{4}w^2p^2}=\sqrt{327.93^2+\tfrac{1}{4}\times 50^2\times 11.00^2}=427.98 \text{ N}.$$

Example 3.7 — Float and Cable

A spherical float used to mark the course for a sailboat race is shown in Figure 3.10(a). A water current from the left to right causes a horizontal drag on the float. The length of the cable between points A and B is 60 m, and the effective mass density of the cable is 2 kg/m when the buoyancy of the cable is accounted for. If the effect of the current on the cable can be neglected, determine the tensions at points A and B.

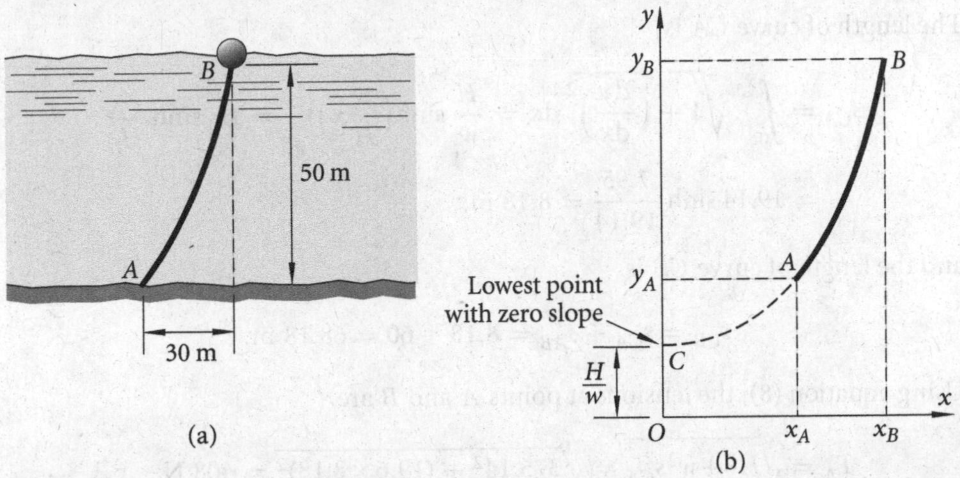

Figure 3.10 A float and a cable.

The weight density of the cable is $w = 2 \times 9.8 = 19.6$ N/m. In this problem, the lowest point with zero slope does not appear between points A and B. To apply the formulation established in this section, add an imaginary segment of cable CA as shown in Figure 3.10(b), and place the origin below the lowest point with zero slope a distance of H/w.

Applying equation (5) to points A and B yields, denoting $H_0 = H/w$,

$$\text{point } A: \quad y_A = \frac{H}{w}\cosh\left(\frac{w}{H}x_A\right) \implies y_A = H_0 \cosh\frac{x_A}{H_0},$$

$$\text{point } B: \quad y_B = \frac{H}{w}\cosh\left(\frac{w}{H}x_B\right) \implies y_A + 50 = H_0 \cosh\frac{x_A + 30}{H_0}.$$

Subtracting these two equations leads to

$$50 = H_0\left(\cosh\frac{x_A + 30}{H_0} - \cosh\frac{x_A}{H_0}\right). \qquad (*)$$

Following the same procedure as in deriving equation (7), the length of the cable is

$$p = \int_{x_A}^{x_B} \sqrt{1 + \left(\frac{dy}{dx}\right)^2}\, dx = \frac{H}{w}\sinh\left(\frac{w}{H}x\right)\Big|_{x_A}^{x_B} = H_0\left(\sinh\frac{x_B}{H_0} - \sinh\frac{x_A}{H_0}\right).$$

Hence

$$60 = H_0\Big(\sinh\frac{x_A+30}{H_0} - \sinh\frac{x_A}{H_0}\Big). \qquad (**)$$

Equations ($*$) and ($**$) give two transcendental equations for two unknowns H_0 and x_A. The equations have to be solved numerically, e.g. using `fsolve` in *Maple*, to yield

$$x_A = 7.95\text{ m},\quad H_0 = \frac{H}{w} = 19.14 \Longrightarrow H = H_0 w = 19.14\times 19.6 = 375.14\text{ N}.$$

The length of curve *CA* is

$$s_{CA} = \int_0^{x_A}\sqrt{1+\Big(\frac{dy}{dx}\Big)^2}\,dx = \frac{H}{w}\sinh\Big(\frac{w}{H}x\Big)\Big|_0^{x_A} = H_0\sinh\frac{x_A}{H_0}$$

$$= 19.14\sinh\frac{7.95}{19.14} = 8.18\text{ m},$$

and the length of curve *CB* is

$$s_{CB} = s_{CA} + s_{AB} = 8.18 + 60 = 68.18\text{ m}.$$

Using equation (8), the tensions at points *A* and *B* are

$$T_A = \sqrt{H^2+w^2 s_{CA}^2} = \sqrt{375.14^2 + (19.6\times 8.18)^2} = 408\text{ N},$$

$$T_B = \sqrt{H^2+w^2 s_{CB}^2} = \sqrt{375.14^2 + (19.6\times 68.18)^2} = 1,388\text{ N}.$$

3.4 Electric Circuits

There are three basic passive electric elements: resistors, capacitors, and inductors.

Resistance R is the capacity of materials to impede the flow of current, which is modeled by a resistor.

Basic Laws

Ohm's Law: $v = iR$, or $i = \dfrac{v}{R}$, where v is the voltage, i is the current.

Kirchhoff's Current Law (KCL): The algebraic sum of all the currents at any node in a circuit equals zero.

Kirchhoff's Voltage Law (KVL): The algebraic sum of all the voltages around any closed path in a circuit equals zero.

A *capacitor* is an electrical component consisting of two conductors separated by an insulator or dielectric material. If the voltage varies with time, the electric field varies with time, which produces a displacement current in the space occupied by the field. The circuit parameter *capacitance* C relates the displacement current to the voltage

$$i(t) = C\frac{dv_C(t)}{dt}, \quad \text{or} \quad v_C(t) = \frac{1}{C}\int_{-\infty}^{t} i(t)\,dt = \frac{1}{C}\int_{t_0}^{t} i(t)\,dt + v_C(t_0).$$

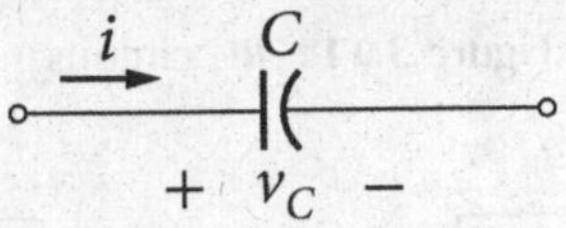

A capacitor behaves as an open circuit in the presence of a constant voltage. Voltage cannot change abruptly across the terminals of a capacitor.

An *inductor* is an electrical component that opposes any change in electrical current. It is composed of a coil of wire wound around a supporting core. If the current varies with time, the magnetic field varies with time, which induces a voltage in the conductor linked by the field. The circuit parameter *inductance* L relates the induced voltage to the current

$$v_L(t) = L\frac{di(t)}{dt}, \quad \text{or} \quad i(t) = \frac{1}{L}\int_{-\infty}^{t} v_L(t)\,dt = \frac{1}{L}\int_{t_0}^{t} v_L(t)\,dt + i(t_0).$$

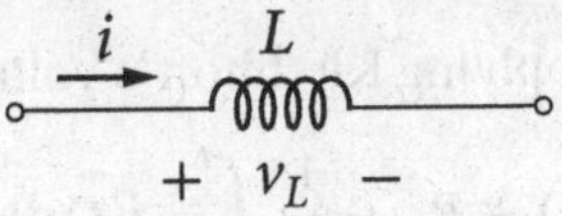

An inductor behaves as a short circuit in the presence of a constant current. Current cannot change abruptly in an inductor.

Four types of simple circuits (see Figures 3.11 and 3.12), a circuit comprising a resistor and a capacitor (*RC* circuit) and a circuit comprising a resistor and inductor (*RL* circuit), either in series or parallel connection, all lead to the first-order linear ordinary differential equation of the form

$$\frac{dx}{dt} + \frac{1}{\tau}x = Q(t).$$

The solution is given by, with $P(t) = 1/\tau$,

$$x(t) = e^{-\int P(t)\,dt}\left[\int Q(t)\,e^{\int P(t)\,dt}dx + B\right] = Be^{-t/\tau} + e^{-t/\tau}\int Q(s)\,e^{s/\tau}ds,$$

where the constant B can be determined using the initial condition: $x(t) = x_0$ when $t = 0 \Longrightarrow B = x_0$. Thus, the solution is

$$x(t) = x_0 e^{-t/\tau} + e^{-t/\tau}\int Q(s)\,e^{s/\tau}ds.$$

(a) Series RC Circuit   (b) Parallel RC Circuit

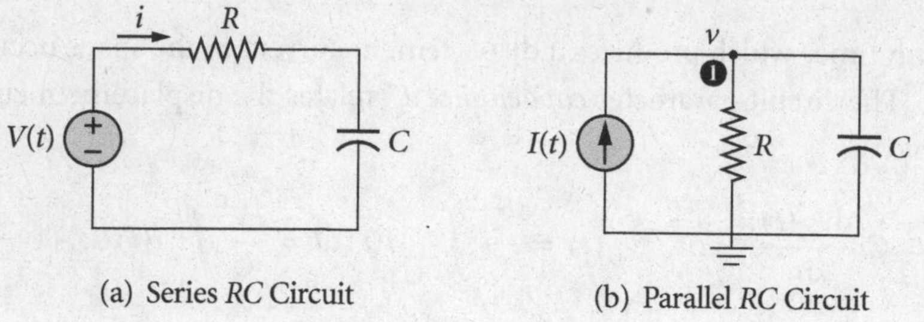

Figure 3.11 RC circuits.

(a) Series RL Circuit   (b) Parallel RL Circuit

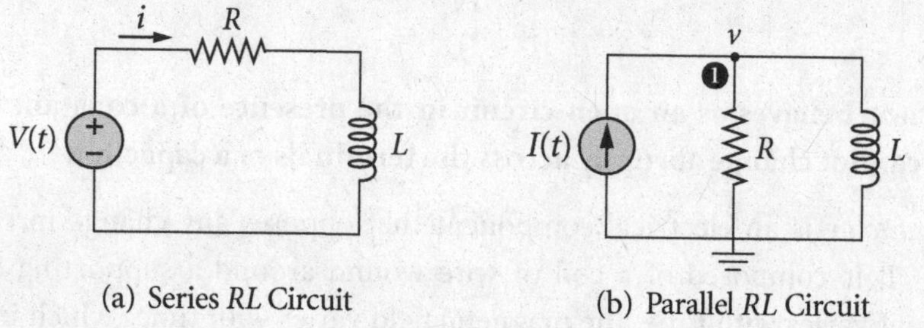

Figure 3.12 RL circuits.

If $Q(t) = Q_0$, the solution becomes

$$x(t) = Q_0\tau + (x_0 - Q_0\tau)e^{-t/\tau}.$$

Series RC Circuit

Referring to Figure 3.11(a), applying Kirchhoff's Voltage Law yields

$$-V(t) + Ri + \frac{1}{C}\int_{-\infty}^{t} i(t)\,dt = 0.$$

Differentiating with respect to t gives

$$R\frac{di}{dt} + \frac{1}{C}i = \frac{dV(t)}{dt} \quad\Longrightarrow\quad \frac{di}{dt} + \frac{1}{RC}i = \frac{1}{R}\frac{dV(t)}{dt},$$

in which $x(t) = i(t)$, $\tau = RC$, $Q(t) = \dfrac{1}{R}\dfrac{dV(t)}{dt}$.

Parallel RC Circuit

Referring to Figure 3.11(b), applying Kirchhoff's Current Law at node 1 yields

$$I(t) - \frac{v}{R} - C\frac{dv}{dt} = 0 \quad\Longrightarrow\quad \frac{dv}{dt} + \frac{1}{RC}v = \frac{I(t)}{C},$$

in which $x(t) = v(t)$, $\tau = RC$, $Q(t) = \dfrac{I(t)}{C}$.

Series RL Circuit

Referring to Figure 3.12(a), applying Kirchhoff's Voltage Law yields

$$-V(t) + Ri + L\frac{di}{dt} = 0 \quad\Longrightarrow\quad \frac{di}{dt} + \frac{R}{L}i = \frac{V(t)}{L},$$

in which $x(t)=i(t)$, $\tau=\dfrac{L}{R}$, $Q(t)=\dfrac{V(t)}{L}$.

Parallel *RL* Circuit

Referring to Figure 3.12(b), applying Kirchhoff's Current Law at node 1 yields

$$I(t)-\frac{v}{R}-\frac{1}{L}\int_{-\infty}^{t} v(t)\,\mathrm{d}t=0.$$

Differentiating with respect to t gives

$$\frac{1}{R}\frac{\mathrm{d}v}{\mathrm{d}t}+\frac{1}{L}v=\frac{\mathrm{d}I(t)}{\mathrm{d}t} \implies \frac{\mathrm{d}v}{\mathrm{d}t}+\frac{R}{L}v=R\frac{\mathrm{d}I(t)}{\mathrm{d}t},$$

in which $x(t)=v(t)$, $\tau=\dfrac{L}{R}$, $Q(t)=R\dfrac{\mathrm{d}I(t)}{\mathrm{d}t}$.

Example 3.8 — First-Order Circuit

For the electric circuit shown in the following figure, determine v_L for $t>0$.

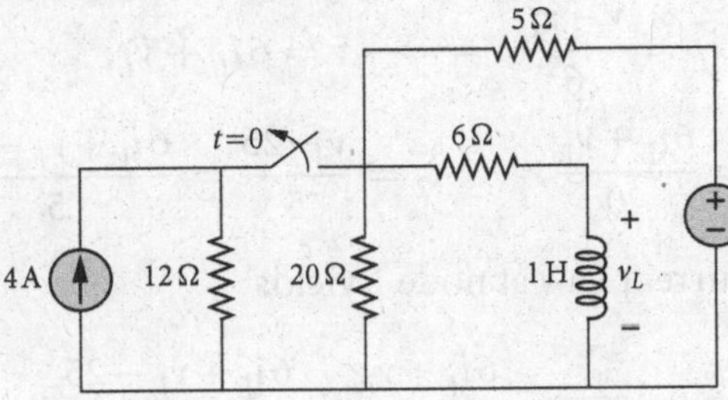

❧ For $t<0$, the switch is closed and the inductor behaves as a short circuit.

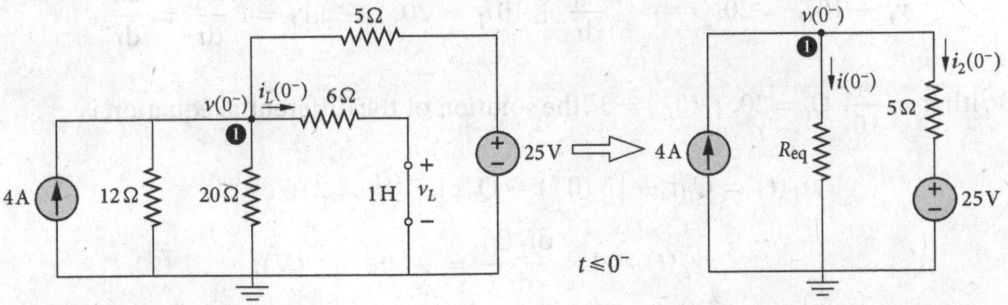

The three resistors of 12 Ω, 20 Ω, and 6 Ω are in parallel connection and can be combined as an equivalent resistor

$$\frac{1}{R_{\text{eq}}}=\frac{1}{12}+\frac{1}{20}+\frac{1}{6} \implies R_{\text{eq}}=\frac{10}{3}.$$

Applying Kirchhoff's Current Law at node 1 yields

$$4=i(0^-)+i_2(0^-)=\frac{v(0^-)}{R_{\text{eq}}}+\frac{v(0^-)-25}{5}=\frac{3\,v(0^-)}{10}+\frac{v(0^-)}{5}-5,$$

$$\therefore \quad v(0^-) = 18\text{ V} \implies i_L(0^-) = \frac{v(0^-)}{6} = 3\text{ A}.$$

❧ At $t = 0$, the switch is open. Since the current in an inductor cannot change abruptly, $i_L(0^-) = i_L(0^+) = 3$ A.

❧ For $t > 0$, the switch is open and the circuit becomes

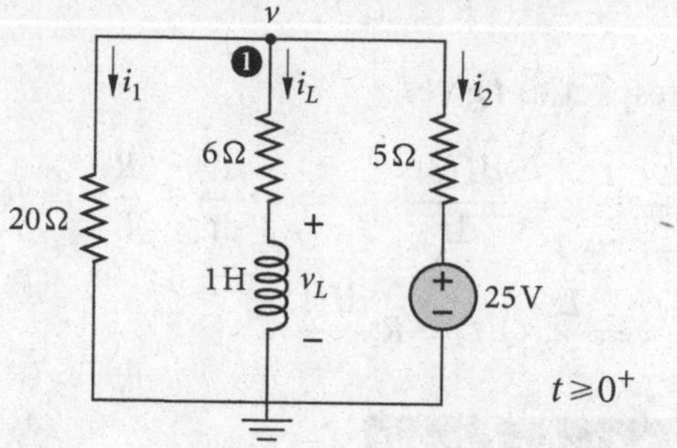

It is easy to evaluate that

$$i_L = \frac{v - v_L}{6} \implies v = 6i_L + v_L,$$

$$i_1 = \frac{v}{20} = \frac{6i_L + v_L}{20}, \qquad i_2 = \frac{v - 25}{5} = \frac{6i_L + v_L - 25}{5}.$$

Applying Kirchhoff's Current Law at node 1 yields

$$i_1 + i_2 + i_L = 0 \implies \frac{6i_L + v_L}{20} + \frac{6i_L + v_L - 25}{5} + i_L = 0,$$

$$v_L + 10i_L = 20 \implies \frac{di_L}{dt} + 10i_L = 20. \quad ✍\ v_L = L\frac{di_L}{dt} = \frac{di_L}{dt}$$

With $\tau = \frac{1}{10}$, $Q_0 = 20$, $i_L(0^+) = 3$, the solution of the differential equation is

$$i_L(t) = Q_0\tau + \left[i_L(0^+) - Q_0\tau\right]e^{-t/\tau} = 2 + e^{-10t},$$

$$\therefore \quad v_L(t) = 1 \times \frac{di_L(t)}{dt} = -10e^{-10t}\ \text{(V)}.$$

Example 3.9 — First-Order Circuit

For the electric circuit shown in the following figure, determine i_1 for $t > 0$.

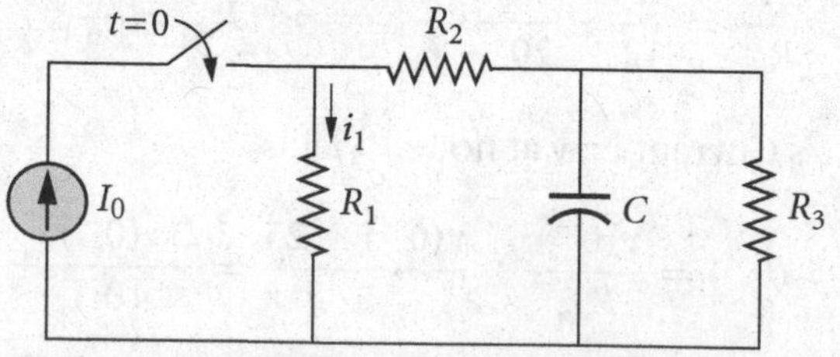

❧ For $t<0$, the switch is open and the current source is disconnected. The capacitor behaves as an open circuit. Hence $i_1(0^-)=0$ and $v_2(0^-)=0$.

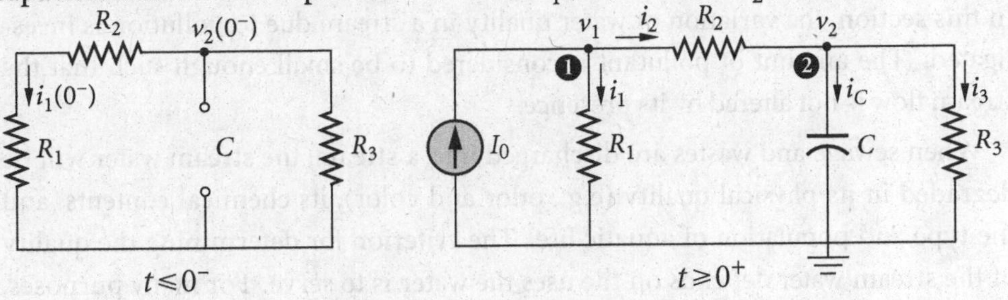

❧ At $t=0$, the switch is closed. Since the capacitor voltage cannot change abruptly, $v_2(0^+)=v_2(0^-)=0$. Applying Kirchhoff's Current Law at node 1 yields

$$I_0=i_1(0^+)+i_2(0^+)=\frac{v_1(0^+)}{R_1}+\frac{v_1(0^+)-v_2(0^+)}{R_2}=\frac{v_1(0^+)}{R_1}+\frac{v_1(0^+)}{R_2},$$

$$\therefore\quad v_1(0^+)=\frac{R_1R_2}{R_1+R_2}I_0 \implies i_1(0^+)=\frac{v_1(0^+)}{R_1}=\frac{R_2}{R_1+R_2}I_0.$$

❧ For $t>0$, $i_1=\dfrac{v_1}{R_1} \implies v_1=R_1 i_1$,

$$i_2=\frac{v_1-v_2}{R_2}=\frac{R_1 i_1-v_2}{R_2},\qquad i_3=\frac{v_2}{R_3},\qquad i_C=C\frac{dv_2}{dt}.$$

Applying Kirchhoff's Current Law at node 1 gives

$$I_0=i_1+i_2 \implies i_2=I_0-i_1=\frac{R_1 i_1-v_2}{R_2} \implies v_2=(R_1+R_2)\,i_1-R_2 I_0.$$

Applying Kirchhoff's Current Law at node 2 leads to

$$i_2=i_C+i_3 \implies I_0-i_1=C\frac{dv_2}{dt}+\frac{v_2}{R_3},$$

$$\therefore\quad I_0-i_1=C\cdot(R_1+R_2)\frac{di_1}{dt}+\frac{(R_1+R_2)\,i_1-R_2 I_0}{R_3},$$

$$\therefore\quad \frac{di_1}{dt}+\frac{1}{\tau}i_1=Q_0,\quad \tau=\frac{C\,(R_1+R_2)\,R_3}{R_1+R_2+R_3},\quad Q_0=\frac{R_2+R_3}{C\,(R_1+R_2)\,R_3}I_0.$$

The solution is given by

$$\begin{aligned} i_1(t)&=Q_0\tau+\big[i_1(0^+)-Q_0\tau\big]e^{-t/\tau}\\ &=\frac{R_2+R_3}{R_1+R_2+R_3}I_0+\left(\frac{R_2}{R_1+R_2}-\frac{R_2+R_3}{R_1+R_2+R_3}\right)I_0\,e^{-t/\tau}. \end{aligned}$$

Remarks: Since these circuits are characterized by first-order differential equations, they are called first-order circuits. They consist of resistors and the equivalent of one energy storage element, such as capacitors and inductors.

3.5 Natural Purification in a Stream

In this section, the variation of water quality in a stream due to pollution is investigated. The amount of pollutant is considered to be small enough such that the stream flow is not altered by its presence.

When sewage and wastes are discharged into a stream, the stream water will be degraded in its physical quality (e.g., odor and color), its chemical contents, and the type and population of aquatic life. The criterion for determining the quality of the stream water depends on the uses the water is to serve. For many purposes, engineers use the concentration of *Dissolved Oxygen* (DO) and decomposable organic matter in the water as indicators of its quality. The DO measures the capacity of the water to assimilate many polluting materials and to support aquatic life. The organic matter consumes oxygen in its decomposition. In sewage, the organic matter includes a great variety of compounds, represented by the amount of oxygen required for its biological decomposition (*Biochemical Oxygen Demand*, or BOD).

Clean stream water is usually saturated with DO. As sewage is added and flows in the stream, the DO in the polluted water is consumed as the organic matter is decomposed. In the meantime, oxygen from the atmosphere dissolves into the water, as it is now no longer saturated with DO. Finally, the organic matter is completely decomposed and the stream water becomes saturated with DO again. This natural process of purification takes place within a period of several days. It is necessary to ascertain the variation of DO and BOD along the flow to determine the effect of pollution on the stream.

The BOD added to the stream is assumed to spread across the stream over a distance that is very short in comparison with the length of the stream where deoxygenation by the BOD and reoxygenation by the atmosphere take place, so that the problem can be considered to be one dimensional with DO and BOD assumed to be uniform at a cross-section.

To derive the governing equations, consider the mass balance of BOD during $\mathrm{d}t$ in a volume $A\,\mathrm{d}x$ bounded by two cross-sections $\mathrm{d}x$ apart, as shown in Figure 3.13, in which A is the cross-sectional area of the stream, x is the distance measured along the stream, and t is time.

Employ the following notations:

$Q =$ the discharge,

$b =$ the concentration of BOD in mass per unit volume of water,

$c =$ the concentration of DO in mass per unit volume of water,

$M =$ the mass of BOD added per unit time per unit discharge along the flow,

$N =$ the mass of oxygen added per unit time per unit discharge along the

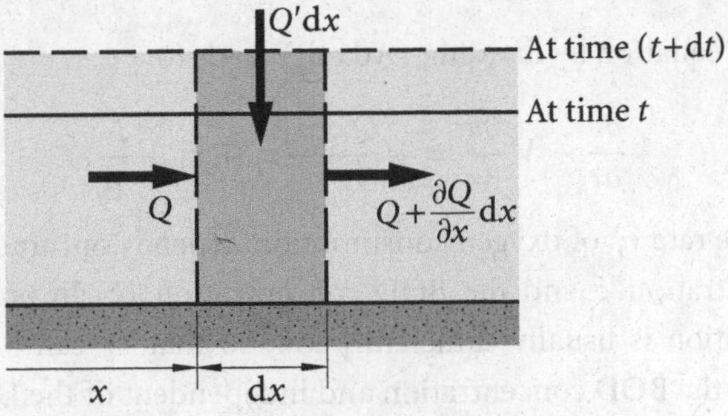

Figure 3.13 Natural purification in a stream.

stream from sources other than the atmosphere (e.g., from photosynthesis of green plants in the stream),

r_1 = the mass of BOD decomposed per unit volume per unit time,

r_2 = the rate of atmospheric reoxygenation in mass per unit volume per unit time.

Any difference between inflow and outflow and between addition and subtraction will cause a change in the mass of BOD contained between these two cross-sections. Thus, during dt,

$$\underbrace{Qb\,dt}_{\text{mass inflow}} - \underbrace{\left(Q+\frac{\partial Q}{\partial x}dx\right)\left(b+\frac{\partial b}{\partial x}dx\right)dt}_{\text{mass outflow}} + \underbrace{M\,dx\,dt}_{\text{added mass}} - \underbrace{r_1 A\,dx\,dt}_{\text{mass decomposed}} = \underbrace{\frac{\partial(bA)}{\partial t}dx\,dt.}_{\text{increase mass in volume}}$$

Simplifying this equation by dividing $(A\,dx\,dt)$ yields

$$\frac{\partial b}{\partial t} + V\frac{\partial b}{\partial x} = -\frac{Q'b}{A} - r_1 + \frac{M}{A}, \tag{1}$$

where $V = \dfrac{Q}{A}$ is the mean velocity at a cross-section, and

$$Q' = \frac{\partial Q}{\partial x} + \frac{\partial A}{\partial t}$$

is the discharge added per unit length of stream. Similarly, from a mass balance for the DO, one has

$$\underbrace{Qc\,dt}_{\text{mass inflow}} - \underbrace{\left(Q+\frac{\partial Q}{\partial x}dx\right)\left(c+\frac{\partial c}{\partial x}dx\right)dt}_{\text{mass outflow}} - \underbrace{r_1 A\,dx\,dt}_{\text{mass consumed by BOD}} + \underbrace{r_2 A\,dx\,dt}_{\text{mass from atmosphere}} + \underbrace{N\,dx\,dt}_{\text{other added mass}}$$

$$= \frac{\partial(cA)}{\partial t}dx\,dt. \quad \text{increase mass in volume}$$

Simplifying the equation by dividing $(A\,dx\,dt)$ leads to

$$\frac{\partial c}{\partial t}+V\frac{\partial c}{\partial x}=-\frac{Q'c}{A}-r_1+r_2+\frac{N}{A}. \tag{2}$$

Generally, the rate r_1 of oxygen consumption depends on, among other factors, the DO concentration c and the BOD concentration b. In practical cases, the BOD concentration is usually sufficiently low, so that r_1 can be assumed to be proportional to the BOD concentration and independent of the DO concentration as long as it is greater than a very small value; that is

$$r_1=k_1 b, \qquad \text{for } c>0^+, \tag{3}$$

where the coefficient k_1 depends on the composition of the sewage and its temperature. For a given sewage, the numerical value of k_1 can be determined in the laboratory.

The rate r_2 of atmospheric reoxygenation is usually assumed to be proportional to the DO deficit (c_s-c):

$$r_2=k_2(c_s-c), \quad \text{for } 0\leqslant c\leqslant c_s, \tag{4}$$

where c_s is the saturation concentration of DO, which depends on the water temperature. The coefficient k_2 depends on the temperature, the area of the air-water interface per unit volume of the stream, and the turbulence of the air and water.

With hydrographical data and sources of BOD and DO of the stream, equation (1) can be solved independently for the BOD distribution $b(x,t)$. The DO distribution $c(x,t)$ can then be obtained from equation (2), as $r_1=k_1 b$ is now a known function of x and t.

For simplicity of analysis, the *steady-state* case is considered, in which the variables, such as $b(x,t)$ and $c(x,t)$, do not change with time t. Hence, equations (1) and (2) become

$$\frac{db}{dx}+\left(\frac{1}{Q}\frac{dQ}{dx}+\frac{k_1}{V}\right)b=\frac{M}{Q}, \tag{5}$$

$$\frac{dc}{dx}+\left[\frac{1}{Q}\frac{dQ}{dx}c+\frac{k_1}{V}b-\frac{k_2}{V}(c_s-c)\right]=\frac{N}{Q},$$

which can be rewritten as

$$\frac{d(c_s-c)}{dx}+\left(\frac{1}{Q}\frac{dQ}{dx}+\frac{k_2}{V}\right)(c_s-c)=\frac{k_1}{V}b+\frac{1}{Q}\frac{dQ}{dx}c_s+\frac{N}{Q}. \tag{6}$$

In the following, various special cases are studied.

1. Consider $b(x)$ as BOD concentration in a stream of constant Q and V. Determine $b(x)$ for the case with $b=b_0$ at $x=0$ and $M=0$. Determine the steady distribution of DO along the stream with $c=c_0$ at $x=0$ and $N=0$.

Since $M=0$, Q and V are constants, $\dfrac{dQ}{dx}=0$, equation (5) becomes

$$\frac{db}{dx}+\frac{k_1}{V}b=0, \qquad \text{✍ Variable separable}$$

the solution of which is

$$\int\frac{db}{b}=-\int\frac{k_1}{V}dx+C \implies \ln b=-\frac{k_1}{V}x+\ln D.$$

Hence

$$b(x)=De^{-k_1x/V},$$

in which the constant D is determined by the initial condition $b=b_0$ at $x=0$:

$$b_0=De^0 \implies D=b_0 \implies b(x)=b_0e^{-k_1x/V}.$$

For $N=0$, equation (6) becomes

$$\frac{d(c_s-c)}{dx}+\frac{k_2}{V}\cdot(c_s-c)=\frac{k_1}{V}b_0e^{-k_1x/V}, \qquad \text{✍ Linear first-order DE}$$

in which the dependent variable is (c_s-c). It is easy to evaluate

$$P(x)=\frac{k_2}{V}, \qquad Q(x)=\frac{k_1}{V}b_0e^{-k_1x/V},$$

$$\int P(x)\,dx=\frac{k_2}{V}x, \quad e^{\int P(x)\,dx}=e^{k_2x/V}, \quad e^{-\int P(x)\,dx}=e^{-k_2x/V},$$

$$\int Q(x)e^{\int P(x)\,dx}dx=\int\frac{k_1}{V}b_0e^{-k_1x/V}\cdot e^{k_2x/V}dx=\frac{k_1b_0}{k_2-k_1}e^{(k_2-k_1)x/V}.$$

The general solution of the differential equation is

$$(c_s-c)=e^{-\int P(x)\,dx}\left[\int Q(x)e^{\int P(x)\,dx}dx+C\right]$$

$$=e^{-k_2x/V}\left(\frac{k_1b_0}{k_2-k_1}e^{(k_2-k_1)x/V}+C\right),$$

in which the constant C is determined by the initial condition $c=c_0$ at $x=0$:

$$(c_s-c_0)=\frac{k_1b_0}{k_2-k_1}+C \implies C=(c_s-c_0)-\frac{k_1b_0}{k_2-k_1}.$$

Hence

$$c_s-c(x)=(c_s-c_0)e^{-k_2x/V}+\frac{k_1b_0}{k_2-k_1}\left(e^{-k_1x/V}-e^{-k_2x/V}\right).$$

2. Determine the BOD distribution $b(x)$ for the case with $Q=Q_0(1+\gamma x)$, constant V, M and k_1, and $b=b_0$ at $x=0$. Determine the DO distribution in the stream with a constant N and $c=c_0$ at $x=0$. From the solution, find the value of c far downstream for the case of a uniform stream ($\gamma=0$).

Since V, M and k_1 are constants, $Q=Q_0(1+\gamma x) \Longrightarrow \dfrac{\mathrm{d}Q}{\mathrm{d}x}=Q_0\gamma$, equation (5) becomes

$$\frac{\mathrm{d}b}{\mathrm{d}x}+\left(\frac{\gamma}{1+\gamma x}+\frac{k_1}{V}\right)b=\frac{M}{Q_0(1+\gamma x)}, \qquad \text{✍ Linear first-order DE}$$

which is of the form $\dfrac{\mathrm{d}b}{\mathrm{d}x}+P(x)\cdot b=Q(x)$, where

$$P(x)=\frac{\gamma}{1+\gamma x}+\frac{k_1}{V}, \quad Q(x)=\frac{M}{Q_0(1+\gamma x)},$$

$$\int P(x)\,\mathrm{d}x=\int\left(\frac{\gamma}{1+\gamma x}+\frac{k_1}{V}\right)\mathrm{d}x=\ln(1+\gamma x)+\frac{k_1}{V}x,$$

$$\mathrm{e}^{\int P(x)\,\mathrm{d}x}=\mathrm{e}^{\ln(1+\gamma x)}\mathrm{e}^{k_1x/V}=(1+\gamma x)\,\mathrm{e}^{k_1x/V}, \quad \mathrm{e}^{-\int P(x)\,\mathrm{d}x}=\frac{1}{1+\gamma x}\mathrm{e}^{-k_1x/V},$$

$$\int Q(x)\,\mathrm{e}^{\int P(x)\,\mathrm{d}x}\mathrm{d}x=\int\frac{M}{Q_0(1+\gamma x)}(1+\gamma x)\,\mathrm{e}^{k_1x/V}\,\mathrm{d}x=\frac{MV}{Q_0k_1}\mathrm{e}^{k_1x/V}.$$

The general solution of the differential equation is

$$\begin{aligned}
b(x)&=\mathrm{e}^{-\int P(x)\,\mathrm{d}x}\left[\int Q(x)\,\mathrm{e}^{\int P(x)\,\mathrm{d}x}\mathrm{d}x+C\right]\\
&=\frac{1}{1+\gamma x}\mathrm{e}^{-k_1x/V}\left(\frac{MV}{Q_0k_1}\mathrm{e}^{k_1x/V}+C\right)=\frac{MV}{Q_0k_1}\frac{1}{1+\gamma x}+\frac{C}{1+\gamma x}\mathrm{e}^{-k_1x/V},
\end{aligned}$$

in which the constant C is determined by the initial condition $b=b_0$ at $x=0$:

$$b_0=\frac{MV}{Q_0k_1}+C \Longrightarrow C=b_0-\frac{MV}{Q_0k_1}.$$

Hence

$$\begin{aligned}
b(x)&=\frac{1}{1+\gamma x}\left[\frac{MV}{Q_0k_1}+\left(b_0-\frac{MV}{Q_0k_1}\right)\mathrm{e}^{-k_1x/V}\right]\\
&=\frac{MV}{Q_0k_1}\frac{1}{1+\gamma x}\left(1-\mathrm{e}^{-k_1x/V}\right)+\frac{b_0}{1+\gamma x}\mathrm{e}^{-k_1x/V}.
\end{aligned}$$

Equation (6) becomes

$$\frac{\mathrm{d}(c_s-c)}{\mathrm{d}x}+\left(\frac{\gamma}{1+\gamma x}+\frac{k_2}{V}\right)\cdot(c_s-c)$$

$$= \frac{k_1}{V}\left[\frac{MV}{Q_0 k_1}\frac{1}{1+\gamma x}\left(1-e^{-k_1 x/V}\right)+\frac{b_0}{1+\gamma x}e^{-k_1 x/V}\right]+\frac{\gamma c_s}{1+\gamma x}+\frac{N}{Q_0(1+\gamma x)},$$

which is a linear first-order differential equation with dependent variable being $(c_s - c)$, and

$$P(x)=\frac{\gamma}{1+\gamma x}+\frac{k_2}{V}, \qquad Q(x)=\frac{\alpha}{1+\gamma x}+\frac{\beta}{1+\gamma x}e^{-k_1 x/V},$$

where

$$\alpha=\frac{k_1}{V}\frac{MV}{Q_0 k_1}+\gamma c_s+\frac{N}{Q_0}=\gamma c_s+\frac{M+N}{Q_0}, \qquad \beta=\frac{k_1 b_0}{V}-\frac{M}{Q_0}.$$

Referring to the differential equation for $b(x)$, one has

$$e^{\int P(x)\,dx}=(1+\gamma x)\,e^{k_2 x/V}, \quad e^{-\int P(x)\,dx}=\frac{1}{1+\gamma x}e^{-k_2 x/V},$$

$$\int Q(x)\,e^{\int P(x)\,dx}dx=\int\left(\frac{\alpha}{1+\gamma x}+\frac{\beta}{1+\gamma x}e^{-k_1 x/V}\right)\cdot(1+\gamma x)\,e^{k_2 x/V}\,dx$$

$$=\int\left[\alpha\, e^{k_2 x/V}+\beta\, e^{(k_2-k_1)x/V}\right]dx=\frac{\alpha V}{k_2}e^{k_2 x/V}+\frac{\beta V}{k_2-k_1}e^{(k_2-k_1)x/V}.$$

The general solution of the differential equation is

$$(c_s-c)=e^{-\int P(x)\,dx}\left[\int Q(x)\,e^{\int P(x)\,dx}dx+C\right]$$

$$=\frac{1}{1+\gamma x}e^{-k_2 x/V}\left[\frac{\alpha V}{k_2}e^{k_2 x/V}+\frac{\beta V}{k_2-k_1}e^{(k_2-k_1)x/V}+C\right],$$

in which the constant C is determined by the initial condition $c=c_0$ at $x=0$:

$$(c_s-c_0)=\frac{\alpha V}{k_2}+\frac{\beta V}{k_2-k_1}+C \implies C=(c_s-c_0)-\frac{\alpha V}{k_2}-\frac{\beta V}{k_2-k_1}.$$

Hence

$$c_s-c(x)=\frac{e^{-k_2 x/V}}{1+\gamma x}\left\{(c_s-c_0)+\frac{\alpha V}{k_2}\left(e^{k_2 x/V}-1\right)+\frac{\beta V}{k_2-k_1}\left[e^{(k_2-k_1)x/V}-1\right]\right\}.$$

For a uniform stream, $\gamma=0$, and for far downstream, $x\to\infty$. By taking the limit as $x\to\infty$, one obtains,

$$\lim_{x\to\infty}\left[c_s-c(x)\right]=\lim_{x\to\infty}\left[(c_s-c_0)\,e^{-k_2 x/V}+\frac{\alpha V}{k_2}\left(1-e^{-k_2 x/V}\right)\right.$$

$$\left.+\frac{\beta V}{k_2-k_1}\left(e^{-k_1 x/V}-e^{-k_2 x/V}\right)\right]=\frac{\alpha V}{k_2} \implies \lim_{x\to\infty}c(x)=c_s-\frac{\alpha V}{k_2}.$$

Procedure for Solving an Application Problem

1. Establish the governing differential equations based on physical principles and geometrical properties underlying the problem.
2. Identify the type of these differential equations and then solve them.
3. Determine the arbitrary constants in the general solutions using the initial or boundary conditions.

3.6 Various Application Problems

Example 3.10 — Ferry Boat

A ferry boat is crossing a river of width a from point A to point O as shown in the following figure. The boat is always aiming toward the destination O. The speed of the river flow is constant $v_{\rm R}$ and the speed of the boat is constant $v_{\rm B}$. Determine the equation of the path traced by the boat.

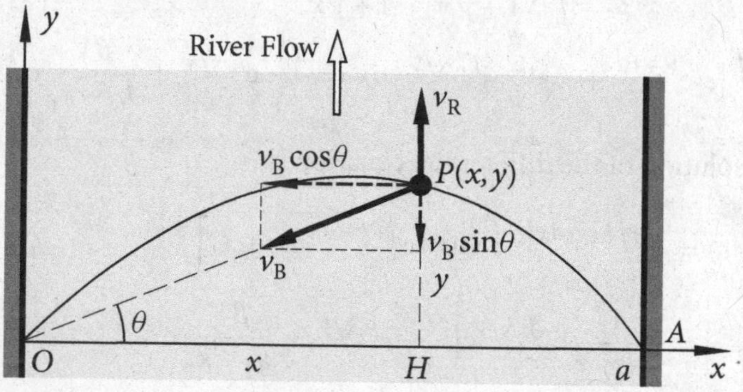

Suppose that, at time t, the boat is at point P with coordinates (x, y). The velocity of the boat has two components: the velocity of the boat $v_{\rm B}$ relative to the river flow (as if the river is not flowing), which is pointing toward the origin O or along line PO, and the velocity of the river $v_{\rm R}$ in the y direction.

Decompose the velocity components v_B and v_R in the x- and y-directions

$$v_x = -v_B\cos\theta, \qquad v_y = v_R - v_B\sin\theta.$$

From ΔOHP, it is easy to see

$$\cos\theta = \frac{OH}{OP} = \frac{x}{\sqrt{x^2+y^2}}, \qquad \sin\theta = \frac{PH}{OP} = \frac{y}{\sqrt{x^2+y^2}}.$$

Hence, the equations of motion are given by

$$v_x = \frac{{\rm d}x}{{\rm d}t} = -v_{\rm B}\frac{x}{\sqrt{x^2+y^2}}, \qquad v_y = \frac{{\rm d}y}{{\rm d}t} = v_{\rm R} - v_{\rm B}\frac{y}{\sqrt{x^2+y^2}}.$$

Since only the equation between x and y is sought, variable t can be eliminated by dividing these two equations

$$\frac{\mathrm{d}y}{\mathrm{d}x}=\frac{\dfrac{\mathrm{d}y}{\mathrm{d}t}}{\dfrac{\mathrm{d}x}{\mathrm{d}t}}=\frac{v_{\mathrm{R}}-v_{\mathrm{B}}\dfrac{y}{\sqrt{x^2+y^2}}}{-v_{\mathrm{B}}\dfrac{x}{\sqrt{x^2+y^2}}}=-\frac{k\sqrt{x^2+y^2}-y}{x},\qquad k=\frac{v_{\mathrm{R}}}{v_{\mathrm{B}}}$$

$$=-k\sqrt{1+\left(\frac{y}{x}\right)^2}+\frac{y}{x}.$$ ✍ Homogeneous DE

Let $u=\dfrac{y}{x}$ or $y=xu$, $\dfrac{\mathrm{d}y}{\mathrm{d}x}=u+x\dfrac{\mathrm{d}u}{\mathrm{d}x}$. Hence, the equation becomes

$$u+x\frac{\mathrm{d}u}{\mathrm{d}x}=-k\sqrt{1+u^2}+u,$$

$$\therefore\quad x\frac{\mathrm{d}u}{\mathrm{d}x}=-k\sqrt{1+u^2}.$$ ✍ Variable separable

The general solution is

$$\int\frac{\mathrm{d}u}{\sqrt{1+u^2}}=-k\int\frac{\mathrm{d}x}{x}+D\implies \ln\left(u+\sqrt{1+u^2}\right)=-k\ln x+\ln C,$$

$$\therefore\quad u+\sqrt{1+u^2}=Cx^{-k}.$$

Replacing u by the original variables yields

$$\frac{y}{x}+\sqrt{1+\left(\frac{y}{x}\right)^2}=Cx^{-k}\implies \sqrt{x^2+y^2}=Cx^{1-k}-y.$$

Squaring both sides leads to

$$x^2+y^2=C^2x^{2(1-k)}-2Cx^{1-k}y+y^2\implies x^2=C^2x^{2(1-k)}-2Cx^{1-k}y.$$

The constant C is determined by the initial condition $t=0$, $x=a$, $y=0$:

$$a^2=C^2a^{2(1-k)}-0\implies C=a^k.$$

Hence, the equation of the path is

$$y=\frac{1}{2Cx^{1-k}}\left[C^2x^{2(1-k)}-x^2\right]=\frac{1}{2}\left(a^kx^{1-k}-a^{-k}x^{1+k}\right),$$

$$\therefore\quad y=\frac{a}{2}\left[\left(\frac{x}{a}\right)^{1-k}-\left(\frac{x}{a}\right)^{1+k}\right].$$

Example 3.11 — Bar with Variable Cross-Section

A bar with circular cross-sections is supported at the top end and is subjected to a load of P as shown in Figure 3.14(a). The length of the bar is L. The weight density of the materials is ρ per unit volume. It is required that the stress at every point is constant σ_a. Determine the equation for the cross-section of the bar.

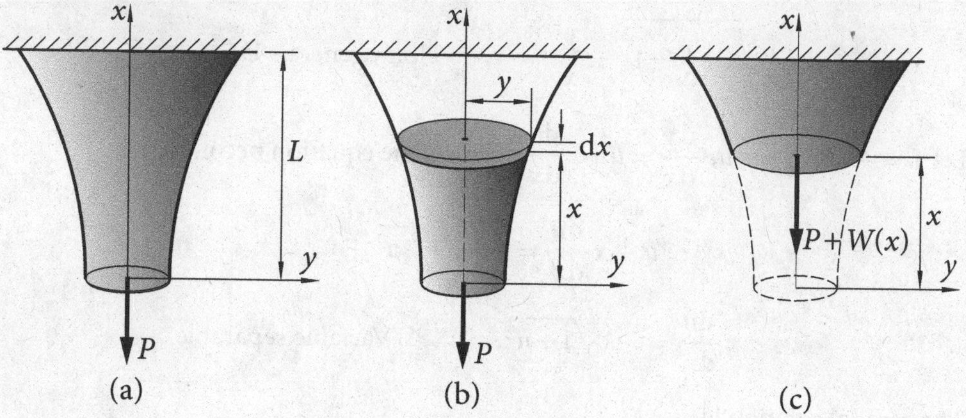

Figure 3.14 A bar under axial load.

Consider a cross-section at level x as shown in Figure 3.14(b). The corresponding radius is y. The volume of a circular disk of thickness $\mathrm{d}x$ is $\mathrm{d}V = \pi y^2 \mathrm{d}x$. The volume of the segment of bar between 0 and x is

$$V(x) = \int_0^x \pi y^2 \mathrm{d}x,$$

and the weight of this segment is

$$W(x) = \rho V(x) = \rho \int_0^x \pi y^2 \mathrm{d}x.$$

The load applied on cross-section at level x is equal to the sum of the externally applied load P and the weight of the segment between 0 and x, i.e.,

$$F(x) = W(x) + P = \rho \int_0^x \pi y^2 \mathrm{d}x + P.$$

The normal stress is

$$\sigma(x) = \frac{F(x)}{A(x)} = \frac{1}{\pi y^2}\left(\rho \int_0^x \pi y^2 \mathrm{d}x + P\right) = \sigma_a \implies \rho \int_0^x \pi y^2 \mathrm{d}x + P = \sigma_a \pi y^2.$$

Differentiating with respect to x yields

$$\rho \pi y^2 = \sigma_a \pi \cdot 2y \frac{\mathrm{d}y}{\mathrm{d}x}.$$

✍ Variable separable

Since $y \neq 0$, the equation can be written as

$$\frac{\rho}{2\sigma_a}\,\mathrm{d}x = \frac{1}{y}\,\mathrm{d}y,$$

and the general solution is given by

$$\frac{\rho}{2\sigma_a}\int \mathrm{d}x = \int \frac{1}{y}\,\mathrm{d}y + C \quad\Longrightarrow\quad \frac{\rho}{2\sigma_a}x = \ln y + C,$$

or

$$y = C\exp\Big(\frac{\rho}{2\sigma_a}x\Big) \quad\Longrightarrow\quad A(x) = \pi y^2 = \pi C^2 \exp\Big(\frac{\rho}{\sigma_a}x\Big).$$

The constant C is determined by the initial condition: $x=0$, $W(0)=0$, $F(0)=P$,

$$\sigma(0) = \frac{P}{A(0)} = \frac{P}{\pi C^2 \exp\Big(\dfrac{\rho}{\sigma_a}\cdot 0\Big)} = \frac{P}{\pi C^2} = \sigma_a \quad\Longrightarrow\quad C^2 = \frac{P}{\pi\sigma_a}.$$

Hence

$$A(x) = \pi\cdot\frac{P}{\pi\sigma_a}\cdot\exp\Big(\frac{\rho}{\sigma_a}x\Big) = \frac{P}{\sigma_a}\exp\Big(\frac{\rho}{\sigma_a}x\Big), \qquad 0 \leqslant x \leqslant L.$$

Example 3.12 — Chain Moving

A uniform chain of length L with mass density per unit length ρ is laid on a smooth horizontal table with an initial hang of length l as shown in Figure 3.15(a). The chain is released from rest at time $t=0$. Show that the time it takes for the chain to leave the table is given by

$$T = \sqrt{\frac{L}{g}}\,\ln\frac{L+\sqrt{L^2-l^2}}{l}.$$

At time t, the length of the chain hanging off the table is $y(t)$ as shown in Figure 3.15(b). The chain is subjected to a downward force $F(t)=(\rho y)g$, which is the weight of the segment of the chain hanging off the table. Apply Newton's Second Law to the chain

$$\downarrow\ ma = \sum F: \quad (\rho L)\ddot{y} = (\rho y)g \quad\Longrightarrow\quad \ddot{y} - \frac{g}{L}y = 0,$$

or

$$\ddot{y} - k^2 y = 0, \qquad k = \sqrt{\frac{g}{L}}.$$

The initial conditions are $t=0$, $y=l$, $\dot{y}=0$.

Remarks: Since the problem is equivalent to the entire chain moving in the vertical direction under gravity $(\rho y)g$ as shown in Figure 3.15(c), the mass m is for the entire chain, not just the segment that is hanging off the table.

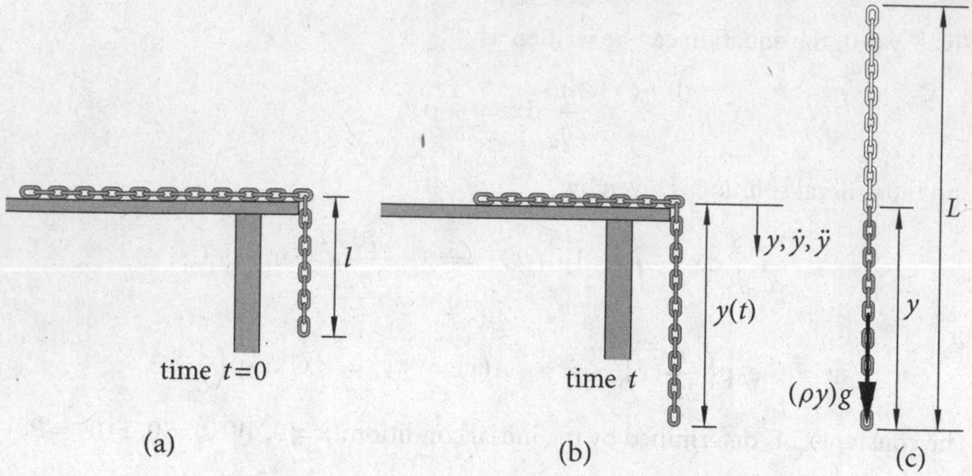

Figure 3.15 A chain moving off a smooth table.

The equation of motion is a second-order differential equation with the independent variable t absent. Let y be the new independent variable and $u = \dfrac{dy}{dt}$ be the new dependent variable, $\dfrac{d^2y}{dt^2} = u\dfrac{du}{dy}$. Hence

$$u\frac{du}{dy} - k^2 y = 0. \quad \text{✍ Variable separable}$$

The general solution is given by

$$\int u\,du = \int k^2 y\,dy + C \implies \frac{1}{2}u^2 = \frac{k^2}{2}y^2 + C.$$

The constant of integration C is determined by the initial condition $t=0$, $y=l$, $u=\dot{y}=0$:

$$\frac{1}{2}\cdot 0^2 = \frac{k^2}{2}\cdot l^2 + C \implies C = -\frac{k^2}{2}l^2.$$

Hence,

$$u^2 = k^2(y^2 - l^2) \implies u = \frac{dy}{dt} = k\sqrt{y^2 - l^2}. \quad \text{✍ Variable separable}$$

The general solution is

$$\int \frac{dy}{\sqrt{y^2 - l^2}} = \int k\,dt + D \implies \ln\left(y + \sqrt{y^2 - l^2}\right) = kt + D.$$

Using the initial condition $t=0$, $y=l$, one obtains

$$\ln l = k\cdot 0 + D \implies D = \ln l.$$

The solution of the equation of motion is

$$\ln\big(y+\sqrt{y^2-l^2}\big) = kt + \ln l,$$

or

$$t = \frac{1}{k}\Big[\ln\big(y+\sqrt{y^2-l^2}\big) - \ln l\Big] = \frac{1}{k}\ln\frac{y+\sqrt{y^2-l^2}}{l}.$$

When the chain leaves the table, $t=T,\ y=L$:

$$T = \sqrt{\frac{L}{g}}\,\ln\frac{L+\sqrt{L^2-l^2}}{l}.$$

Example 3.13 — Chain Moving

One end of a pile of uniform chain falls through a hole in its support and pulls the remaining links after it as shown. The links, which are initially at rest, acquire the velocity of the chain suddenly without any frictional resistance or interference from the support and adjacent links. At $t=0,\ y(t)=0$ and $v(t)=\dot{y}(t)=0$. Determine the length $y(t)$ and the velocity $v(t)$ of the chain.

At time t, the length of the chain hanging off the support is $y(t)$ and the velocity of the chain is $v(t)=\dot{y}(t)$. The chain is subjected to a downward force $F(t)=(\rho y)g$, which is the weight of the segment of the chain hanging off the support.

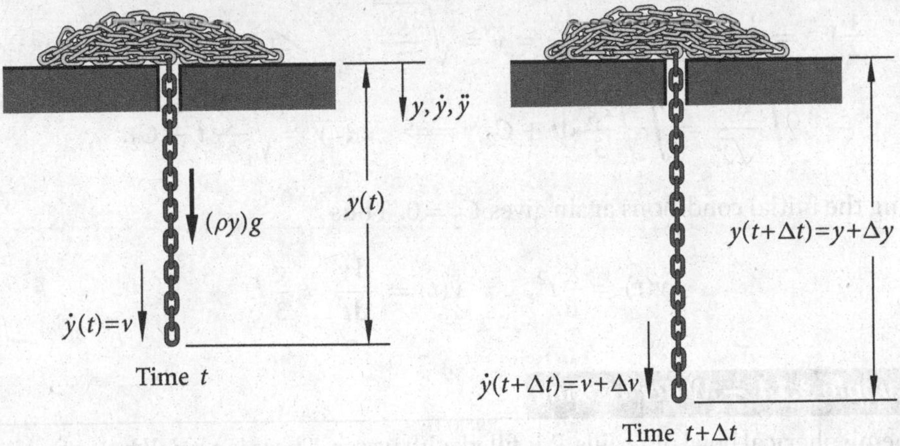

To set up the equation of motion, apply the Impulse-Momentum Principle:

$$(\text{Momentum at time } t) + (\text{Impulse during } \Delta t) = (\text{Momentum at time } t+\Delta t),$$

where

$$\text{Momentum at time } t = (\rho y)v, \qquad \rho = \text{mass density of the chain},$$

$$\text{Momentum at time } t+\Delta t = \big[\rho(y+\Delta y)\big](v+\Delta v),$$

$$\text{Impulse during } \Delta t = \big[(\rho y)g\big]\Delta t.$$

Hence,

$$\rho y v + \rho g y \Delta t = \rho(y+\Delta y)(v+\Delta v).$$

Dividing the equation by Δt and taking the limit as $\Delta t \to 0$ result in the equation of motion

$$y\frac{dv}{dt} + v\frac{dy}{dt} = gy \quad \Longrightarrow \quad \frac{d(yv)}{dt} = gy.$$

Noting that $v = \dfrac{dy}{dt} \Rightarrow dt = \dfrac{dy}{v}$, and letting $V = yv$, one has

$$\frac{dV}{\dfrac{dy}{v}} = gy \quad \Longrightarrow \quad v\,dV = gy\,dy.$$

Multiplying the equation by y yields

$$V\,dV = gy^2 dy. \qquad \text{✍ Variable separable}$$

Integrating both sides gives

$$\tfrac{1}{2}V^2 = \tfrac{1}{3}gy^3 + C_1.$$

Using the initial conditions $y=0$ and $v=0$ when $t=0$, one has $C_1 = 0$. Hence

$$\tfrac{1}{2}V^2 = \tfrac{1}{3}gy^3 \quad \Longrightarrow \quad \frac{dy}{dt} = v = \sqrt{\frac{2gy}{3}}. \qquad \text{✍ Variable separable}$$

$$\int \frac{dy}{\sqrt{y}} = \int \sqrt{\frac{2g}{3}}\,dt + C_2 \quad \Longrightarrow \quad 2\sqrt{y} = \sqrt{\frac{2g}{3}}\,t + C_2.$$

Using the initial conditions again gives $C_2 = 0$. Thus

$$y(t) = \frac{g}{6}t^2, \qquad v(t) = \frac{dy}{dt} = \frac{g}{3}t.$$

Example 3.14 — Water Leaking

A hemispherical bowl of radius R is filled with water. There is a small hole of radius r at the bottom of the convex surface as shown in Figure 3.16(a). Assume that the velocity of efflux of the water when the water level is at height h is $v = c\sqrt{2gh}$, where c is the discharge coefficient. The volume of the cap of the sphere of height h, shown as shaded volume in Figure 3.16(a), is given by

$$V = \frac{\pi}{3}h^2(3R - h).$$

Determine the time taken for the bowl to empty.

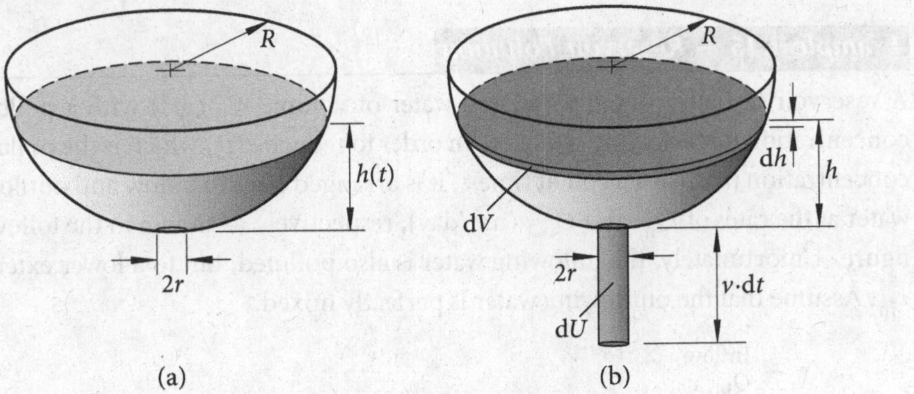

Figure 3.16 A hemispherical bowl with a hole.

At time t, the water level is $h(t)$ and the volume of the water is

$$V = \frac{\pi}{3} h^2 (3R - h).$$

Considering a small time interval dt, the water level drops dh as shown in Figure 3.16(b). The water lost is

$$dV = \frac{dV}{dh} \cdot dh = \frac{\pi}{3}\big[2h(3R-h) + h^2(-1)\big]dh = \pi(2Rh - h^2)dh.$$

The water lost is leaked from the hole at the bottom. The water level drops, i.e., $dh < 0$, which leads to a negative volume change, i.e., $dV < 0$.

Since the velocity of efflux of water is v, the amount of water leaked during time dt is

$$dU = \pi r^2 \cdot v\, dt = \pi r^2 \cdot c\sqrt{2gh}\, dt,$$

which is indicated by the small shaded cylinder at the bottom of the bowl.

From the conservation of water volume, $dV + dU = 0$, i.e.,

$$\pi(2Rh - h^2)\, dh + \pi r^2 \cdot c\sqrt{2gh}\, dt = 0 \implies (2R - h)\sqrt{h}\, dh = -r^2 c\sqrt{2g}\, dt.$$

The equation is variable separable and the general solution is

$$\int \big(2Rh^{\frac{1}{2}} - h^{\frac{3}{2}}\big)\, dh = -r^2 c\sqrt{2g} \int dt + D,$$

$$2R\,\frac{h^{\frac{3}{2}}}{\frac{3}{2}} - \frac{h^{\frac{5}{2}}}{\frac{5}{2}} = -r^2 c\sqrt{2g}\, t + D. \qquad \int x^n\, dx = \frac{x^{n+1}}{n+1}$$

The constant D is determined from the initial condition $t = 0$, $h = R$:

$$\frac{4}{3} R^{\frac{5}{2}} - \frac{2}{5} R^{\frac{5}{2}} = D \implies D = \frac{14}{15} R^{\frac{5}{2}}.$$

When the bowl is empty, $t = T$, $h = 0$:

$$0 = -r^2 c\sqrt{2g}\, T + \frac{14}{15} R^{\frac{5}{2}} \implies T = \frac{14}{15}\frac{R^{\frac{5}{2}}}{r^2 c\sqrt{2g}} = \frac{14}{15c}\Big(\frac{R}{r}\Big)^2 \sqrt{\frac{R}{2g}}.$$

Example 3.15 — Reservoir Pollution

A reservoir initially contains polluted water of volume V_0 (m^3) with a pollutant concentration in *percentage* being c_0. In order to reduce $c(t)$, which is the pollutant concentration in the reservoir at time t, it is arranged to have inflow and outflow of water at the rates of Q_{in} and Q_{out} (m^3/day), respectively, as shown in the following figure. Unfortunately, the inflowing water is also polluted, but to a lower extent of c_{in}. Assume that the outflowing water is perfectly mixed.

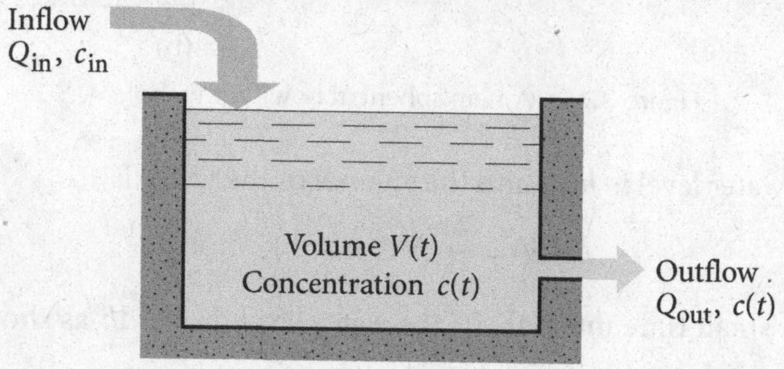

1. Set up the differential equation governing the pollutant $c(t)$.
2. Considering the case with the following parameters

$$V_0 = 500 \text{ m}^3, \quad Q_{in} = 200 \text{ m}^3, \quad Q_{out} = 195 \text{ m}^3, \quad c_0 = 0.05\%, \quad c_{in} = 0.01\%,$$

find the time (in days) it will take to reduce the pollutant concentration to the acceptable level of 0.02%.

---

1. At time t,

$$\text{Volume } V(t) = V_0 + (Q_{in} - Q_{out})\, t,$$

$$\text{Pollutant concentration } c(t),$$

$$\text{Amount of pollutant} = V(t)\,c(t) = \big[V_0 + (Q_{in} - Q_{out})\, t\big]\, c.$$

At time $t + \Delta t$,

$$\text{Volume } V(t+\Delta t) = V_0 + (Q_{in} - Q_{out})\,(t + \Delta t),$$

$$\text{Pollutant concentration } c(t+\Delta t) = c(t) + \Delta c,$$

$$\begin{aligned}\text{Amount of pollutant} &= V(t+\Delta t)\,c(t+\Delta t)\\ &= \big[V_0 + (Q_{in} - Q_{out})(t+\Delta t)\big]\,(c + \Delta c),\end{aligned}$$

$$\text{Inflow pollutant} = Q_{in}\,\Delta t \cdot c_{in},$$

$$\text{Outflow pollutant} = Q_{out}\,\Delta t \cdot c.$$

Since

$$\text{(Amount of pollutant at } t+\Delta t) = \text{(Amount of pollutant at } t) + \big[\text{(Inflow pollutant)} - \text{(Outflow pollutant)}\big],$$

$$\therefore\ \big[V_0 + (Q_{\text{in}} - Q_{\text{out}})(t+\Delta t)\big](c+\Delta c) = \big[V_0 + (Q_{\text{in}} - Q_{\text{out}})\,t\big]c + Q_{\text{in}}\Delta t\cdot c_{\text{in}} - Q_{\text{out}}\Delta t\cdot c,$$

expanding yields

$$\big[V_0 + (Q_{\text{in}} - Q_{\text{out}})t\big]c + \big[V_0 + (Q_{\text{in}} - Q_{\text{out}})t\big]\Delta c + (Q_{\text{in}} - Q_{\text{out}})\Delta t\cdot c + (Q_{\text{in}} - Q_{\text{out}})\,\Delta t\cdot\Delta c = \big[V_0 + (Q_{\text{in}} - Q_{\text{out}})t\big]c + Q_{\text{in}}\Delta t\cdot c_{\text{in}} - Q_{\text{out}}\Delta t\cdot c.$$

Neglecting higher order term $\Delta t\cdot\Delta c$, dividing by Δt, and simplifying lead to

$$\big[V_0 + (Q_{\text{in}} - Q_{\text{out}})\,t\big]\frac{\Delta c}{\Delta t} + Q_{\text{in}}\,c = Q_{\text{in}}\,c_{\text{in}}.$$

Taking the limit $\Delta t \to 0$ results in the differential equation

$$\big[V_0 + (Q_{\text{in}} - Q_{\text{out}})\,t\big]\frac{\mathrm{d}c}{\mathrm{d}t} + Q_{\text{in}}\,c = Q_{\text{in}}\,c_{\text{in}}.$$

2. For the parameters

$$V_0 = 500, \quad Q_{\text{in}} = 200, \quad Q_{\text{out}} = 195, \quad c_0 = 0.05, \quad c_{\text{in}} = 0.01,$$

the differential equation becomes

$$\big[500 + (200-195)\,t\big]\frac{\mathrm{d}c}{\mathrm{d}t} + 200\,c = 200\times 0.01 \implies (100+t)\frac{\mathrm{d}c}{\mathrm{d}t} = 0.4 - 40\,c.$$

The equation is variable separable and the general solution is

$$\int \frac{\mathrm{d}c}{0.4-40c} = \int \frac{\mathrm{d}t}{100+t} + C \implies -\frac{1}{40}\ln(0.4-40c) = \ln(100+t) + \ln D,$$

$$\therefore\quad (0.4-40c)^{-\frac{1}{40}} = D(100+t). \qquad ✍\ C = \ln D$$

The constant D is determined from the initial condition $t=0$, $c=c_0=0.05$:

$$(0.4-40c_0)^{-\frac{1}{40}} = D\cdot 100 \implies D = \frac{(0.4-40c_0)^{-\frac{1}{40}}}{100}.$$

The solution becomes

$$(0.4-40c)^{-\frac{1}{40}} = \frac{(0.4-40c_0)^{-\frac{1}{40}}}{100}(100+t) \implies t = 100\left[\left(\frac{0.4-40c}{0.4-40c_0}\right)^{-\frac{1}{40}} - 1\right].$$

When the pollutant concentration is reduced to $c=0.02$, the time required is

$$t = 100\left[\left(\frac{0.4-40\times 0.02}{0.4-40\times 0.05}\right)^{-\frac{1}{40}} - 1\right] = 3.53 \text{ days}.$$

Problems

3.1 At a crime scene, a forensic technician found the body temperature of the victim was 33°C at 6:00 p.m. One hour later, the coroner arrived and found the body temperature of the victim fell to 31.5°C. The forensic technician determined that the change in the atmospheric temperature could be modeled satisfactorily as $20e^{-0.02t}$ in the time window of ± 3 hours starting at 6:00 p.m. It is known that the body temperature of a live person is 37°C. When was the victim murdered?
ANS 3:42 p.m.

3.2 The value of proportionality of cooling of a large workshop is k (1/hr) due to its ventilating system. The atmospheric temperature fluctuates sinusoidally with a period of 24 hours, reaching a minimum of 15°C at 2:00 a.m. and a maximum of 35°C at 2:00 p.m. Let t denote the time in hours starting with $t=0$ at 8:00 a.m.

1. By applying Newton's Law of Cooling, set up the differential equation governing the temperature of the workshop $T(t)$.

   ANS $$\frac{dT}{dt} + kT = k\left(25 + 10\sin\frac{\pi t}{12}\right)$$

2. Determine the *steady-state* solution of the differential equation, which is solution for time t large or the solution due to the atmospherical temperature change.

   ANS $$T_{\text{steady-state}} = 25 + \frac{120k}{\sqrt{144k^2+\pi^2}}\sin\left[\frac{\pi t}{12} - \tan^{-1}\left(\frac{\pi}{12k}\right)\right]$$

3. If $k=0.2$ (1/hr), what are the maximum and minimum temperatures that the workshop will reach? ANS $T_{\min}=18.9$°C, $T_{\max}=31.1$°C

3.3 Suppose that the air resistance on a parachute is proportional to the effective area A of the parachute, which is the area of the parachute projected in the horizontal plane, and to the square of its velocity v, i.e., $R=kAv^2$, where k is a constant. A parachute of mass m falls with zero initial velocity, i.e., $x(t)=0$ and $v(t)=0$ when $t=0$.

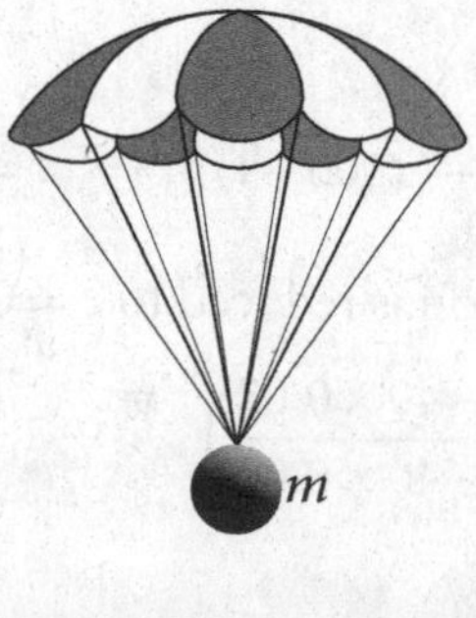

1. Show that the terminal velocity v_T of the parachute is

$$v_T = \lim_{t\to\infty} v(t) = \sqrt{\frac{mg}{kA}}.$$

2. Show that the velocity and the displacement of the mass are given by

$$v(t) = v_T \tanh\left(\frac{gt}{v_T}\right), \quad x(t) = \frac{v_T^2}{g}\ln\cosh\left(\frac{gt}{v_T}\right). \quad \tanh z = \frac{e^z - e^{-z}}{e^z + e^{-z}}$$

3.4 A mass m falls from a height of $H = 3200$ km with zero initial velocity as shown in the following figure. The gravity on the mass changes as

$$F = \frac{mgR^2}{(R+H-x)^2},$$

where $R = 6400$ km is the radius of the earth. Neglect the air resistance. Determine the time it takes for the mass to reach the ground and the velocity of the mass.

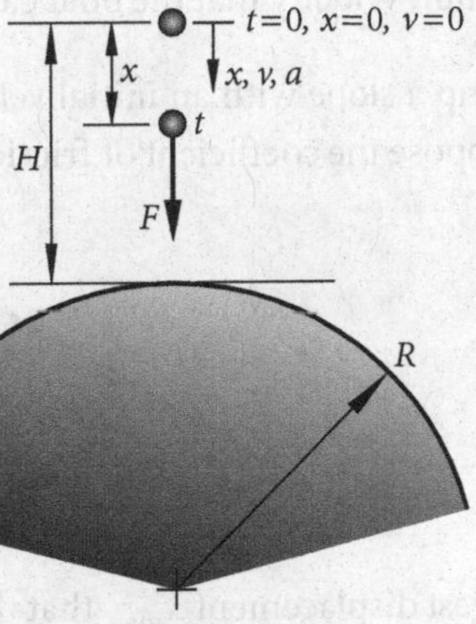

Hint: Use the formulas $a = \frac{dv}{dt} = \frac{dv}{dx}\frac{dx}{dt} = v\frac{dv}{dx}$, and

$$\int \sqrt{\frac{k-x}{x}}\,dx = \sqrt{x(k-x)} + k\sin^{-1}\sqrt{\frac{x}{k}} + C.$$

ANS $v = \sqrt{\frac{2gRH}{R+H}} = 6.47$ km/sec

$$t = \frac{1}{R}\sqrt{\frac{R+H}{2g}}\left[\sqrt{RH} + (R+H)\sin^{-1}\sqrt{\frac{H}{R+H}}\right] = 1,141 \text{ sec}$$

3.5 An undersea explorer traveling along a straight line in a horizontal direction is propelled by a constant force T. Suppose the resistant force is $R = a + bv^2$, where $0 < a < T$ and $b > 0$ are constants, and v is the velocity of the explorer.

1. If the explorer is at rest at time $t=0$, show that the velocity of the explorer is

$$v(t) = \alpha \tanh \beta t, \qquad \alpha = \sqrt{\frac{T-a}{b}}, \quad \beta = \frac{\alpha b}{m} = \frac{1}{m}\sqrt{b(T-a)}.$$

2. Determine the distance $x(t)$ traveled at time t. ANS $x = \dfrac{\alpha}{\beta} \ln \cosh \beta t$

3.6 A boat starting from rest at time $t=0$ is propelled by a constant force F. The resistance from air is proportional to the velocity, i.e., $R_{\text{air}} = -\alpha v$, and the resistance from water is proportional to the square of the velocity, i.e., $R_{\text{water}} = -\beta v^2$, in which α, β are positive constants. Assume that $F > \alpha v + \beta v^2$.

1. Show that the velocity of the boat is given by

$$v(t) = a \tanh\left(\frac{\beta a t}{m} + \tanh^{-1}\frac{b}{a}\right) - b, \quad \text{where } a^2 = \frac{F}{\beta} + \frac{\alpha^2}{4\beta^2}, \quad b = \frac{\alpha}{2\beta}.$$

2. What is the maximum velocity that the boat can reach? ANS $v_{\text{max}} = a - b$

3.7 A mass m moves up a slope with an initial velocity of 10 m/sec as shown in the following figure. Suppose the coefficient of friction between the mass and slope is 0.25.

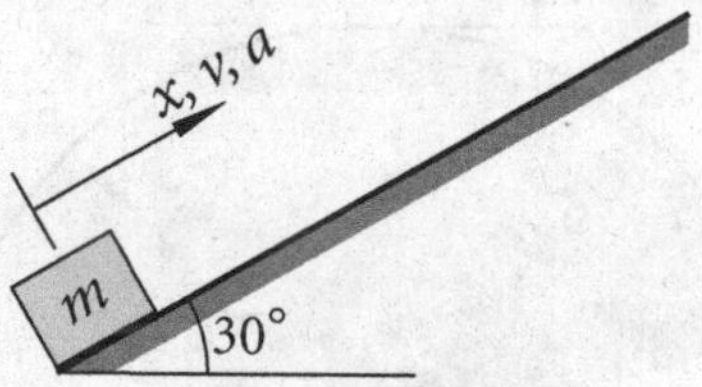

1. Determine the largest displacement x_{max} that the mass can reach and the time it takes. ANS $x_{\text{max}} = 7.12$ m, $t = 1.42$ sec

2. Determine the time it takes for the mass to return to its original position and the corresponding velocity. ANS $v = 6.29$ m/sec, $t = 3.69$ sec

3.8 A skier skis from rest on a slope with $\theta = 30°$ from point A at $t=0$ as shown in the following figure. The skier is clocked $t_1 = 3.61$ sec at the 25-m checkpoint and $t_2 = 5.30$ sec at the 50-m checkpoint. The length of the slope is $AB = L = 100$ m. The wind resistance is proportional to the speed of the skier, i.e., $R = kv$.

1. Find the coefficient of wind resistance k and the coefficient of kinetic friction μ between the snow and the skis. ANS $k = 10.395$ N·sec/m, $\mu = 0.041$

2. Determine the time t_3 that it takes for the skier to reach the bottom of the slope B and the corresponding speed v_3. ANS $t_3 = 7.92$ sec, $v_3 = 21.20$ m/sec

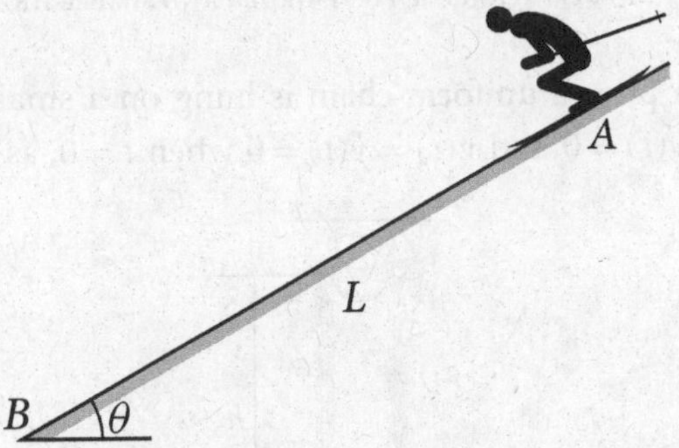

3.9 A uniform chain of length $2L$ with mass density per unit length ρ is laid on a rough inclined surface with $y=L$ at $t=0$ as shown in the following figure.

The coefficients of static and kinetic friction between the chain and the surface have inclined the same value μ. The chain starts to drop from rest at time $t=0$. Show that the relationship between the velocity of the chain v and y is given by

$$v=\sqrt{\frac{(1+\sin\theta+\mu\cos\theta)g}{2L}(y^2-L^2)-2(\sin\theta+\mu\cos\theta)g\,(y-L)}.$$

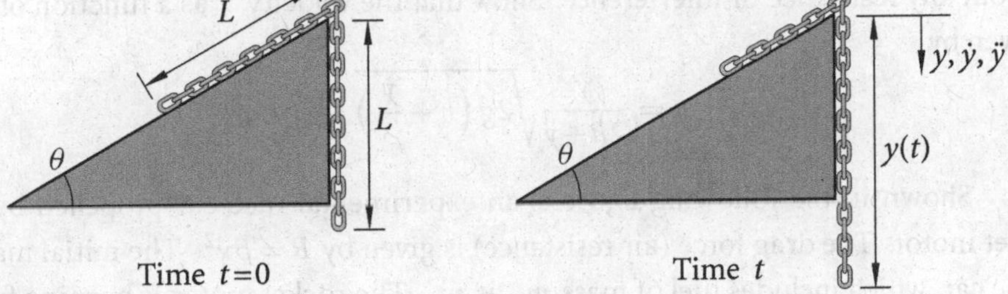

3.10 A uniform chain of length L with mass density per unit length ρ is laid on a rough inclined surface with $y=0$ at $t=0$ as shown in the following figure.

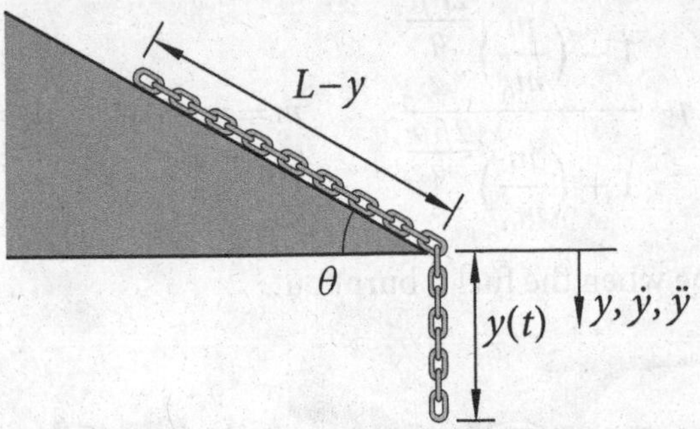

The coefficients of static and kinetic friction between the chain and the surface have the same value μ. The chain is release from rest at time $t=0$. Show that the relationship between the velocity of the chain v and y is given by

$$v=\sqrt{2gy\left[(\sin\theta-\mu\cos\theta)+\frac{1-\sin\theta+\mu\cos\theta}{2L}y\right]}.$$

3.11 One end of a pile of uniform chain is hung on a small smooth pulley of negligible size with $y(t)=0$ and $v(t)=\dot{y}(t)=0$ when $t=0$, as shown here.

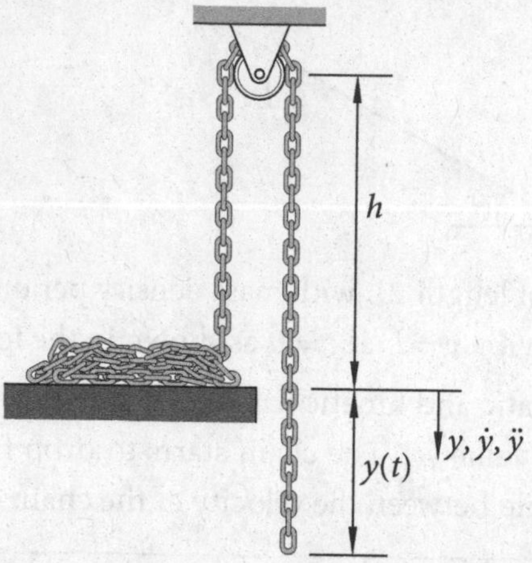

The chain starts to fall at time $t=0$ and pulls the remaining links. The links on the support, which are initially at rest, acquire the velocity of the chain suddenly without any resistance or interference. Show that the velocity v as a function of y is given by

$$v = \frac{y}{2h+y}\sqrt{2g\left(h+\frac{y}{3}\right)}.$$

3.12 Shown in the following figure is an experimental race car propelled by a rocket motor. The drag force (air resistance) is given by $R=\beta v^2$. The initial mass of the car, which includes fuel of mass m_f, is m_0. The rocket motor is burning fuel at the rate of q with an exhaust velocity of u relative to the car. The car is at rest at $t=0$. Show that the velocity of the car is given by, for $0 \leqslant t \leqslant T$,

$$v(t) = \mu \cdot \frac{1-\left(\dfrac{m}{m_0}\right)^{\frac{2\beta\mu}{q}}}{1+\left(\dfrac{m}{m_0}\right)^{\frac{2\beta\mu}{q}}}, \qquad m=m_0-qt, \quad \mu^2=\frac{qu}{\beta},$$

$T=m_f/q$ is the time when the fuel is burnt out.

For $m_0=900$ kg, $m_f=450$ kg, $q=15$ kg/sec, $u=500$ m/sec, $\beta=0.3$, what is the burnout velocity of the car? ANS $v_f=555.2$ km/hr

3.13 Cable OA supports a structure of length L as shown in the following figure. The structure applies a trapezoidally distributed load on the cable through the hangers. The load intensity is w_0 at the left end and w_1 at the right end. Cable OA has zero slope at its lowest point O and has sag h. Determine the tensions $T_O = H$ at O and T_A at A.

ANS $H = \dfrac{L^2}{6h}(2w_0 + w_1), \quad T_A = \sqrt{H^2 + W(L)^2}, \quad W(L) = \dfrac{w_0 + w_1}{2}L$

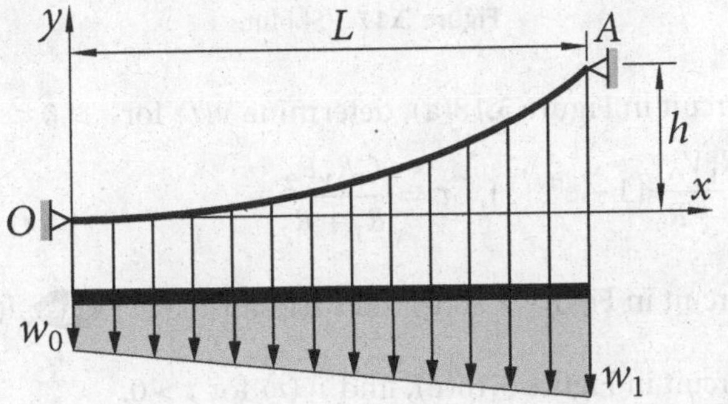

3.14 As a structural engineer, you are asked to design a footbridge across the Magnificent Gorge, which is 50 m across. The configuration of the bridge is shown in the following figure; the lowest point of the cable is 20 m below the left support and 10 m below the right support. Both supports are anchored on the cliffs. The effective load of the footbridge is assumed to be uniform with an intensity of 500 N/m. Determine the maximum and minimum tensions in the cable, and the length of the cable. ANS $T_{min} = H = 10723$ N, $T_{max} = T_A = 18151$ N, Length $= 60.36$ m

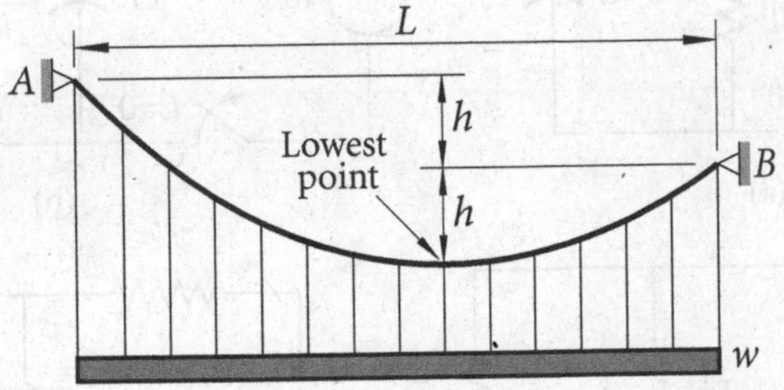

3.15 Consider the moving cable AB of a ski lift between two supporting towers as shown in Figure 3.17. The cable has a mass of 10 kg/m and carries equally spaced chairs and passengers, which result in an added mass of 20 kg/m when averaged over the length of the cable. The cable leads horizontally from the supporting guide wheel at A. Determine the tensions in the cable at A and B, and the length of the cable between A and B.

ANS $T_A = H = 27387$ N, $T_B = 33267$ N, $s = 64.24$ m

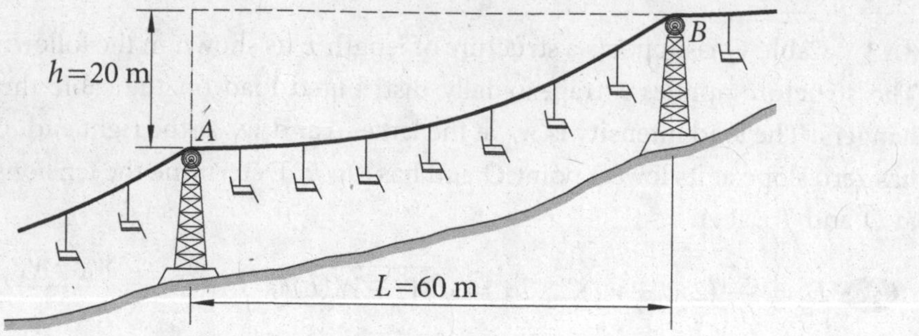

Figure 3.17 Ski lift.

3.16 For the circuit in Figure 3.18(a), determine $v(t)$ for $t>0$.

ANS $v(t)=\dfrac{R_1V_0}{R_1+R_2}\left(1-e^{-t/\tau}\right),\quad \tau=\dfrac{CR_1R_2}{R_1+R_2}$

3.17 For the circuit in Figure 3.18(b), find $i(t)$ for $t>0$. ANS $i(t)=e^{-t/40}$ (A)

3.18 For the circuit in Figure 3.18(c), find $i_L(t)$ for $t>0$.

ANS $i_L(t)=6+e^{-7t}$ (A)

3.19 For the circuit in Figure 3.18(d), find $i_L(t)$ for $t>0$. ANS $i_L(t)=\dfrac{V_0}{R_1}$

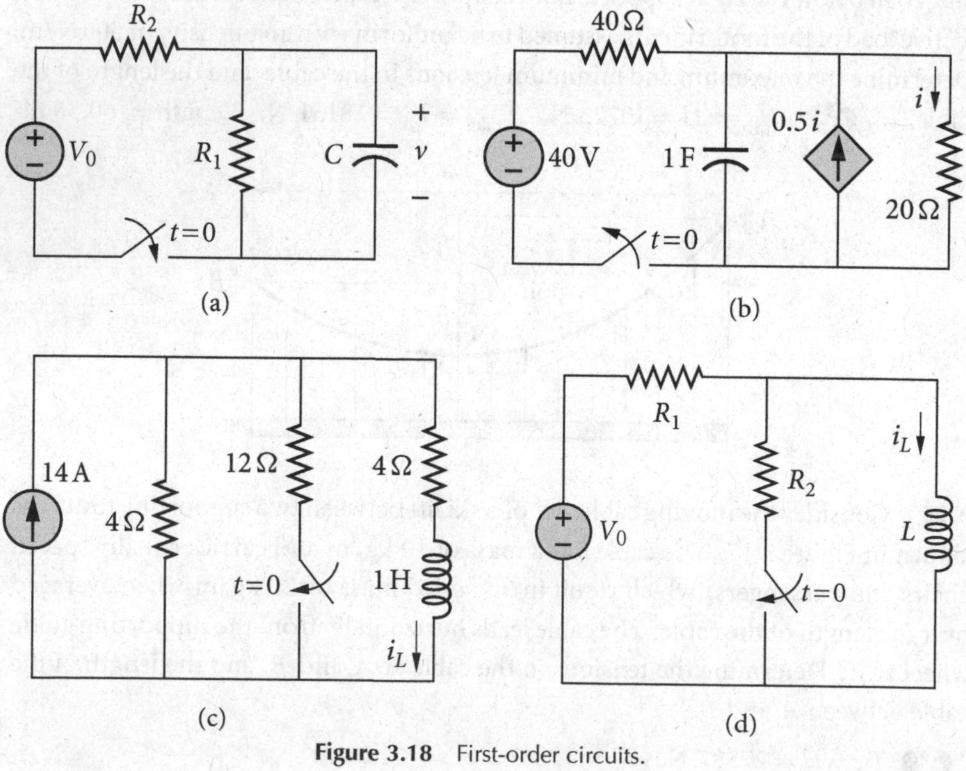

Figure 3.18 First-order circuits.

3.20 Consider $b(x)$ as the concentration of an inert pollutant ($k_1=0$) which is added to the stream at an outfall such that $b=b_0$ at $x=0$ and $M=0$. Due to the increase of drainage area, the discharge increases along the stream as $Q=Q_0(1+\gamma x)$, where Q_0 and γ are constants. Determine the BOD distribution $b(x)$ and the DO distribution $c(x)$ in the stream with $c=c_0$ at $x=0$ and $N=0$.

ANS $$b(x)=\frac{b_0}{1+\gamma x}$$
$$c(x)=\frac{1}{1+\gamma x}\left[\frac{(k_2-\gamma V)c_s}{k_2}\left(1-e^{-k_2x/V}\right)+c_0e^{-k_2x/V}+\gamma c_s x\right]$$

3.21 In a stream of constant Q and V, BOD is added uniformly along the stream, starting at $x=0$ where $b=0$. Determine the BOD distribution $b(x)$ and the DO distribution $c(x)$ in the stream with $c=c_0$ at $x=0$ and $N=0$.

ANS $$b(x)=\frac{MV}{Qk_1}\left(1-e^{-k_1x/V}\right)$$
$$c(x)=c_s+(c_0-c_s)e^{-k_2x/V}+\frac{MV}{Q}\left(\frac{e^{-k_2x/V}-1}{k_2}+\frac{e^{-k_1x/V}-e^{-k_2x/V}}{k_2-k_1}\right)$$

3.22 A drop of liquid of initial mass of m_0 falls vertically in air from rest. The liquid evaporates uniformly, losing mass m_1 in unit time. Suppose the resistance from air is proportional to the velocity of the drop, i.e., $R=kv$. Show that the velocity of the drop $v(t)$ is

$$v(t)=\frac{g}{k-m_1}\left[(m_0-m_1t)-m_0\left(1-\frac{m_1}{m_0}t\right)^{k/m_1}\right],\quad t<\frac{m_0}{m_1}.$$

3.23 As an engineer, you are asked to design a bridge pier with circular horizontal cross-sections as shown in the following figure.

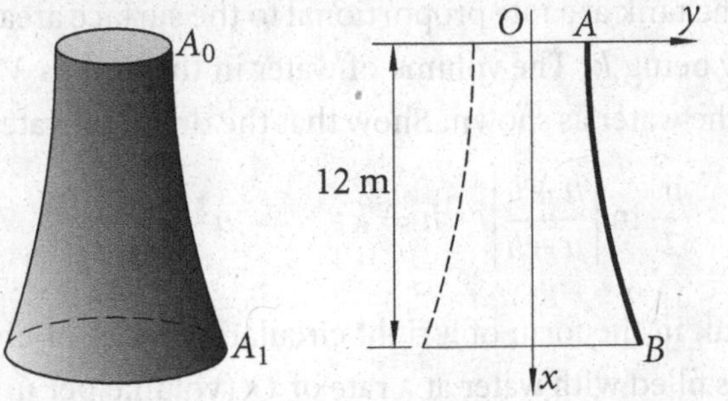

The height of the pier is 12 m. The top cross-section is subjected to a uniformly distributed pressure with a resultant $P=3\times10^5$ N. It is known that the material of the pier has a density $\rho=2.5\times10^4$ N/m^3 and an allowable pressure $P_A=3\times10^5$ N/m^2. Design the bridge pier with a minimum amount of material and include the weight of the bridge pier in the calculations.

1. Note that the surface of the pier is obtained by rotating curve AB about the x-axis. Determine the equation of curve AB.
2. Determine the areas of the top and bottom cross-sections A_0 and A_1.

ANS $y=\dfrac{1}{\sqrt{\pi}}\exp\left(\dfrac{x}{24}\right)$, $A_0=1\ \text{m}^2$, $A_1=2.718\ \text{m}^2$

3.24 A tank in the form of a right circular cone of height H, radius R, with its vertex below the base is filled with water as shown in the following figure. A hole, having a cross-sectional area a at the vertex, causes the water to leak out. Assume that the leaking flow velocity is $v(t)=c\sqrt{2gh(t)}$, where c is the discharge coefficient.

1. Show that the differential equation governing the height of water level $h(t)$ is

$$h^{\frac{3}{2}}\frac{dh}{dt}=-\frac{acH^2}{A}\sqrt{2g}, \qquad A=\pi R^2.$$

2. Show that the time for the cone to empty is $T=\dfrac{2A}{5ac}\sqrt{\dfrac{H}{2g}}$.

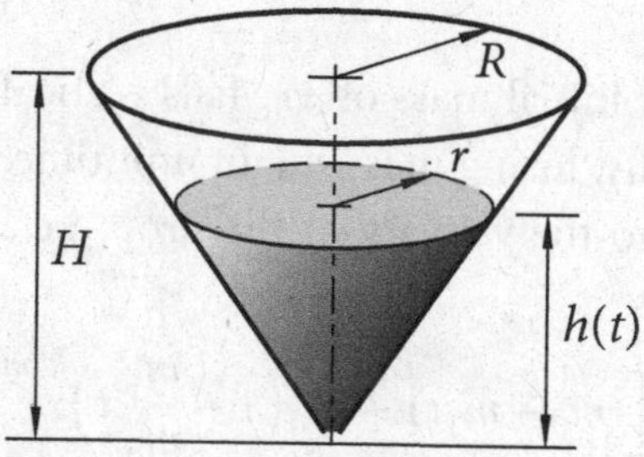

3.25 A conical tank with an open top of radius R has a depth H (Figure 3.19) is initially empty. Water is added at a rate of q (volume per unit time). Water evaporates from the tank at a rate proportional to the surface area with the constant of proportionality being k. The volume of water in the tank is $V=\frac{1}{3}\pi r^2 h$, where h is the depth of the water as shown. Show that the depth of water $h(t)$ is given by

$$\frac{\mu}{2}\ln\left|\frac{\mu+h}{\mu-h}\right|-h=kt, \qquad \mu^2=\frac{H^2 q}{\pi k R^2}.$$

3.26 A water tank in the form of a right circular cylinder of cross-sectional area A and height H is filled with water at a rate of Q (volume per unit time) as shown in Figure 3.20. A hole, having a cross-sectional area a at the base, causes the water to leak out. The leaking flow velocity is $v(t)=c\sqrt{2gh(t)}$, where c is the discharge coefficient. The tank is initially empty. Show that the time T to fill the tank is

$$T=\frac{2A}{b^2}\left(-b\sqrt{H}+Q\ln\frac{Q}{Q-b\sqrt{H}}\right), \qquad b=ac\sqrt{2g}.$$

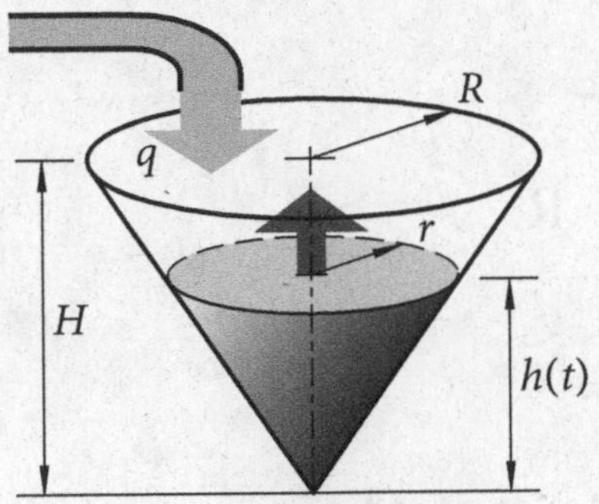

Figure 3.19

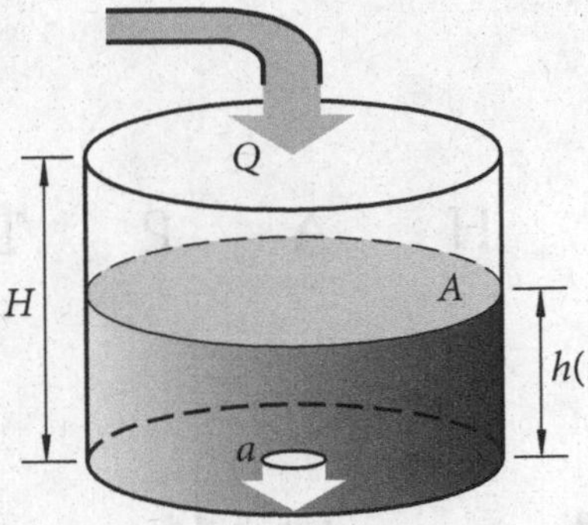

Figure 3.20

3.27 Lake Ontario has a volume of approximately $V=1600$ km^3. It is heavily polluted by a certain pollutant with a concentration of $c_0=0.1\%$. As a part of pollution control, the pollutant concentration of the inflow is reduced to $c_{\text{in}}=0.02\%$. The inflow and outflow rates are $Q_{\text{in}}=Q_{\text{out}}=Q=500$ km^3 per year. Suppose that pollutant from the inflow is well mixed with the lake water before leaving the lake. How long will it take for the pollutant concentration in the lake to reduce to 0.05%?

ANS $t=3.14$ years

3.28 A mouse Q runs along the positive y-axis at the constant speed v_1 starting at the origin. A cat P chases the mouse along curve C at the constant speed v_2 starting at point (1, 0) as shown in the following figure. At any time instance, line PQ is tangent to curve C. Determine the equation of curve C.

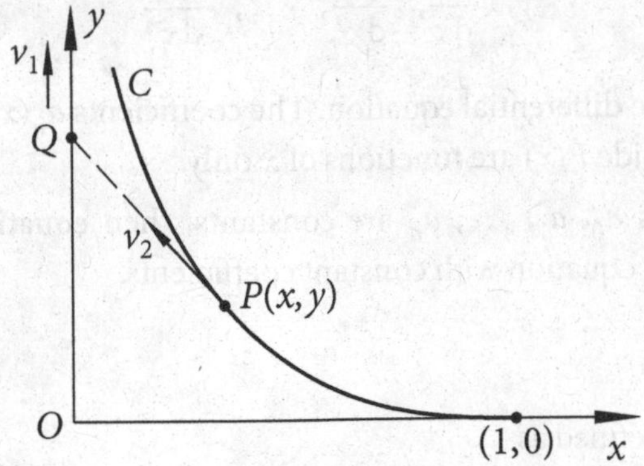

ANS $y=-\dfrac{1}{2}\left[\dfrac{x^{\mu+1}}{\mu+1}+\dfrac{x^{-(\mu-1)}}{\mu-1}\right]+\dfrac{\mu}{\mu^2-1}, \quad \mu=\dfrac{v_1}{v_2}\neq 1$

$y=\dfrac{1}{4}(1-x^2)+\dfrac{1}{2}\ln x, \quad v_1=v_2$

CHAPTER 4

Linear Differential Equations

4.1 General Linear Ordinary Differential Equations

In general, an nth-order linear ordinary differential equation is of the form

$$a_n(x)\frac{d^n y}{dx^n} + a_{n-1}(x)\frac{d^{n-1}y}{dx^{n-1}} + \cdots + a_1(x)\frac{dy}{dx} + a_0(x)y = F(x), \qquad (1)$$

in which the dependent variable y and its derivatives of various orders

$$y, \quad \frac{dy}{dx}, \quad \frac{d^2y}{dx^2}, \quad \ldots, \quad \frac{d^n y}{dx^n}$$

appear linearly in the differential equation. The coefficients $a_0(x), a_1(x), \ldots, a_n(x)$ and the right-hand side $F(x)$ are functions of x only.

If the coefficients a_0, a_1, ..., a_n are constants, then equation (1) is a linear ordinary differential equation with constant coefficients.

The D-Operator

The D-operator is defined as

$Dy \equiv \dfrac{dy}{dx}$, ✍ Dy is taking first-order derivative of y w.r.t. x.

$D^2y = D(Dy) = \dfrac{d^2y}{dx^2}$, ✍ D^2y is taking second-order derivative of y w.r.t. x.

... ...

$D^n y = \dfrac{d^n y}{dx^n}$, n is a positive integer. ✍ $D^n y$ is taking nth-order derivative of y w.r.t. x.

Hence the D-operator is a differential operator; applying the D-operator on function $f(x)$ means differentiating $f(x)$ with respect to x, i.e.,

$$Df(x) = \frac{df(x)}{dx}.$$

Properties of the D-Operator

The following properties of the D-operator can be easily verified:

(1) $D[y_1(x)+y_2(x)] = \frac{d}{dx}(y_1+y_2) = \frac{dy_1}{dx}+\frac{dy_2}{dx} = Dy_1 + Dy_2;$

(2) $D[cy(x)] = \frac{d}{dx}(cy) = c\frac{dy}{dx} = c\,Dy, \quad c = \text{constant};$

(3) $D[c_1y_1(x)+c_2y_2(x)] = c_1\,Dy_1 + c_2\,Dy_2, \quad c_1,\ c_2 = \text{constants}.$

Hence, D is a linear operator. Using the D-operator, the general nth-order linear differential equation (1) can be rewritten as

$$[a_n(x)D^n + a_{n-1}(x)D^{n-1} + \cdots + a_1(x)D + a_0(x)]y = F(x) \implies \phi(D)y = F(x),$$

where $\phi(D)$ is an operator given by

$$\phi(D) = a_n(x)D^n + a_{n-1}(x)D^{n-1} + \cdots + a_1(x)D + a_0(x) = \sum_{r=0}^{n} a_r(x)D^r.$$

The operator $\phi(D)$ is a linear operator, since

$$\phi(D)(y_1+y_2) = \phi(D)y_1 + \phi(D)y_2,$$

$$\phi(D)(cy) = c\,\phi(D)y, \quad c = \text{constant},$$

$$\phi(D)(c_1y_1+c_2y_2) = c_1\phi(D)y_1 + c_2\phi(D)y_2, \quad c_1,\ c_2 = \text{constants}.$$

Example 4.1

Rewrite the following differential equations using the D-operator:

(1) $6x^2\frac{d^2y}{dx^2} + 2x\frac{dy}{dx} - 3y = x^3e^{2x};$

(2) $5\frac{d^3x}{dt^3} + 2\frac{d^2x}{dt^2} - \frac{dx}{dt} + 7x = 3\sin 8t.$

---

(1) $(6x^2D^2+2xD-3)y = x^3e^{2x}, \quad D \equiv \frac{d}{dx};$ ✍ The independent variable is x.

(2) $(5D^3+2D^2-D+7)x = 3\sin 8t, \quad D \equiv \frac{d}{dt}.$ ✍ The independent variable is t.

Fundamental Theorem

Let $y = u(x)$ be **any** solution of the differential equation

$$\phi(D)\,y = F(x), \tag{2}$$

where $\phi(D) = a_n(x)D^n + a_{n-1}(x)D^{n-1} + \cdots + a_1(x)D + a_0(x)$, and $y = v(x)$ be **a** solution of the *complementary differential equation*, or the homogeneous differential equation,

$$\phi(D)\,y = 0, \tag{3}$$

which is obtained by setting the right-hand side of equation (2) to zero. Then $y = u(x) + v(x)$ is also a solution of equation (2).

Proof: Since $u(x)$ and $v(x)$ are solutions of differential equations (2) and (3), respectively,

$$\phi(D)u(x) = F(x), \qquad \phi(D)v(x) = 0.$$

Adding these two equations yields

$$\phi(D)u(x) + \phi(D)v(x) = F(x) + 0.$$

Since $\phi(D)$ is a linear operator, one has

$$\phi(D)\big[u(x) + v(x)\big] = \phi(D)u(x) + \phi(D)v(x).$$

Hence,

$$\phi(D)\big[u(x) + v(x)\big] = F(x),$$

i.e., $y = u(x) + v(x)$ is a solution of differential equation (2). ■

Procedure for Finding the General Solution

1. Find *a particular solution* $y_P(x)$ of the original equation (2), i.e.,

   $$\phi(D)\,y_P = F(x), \qquad \text{✍ Particular solution}$$

   where the subscript "P" stands for *particular* solution.
2. Find *the general solution* $y_C(x)$ of the complementary differential equation (3), i.e.,

   $$\phi(D)\,y_C = 0, \qquad \text{✍ Complementary solution}$$

   where the subscript "C" stands for *complementary* solution.
3. Add $y_P(x)$ and $y_C(x)$ to obtain the general solution $y(x)$, i.e.,

   $$y(x) = y_C(x) + y_P(x). \qquad \text{✍ General solution}$$

In the following sections, various methods for determining complementary solutions and particular solutions for linear differential equations with constant coefficients are studied in detail.

4.2 Complementary Solutions

The complementary differential equation is obtained by setting the right-hand side of the differential equation to zero, i.e.,

$$\phi(D)y = 0, \qquad \phi(D) = a_n D^n + a_{n-1} D^{n-1} + \cdots + a_1 D + a_0,$$

where $a_0, a_1, \ldots, a_n$ are constants.

4.2.1 *Characteristic Equation Having Real Distinct Roots*

Motivating Example

Consider a linear first-order differential equation of the form

$$\frac{dy}{dx} + ay = 0 \quad \text{or} \quad (D+a)y = 0, \qquad a = \text{constant.}$$

The equation is of the form

$$\frac{dy}{dx} + P(x) \cdot y = Q(x), \qquad P(x) = a, \; Q(x) = 0,$$

and, using the result in Section 2.4.1, the solution is

$$y = e^{-\int P(x)\,dx} \left[\int Q(x)\, e^{\int P(x)\,dx} dx + C \right] = Ce^{-ax},$$

i.e., the solution is of the form $e^{\lambda x}$ with $\lambda = -a$.

Since the solution of the linear first-order differential equation in the above example is $y = Ce^{-ax}$, one is tempted to try a solution of the form $y = e^{\lambda x}$, where λ is a constant to be determined, for the general nth-order differential equation $\phi(D)y = 0$. It is easy to verify that

$$Dy = De^{\lambda x} = \frac{d}{dx}(e^{\lambda x}) = \lambda e^{\lambda x},$$

$$D^2 y = D^2 e^{\lambda x} = D(De^{\lambda x}) = \lambda^2 e^{\lambda x},$$

$$\cdots \cdots$$

$$D^n y = D^n e^{\lambda x} = \lambda^n e^{\lambda x}.$$

Substituting into the differential equation $\phi(D)y=0$ results in

$$(a_n\lambda^n + a_{n-1}\lambda^{n-1} + \cdots + a_1\lambda + a_0)\,e^{\lambda x} = 0.$$

Since $e^{\lambda x} \neq 0$, one must have

$$a_n\lambda^n + a_{n-1}\lambda^{n-1} + \cdots + a_1\lambda + a_0 = 0.$$ ✍ Characteristic equation

This algebraic equation, called the *characteristic equation* or *auxiliary equation*, will give n roots $\lambda_1, \lambda_2, \ldots, \lambda_n$, which are called *characteristic numbers* and can be real or complex.

If the roots $\lambda_1, \lambda_2, \ldots, \lambda_n$ are distinct, there will be n solutions of the form

$$e^{\lambda_1 x}, \quad e^{\lambda_2 x}, \quad \ldots, \quad e^{\lambda_n x}.$$

Since $\phi(D)y=0$ is a linear differential equation, the sum of two solutions is also a solution. Hence

$$y = C_1 e^{\lambda_1 x} + C_2 e^{\lambda_2 x} + \cdots + C_n e^{\lambda_n x}$$

is also a solution, where $C_1, C_2, \ldots, C_n$ are arbitrary constants.

It is known that the general solution of an nth-order differential equation $\phi(D)y=0$ should contain n arbitrary constants. Since the equation above contains n arbitrary constants, the solution is therefore the general solution of the complementary equation $\phi(D)y=0$, i.e.,

$$y_C = C_1 e^{\lambda_1 x} + C_2 e^{\lambda_2 x} + \cdots + C_n e^{\lambda_n x},$$

where $\lambda_1, \lambda_2, \ldots, \lambda_n$ are distinct.

Procedure for Finding the Complementary Solution

1. For the nth-order linear differential equation

$$\phi(D)y = F(x) \quad \text{or} \quad (a_n D^n + a_{n-1}D^{n-1} + \cdots + a_1 D + a_0)\,y = F(x),$$

set the right-hand side to zero to obtain the *complementary equation*

$$\phi(D)y = 0 \quad \text{or} \quad (a_n D^n + a_{n-1}D^{n-1} + \cdots + a_1 D + a_0)\,y = 0.$$

2. Replace D by λ to obtain the *characteristic equation*

$$\phi(\lambda) = 0 \quad \text{or} \quad a_n\lambda^n + a_{n-1}\lambda^{n-1} + \cdots + a_1\lambda + a_0 = 0.$$

3. Solve the characteristic equation, which is an algebraic equation, to find the *characteristic numbers* (roots) $\lambda_1, \lambda_2, \ldots, \lambda_n$.
4. Write the complementary solution y_C using the characteristic numbers $\lambda_1, \lambda_2, \ldots, \lambda_n$.

Hyperbolic Functions

The hyperbolic functions are defined as

$$\sinh x = \frac{e^x - e^{-x}}{2} = -\sinh(-x), \quad \cosh x = \frac{e^x + e^{-x}}{2} = \cosh(-x),$$

$$\tanh x = \frac{\sinh x}{\cosh x}, \quad \coth x = \frac{\cosh x}{\sinh x}, \quad \operatorname{csch} x = \frac{1}{\sinh x}, \quad \operatorname{sech} x = \frac{1}{\cosh x}.$$

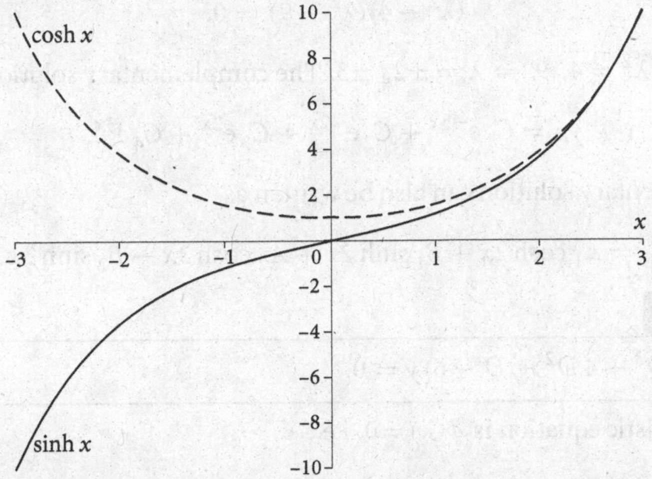

Analogous to the trigonometric identities, the hyperbolic functions satisfy many identities

$$\cosh^2 x - \sinh^2 x = 1, \qquad \sinh 2x = 2 \sinh x \cosh x,$$

$$\cosh 2x = \cosh^2 x + \sinh^2 x = 2\cosh^2 x - 1 = 1 + 2\sinh^2 x,$$

$$\cosh x + \sinh x = e^x, \qquad \cosh x - \sinh x = e^{-x}.$$

The derivatives and the integrals of hyperbolic sine and cosine functions are

$$\frac{d}{dx}(\sinh x) = \cosh x, \qquad \frac{d}{dx}(\cosh x) = \sinh x,$$

$$\int \sinh x \, dx = \cosh x, \qquad \int \cosh x \, dx = \sinh x.$$

An important application of hyperbolic functions is in the complementary solutions of linear ordinary differential equations. If the characteristic equation has roots $\lambda = \pm\beta$, the corresponding complementary solution is

$$y_C = C_1 e^{-\beta x} + C_2 e^{\beta x} = C_1 (\cosh \beta x - \sinh \beta x) + C_2 (\cosh \beta x + \sinh \beta x)$$

$$= (C_1 + C_2) \cosh \beta x + (C_2 - C_1) \sinh \beta x,$$

$$\lambda = \pm\beta \implies y_C = A \cosh \beta x + B \sinh \beta x.$$

Example 4.2

Solve $(D^4 - 13D^2 + 36)y = 0.$

The characteristic equation is $\phi(\lambda) = 0$, i.e.,

$$\lambda^4 - 13\lambda^2 + 36 = 0,$$

which is a quadratic equation in λ^2 and can be factorized as

$$(\lambda^2 - 4)(\lambda^2 - 9) = 0.$$

The roots are $\lambda^2 = 4,\ 9 \Longrightarrow \lambda = \pm 2,\ \pm 3$. The complementary solution is

$$y_C = C_1 e^{-3x} + C_2 e^{-2x} + C_3 e^{2x} + C_4 e^{3x}.$$

The complementary solution can also be written as

$$y_C = A_1 \cosh 2x + B_1 \sinh 2x + A_2 \cosh 3x + B_2 \sinh 3x.$$

Example 4.3

Solve $(D^3 + 4D^2 + D - 6)y = 0.$

The characteristic equation is $\phi(\lambda) = 0$, i.e.,

$$\lambda^3 + 4\lambda^2 + \lambda - 6 = 0.$$

By trial-and-error, find a root of the characteristic equation

$$1^3 + 4 \times 1^2 + 1 - 6 = 0 \Longrightarrow \lambda = 1 \text{ is a root} \Longrightarrow (\lambda - 1) \text{ is a factor.}$$

Use long division to find the other factor as follows

$$\begin{array}{r|llll}
 & \lambda^2 & + 5\lambda & + 6 & \\
\hline
\lambda - 1 & \lambda^3 + & 4\lambda^2 + & \lambda - 6 & \\
 & \lambda^3 - & \lambda^2 & & (- \\
\hline
 & & 5\lambda^2 + & \lambda - 6 & \\
 & & 5\lambda^2 - & 5\lambda & (- \\
\hline
 & & & 6\lambda - 6 & \\
 & & & 6\lambda - 6 & (- \\
\hline
 & & & 0 &
\end{array}$$

The characteristic equation becomes

$$(\lambda - 1)(\lambda^2 + 5\lambda + 6) = 0 \Longrightarrow (\lambda - 1)(\lambda + 2)(\lambda + 3) = 0.$$

The three roots are $\lambda = -3, -2, +1$, and the complementary solution is

$$y_C = C_1 e^{-3x} + C_2 e^{-2x} + C_3 e^{x}.$$

4.2.2 *Characteristic Equation Having Complex Roots*

The following example is considered to derive the general result.

Example 4.4

Solve $\quad (D^2+2D+3)y=0.$

The characteristic equation is $\phi(\lambda)=0$, i.e., $\lambda^2+2\lambda+3=0$, and its roots are

$$\lambda=\frac{-2\pm\sqrt{2^2-4\times1\times3}}{2}=-1\pm\sqrt{-2}=-1\pm \mathrm{i}\sqrt{2}. \qquad \text{✍ } \mathrm{i}=\sqrt{-1}$$

The complementary solution is

$$y_C=C_1\mathrm{e}^{(-1-\mathrm{i}\sqrt{2})x}+C_2\mathrm{e}^{(-1+\mathrm{i}\sqrt{2})x}=\mathrm{e}^{-x}\big(C_1\mathrm{e}^{-\mathrm{i}\sqrt{2}x}+C_2\mathrm{e}^{\mathrm{i}\sqrt{2}x}\big)$$

Using Euler's formula

$$\mathrm{e}^{\mathrm{i}\theta}=\cos\theta+\mathrm{i}\sin\theta, \qquad \mathrm{e}^{-\mathrm{i}\theta}=\cos\theta-\mathrm{i}\sin\theta,$$

the solution y_C can be written as

$$\begin{aligned} y_C&=\mathrm{e}^{-x}\big[C_1\big(\cos\sqrt{2}x-\mathrm{i}\sin\sqrt{2}x\big)+C_2\big(\cos\sqrt{2}x+\mathrm{i}\sin\sqrt{2}x\big)\big]\\ &=\mathrm{e}^{-x}\big[(C_1+C_2)\cos\sqrt{2}x+\mathrm{i}\,(-C_1+C_2)\sin\sqrt{2}x\big]. \end{aligned}$$

Since the differential equation is real, its solution must be real. Hence the coefficients of $(\cos\sqrt{2}x)$ and $(\sin\sqrt{2}x)$ must be real:

$$C_1+C_2=2a, \quad \mathrm{i}\,(-C_1+C_2)=2b \Longrightarrow -C_1+C_2=-\mathrm{i}\,2b, \quad a, b \text{ real constants,}$$

or

$$C_1=a+\mathrm{i}\,b, \qquad C_2=a-\mathrm{i}\,b,$$

i.e., C_1 and C_2 are complex conjugate, $C_1=\bar{C}_2$.

The complementary solution becomes

$$\begin{aligned} y_C&=\mathrm{e}^{-x}\big[2a\cos\sqrt{2}x+2b\sin\sqrt{2}x\big]\\ &=\mathrm{e}^{-x}\big(A\cos\sqrt{2}x+B\sin\sqrt{2}x\big), \qquad A=2a,\ \ B=2b, \end{aligned}$$

where A and B are real constants.

In general, if the characteristic equation has a pair of complex roots $\lambda=\alpha\pm\mathrm{i}\beta$, where α and β are real, then the complementary solution is

$$\lambda=\alpha\pm\mathrm{i}\beta \Longrightarrow y_C(x)=\mathrm{e}^{\alpha x}(A\cos\beta x+B\sin\beta x),$$

where A and B are real constants.

Example 4.5

Solve $(D^3+2D^2+9D+18)y=0.$

The characteristic equation is $\phi(\lambda)=0$, i.e.,

$$\lambda^3+2\lambda^2+9\lambda+18=0.$$

Try to find a root of the characteristic equation

$$\because \quad (-2)^3+2\times(-2)^2+9\times(-2)+18=-8+8-18+18=0,$$

$$\therefore \quad \lambda=-2 \text{ is a root} \implies (\lambda+2) \text{ is a factor.}$$

Use long division to find the other factor

$$\begin{array}{r|rrrrl} & \lambda^2 & & + & 9 & \\ \hline \lambda+2 & \lambda^3 + 2\lambda^2 & + & 9\lambda & + 18 & \\ & \lambda^3 + 2\lambda^2 & & & & (- \\ \hline & & & 9\lambda & + 18 & \\ & & & 9\lambda & + 18 & (- \\ \hline & & & & 0 & \end{array}$$

The characteristic equation becomes

$$(\lambda+2)(\lambda^2+9)=0.$$

The three roots are $\lambda=-2, \pm\mathrm{i}\,3$, and the complementary solution is

$$\begin{aligned} y_C &= C\mathrm{e}^{-2x}+\mathrm{e}^{0\cdot x}(A\cos 3x+B\sin 3x)\\ &= C\mathrm{e}^{-2x}+A\cos 3x+B\sin 3x. \end{aligned}$$

Before more challenging problems can be studied, the following formula for evaluating the nth roots of complex numbers is reviewed.

Review of Complex Numbers

Given that $\lambda^n=a\pm\mathrm{i}\,b$, n is a positive integer, it is required to find λ. The complex numbers can be converted from the rectangular form to the polar form:

$$\lambda^n = a\pm\mathrm{i}\,b \qquad \text{✍ Rectangular form of complex numbers}$$

$$= \underbrace{\sqrt{a^2+b^2}}_{r}\left(\underbrace{\frac{a}{\sqrt{a^2+b^2}}}_{\cos\theta}\pm\mathrm{i}\,\underbrace{\frac{b}{\sqrt{a^2+b^2}}}_{\sin\theta}\right)$$

$$= r(\cos\theta \pm \mathrm{i}\sin\theta), \qquad r = \sqrt{a^2+b^2}, \quad \theta = \tan^{-1}\frac{b}{a}$$ ✍ Polar form

$$= r\mathrm{e}^{\pm \mathrm{i}\theta}.$$ ✍ Euler's formula

Since $\cos\theta$ and $\sin\theta$ are periodic functions of period 2π,

$$\lambda^n = r\big[\cos(2k\pi+\theta) \pm \mathrm{i}\sin(2k\pi+\theta)\big] = r\mathrm{e}^{\pm \mathrm{i}(2k\pi+\theta)}.$$

Hence

$$\lambda = \sqrt[n]{r}\,\mathrm{e}^{\pm \mathrm{i}(2k\pi+\theta)/n},$$

or

$$\lambda = \sqrt[n]{r}\left(\cos\frac{2k\pi+\theta}{n} \pm \mathrm{i}\sin\frac{2k\pi+\theta}{n}\right), \qquad k = 0, 1, \ldots, n-1.$$

It can be easily shown that when $k = pn+q$, where p and q are positive integers, the value of λ repeats that when $k=q$. Therefore only n values of $k = 0, 1, \ldots, n-1$ are taken.

Example 4.6

Solve $\quad (D^5 - 1)y = 0.$

The characteristic equation is $\phi(\lambda) = 0$, i.e.,

$$\lambda^5 - 1 = 0 \implies \lambda^5 = 1 = 1 + \mathrm{i}\,0 = \cos 0 + \mathrm{i}\sin 0.$$

The five roots are given by

$$\lambda = \cos\frac{2k\pi+0}{5} + \mathrm{i}\sin\frac{2k\pi+0}{5}, \qquad k = 0, 1, \ldots, 4,$$

$k=0$: $\quad \lambda = \cos 0 + \mathrm{i}\sin 0 = 1,$

$k=1$: $\quad \lambda = \cos\frac{2\pi}{5} + \mathrm{i}\sin\frac{2\pi}{5},$

$k=2$: $\quad \lambda = \cos\frac{4\pi}{5} + \mathrm{i}\sin\frac{4\pi}{5} = \cos\left(\pi - \frac{\pi}{5}\right) + \mathrm{i}\sin\left(\pi - \frac{\pi}{5}\right)$

$$= -\cos\frac{\pi}{5} + \mathrm{i}\sin\frac{\pi}{5},$$

$k=3$: $\quad \lambda = \cos\frac{6\pi}{5} + \mathrm{i}\sin\frac{6\pi}{5} = \cos\left(\pi + \frac{\pi}{5}\right) + \mathrm{i}\sin\left(\pi + \frac{\pi}{5}\right)$

$$= -\cos\frac{\pi}{5} - \mathrm{i}\sin\frac{\pi}{5},$$

$k=4$: $\quad \lambda = \cos\frac{8\pi}{5} + \mathrm{i}\sin\frac{8\pi}{5} = \cos\left(2\pi - \frac{2\pi}{5}\right) + \mathrm{i}\sin\left(2\pi - \frac{2\pi}{5}\right)$

$$= \cos\frac{2\pi}{5} - \mathrm{i}\sin\frac{2\pi}{5}.$$

Hence, the five roots are

$$\lambda = 1, \quad \cos\frac{2\pi}{5} \pm \mathrm{i}\sin\frac{2\pi}{5}, \quad -\cos\frac{\pi}{5} \pm \mathrm{i}\sin\frac{\pi}{5}.$$

The complementary solution is

$$y_C = C\mathrm{e}^x + \exp\Bigl(\cos\frac{2\pi}{5}\cdot x\Bigr)\Bigl[A_1\cos\Bigl(\sin\frac{2\pi}{5}\cdot x\Bigr) + B_1\sin\Bigl(\sin\frac{2\pi}{5}\cdot x\Bigr)\Bigr]$$
$$+ \exp\Bigl(-\cos\frac{\pi}{5}\cdot x\Bigr)\Bigl[A_2\cos\Bigl(\sin\frac{\pi}{5}\cdot x\Bigr) + B_2\sin\Bigl(\sin\frac{\pi}{5}\cdot x\Bigr)\Bigr].$$

Example 4.7

Solve $\quad (D^4 - 16D^2 + 100)\,y = 0.$

The characteristic equation is $\phi(\lambda)=0$, i.e., $\lambda^4 - 16\lambda^2 + 100 = 0$, a quadratic equation in λ^2. Hence

$$\lambda^2 = \frac{16 \pm \sqrt{16^2 - 4\times1\times100}}{2} = \frac{16 \pm \sqrt{-144}}{2} = 8 \pm \mathrm{i}6 = 10\Bigl(\frac{4}{5} \pm \mathrm{i}\frac{3}{5}\Bigr)$$
$$= 10\,(\cos\theta \pm \mathrm{i}\sin\theta), \quad \cos\theta = \frac{4}{5}, \quad \sin\theta = \frac{3}{5}.$$

The roots are given by

$$\lambda = \sqrt{10}\Bigl(\cos\frac{2k\pi+\theta}{2} \pm \mathrm{i}\sin\frac{2k\pi+\theta}{2}\Bigr), \qquad k = 0,\ 1,$$

$$k=0:\quad \lambda = \sqrt{10}\Bigl(\cos\frac{\theta}{2} \pm \mathrm{i}\sin\frac{\theta}{2}\Bigr),$$

$$k=1:\quad \lambda = \sqrt{10}\Bigl[\cos\Bigl(\pi+\frac{\theta}{2}\Bigr) \pm \mathrm{i}\sin\Bigl(\pi+\frac{\theta}{2}\Bigr)\Bigr] = \sqrt{10}\Bigl(-\cos\frac{\theta}{2} \mp \mathrm{i}\sin\frac{\theta}{2}\Bigr).$$

The four roots are

$$\lambda = \sqrt{10}\Bigl(\pm\cos\frac{\theta}{2} \pm \mathrm{i}\sin\frac{\theta}{2}\Bigr).$$

Using the half-angle formula, one obtains

$$\cos\frac{\theta}{2} = \sqrt{\frac{1+\cos\theta}{2}} = \sqrt{\frac{1+\frac{4}{5}}{2}} = \frac{3}{\sqrt{10}},$$

$$\sin\frac{\theta}{2} = \sqrt{\frac{1-\cos\theta}{2}} = \sqrt{\frac{1-\frac{4}{5}}{2}} = \frac{1}{\sqrt{10}},$$

and hence

$$\lambda = \sqrt{10}\Bigl(\pm\frac{3}{\sqrt{10}} \pm \mathrm{i}\frac{1}{\sqrt{10}}\Bigr) = \pm3 \pm \mathrm{i}1.$$

The complementary solution is

$$y_C = \mathrm{e}^{3x}(A_1\cos x + B_1\sin x) + \mathrm{e}^{-3x}(A_2\cos x + B_2\sin x).$$

4.2.3 Characteristic Equation Having Repeated Roots

Example 4.8

Solve $(D^2 - 4D + 4)y = 0.$

The characteristic equation is $\phi(\lambda)=0$, i.e.,

$$\lambda^2 - 4\lambda + 4 = 0 \implies (\lambda - 2)^2 = 0 \implies \lambda = 2,\ 2. \quad \text{A double root}$$

If the complementary solution is written as

$$y = C_1 e^{2x} + C_2 e^{2x} = (C_1+C_2)\, e^{2x} = C e^{2x}, \qquad C = C_1 + C_2,$$

then there is only one arbitrary constant. But for a *second-order* differential equation, the solution must contain *two* arbitrary constants. The problem is due to the fact that the two solutions $C_1 e^{2x}$ and $C_2 e^{2x}$ are linearly dependent.

Hence, one must seek a second linearly independent solution. Try a solution of the form $y(x) = v(x)\, e^{2x}$. Since

$$Dy = D(ve^{2x}) = e^{2x}\, Dv + 2ve^{2x},$$

$$D^2y = D(e^{2x}\, Dv + 2ve^{2x}) = e^{2x}(D^2v + 4\,Dv + 4v),$$

substituting in the original equation yields

$$e^{2x}\big[(D^2v + 4\,Dv + 4v) - 4(Dv + 2v) + 4v\big] = 0 \implies e^{2x}\, D^2v = 0.$$

Hence $v(x)$ satisfies the differential equation $D^2v = 0$ or $\dfrac{d^2v}{dx^2} = 0$. Integrating the equation twice leads to

$$v(x) = C_0 + C_1 x.$$

The solution is then

$$y(x) = v(x)\, e^{2x} = \underbrace{(C_0 + C_1 x)}_{\text{polynomial of degree 1}}\, e^{2x},$$

in which there are two arbitrary constants.

It is seen that, when the characteristic equation has a double root $\lambda = 2$, the coefficient of e^{2x} is a polynomial of degree 1 instead of a constant. In general, the following results can be obtained.

Summary on the Forms of Complementary Solution

Having obtained the roots of the characteristic equation or the characteristic numbers, the complementary solution can be easily written.

1. If the characteristic equation $\phi(\lambda)=0$ has a p-fold root $\lambda=a$, the corresponding solution is

$$\underbrace{\lambda=a,\ \ldots,\ a}_{p\text{ times}} \implies y_C = \underbrace{\left(C_0+C_1x+\ldots+C_{p-1}x^{p-1}\right)}_{\text{polynomial of degree } p-1}e^{ax}$$

2. If the characteristic equation $\phi(\lambda)=0$ has a pair of complex roots $\lambda=\alpha\pm i\beta$ of p-fold, the corresponding solution is

$$\underbrace{\lambda=\alpha\pm i\beta,\ \ldots,\ \alpha\pm i\beta}_{p\text{ times}} \Rightarrow y_C = e^{\alpha x}\Big[\left(A_0+A_1x+\ldots+A_{p-1}x^{p-1}\right)\cos\beta x + \underbrace{\left(B_0+B_1x+\ldots+B_{p-1}x^{p-1}\right)}_{\text{polynomials of degree } p-1}\sin\beta x\Big]$$

Example 4.9

Given that the characteristic equations have the following roots, write the complementary solutions:

1. $-3\pm i2,\ -1,\ 0,\ 2$;
2. $3,\ 3,\ 3,\ 0,\ 0,\ 1\pm i3$;
3. $2\pm i5,\ 2\pm i5,\ 1,\ 1,\ 1,\ 1,\ 3$.

---

1. $y_C(x) = e^{-3x}(A\cos 2x + B\sin 2x) + C_1e^{-x} + C_2 + C_3e^{2x}$.
2. $y_C(x) = (C_0 + C_1x + C_2x^2)e^{3x}$ ✍ Corresponding to $\lambda=3,\ 3,\ 3$
   $+ D_0 + D_1x$ ✍ Corresponding to $\lambda=0,\ 0$
   $+ e^{x}(A\cos 3x + B\sin 3x)$. ✍ Corresponding to $\lambda=1\pm i3$
3. $y_C(x) = e^{2x}\left[(A_0+A_1x)\cos 5x+(B_0+B_1x)\sin 5x\right]$ ✍ Corresponding to $\lambda=2\pm i5,\ 2\pm i5$
   $+ (C_0+C_1x+C_2x^2+C_3x^3)e^{x}$ ✍ Corresponding to $\lambda=1,1,1,1$
   $+ De^{3x}$. ✍ Corresponding to $\lambda=3$

Example 4.10

Solve $\quad (D^8 + 18D^6 + 81D^4)y = 0.$

The characteristic equation is $\phi(\lambda)=0$, i.e.,

$$\lambda^8 + 18\lambda^6 + 81\lambda^4 = 0,$$

which can be factorized as $\lambda^4(\lambda^2+9)^2=0$. Hence the roots are

$$\lambda = 0,\ 0,\ 0,\ 0,\ \pm i3,\ \pm i3,$$

and the complementary solution is

$$y_C(x) = (C_0 + C_1x + C_2x^2 + C_3x^3)\,e^{0\cdot x} \quad \text{✍ Corresponding to } \lambda=0,\ 0,\ 0,\ 0$$

$$+\,e^{0\cdot x}\big[(A_0+A_1x)\cos 3x + (B_0+B_1x)\sin 3x\big] \quad \text{✍ Corresponding to } \lambda=\pm i3,\ \pm i3$$

$$= C_0 + C_1x + C_2x^2 + C_3x^3 + (A_0+A_1x)\cos 3x + (B_0+B_1x)\sin 3x.$$

Example 4.11

Solve $\quad (D^7 - 3D^6 + 4D^5 - 4D^4 + 3D^3 - D^2)y = 0.$

The characteristic equation is $\phi(\lambda)=0$, i.e.,

$$\lambda^7 - 3\lambda^6 + 4\lambda^5 - 4\lambda^4 + 3\lambda^3 - \lambda^2 = 0.$$

Factorizing the left-hand side yields

$$\begin{aligned}
&\lambda^7 - 3\lambda^6 + 4\lambda^5 - 4\lambda^4 + 3\lambda^3 - \lambda^2\\
&\quad = \lambda^2(\lambda^5 - 3\lambda^4 + 4\lambda^3 - 4\lambda^2 + 3\lambda - 1)\\
&\quad = \lambda^2\big[(\lambda^5 - 3\lambda^4 + 3\lambda^3 - \lambda^2) + (\lambda^3 - 3\lambda^2 + 3\lambda - 1)\big]\\
&\quad = \lambda^2\big[\lambda^2(\lambda-1)^3 + (\lambda-1)^3\big] = \lambda^2(\lambda-1)^3(\lambda^2+1).
\end{aligned}$$

The roots are $\lambda = 0,\ 0,\ 1,\ 1,\ 1,\ \pm i$, and the complementary solution is

$$y_C(x) = C_0 + C_1x \quad \text{✍ Corresponding to } \lambda=0,\ 0$$

$$+\,(D_0 + D_1x + D_2x^2)\,e^x \quad \text{✍ Corresponding to } \lambda=1,\ 1,\ 1$$

$$+\,A\cos x + B\sin x. \quad \text{✍ Corresponding to } \lambda=\pm i$$

4.3 Particular Solutions

In this section, three methods, i.e., the *method of undetermined coefficients*, the *D-operator method*, and the *method of variation of parameters*, will be introduced for finding particular solutions of linear ordinary differential equations.

4.3.1 *Method of Undetermined Coefficients*

The method is illustrated through specific examples.

Example 4.12

Solve $(D^2+9)y = 3e^{2x}$.

The characteristic equation is $\lambda^2+9=0 \implies \lambda = \pm i3$. The complementary solution is

$$y_C = A\cos 3x + B\sin 3x.$$

To find a particular solution y_P, look at the right-hand side of the differential equation, i.e., $F(x)=3e^{2x}$. Since the derivatives of an exponential function are exponential functions, one is tempted to try

$$y_P = Ce^{2x} \implies Dy_P = 2Ce^{2x}, \quad D^2y_P = 4Ce^{2x},$$

where C is a constant to be determined.

Substituting into the equation yields

$$4Ce^{2x} + 9\cdot Ce^{2x} = 3e^{2x} \implies 13Ce^{2x} = 3e^{2x}.$$

Comparing the coefficients of e^{2x} leads to

$$13C = 3 \implies C = \frac{3}{13} \implies y_P = \frac{3}{13}e^{2x}.$$

The general solution is

$$y(x) = y_C + y_P = A\cos 3x + B\sin 3x + \frac{3}{13}e^{2x}.$$

Table 4.1 Method of undetermined coefficients.

	Corresponding to Right-hand Side $F(x)$	**Assumed Form of y_P**
(1)	Polynomial of degree k	Polynomial of degree k
(2)	$e^{\alpha x}$	$Ce^{\alpha x}$
(3)	$\sin\beta x,\ \cos\beta x$	$A\cos\beta x + B\sin\beta x$
(4)	$e^{\alpha x}\,(a_0+a_1x+\cdots+a_kx^k)\cos\beta x,$ $e^{\alpha x}\,(b_0+b_1x+\cdots+b_kx^k)\sin\beta x$ (bracketed: Polynomial of degree k)	$e^{\alpha x}\big[(A_0+A_1x+\cdots+A_kx^k)\cos\beta x$ $+(B_0+B_1x+\cdots+B_kx^k)\sin\beta x\big]$ (bracketed: Polynomial of degree k)

❧ The essence of the method of undetermined coefficients is to assume a form for a particular solution, with coefficients to be determined, according to the form of the right-hand side of the differential equation. The coefficients are then determined by substituting the assumed particular solution into the differential equation. In general, one uses the results given in Table 4.1 to assume the form of a particular solution.

Remarks: For Cases (3) and (4), even if the right-hand side has only sine or cosine function, the assumed form of a particular solution must contain both the sine and cosine functions.

Example 4.13

Solve $(D^2+3D+2)\,y = 42\,e^{5x} + 390\sin 3x + 8x^2 - 2.$

The characteristic equation is $\lambda^2+3\lambda+2=0$. Factorizing the left-hand side gives

$$(\lambda+1)(\lambda+2)=0 \implies \lambda=-2,-1.$$

The complementary solution is

$$y_C = C_1\,e^{-2x} + C_2\,e^{-x}.$$

Use the method of undetermined coefficients to find y_P

Corresponding to right-hand side $F(x)$	Assumed form for y_P
$42\,e^{5x}$	$C\,e^{5x}$
$390\sin 3x$	$A\cos 3x + B\sin 3x$
$8x^2-2$	$D_2x^2+D_1x+D_0$

Hence, assume a particular solution of the form

$$y_P = C\,e^{5x} + A\cos 3x + B\sin 3x + D_2x^2 + D_1x + D_0.$$

Differentiating y_P yields

$$D\,y_P = 5C\,e^{5x} - 3A\sin 3x + 3B\cos 3x + 2D_2x + D_1,$$

$$D^2y_P = 25C\,e^{5x} - 9A\cos 3x - 9B\sin 3x + 2D_2.$$

Substituting into the differential equation leads to

$$\begin{aligned}(D^2+3D+2)\,y_P &= (25C\,e^{5x} - 9A\cos 3x - 9B\sin 3x + 2D_2) \quad \text{✍ } D^2y_P\\ &\quad + 3(5C\,e^{5x} - 3A\sin 3x + 3B\cos 3x + 2D_2x + D_1) \quad \text{✍ } 3D\,y_P\\ &\quad + 2(C\,e^{5x} + A\cos 3x + B\sin 3x + D_2x^2 + D_1x + D_0) \quad \text{✍ } 2y_P\\ &= 42C\,e^{5x} + (-7A+9B)\cos 3x + (-9A-7B)\sin 3x\\ &\quad + 2D_2x^2 + (2D_1+6D_2)x + (2D_0+3D_1+2D_2) \quad \text{✍ Collecting similar terms}\\ &= 42\,e^{5x} + 390\sin 3x + 8x^2 - 2. \quad \text{✍ Right-hand side of the DE}\end{aligned}$$

Equating coefficients of corresponding terms results in

$$\begin{aligned}
&e^{5x}: && 42C = 42 \implies C = 1,\\
&\cos 3x: && \left.\begin{array}{l} -7A + 9B = 0 \\ -9A - 7B = 390 \end{array}\right\} \implies \left\{\begin{array}{l} A = -27, \\ B = -21, \end{array}\right.\\
&\sin 3x: && \\
&x^2: && 2D_2 = 8 \implies D_2 = 4,\\
&x: && 2D_1 + 6D_2 = 0 \implies D_1 = -3D_2 = -12,\\
&1: && 2D_0 + 3D_1 + 2D_2 = -2 \implies D_0 = \tfrac{1}{2}(-2 - 3D_1 - 2D_2) = 13.
\end{aligned}$$

Hence, a particular solution is

$$y_P = e^{5x} - 27\cos 3x - 21\sin 3x + 4x^2 - 12x + 13.$$

Example 4.14

Solve $\quad (D^2 + 2D + 3)y = 34e^x \cos 2x + 1331x^2 e^{2x}.$

The characteristic equation is

$$\lambda^2 + 2\lambda + 3 = 0 \implies \lambda = \frac{-2 \pm \sqrt{2^2 - 4\times 1\times 3}}{2} = -1 \pm i\sqrt{2}.$$

The complementary solution is

$$y_C = e^{-x}(C_1 \cos\sqrt{2}x + C_2 \sin\sqrt{2}x).$$

Use the method of undetermined coefficients to find y_P

Corresponding to right-hand side $F(x)$	Assumed form for y_P
$34e^x \cos 2x$	$e^x(A\cos 2x + B\sin 2x)$
$1331x^2 e^{2x}$	$(D_2x^2 + D_1x + D_0)e^{2x}$

Hence, assume a particular solution of the form

$$y_P = e^x(A\cos 2x + B\sin 2x) + (D_2x^2 + D_1x + D_0)e^{2x}.$$

Differentiating y_P with respect to x yields

$$\begin{aligned}
Dy_P &= e^x\big[(A + 2B)\cos 2x + (-2A + B)\sin 2x\big]\\
&\quad + \big[2D_2x^2 + (2D_1 + 2D_2)x + (2D_0 + D_1)\big]e^{2x},\\
D^2y_P &= e^x\big[(-3A + 4B)\cos 2x + (-4A - 3B)\sin 2x\big]\\
&\quad + \big[4D_2x^2 + (4D_1 + 8D_2)x + (4D_0 + 4D_1 + 2D_2)\big]e^{2x}.
\end{aligned}$$

Substituting into the differential equation leads to

$$
\begin{aligned}
&(D^2+2D+3)\,y_P \\
&= e^x\big[(-3A+4B)\cos 2x+(-4A-3B)\sin 2x\big] \\
&\qquad +\big[4D_2x^2+(4D_1+8D_2)x+(4D_0+4D_1+2D_2)\big]e^{2x} \qquad \text{✍ } D^2y_P \\
&\quad +2\big\{e^x\big[(A+2B)\cos 2x+(-2A+B)\sin 2x\big] \\
&\qquad +\big[2D_2x^2+(2D_1+2D_2)x+(2D_0+D_1)\big]e^{2x}\big\} \qquad \text{✍ } 2Dy_P \\
&\quad +3\big\{e^x(A\cos 2x+B\sin 2x)+(D_2x^2+D_1x+D_0)\,e^{2x}\big\} \qquad \text{✍ } 3y_P \\
&= e^x\big[(2A+8B)\cos 2x+(-8A+2B)\sin 2x\big] \qquad \text{✍ Collecting similar terms} \\
&\qquad +\big[11D_2x^2+(11D_1+12D_2)x+(11D_0+6D_1+2D_2)\big]e^{2x} \\
&= 34e^x\cos 2x+1331x^2e^{2x}. \qquad \text{✍ Right-hand side of the DE}
\end{aligned}
$$

Equating coefficients of corresponding terms results in

$$
\left.\begin{aligned}
e^x\cos 2x:&\quad 2A+8B=34 \\
e^x\sin 2x:&\quad -8A+2B=0
\end{aligned}\right\}
\implies
\left\{\begin{aligned}
A&=1, \\
B&=4,
\end{aligned}\right.
$$

$$
\begin{aligned}
x^2e^{2x}:&\quad 11D_2=1331 \implies D_2=121, \\
xe^{2x}:&\quad 11D_1+12D_2=0 \implies D_1=-\tfrac{12}{11}D_2=-132, \\
e^{2x}:&\quad 11D_0+6D_1+2D_2=0 \implies D_0=-\tfrac{1}{11}(6D_1+2D_2)=50.
\end{aligned}
$$

Hence, a particular solution is

$$y_P = e^x(\cos 2x+4\sin 2x)+(121x^2-132x+50)\,e^{2x}.$$

Example 4.15

Solve $\quad (D^2+8D+25)\,y = 4\cos^3 2x.$

The characteristic equation is $\lambda^2+8\lambda+25=0$. The roots are

$$\lambda = \frac{-8\pm\sqrt{8^2-4\times1\times25}}{2} = -4\pm i3.$$

The complementary solution is

$$y_C = e^{-4x}(A\cos 3x+B\sin 3x).$$

The right-hand side of the equation is converted to the standard form using trigonometric identities:

$$4\cos^3 2x = 4\cos 2x\cos^2 2x$$

$$= 2\cos 2x\,(1+\cos 4x) = 2\cos 2x + 2\cos 2x\cos 4x \quad ✍\ \cos^2 A = \frac{1+\cos 2A}{2}$$

$$= 2\cos 2x + \cos 6x + \cos 2x \quad ✍\ 2\cos A\cos B = \cos(A+B)+\cos(A-B)$$

$$= \cos 6x + 3\cos 2x.$$

Use the method of undetermined coefficients to find y_P

Corresponding to right-hand side $F(x)$	**Assumed form for y_P**
$\cos 6x$	$A_1\cos 6x + B_1\sin 6x$
$3\cos 2x$	$A_2\cos 2x + B_2\sin 2x$

Hence, assume a particular solution of the form

$$y_P = A_1\cos 6x + B_1\sin 6x + A_2\cos 2x + B_2\sin 2x.$$

Differentiating y_P with respect to x yields

$$\mathcal{D}y_P = -6A_1\sin 6x + 6B_1\cos 6x - 2A_2\sin 2x + 2B_2\cos 2x$$

$$\mathcal{D}^2 y_P = -36A_1\cos 6x - 36B_1\sin 6x - 4A_2\cos 2x - 4B_2\sin 2x.$$

Substituting into the differential equation leads to

$$\begin{aligned}
&(\mathcal{D}^2+8\mathcal{D}+25)\,y_P \\
&= -36A_1\cos 6x - 36B_1\sin 6x - 4A_2\cos 2x - 4B_2\sin 2x \quad ✍\ \mathcal{D}^2 y_P \\
&\quad + 8(-6A_1\sin 6x + 6B_1\cos 6x - 2A_2\sin 2x + 2B_2\cos 2x) \quad ✍\ 8\mathcal{D}y_P \\
&\quad + 25(A_1\cos 6x + B_1\sin 6x + A_2\cos 2x + B_2\sin 2x) \quad ✍\ 25y_P \\
&= (-11A_1+48B_1)\cos 6x + (-48A_1-11B_1)\sin 6x \\
&\quad + (21A_2+16B_2)\cos 2x + (-16A_2+21B_2)\sin 2x \quad ✍\ \text{Collecting similar terms} \\
&= \cos 6x + 3\cos 2x. \quad ✍\ \text{Right-hand side of the DE}
\end{aligned}$$

Equating coefficients of corresponding terms results in

$$\left.\begin{aligned}\cos 6x:&\quad -11A_1+48B_1=1\\ \sin 6x:&\quad -48A_1-11B_1=0\end{aligned}\right\} \Longrightarrow \left\{\begin{aligned}A_1&=-\frac{11}{2425},\\ B_1&=\frac{48}{2425},\end{aligned}\right.$$

$$\left.\begin{aligned}\cos 2x:&\quad 21A_2+16B_2=3\\ \sin 2x:&\quad -16A_2+21B_2=0\end{aligned}\right\} \Longrightarrow \left\{\begin{aligned}A_2&=\frac{63}{697},\\ B_2&=\frac{48}{697}.\end{aligned}\right.$$

Hence, a particular solution is

$$y_P = -\frac{11}{2425}\cos 6x + \frac{48}{2425}\sin 6x + \frac{63}{697}\cos 2x + \frac{48}{697}\sin 2x.$$

Exceptions in the Method of Undetermined Coefficients

The method fails if the right-hand side of the differential equation is already contained in the complementary solution y_C.

Example 4.16

Solve $(D^2+2D)y = 4x^2+2x+3.$

The characteristic equation is $\lambda^2+2\lambda=0 \implies \lambda = 0,\,-2$. The complementary solution is

$$y_C = C_0 + C_1\,e^{-2x}.$$

Apply the method of undetermined coefficients to find a particular solution y_P. Corresponding to the right-hand side $4x^2+2x+3$, the form of y_P would normally be assumed as $D_2x^2+D_1x+D_0$. However

$$(D^2+2D)(D_2x^2+D_1x+D_0) = 2D_2+2(2D_2x+D_1) \neq 4x^2+2x+3.$$

This is because the constant "3" in the right-hand side is already contained in the complementary solution C_0. Hence, one has to assume the form of a particular solution to be

$$y_P = x(D_2x^2+D_1x+D_0).$$

Differentiating y_P with respect to x yields

$$Dy_P = 3D_2x^2+2D_1x+D_0, \qquad D^2y_P = 6D_2x+2D_1.$$

Substituting into the differential equation leads to

$$\begin{aligned}(D^2+2D)y_P &= 6D_2x+2D_1+2(3D_2x^2+2D_1x+D_0)\\ &= 6D_2x^2+(4D_1+6D_2)x+(2D_0+2D_1)\\ &= 4x^2+2x+3. \quad \text{✍ Right-hand side of the DE}\end{aligned}$$

Equating coefficients of corresponding terms results in

$$\begin{aligned}x^2:&\quad 6D_2=4 \implies D_2=\tfrac{2}{3},\\ x:&\quad 4D_1+6D_2=2 \implies D_1=\tfrac{1}{4}(2-6D_2)=-\tfrac{1}{2},\\ 1:&\quad 2D_0+2D_1=3 \implies D_0=\tfrac{1}{2}(3-2D_1)=2.\end{aligned}$$

Hence, a particular solution is

$$y_P = x\big(\tfrac{2}{3}x^2-\tfrac{1}{2}x+2\big).$$

Exceptions in the Method of Undetermined Coefficients

In general, if any of the normally assumed terms of a particular solution occurs in the complementary solution, one must **multiply these assumed terms by a power of x** which is **sufficiently high, but not higher,** so that **none of these assumed terms occur in the complementary solution.**

Example 4.17

Given the complementary solution y_C and the right-hand side $F(x)$ of the differential equation, specify the form of a particular solution y_P using the method of undetermined coefficients.

(1) $y_C = c_1 e^{-x} + c_2 e^{3x} + (d_0 + d_1 x + d_2 x^2) e^{5x}$,

$F(x) = 3e^{-x} + 6e^{2x} - 4e^{5x}$;

(2) $y_C = e^{2x}(a \cos 3x + b \sin 3x) + c_0 + c_1 x + c_2 x^2$,

$F(x) = 5x e^{2x} \cos 3x + 3x + e^{2x}$;

(3) $y_C = (c_0 + c_1 x) e^x + d_0 + a \sin 2x + b \cos 2x$,

$F(x) = 2x e^x + 3x^2 + \cos 3x$.

---

(1)

$F(x)$	Normally Assumed Form for y_P	Contained in y_C	Modification
$3e^{-x}$	$C_1 e^{-x}$	$c_1 e^{-x}$	$x \cdot C_1 e^{-x}$
$6e^{2x}$	$C_2 e^{2x}$	———	———
$-4e^{5x}$	$C_3 e^{5x}$	$(d_0 + d_1 x + d_2 x^2) e^{5x}$	$x^3 \cdot C_3 e^{5x}$

$$y_P = x \cdot C_1 e^{-x} + C_2 e^{2x} + x^3 \cdot C_3 e^{5x}.$$

(2)

$F(x)$	Normally Assumed Form for y_P	Contained in y_C	Modification
$5x e^{2x} \cos 3x$	$e^{2x}[(A_0 + A_1 x) \cos 3x + (B_0 + B_1 x) \sin 3x]$	$e^{2x}(a \cos 3x + b \sin 3x)$	$x \cdot e^{2x}[(A_0 + A_1 x) \cos 3x + (B_0 + B_1 x) \sin 3x]$
$3x$	$C_0 + C_1 x$	$c_0 + c_1 x + c_2 x^2$	$x^3 \cdot (C_0 + C_1 x)$
e^{2x}	$D e^{2x}$	———	———

$$y_P = x \cdot e^{2x}[(A_0 + A_1 x) \cos 3x + (B_0 + B_1 x) \sin 3x] + x^3 \cdot (C_0 + C_1 x) + D e^{2x}.$$

(3)

$F(x)$	Normally Assumed Form for y_P	Contained in y_C	Modification
$x\,e^x$	$(C_0+C_1x)\,e^x$	$(c_0+c_1x)\,e^x$	$x^2\cdot(C_0+C_1x)\,e^x$
$3x^2$	$D_0+D_1x+D_2x^2$	d_0	$x\cdot(D_0+D_1x+D_2x^2)$
$\cos 3x$	$A\cos 3x+B\sin 3x$	———	———

$$y_P = x^2\cdot(C_0+C_1x)\,e^x + x\cdot(D_0+D_1x+D_2x^2) + A\cos 3x + B\sin 3x.$$

Example 4.18

Solve $\quad (D^4-4D^2)y = \sinh 2x + 2x^2.$

The characteristic equation is $\phi(\lambda)=0$, i.e.,

$$\lambda^4-4\lambda^2=\lambda^2(\lambda^2-4)=0 \implies \lambda=\pm 2,\ 0,\ 0.$$

The complementary solution is $y_C=c_1e^{2x}+c_2e^{-2x}+d_0+d_1x$, which can be expressed using hyperbolic functions:

$$y_C = a\cosh 2x + b\sinh 2x + d_0 + d_1x.$$

Remarks: When using the method of undetermined coefficients to find a particular solution, the hyperbolic functions $(\cosh\beta x)$ and $(\sinh\beta x)$ can be treated similar to the sinusoidal functions $(\cos\beta x)$ and $(\sin\beta x)$.

$F(x)$	Normally Assumed Form for y_P	Contained in y_C	Modification
$\sinh 2x$	$A\cosh 2x+B\sinh 2x$	$a\cosh 2x+b\sinh 2x$	$x\cdot(A\cosh 2x+B\sinh 2x)$
$2x^2$	$D_0+D_1x+D_2x^2$	d_0+d_1x	$x^2\cdot(D_0+D_1x+D_2x^2)$

Hence, assume a particular solution of the form

$$y_P = x\cdot(A\cosh 2x+B\sinh 2x) + x^2\cdot(D_0+D_1x+D_2x^2).$$

Differentiating y_P with respect to x yields

$$\begin{aligned}
Dy_P &= (A\cosh 2x+B\sinh 2x) + 2x\,(A\sinh 2x+B\cosh 2x)\\
&\quad + 2D_0x+3D_1x^2+4D_2x^3,\\
D^2y_P &= 4\,(A\sinh 2x+B\cosh 2x) + 4x\,(A\cosh 2x+B\sinh 2x)\\
&\quad + 2D_0+6D_1x+12D_2x^2,\\
D^3y_P &= 12\,(A\cosh 2x+B\sinh 2x) + 8x\,(A\sinh 2x+B\cosh 2x) + 6D_1+24D_2x,
\end{aligned}$$

$$D^4 y_P = 32\,(A \sinh 2x + B \cosh 2x) + 16x\,(A \cosh 2x + B \sinh 2x) + 24D_2.$$

Substituting into the differential equation leads to

$$\begin{aligned}(D^4 - 4D^2)\,y_P\\ &= 32\,(A \sinh 2x + B \cosh 2x) + 16x\,(A \cosh 2x + B \sinh 2x) + 24D_2 \quad ✍\ D^4 y_P\\ &\quad - 4\big[4\,(A \sinh 2x + B \cosh 2x) + 4x\,(A \cosh 2x + B \sinh 2x)\\ &\qquad + 2D_0 + 6D_1 x + 12D_2 x^2\big] \quad ✍\ 4D^2 y_P\\ &= 16\,(A \sinh 2x + B \cosh 2x) + (24D_2 - 8D_0) - 24D_1 x - 48D_2 x^2\\ &= \sinh 2x + 2x^2. \quad ✍\ \text{Right-hand side of the DE}\end{aligned}$$

Equating coefficients of corresponding terms results in

$$\begin{aligned}\cosh 2x:&\quad 16B = 0 \implies B = 0,\\ \sinh 2x:&\quad 16A = 1 \implies A = \tfrac{1}{16},\\ x^2:&\quad -48D_2 = 2 \implies D_2 = -\tfrac{1}{24},\\ x:&\quad -24D_1 = 0 \implies D_1 = 0,\\ 1:&\quad 24D_2 - 8D_0 = 0 \implies D_0 = 3D_2 = -\tfrac{1}{8}.\end{aligned}$$

Hence, a particular solution is

$$y_P = \frac{x}{16} \cosh 2x - \frac{x^2}{8} - \frac{x^4}{24}.$$

4.3.2 Method of Operators

In Section 4.1, the D-operator is defined as the differential operator, i.e.,

$$D(\cdot) \equiv \frac{d(\cdot)}{dx}.$$

Define the inverse operator of D, denoted as D^{-1}, by $(D^{-1}D)y \equiv y$. Hence

$$D^{-1}(\cdot) = \frac{1}{D}(\cdot) \equiv \int (\cdot)\,dx,$$

i.e., the inverse operator D^{-1} is the integral operator.

Similarly, for differential equation $\phi(D)\,y = F(x)$, define the inverse operator $\phi^{-1}(D)$ by

$$[\phi^{-1}(D)\,\phi(D)]\,y = y \quad \text{or} \quad \left[\frac{1}{\phi(D)}\phi(D)\right] y = y.$$

For the operator $\phi(D)$ of the form

$$\phi(D) = a_n D^n + a_{n-1} D^{n-1} + \cdots + a_1 D + a_0,$$

where $a_0, a_1, \ldots, a_n$ are constant coefficients, a solution of the differential equation $\phi(D)\,y = F(x)$ can be rewritten using the inverse operator

$$\phi(D)\,y = F(x) \implies y = \frac{1}{\phi(D)} F(x),$$

where

$$\phi^{-1}(D) = \frac{1}{\phi(D)} = \frac{1}{a_n D^n + a_{n-1} D^{n-1} + \cdots + a_1 D + a_0}.$$

The inverse operator $\phi^{-1}(D)$ has the following properties:

1. $\dfrac{1}{\phi_1(D)\,\phi_2(D)} F(x) = \dfrac{1}{\phi_1(D)}\left[\dfrac{1}{\phi_2(D)} F(x)\right] = \dfrac{1}{\phi_2(D)}\left[\dfrac{1}{\phi_1(D)} F(x)\right];$

2. $\dfrac{1}{\phi(D)}\big[F_1(x) + F_2(x)\big] = \dfrac{1}{\phi(D)} F_1(x) + \dfrac{1}{\phi(D)} F_2(x).$

Remarks: For the given differential equation $\phi(D)\,y = F(x)$, the operator $\phi(D)$ and function $F(x)$ are unique. However, the function $\phi^{-1}(D)F(x)$ is not uniquely determined. The difference between any two results is a solution of the complementary equation $\phi(D)\,y = 0$. But this is not important because only **one** particular solution is to be determined. In fact, one always tries to find a simple result of $\phi^{-1}(D)F(x)$.

Theorem 1

$$\frac{1}{\phi(D)} e^{ax} = \frac{1}{\phi(a)} e^{ax}, \quad \text{provided } \phi(a) \neq 0.$$

✍ Replace D by a.

☞ If $\phi(a) = 0$, use Theorem 4.

Proof: Since $De^{ax} = a\,e^{ax}$, $D^2 e^{ax} = a^2 e^{ax}$, $\ldots$, $D^n e^{ax} = a^n e^{ax}$, then

$$\begin{aligned}\phi(D)e^{ax} &= (a_n D^n + a_{n-1} D^{n-1} + \cdots + a_1 D + a_0)\,e^{ax} \\ &= (a_n a^n + a_{n-1} a^{n-1} + \cdots + a_1 a + a_0)\,e^{ax} = \phi(a)e^{ax}.\end{aligned}$$

Applying $\phi^{-1}(D)$ on both sides yields:

$$\phi^{-1}(D)\,\phi(D)e^{ax} = \phi^{-1}(D)\big[\phi(a)e^{ax}\big] \implies e^{ax} = \phi(a)\,\phi^{-1}(D)e^{ax},$$

$$\therefore \quad \frac{1}{\phi(D)} e^{ax} = \frac{1}{\phi(a)} e^{ax}, \qquad \phi(a) \neq 0. \qquad \blacksquare$$

Theorem 2 (Shift Theorem)

$$\frac{1}{\phi(D)}\big[e^{ax}f(x)\big]=e^{ax}\frac{1}{\phi(D+a)}f(x).$$ Move e^{ax} out of $\phi^{-1}(D)$, shift the operator D by a.

Proof: Consider the differential equation $\phi(D)y=F(x)$. Since

$$(D+a)\big(e^{-ax}y\big)=D\big(e^{-ax}y\big)+ae^{-ax}y=e^{-ax}Dy,$$
$$(D+a)^2\big(e^{-ax}y\big)=(D+a)\big(e^{-ax}Dy\big)=e^{-ax}D^2y,$$
$$\cdots\ \cdots$$
$$(D+a)^n\big(e^{-ax}y\big)=e^{-ax}D^ny,$$

one obtains

$$\big[a_n(D+a)^n+a_{n-1}(D+a)^{n-1}+\cdots+a_1(D+a)+a_0\big]\big(e^{-ax}y\big)$$
$$=e^{-ax}(a_nD^n+a_{n-1}D^{n-1}+\cdots+a_1D+a_0)y,$$

$$\therefore\quad \phi(D+a)\big(e^{-ax}y\big)=e^{-ax}\phi(D)y \implies y=e^{ax}\frac{1}{\phi(D+a)}\big[e^{-ax}\phi(D)y\big].$$

Using $\phi(D)y=e^{ax}f(x)$ and $y=\phi^{-1}(D)\big[e^{ax}f(x)\big]$ leads to

$$\frac{1}{\phi(D)}\big[e^{ax}f(x)\big]=e^{ax}\frac{1}{\phi(D+a)}f(x).$$ ∎

Example 4.19

Solve $(D^2+5D+4)y=e^{2x}+x^2e^{-2x}$.

The characteristic equation is

$$\lambda^2+5\lambda+4=0 \implies (\lambda+4)(\lambda+1)=0 \implies \lambda=-4,\ -1.$$

The complementary solution is $y_C=C_1e^{-4x}+C_2e^{-x}$. For a particular solution

$$y_P=\frac{1}{D^2+5D+4}(e^{2x}+x^2e^{-2x}).$$

$$y_{P1}=\frac{1}{D^2+5D+4}e^{2x}=\frac{1}{2^2+5\times 2+4}e^{2x}=\frac{1}{18}e^{2x}.$$ Theorem 1: $a=2$

$$y_{P2}=\frac{1}{D^2+5D+4}(x^2e^{-2x})$$ Apply Shift Theorem to $\frac{1}{\phi(D)}\big[e^{ax}f(x)\big]$.

$$=e^{-2x}\frac{1}{(D-2)^2+5\,(D-2)+4}x^2$$ Take e^{-2x} out of the operator and shift operator D by -2.

$$= e^{-2x} \frac{1}{D^2+D-2} x^2 = -\frac{e^{-2x}}{2} \frac{1}{1-\frac{D}{2}-\frac{D^2}{2}} x^2$$

Write the operator in ascending order of D. Set the constant to 1.

$$= -\frac{e^{-2x}}{2}\left(1+\frac{D}{2}+\frac{3D^2}{4}+\cdots\right)x^2$$

Use long division to expand the operator and stop at D^2.

$$= -\frac{e^{-2x}}{2}\left(x^2+x+\frac{3}{2}+0+\cdots\right)$$

$D^k x^2 = 0$, for $k > 2$

$$= -\frac{e^{-2x}}{2}\left(x^2+x+\frac{3}{2}\right).$$

Stop at D^2.

$$\begin{array}{r|l}
 & 1 + \frac{D}{2} + \frac{3D^2}{4} \\
\hline
1-\frac{D}{2}-\frac{D^2}{2} & 1 \\
 & 1 - \frac{D}{2} - \frac{D^2}{2} \qquad (- \\
\hline
 & \frac{D}{2} + \frac{D^2}{2} \\
 & \frac{D}{2} - \frac{D^2}{4} - \frac{D^3}{4} \qquad (- \\
\hline
 & \frac{3D^2}{4} + \frac{D^3}{4}
\end{array}$$

Alternatively, one can use the series expansion

$$\frac{1}{1-u} = 1 + u + u^2 + \cdots + u^k + \cdots$$

to expand the operator

$$\frac{1}{1-\frac{D}{2}-\frac{D^2}{2}} = \frac{1}{1-\left(\frac{D}{2}+\frac{D^2}{2}\right)}$$

$\frac{1}{1-u} = 1 + u + u^2 + u^3 + \cdots$

$$= 1+\left(\frac{D}{2}+\frac{D^2}{2}\right)+\left(\frac{D}{2}+\frac{D^2}{2}\right)^2+\left(\frac{D}{2}+\frac{D^2}{2}\right)^3+\cdots$$

$$= 1+\left(\frac{D}{2}+\frac{D^2}{2}\right)+\left(\frac{D^2}{4}+\cdots\right)+\cdots$$

Keep terms up to D^2.

$$= 1+\frac{D}{2}+\frac{3D^2}{4}+\cdots.$$

Hence, a particular solution is

$$y_P = y_{P1} + y_{P2} = \frac{1}{18}e^{2x} - \frac{e^{-2x}}{2}\left(x^2+x+\frac{3}{2}\right).$$

Technique for Evaluating a Polynomial

$$\frac{1}{a_n D^n + a_{n-1} D^{n-1} + \cdots + a_1 D + a_0} \underbrace{(C_0 + C_1 x + \cdots + C_k x^k)}_{\text{Polynomial of degree } k}$$

$$= \frac{1}{a_0} \underbrace{\frac{1}{1 + \frac{a_1}{a_0} D + \frac{a_2}{a_0} D^2 + \cdots + \frac{a_n}{a_0} D^n}}_{\substack{\text{Rewrite the operator in ascending order of } D. \\ \text{Set the constant to 1.}}} (C_0 + C_1 x + \cdots + C_k x^k)$$

$$= \frac{1}{a_0} \underbrace{(1 + b_1 D + b_2 D^2 + \cdots + b_k D^k + \cdots)}_{\text{Expand the operator and stop at } D^k.} (C_0 + C_1 x + \cdots + C_k x^k),$$

which can be easily evaluated using

$$D x^q = q x^{q-1}, \quad D^2 x^q = q(q-1) x^{q-2}, \quad \ldots, \quad D^q x^q = q(q-1) \cdots 1,$$

$$D^p x^q = 0, \quad \text{for } p > q.$$

Expansion of the operator can be obtained using

1. Long division, or
2. Series $\quad \dfrac{1}{1-u} = 1 + u + u^2 + \cdots + u^k + \cdots.$

☞ This method of expanding operators using long division or series is applicable only when evaluating **polynomials.**

Special Case: $\dfrac{1}{a_n D^n + a_{n-1} D^{n-1} + \cdots + a_1 D + a_0} C_0 = \dfrac{C_0}{a_0}$ ✍ Constant

Example 4.20

Find a particular solution of $(D^2 + 6D + 9) y = (x^3 + 2x) e^{-3x}$.

Using the method of operators, a particular solution is given by

$$y_P = \frac{1}{D^2 + 6D + 9} \left[(x^3 + 2x) e^{-3x} \right]$$ ✍ Shift Theorem: $a = -3$.

$$= e^{-3x} \frac{1}{(D-3)^2 + 6(D-3) + 9} (x^3 + 2x)$$ ✍ Take e^{-3x} out of the operator and shift operator D by -3.

$$= e^{-3x} \frac{1}{D^2} (x^3 + 2x).$$

Since D^{-1} is the integral operator, $D^{-2}f(x)$ means integrating $f(x)$ twice:

$$x^3+2x \xrightarrow{\text{Integrate}} \tfrac{1}{4}x^4+x^2 \xrightarrow{\text{Integrate}} \tfrac{1}{20}x^5+\tfrac{1}{3}x^3.$$

Hence,

$$y_P = e^{-3x}\left(\tfrac{1}{20}x^5+\tfrac{1}{3}x^3\right).$$

Example 4.21

Find a particular solution of $(D^2-2D+2)y = x^2 e^x \cos 2x.$

From Euler's formula

$$e^{i\theta}=\cos\theta+i\sin\theta \implies \cos 2x = \mathcal{R}e(e^{i2x}) \quad\text{and}\quad e^x\cos 2x=\mathcal{R}e\left[e^{(1+i2)x}\right].$$

Applying the method of operators, a particular solution is

$$y_P = \mathcal{R}e\left\{\frac{1}{D^2-2D+2}\left[x^2 e^{(1+i2)x}\right]\right\}$$

✍ Theorem 2: $a=1+i2$. Take $e^{(1+i2)x}$ out of the operator, shift D by $(1+i2)$.

$$= \mathcal{R}e\left\{ e^{(1+i2)x}\frac{1}{[D+(1+i2)]^2-2[D+(1+i2)]+2}x^2\right\}$$

$$= \mathcal{R}e\left[e^{(1+i2)x}\frac{1}{D^2+i4D-3}x^2\right]$$

$$= \mathcal{R}e\left[-\frac{1}{3}e^{(1+i2)x}\frac{1}{1-\left(\frac{i4}{3}D+\frac{1}{3}D^2\right)}x^2\right]$$

✍ Expand the operator using series. Stop at D^2.

$$= \mathcal{R}e\left\{ -\frac{1}{3}e^{(1+i2)x}\left[1+\left(\frac{i4}{3}D+\frac{1}{3}D^2\right)+\left(\frac{i4}{3}D+\frac{1}{3}D^2\right)^2+\cdots\right]x^2\right\}$$

$$= \mathcal{R}e\left[-\frac{1}{3}e^{(1+i2)x}\left(x^2+\frac{i4}{3}Dx^2-\frac{13}{9}D^2x^2+\cdots\right)\right]$$

$$= \mathcal{R}e\left[-\frac{1}{3}e^{x}(\cos 2x+i\sin 2x)\left(x^2+\frac{i8}{3}x-\frac{26}{9}\right)\right]$$

✍ Apply Euler's formula to $e^{(1+i2)x}$.

$$= -\frac{1}{3}e^x\left(x^2\cos 2x-\frac{8}{3}x\sin 2x-\frac{26}{9}\cos 2x\right).$$

✍ Expand the product and take the real part.

Example 4.22

Find a particular solution of $(D^2+3D-4)y = 6\sin 3x.$

Using the method of operators, a particular solution is given by

$$y_P = \frac{6}{D^2+3D-4}\sin 3x.$$

In order to use Theorem 1, sin $3x$ must be converted to an exponential function using Euler's formula $e^{i\theta} = \cos\theta + i\sin\theta \implies \cos\theta = \mathcal{R}e(e^{i\theta}),\ \sin\theta = \mathcal{I}m(e^{i\theta})$:

$$y_P = \mathcal{I}m\left(\frac{6}{D^2+3D-4}e^{i3x}\right) = \mathcal{I}m\left[\frac{6}{(i3)^2+3(i3)-4}e^{i3x}\right]$$ ✍ Theorem 1: $a = i3$, replace D by i3.

$$= \mathcal{I}m\left(\frac{6}{-13+i9}e^{i3x}\right) = \mathcal{I}m\left[\frac{6(-13-i9)}{(-13+i9)(-13-i9)}e^{i3x}\right]$$

✍ Multiply both the numerator and denominator by $(-13-i9)$.

$$= \mathcal{I}m\left[\frac{6(-13-i9)}{13^2+9^2}(\cos 3x + i\sin 3x)\right]$$ ✍ Expand e^{i3x} using Euler's formula.

$$= \frac{3}{125}\mathcal{I}m\big[\underbrace{(-13\cos 3x + 9\sin 3x)}_{\text{Real part}} + i\underbrace{(-13\sin 3x - 9\cos 3x)}_{\text{Imaginary part}}\big]$$

$$= -\frac{3}{125}(13\sin 3x + 9\cos 3x).$$

Remarks:

- This approach involves manipulations of complex numbers, which could be tedious. Note that $i^2 = -1,\ i^4 = 1,\ i^6 = -1,\ \ldots$. When one deals with only the even power terms of D, one can avoid complex numbers.
- Since $\cos\beta x$ and $\sin\beta x$ are related to $e^{i\beta x}$, when applying Theorem 1, $a = i\beta$ and $a^2 = (i\beta)^2 = -\beta^2$. Hence, by dealing with only even power terms of D, a simpler method can be devised as follows.

$$y_P = \frac{6}{D^2+3D-4}\sin 3x$$ ✍ Theorem 1: $\sin 3x = \mathcal{I}m(e^{i3x})$; replace D^2 by $(i3)^2 = -3^2$.

$$= \frac{6}{(-3^2)+3D-4}\sin 3x = \frac{6}{-13+3D}\sin 3x$$

$$= \frac{6\,(-13-3D)}{(-13+3D)\,(-13-3D)}\sin 3x$$ ✍ Multiply both the numerator and denominator by $(-13-3D)$.

$$= \frac{-6(13+3D)}{169-9\,D^2}\sin 3x$$ ✍ Apply Theorem 1. Replace D^2 in the denominator by -3^2.

$$= -\frac{6}{169-9(-3^2)}(13\sin 3x + 3\,D\sin 3x)$$ ✍ Expand the numerator, $D\sin 3x = (\sin 3x)' = 3\cos 3x$.

$$= -\frac{3}{125}(13\sin 3x + 9\cos 3x).$$

This procedure can be summarized as Theorem 3. Note that $\phi(D)$ can always be written as $\phi_1(D^2) + \phi_2(D^2)D$, e.g.

$$\phi(D) = 3D^3 + 2D^2 + D + 1 = (2D^2+1) + (3D^2+1)D.$$

Theorem 3

$$\frac{1}{\phi(D)}\begin{Bmatrix}\sin\beta x\\ \cos\beta x\end{Bmatrix}=\frac{1}{\phi_1(D^2)+\phi_2(D^2)D}\begin{Bmatrix}\sin\beta x\\ \cos\beta x\end{Bmatrix}$$

✍ $\sin\beta x$ and $\cos\beta x$ are related to $e^{i\beta x}$.

$$=\frac{1}{\phi_1(-\beta^2)+\phi_2(-\beta^2)D}\begin{Bmatrix}\sin\beta x\\ \cos\beta x\end{Bmatrix}$$

✍ Replace D^2 by $(i\beta)^2=-\beta^2$.

$$=\frac{[\phi_1(-\beta^2)-\phi_2(-\beta^2)D]}{[\phi_1(-\beta^2)+\phi_2(-\beta^2)D][\phi_1(-\beta^2)-\phi_2(-\beta^2)D]}\begin{Bmatrix}\sin\beta x\\ \cos\beta x\end{Bmatrix}$$

✍ Multiply both numerator and denominator by $[\phi_1(-\beta^2)-\phi_2(-\beta^2)D]$.

$$=\frac{\phi_1(-\beta^2)-\phi_2(-\beta^2)D}{[\phi_1(-\beta^2)]^2-[\phi_2(-\beta^2)]^2D^2}\begin{Bmatrix}\sin\beta x\\ \cos\beta x\end{Bmatrix}$$

✍ $(a-b)(a+b)=a^2-b^2$

$$=\frac{1}{[\phi_1(-\beta^2)]^2-[\phi_2(-\beta^2)]^2(-\beta^2)}\begin{Bmatrix}\phi_1(-\beta^2)\sin\beta x-\phi_2(-\beta^2)D\sin\beta x\\ \phi_1(-\beta^2)\cos\beta x-\phi_2(-\beta^2)D\cos\beta x\end{Bmatrix}$$

✍ Replace D^2 in the denominator by $-\beta^2$; expand the numerator.

$$=\frac{1}{[\phi_1(-\beta^2)]^2+\beta^2[\phi_2(-\beta^2)]^2}\begin{Bmatrix}\phi_1(-\beta^2)\sin\beta x-\beta\,\phi_2(-\beta^2)\cos\beta x\\ \phi_1(-\beta^2)\cos\beta x+\beta\,\phi_2(-\beta^2)\sin\beta x\end{Bmatrix}.$$

☞ If $[\phi_1(-\beta^2)]^2+\beta^2[\phi_2(-\beta^2)]^2=0$, use Theorem 4.

It is not advisable to memorize the result of Theorem 3, but to treat the theorem as a technique for finding a particular solution corresponding to a *sinusoidal function*.

Example 4.23

Evaluate $\quad y_P=\dfrac{1}{D^2-4D+3}\big[e^x(2\sin 3x-3\cos 2x)\big].$

$$y_P=e^x\frac{1}{(D+1)^2-4(D+1)+3}(2\sin 3x-3\cos 2x)$$

✍ Theorem 2: take e^x out of operator and shift D by $+1$.

$$=e^x\frac{1}{D^2-2D}(2\sin 3x-3\cos 2x)$$

$$=e^x\Big[\frac{1}{(-3^2)-2D}(2\sin 3x)$$

✍ Replace D^2 by -3^2.

$$+\frac{1}{(-2^2)-2D}(-3\cos 2x)\Big]$$

✍ Replace D^2 by -2^2.

$$= e^x\left[\frac{-2\,(9-2D)}{(9+2D)\,(9-2D)}\sin 3x + \frac{3\,(4-2D)}{(4+2D)\,(4-2D)}\cos 2x\right]$$

$$= e^x\left[\frac{-2(9-2D)}{81-4\,D^2}\sin 3x + \frac{3(4-2D)}{16-4\,D^2}\cos 2x\right]$$

$$= e^x\left[\frac{-2}{81-4\,(-3^2)}(9\sin 3x - 2D\sin 3x)\right.$$ ✍ Replace D^2 by -3^2, expand the numerator.

$$\left. + \frac{3}{16-4\,(-2^2)}(4\cos 2x - 2D\cos 2x)\right]$$ ✍ Replace D^2 by -2^2, expand the numerator.

$$= e^x\left[-\tfrac{2}{39}(3\sin 3x - 2\cos 3x) + \tfrac{3}{8}(\cos 2x + \sin 2x)\right].$$

Example 4.24

Solve $(D^2+6D+9)y = 72\sin^4 3x.$

The characteristic equation is $\lambda^2+6\lambda+9=0 \Longrightarrow (\lambda+3)^2=0 \Longrightarrow \lambda=-3,-3.$ The complementary solution is

$$y_C = (C_0+C_1x)\,e^{-3x}.$$

The right-hand side of the differential equation can be converted to the standard form using trigonometric identities:

$$72\sin^4 3x = 72(\sin^2 3x)^2 = 72\left(\frac{1-\cos 6x}{2}\right)^2 = 18(1-2\cos 6x+\cos^2 6x)$$

$$= 18\left(1-2\cos 6x+\frac{1+\cos 12x}{2}\right) = 27-36\cos 6x+9\cos 12x.$$

$$y_P = \frac{1}{D^2+6D+9}\,27$$ ✍ Special case of a polynomial: constant

$$-\,36\frac{1}{D^2+6D+9}\cos 6x$$ ✍ Theorem 3: replace D^2 by -6^2.

$$+\,9\frac{1}{D^2+6D+9}\cos 12x$$ ✍ Theorem 3: replace D^2 by -12^2.

$$= 3 - 36\frac{1}{(-6^2)+6D+9}\cos 6x + 9\frac{1}{(-12^2)+6D+9}\cos 12x$$

$$= 3 - 12\frac{(2D+9)}{(2D-9)(2D+9)}\cos 6x + 3\frac{(2D+45)}{(2D-45)(2D+45)}\cos 12x$$

$$= 3 - \frac{12(2D+9)}{4D^2-81}\cos 6x + \frac{3(2D+45)}{4D^2-2025}\cos 12x$$

$$= 3 - \frac{12}{4(-6^2)-81}(-12\sin 6x+9\cos 6x)$$ ✍ Replace D^2 by -6^2, expand the numerator.

$$+\frac{3}{4(-12^2)-2025}(-24\sin 12x+45\cos 12x)$$ ✍ Replace D^2 by -12^2, expand the numerator.

$$=3-\frac{16}{25}\sin 6x+\frac{12}{25}\cos 6x+\frac{8}{289}\sin 12x-\frac{15}{289}\cos 12x.$$

Theorem 4

If $\phi(a)=0,\ \phi'(a)=0,\ \phi''(a)=0,\ \ldots,\ \phi^{(p-1)}(a)=0,\ \phi^{(p)}(a)\neq 0$, i.e., a is a p-fold root of the characteristic equation $\phi(\lambda)=0$, then

$$\frac{1}{\phi(D)}e^{ax}=\frac{1}{\phi^{(p)}(a)}x^p e^{ax}.$$ ✍ Take pth-order derivative of $\phi(D)$; replace D by a; multiply the result by x^p.

Proof: Since a is a p-fold root of the characteristic equation $\phi(\lambda)=0$, $\phi(D)$ can be written as $\phi(D)=(D-a)^p\phi_1(D)$, $\phi_1(a)\neq 0$. Hence

$$\frac{1}{\phi(D)}e^{ax}=\frac{1}{(D-a)^p\phi_1(D)}(e^{ax}\cdot 1)$$ ✍ Rewrite e^{ax} as $(e^{ax}\cdot 1)$.

$$=e^{ax}\frac{1}{D^p\phi_1(D+a)}(1)$$ ✍ Apply Theorem 2. Take e^{ax} out of the operator and shift D by a.

$$=e^{ax}\frac{1}{D^p}\left\{\frac{1}{\phi_1(D+a)}e^{0\cdot x}\right\}$$ ✍ Rewrite 1 as $e^{0\cdot x}$.

$$=e^{ax}\frac{1}{D^p}\frac{1}{\phi_1(a)}(1).$$ ✍ Apply Theorem 1 on the shaded part, $a=0$.

Since $\frac{1}{D}$ is the integral operator, $\frac{1}{D^p}$ means integrating 1 with respect to x for p times, which gives

$$\frac{1}{D^p}(1)=\frac{1}{p!}x^p \quad\text{and}\quad \frac{1}{\phi(D)}e^{ax}=\frac{1}{p!\,\phi_1(a)}x^p e^{ax}.$$

On the other hand, differentiating $\phi(D)$ yields

$$\phi'(D)=p(D-a)^{p-1}\phi_1(D)+(D-a)^p\phi_1'(D),$$

$$\phi''(D)=p(p-1)(D-a)^{p-2}\phi_1(D)+2p(D-a)^{p-1}\phi_1'(D)+(D-a)^p\phi_1''(D),$$

$$\phi'''(D)=p(p-1)(p-2)(D-a)^{p-3}\phi_1(D)+3p(p-1)(D-a)^{p-2}\phi_1'(D)$$
$$+3p(D-a)^{p-1}\phi_1''(D)+(D-a)^p\phi_1'''(D),$$

$\cdots\ \cdots$

From the shaded terms, it can be easily seen that $\phi^{(p)}(a)=p!\,\phi_1(a)$. Hence,

$$\frac{1}{\phi(D)}e^{ax}=\frac{1}{\phi^{(p)}(a)}x^p e^{ax}.$$ ■

Example 4.25

Evaluate $y_P = \dfrac{1}{(D-2)^3}\,e^{2x}$.

Use Theorem 4:

$$\phi(D) = (D-2)^3, \qquad \phi(2) = 0,$$
$$\phi'(D) = 3(D-2)^2, \qquad \phi'(2) = 0,$$
$$\phi''(D) = 6(D-2), \qquad \phi''(2) = 0,$$
$$\phi'''(D) = 6, \qquad \phi'''(2) = 6 \neq 0.$$
$$\therefore \quad y_P = \frac{1}{\phi'''(2)}x^3\,e^{2x} = \frac{1}{6}x^3\,e^{2x}.$$

Example 4.26

Solve $(D^2 + 4D + 13)\,y = e^{-2x}\sin 3x$.

The characteristic equation is $\lambda^2+4\lambda+13=0$, which gives

$$\lambda = \frac{-4 \pm \sqrt{4^2 - 4\times 13}}{2} = -2 \pm i3.$$

Hence the complementary solution is $y_C = e^{-2x}(A\cos 3x + B\sin 3x)$.

Remarks: Note that the right-hand side of the differential equation is contained in the complementary solution. Using the method of undetermined coefficient, the assumed form of a particular solution is $x\cdot e^{-2x}\,(a\cos 3x + b\sin 3x)$.

A particular solution is given by

$$y_P = \frac{1}{D^2+4D+13}\left(e^{-2x}\sin 3x\right) = e^{-2x}\,\frac{1}{(D-2)^2+4(D-2)+13}\,\sin 3x$$

✍ Theorem 2: take e^{-2x} out of the operator, shift D by -2.

$$= e^{-2x}\frac{1}{D^2+9}\,\sin 3x = e^{-2x}\,\mathcal{Im}\left[\frac{1}{D^2+9}\,e^{i3x}\right].$$

This can be evaluated using Theorem 4:

$$\phi(D) = D^2+9, \qquad \phi(i3) = (i3)^2+9 = 0,$$
$$\phi'(D) = 2D, \qquad \phi'(i3) = 2(i3) = i6 \neq 0.$$

Hence,

$$y_P = e^{-2x}\,\mathcal{Im}\left[\frac{1}{\phi'(i3)}\,x\,e^{i3x}\right] \qquad \text{✍ Theorem 4}$$
$$= e^{-2x}\,\mathcal{Im}\left[\frac{1}{i6}\,x\,(\cos 3x + i\sin 3x)\right] = e^{-2x}\,\mathcal{Im}\left[-\frac{i}{6}\,x\,(\cos 3x + i\sin 3x)\right]$$
$$= -\frac{1}{6}\,x\,e^{-2x}\cos 3x.$$

4.3.3 *Method of Variation of Parameters*

When the right-hand side of a linear ordinary differential equation with constant coefficients is a combination of polynomials, exponential functions, and sinusoidal functions in the following form

$$e^{\alpha x}\big[(a_0+a_1x+\cdots+a_kx^k)\cos\beta x+(b_0+b_1x+\cdots+b_kx^k)\sin\beta x\big],$$

the *method of undetermined coefficients* or the *method of D-operator* can be applied to find a particular solution. Otherwise, the *method of variation of parameters* must be employed, which is the *most general* method.

The method of variation of parameters is illustrated using specific examples.

Example 4.27

Solve the differential equation $\quad y''+y=\csc^3 x. \quad (1)$

The characteristic equation is $\lambda^2+1=0$, which gives $\lambda=\pm\mathrm{i}$. The complementary solution is

$$y_C = A\cos x + B\sin x. \qquad (2a)$$

Differentiating y_C with respect to x yields

$$y_C' = -A\sin x + B\cos x. \qquad (2b)$$

Note that in equations (2), A and B are constants.

Remarks: In general, for an nth-order linear differential equation, the complementary solution y_C is differentiated $(n-1)$ times to obtain equations for $y_C', y_C'', \ldots, y_C^{(n-1)}$.

To find a particular solution y_P, vary the constants A and B in equations (2) to make them functions of x, i.e., $A \Rightarrow a(x)$, $B \Rightarrow b(x)$. Thus the expressions for y_P and y_P' are obtained

$$y_P = a(x)\cos x + b(x)\sin x, \qquad (3a)$$

$$y_P' = -a(x)\sin x + b(x)\cos x. \qquad (3b)$$

Differentiating equation (3a) with respect to x yields

$$y_P' = a'(x)\cos x - a(x)\sin x + b'(x)\sin x + b(x)\cos x$$ ✍ Differentiate using the product rule.

$$= -a(x)\sin x + b(x)\cos x,$$ ✍ Compare with equation (3b).

which leads to

$$a'(x)\cos x + b'(x)\sin x = 0. \qquad (4a)$$

Differentiating equation (3b) with respect to x and substituting into equation (1) result in

$$
\begin{aligned}
y_P'' + y_P &= \left[-a'(x)\sin x - a(x)\cos x + b'(x)\cos x - b(x)\sin x\right] \\
&\quad + \left[a(x)\cos x + b(x)\sin x\right] \\
&= \csc^3 x,
\end{aligned}
$$

which yields

$$-a'(x)\sin x + b'(x)\cos x = \csc^3 x. \tag{4b}$$

Equations (4) give two linear algebraic equations for two unknowns $a'(x)$ and $b'(x)$, which can be solved easily

$$
\begin{aligned}
&\text{Eq(4a)} \times \cos x: \qquad && a'(x)\cos^2 x + b'(x)\sin x\cos x = 0, \\
&\text{Eq(4b)} \times \sin x: \qquad && -a'(x)\sin^2 x + b'(x)\sin x\cos x = \csc^2 x.
\end{aligned}
$$

Subtracting these two equations leads to

$$a'(x) = -\csc^2 x. \qquad ✍ \sin^2 x + \cos^2 x = 1. \tag{5a}$$

Similarly,

$$\text{Eq(4a)} \times \sin x + \text{Eq(4b)} \times \cos x: \qquad b'(x) = \csc^3 x\cos x. \tag{5b}$$

Integrating equation (5a) yields

$$a(x) = -\int \csc^2 x\,dx = \cot x,$$

and integrating equation (5b) gives

$$b(x) = \int \csc^3 x\cos x\,dx = \int \frac{d(\sin x)}{\sin^3 x} = -\frac{1}{2\sin^2 x}.$$

The general solution is then given by

$$
\begin{aligned}
y = y_C + y_P &= A\cos x + B\sin x + a(x)\cos x + b(x)\sin x \\
&= A\cos x + B\sin x + \cot x \cdot \cos x - \frac{1}{2\sin^2 x} \cdot \sin x \\
&= A\cos x + B\sin x + \frac{\cos^2 x}{\sin x} - \frac{1}{2\sin x}.
\end{aligned}
$$

Example 4.28

Solve the differential equation $\qquad y'' - 2y' + y = \ln x. \qquad (1)$

The characteristic equation is $\lambda^2-2\lambda+1=0 \implies \lambda=1,1$. The complementary solution is

$$y_C=(C_0+C_1x)\,e^x, \tag{2a}$$

where C_0 and C_1 are constants. Differentiating y_C with respect to x yields

$$y_C'=C_1e^x+(C_0+C_1x)\,e^x=\big[(C_0+C_1)+C_1x\big]e^x. \tag{2b}$$

To find a particular solution y_P, vary the constants C_0 and C_1 in equations (2) to make them functions of x, i.e., $C_0 \Rightarrow c_0(x)$, $C_1 \Rightarrow c_1(x)$, to obtain the expressions for y_P and y_P'

$$y_P=\big[c_0(x)+c_1(x)\cdot x\big]e^x, \tag{3a}$$

$$y_P'=\big\{\big[c_0(x)+c_1(x)\big]+c_1(x)\cdot x\big\}e^x. \tag{3b}$$

Differentiating equation (3a) with respect to x yields:

$$y_P'=\big[c_0'(x)+c_1'(x)\cdot x+c_1(x)\big]e^x+\big[c_0(x)+c_1(x)\cdot x\big]e^x$$ ✍ Differentiate using the product rule.

$$=\big\{\big[c_0(x)+c_1(x)\big]+c_1(x)\cdot x\big\}e^x,$$ ✍ Compare with equation (3b).

which leads to

$$c_0'(x)+c_1'(x)\cdot x=0. \tag{4a}$$

Differentiating equation (3b) with respect to x and substituting into equation (1) result in

$$\begin{aligned} &y_P''-2y_P'+y_P\\ &\quad=\big\{\big[c_0'(x)+c_1'(x)\big]+c_1'(x)\cdot x+c_1(x)\big\}e^x+\big\{\big[c_0(x)+c_1(x)\big]+c_1(x)\cdot x\big\}e^x\\ &\qquad-2\big\{\big[c_0(x)+c_1(x)\big]+c_1(x)\cdot x\big\}e^x+\big[c_0(x)+c_1(x)\cdot x\big]e^x\\ &\quad=\big[c_0'(x)+c_1'(x)+c_1'(x)\cdot x\big]e^x=\ln x. \end{aligned}$$ ✍ Right-hand side of equation (1)

Hence,

$$c_0'(x)+c_1'(x)+c_1'(x)\cdot x=e^{-x}\ln x. \tag{4b}$$

Equations (4) give two linear algebraic equations for two unknowns $c_0'(x)$ and $c_1'(x)$. Subtracting equation (4a) from (4b) yields

$$c_1'(x)=e^{-x}\ln x. \tag{5a}$$

From equation (4a), one has

$$c_0'(x)=-c_1'(x)\cdot x=-xe^{-x}\ln x. \tag{5b}$$

Integrating equations (5) yields

$$c_1(x) = \int e^{-x} \ln x \, dx = -\int \ln x \, d(e^{-x}) \quad \text{✍ Integration by parts}$$

$$= -\left[e^{-x} \ln x - \int \frac{e^{-x}}{x} dx\right].$$

$$c_0(x) = -\int x e^{-x} \ln x \, dx = \int x \ln x \, d(e^{-x}) \quad \text{✍ Integration by parts}$$

$$= x e^{-x} \ln x - \int e^{-x}(\ln x + 1) \, dx = x e^{-x} \ln x + e^{-x} - \int e^{-x} \ln x \, dx$$

$$= x e^{-x} \ln x + e^{-x} + e^{-x} \ln x - \int \frac{e^{-x}}{x} dx,$$

A particular solution is then given by

$$y_P = \left[c_0(x) + c_1(x) \cdot x\right] e^x$$

$$= \left\{x e^{-x} \ln x + e^{-x} + e^{-x} \ln x - \int \frac{e^{-x}}{x} dx + \left[-e^{-x} \ln x + \int \frac{e^{-x}}{x} dx\right] \cdot x\right\} e^x$$

$$= 1 + \ln x + e^x (x-1) \int \frac{e^{-x}}{x} dx.$$

Example 4.29

Solve the differential equation $\quad y''' - y' = \dfrac{e^x}{1+e^x}.$ (1)

The characteristic equation is $\lambda^3 - \lambda = 0 \implies \lambda(\lambda+1)(\lambda-1) = 0 \implies \lambda = -1, 1, 0.$ The complementary solution is

$$y_C = +C_1 e^{-x} + C_2 e^x + C_3, \tag{2a}$$

where C_1, C_2, and C_3 are constants. Differentiating y_C with respect to x yields

$$y_C' = -C_1 e^{-x} + C_2 e^x, \tag{2b}$$

$$y_C'' = +C_1 e^{-x} + C_2 e^x. \tag{2c}$$

Apply the method of variation of parameters and vary the constants $C_1 \Longrightarrow c_1(x)$, $C_2 \Longrightarrow c_2(x)$, $C_3 \Longrightarrow c_3(x)$ to express a particular solution and its first- and second-order derivatives as

$$y_P = +c_1(x) e^{-x} + c_2(x) e^x + c_3(x), \tag{3a}$$

$$y_P' = -c_1(x) e^{-x} + c_2(x) e^x, \tag{3b}$$

$$y_P'' = +c_1(x) e^{-x} + c_2(x) e^x. \tag{3c}$$

Differentiating equation (3a) with respect to x yields:

$$\begin{aligned} y_P' &= c_1'(x)\,\mathrm{e}^{-x} - c_1(x)\,\mathrm{e}^{-x} + c_2'(x)\,\mathrm{e}^{x} + c_2(x)\,\mathrm{e}^{x} + c_3'(x) \\ &= -c_1(x)\,\mathrm{e}^{-x} + c_2(x)\,\mathrm{e}^{x}, \quad \text{✍ Compare with equation (3b).} \end{aligned}$$

which leads to

$$c_1'(x)\,\mathrm{e}^{-x} + c_2'(x)\,\mathrm{e}^{x} + c_3'(x) = 0. \tag{4a}$$

Differentiating equation (3b) with respect to x yields:

$$\begin{aligned} y_P'' &= -c_1'(x)\,\mathrm{e}^{-x} + c_1(x)\,\mathrm{e}^{-x} + c_2'(x)\,\mathrm{e}^{x} + c_2(x)\,\mathrm{e}^{x} \\ &= c_1(x)\,\mathrm{e}^{-x} + c_2(x)\,\mathrm{e}^{x}, \quad \text{✍ Compare with equation (3c).} \end{aligned}$$

which leads to

$$-c_1'(x)\,\mathrm{e}^{-x} + c_2'(x)\,\mathrm{e}^{x} = 0. \tag{4b}$$

Differentiating equation (3c) and substituting into equation (1) result in

$$\begin{aligned} y_P''' - y_P' &= \big[\,c_1'(x)\,\mathrm{e}^{-x} - c_1(x)\,\mathrm{e}^{-x} + c_2'(x)\,\mathrm{e}^{x} + c_2(x)\,\mathrm{e}^{x}\big] - \big[-c_1(x)\,\mathrm{e}^{-x} + c_2(x)\,\mathrm{e}^{x}\big] \\ &= \frac{\mathrm{e}^{x}}{1+\mathrm{e}^{x}}, \quad \text{✍ Right-hand side of equation (1)} \end{aligned}$$

which leads to

$$c_1'(x)\,\mathrm{e}^{-x} + c_2'(x)\,\mathrm{e}^{x} = \frac{\mathrm{e}^{x}}{1+\mathrm{e}^{x}}. \tag{4c}$$

Equations (4) give three linear algebraic equations for three unknowns $c_1'(x)$, $c_2'(x)$, and $c_3'(x)$, which can be solved using Cramer's Rule or Gaussian elimination:

$$\text{Eq(4c)} - \text{Eq(4b)}: \quad 2c_1'(x)\,\mathrm{e}^{-x} = \frac{\mathrm{e}^{x}}{1+\mathrm{e}^{x}} \implies c_1'(x) = \frac{\mathrm{e}^{2x}}{2(1+\mathrm{e}^{x})}, \tag{5a}$$

$$\text{Eq(4c)} + \text{Eq(4b)}: \quad 2c_2'(x)\,\mathrm{e}^{x} = \frac{\mathrm{e}^{x}}{1+\mathrm{e}^{x}} \implies c_2'(x) = \frac{1}{2(1+\mathrm{e}^{x})}, \tag{5b}$$

$$\text{Eq(4a)} - \text{Eq(4c)}: \quad c_3'(x) = -\frac{\mathrm{e}^{x}}{1+\mathrm{e}^{x}}. \tag{5c}$$

Integrating equations (5) yields

$$c_3(x) = -\int \frac{\mathrm{e}^{x}}{1+\mathrm{e}^{x}}\,\mathrm{d}x = -\int \frac{1}{1+\mathrm{e}^{x}}\,\mathrm{d}(1+\mathrm{e}^{x}) = -\ln(1+\mathrm{e}^{x}).$$

$$\begin{aligned} c_1(x) &= \frac{1}{2}\int \frac{\mathrm{e}^{2x}}{1+\mathrm{e}^{x}}\,\mathrm{d}x = \frac{1}{2}\int \frac{(1+\mathrm{e}^{x})-1}{1+\mathrm{e}^{x}}\,\mathrm{e}^{x}\,\mathrm{d}x \\ &= \frac{1}{2}\int \Big(1 - \frac{1}{1+\mathrm{e}^{x}}\Big)\mathrm{e}^{x}\,\mathrm{d}x = \frac{1}{2}\big[\mathrm{e}^{x} - \ln(1+\mathrm{e}^{x})\big], \end{aligned}$$

$$c_2(x) = \frac{1}{2}\int \frac{1}{1+e^x}dx = \frac{1}{2}\int \frac{e^{-x}}{e^{-x}+1}dx = -\frac{1}{2}\int \frac{1}{e^{-x}+1}d(e^{-x}+1)$$
$$= -\tfrac{1}{2}\ln(e^{-x}+1) = \tfrac{1}{2}\big[-\ln(1+e^x)+x\big],$$

A particular solution is then given by

$$y_P = c_1(x)\,e^{-x} + c_2(x)\,e^x + c_3(x)$$
$$= \tfrac{1}{2}\big[e^x - \ln(1+e^x)\big]\cdot e^{-x} + \tfrac{1}{2}\big[-\ln(1+e^x)+x\big]\cdot e^x - \ln(1+e^x)$$
$$= \tfrac{1}{2}\big[1 + xe^x - (e^{-x}+e^x+2)\ln(1+e^x)\big].$$

4.4 Euler Differential Equations

An Euler differential equation is a linear ordinary differential equation with variable coefficients of the form

$$a_n x^n \frac{d^n y}{dx^n} + a_{n-1}x^{n-1}\frac{d^{n-1}y}{dx^{n-1}} + \cdots + a_1 x\frac{dy}{dx} + a_0 y = f(x),$$

where a_0, a_1, $\ldots$, a_n are constants. Using the D-operator, $D(\cdot)\equiv d(\cdot)/dx$, an Euler differential equation can be written as

$$\big(a_n x^n D^n + a_{n-1}x^{n-1}D^{n-1} + \cdots + a_1 x D + a_0\big)y = f(x).$$

To solve the equation, use the substitution $x=e^z$ to convert it to one with constant coefficients.

Assuming the independent variable $x>0$, let $x=e^z$ or $z=\ln x$ and adopt the script $\mathscr{D}$ notation, $\mathscr{D}(\cdot)\equiv d(\cdot)/dz$, to denote differentiation with respect to z. Hence, using the chain rule,

$$\frac{dy}{dx} = \frac{dy}{dz}\frac{dz}{dx} = \frac{dy}{dz}\frac{1}{x}, \qquad \frac{dz}{dx} = \frac{d}{dx}(\ln x) = \frac{1}{x}$$

which yields

$$x\frac{dy}{dx} = \frac{dy}{dz} \implies xD(\cdot) = \mathscr{D}(\cdot).$$

Similarly,

$$\frac{d^2y}{dx^2} = \frac{d}{dx}\Big(\frac{dy}{dx}\Big) = \frac{d}{dx}\Big(\frac{1}{x}\frac{dy}{dz}\Big)$$
$$= -\frac{1}{x^2}\frac{dy}{dz} + \frac{1}{x}\frac{d}{dz}\Big(\frac{dy}{dz}\Big)\frac{dz}{dx} = -\frac{1}{x^2}\frac{dy}{dz} + \frac{1}{x^2}\frac{d^2y}{dz^2},$$

leading to

$$x^2\frac{d^2y}{dx^2} = \frac{d^2y}{dz^2} - \frac{dy}{dz} \implies x^2D^2(\cdot) = (\mathscr{D}^2-\mathscr{D})(\cdot) = \mathscr{D}(\mathscr{D}-1)(\cdot).$$

It can be proved, by mathematical induction, that

$$x^n D^n(\cdot) = \mathscr{D}(\mathscr{D}-1)(\mathscr{D}-2)\cdots(\mathscr{D}-n+1)(\cdot), \quad n = \text{positive integer.}$$

Hence, in terms of the script $\mathscr{D}$ operator, $\mathscr{D}(\cdot) \equiv d(\cdot)/dz$, an Euler's equation becomes an equation with constant coefficients.

For example, consider

$$(a_2 x^2 D^2 + a_1 x D + a_0)y = f(x), \qquad D(\cdot) = d(\cdot)/dx.$$

Applying the variable substitution $x = e^z$, the differential equation becomes

$$\big[a_2 \mathscr{D}(\mathscr{D}-1) + a_1 \mathscr{D} + a_0\big]y = f(e^z), \qquad \mathscr{D}(\cdot) = d(\cdot)/dz,$$

or

$$\big[a_2 \mathscr{D}^2 + (a_1 - a_2)\mathscr{D} + a_0\big]y = f(e^z).$$

Example 4.30

Solve $\quad (x^2 D^2 - xD + 2)y = x(\ln x)^3, \quad D(\cdot) \equiv d(\cdot)/dx.$

Letting $x = e^z$, $z = \ln x$, and $\mathscr{D}(\cdot) \equiv d(\cdot)/dz$, the differential equation becomes

$$\big[\mathscr{D}(\mathscr{D}-1) - \mathscr{D} + 2\big]y = e^z z^3 \implies (\mathscr{D}^2 - 2\mathscr{D} + 2)y = e^z z^3.$$

The characteristic equation is $\lambda^2 - 2\lambda + 2 = 0$, which gives

$$\lambda = \frac{-(-2) \pm \sqrt{(-2)^2 - 4 \times 1 \times 2}}{2} = 1 \pm i.$$

The complementary solution is

$$y_C = e^z(A\cos z + B\sin z).$$

A particular solution is given by

$$y_P = \frac{1}{\mathscr{D}^2 - 2\mathscr{D} + 2}(e^z \cdot z^3)$$ ✍ Theorem 2: take e^z out of the operator and shift operator D by $+1$.

$$= e^z \frac{1}{(\mathscr{D}+1)^2 - 2(\mathscr{D}+1) + 2} z^3 = e^z \frac{1}{\mathscr{D}^2 + 1} z^3$$

$$= e^z\big[1 - \mathscr{D}^2 + (\mathscr{D}^2)^2 - \cdots\big]z^3$$ ✍ Expand the operator in series, stop at $\mathscr{D}^3$.

$$= e^z(z^3 - 6z).$$

Hence, the general solution is

$$y = y_C + y_P = e^z(A\cos z + B\sin z) + e^z(z^3 - 6z)$$

$$= x\big[A\cos(\ln x) + B\sin(\ln x) + (\ln x)^3 - 6\ln x\big].$$ ✍ Change back to the original variable x.

Example 4.31

Solve $(x^3 D^3 + xD - 1)y = 4x^5, \quad D(\cdot) \equiv d(\cdot)/dx.$

Let $x = e^z$, $z = \ln x$, and $\mathscr{D}(\cdot) \equiv d(\cdot)/dz$. The differential equation becomes

$$\big[\mathscr{D}(\mathscr{D}-1)(\mathscr{D}-2) + \mathscr{D} - 1\big]y = 4e^{5z} \implies (\mathscr{D}-1)^3 y = 4e^{5z}.$$

The characteristic equation is $(\lambda-1)^3 = 0 \implies \lambda = 1, 1, 1.$ The complementary solution is

$$y_C = (C_0 + C_1 z + C_2 z^2)e^z.$$

A particular solution is given by

$$y_P = 4\frac{1}{(\mathscr{D}-1)^3}e^{5z} = 4\frac{1}{(5-1)^3}e^{5z} = \frac{e^{5z}}{16}.$$ ✍ Theorem 1: $a=5$

The general solution is

$$y = y_C + y_P = (C_0 + C_1 z + C_2 z^2)e^z + \tfrac{1}{16}e^{5z}$$

$$= \big[C_0 + C_1 \ln x + C_2(\ln x)^2\big]x + \tfrac{1}{16}x^5.$$ ✍ Change back to the original variable x.

4.5 Summary

Consider an nth-order ordinary differential equation with constant coefficients

$$\phi(D)y = F(x), \quad \phi(D) = a_n D^n + a_{n-1}D^{n-1} + \cdots + a_1 D + a_0, \quad D(\cdot) \equiv d(\cdot)/dx.$$

General solution $y(x) = y_C(x) + y_P(x)$.

Complementary Solution

Complementary differential equation: $\phi(D)y = 0$. ✍ R.H.S. set to 0.

Characteristic equation: $\phi(\lambda) = 0 \implies$ characteristic numbers $\lambda_1, \lambda_2, \ldots, \lambda_n$.

- $\phi(\lambda) = 0$ has a real root $\lambda = a$ of p-fold

$$y_C = \underbrace{\left(C_0 + C_1 x + \ldots + C_{p-1}x^{p-1}\right)}_{\text{polynomial of degree } p-1} e^{ax}$$

- $\phi(\lambda) = 0$ has a pair of complex roots $\lambda = \alpha \pm i\beta$ of p-fold

$$y_C = e^{\alpha x}\big[(A_0 + A_1 x + \ldots + A_{p-1}x^{p-1})\cos\beta x + \underbrace{(B_0 + B_1 x + \ldots + B_{p-1}x^{p-1})}_{\text{polynomials of degree } p-1}\sin\beta x\big]$$

Particular Solution

1. The *method of variation of parameters* is the most general method that can be used for *any functions* on the right-hand side.
2. When the right-hand side of the differential equation is not of the form

$$e^{\alpha x}\big[(a_0+a_1x+\cdots+a_kx^k)\cos\beta x+(b_0+b_1x+\cdots+b_kx^k)\sin\beta x\big],$$

the *method of variation of parameters* ***must*** be applied.
3. When the right-hand side is of this form, the method of undetermined coefficients, the method of D-operator, or the general method of variation of parameters can be applied. The *method of D-operator* is usually the easiest and the most efficient; hence it is the preferred method.

1. Method of Undetermined Coefficients

	Corresponding to Right-hand Side $F(x)$	Assumed Form of y_P
(1)	Polynomial of degree k	Polynomial of degree k
(2)	$\sin\beta x,\ \cos\beta x$	$A\cos\beta x+B\sin\beta x$
(3)	$e^{\alpha x}$	$Ce^{\alpha x}$
(4)	$e^{\alpha x}(a_0+a_1x+\cdots+a_kx^k)\cos\beta x$, $e^{\alpha x}\underbrace{(b_0+b_1x+\cdots+b_kx^k)}_{\text{Polynomial of degree } k}\sin\beta x$	$e^{\alpha x}\big[(A_0+A_1x+\cdots+A_kx^k)\cos\beta x$ $+\underbrace{(B_0+B_1x+\cdots+B_kx^k)}_{\text{Polynomial of degree } k}\sin\beta x\big]$

If a normally assumed term of a particular solution occurs in the complementary solution, it must be multiplied by a power of x, which is *sufficiently high but not higher*, so that it does not occur in the complementary solution.

2. Method of D-Operator

❧ **Polynomial:** Used when $\phi^{-1}(D)$ operates on a polynomial.

$$\frac{1}{a_nD^n+a_{n-1}D^{n-1}+\cdots+a_1D+a_0}\underbrace{(C_0+C_1x+\cdots+C_kx^k)}_{\text{Polynomial of degree } k}$$

$$=\frac{1}{a_0}\underbrace{\frac{1}{1+\dfrac{a_1}{a_0}D+\dfrac{a_2}{a_0}D^2+\cdots+\dfrac{a_n}{a_0}D^n}}_{\text{Rewrite the operator in ascending order of } D.\text{ Set the constant to 1.}}(C_0+C_1x+\cdots+C_kx^k)$$

$$=\frac{1}{a_0}\underbrace{(1+b_1D+b_2D^2+\cdots+b_kD^k)}_{\text{Expand the operator using series or long division. Stop at } D^k.}(C_0+C_1x+\cdots+C_kx^k).$$

❧ **Shift Theorem (Theorem 2):** Used when $\phi^{-1}(D)$ operates on $e^{ax} f(x)$. The Theorem takes the exponential function e^{ax} out of the operation

$$\frac{1}{\phi(D)}\big[e^{ax} f(x)\big] = e^{ax}\frac{1}{\phi(D+a)} f(x).$$ ✍ Take e^{ax} out of the operator and shift operator D by a.

❧ **Theorem 1:** Used when $\phi^{-1}(D)$ operates on exponential function e^{ax}.

$$\frac{1}{\phi(D)} e^{ax} = \frac{1}{\phi(a)} e^{ax}, \qquad \phi(a) \neq 0.$$ ✍ Replace D by a.

☞ If $\phi(a) = 0$, use Theorem 4.

❧ **Theorem 3:** Used when $\phi^{-1}(D)$ operates on $\sin\beta x$ or $\cos\beta x$.

$$\frac{1}{\phi(D)}\begin{Bmatrix}\sin\beta x\\ \cos\beta x\end{Bmatrix} = \frac{1}{\phi_1(D^2)+\phi_2(D^2)D}\begin{Bmatrix}\sin\beta x\\ \cos\beta x\end{Bmatrix}$$ ✍ Replace D^2 by $-\beta^2$.

$$= \frac{1}{\phi_1(-\beta^2)+\phi_2(-\beta^2)D}\begin{Bmatrix}\sin\beta x\\ \cos\beta x\end{Bmatrix}$$

$$= \frac{\phi_1(-\beta^2)-\phi_2(-\beta^2)D}{\big[\phi_1(-\beta^2)\big]^2-\big[\phi_2(-\beta^2)\big]^2 D^2}\begin{Bmatrix}\sin\beta x\\ \cos\beta x\end{Bmatrix}$$ ✍ Replace D^2 by $-\beta^2$. Expand the numerator.

$$= \frac{1}{\big[\phi_1(-\beta^2)\big]^2+\beta^2\big[\phi_2(-\beta^2)\big]^2}\begin{Bmatrix}\phi_1(-\beta^2)\sin\beta x-\beta\phi_2(-\beta^2)\cos\beta x\\ \phi_1(-\beta^2)\cos\beta x+\beta\phi_2(-\beta^2)\sin\beta x\end{Bmatrix}.$$

☞ If $\big[\phi_1(-\beta^2)\big]^2+\beta^2\big[\phi_2(-\beta^2)\big]^2 = 0$, use Theorem 4.

❧ **Theorem 4:** Used when the denominator is 0 when Theorem 1 or 3 is applied.

$$\frac{1}{\phi(D)} e^{ax} = \frac{1}{\phi^{(p)}(a)} x^p e^{ax}, \qquad \phi(a) = \phi'(a) = \cdots = \phi^{(p-1)}(a) = 0, \quad \phi^{(p)}(a) \neq 0.$$

Euler's Formula: $e^{i\beta x} = \cos\beta x + i\sin\beta x$,

$$\cos\beta x = \mathcal{Re}(e^{i\beta x}), \quad \sin\beta x = \mathcal{Im}(e^{i\beta x}).$$

3. Method of Variation of Parameters

1. For an nth-order linear differential equation $\phi(D)y = F(x)$, determine the complementary solution

$$y_C = f(x; C_1, C_2, \ldots, C_n),$$

where $C_1, C_2, \ldots, C_n$ are n arbitrary constants.

2. Differentiate y_C with respect to x to yield the equations $y_C', y_C'', \ldots, y_C^{(n-1)}$.

3. Vary the constants $C_1, C_2, \ldots, C_n$ to make them functions of x, and obtain n equations $y_P^{(k)}$, $k=0,1,\ldots,n-1$,

$$y_C^{(k)}(x;C_1,C_2,\ldots,C_n) \xrightarrow[i=1,2,\ldots,n]{C_i \Rightarrow c_i(x)} y_P^{(k)}\big(x;c_1(x),c_2(x),\ldots,c_n(x)\big).$$

4. Differentiating $y_P^{(k)}$ with respect to x and comparing with the expressions $y_P^{(k+1)}$, $k=0,1,\ldots,n-2$, and substituting y_P and the derivatives into the original differential equation yield n linear algebraic equations for $c_1'(x)$, $c_2'(x)$, $\ldots$, $c_n'(x)$.
5. Solve these equations for $c_1'(x)$, $c_2'(x)$, $\ldots$, $c_n'(x)$.
6. Integrate to obtain the functions $c_1(x)$, $c_2(x)$, $\ldots$, $c_n(x)$. A particular solution $y_P\big(x;c_1(x),c_2(x),\ldots,c_n(x)\big)$ is then obtained.

Euler Differential Equations

$$(a_n x^n D^n + a_{n-1}x^{n-1}D^{n-1} + \cdots + a_1 x D + a_0)y = f(x), \qquad D(\cdot)\equiv \mathrm{d}(\cdot)/\mathrm{d}x.$$

Letting $x=\mathrm{e}^z$ or $z=\ln x$, $x>0$, $\mathscr{D}(\cdot)\equiv \mathrm{d}(\cdot)/\mathrm{d}z$, then

$$x^n D^n(\cdot) = \mathscr{D}(\mathscr{D}-1)(\mathscr{D}-2)\cdots(\mathscr{D}-n+1)(\cdot), \qquad n=\text{positive integer}.$$

The Euler differential equation is converted to a differential equation with constant coefficients.

Problems

Complementary Solutions

4.1 $(D^3-2D^2+D-2)y=0$ ANS $y_C = A\cos x + B\sin x + Ce^{2x}$

4.2 $(D^3+D^2+9D+9)y=0$ ANS $y_C = A\cos 3x + B\sin 3x + Ce^{-x}$

4.3 $(D^3+D^2-D-1)y=0$ ANS $y_C = (C_0+C_1x)\mathrm{e}^{-x} + C\mathrm{e}^{x}$

4.4 $(D^3+8)y=0$ ANS $y_C = \mathrm{e}^{x}\big[A\cos(\sqrt{3}x) + B\sin(\sqrt{3}x)\big] + Ce^{-2x}$

4.5 $(D^3-8)y=0$ ANS $y_C = \mathrm{e}^{-x}\big[A\cos(\sqrt{3}x) + B\sin(\sqrt{3}x)\big] + Ce^{2x}$

4.6 $(D^4+4)y=0$ ANS $y_C = \mathrm{e}^{x}(A_1\cos x + B_1\sin x) + \mathrm{e}^{-x}(A_2\cos x + B_2\sin x)$

4.7 $(D^4+18D^2+81)y=0$ ANS $y_C = (A_0+A_1x)\cos 3x + (B_0+B_1x)\sin 3x$

4.8 $(D^4-4D^2+16)y=0$

ANS $y_C = \mathrm{e}^{\sqrt{3}x}(A_1\cos x + B_1\sin x) + \mathrm{e}^{-\sqrt{3}x}(A_2\cos x + B_2\sin x)$

4.9 $(D^4-2D^3+2D^2-2D+1)y=0$ ANS $y_C=(C_0+C_1x)e^x+A\cos x+B\sin x$

4.10 $(D^4-5D^3+5D^2+5D-6)y=0$

ANS $y_C=C_1e^{-x}+C_2e^x+C_3e^{2x}+C_4e^{3x}$

4.11 $(D^5-6D^4+9D^3)y=0$ ANS $y_C=C_0+C_1x+C_2x^2+(D_0+D_1x)e^{3x}$

4.12 $(D^6-64)y=0$ ANS $y_C=C_1e^{-2x}+C_2e^{2x}$
$+e^{-x}\left[A_1\cos(\sqrt{3}x)+B_1\sin(\sqrt{3}x)\right]+e^x\left[A_2\cos(\sqrt{3}x)+B_2\sin(\sqrt{3}x)\right]$

Particular Solutions — Method of Undetermined Coefficients

For the following differential equations, specify the form of a particular solution using the method of undetermined coefficients.

4.13 $(D^2+6D+10)y=3xe^{-3x}-2e^{3x}\cos x$

4.14 $(D^2-8D+17)y=e^{4x}(x^2-3x\sin x)$

4.15 $(D^2-2D+2)y=(x+e^x)\sin x$

4.16 $(D^2+4)y=\sinh x\sin 2x$

4.17 $(D^2+2D+2)y=\cosh x\sin x$

4.18 $(D^3+D)y=\sin x+x\cos x$

4.19 $(D^3-2D^2+4D-8)y=e^{2x}\sin 2x+2x^2$

4.20 $(D^3-4D^2+3D)y=x^2+xe^{2x}$

4.21 $(D^4+D^2)y=7x-3\cos x$

4.22 $(D^4+5D^2+4)y=\sin x\cos 2x$

Particular Solutions — D-Operator Method

4.23 $(D^5-3D^3+1)y=9e^{2x}$ ANS $y_P=e^{2x}$

4.24 $(D-1)^3y=48xe^x$ ANS $y_P=2x^4e^x$

4.25 $(D^3-3D)y=9x^2$ ANS $y_P=-x^3-2x$

4.26 $(D^5+4D^3)y=7+x$ ANS $y_P=\frac{1}{96}x^3(28+x)$

4.27 $(D^2-D-2)y=36xe^{2x}$ ANS $y_P=2e^{2x}(3x^2-2x)$

4.28 $(D^4+16)y=64\cos 2x$ ANS $y_P=2\cos 2x$

4.29 $(D^4+4D^2-1)y=44\sin 3x$ ANS $y_P=\sin 3x$

4.30 $(D^3 + D^2 + 5D + 5)y = 5\cos 2x$ ANS $y_P = 2\sin 2x + \cos 2x$

4.31 $(D^2 + 3D + 5)y = 5e^{-x}\sin 2x$ ANS $y_P = -e^{-x}(2\cos 2x + \sin 2x)$

4.32 $(D^4 - 1)y = 4e^{-x}$ ANS $y_P = -xe^{-x}$

4.33 $(D^2 + 4)y = 8\sin^2 x$ ANS $y_P = 1 - x\sin 2x$

4.34 $(D^3 - D^2 + D - 1)y = 4\sin x$ ANS $y_P = x(\cos x - \sin x)$

4.35 $(D^4 - D^2)y = 2e^x$ ANS $y_P = xe^x$

General Solutions

4.36 $y'' - 4y' + 4y = (1 + x)e^x + 2e^{2x} + 3e^{3x}$

ANS $y = (C_0 + C_1 x)e^{2x} + (x+3)e^x + x^2e^{2x} + 3e^{3x}$

4.37 $(D^2 - 2D + 5)y = 4e^x\cos 2x$

ANS $y = e^x(A\cos 2x + B\sin 2x) + xe^x\sin 2x$

4.38 $(D^2 + 4)y = 4\sin 2x$ ANS $y = A\cos 2x + B\sin 2x - x\cos 2x$

4.39 $(D^2 - 1)y = 12x^2e^x + 3e^{2x} + 10\cos 3x$

ANS $y_P = C_1e^{-x} + C_2e^x + e^x(2x^3 - 3x^2 + 3x) + e^{2x} - \cos 3x$

4.40 $y'' + y = 2\sin x - 3\cos 2x$ ANS $y = A\cos x + B\sin x - x\cos x + \cos 2x$

4.41 $y'' - y' = e^x(10 + x^2)$ ANS $y = C_1 + C_2e^x + e^x\left(\frac{1}{3}x^3 - x^2 + 12x\right)$

4.42 $(D^2 - 4)y = 96x^2e^{2x} + 4e^{-2x}$

ANS $y = C_1e^{-2x} + C_2e^{2x} + e^{2x}(8x^3 - 6x^2 + 3x) - xe^{-2x}$

4.43 $(D^2 + 2D + 2)y = 5\cos x + 10\sin 2x$

ANS $y = e^{-x}(A\cos x + B\sin x) + \cos x + 2\sin x - 2\cos 2x - \sin 2x$

4.44 $(D^2 - 2D + 2)y = 4x - 2 + 2e^x\sin x$

ANS $y = e^x(A\cos x + B\sin x) + 2x + 1 - xe^x\cos x$

4.45 $(D^2 - 4D + 4)y = 4xe^{2x}\sin 2x$

ANS $y = (C_0 + C_1x)e^{2x} - e^{2x}(x\sin 2x + \cos 2x)$

4.46 $(D^3 - D^2 + D - 1)y = 15\sin 2x$

ANS $y = Ce^x + A\cos x + B\sin x + 2\cos 2x + \sin 2x$

4.47 $(D^3 + 3D^2 - 4)y = 40\sin 2x$

ANS $y = (C_0 + C_1x)e^{-2x} + C_2e^x + \cos 2x - 2\sin 2x$

4.48 $y''' - y'' + y' - y = 2e^x + 5e^{2x}$

ANS $y = A\cos x + B\sin x + Ce^x + xe^x + e^{2x}$

4.49 $(D^3 - 6D^2 + 11D - 6)y = 10e^x \sin x$

ANS $y = C_1e^x + C_2e^{2x} + C_3e^{3x} + e^x(3\sin x - \cos x)$

4.50 $(D^3 - 2D - 4)y = 50(\sin x + e^{2x})$

ANS $y = Ce^{2x} + e^{-x}(A\cos x + B\sin x) + 6\cos x - 8\sin x + 5xe^{2x}$

4.51 $y''' - 3y'' + 4y = 12e^{2x} + 4e^{3x}$

ANS $y = (C_0 + C_1x)e^{2x} + C_3e^{-x} + 2x^2e^{2x} + e^{3x}$

4.52 $(D^4 - 8D^2 + 16)y = 32e^{2x} + 16x^3$

ANS $y = (C_0 + C_1x)e^{-2x} + (D_0 + D_1x)e^{2x} + x^2e^{2x} + x^3 + 3x$

4.53 $(D^4 - 18D^2 + 81)y = 72e^{3x} + 729x^2$

ANS $y = (C_0 + C_1x)e^{-3x} + (D_0 + D_1x)e^{3x} + x^2e^{3x} + 9x^2 + 4$

Method of Variation of Parameters

4.54 $y'' - y = x^{-1} - 2x^{-3}$ ANS $y = C_1e^{-x} + C_2e^x - x^{-1}$

4.55 $y'' - y = \dfrac{1}{\sinh x}$ ANS $y = C_1e^{-x} + C_2e^x - xe^{-x} + \sinh x \ln|1 - e^{-2x}|$

4.56 $y'' - 2y' + y = \dfrac{e^x}{x}$ ANS $y = (C_0 + C_1x + x\ln|x|)e^x$

4.57 $y'' + 3y' + 2y = \sin e^x$ ANS $y = C_1e^{-2x} + C_2e^{-x} - e^{-2x}\sin e^x$

4.58 $y'' - 3y' + 2y = \sin e^{-x}$ ANS $y = C_1e^x + C_2e^{2x} - e^{2x}\sin e^{-x}$

4.59 $y'' + y = \sec^3 x$ ANS $y = A\cos x + B\sin x + \frac{1}{2}\sec x$

4.60 $y'' - y = (1 - e^{2x})^{-\frac{1}{2}}$ ANS $y = C_1e^{-x} + C_2e^x - \frac{1}{2}e^{-x}\sin^{-1}e^x - \frac{1}{2}\sqrt{1 - e^{2x}}$

4.61 $y'' - y = e^{-2x}\sin e^{-x}$ ANS $y = C_1e^{-x} + C_2e^x - \sin e^{-x} - e^x\cos e^{-x}$

4.62 $y'' + 2y' + y = 15e^{-x}\sqrt{x+1}$ ANS $y = e^{-x}\left[C_0 + C_1x + 4(x+1)^{\frac{5}{2}}\right]$

4.63 $y'' + 4y = 2\tan x$

ANS $y = A\cos 2x + B\sin 2x + \sin 2x \ln|\cos x| - x\cos 2x$

4.64 $y'' - 2y' + y = \dfrac{e^{2x}}{(e^x + 1)^2}$ ANS $y = (C_0 + C_1x)e^x + e^x\ln(1 + e^x)$

4.65 $y'' + y' = \dfrac{1}{1 + e^x}$ ANS $y = C_1 + C_2e^{-x} - \ln(e^{-x} + 1) - e^{-x}\ln(e^x + 1)$

Euler Differential Equations

4.66 $(x^2D^2 - xD + 1)y = \ln x$ ANS $y = (C_0 + C_1 \ln x)x + 2 + \ln x$

4.67 $x^2y'' + 3xy' + 5y = \dfrac{5}{x^2}\ln x$

ANS $y = x^{-1}\big[A\cos(2\ln x) + B\sin(2\ln x)\big] + x^{-2}\left(\frac{2}{5} + \ln x\right)$

4.68 $(x^3D^3 + 2x^2D^2 - xD + 1)y = 9x^2 \ln x$

ANS $y_C = C_1x^{-1} + (C_2 + C_3 \ln x)x + (3\ln x - 7)x^2$

4.69 $\big[(x-2)^2D^2 - 3(x-2)D + 4\big]y = x$

ANS $y_C = (x-2)^2\big(C_0 + C_1 \ln|x-2|\big) + x - \frac{3}{2}$

4.70 $x^3y''' + 3x^2y'' + xy' - y = x^2$

ANS $y = Cx + \dfrac{1}{\sqrt{x}}\left[A\cos\left(\dfrac{\sqrt{3}}{2}\ln x\right) + B\sin\left(\dfrac{\sqrt{3}}{2}\ln x\right)\right] + \dfrac{x^2}{7}$

CHAPTER 5

Applications of Linear Differential Equations

5.1 Vibration of a Single Degree-of-Freedom System

5.1.1 Formulation—Equation of Motion

In this section, the vibration of a single story shear building as shown in Figure 5.1, which is considered as a model of a single degree-of-freedom (DOF) system, is studied.

A single story shear building consists of a rigid girder with mass m, which is supported by columns with combined stiffness k. The columns are assumed to be weightless, inextensible in the axial (vertical) direction, and they can only take shear forces but not bending moments. In the horizontal direction, the columns act as a spring of stiffness k. As a result, the girder can only move in the horizontal direction, and its motion can be described by a single variable $x(t)$; hence the system is called a single degree-of-freedom (DOF) system. The number of degrees-of-freedom is the total number of variables required to describe the motion of a system.

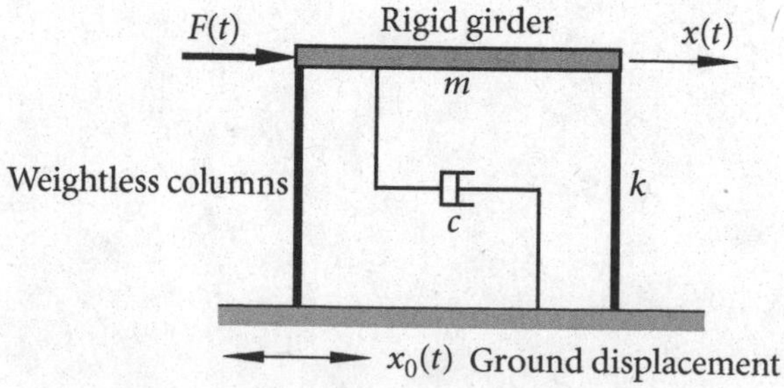

Figure 5.1 A single-story shear building.

The combined stiffness k of the columns can be determined as follows. Apply a horizontal static force P on the girder. If the displacement of the girder is Δ as shown in Figure 5.2, then the combined stiffness of the columns is $k = P/\Delta$.

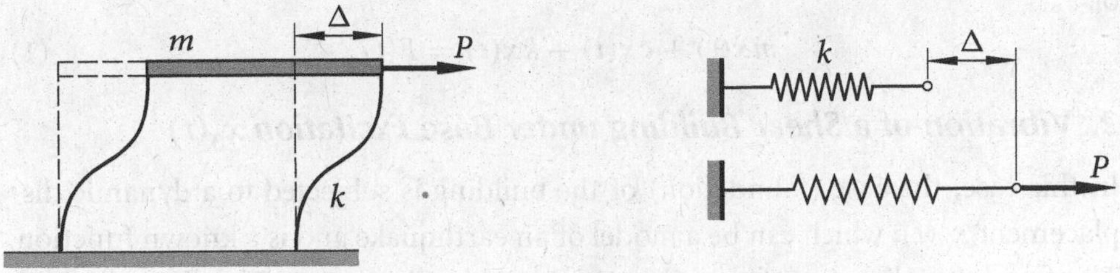

Figure 5.2 Determination of column stiffness.

The internal friction between the girder and the columns is described by a viscous dashpot damper with damping coefficient c. A dashpot damper is shown schematically in Figure 5.3 and provides a damping force $-c(v_B - v_A)$, where v_A and v_B are the velocities of points A and B, respectively, and $(v_B - v_A)$ is the relative velocity between points B and A. The damping force is opposite to the direction of the relative velocity.

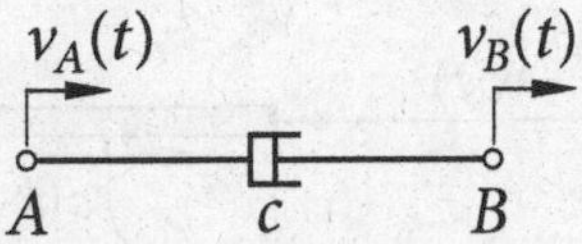

Figure 5.3 A dashpot damper.

1. Vibration of a Shear Building under Externally Applied Force $F(t)$

In this case, the girder is subjected to an externally applied force $F(t)$, which can be a model of wind load. Consider the vibration of the girder; its free-body diagram is drawn in Figure 5.4.

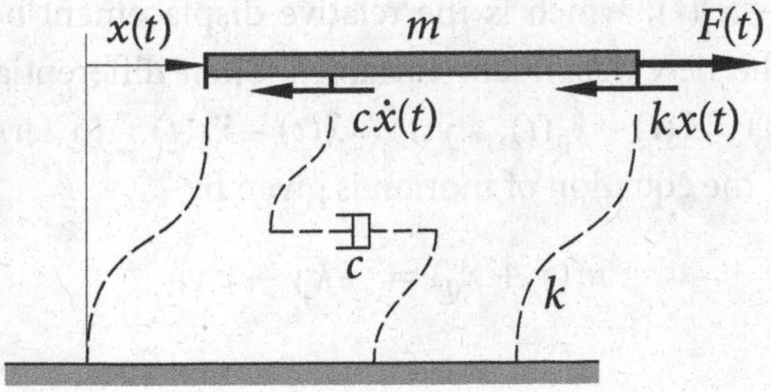

Figure 5.4 Free-body diagram of the building under externally applied force.

The girder is subjected to the shear force (elastic force) $kx(t)$, the viscous damping force $c\dot{x}(t)$, and the externally applied load $F(t)$. The equation of motion is governed by Newton's Second Law

$$\rightarrow ma = \sum F \implies m\ddot{x}(t) = -kx(t) - c\dot{x}(t) + F(t),$$

or

$$m\ddot{x}(t) + c\dot{x}(t) + kx(t) = F(t). \tag{1}$$

2. Vibration of a Shear Building under Base Excitation $x_0(t)$

In this case, the base (foundation) of the building is subjected to a dynamic displacement $x_0(t)$, which can be a model of an earthquake and is a known function. The free-body diagram of the girder is shown in Figure 5.5. The shear (elastic) force and the damping force applied on the girder are given by

Shear force $= k \cdot$ (Relative displacement between girder and base) $= k(x - x_0)$,

Damping force $= c \cdot$ (Relative velocity between girder and base) $= c(\dot{x} - \dot{x}_0)$.

Newton's Second Law requires that

$$\rightarrow ma = \sum F \implies m\ddot{x}(t) = -k\big[x(t) - x_0(t)\big] - c\big[\dot{x}(t) - \dot{x}_0(t)\big].$$

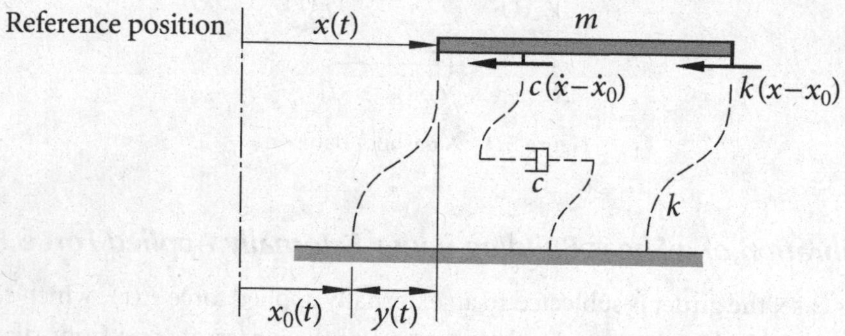

Figure 5.5 Free-body diagram of the building under base excitation.

Let $y(t) = x(t) - x_0(t)$, which is the relative displacement between the girder and the base, be the new dependent variable. Then differentiating with respect to t results in $\dot{y}(t) = \dot{x}(t) - \dot{x}_0(t)$, $\ddot{y}(t) = \ddot{x}(t) - \ddot{x}_0(t)$. In terms of the relative displacement $y(t)$, the equation of motion is given by

$$m(\ddot{y} + \ddot{x}_0) = -ky - c\dot{y},$$

i.e.,

$$m\ddot{y}(t) + c\dot{y}(t) + ky(t) = -m\ddot{x}_0(t), \tag{2}$$

where $\ddot{x}_0(t)$ is the base or ground acceleration. The loading on the girder created from ground excitation (earthquake) is $F(t) = -m\ddot{x}_0(t)$, which is proportional to the mass of the girder and the ground acceleration.

Both systems (1) and (2) are second-order linear ordinary differential equations with constant coefficients.

In general, a linear single degree-of-freedom system can be modeled by a mechanical mass-damper-spring system. The example of a single story shear building under externally applied load or base excitation can be described using the following equivalent mass-damper-spring system.

1. Single Degree-of-Freedom System under Externally Applied Force

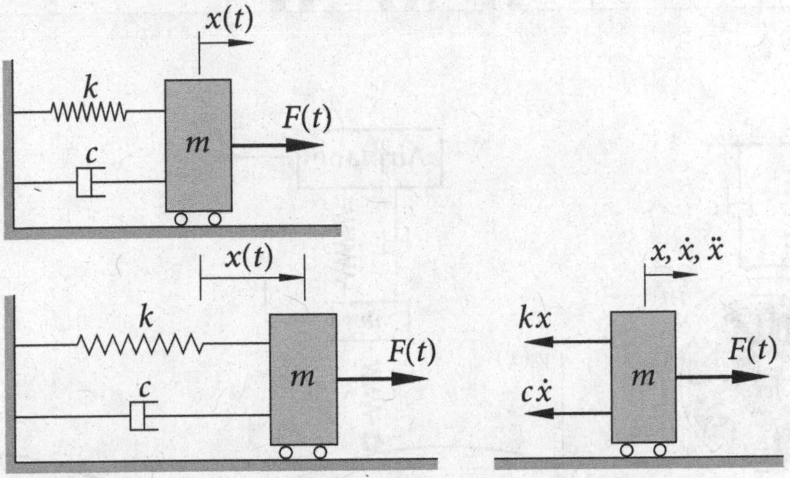

Figure 5.6 A mass-damper-spring system under externally applied force.

2. Single Degree-of-Freedom System under Base Excitation

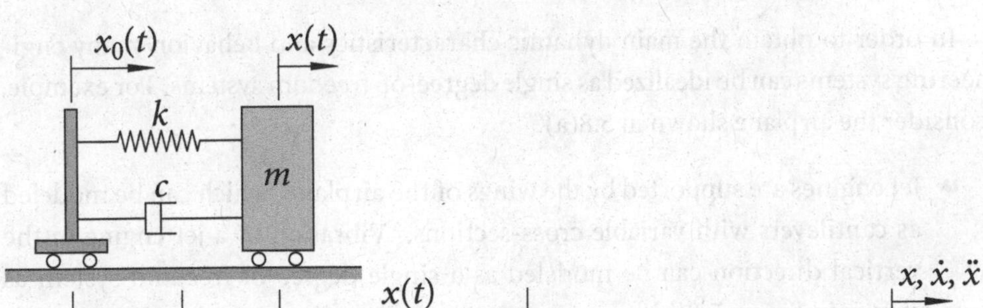

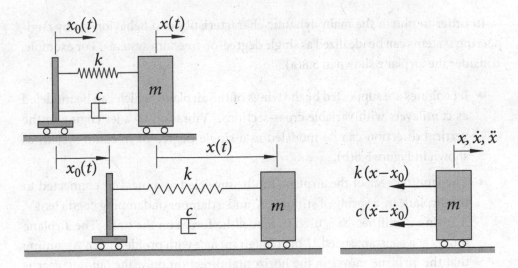

Figure 5.7 A mass-damper-spring system under base excitation.

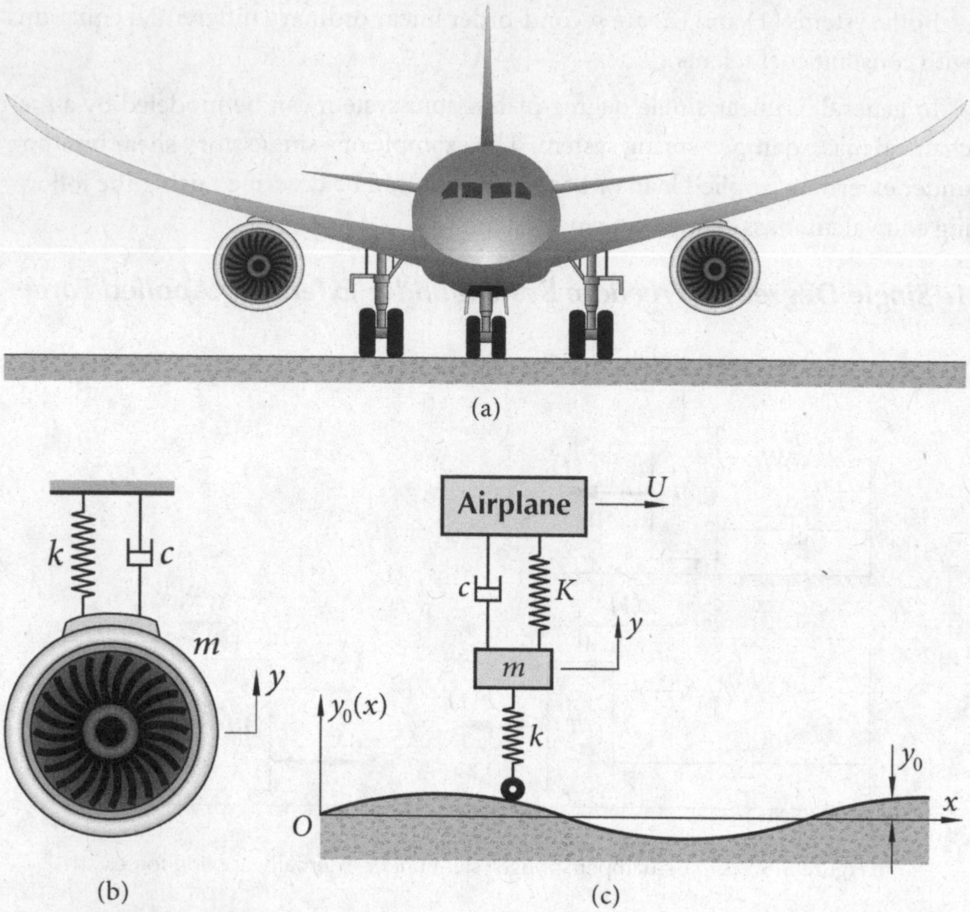

Figure 5.8 Mathematical modeling of jet engine and landing gear.

In order to obtain the main dynamic characteristics and behavior, many engineering systems can be idealized as single degree-of-freedom systems. For example, consider the airplane shown in 5.8(a).

- Jet engines are supported by the wings of the airplane, which can be modeled as cantilevers with variable cross-sections. Vibration of a jet engine in the vertical direction can be modeled as a single degree-of-freedom system as shown in Figure 5.8(b).
- The landing gear of the airplane can be modeled as a mass m connected to the airplane by a spring of stiffness K and a damper of damping coefficient c. A spring of stiffness k is used to model the forces on the tires. The airplane moves at a constant speed U on a rough surface with profile $y_0(x)$. Assuming that the airplane moves in the horizontal direction only, the landing gear is modeled as a single degree-of-freedom system as shown in Figure 5.8(c).

5.1.2 Response of a Single Degree-of-Freedom System

From Section 5.1.1, the equation of motion of a single degree-of-freedom system is given by

$$m\ddot{x} + c\dot{x} + kx = F(t),$$

where $F(t)$ is a known forcing function. It is a second-order linear differential equation with constant coefficients, and can be rewritten in the standard form in the theory of vibration by dividing both sides by m

$$\ddot{x} + \frac{c}{m}\dot{x} + \frac{k}{m}x = \frac{F(t)}{m}.$$

Denoting

$$\frac{k}{m} = \omega_0^2, \qquad \omega_0 = \text{natural circular frequency},$$

$$\frac{c}{m} = 2\zeta\omega_0, \qquad \zeta = \text{nondimensional damping coefficient},$$

the equation of motion becomes

$$\ddot{x} + 2\zeta\omega_0\dot{x} + \omega_0^2 x = \frac{F(t)}{m}.$$

The solution of the system $x(t)$ consists of two parts: complementary solution and particular solution.

- The complementary solution $x_C(t)$ is obtained when the right-hand side of the equation is set to zero, i.e., $F(t)=0$, implying that the system is not subjected to loading. In the terminology of vibration, the system is in free (not forced) vibration, and the solution of the equation is the response of *free vibration.*
- The particular solution $x_P(t)$ corresponds to the right-hand side of the equation or the forcing term $F(t)$, hence the response of *forced vibration.*

5.1.2.1 Free Vibration—Complementary Solution

The equation of motion is the complementary differential equation given by

$$\ddot{x} + 2\zeta\omega_0\dot{x} + \omega_0^2 x = 0.$$

The characteristic equation is $\lambda^2 + 2\zeta\omega_0\lambda + \omega_0^2 = 0$, which gives

$$\lambda = \frac{-2\zeta\omega_0 \pm \sqrt{(2\zeta\omega_0)^2 - 4\times 1\times\omega_0^2}}{2} = \omega_0\left(-\zeta \pm \sqrt{\zeta^2-1}\right).$$

Whether the characteristic equation has distinct real roots, double roots, or complex roots depends on the value of the nondimensional damping coefficient ζ.

Case 1: Underdamped System $0 \leqslant \zeta < 1$

Most engineering structures fall in this category with damping coefficient ζ usually less than 10%. The roots of the characteristic equation are

$$\lambda = \omega_0\big(-\zeta \pm \mathrm{i}\sqrt{1-\zeta^2}\big) = -\zeta\omega_0 \pm \mathrm{i}\omega_d,$$

where $\omega_d = \omega_0\sqrt{1-\zeta^2}$ is the damped natural circular frequency. The complementary solution is

$$x_C(t) = \mathrm{e}^{-\zeta\omega_0 t}(A\cos\omega_d t + B\sin\omega_d t),$$

where constants A and B are determined from the initial conditions $x(0) = x_0$ and $\dot{x}(0) = v_0$. Since

$$\begin{aligned}
\dot{x}_C(t) &= -\zeta\omega_0\mathrm{e}^{-\zeta\omega_0 t}(A\cos\omega_d t + B\sin\omega_d t)\\
&\quad + \mathrm{e}^{-\zeta\omega_0 t}(-A\omega_d\sin\omega_d t + B\omega_d\cos\omega_d t)\\
&= \mathrm{e}^{-\zeta\omega_0 t}\big[(-\zeta\omega_0 A + \omega_d B)\cos\omega_d t + (-\zeta\omega_0 B - \omega_d A)\sin\omega_d t\big],
\end{aligned}$$

substituting in the initial conditions yields

$$x_C(0) = A = x_0,$$

$$\dot{x}_C(0) = -\zeta\omega_0 A + \omega_d B = v_0 \quad\Longrightarrow\quad B = \frac{v_0 + \zeta\omega_0 x_0}{\omega_d}.$$

Hence the response of free vibration is

$$x_C(t) = \mathrm{e}^{-\zeta\omega_0 t}\Big(x_0\cos\omega_d t + \frac{v_0 + \zeta\omega_0 x_0}{\omega_d}\sin\omega_d t\Big), \qquad 0 \leqslant \zeta < 1.$$

Special Case: Undamped System $\zeta = 0$, $\omega_d = \omega_0$

The response becomes

$$\begin{aligned}
x_C(t) &= x_0\cos\omega_0 t + \frac{v_0}{\omega_0}\sin\omega_0 t\\
&= \underbrace{\sqrt{x_0^2 + \Big(\frac{v_0}{\omega_0}\Big)^2}}_{a}\Bigg[\underbrace{\frac{x_0}{\sqrt{x_0^2 + (v_0/\omega_0)^2}}}_{\cos\varphi}\cos\omega_0 t + \underbrace{\frac{v_0/\omega_0}{\sqrt{x_0^2 + (v_0/\omega_0)^2}}}_{\sin\varphi}\sin\omega_0 t\Bigg]\\
&= a\cos(\omega_0 t - \varphi),
\end{aligned}$$

which is a harmonic function shown in Figure 5.9 with amplitude a and phase angle φ given by

$$a = \sqrt{x_0^2 + \Big(\frac{v_0}{\omega_0}\Big)^2}, \qquad \varphi = \tan^{-1}\Big(\frac{v_0}{\omega_0 x_0}\Big).$$

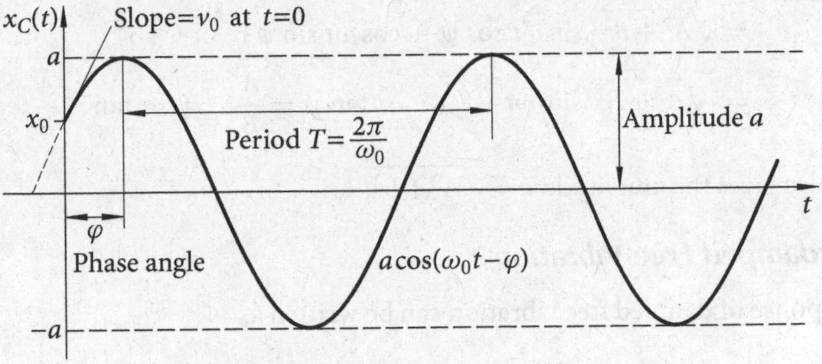

Figure 5.9 Response of undamped free vibration.

The harmonic solution has period $T=2\pi/\omega_0$; hence ω_0 is called the *natural circular frequency* of the system, with unit rad/sec. The *natural frequency* is $f=\omega_0/(2\pi)=1/T$, in cycles/sec or Hz. The maximum displacement is

$$\max x_C(t) = a = \sqrt{x_0^2+\left(\frac{v_0}{\omega_0}\right)^2} = \text{amplitude of the motion.}$$

Remarks: Using the trigonometric identities

$$\sin(\alpha+\beta) = \sin\alpha\cos\beta + \cos\alpha\sin\beta, \quad \cos(\alpha-\beta) = \cos\alpha\cos\beta + \sin\alpha\sin\beta,$$

the amplitude of

$$x(t) = A\cos\omega t + B\sin\omega t$$

can be determined as follows.

Rewrite the expression of $x(t)$

$$\begin{aligned} x(t) &= \sqrt{A^2+B^2}\Bigg(\underbrace{\frac{A}{\sqrt{A^2+B^2}}}_{\cos\varphi}\cos\omega t + \underbrace{\frac{B}{\sqrt{A^2+B^2}}}_{\sin\varphi}\sin\omega t\Bigg) \\ &= \sqrt{A^2+B^2}\,(\cos\omega t\cos\varphi + \sin\omega t\sin\varphi) \\ &= \sqrt{A^2+B^2}\,\cos(\omega t-\varphi), \qquad \tan\varphi = \frac{B}{A}, \quad \varphi = \tan^{-1}\frac{B}{A}, \end{aligned}$$

which gives the amplitude $a=\sqrt{A^2+B^2}$.

Alternatively,

$$x(t) = \sqrt{A^2+B^2}\Bigg(\underbrace{\frac{B}{\sqrt{A^2+B^2}}}_{\cos\psi}\sin\omega t + \underbrace{\frac{A}{\sqrt{A^2+B^2}}}_{\sin\psi}\cos\omega t\Bigg)$$

$$= \sqrt{A^2+B^2}\,(\sin\omega t\cos\psi + \cos\omega t\sin\psi)$$

$$= \sqrt{A^2+B^2}\,\sin(\omega t+\psi), \qquad \tan\psi = \frac{A}{B}, \quad \psi = \tan^{-1}\frac{A}{B},$$

which also gives the amplitude $a=\sqrt{A^2+B^2}$.

Underdamped Free Vibration

The response of damped free vibration can be written as

$$x_C(t) = a\mathrm{e}^{-\zeta\omega_0 t}\cos(\omega_d t-\varphi),$$

where

$$a = \sqrt{x_0^2 + \left(\frac{v_0+\zeta\omega_0 x_0}{\omega_d}\right)^2}, \qquad \varphi = \tan^{-1}\left(\frac{v_0+\zeta\omega_0 x_0}{\omega_d x_0}\right).$$

To sketch the response of damped free vibration, compare it with the response of undamped free vibration. The difference lies in the amplitude: instead of having a constant amplitude a, the amplitude $a\mathrm{e}^{-\zeta\omega_0 t}$ decays exponentially with time. The following steps are followed in sketching the response $x_C(t)$ (Figure 5.10).

1. Sketch the sinusoidal function $\cos(\omega_d t-\varphi)$, in which ω_d is the *damped natural circular frequency* (rad/sec) and φ is the phase angle. The period is $T_d=2\pi/\omega_d$ and $f_d=\omega_d/(2\pi)=1/T_d$ is the *damped natural frequency* in cycles/sec or Hz.
2. Sketch the amplitude $a\mathrm{e}^{-\zeta\omega_0 t}$ and its mirror image $-a\mathrm{e}^{-\zeta\omega_0 t}$ in dashed lines. These two lines form the envelope of the response.
3. Fit the sinusoidal function $\cos(\omega_d t-\varphi)$ inside the envelope to obtain the response of damped free vibration.

The response of damped free vibration can be used to determine the damping coefficient ζ as follows:

At time t, the response is

$$x_C(t) = \mathrm{e}^{-\zeta\omega_0 t}\cos(\omega_d t-\varphi).$$

After a period T_d, the response becomes

$$x_C(t+T_d) = \mathrm{e}^{-\zeta\omega_0(t+T_d)}\cos\left[\omega_d(t+T_d)-\varphi\right] = \mathrm{e}^{-\zeta\omega_0(t+T_d)}\cos(\omega_d t-\varphi),$$

which leads to

$$\frac{x_C(t+T_d)}{x_C(t)} = \frac{\mathrm{e}^{-\zeta\omega_0(t+T_d)}\cos(\omega_d t-\varphi)}{\mathrm{e}^{-\zeta\omega_0 t}\cos(\omega_d t-\varphi)} = \mathrm{e}^{-\zeta\omega_0 T_d}.$$

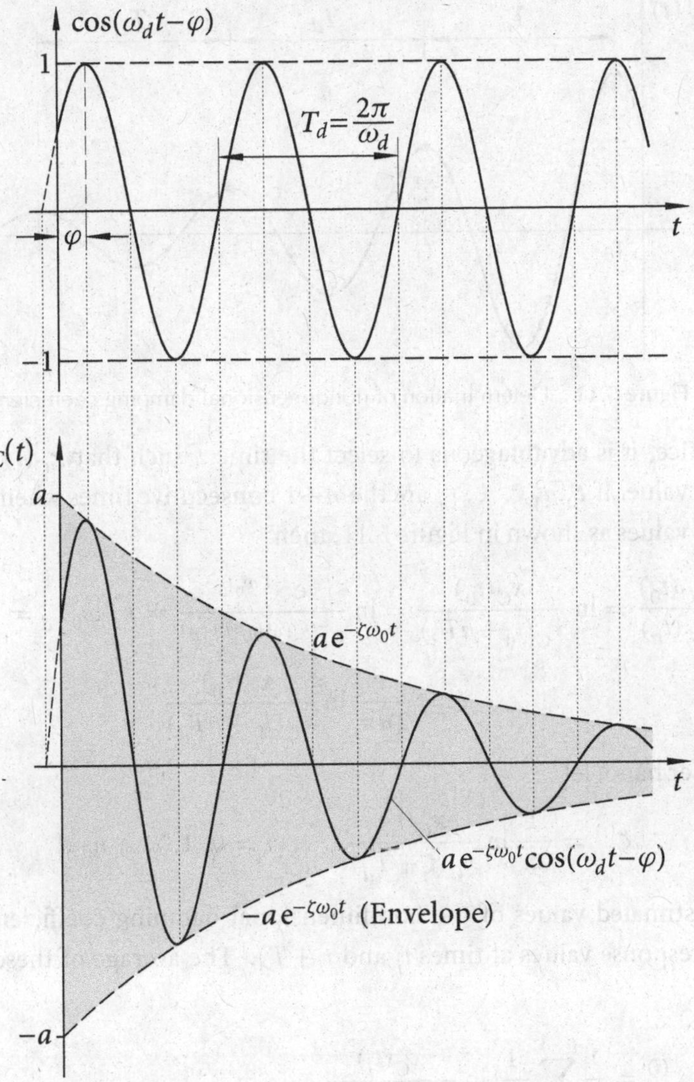

Figure 5.10 Response of underdamped free vibration.

Taking logarithm yields

$$\ln \frac{x_C(t+T_d)}{x_C(t)} = -\zeta\omega_0 T_d,$$

in which $\zeta\omega_0 T_d$ is called the logarithmic decrement δ. For lightly damped structures with $0 < \zeta \ll 1$,

$$\delta = \zeta\omega_0 T_d = \zeta\omega_0 \frac{2\pi}{\omega_d} = \zeta\omega_0 \frac{2\pi}{\omega_0\sqrt{1-\zeta^2}} = \frac{2\pi\zeta}{\sqrt{1-\zeta^2}} \approx 2\pi\zeta.$$

Hence, the nondimensional damping coefficient ζ is given by

$$\ln \frac{x_C(t+T_d)}{x_C(t)} = -2\pi\zeta \implies \zeta = \frac{1}{2\pi} \ln \frac{x_C(t)}{x_C(t+T_d)}.$$

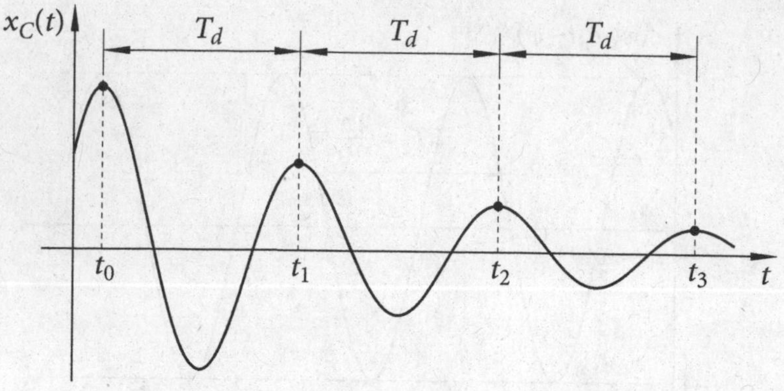

Figure 5.11 Determination of nondimensional damping coefficient.

In practice, it is advantageous to select the time t such that $x_C(t)$ reaches the maximum value. If $t_0, t_1, \ldots, t_n$ are the $n+1$ consecutive times when $x_C(t)$ takes maximum values as shown in Figure 5.11, then

$$\ln\frac{x_C(t_0)}{x_C(t_n)}=\ln\frac{x_C(t_0)}{x_C(t_0+nT_d)}=\ln\frac{\mathrm{e}^{-\zeta\omega_0 t_0}}{\mathrm{e}^{-\zeta\omega_0(t_0+nT_d)}}=n\zeta\omega_0T_d=2n\pi\zeta$$

$$\therefore\quad \zeta=\frac{1}{2n\pi}\ln\frac{x_C(t_0)}{x_C(t_0+nT_d)}.$$

On the other hand, let

$$\zeta^{(i)}=\frac{1}{2\pi}\ln\frac{x_C(t_i)}{x_C(t_i+T_d)},\qquad i=0,1,\ldots,n-1,$$

be the n estimated values of the nondimensional damping coefficient estimated using the response values at times t_i and t_i+T_d. The average of these n values is given by

$$\bar{\zeta}=\frac{1}{n}\sum_{i=0}^{n-1}\zeta^{(i)}=\frac{1}{n}\sum_{i=0}^{n-1}\frac{1}{2\pi}\ln\frac{x_C(t_i)}{x_C(t_i+T_d)}$$

$$=\frac{1}{2n\pi}\left[\ln\frac{x_C(t_0)}{x_C(t_0+T_d)}+\ln\frac{x_C(t_1)}{x_C(t_1+T_d)}+\cdots+\ln\frac{x_C(t_{n-1})}{x_C(t_{n-1}+T_d)}\right]$$

$$=\frac{1}{2n\pi}\ln\left[\frac{x_C(t_0)}{x_C(t_0+T_d)}\cdot\frac{x_C(t_1)}{x_C(t_1+T_d)}\cdots\frac{x_C(t_{n-1})}{x_C(t_{n-1}+T_d)}\right]$$

✍ $x_C(t_i+T_d)=x_C(t_{i+1}),\ i=0,1,\ldots,n-1$

$$=\frac{1}{2n\pi}\ln\frac{x_C(t_0)}{x_C(t_n)}.$$

From the Central Limit Theorem in the theory of probability, it is known that $\bar{\zeta}$ approaches the true value of the nondimensional damping coefficient when $n\to\infty$. Hence, by using a larger value of n in the above equation, a better estimation of ζ is achieved.

Case 2: Critically Damped System $\zeta = 1$

When $\zeta = 1$, the system is called critically damped. The characteristic equation has a double root $\lambda = -\omega_0, -\omega_0$. The complementary solution is given by

$$x_C(t) = (C_0 + C_1 t)\,\mathrm{e}^{-\omega_0 t},$$

in which the constants C_0 and C_1 are determined from the initial conditions $x(0) = x_0$, $\dot{x}(0) = v_0$. Since

$$\dot{x}_C(t) = C_1\,\mathrm{e}^{-\omega_0 t} - (C_0 + C_1 t)\,\omega_0\,\mathrm{e}^{-\omega_0 t},$$

one has

$$x_C(0) = C_0 = x_0, \qquad \dot{x}_C(0) = C_1 - C_0\omega_0 = v_0 \implies C_1 = v_0 + \omega_0 x_0.$$

Hence, the response is

$$x_C(t) = \big[x_0 + (v_0 + \omega_0 x_0)\,t\big]\,\mathrm{e}^{-\omega_0 t}, \qquad \zeta = 1.$$

A typical response of the free vibration of a critically damped system is shown in Figure 5.12, which is not oscillatory and decays exponentially.

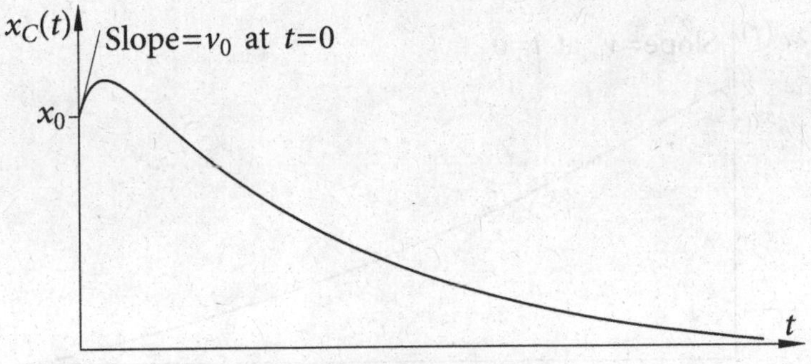

Figure 5.12 Response of critically damped free vibration.

Case 3: Overdamped System $\zeta > 1$

The system is called overdamped when the nondimensional damping coefficient $\zeta > 1$. The characteristic equation has two distinct real roots $\lambda = \omega_0\big(-\zeta \pm \sqrt{\zeta^2 - 1}\big)$. The complementary solution is given by

$$x_C(t) = C_1\,\mathrm{e}^{-\omega_0(\zeta - \sqrt{\zeta^2-1})t} + C_2\,\mathrm{e}^{-\omega_0(\zeta + \sqrt{\zeta^2-1})t},$$

in which the constants C_1 and C_2 are determined from the initial conditions $x(0) = x_0$, $\dot{x}(0) = v_0$. Since

$$\dot{x}_C(t) = -\omega_0\Big[C_1\big(\zeta - \sqrt{\zeta^2-1}\big)\,\mathrm{e}^{-\omega_0(\zeta - \sqrt{\zeta^2-1})t} + C_2\big(\zeta + \sqrt{\zeta^2-1}\big)\,\mathrm{e}^{-\omega_0(\zeta + \sqrt{\zeta^2-1})t}\Big],$$

one obtains

$$x_C(0) = C_1 + C_2 = x_0,$$

$$\dot{x}_C(0) = -\omega_0 \Big[C_1\big(\zeta - \sqrt{\zeta^2-1}\big) + C_2\big(\zeta + \sqrt{\zeta^2-1}\big) \Big] = v_0$$

$$\Longrightarrow \begin{cases} C_1 = \dfrac{1}{2\omega_0\sqrt{\zeta^2-1}} \big[v_0 + \big(\zeta + \sqrt{\zeta^2-1}\big)\,\omega_0 x_0\big], \\[2ex] C_2 = -\dfrac{1}{2\omega_0\sqrt{\zeta^2-1}} \big[v_0 + \big(\zeta - \sqrt{\zeta^2-1}\big)\,\omega_0 x_0\big]. \end{cases}$$

Hence, the response of free vibration of an overdamped system is

$$x_C(t) = \frac{1}{2\omega_0\sqrt{\zeta^2-1}} \Big\{ \big[v_0 + \big(\zeta + \sqrt{\zeta^2-1}\big)\,\omega_0 x_0\big] \mathrm{e}^{-\omega_0(\zeta - \sqrt{\zeta^2-1})t} - \big[v_0 + \big(\zeta - \sqrt{\zeta^2-1}\big)\,\omega_0 x_0\big] \mathrm{e}^{-\omega_0(\zeta + \sqrt{\zeta^2-1})t} \Big\}, \qquad \zeta > 1.$$

A typical plot of the response is shown in Figure 5.13, which is not oscillatory and decays exponentially.

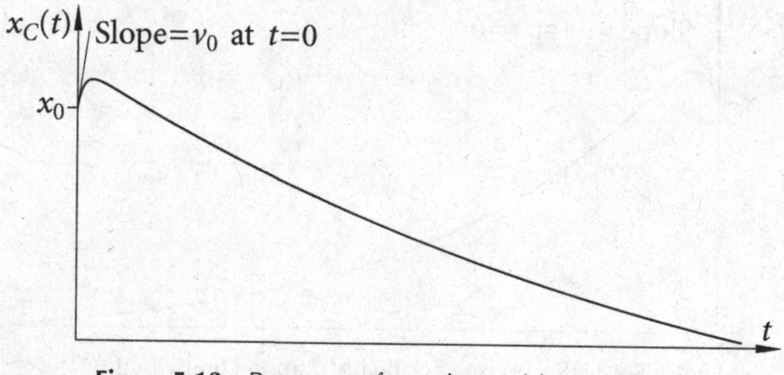

Figure 5.13 Response of overdamped free vibration.

5.1.2.2 Forced Vibration—Particular Solution

For an underdamped system with $0 < \zeta < 1$, the complementary solution or the response of free vibration is given by

$$x_C(t) = \mathrm{e}^{-\zeta\omega_0 t}\Big(x_0 \cos\omega_d t + \frac{v_0 + \zeta\omega_0 x_0}{\omega_d} \sin\omega_d t\Big),$$

which decays exponentially and approaches zero as $t \to \infty$, as shown in Figure 5.10. Hence the complementary solution is called the *transient* solution. Because its value becomes negligible after some time, its effect is small and is not important in practice.

The particular solution $x_P(t)$ is associated with the right-hand side of the differential equation and hence corresponds to forced vibration. The particular solution is called the *steady-state* solution, because it is the solution that persists when time is large. Suppose $F(t)$ is periodic and of the form $F(t)=F_0 \sin \Omega t$. Then the particular solution satisfies

$$\ddot{x}_P + 2\zeta\omega_0\dot{x}_P + \omega_0^2 x_P = \frac{F_0}{m}\sin\Omega t,$$

or, in the D-operator notation,

$$(D^2 + 2\zeta\omega_0 D + \omega_0^2)x_P = \frac{F_0}{m}\sin\Omega t.$$

Hence

$$\begin{aligned}
x_P &= \frac{F_0}{m}\frac{1}{D^2+2\zeta\omega_0 D+\omega_0^2}\sin\Omega t \\
&= \frac{F_0}{m}\frac{1}{-\Omega^2+2\zeta\omega_0 D+\omega_0^2}\sin\Omega t \qquad \text{Theorem 3 of Chapter 4: replace } D^2 \text{ by } -\Omega^2. \\
&= \frac{F_0}{m}\frac{(\omega_0^2-\Omega^2)-2\zeta\omega_0 D}{\left[(\omega_0^2-\Omega^2)+2\zeta\omega_0 D\right]\left[(\omega_0^2-\Omega^2)-2\zeta\omega_0 D\right]}\sin\Omega t \\
&= \frac{F_0}{m}\frac{(\omega_0^2-\Omega^2)-2\zeta\omega_0 D}{(\omega_0^2-\Omega^2)^2-(2\zeta\omega_0 D)^2}\sin\Omega t \\
&= \frac{F_0}{m}\frac{(\omega_0^2-\Omega^2)\sin\Omega t-2\zeta\omega_0\Omega\cos\Omega t}{(\omega_0^2-\Omega^2)^2+(2\zeta\omega_0\Omega)^2}. \qquad \text{Replace } D^2 \text{ by } -\Omega^2 \text{ in denominator, and evaluate numerator.}
\end{aligned}$$

Defining angle φ by

$$\cos\varphi = \frac{\omega_0^2-\Omega^2}{\sqrt{(\omega_0^2-\Omega^2)^2+(2\zeta\omega_0\Omega)^2}}, \qquad \sin\varphi = \frac{2\zeta\omega_0\Omega}{\sqrt{(\omega_0^2-\Omega^2)^2+(2\zeta\omega_0\Omega)^2}},$$

the response of forced vibration becomes

$$x_P(t) = \frac{F_0}{m}\frac{1}{\sqrt{(\omega_0^2-\Omega^2)^2+(2\zeta\omega_0\Omega)^2}}\sin(\Omega t-\varphi).$$

The applied force $F(t)$ is maximum at $\Omega t = \frac{1}{2}\pi, \frac{3}{2}\pi, \ldots$; the response $x_P(t)$ is maximum at $\Omega t - \varphi = \frac{1}{2}\pi, \frac{3}{2}\pi, \ldots$. Hence the response $x_P(t)$ lags behind the forcing by a time φ/Ω. The angle φ is called a phase angle or phase lag.

The amplitude of the response of forced vibration is

$$\left|x_P(t)\right|_{\max} = \frac{F_0}{m}\frac{1}{\sqrt{(\omega_0^2-\Omega^2)^2+(2\zeta\omega_0\Omega)^2}}.$$

Denoting

$$r = \frac{\Omega}{\omega_0} = \frac{\text{Excitation frequency}}{\text{Undamped natural frequency}} = \text{Frequency ratio},$$

one has

$$\left|x_P(t)\right|_{\max} = \frac{\dfrac{F_0}{m\omega_0^2}}{\sqrt{\left[1-\left(\dfrac{\Omega}{\omega_0}\right)^2\right]^2+\left(2\zeta\dfrac{\Omega}{\omega_0}\right)^2}} = \frac{\dfrac{F_0}{m\omega_0^2}}{\sqrt{(1-r^2)^2+(2\zeta r)^2}}$$

$$= \frac{F_0}{k}\,\frac{1}{\sqrt{(1-r^2)^2+(2\zeta r)^2}}, \qquad \omega_0^2 = \frac{k}{m} \implies k = m\omega_0^2.$$

If dynamic effect is not considered, i.e., if only static terms are considered and the shaded dynamic terms in the following equation are dropped, one obtains

$$\underbrace{m\ddot{x}}_{\substack{\text{Dynamic}\\ \text{inertia}\\ \text{force}}} + \underbrace{c\dot{x}}_{\substack{\text{Dynamic}\\ \text{damping}\\ \text{force}}} + \underbrace{kx}_{\substack{\text{Static}\\ \text{elastic}\\ \text{force}}} = F(t) = F_0 \underbrace{\sin\Omega t}_{\substack{\text{Dynamic}\\ \text{variation}}} \implies x_{\text{static}} = \frac{F_0}{k},$$

which is the static displacement of the system under the static force F_0.

Define

$$\frac{\left|x_P(t)\right|_{\max}}{x_{\text{static}}} = \frac{1}{\sqrt{(1-r^2)^2+(2\zeta r)^2}} = \text{Dynamic Magnification Factor}.$$

The Dynamic Magnification Factor (DMF) is plotted in Figure 5.14 for various values of the nondimensional damping coefficient ζ. It is one of the most important quantities describing the dynamic behavior of an underdamped single degree-of-freedom system under sinusoidal excitation.

- When $r \to 0$, i.e., when the excitation frequency $\Omega \to 0$, DMF$\to 1$. In this case, the dynamic excitation approaches a static force and the amplitude of dynamic response approaches the static displacement.
- When $r \to \infty$, i.e., when the excitation frequency Ω is large compared to the natural frequency ω_0, DMF$\to 0$, which means that the dynamic response approaches zero. This result can be understood intuitively as follows: the system is excited (pushed and pulled) at such a high frequency that it does not "know" which way to move so that it just "stands" still or there is no response.
- When $r \approx 1$ or the excitation frequency Ω is close to the natural frequency ω_0, $\Omega \approx \omega_0$, DMF tends to large values for small damping.

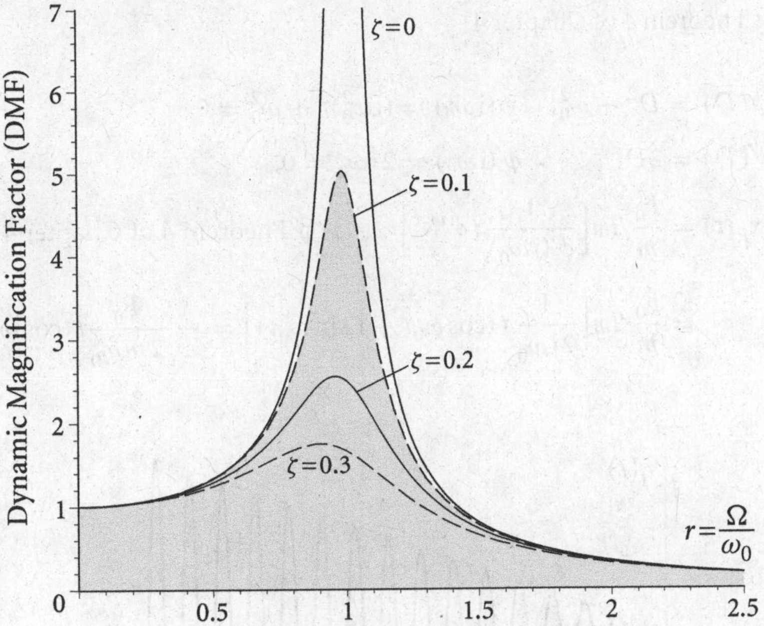

Figure 5.14 Dynamic magnification factor (DMF).

- Maximum value of DMF occurs when $\mathrm{d(DMF)}/\mathrm{d}r=0$, or

$$\frac{\mathrm{d(DMF^2)}}{\mathrm{d}(r^2)}=\frac{\mathrm{d}}{\mathrm{d}\rho}\left[\frac{1}{(1-\rho)^2+4\zeta^2\rho}\right]=-\frac{2(1-\rho)(-1)+4\zeta^2}{[(1-\rho)^2+4\zeta^2\rho]^2}=0,\ \rho=r^2,$$

$$\rho-1+2\zeta^2=0 \implies r^2=1-2\zeta^2,$$

$$\therefore\quad r=\sqrt{1-2\zeta^2}\approx 1-\zeta^2\approx 1,\quad \text{if } \zeta\ll 1.$$

The maximum value of DMF is approximately

$$\mathrm{DMF_{max}}\approx \mathrm{DMF}\Big|_{r=1}=\frac{1}{\sqrt{(1-r^2)^2+(2\zeta r)^2}}\Bigg|_{r=1}=\frac{1}{2\zeta}.$$

Hence, the smaller the damping coefficient, the larger the DMF values or the amplitudes of dynamic response. When $\zeta=0$ and $\Omega=\omega_0$, a response of unbound amplitude occurs and the system is in resonance.

Resonance

When $\zeta=0$ and $\Omega=\omega_0$, the equation of motion becomes

$$\ddot{x}_P+\omega_0^2\,x_P=\frac{F_0}{m}\sin\omega_0 t,$$

and the particular solution is given by

$$x_P(t)=\frac{F_0}{m}\frac{1}{D^2+\omega_0^2}\sin\omega_0 t=\frac{F_0}{m}\,\mathcal{Im}\left(\frac{1}{D^2+\omega_0^2}\,\mathrm{e}^{\mathrm{i}\,\omega_0 t}\right).$$

Applying Theorem 4 of Chapter 4,

$$\phi(D) = D^2 + \omega_0^2, \quad \phi(i\omega_0) = (i\omega_0^2)^2 + \omega_0^2 = 0,$$

$$\phi'(D) = 2D, \qquad \phi'(i\omega_0) = 2i\omega_0 \neq 0,$$

$$\therefore\ x_P(t) = \frac{F_0}{m}\,\mathcal{I}m\Big[\frac{1}{\phi'(i\omega_0)}\,t\,e^{i\omega_0 t}\Big] \qquad \text{Theorem 4 of Chapter 4}$$

$$= \frac{F_0}{m}\,\mathcal{I}m\Big[\frac{1}{2i\omega_0}\,t\,(\cos\omega_0 t + i\sin\omega_0 t)\Big] = -\frac{F_0}{2m\omega_0}\,t\cos\omega_0 t.$$

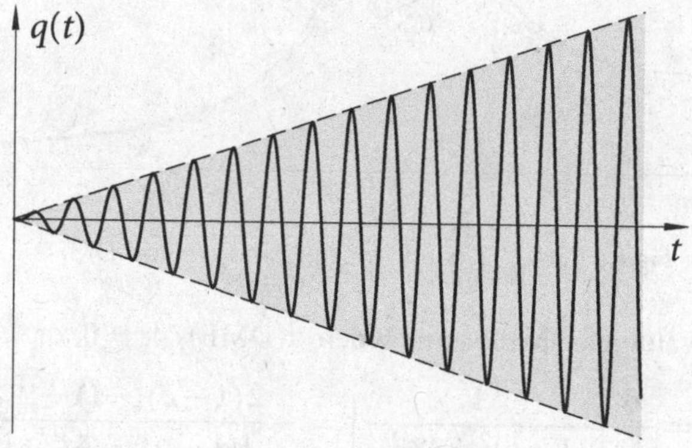

Figure 5.15 Response of a system in resonance.

To sketch the response, $F_0 t/(2m\omega_0)$, which is a straight line, is regarded as the amplitude of the response. This straight line and its mirror image $-F_0 t/(2m\omega_0)$ form the envelop of the sinusoidal function $-\cos\omega_0 t$. Fitting the sinusoidal function $-\cos\omega_0 t$ inside the envelop results in the response as shown in Figure 5.15.

Hence, when the system is undamped and the excitation frequency is equal to the natural frequency, the system is in resonance and the amplitude of the response of the system grows linearly with time.

Undamped System under Sinusoidal Excitation

For a *damped* system, the complementary solution or the response of free vibration (due to initial conditions) decays exponentially to zero; hence its effect diminishes when time increases. It is therefore important to focus on the particular solution or the response of forced vibration (due to externally applied forcing).

However, for an *undamped* system, the effect of response due to free vibration is significant. Furthermore, it interacts with the response due to forced vibration to produce a response that is different from either the complementary solution or the particular solution alone.

The equation of motion of an undamped system under sinusoidal excitation is

$$\ddot{x} + \omega_0^2 x = \frac{F_0}{m} \sin \Omega t,$$

and the complementary solution and a particular solution are

$$x_C(t) = A \cos \omega_0 t + B \sin \omega_0 t, \qquad x_P(t) = \frac{F_0}{m} \frac{\sin \Omega t}{\omega_0^2 - \Omega^2}.$$

Hence, the general solution is given by

$$x(t) = x_C(t) + x_P(t) = A \cos \omega_0 t + B \sin \omega_0 t + \frac{F_0}{m(\omega_0^2 - \Omega^2)} \sin \Omega t,$$

where the constants A and B are determined by the initial conditions $x(0) = x_0$ and $\dot{x}(0) = v_0$. Since

$$\dot{x}(t) = -A\omega_0 \sin \omega_0 t + B\omega_0 \cos \omega_0 t + \frac{F_0 \Omega}{m(\omega_0^2 - \Omega^2)} \cos \Omega t,$$

one has $x(0) = A = x_0$, and

$$\dot{x}(0) = B\omega_0 + \frac{F_0 \Omega}{m(\omega_0^2 - \Omega^2)} = v_0 \implies B = \frac{1}{\omega_0}\left[v_0 - \frac{F_0 \Omega}{m(\omega_0^2 - \Omega^2)}\right].$$

The response of the system is

$$\begin{aligned} x(t) &= x_0 \cos \omega_0 t + \frac{1}{\omega_0}\left[v_0 - \frac{F_0 \Omega}{m(\omega_0^2 - \Omega^2)}\right] \sin \omega_0 t + \frac{F_0}{m(\omega_0^2 - \Omega^2)} \sin \Omega t \\ &= x_0 \cos \omega_0 t + \frac{1}{\omega_0}\left[v_0 + \frac{F_0(\omega_0 - \Omega)}{m(\omega_0^2 - \Omega^2)}\right] \sin \omega_0 t - \frac{F_0(\sin \omega_0 t - \sin \Omega t)}{m(\omega_0^2 - \Omega^2)} \\ &= x_0 \cos \omega_0 t + \frac{1}{\omega_0}\left[v_0 + \frac{F_0}{m(\omega_0 + \Omega)}\right] \sin \omega_0 t - \underbrace{\frac{F_0(\sin \omega_0 t - \sin \Omega t)}{m(\omega_0^2 - \Omega^2)}}_{X(t)}, \end{aligned}$$

in which the first two terms are sinusoidal functions of frequency ω_0. The last term can be written as

$$X(t) = \frac{F_0(\sin \omega_0 t - \sin \Omega t)}{m(\omega_0^2 - \Omega^2)} = \underbrace{\frac{2F_0}{m(\omega_0^2 - \Omega^2)} \sin\left(\frac{\omega_0 - \Omega}{2} t\right)}_{a(t)} \cos\left(\frac{\omega_0 + \Omega}{2} t\right).$$

✍ $\sin A - \sin B = 2 \cos \frac{A+B}{2} \sin \frac{A-B}{2}$

It has been shown that the response due to forced vibration is of large amplitude when the excitation frequency is close to the natural frequency.

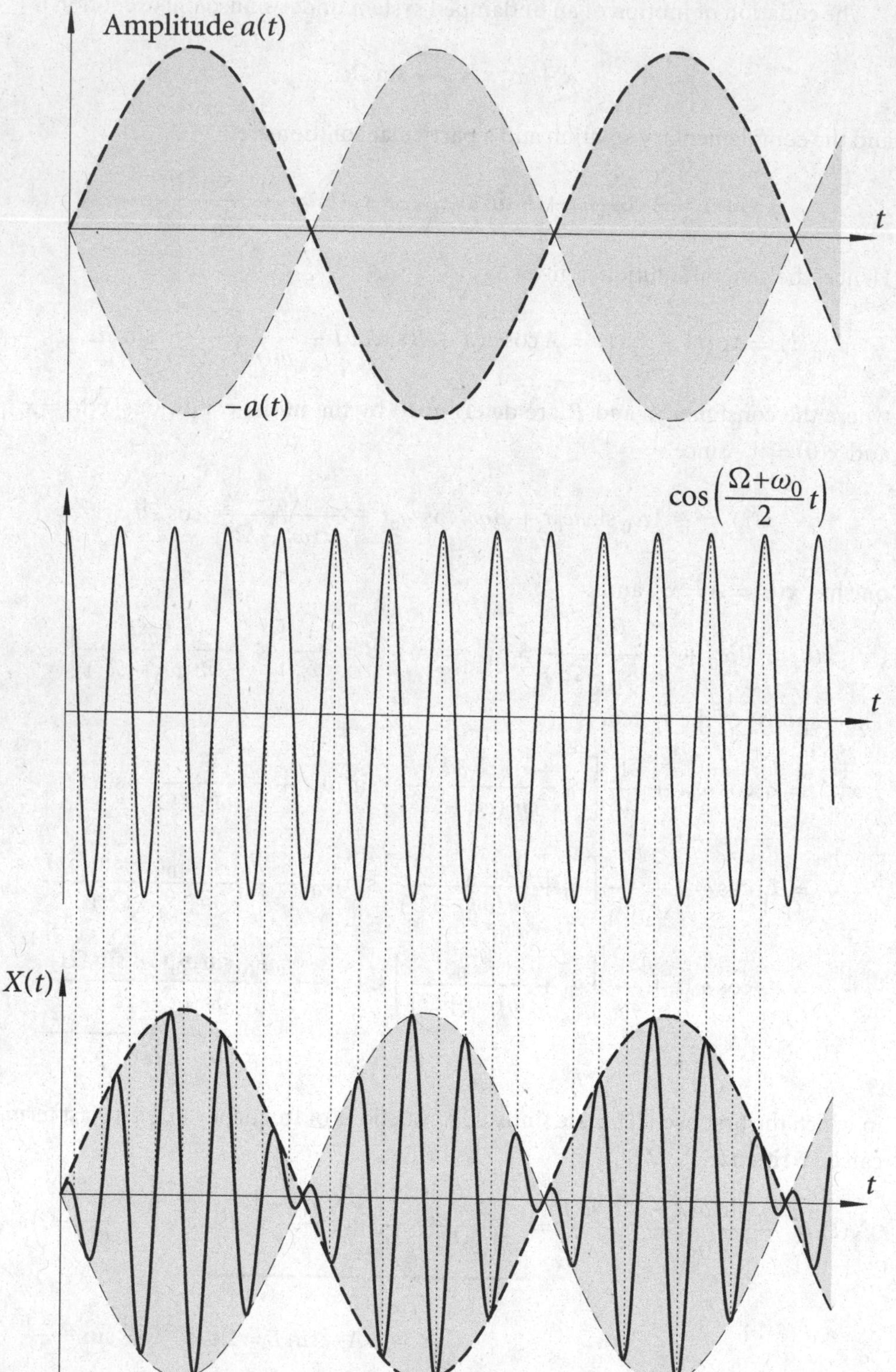

Figure 5.16 Plotting of $X(t)$.

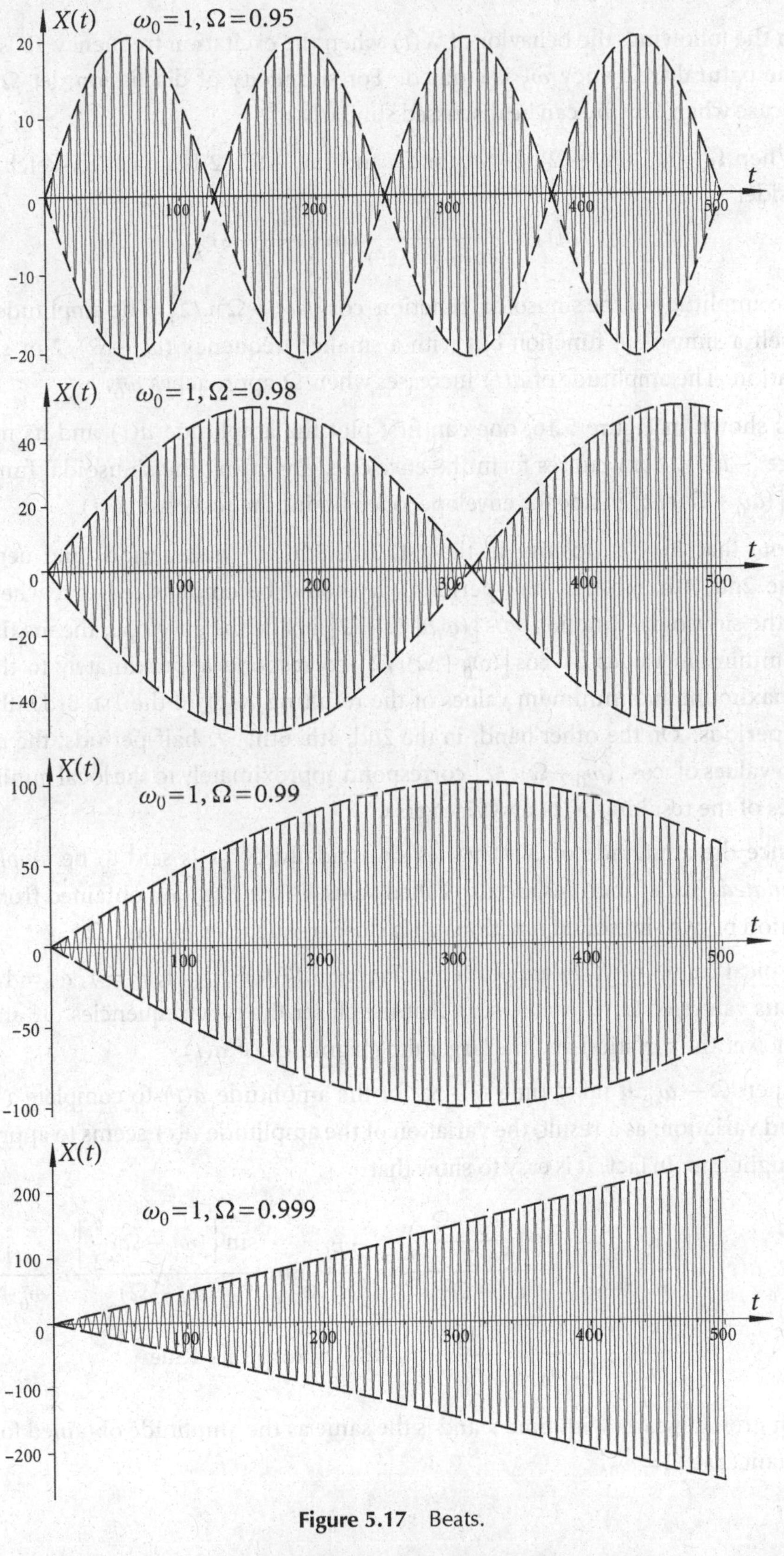

Figure 5.17 Beats.

In the following, the behavior of $X(t)$ when the excitation frequency Ω is close to the natural frequency ω_0 is studied. For simplicity of discussion, let $\Omega < \omega_0$. The case when $\Omega > \omega_0$ can be discussed similarly.

When $\Omega \approx \omega_0$, $(\omega_0+\Omega)/2 \approx \omega_0$ or Ω, and $(\omega_0-\Omega)/2$ is small. To sketch $X(t)$, consider

$$a(t) = \frac{2F_0}{m(\omega_0^2-\Omega^2)} \sin\left(\frac{\omega_0-\Omega}{2}t\right)$$

as the amplitude of the sinusoidal function $\cos\left[(\omega_0+\Omega)t/2\right]$. The amplitude $a(t)$ is itself a sinusoidal function but with a smaller frequency $(\omega_0-\Omega)/2$ or slower variation. The amplitude of $a(t)$ increases when Ω approaches ω_0.

As shown in Figure 5.16, one can first plot the amplitude $a(t)$ and its mirror image $-a(t)$; both curves form the envelope. By fitting the sinusoidal function $\cos\left[(\omega_0+\Omega)t/2\right]$ inside the envelope, one obtains the response $X(t)$.

Note that $a(t)$ is positive in the 1st, 3rd, 5th, ... half-periods and negative in the 2nd, 4th, 6th, ... half-periods. It should be emphasized that when fitting the sinusoidal function $\cos\left[(\omega_0+\Omega)t/2\right]$ inside the envelope, the maximum and minimum values of $\cos\left[(\omega_0+\Omega)t/2\right]$ correspond approximately to the local maximum and minimum values of the resulting $X(t)$ in the 1st, 3rd, 5th, ... half-periods. On the other hand, in the 2nd, 4th, 6th, ... half-periods, the maximum values of $\cos\left[(\omega_0+\Omega)t/2\right]$ correspond approximately to the local minimum values of the resulting $X(t)$ and vice versa.

Since the amplitude of $X(t)$ varies sinusoidally, $X(t)$ is said to be *amplitude modulated*. Such functions are also called *beats*. Beats $X(t)$ are obtained from the addition of two sinusoidal functions with close frequencies ω_0 and Ω.

Typical results of $X(t)$ are plotted in Figure 5.17 for $F_0=1$, $m=1$, $\omega_0=1$, and various values of Ω. It can be seen that the closer the two frequencies ω_0 and Ω, the slower the variation and the larger the magnitude of $a(t)$.

When $\Omega \to \omega_0$, it takes time $t \to \infty$ for the amplitude $a(t)$ to complete a half-period variation; as a result, the variation of the amplitude $a(t)$ seems to approach a straight line. In fact, it is easy to show that

$$\lim_{\Omega\to\omega_0} a(t) = \frac{2F_0}{m} \lim_{\Omega\to\omega_0} \frac{\sin\left(\dfrac{\omega_0-\Omega}{2}t\right)}{\omega_0^2-\Omega^2} = \frac{2F_0}{m} \lim_{\Omega\to\omega_0} \frac{\sin\left[(\omega_0-\Omega)\dfrac{t}{2}\right]}{(\omega_0-\Omega)} \cdot \frac{1}{(\omega_0+\Omega)}$$

$$= \frac{2F_0}{m} \cdot \frac{t}{2} \cdot \frac{1}{2\omega_0} = \frac{F_0}{2m\omega_0} t, \qquad \text{✍ L'Hospital's Rule}$$

which grows linearly with time t and is the same as the amplitude obtained for the resonant case $\Omega = \omega_0$.

5.2 Electric Circuits

Series RLC Circuit

A circuit consisting of a resistor R, an inductor L, a capacitor C, and a voltage source $V(t)$ connected in series, shown in Figure 5.18, is called the series RLC circuit. Applying Kirchhoff's Voltage Law, one has

$$-V(t) + Ri + L\frac{di}{dt} + \frac{1}{C}\int_{-\infty}^{t} i\,dt = 0.$$

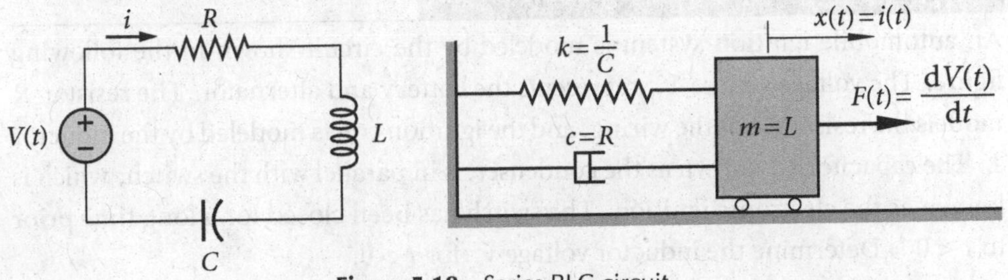

Figure 5.18 Series RLC circuit.

Differentiating with respect to t yields

$$L\frac{d^2i}{dt^2} + R\frac{di}{dt} + \frac{1}{C}i = \frac{dV(t)}{dt},$$

or, in the standard form,

$$\frac{d^2i}{dt^2} + 2\zeta\omega_0\frac{di}{dt} + \omega_0^2 i = \frac{1}{L}\frac{dV(t)}{dt}, \quad \omega_0^2 = \frac{1}{LC}, \quad \zeta\omega_0 = \frac{R}{2L}.$$

The series RLC circuit is equivalent to a mass-damper-spring system as shown.

Parallel RLC Circuit

A circuit consisting of a resistor R, an inductor L, a capacitor C, and a current source $I(t)$ connected in parallel, as shown in Figure 5.19, is called the parallel RLC circuit. Applying Kirchhoff's Current Law at node 1, one has

$$I(t) = C\frac{dv}{dt} + \frac{1}{L}\int_{-\infty}^{t} v\,dt + \frac{v}{R}.$$

Differentiating with respect to t yields

$$C\frac{d^2v}{dt^2} + \frac{1}{R}\frac{dv}{dt} + \frac{1}{L}v = \frac{dI(t)}{dt},$$

or, in the standard form,

$$\frac{d^2v}{dt^2} + 2\zeta\omega_0\frac{dv}{dt} + \omega_0^2 v = \frac{1}{C}\frac{dI(t)}{dt}, \quad \omega_0^2 = \frac{1}{LC}, \quad \zeta\omega_0 = \frac{1}{2RC}.$$

The parallel RLC circuit is equivalent to a mass-damper-spring system as shown.

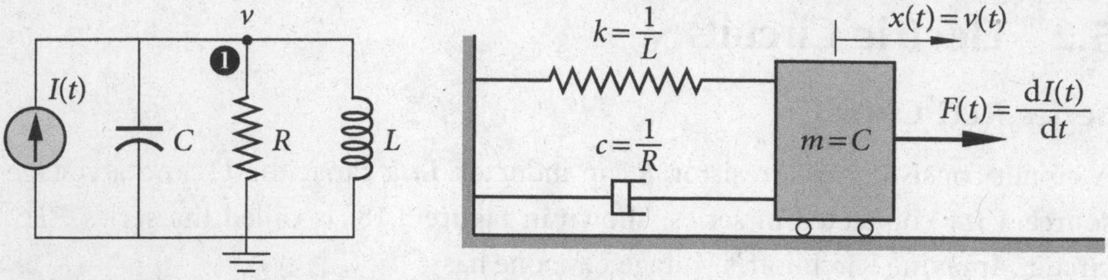

Figure 5.19 Parallel RLC circuit.

Example 5.1 — Automobile Ignition Circuit

An automobile ignition system is modeled by the circuit shown in the following figure. The voltage source V_0 represents the battery and alternator. The resistor R models the resistance of the wiring, and the ignition coil is modeled by the inductor L. The capacitor C, known as the condenser, is in parallel with the switch, which is known as the electronic ignition. The switch has been closed for a long time prior to $t<0^-$. Determine the inductor voltage v_L for $t>0$.

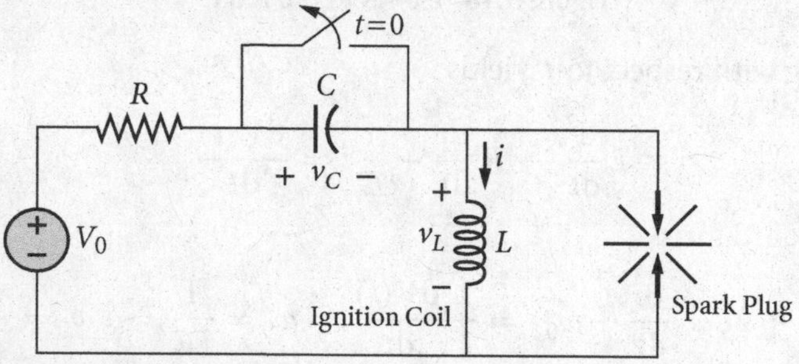

For $V_0=12$ V, $R=4\,\Omega$, $C=1\,\mu$F, $L=8$ mH, determine the maximal inductor voltage and the time when it is reached.

❧ For $t<0$, the switch is closed, the capacitor behaves as an open circuit and the inductor behaves as a short circuit as shown. Hence $i(0^-)=V_0/R$, $v_C(0^-)=0$.

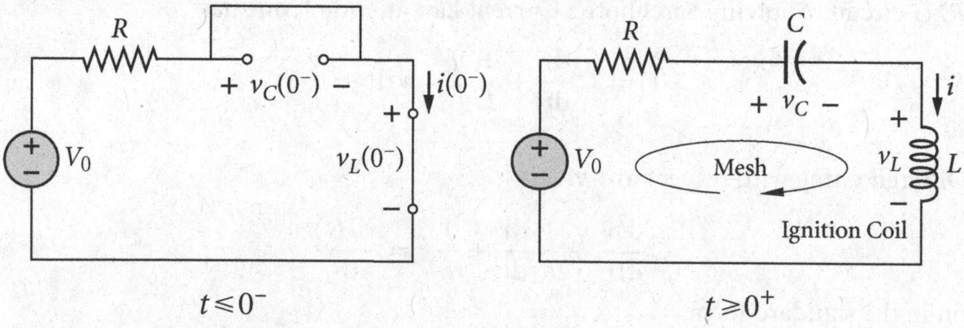

❧ At $t=0$, the switch is opened. Since the current in an inductor and the voltage across a capacitor cannot change abruptly, one has $i(0^+)=i(0^-)=V_0/R$, $v_C(0^+)=v_C(0^-)=0$. The derivative $i'(0^+)$ is obtained from $v_L(0^+)$, which is determined by applying Kirchhoff's Voltage Law to the mesh at $t=0^+$:

$$-V_0+Ri(0^+)+v_C(0^+)+v_L(0^+)=0 \implies v_L(0^+)=V_0-Ri(0^+)=0,$$

$$v_L(0^+)=L\frac{di(0^+)}{dt} \implies i'(0^+)=\frac{v_L(0^+)}{L}=0.$$

✍ A mesh is a loop which does not contain any other loops within it.

❧ For $t>0$, applying Kirchhoff's Voltage Law to the mesh leads to

$$-V_0+Ri+\frac{1}{C}\int_{-\infty}^{t} i\,dt+L\frac{di}{dt}=0.$$

Differentiating with respect to t yields

$$R\frac{di}{dt}+\frac{i}{C}+L\frac{d^2i}{dt^2}=0 \implies \frac{d^2i}{dt^2}+2\zeta\omega_0\frac{di}{dt}+\omega_0^2 i=0,\ \omega_0^2=\frac{1}{LC},\ \zeta\omega_0=\frac{R}{2L}.$$

If $\zeta<1$, the system is underdamped and the solution of the differential equation is

$$i(t)=e^{-\zeta\omega t}\left[i(0^+)\cos\omega_d t+\frac{i'(0^+)+\zeta\omega_0 i(0^+)}{\omega_d}\sin\omega_d\right]$$

$$=i(0^+)e^{-\zeta\omega_0 t}\left(\cos\omega_d t+\frac{\zeta\omega_0}{\omega_d}\sin\omega_d t\right),$$

where $\omega_d=\omega_0\sqrt{1-\zeta^2}$ is the damped natural frequency.

The voltage across the inductor is

$$v_L(t)=L\frac{di(t)}{dt}=-Li(0^+)\frac{\omega_0^2}{\omega_d}e^{-\zeta\omega_0 t}\sin\omega_d t.$$

For $V_0=12$ V, $R=4\,\Omega$, $C=1\,\mu\text{F}=1\times10^{-6}$ F, $L=8\,\text{mH}=8\times10^{-3}$ H,

$$\omega_0=\frac{1}{\sqrt{LC}}=\frac{1}{\sqrt{1\times10^{-6}\times8\times10^{-3}}}=1.118\times10^4,$$

$$\zeta\omega_0=\frac{R}{2L}=\frac{4}{2\times8\times10^{-3}}=250,\quad \zeta=\frac{250}{\omega_0}=0.02236,$$

$$\omega_d=\omega_0\sqrt{1-\zeta^2}\approx\omega_0=1.118\times10^4,\quad i(0^+)=\frac{V_0}{R}=\frac{12}{4}=3,$$

$$\therefore\quad v_L=-8\times10^{-3}\times3\times1.118\times10^4 e^{-250t}\sin(1.118\times10^4 t)$$

$$=-268.32e^{-250t}\sin(1.118\times10^4 t),$$

which is maximum when $1.118\times10^4 t=\pi/2$ or $t=1.405\times10^{-4}$ sec $=140.5\,\mu$s and is given by

$$v_{L,\max}(t)=-268.32e^{-250\times1.405\times10^{-4}}=-259\,\text{V}.$$

A device known as a transformer is then used to step up the inductor voltage to the range of 6000 to 10,000 V required to fire the spark plug in a typical automobile.

Example 5.2 — Second-Order Circuit

For the electric circuit shown in the following figure, derive the differential equation governing $i(t)$ for $t>0$.

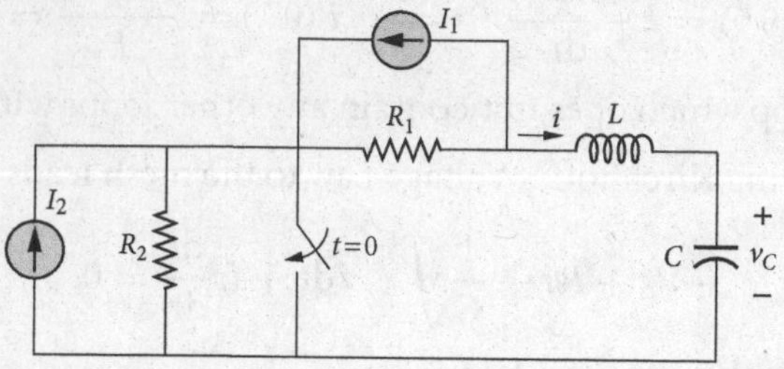

For $R_1=6\,\Omega$, $R_2=2\,\Omega$, $C=0.04\,$F, $L=1\,$H, $I_1=I_2=2\,$A, find $i(t)$ and $v_C(t)$.

❧ For $t<0$, the switch is open. The inductor behaves as a short circuit and the capacitor behaves as an open circuit. Hence, $i(0^-)=0$, $i_1(0^-)=-I_1$, $i_2(0^-)=I_2$, and

$$v_{R2}(0^-)=v_{R1}(0^-)+v_C(0^-) \implies v_C(0^-)=R_2I_2-R_1I_1.$$

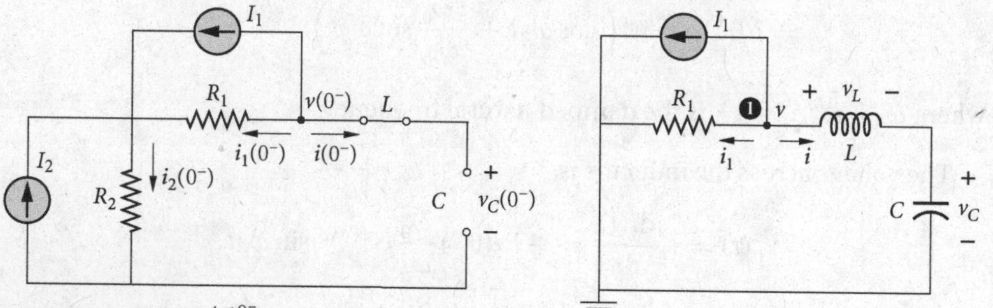

$t\leq 0^-$ $t\geq 0^+$

❧ For $t=0$, the switch is closed, the current source I_2 and the resistor R_2 are short-circuited. The circuit becomes as shown in the figure. Since the current in an inductor cannot change abruptly, $i(0^+)=i(0^-)=0$. Since the voltage across a capacitor cannot change abruptly, $v_C(0^-)=v_C(0^+)=R_2I_2-R_1I_1$.

Note that $v(0^+)=v_L(0^+)+v_C(0^+)=v_L(0^+)+R_2I_2-R_1I_1$,

$$i_1(0^+)=\frac{v(0^+)}{R_1}=\frac{v_L(0^+)+R_2I_2-R_1I_1}{R_1}.$$

Applying Kirchhoff's Current Law at node 1 yields

$$I_1+i(0^+)+i_1(0^+)=0 \implies I_1+0+\frac{v_L(0^+)+R_2I_2-R_1I_1}{R_1}=0,$$

$$\therefore\quad V_0(0^+)=-R_2I_2=L\frac{di(0^+)}{dt} \implies i'(0^+)=-\frac{R_2I_2}{L}.$$

❧ For $t > 0$,

$$i_1 = \frac{V}{R_1}, \quad v = v_L + v_C = L\frac{di}{dt} + \frac{1}{C}\int_{-\infty}^{t} i\,dt.$$

Applying Kirchhoff's Current Law at node 1 leads to

$$I_1 + i + i_1 = 0 \implies I_1 + i + \frac{1}{R_1}\left(L\frac{di}{dt} + \frac{1}{C}\int_{-\infty}^{t} i\,dt\right) = 0.$$

Differentiating with respect to t yields

$$\frac{d^2 i}{dt^2} + \frac{R_1}{L}\frac{di}{dt} + \frac{1}{CL}i = 0, \qquad i(0^+) = 0, \quad i'(0^+) = -\frac{R_2 I_2}{L}.$$

For $R_1 = 6\,\Omega$, $R_2 = 2\,\Omega$, $C = 0.04\,\text{F}$, $L = 1\,\text{H}$, $I_1 = I_2 = 2\,\text{A}$, the equation becomes

$$\frac{d^2 i}{dt^2} + 6\frac{di}{dt} + 25i = 0, \qquad i(0^+) = 0, \quad i'(0^+) = -4,$$

$$\omega_0 = 5, \quad 2\zeta\omega_0 = 6 \implies \zeta\omega_0 = 3, \ \zeta = \frac{3}{5} < 1, \quad \omega_d = \omega\sqrt{1-\zeta^2} = 4.$$

The system is underdamped and the solution is given by

$$i(t) = e^{-\zeta\omega_0 t}\left[i(0^+)\cos\omega_d t + \frac{i'(0^+) + \zeta\omega_0 i(0^+)}{\omega_d}\sin\omega_d t\right] = -e^{-3t}\sin 4t.$$

The voltage across the capacitor is, with $v_C(0^+) = -8$,

$$\begin{aligned} v_C &= \frac{1}{C}\int_0^t i(t)\,dt + v_C(0^+) = 25\int_0^t -e^{-3t}\sin 4t - 8 \\ &= -12 + e^{-3t}(3\sin 4t + 4\cos 4t)\ (\text{V}). \end{aligned}$$

Remarks: Circuits consisting of resistors and the equivalent of two energy storage elements, such as capacitors and inductors, are called second-order circuits, because they are characterized by second-order linear ordinary differential equations. The governing equations are of the same form as that of a single degree-of-freedom system. As a result, the solutions and the behavior of second-order circuits are the same as those of a single degree-of-freedom system.

5.3 Vibration of a Vehicle Passing a Speed Bump

As an application of single degree-of-freedom system, the vibration of a vehicle passing a speed bump is studied in this section.

The vehicle is modeled by a damped single degree-of-freedom system with mass m, spring stiffness k, and damping coefficient c as shown in Figure 5.20. The vehicle has been moving at a constant speed U on a smooth surface.

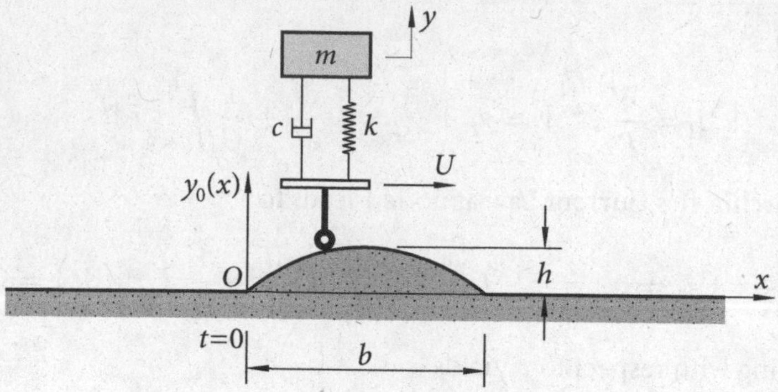

Figure 5.20 A vehicle passing a speed bump.

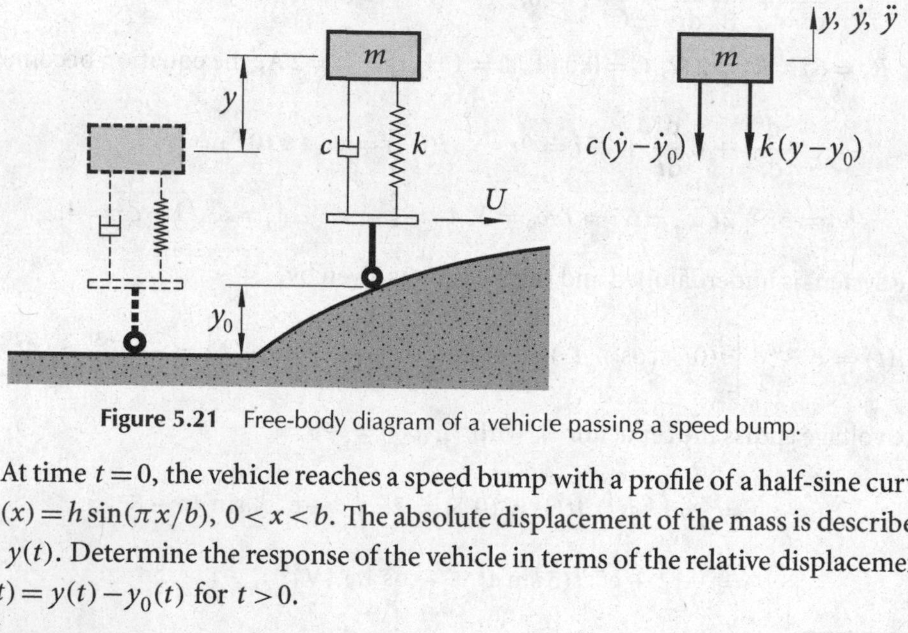

Figure 5.21 Free-body diagram of a vehicle passing a speed bump.

At time $t=0$, the vehicle reaches a speed bump with a profile of a half-sine curve $y_0(x)=h\sin(\pi x/b)$, $0<x<b$. The absolute displacement of the mass is described by $y(t)$. Determine the response of the vehicle in terms of the relative displacement $z(t)=y(t)-y_0(t)$ for $t>0$.

To set up the equation of motion of the vehicle, consider the free-body diagram of the mass m as shown in Figure 5.21. Newton's Second Law requires that

$$\uparrow ma=\sum F: \quad m\ddot{y}=-c(\dot{y}-\dot{y}_0)-k(y-y_0).$$

Letting the relative displacement of the mass be $z(t)=y(t)-y_0(t)$, the equation of motion becomes

$$m(\ddot{z}+\ddot{y}_0)=-c\dot{z}-kz \implies m\ddot{z}+c\dot{z}+kz=-m\ddot{y}_0. \tag{1}$$

Phase 1: On the Speed Bump $0\leqslant t\leqslant T,\ \ T=b/U$

Since the vehicle is moving at the constant speed U, one has $x=Ut$. The speed bump is of the half-sine shape

$$y_0(x)=h\sin\Big(\frac{\pi}{b}x\Big),\quad 0\leqslant x\leqslant b \implies y_0(t)=h\sin\Big(\frac{\pi U}{b}t\Big),\quad 0\leqslant t\leqslant T.$$

The equation of motion can be written as

$$m\ddot{z}+c\dot{z}+kz = mh\Omega^2 \sin\Omega t \implies (D^2+2\zeta\omega_0 D+\omega_0^2)z = h\Omega^2 \sin\Omega t, \quad (2)$$

where

$$\omega_0=\sqrt{\frac{k}{m}}, \qquad 2\zeta\omega_0 = \frac{c}{m}, \qquad \Omega = \frac{\pi U}{b}.$$

The characteristic equation is $\lambda^2+2\zeta\omega_0\lambda+\omega_0^2=0$, which gives, assuming $\zeta<1$,

$$\lambda = -\zeta\omega_0 \pm \omega_0\sqrt{\zeta^2-1} = -\zeta\omega_0 \pm \mathrm{i}\omega_d, \quad \omega_d = \omega_0\sqrt{1-\zeta^2}.$$

The complementary solution is

$$z_C(t) = \mathrm{e}^{-\zeta\omega_0 t}(A_1\cos\omega_d t + B_1 \sin\omega_d t).$$

A particular solution can be obtained using the D-operator method

$$z_P(t) = \frac{1}{D^2+2\zeta\omega_0 D+\omega_0^2}(h\Omega^2\sin\Omega t)$$

$$= h\Omega^2\frac{1}{-\Omega^2+2\zeta\omega_0 D+\omega_0^2}\sin\Omega t$$

✍ Theorem 3 of Chapter 4: replace D^2 by $-\Omega^2$.

$$= h\Omega^2\frac{(\omega_0^2-\Omega^2)-2\zeta\omega_0 D}{(\omega_0^2-\Omega^2)^2-(2\zeta\omega_0 D)^2}\sin\Omega t$$

$$= \frac{h\Omega^2}{(\omega_0^2-\Omega^2)^2+(2\zeta\omega_0\Omega)^2}\big[(\omega_0^2-\Omega^2)\sin\Omega t - 2\zeta\omega_0\Omega\cos\Omega t\big]$$

✍ Replace D^2 by $-\Omega^2$ in denominator and evaluate numerator.

$$= a\sin(\Omega t-\varphi),$$

where

$$a = \frac{h\Omega^2}{\sqrt{(\omega_0^2-\Omega^2)^2+(2\zeta\omega_0\Omega)^2}}, \qquad \varphi = \tan^{-1}\frac{2\zeta\omega_0\Omega}{\omega_0^2-\Omega^2}.$$

✍ Use $A\sin\theta - B\cos\theta = \sqrt{A^2+B^2}\sin(\theta-\varphi)$, $\varphi=\tan^{-1}(B/A)$.

The general solution is, for $0\leqslant t\leqslant T$,

$$z(t) = z_C(t)+z_P(t) = \mathrm{e}^{-\zeta\omega_0 t}(A_1\cos\omega_d t + B_1\sin\omega_d t) + a\sin(\Omega t-\varphi), \quad (3)$$

where the constants are determined by the initial conditions $z(0)$ and $\dot{z}(0)$. Since the vehicle has been traveling on a smooth surface, $y(0)=0$ and $\dot{y}(0)=0$; hence $z(0)=0$ and $\dot{z}(0)=0$. Since

$$\dot{z}(t) = -\zeta\omega_0\mathrm{e}^{-\zeta\omega_0 t}(A_1\cos\omega_d t + B_1\sin\omega_d t)$$
$$+\mathrm{e}^{-\zeta\omega_0 t}(-A_1\omega_d\sin\omega_d t + B_1\omega_d\cos\omega_d t) + a\Omega\cos(\Omega t-\varphi), \quad (4)$$

one has

$$z(0) = A_1 + a\sin(-\varphi) = 0 \implies A_1 = a\sin\varphi,$$

$$\dot{z}(0) = -\zeta\omega_0 A_1 + B_1\omega_d + a\Omega\cos(-\varphi) = 0$$

$$\implies B_1 = \frac{1}{\omega_d}(\zeta\omega_0 A - a\Omega\cos\varphi) = \frac{a}{\omega_d}(\zeta\omega_0\sin\varphi - \Omega\cos\varphi).$$

At time $t = T$, $T = b/U$, the relative displacement and velocity of the vehicle are $z(T)$ and $\dot{z}(T)$, which can be obtained by evaluating equations (3) and (4) at $t = T$.

Phase 2: Passed the Speed Bump $t \geqslant T,\ T = b/U$

For time $t \geqslant T$, $T = b/U$, the vehicle has gone over the speed bump and the surface is smooth again with $y_0(t) = 0$. The equation of motion becomes

$$m\ddot{z} + c\dot{z} + kz = 0 \implies (D^2 + 2\zeta\omega_0 D + \omega_0^2)z = 0, \quad t \geqslant T. \tag{5}$$

The vehicle is not forced and the response is due to free vibration with initial conditions at time $t = T$.

The complementary solution, which is also the general solution, is

$$z(t) = e^{-\zeta\omega_0 t}(A_2\cos\omega_d t + B_2\sin\omega_d t), \qquad t \geqslant T,$$

$$\dot{z}(t) = -\zeta\omega_0 e^{-\zeta\omega_0 t}(A_2\cos\omega_d t + B_2\sin\omega_d t)$$
$$+ e^{-\zeta\omega_0 t}(-A_2\omega_d\sin\omega_d t + B_2\omega_d\cos\omega_d t),$$

where constants A_2 and B_2 are determined from the initial conditions $z(T)$ and $\dot{z}(T)$ obtained in Phase 1:

$$z(T) = e^{-\zeta\omega_0 T}\big[A_2\cos(\omega_d T) + B_2\sin(\omega_d T)\big],$$

$$\dot{z}(T) = e^{-\zeta\omega_0 T}\big\{\big[-\zeta\omega_0\cos(\omega_d T) - \omega_d\sin(\omega_d T)\big]A_2$$
$$+\big[-\zeta\omega_0\sin(\omega_d T) + \omega_d\cos(\omega_d T)\big]B_2\big\},$$

or

$$\alpha_{11}A_2 + \alpha_{12}B_2 = \beta_1, \qquad \alpha_{21}A_2 + \alpha_{22}B_2 = \beta_2,$$

where

$$\alpha_{11} = \cos(\omega_d T), \qquad \alpha_{21} = -\zeta\omega_0\cos(\omega_d T) - \omega_d\sin(\omega_d T),$$

$$\alpha_{12} = \sin(\omega_d T), \qquad \alpha_{22} = -\zeta\omega_0\sin(\omega_d T) + \omega_d\cos(\omega_d T),$$

$$\beta_1 = e^{\zeta\omega_0 T}z(T), \qquad \beta_2 = e^{\zeta\omega_0 T}\dot{z}(T).$$

Using Gaussian elimination or Cramer's Rule, it is easy to solve for A_2 and B_2 as

$$A_2 = \frac{\alpha_{22}\beta_1 - \alpha_{12}\beta_2}{\alpha_{11}\alpha_{22} - \alpha_{21}\alpha_{12}}, \qquad B_2 = \frac{\alpha_{11}\beta_2 - \alpha_{21}\beta_1}{\alpha_{11}\alpha_{22} - \alpha_{21}\alpha_{12}}.$$

✍ See page 307 for a brief review on Cramer's Rule.

Numerical Results

As a numerical example, use the following parameters

$$f_0 = 3\text{ Hz} \implies \omega_0 = 2\pi f_0 = 6\pi \text{ rad/sec}, \quad \zeta = 0.1,$$

$$U = 1.8\text{ km/hr} = 0.5\text{ m/sec}, \quad b = 0.5\text{ m}, \quad h = 0.1\text{ m}, \quad T = b/U = 1\text{ sec}.$$

When $0 \leqslant t \leqslant 1$ sec, the vehicle is on the speed bump; whereas when $t \geqslant 1$ sec, the vehicle has passed over the speed bump. It is easy to evaluate that

$$\omega_d = 18.755\text{ rad/sec}, \quad a = 0.0028555\text{ m}, \quad \varphi = 0.034272\text{ rad},$$

$$A_1 = 0.000097844\text{ m}, \quad B_1 = -0.00046819\text{ m},$$

$$\therefore \quad z(t) = \mathrm{e}^{-1.8850t}\big[0.000097844\cos(18.755t) - 0.00046819\sin(18.755t)\big]$$
$$+\, 0.0028555\sin(3.1416t - 0.0343) \text{ (m)}, \quad 0 \leqslant t \leqslant 1.$$

At $t = 1$ sec, $z(1) = 0.00011934$ m, $\dot{z}(1) = -0.010307$ m/sec. In the second phase, $t \geqslant 1$,

$$\alpha_{11} = 0.99554, \quad \alpha_{12} = -0.094344,$$

$$\alpha_{21} = -0.10712 \text{ rad/sec}, \quad \alpha_{12} = 18.849 \text{ rad/sec},$$

$$\beta_1 = 0.00078599\text{ m}, \quad \beta_2 = -0.067883\text{ m/sec},$$

$$A_2 = 0.00044846\text{ m}, \quad B_2 = -0.0035988\text{ m},$$

$$\therefore \quad z(t) = \mathrm{e}^{-1.8850t}\big[0.00044846\cos(18.755t) - 0.0035988\sin(18.755t)\big] \text{ (m)}.$$

The relative displacement response $z(t)$ is shown in Figure 5.22 for $\zeta = 0.1$ and 0.01. It can be seen that the response decays rapidly for large values of ζ.

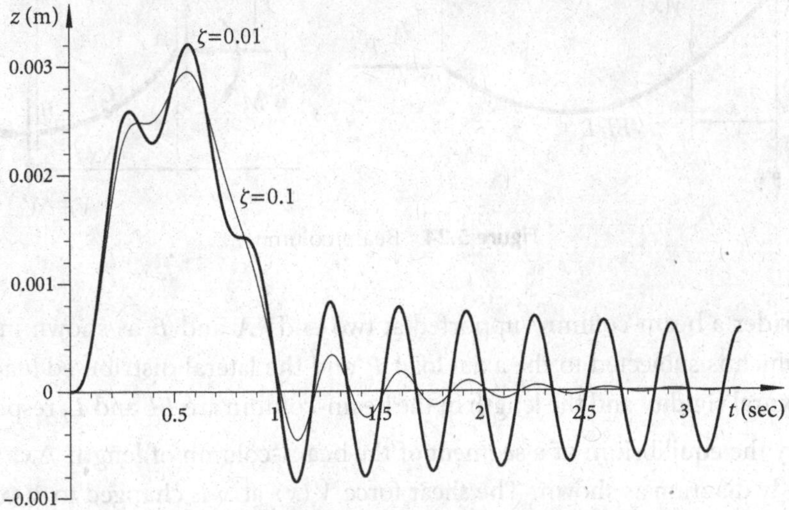

Figure 5.22 Response of a vehicle passing a speed bump.

5.4 Beam-Columns

A beam-column is a structural member subjected simultaneously to axial load and bending moments produced by lateral forces or eccentricity of the axial load. Beam-columns are found in many engineering structures. For example, a water tower as shown in Figure 5.23 is a beam-column; the supporting column is subjected to the axial load due to the weight of the water tank and the lateral load due to wind.

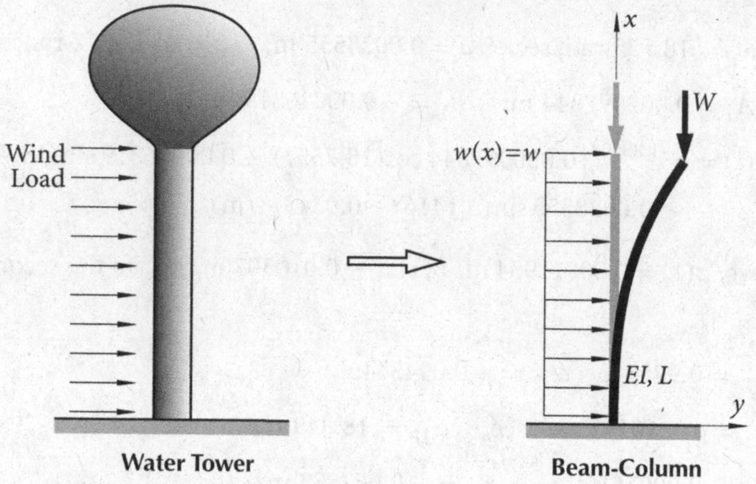

Figure 5.23 Water tower.

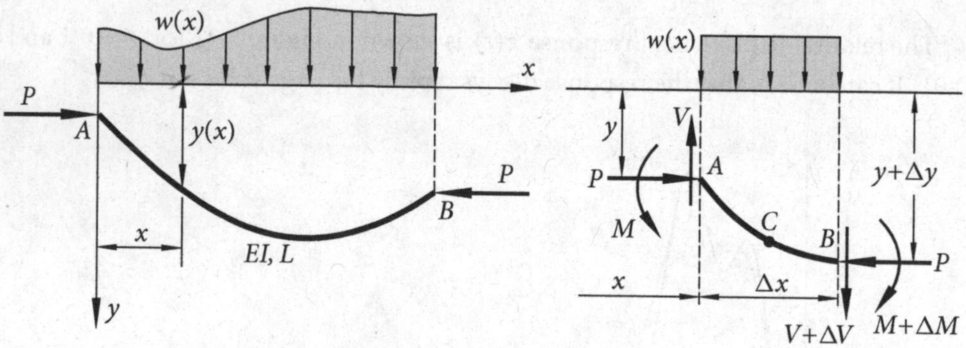

Figure 5.24 Beam-column.

Consider a beam-column supported at two ends A and B as shown in Figure 5.24, which is subjected to the axial load P and the lateral distributed load $w(x)$. The flexural rigidity and the length of the beam-column are EI and L, respectively.

Study the equilibrium of a segment of the beam-column of length Δx, with its free-body diagram as shown. The shear force $V(x)$ at x is changed to $V(x)+\Delta V$ at $x+\Delta x$; the bending moment $M(x)$ at x is changed to $M(x)+\Delta M$ at $x+\Delta x$.

For small Δx, the distributed load is approximated by the uniformly distributed load of intensity $w(x)$ over the beam segment.

Summing up the forces in the y direction and taking the limit as $\Delta x \to 0$ yield

$$\downarrow \sum F_y = 0: \quad [V(x)+\Delta V] - V(x) + w(x)\,\Delta x = 0 \implies \frac{dV}{dx} = -w(x).$$

Summing up the moments about point C, which is the midpoint of the beam segment, gives

$$\curvearrowright \sum M_C = 0: \quad [M(x)+\Delta M] - M(x) + V(x)\cdot\frac{\Delta x}{2} + [V(x)+\Delta V]\cdot\frac{\Delta x}{2}$$

$$+ P\cdot\Delta y = 0 \implies \frac{dM}{dx} + P\frac{dy}{dx} + V = 0.$$

Table 5.1 Boundary conditions.

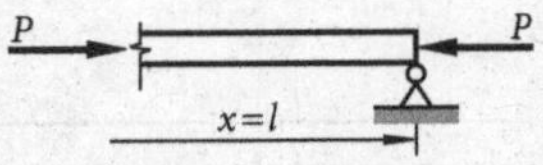	**Pinned End:** Deflection $= 0 \implies y(l) = 0$ Moment $= 0 \implies y''(l) = 0$
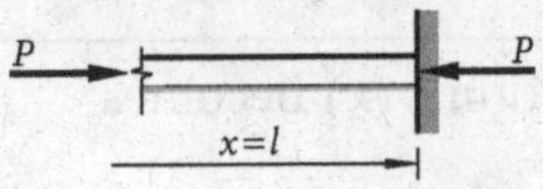	**Clamped End:** Deflection $= 0 \implies y(l) = 0$ Slope $= 0 \implies y'(l) = 0$
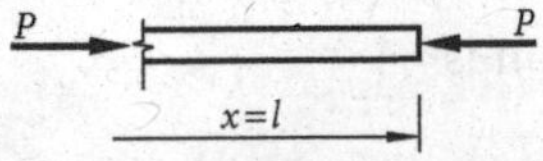	**Free End:** Moment $= 0 \implies y''(l) = 0$ Shear Force $= 0 \implies y'''(l) + \alpha^2 y'(l) = 0$
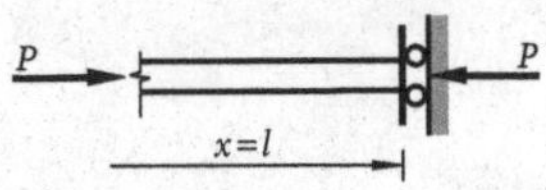	**Sliding End:** Slope $= 0 \implies y'(l) = 0$ Shear Force $= 0 \implies y'''(l) + \alpha^2 y'(l) = 0$ $\implies y'''(l) = 0$

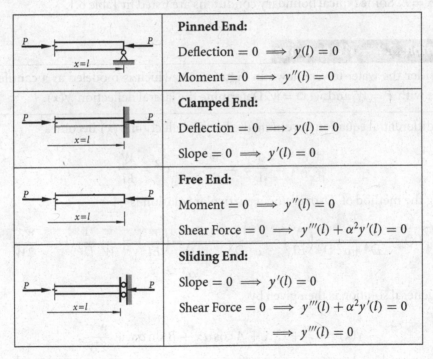

Eliminating V between these two equations leads to

$$\frac{d^2M}{dx^2} + P\frac{d^2y}{dx^2} = w(x).$$

Using the moment-curvature relationship

$$M(x) = EI\,y''(x)$$

results in a fourth-order linear ordinary differential equation governing the lateral deflection $y(x)$

$$\frac{d^4y}{dx^4}+\alpha^2\frac{d^2y}{dx^2}=\frac{w(x)}{EI}, \qquad \alpha^2=\frac{P}{EI}.$$

Note that the shear force becomes

$$V(x)=-\frac{dM}{dx}-P\frac{dy}{dx}=-EI\,y'''(x)-P\,y'(x)=-EI\big[y'''(x)+\alpha^2 y'(x)\big].$$

The characteristic equation is $\lambda^4+\alpha^2\lambda^2=0 \implies \lambda=0, 0, \pm i\alpha$. Hence, the general solution is

$$y(x)=C_0+C_1x+A\cos\alpha x+B\sin\alpha x+y_P(x),$$

where $y_P(x)$ is a particular solution which depends on $w(x)$. The constants A, B, C_0, and C_1 are determined using the boundary conditions at ends $x=0$ and $x=L$. Some typical boundary conditions are listed in Table 6.1.

Example 5.3 — Water Tower

Consider the water tower shown in Figure 5.23, which is modeled as a cantilever beam with $P=W$ and $w(x)=w$. Determine the lateral deflection $y(x)$.

The differential equation governing the lateral deflection $y(x)$ becomes

$$\frac{d^4y}{dx^4}+\alpha^2\frac{d^2y}{dx^2}=\frac{w}{EI}, \qquad \alpha^2=\frac{W}{EI}.$$

Using the method of D-operator, a particular solution is

$$y_P(x)=\frac{1}{D^4+\alpha^2D^2}\left(\frac{w}{EI}\right)=\frac{1}{\alpha^2D^2}\frac{1}{1+\dfrac{D^2}{\alpha^2}}\left(\frac{w}{EI}\right)=\frac{w}{W}\frac{1}{D^2}(1)=\frac{wx^2}{2W}.$$

The general solution is then given by

$$y(x)=C_0+C_1x+A\cos\alpha x+B\sin\alpha x+\frac{wx^2}{2W},$$

where A, B, C_0, and C_1 are determined from the boundary conditions:

$$x=0 \text{ (clamped)}: \qquad y(0)=0, \qquad y'(0)=0,$$

$$x=L \text{ (free)}: \qquad y''(L)=0, \qquad y'''(L)+\alpha^2y'(L)=0.$$

Since

$$y'(x)=C_1-A\alpha\sin\alpha x+B\alpha\cos\alpha x+\frac{wx}{W},$$

$$y''(x) = -A\alpha^2 \cos\alpha x - B\alpha^2 \sin\alpha x + \frac{w}{W},$$

$$y'''(x) = A\alpha^3 \sin\alpha x - B\alpha^3 \cos\alpha x,$$

one has

$$y'''(L) + \alpha^2 y'(L) = 0:$$

$$(A\alpha^3 \sin\alpha x - B\alpha^3 \cos\alpha x) + \alpha^2\Big(C_1 - A\alpha \sin\alpha x + B\alpha \cos\alpha x + \frac{wx}{W}\Big) = 0,$$

$$\therefore \quad C_1 = -\frac{wL}{W},$$

$$y'(0) = C_1 + B\alpha = 0 \implies B = -\frac{C_1}{\alpha} = \frac{wL}{\alpha W},$$

$$y''(L) = -A\alpha^2 \cos\alpha L - B\alpha^2 \sin\alpha L + \frac{w}{W} = 0 \implies A = \frac{w(1-\alpha L \sin\alpha L)}{\alpha^2 W \cos\alpha L},$$

$$y(0) = C_0 + A = 0 \implies C_0 = -A = -\frac{w(1-\alpha L \sin\alpha L)}{\alpha^2 W \cos\alpha L}.$$

Hence, the general solution becomes

$$y(x) = \frac{w(1-\alpha L \sin\alpha L)}{\alpha^2 W \cos\alpha L}(\cos\alpha x - 1) + \frac{wL}{\alpha W}(\sin\alpha x - \alpha x) + \frac{wx^2}{2W}.$$

Remarks: When the lateral load $w(x)$ includes distributed load over only a portion of the beam-column or concentrated loads, it can be better expressed using the Heaviside step function or the Dirac delta function. The differential equation can be easily solved using the Laplace transform as illustrated in Section 6.6.2.

Example 5.4 — Buckling of a Column

Consider the simply supported column AB subjected to axial compressive load. A rotational spring of stiffness κ provides resistance to rotation of end B. Set up the buckling equation for the column.

Since the lateral load $w(x) = 0$, the differential equation governing the lateral deflection $y(x)$ becomes

$$\frac{d^4 y}{dx^4} + \alpha^2 \frac{d^2 y}{dx^2} = 0, \qquad \alpha^2 = \frac{P}{EI}.$$

The solution is

$$y(x) = C_0 + C_1 x + A \cos\alpha x + B \sin\alpha x,$$

where A, B, C_0, and C_1 are determined from the boundary conditions:

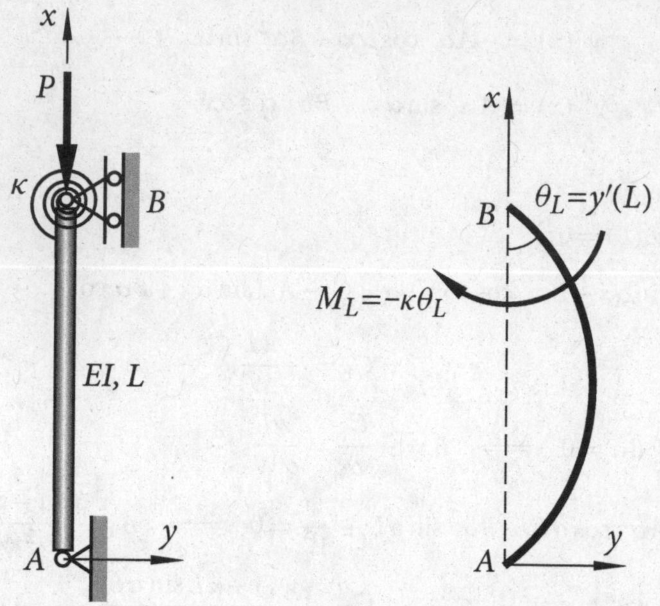

$$\text{support } A,\quad x=0:\qquad y(0)=0,\qquad y''(0)=0,$$

$$\text{support } B,\quad x=L:\qquad y(L)=0,\qquad M(L)=EI\,y''(L)=-\kappa\,y'(L).$$

Since

$$y'(x) = C_1 - A\alpha\sin\alpha x + B\alpha\cos\alpha x,$$
$$y''(x) = -A\alpha^2\cos\alpha x - B\alpha^2\sin\alpha x,$$

one has

$$y''(0) = -A\alpha^2 = 0 \quad\Longrightarrow\quad A=0,$$

$$y(0) = C_0 + A = 0 \quad\Longrightarrow\quad C_0 = -A = 0,$$

$$y(L) = C_1 L + B\sin\alpha L = 0 \quad\Longrightarrow\quad C_1 = -\frac{\sin\alpha L}{L}B,$$

$$EI\,y''(L) = -\kappa\,y'(L) \quad\Longrightarrow\quad EI(-B\alpha^2\sin\alpha L) = -\kappa\,(C_1 + B\alpha\cos\alpha L),$$

$$\left\{\kappa\left(-\frac{\sin\alpha L}{L} + \alpha\cos\alpha L\right) - EI\,\alpha^2\sin\alpha L\right\}B = 0.$$

When the column buckles, it has nonzero deflection $y(x)$, which means $B\neq 0$. Note that $\sin\alpha L\neq 0$, otherwise, $B=0$. Hence, the buckling equation is given by

$$\kappa\left(-\frac{\sin\alpha L}{L} + \alpha\cos\alpha L\right) - EI\,\alpha^2\sin\alpha L = 0,$$

$$\therefore\quad \frac{\kappa L}{EI}(\alpha L\cot\alpha L - 1) - (\alpha L)^2 = 0,\qquad \alpha^2 = \frac{P}{EI}.$$

5.5 Various Application Problems

Example 5.5 — Jet Engine Vibration

As shown in Figure 5.8, jet engines are supported by the wings of the airplane. To study the horizontal motion of a jet engine, it is modeled as a rigid body supported by an elastic beam. The mass of the engine is m and the moment of inertia about its centroidal axis C is J. The elastic beam is further modeled as a massless bar hinged at A, with the rotational spring κ providing restoring moment equal to $\kappa\theta$, where θ is the angle between the bar and the vertical line as shown in Figure 5.25.

For small rotations, i.e., $|\theta| \ll 1$, set up the equation of motion for the jet engine in term of θ. Find the natural frequency of oscillation.

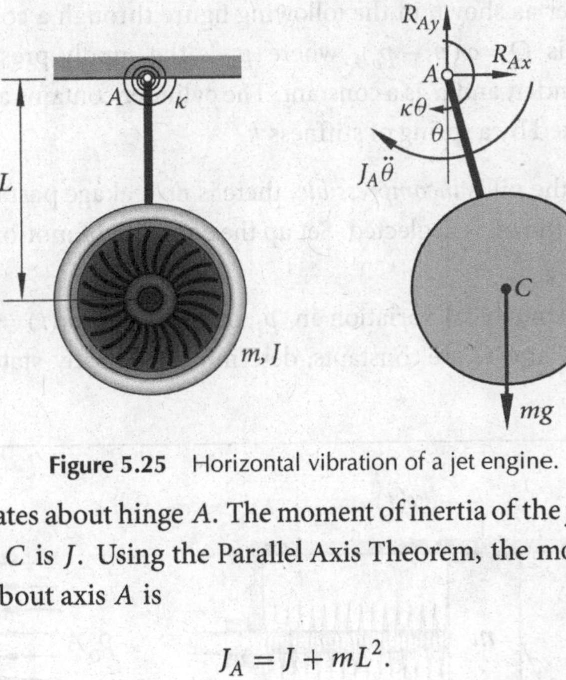

Figure 5.25 Horizontal vibration of a jet engine.

The system rotates about hinge A. The moment of inertia of the jet engine about its centroidal axis C is J. Using the Parallel Axis Theorem, the moment of inertia of the jet engine about axis A is

$$J_A = J + mL^2.$$

Draw the free-body diagram of the jet engine and the supporting bar as shown. The jet engine is subjected to gravity mg. Remove the hinge at A and replace it by two reaction force components R_{Ax} and R_{Ay}. Since the bar rotates an angle θ counterclockwise, the rotational spring provides a clockwise restoring moment $\kappa\theta$.

Since the angular acceleration of the system is $\ddot{\theta}$ counterclockwise, the inertia moment is $J_A\ddot{\theta}$ clockwise.

Applying D'Alembert's Principle, the free-body as shown in Figure 5.25 is in dynamic equilibrium. Hence,

$$\curvearrowleft \sum M_A = 0: \quad J_A\ddot{\theta} + \kappa\theta + mg \cdot L\sin\theta = 0.$$

For small rotations $|\theta| \ll 1$, $\sin\theta \approx \theta$, the equation of motion is

$$(J+mL^2)\ddot{\theta} + (\kappa + mgL)\theta = 0.$$

Since

$$\ddot{\theta} + \frac{\kappa+mgL}{J+mL^2}\theta = 0 \implies \ddot{\theta} + \omega_0^2\theta = 0,$$

the natural circular frequency ω_0 of oscillation is given by

$$\omega_0 = \sqrt{\frac{\kappa+mgL}{J+mL^2}}.$$

Example 5.6 — Piston Vibration

Oil enters a cylinder as shown in the following figure through a constriction such that the flow rate is $Q=\alpha(p_i - p_o)$, where p_i is the supply pressure, p_o is the pressure in the cylinder, and α is a constant. The cylinder contains a piston of mass m and area A backed by a spring of stiffness k.

1. Assume that the oil is *incompressible*, there is no leakage past the piston, and the inertia of the oil is neglected. Set up the equation of motion for the piston displacement x.
2. If there is a sinusoidal variation in p_i of the form $p_i(t) = P_0 + P_1 \sin\Omega t$, where P_0, P_1, and Ω are constants, determine the steady-state displacement $x_P(t)$.

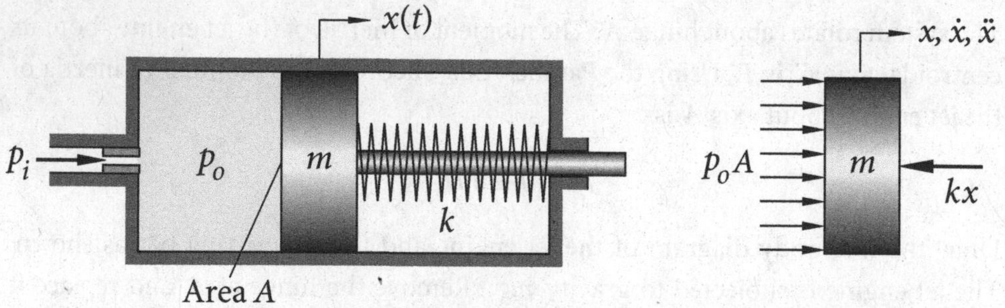

1. To set up the equation of motion, consider the free-body of the piston as shown. The piston is subjected to two forces: the force due to the internal oil pressure p_oA and the spring force kx. Newton's Second Law requires that

$$\rightarrow ma = \sum F: \qquad m\ddot{x} = p_oA - kx.$$

Since the oil is incompressible, in time Δt,

$$\text{Inflow} = Q\,\Delta t = \alpha(p_i - p_o)\,\Delta t = A\,\Delta x,$$

where Δx is the displacement of the piston displaced by the oil inflow. Solving for p_o yields

$$p_o = p_i - \frac{A}{\alpha}\frac{\Delta x}{\Delta t}.$$

Taking the limit as $\Delta t \to 0$ leads to

$$p_o = p_i - \frac{A}{\alpha}\frac{\mathrm{d}x}{\mathrm{d}t}.$$

Hence, the equation of motion becomes

$$m\ddot{x} = \Big(p_i - \frac{A}{\alpha}\dot{x}\Big)A - kx \implies m\ddot{x} + \frac{A^2}{\alpha}\dot{x} + kx = p_i A,$$

$$\therefore \quad \ddot{x} + \frac{A^2}{\alpha m}\dot{x} + \frac{k}{m}x = \frac{A}{m}p_i.$$

2. Since $p_i(t) = P_0 + P_1 \sin \Omega t$, using the D-operator, the equation of motion can be written as

$$(D^2 + cD + \omega_0^2)x = \frac{A}{m}(P_0 + P_1 \sin \Omega t), \qquad c = \frac{A^2}{\alpha m}, \quad \omega_0^2 = \frac{k}{m}.$$

A particular solution is given by

$$\begin{aligned}
x_P(t) &= \frac{A}{m}\frac{1}{D^2 + cD + \omega_0^2}(P_0 + P_1 \sin \Omega t) \\
&= \frac{AP_0}{m\omega_0^2} + \frac{AP_1}{m}\frac{1}{-\Omega^2 + cD + \omega_0^2}\sin \Omega t, \quad \text{✍ Theorem 3 of Chapter 4: replace } D^2 \text{ by } -\Omega^2. \\
&= \frac{AP_0}{m\omega_0^2} + \frac{AP_1}{m}\frac{(\omega_0^2 - \Omega^2) - cD}{[(\omega_0^2 - \Omega^2) + cD][(\omega_0^2 - \Omega^2) - cD]}\sin \Omega t \\
&= \frac{AP_0}{m\omega_0^2} + \frac{AP_1}{m}\frac{(\omega_0^2 - \Omega^2) - cD}{(\omega_0^2 - \Omega^2)^2 - c^2D^2}\sin \Omega t \\
&= \frac{AP_0}{m\omega_0^2} + \frac{AP_1}{m}\frac{(\omega_0^2 - \Omega^2)\sin \Omega t - c\Omega \cos \Omega t}{(\omega_0^2 - \Omega^2)^2 + c^2\Omega^2} \quad \text{✍ Replace } D^2 \text{ by } -\Omega^2 \text{ and evaluate the numerator.} \\
&= \frac{AP_0}{m\omega_0^2} + \frac{AP_1}{m}\frac{\sin(\Omega t - \varphi)}{\sqrt{(\omega_0^2 - \Omega^2)^2 + c^2\Omega^2}}, \quad \varphi = \tan^{-1}\frac{c\Omega}{\omega_0^2 - \Omega^2}.
\end{aligned}$$

✍ Use $A \sin\theta - B\cos\theta = \sqrt{A^2 + B^2}\sin(\theta - \varphi)$, $\varphi = \tan^{-1}(B/A)$.

Example 5.7 — Single Degree-of-Freedom System

The single degree-of-freedom system described by $x(t)$, as shown in Figure 5.26(a), is subjected to a sinusoidal load $F(t)=F_0 \sin \Omega t$. Assume that the mass m, the spring stiffnesses k_1 and k_2, the damping coefficient c, and F_0 and Ω are known. Determine the *steady-state amplitude* of the response of $x_P(t)$.

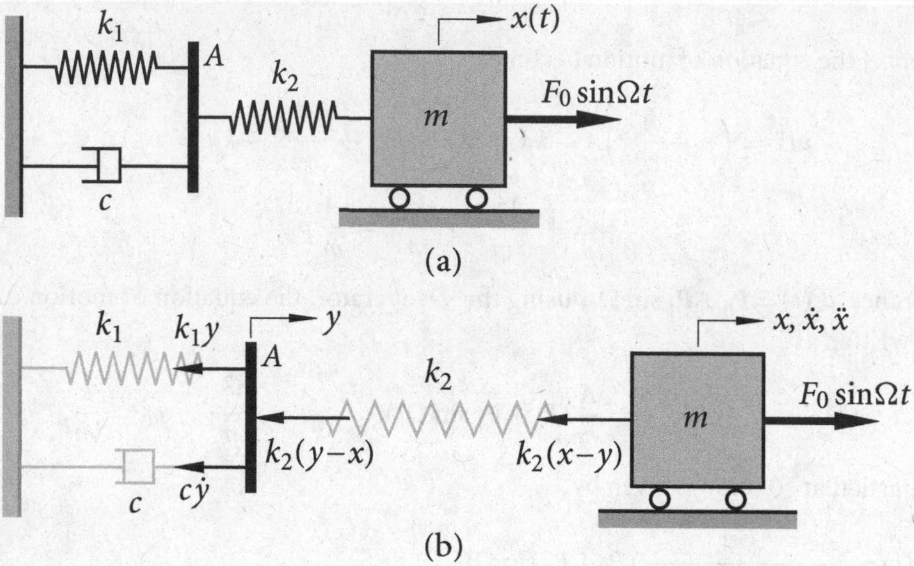

Figure 5.26 A vibrating system.

Introduce a displacement $y(t)$ at A as shown in Figure 5.26(b). Consider the free-body of A. The extension of spring k_1 is y and the compression of spring k_2 is $y-x$. Body A is subjected to three forces: spring force $k_1 y$, damping force $c\dot{y}$, and spring force $k_2(y-x)$. Newton's Second Law requires

$$\rightarrow m_A\ddot{y} = \sum F: \qquad m_A\ddot{y} = -k_1 y - c\dot{y} - k_2(y-x).$$

Since the mass of A is zero, i.e., $m_A=0$, one has

$$x = \frac{(k_1+k_2)y + c\dot{y}}{k_2}. \tag{1}$$

Consider the free-body of mass m. The extension of spring k_2 is $x-y$. The mass is subjected to two forces: spring force $k_2(x-y)$ and the externally applied load $F_0 \sin \Omega t$. Applying Newton's Second Law gives

$$\rightarrow m\ddot{x} = \sum F: \qquad m\ddot{x} = F_0 \sin \Omega t - k_2(x-y).$$

Substituting equation (1) yields the equation of motion

$$m\frac{(k_1+k_2)\,\ddot{y} + c\,\dddot{y}}{k_2} = F_0 \sin \Omega t - k_2\Big[\frac{(k_1+k_2)y + c\dot{y}}{k_2} - y\Big],$$

$$\frac{cm}{k_2}\dddot{y}+\frac{m(k_1+k_2)}{k_2}\ddot{y}+c\dot{y}+k_1 y=F_0\sin\Omega t,$$

or, using the D-operator

$$\left[\frac{cm}{k_2}D^3+\frac{m(k_1+k_2)}{k_2}D^2+cD+k_1\right]y=F_0\sin\Omega t.$$

A particular solution is given by

$$y_P=\frac{1}{\dfrac{cm}{k_2}D^3+\dfrac{m(k_1+k_2)}{k_2}D^2+cD+k_1}F_0\sin\Omega t$$

$$=k_2F_0\frac{1}{[m(k_1+k_2)D^2+k_1k_2]+c(mD^2+k_2)D}\sin\Omega t$$

$$=k_2F_0\frac{1}{[-m(k_1+k_2)\Omega^2+k_1k_2]+c(-m\Omega^2+k_2)D}\sin\Omega t$$

✍ Theorem 3 of Chapter 4: replace D^2 by $-\Omega^2$.

$$=k_2F_0\frac{[k_1k_2-m(k_1+k_2)\Omega^2]-c(k_2-m\Omega^2)D}{[k_1k_2-m(k_1+k_2)\Omega^2]^2-c^2(k_2-m\Omega^2)^2D^2}\sin\Omega t$$

$$=k_2F_0\frac{[k_1k_2-m(k_1+k_2)\Omega^2]\sin\Omega t-c\Omega(k_2-m\Omega^2)\cos\Omega t}{[k_1k_2-m(k_1+k_2)\Omega^2]^2+c^2(k_2-m\Omega^2)^2\Omega^2}$$

✍ Replace D^2 by $-\Omega^2$ and evaluate the numerator.

$$=k_2F_0\frac{\sqrt{[k_1k_2-m(k_1+k_2)\Omega^2]^2+c^2\Omega^2(k_2-m\Omega^2)^2}\,\sin(\Omega t-\varphi)}{[k_1k_2-m(k_1+k_2)\Omega^2]^2+c^2\Omega^2(k_2-m\Omega^2)^2}$$

✍ Use $A\sin\theta-B\cos\theta=\sqrt{A^2+B^2}\sin(\theta-\varphi),\quad \varphi=\tan^{-1}(B/A)$.

$$=k_2a\sin(\Omega t-\varphi),$$

where

$$a=\frac{F_0}{\sqrt{[k_1k_2-m(k_1+k_2)\Omega^2]^2+c^2\Omega^2(k_2-m\Omega^2)^2}}.$$

Using equation (1), one obtains

$$x_P(t)=\frac{(k_1+k_2)y_P+c\dot{y}_P}{k_2}=\frac{(k_1+k_2)\cdot k_2a\sin(\Omega t-\varphi)+c\cdot k_2a\Omega\cos(\Omega t-\varphi)}{k_2}$$

$$=a\big[(k_1+k_2)\sin(\Omega t-\varphi)+c\Omega\cos(\Omega t-\varphi)\big].$$

The amplitude of $x_P(t)$ is ✍ $x=A\sin\theta\pm B\cos\theta\Longrightarrow x_{\text{amplitude}}=\sqrt{A^2+B^2}$

$$x_{P,\,\text{amplitude}}=a\sqrt{(k_1+k_2)^2+c^2\Omega^2}.$$

Example 5.8 — Flywheel Vibration

A pair of uniform parallel bars AB of length L and together of mass m are hinged at A and supported at B by a spring of stiffness k as shown. The bars carry a uniform flywheel D of mass M and radius r supported in bearings, with $AD = l$. When the system is in static equilibrium, AB is horizontal. The flywheel rotates at angular velocity Ω and has a small eccentricity e, i.e., $DG = e$, where G is the center of gravity of the flywheel. Determine the natural frequency of vibration of the system and the total vertical movement of B (assuming no resonance).

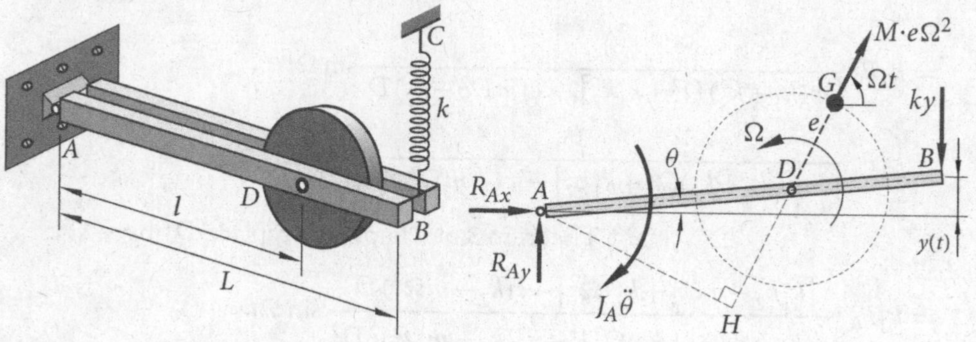

The moment of inertia of the bars AB about hinge A is

$$J_A^{\text{Bars}} = \frac{1}{3} mL^2.$$

The moment of inertia of the flywheel about its axis of rotation D is

$$J_D^{\text{Flywheel}} = \frac{1}{2} Mr^2.$$

Using the Parallel Axis Theorem, the moment of inertia of the flywheel about hinge A is

$$J_A^{\text{Flywheel}} = J_D^{\text{Flywheel}} + Ml^2 = \frac{1}{2} Mr^2 + Ml^2.$$

Hence, the moment of inertia of the system about hinge A is

$$J_A = J_A^{\text{Bars}} + J_A^{\text{Flywheel}} = \frac{1}{3} mL^2 + \frac{1}{2} M(r^2 + 2l^2).$$

Suppose bars AB has a *small* angular rotation $\theta(t)$ about hinge A as shown. The upward vertical displacement of end B is $y(t) = L\theta(t)$, $\theta \ll 1$. Hence, the downward spring force applied on bars AB at B is $ky = kL\theta$.

When the flywheel rotates at angular velocity Ω, the center of gravity G moves on a circle of radius e with angular velocity Ω, resulting in a centrifugal acceleration $e\Omega^2$. Hence, the centrifugal force is $M \cdot e\Omega^2$. The moment of the centrifugal force about point A is $(Me\Omega^2) \times AH$, where AH is the moment arm given by

$$AH = AD \cdot \sin \angle ADH \approx l \sin \Omega t. \qquad ✍ \angle ADH = \Omega t - \theta \approx \Omega t, \text{ for } \theta \ll 1$$

Consider the free-body diagram of the bars AB and the flywheel. The inertia moment of the system about point A is $J_A\ddot{\theta}$, where $\ddot{\theta}$ is the angular acceleration of the system about point A. From D'Alembert's Principle, the system is in dynamic equilibrium under the inertia moment $J_A\ddot{\theta}$ and the externally applied forces, i.e., the spring force $kL\theta$, the centrifugal force $Me\Omega^2$, and the two components R_{Ax}, R_{Ay} of the reaction force at hinge A. Summing up the moments about A yields

$$\curvearrowleft \sum M_A = 0: \quad -J_A\ddot{\theta} + Me\Omega^2 \cdot l \sin\Omega t - kL\theta \cdot L = 0,$$

$$\therefore \quad J_A\ddot{\theta} + kL^2\theta = Mel\Omega^2 \sin\Omega t \quad \Longrightarrow \quad \ddot{\theta} + \omega_0^2\theta = \frac{Mel\Omega^2}{J_A}\sin\Omega t,$$

where ω_0 is the natural circular frequency of the system given by

$$\omega_0^2 = \frac{kL^2}{J_A} = \frac{kL^2}{\frac{1}{3}mL^2 + \frac{1}{2}M(r^2+2l^2)}.$$

The response of the forced vibration is, assuming no resonance, i.e., $\Omega \neq \omega_0$,

$$\theta_P(t) = \frac{1}{D^2+\omega_0^2}\left(\frac{Mel\Omega^2}{J_A}\sin\Omega t\right)$$

$$= \frac{Mel\Omega^2}{J_A}\frac{1}{\omega_0^2-\Omega^2}\sin\Omega t.$$

Theorem 3 of Chapter 4: replace D^2 by $-\Omega^2$.

The amplitude of the forced vibration is

$$a = \frac{Mel\Omega^2}{J_A\left|\omega_0^2-\Omega^2\right|},$$

and the total vertical movement of B is $2a$.

Example 5.9 — Displacement Meter

The following figure shows the configuration of a displacement meter used for measuring the vibration of the structure upon which the meter is mounted. The structure undergoes vertical displacement, which may be modeled as $y_0 = a_0 \sin\Omega t$ with a_0 and Ω to be measured by the displacement meter. The structure excites the mass-spring-damper system of the displacement meter. The displacement of tip D of rod AD is recorded on the rotating drum. The record shows that the steady-state displacement of tip D has a peak-to-peak amplitude of $2a$ and the distance between two adjacent peaks is d. The rotating drum has a radius r and rotates at a constant speed of ν rpm. Determine the amplitude a_0 and the circular frequency Ω of the displacement of the structure.

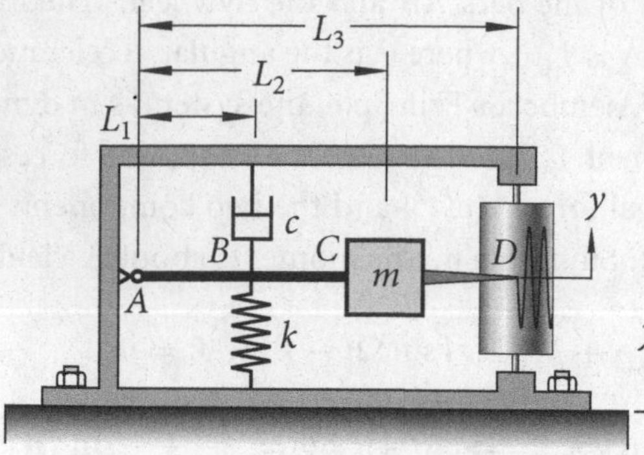

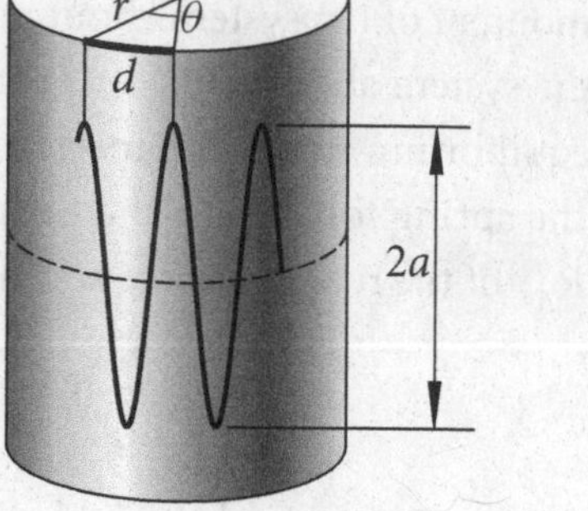

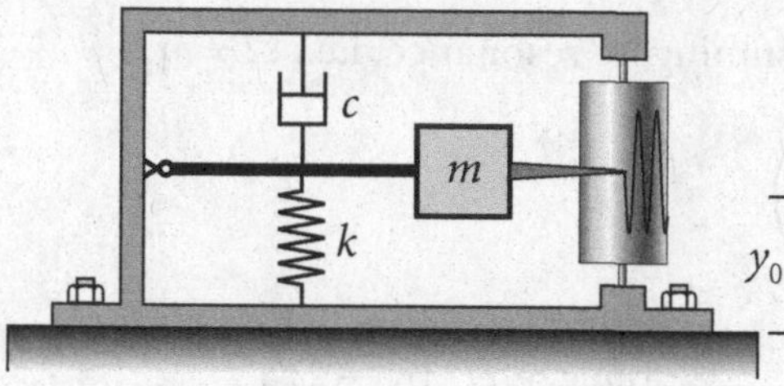

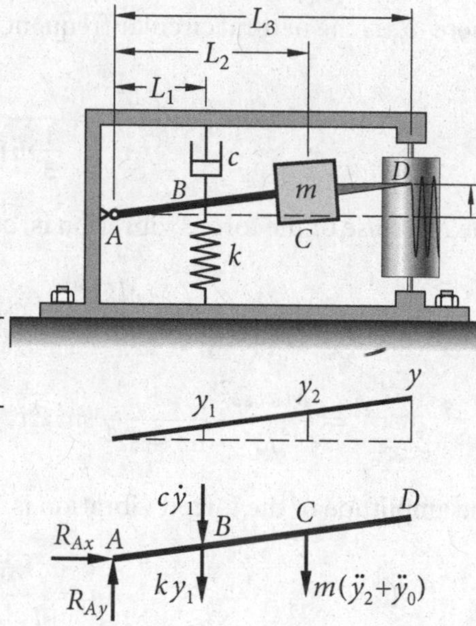

To establish the differential equation governing the displacement y, consider the free-body diagram of rod AD. D'Alembert's Principle is applied to set up the equation of motion.

Since the relative displacement of tip D is y, the relative displacements at B and C can be determined using similar triangles

$$y_1 = \frac{L_1}{L_3}\, y, \qquad y_2 = \frac{L_2}{L_3}\, y.$$

The extension of the spring is y_1, resulting in the spring force $k y_1$ applied on rod AD at point B. Similarly, the damping force applied on rod AD at point B is $c\dot{y}_1$. The inertia force applied at point C depends on the *absolute* acceleration of point C, i.e., $\ddot{y}_2+\ddot{y}_0$; hence, the inertia force is $m(\ddot{y}_2+\ddot{y}_0)$, opposite to the direction of the absolute acceleration. Because the rod is supported by a hinge at

A, two reaction force components R_{Ax} and R_{Ay} are applied at A when the hinge is removed.

Rod AD is in dynamic equilibrium under the spring force ky_1, the damping force $c\dot{y}_1$, the inertia force $m(\ddot{y}_2+\ddot{y}_0)$, and the support reaction forces R_{Ax}, R_{Ay}. Summing up the moments about point A yields:

$$\curvearrowright \sum M_A = 0: \qquad m(\ddot{y}_2+\ddot{y}_0)\cdot L_2 + (c\dot{y}_1 + ky_1)\cdot L_1 = 0,$$

or

$$m\cdot\frac{L_2}{L_3}\ddot{y}\cdot L_2 + \left(c\cdot\frac{L_1}{L_3}\dot{y} + k\cdot\frac{L_1}{L_3}y\right)\cdot L_1 = -m\ddot{y}_0\cdot L_2.$$

Noting that $y_0 = a_0 \sin\Omega t$, one obtains

$$mL_2^2\,\ddot{y} + cL_1^2\,\dot{y} + kL_1^2\,y = a_0 m L_2 L_3 \Omega^2 \sin\Omega t,$$

$$\therefore \quad (MD^2 + CD + K)\,y = a_s \sin\Omega t,$$

where

$$M = mL_2^2, \quad C = cL_1^2, \quad K = kL_1^2, \quad a_s = a_0 m L_2 L_3 \Omega^2.$$

The steady-state response is given by

$$y_P(t) = a_s\,\frac{1}{MD^2 + CD + K}\sin\Omega t$$

$$= a_s\,\frac{1}{-M\Omega^2 + CD + K}\sin\Omega t, \qquad \text{✍ Theorem 3 of Chapter 4: replace } D^2 \text{ by } -\Omega^2.$$

$$= a_s\,\frac{(K-M\Omega^2) - CD}{\big[(K-M\Omega^2)+CD\big]\big[(K-M\Omega^2)-CD\big]}\sin\Omega t$$

$$= a_s\,\frac{(K-M\Omega^2) - CD}{(K-M\Omega^2)^2 - C^2D^2}\sin\Omega t$$

$$= a_s\,\frac{(K-M\Omega^2)\sin\Omega t - C\Omega\cos\Omega t}{(K-M\Omega^2)^2 + C^2\Omega^2}, \qquad \text{✍ Replace } D^2 \text{ by } -\Omega^2 \text{ and evaluate the numerator.}$$

$$= \frac{a_s}{\sqrt{(K-M\Omega^2)^2 + C^2\Omega^2}}\sin(\Omega t - \varphi).$$

✍ Use $A\sin\theta - B\cos\theta = \sqrt{A^2+B^2}\,\sin(\theta-\varphi), \quad \varphi = \tan^{-1}(B/A)$.

Hence, the amplitude of the steady-state displacement of tip D is

$$a = \frac{a_s}{\sqrt{(K-M\Omega^2)^2 + C^2\Omega^2}} = \frac{a_0 m L_2 L_3 \Omega^2}{\sqrt{(kL_1^2 - mL_2^2\Omega^2)^2 + (cL_1^2\Omega)^2}},$$

which leads to

$$a_0 = \frac{\sqrt{(kL_1^2 - mL_2^2\Omega^2)^2 + (cL_1^2\Omega)^2}}{mL_2L_3\Omega^2}\,a.$$

On the record paper, the distance d between two adjacent peaks is measured in length, which needs to be changed to time to yield the period T of the response. Since the drum rotates at a speed of ν rpm, i.e., it rotates an angle of $2\pi\nu$ in 60 seconds, hence the time T it takes to rotate an angle θ, as shown in the figure, is given by

$$\frac{T}{60} = \frac{\theta}{2\pi\nu} \implies T = \frac{30\theta}{\pi\nu}.$$

Furthermore, since $d = r\theta$, which is the arc length corresponding to angle θ, one has

$$T = \frac{30}{\pi\nu}\,\frac{d}{r} = \frac{30d}{\pi r\nu}.$$

The frequency of vibration of tip D is

$$f = \frac{1}{T} = \frac{\pi r\nu}{30d}.$$

Since the steady-state response and the excitation have the same frequency, one obtains

$$\Omega = 2\pi f = \frac{\pi^2 r\nu}{15d}.$$

Problems

5.1 A circular cylinder of radius r and mass m is supported by a spring of stiffness k and partially submerges in a liquid of density γ. Suppose that, during vibration, the cylinder does not completely submerge in the liquid. Set up the equation of motion of the cylinder for the oscillation about the equilibrium position and determine the period of the oscillation.

ANS $m\ddot{y} + (k+\gamma\pi r^2)y = 0, \quad T = 2\pi\sqrt{\dfrac{m}{k+\gamma\pi r^2}}$

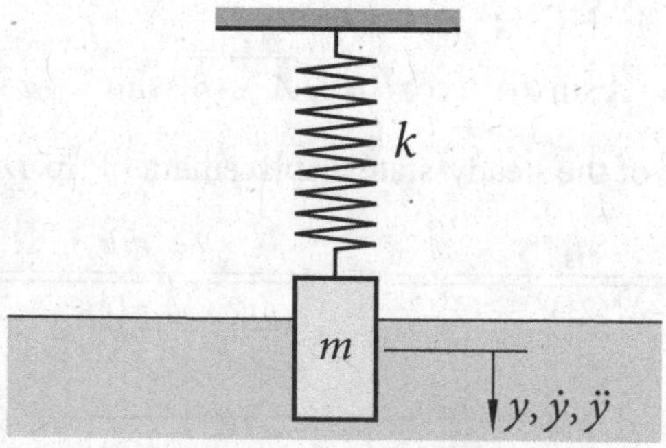

5.2 A cylinder of radius r, height h, and mass m floats with its axis vertical in a liquid of density ρ as shown in the following figure.

☞ **Archimedes' Principle:** An object partially or totally submerged in a fluid is buoyed up by a force equal to the weight of the fluid displaced.

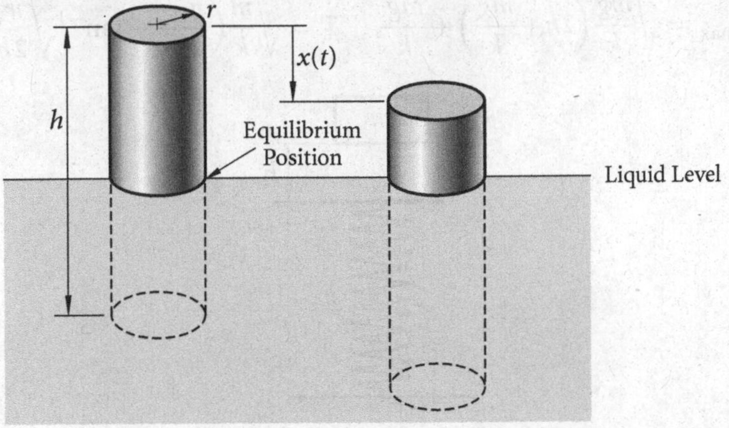

1. Set up the differential equation governing the displacement $x(t)$, measured relative to the equilibrium position, and determine the period of oscillation.

2. If the cylinder is set into oscillation by being pushed down a displacement x_0 at $t=0$ and then released, determine the response $x(t)$.

ANS $m\ddot{x} + \rho\pi r^2 x = 0,\ \ T = \dfrac{2}{r}\sqrt{\dfrac{\pi m}{\rho}};\ \ x(t) = x_0 \cos\omega_0 t,\ \ \omega_0 = r\sqrt{\dfrac{\rho\pi}{m}}$

5.3 A cube of mass m is immersed in a liquid as shown. The length of each side of the cube is L. At time $t=0$, the top surface of the cube is leveled with the surface of the liquid due to buoyancy. The cube is lifted by a constant force F. Show that the time T when the bottom surface is leveled with the liquid surface is given by

$$T = \sqrt{\frac{L}{g}}\cos^{-1}\left(1 - \frac{mg}{F}\right).$$

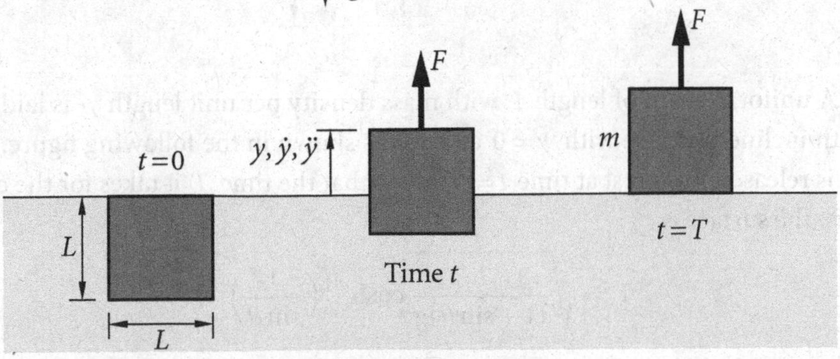

5.4 A mass m is dropped with zero initial velocity from a height of h above a spring of stiffness k as shown in the following figure. Determine the maximum compression of the spring and the duration between the time when the mass contacts the spring and the time when the spring reaches maximum compression.

ANS $y_{\max}=\sqrt{\frac{mg}{k}\left(2h+\frac{mg}{k}\right)}+\frac{mg}{k},\quad T=\sqrt{\frac{m}{k}}\left(\frac{\pi}{2}+-\tan^{-1}\sqrt{\frac{mg}{2hk}}\right)$

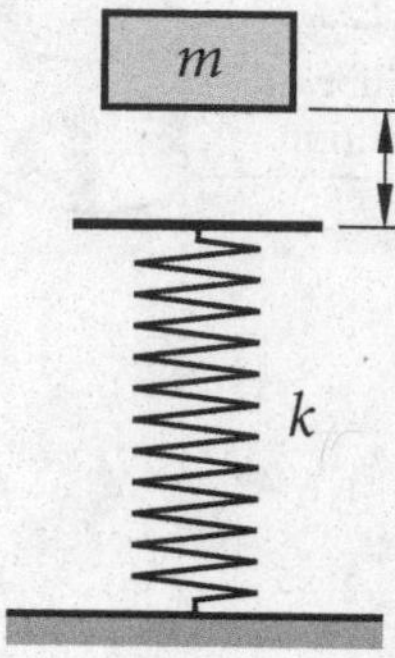

5.5 A uniform chain of length L with mass density per unit length ρ is laid on a rough horizontal table with an initial hang of length l, i.e., $y=l$ at $t=0$ as shown in the following figure. The coefficients of static and kinetic friction between the chain and the surface have the same value μ. The chain is released from rest at time $t=0$ and it starts sliding off the table if $(1+\mu)l>\mu L$. Show that the time T it takes for the chain to leave the table is

$$T=\sqrt{\frac{L}{(1+\mu)g}}\cosh^{-1}\left[\frac{L}{(1+\mu)l-\mu L}\right].$$

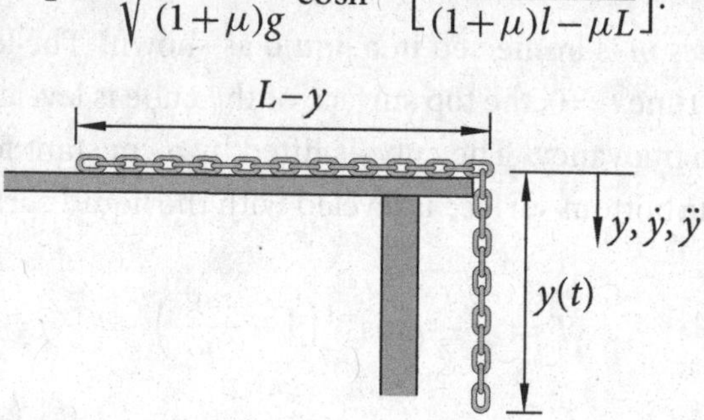

5.6 A uniform chain of length L with mass density per unit length ρ is laid on a smooth inclined surface with $y=0$ at $t=0$ as shown in the following figure. The chain is released from rest at time $t=0$. Show that the time T it takes for the chain to leave the surface is

$$T=\sqrt{\frac{L}{(1-\sin\theta)g}}\cosh^{-1}\left(\frac{1}{\sin\theta}\right).$$

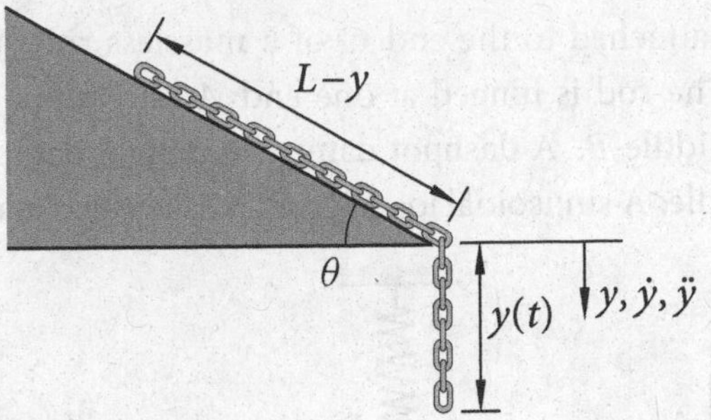

5.7 A uniform chain of length L with mass density per unit length ρ is hung on a small smooth pulley with $y(t)=l$ when $t=0$, $l>L/2$, as shown in the Figure 5.27. The chain is released from rest at time $t=0$. Show that the time T it takes for the chain to leave the pulley is

$$T=\sqrt{\frac{L}{2g}}\,\cosh^{-1}\!\left(\frac{L}{2l-L}\right).$$

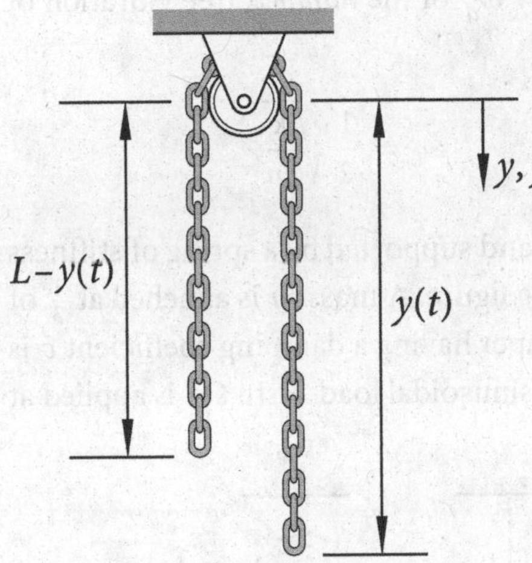

Figure 5.27

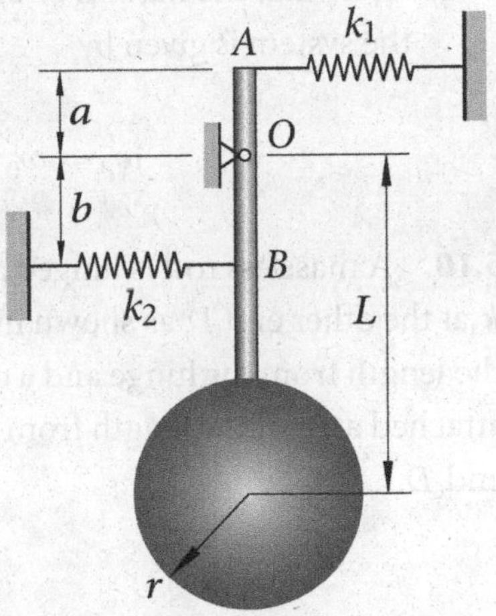

Figure 5.28

5.8 A pendulum as shown in Figure 5.28 consists of a uniform solid sphere of radius r and mass m connected by a weightless bar to hinge O. The bar is further constrained by two linear springs of stiffnesses k_1 and k_2 at A and B, respectively. It is known that the moment of inertia of a solid sphere of radius r and mass m about its diameter is $\frac{2}{5}mr^2$. Show that the equation of motion governing the angle of rotation of the pendulum about O and the natural period of oscillation of the pendulum are given by

$$m\left(\tfrac{2}{5}r^2+L^2\right)\ddot{\theta}+(k_1a^2+k_2b^2+mgL)\,\theta=0,\qquad T=2\pi\sqrt{\frac{m\left(\frac{2}{5}r^2+L^2\right)}{k_1a^2+k_2b^2+mgL}}.$$

5.9 A mass m is attached to the end C of a massless rod AC as shown in the following figure. The rod is hinged at one end A and supported by a spring of stiffness k at the middle B. A dashpot damper having a damping coefficient c is attached at the middle. A sinusoidal load $F \sin \Omega t$ is applied at end C.

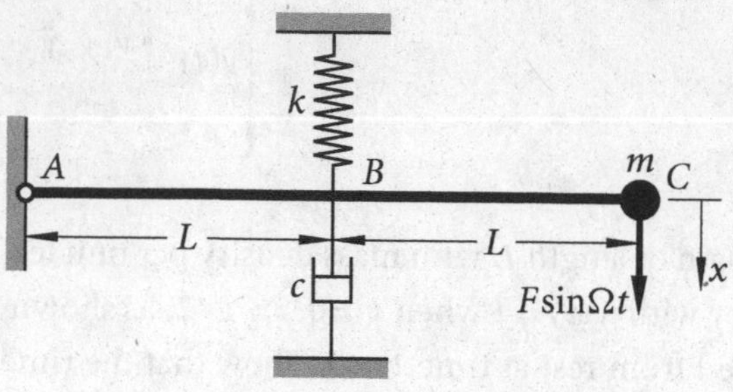

1. Show that the equation of motion governing displacement $x(t)$ of end C is
$$4m\ddot{x} + c\dot{x} + kx = 4F \sin \Omega t.$$
2. Show that the natural circular frequency ω_d of the *damped* free vibration of the system is given by
$$\omega_d = \omega_0\sqrt{1 - \frac{c^2}{16km}}, \qquad \omega_0 = \frac{1}{2}\sqrt{\frac{k}{m}}.$$

5.10 A massless rod is hinged at one end A and supported by a spring of stiffness k at the other end D as shown in the following figure. A mass m is attached at $\frac{1}{3}$ of the length from the hinge and a dash-pot damper having a damping coefficient c is attached at $\frac{2}{3}$ of the length from the hinge. A sinusoidal load $F \sin \Omega t$ is applied at end D.

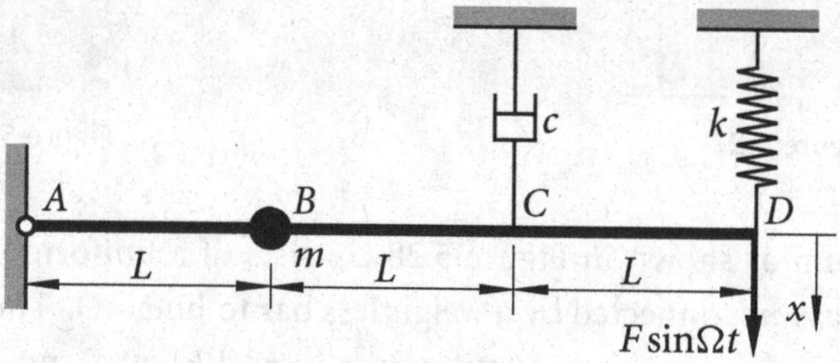

1. Show that the equation of motion governing displacement $x(t)$ of end D is
$$m\ddot{x} + 4c\dot{x} + 9kx = 9F \sin \Omega t.$$
2. Show that the natural circular frequency ω_d of the *damped* free vibration of the system is given by
$$\omega_d = \omega_0\sqrt{1 - \frac{4c^2}{9km}}, \qquad \omega_0 = 3\sqrt{\frac{k}{m}}.$$

5.11 A damped single degree-of-freedom system is shown in the following figure. The displacement of the mass M is described by $x(t)$. The excitation is provided by $x_0(t) = a \sin \Omega t$.

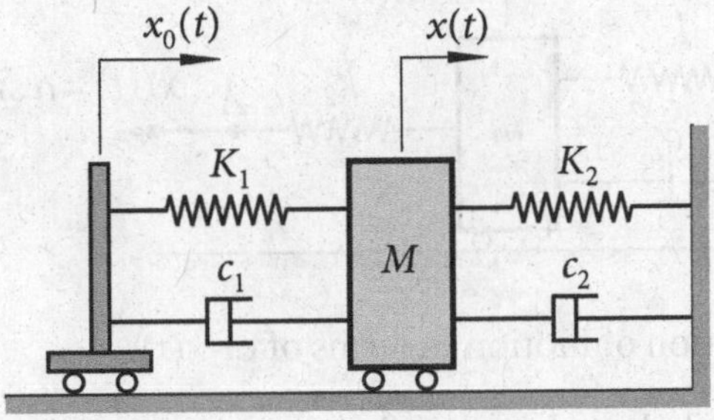

1. Show that the equation of motion governing the displacement of the mass M is given by

$$\ddot{x} + 2\zeta\omega_0\dot{x} + \omega_0^2 x = \alpha \sin \Omega t + \beta \cos \Omega t,$$

where

$$\omega_0 = \sqrt{\frac{K_1 + K_2}{M}}, \quad 2\zeta\omega_0 = \frac{c_1 + c_2}{M}, \quad \alpha = \frac{aK_1}{M}, \quad \beta = \frac{\Omega a c_1}{M}.$$

2. Determine the *amplitude* of the steady-state response $x_P(t)$.

ANS $\sqrt{\dfrac{\left[\alpha(\omega_0^2 - \Omega^2) + 2\zeta\omega_0\beta\Omega\right]^2 + \left[\beta(\omega_0^2 - \Omega^2) - 2\zeta\omega_0\alpha\Omega\right]^2}{(\omega_0^2 - \Omega^2)^2 + (2\zeta\omega_0\Omega)^2}}$

5.12 The single degree-of-freedom system shown in the following figure is subjected to dynamic force $F(t) = F_0 \sin \Omega t$.

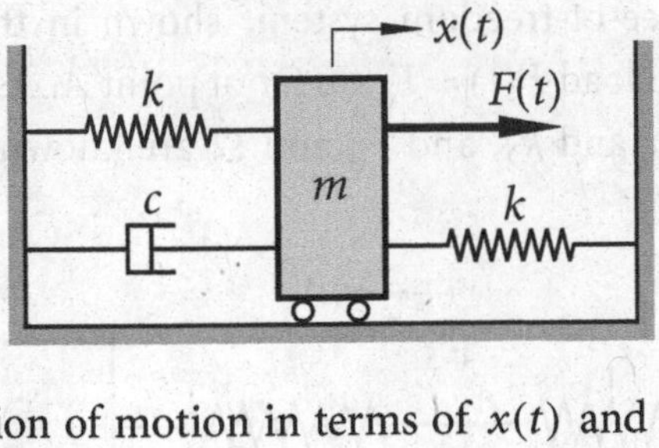

1. Set up the equation of motion in terms of $x(t)$ and determine the damped natural circular frequency.

2. Determine the steady-state response of the system $x_P(t)$.

ANS $m\ddot{x} + c\dot{x} + 2kx = F_0 \sin \Omega t, \quad \omega_d = \sqrt{\dfrac{2k}{m}\left(1 - \dfrac{c^2}{8km}\right)}$

$$x_P(t) = \frac{F_0\left[(2k - m\Omega^2)\sin \Omega t - c\Omega \cos \Omega t\right]}{(2k - m\Omega^2)^2 + c^2\Omega^2}$$

5.13 The single degree-of-freedom system shown in the following figure is subjected to dynamic displacement $x_0(t) = a \sin \Omega t$ at point A.

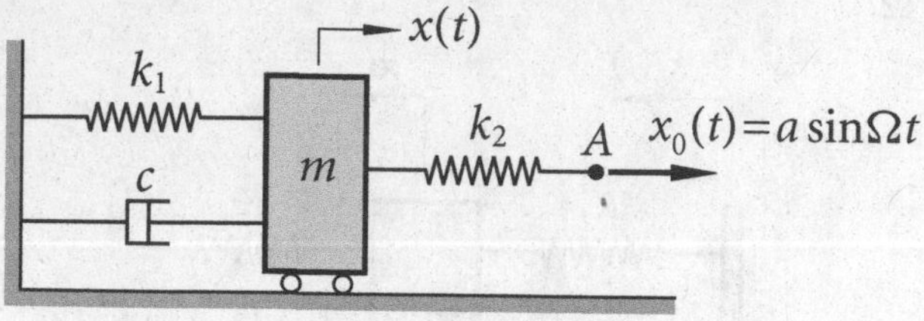

1. Set up the equation of motion in terms of of $x(t)$.
2. If the system is lightly damped, determine the *steady-state* response of the system $x_P(t)$.

ANS $m\ddot{x} + c\dot{x} + (k_1 + k_2)x = ak_2 \sin \Omega t$

$$x_P(t) = ak_2 \frac{(k_1 + k_2 - m\Omega^2) \sin \Omega t - c\Omega \cos \Omega t}{(k_1 + k_2 - m\Omega^2)^2 + c^2\Omega^2}$$

5.14 A precision instrument having a mass of $m = 400$ kg is to be mounted on a floor. It is known that the floor vibrates vertically with a peak-to-peak amplitude of 2 mm and frequency of 5 Hz. To reduce the effect of vibration of the floor on the instrument, four identical springs are placed underneath the instrument. If the peak-to-peak amplitude of vibration of the instrument is to be limited to less than 0.2 mm, determine the stiffness of each spring. Neglect damping.

ANS $k = 8.97$ kN/m

5.15 The single degree-of-freedom system, shown in the following figure, is subjected to a sinusoidal load $F(t) = F_0 \sin \Omega t$ at point A. Assume that the mass m, the spring stiffnesses k_1 and k_2, and F_0 and Ω are known. The system is at rest when $t = 0$.

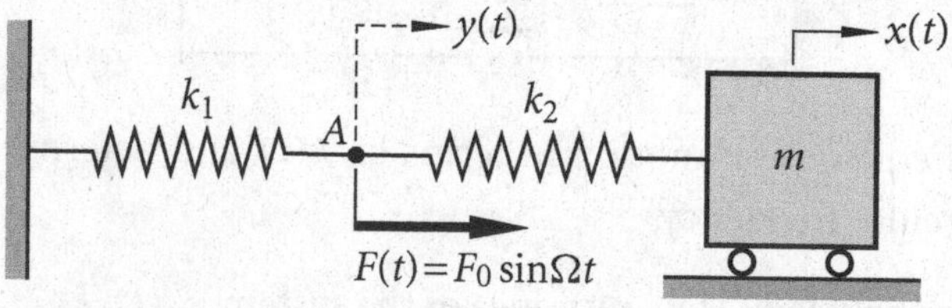

1. Show that the differential equation governing the displacement of the mass $x(t)$ is

$$\ddot{x} + \omega_0^2 x = f \sin \Omega t, \quad \omega_0 = \sqrt{\frac{k_1 k_2}{m(k_1 + k_2)}}, \quad f = \frac{k_2}{m(k_1 + k_2)} F_0.$$

2. For the case $\Omega \neq \omega_0$, determine the response of the system $x(t)$.
3. For the case $\Omega = \omega_0$, determine the response of the system $x(t)$.

ANS 2. $\Omega \neq \omega_0$: $x(t) = \dfrac{f}{\omega_0^2 - \Omega^2}\left(-\dfrac{\Omega}{\omega_0}\sin\omega_0 t + \sin\Omega t\right)$;

3. $\Omega = \omega_0$: $x(t) = -\dfrac{f}{2\omega_0^2}\left(-\sin\omega_0 t + \omega_0 t\cos\omega_0 t\right)$

5.16 A vehicle is modeled by a damped single degree-of-freedom system with mass M, spring stiffness K, and damping coefficient c as shown in the following figure. The absolute displacement of the mass M is described by $y(t)$. The vehicle is moving at a constant speed U on a wavy surface with profile $y_0(x) = \mu \sin \Omega x$. At time $t = 0$, the vehicle is at $x = 0$.

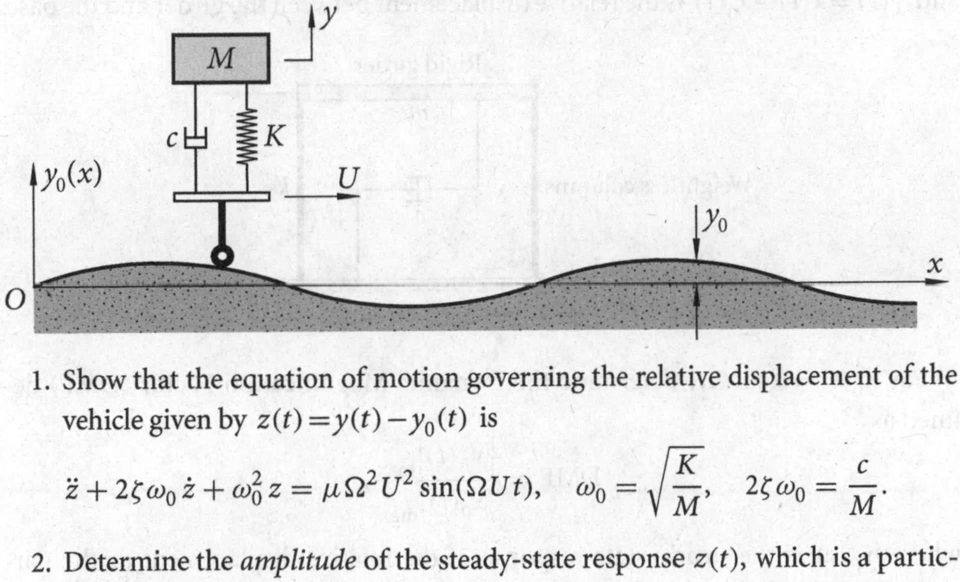

1. Show that the equation of motion governing the relative displacement of the vehicle given by $z(t) = y(t) - y_0(t)$ is

$$\ddot{z} + 2\zeta\omega_0\dot{z} + \omega_0^2 z = \mu\Omega^2U^2\sin(\Omega U t), \quad \omega_0 = \sqrt{\frac{K}{M}}, \quad 2\zeta\omega_0 = \frac{c}{M}.$$

2. Determine the *amplitude* of the steady-state response $z(t)$, which is a particular solution of the equation of motion.
3. Assuming that the damping coefficient $c = 0$, determine the speed U at which resonance occurs.

ANS $\dfrac{\mu\Omega^2U^2}{\sqrt{(\omega_0^2 - \Omega^2U^2)^2 + (2\zeta\omega_0\Omega U)^2}}$; $\quad U = \dfrac{1}{\Omega}\sqrt{\dfrac{K}{M}}$

5.17 The landing gear of an airplane as shown in Figure 5.8 can be modeled as a mass connected to the airplane by a spring of stiffness K and a damper of damping coefficient c. A spring of stiffness k is used to model the forces on the tires. The airplane lands at time $t = 0$ with $x = 0$ and moves at a constant speed U on a wavy surface with profile $y_0(x) = \mu \sin \Omega x$. Assuming that the airplane moves in the horizontal direction only, determine the steady-state response of the absolute displacement $y(t)$ of the mass m.

ANS $$y(t)=\frac{k\mu\sin(\Omega U t-\varphi)}{\sqrt{(K+k-m\Omega^2U^2)^2+(c\Omega U)^2}}, \quad \varphi=\tan^{-1}\frac{c\Omega U}{K+k-m\Omega^2U^2}$$

5.18 In Section 5.1, it is derived that the equation of motion of a single story shear building under the base excitation $x_0(t)$ is given by

$$m\ddot{y}(t)+c\dot{y}(t)+ky=-m\ddot{x}_0(t),$$

or

$$\ddot{y}(t)+2\zeta\omega_0\dot{y}(t)+\omega_0^2y=-\ddot{x}_0(t),$$

where

$$\omega_0^2=\frac{k}{m}, \quad 2\zeta\omega_0=\frac{c}{m},$$

and $y(t)=x(t)-x_0(t)$ is the relative displacement between the girder and the base.

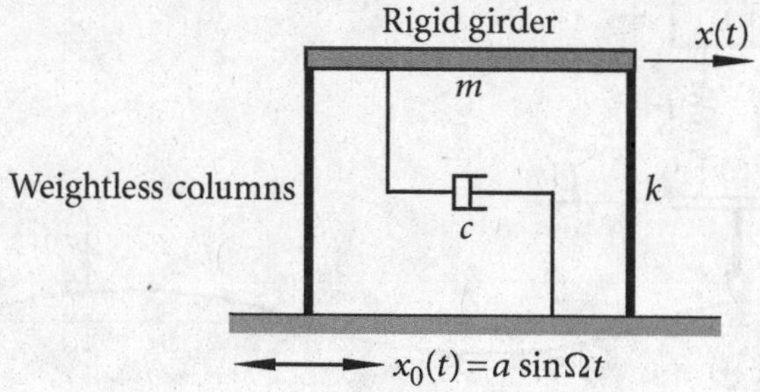

For $x_0(t)=a\sin\Omega t$, determine the Dynamic Magnification Factor (DMF) defined as

$$\text{DMF}=\frac{|y_P(t)|_{\max}}{|x_0(t)|_{\max}},$$

where $y_P(t)$ is the steady-state response of the relative displacement or the particular solution due to the base excitation. Plot DMF versus the frequency ratio $r=\Omega/\omega_0$ for $\zeta=0, 0.1, 0.2$, and 0.3.

ANS $$\text{DMF}=\frac{r^2}{\sqrt{(1-r^2)^2+(2\zeta r)^2}}, \quad r=\frac{\Omega}{\omega_0}$$

5.19 Consider the undamped single degree-of-freedom system with $m=10$ kg, $k=1$ kN/m. The system is subjected to a dynamic load $F(t)$ as shown in the following figure. The system is at rest at time $t=0$.

Determine the analytical expression of the displacement as a function of time up to $t=10$ sec. ANS $x(t)=0.02(10t-\sin 10t), \quad 0\leqslant t\leqslant 5$

$$x(t)=1.005\cos 10(t-5)+0.0007\sin 10(t-5), \quad 5\leqslant t\leqslant 10$$

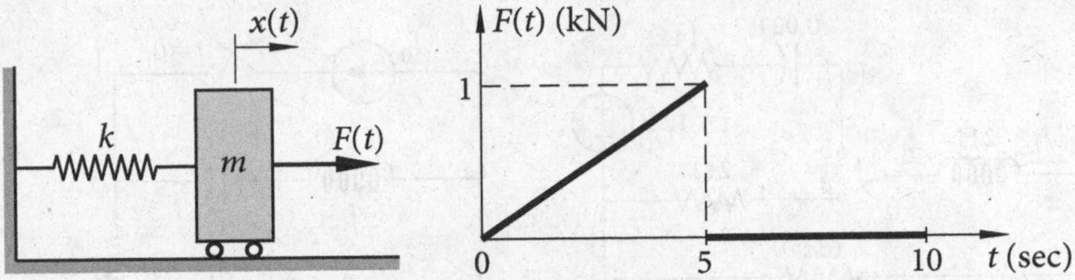

5.20 The following figure shows the configuration of a displacement meter used for measuring the vibration of the structure that the meter is mounted on. The structure undergoes vertical displacement $a_0 \sin \Omega t$ and excites the mass-spring-damper system of the displacement meter. The displacement of the mass is recorded on the rotating drum. It is known that $m = 1$ kg, $k = 1000$ N/m, $c = 5$ N·sec/m, and the steady-state record on the rotating drum shows a sinusoidal function with frequency of 5 Hz and peak-to-peak amplitude of 50 mm. Determine the amplitude a_0 and the frequency $f = \Omega/(2\pi)$ of the displacement of the structure.

ANS $a_0 = 4.0$ mm, $f = 5$ Hz

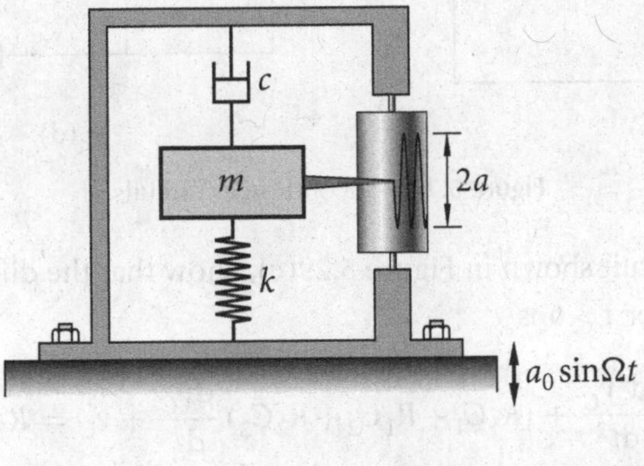

5.21 For the circuit shown in Figure 5.29(a), the switch has been at position a for a long time prior to $t = 0^-$. At $t = 0$, the switch is moved to position b. Determine $i(t)$ for $t > 0$. ANS $i(t) = (3 - 9t)\,e^{-5t}$ (A)

5.22 For the circuit shown in Figure 5.29(b), the switch has been at position a for a long time prior to $t = 0^-$. At $t = 0$, the switch is moved to position b. Show that the differential equation governing $v_C(t)$ for $t > 0$ is

$$\frac{d^2 v_C}{dt^2} + \frac{R}{L}\frac{dv_C}{dt} + \frac{1}{LC} v_C = \frac{V(t)}{LC}, \qquad v_C(0^+) = -RI_0, \quad \frac{dv_C(0^+)}{dt} = 0.$$

For $R = 6\,\Omega$, $C = \frac{1}{25}$ F, $L = 1$ H, $I_0 = 1$ A, $V(t) = 39 \sin 2t$ (V), determine $v_C(t)$ for $t > 0$. ANS $v_C(t) = 7e^{-3t}(2\cos 4t - \sin 4t) + 35 \sin 2t - 20 \cos 2t$ (V)

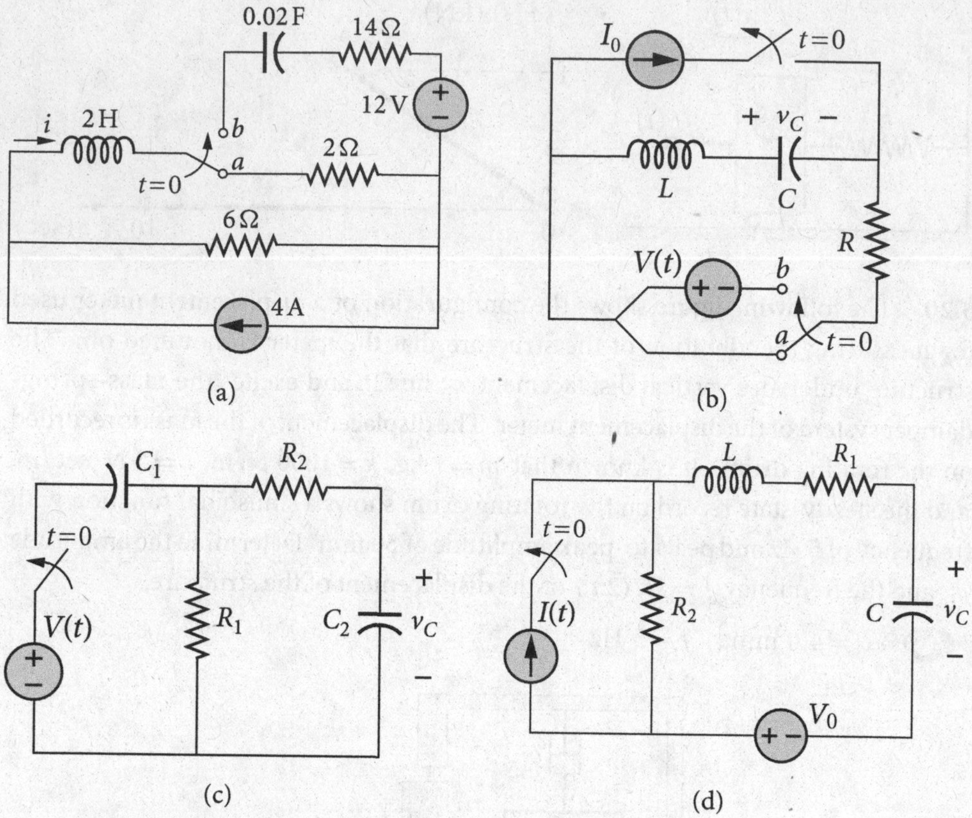

Figure 5.29 Second-order circuits.

5.23 For the circuit shown in Figure 5.29(c), show that the differential equation governing $v_C(t)$ for $t>0$ is

$$R_1C_1R_2C_2\frac{d^2v_C}{dt^2}+(R_1C_1+R_1C_2+R_2C_2)\frac{dv_C}{dt}+v_C=R_1C_1\frac{dV(t)}{dt},$$

with the initial conditions given by $v_C(0^+)=0,\quad \dfrac{dv_C(0^+)}{dt}=\dfrac{V(0^+)}{R_2C_2}.$

For $R_1=1\,\Omega$, $R_2=2\,\Omega$, $C_1=2\,\text{F}$, $C_2=1\,\text{F}$, $V(t)=12e^{-t}$ (V), determine $v_C(t)$ for $t>0$. **Ans** $v_C(t)=-\frac{8}{3}e^{-\frac{t}{4}}+\frac{8}{3}(1+3t)e^{-t}$ (V)

5.24 For the circuit shown in Figure 5.29(d), show that the differential equation governing $v_C(t)$ for $t>0$ is

$$\frac{d^2v_C}{dt^2}+\frac{R_1+R_2}{L}\frac{dv_C}{dt}+\frac{1}{LC}v_C=\frac{V_0+R_2I(t)}{LC},\qquad v_C(0^+)=V_0,\qquad \frac{dv_C(0^+)}{dt}=0.$$

For $R_1=R_2=5\,\Omega$, $C=0.2\,\text{F}$, $L=5\,\text{H}$, $V_0=12\,\text{V}$, $I(t)=2\sin t$ (A), determine $v_C(t)$ for $t>0$. **Ans** $v_C(t)=5(1+t)e^{-t}+12-5\cos t$ (V)

5.25 Consider column AB clamped at the base and pin-supported at the top by an elastic spring of stiffness k. Show that the buckling equation for the column is

$$\frac{kL^3}{EI}\left[\tan(\alpha L) - (\alpha L)\right] + (\alpha L)^3 = 0, \qquad \alpha^2 = \frac{P}{EI}.$$

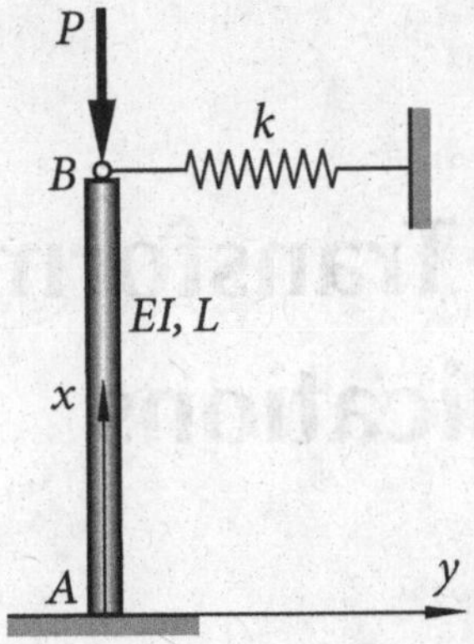

5.26 Consider the beam-column shown in the following figure. Determine the lateral deflection $y(x)$.

ANS $y(x) = -\frac{w}{P}\left[\frac{Lx}{6}\left(1-\frac{x^2}{L^2}\right) + \frac{1}{\alpha^2}\left(\frac{x}{L} - \frac{\sin\alpha x}{\sin\alpha L}\right)\right], \quad \alpha^2 = \frac{P}{EI}.$

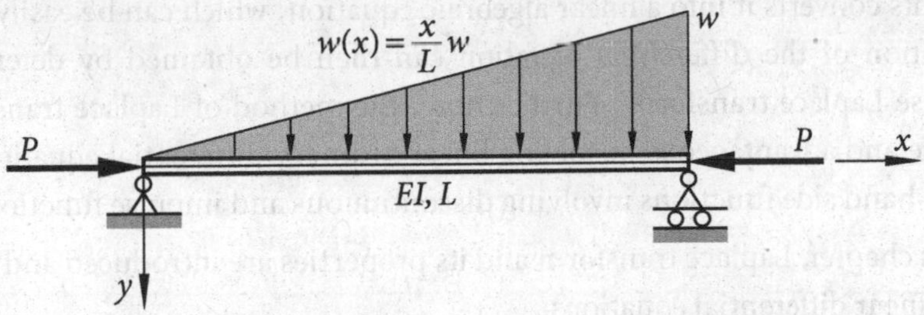

CHAPTER 6

The Laplace Transform and Its Applications

The Laplace transform is one of the most important *integral transforms*. Because of a number of special properties, it is very useful in studying linear differential equations.

Applying the Laplace transform to a linear differential equation with constant coefficients converts it into a linear algebraic equation, which can be easily solved. The solution of the differential equation can then be obtained by determining the inverse Laplace transform. Furthermore, the method of Laplace transform is preferable and advantageous in solving linear ordinary differential equations with the right-hand side functions involving discontinuous and impulse functions.

In this chapter, Laplace transform and its properties are introduced and applied to solve linear differential equations.

6.1 The Laplace Transform

Definition — Laplace Transform

Let $f(t)$, $t>0$, be a given function. The Laplace transform $F(s)$ of function $f(t)$ is defined by

$$F(s) = \mathscr{L}\{f(t)\} = \int_0^{\infty} \mathrm{e}^{-st} f(t)\,\mathrm{d}t, \quad s>0.$$

The integral in the Laplace transform is *improper* because of the unbounded interval of integration and is given by

$$\int_0^{\infty} \mathrm{e}^{-st} f(t)\,\mathrm{d}t = \lim_{M\to\infty} \int_0^{M} \mathrm{e}^{-st} f(t)\,\mathrm{d}t.$$

There are tables of Laplace transforms for various functions $f(t)$, similar to tables of integrals. Some frequently used Laplace transforms are listed in Table of Laplace Transforms (Appendix A.4).

Example 6.1

Determine the Laplace transforms of the following functions

1. $f(t) = 1$; 2. $f(t) = \mathrm{e}^{at}$; 3. $f(t) = \begin{Bmatrix} \sin\omega t \\ \cos\omega t \end{Bmatrix}$.

1. $F(s) = \mathscr{L}\{1\} = \displaystyle\int_0^\infty \mathrm{e}^{-st}\cdot 1\,\mathrm{d}t = -\frac{1}{s}\,\mathrm{e}^{-st}\Big|_{t=0}^{\infty} = \frac{1}{s}, \quad s>0.$

2.
$$F(s) = \mathscr{L}\{\mathrm{e}^{at}\} = \int_0^\infty \mathrm{e}^{-st}\cdot \mathrm{e}^{at}\,\mathrm{d}t = \int_0^\infty \mathrm{e}^{-(s-a)t}\,\mathrm{d}t$$
$$= -\frac{1}{s-a}\,\mathrm{e}^{-(s-a)t}\Big|_{t=0}^{\infty} = \frac{1}{s-a}, \quad s>a.$$

3. Note Euler's formula $\mathrm{e}^{\mathrm{i}\,\omega t} = \cos\omega t + \mathrm{i}\sin\omega t$. To determine the Laplace transforms of $\cos\omega t$ and $\sin\omega t$, consider the Laplace transform of $\mathrm{e}^{\mathrm{i}\,\omega t}$:

$$\mathscr{L}\{\mathrm{e}^{\mathrm{i}\,\omega t}\} = \int_0^\infty \mathrm{e}^{-st}\cdot \mathrm{e}^{\mathrm{i}\,\omega t}\,\mathrm{d}t = \int_0^\infty \mathrm{e}^{-(s-\mathrm{i}\,\omega)t}\,\mathrm{d}t = -\frac{1}{s-\mathrm{i}\,\omega}\,\mathrm{e}^{-(s-\mathrm{i}\,\omega)t}\Big|_{t=0}^{\infty}$$
$$= \frac{1}{s-\mathrm{i}\,\omega} = \frac{s+\mathrm{i}\,\omega}{(s-\mathrm{i}\,\omega)(s+\mathrm{i}\,\omega)} = \frac{s+\mathrm{i}\,\omega}{s^2+\omega^2}, \quad s>0.$$

Since $\cos\omega t = \mathcal{Re}(\mathrm{e}^{\mathrm{i}\,\omega t})$ and $\sin\omega t = \mathcal{Im}(\mathrm{e}^{\mathrm{i}\,\omega t})$, one has

$$\mathscr{L}\{\cos\omega t\} = \mathcal{Re}\big[\mathscr{L}\{\mathrm{e}^{\mathrm{i}\,\omega t}\}\big] = \frac{s}{s^2+\omega^2},$$
$$\mathscr{L}\{\sin\omega t\} = \mathcal{Im}\big[\mathscr{L}\{\mathrm{e}^{\mathrm{i}\,\omega t}\}\big] = \frac{\omega}{s^2+\omega^2}.$$

Properties of the Laplace Transform

1. Note that

$$\mathscr{L}\{c_1 f_1(t) + c_2 f_2(t)\} = \int_0^\infty \mathrm{e}^{-st}\big[c_1 f_1(t) + c_2 f_2(t)\big]\mathrm{d}t, \quad c_1 \text{ and } c_2 \text{ are constants}$$
$$= c_1\int_0^\infty \mathrm{e}^{-st} f_1(t)\,\mathrm{d}t + c_2\int_0^\infty \mathrm{e}^{-st} f_2(t)\,\mathrm{d}t$$
$$= c_1\mathscr{L}\{f_1(t)\} + c_2\mathscr{L}\{f_2(t)\}$$

∴ The Laplace transform $\mathscr{L}\{\cdot\}$ is a linear operator.

2. Laplace Transform of Derivatives

$$\mathscr{L}\{f'(t)\} = \int_0^\infty \mathrm{e}^{-st} f'(t)\,\mathrm{d}t = \int_0^\infty \mathrm{e}^{-st}\,\mathrm{d}[f(t)]$$

$$= \mathrm{e}^{-st} f(t)\Big|_{t=0}^{\infty} - \int_0^\infty f(t)\big(-s\,\mathrm{e}^{-st}\big)\,\mathrm{d}t \qquad \text{✍ Integration by parts}$$

$$= -f(0) + s\int_0^\infty \mathrm{e}^{-st} f(t)\,\mathrm{d}t,$$

$$\mathscr{L}\{f'(t)\} = s\,F(s) - f(0).$$

Using this result, Laplace transforms of higher-order derivatives can be derived:

$$\mathscr{L}\{f''(t)\} = s\,\mathscr{L}\{f'(t)\} - f'(0), \qquad \text{✍ } f''(t) = \big[f'(t)\big]'$$

$$= s\big[s\,F(s) - f(0)\big] - f'(0),$$

$$\mathscr{L}\{f''(t)\} = s^2 F(s) - s\,f(0) - f'(0).$$

In general, one obtains

$$\mathscr{L}\{f^{(n)}(t)\} = s^n F(s) - s^{n-1} f(0) - s^{n-2} f'(0) - \cdots - s\,f^{(n-2)}(0) - f^{(n-1)}(0).$$

Remarks: The Laplace transform of the nth-order derivative $f^{(n)}(t)$ of function $f(t)$ is reduced to $s^n F(s)$, along with terms of the form $s^i f^{(n-i-1)}(0)$, $i=0, 1, \ldots, n-1$.

3. Property of Shifting

Given the Laplace transform $F(s) = \int_0^\infty \mathrm{e}^{-st} f(t)\,\mathrm{d}t$, replacing s by $s-a$ leads to

$$F(s-a) = \int_0^\infty \mathrm{e}^{-(s-a)t} f(t)\,\mathrm{d}t = \int_0^\infty \mathrm{e}^{-st}\big[\mathrm{e}^{at} f(t)\big]\mathrm{d}t = \mathscr{L}\{\mathrm{e}^{at} f(t)\},$$

$$\mathscr{L}\{\mathrm{e}^{at} f(t)\} = F(s-a) = \mathscr{L}\{f(t)\}\Big|_{s\to(s-a)}.$$

The Laplace transform of $f(t)$ multiplied by e^{at} is equal to $F(s-a)$, with s shifted by $-a$.

4. Property of Differentiation

Since $F(s) = \int_0^\infty \mathrm{e}^{-st} f(t)\,\mathrm{d}t$, differentiating with respect to s gives

$$F'(s) = \int_0^\infty (-t)\mathrm{e}^{-st} f(t)\,\mathrm{d}t = -\int_0^\infty \mathrm{e}^{-st}\big[t\,f(t)\big]\mathrm{d}t,$$

$$\mathscr{L}\{t\,f(t)\} = -F'(s).$$

Differentiating with respect to s again leads to

$$F''(s) = -\int_0^\infty (-t)\,\mathrm{e}^{-st}\big[t\,f(t)\big]\mathrm{d}t = \int_0^\infty \mathrm{e}^{-st}\big[t^2 f(t)\big]\mathrm{d}t,$$

$$\mathscr{L}\{t^2 f(t)\} = F''(s).$$

Following this procedure, one obtains

$$\mathscr{L}\{t^n f(t)\} = (-1)^n F^{(n)}(s) = (-1)^n \frac{\mathrm{d}^n F(s)}{\mathrm{d}s^n}, \quad n = 1, 2, \ldots.$$

Example 6.2

Determine the Laplace transform of $f(t) = t^n$, $n = 0, 1, \ldots$.

Solution 1: Using the formula of Laplace transform of derivative

$$\mathscr{L}\{f'(t)\} = s\,\mathscr{L}\{f(t)\} - f(0), \quad \text{with } f(t) = t^n,$$

one obtains

$$\mathscr{L}\{n\,t^{n-1}\} = s\,\mathscr{L}\{t^n\} - 0^n \implies \mathscr{L}\{t^n\} = \frac{n}{s}\,\mathscr{L}\{t^{n-1}\}.$$

Hence, for $n = 1, 2, \ldots$,

$$n = 1: \quad \mathscr{L}\{t\} = \frac{1}{s}\,\mathscr{L}\{1\} = \frac{1}{s^2}, \qquad ✍\ \mathscr{L}\{1\} = \frac{1}{s}$$

$$n = 2: \quad \mathscr{L}\{t^2\} = \frac{2}{s}\,\mathscr{L}\{t\} = \frac{2}{s}\cdot\frac{1}{s^2} = \frac{2\cdot 1}{s^3},$$

$$n = 3: \quad \mathscr{L}\{t^3\} = \frac{3}{s}\,\mathscr{L}\{t^2\} = \frac{3}{s}\cdot\frac{2\cdot 1}{s^3} = \frac{3\cdot 2\cdot 1}{s^4},$$

$\cdots\ \cdots$

in general,

$$\mathscr{L}\{t^n\} = \frac{n!}{s^{n+1}}, \quad n = 0, 1, \ldots. \qquad ✍\ 0! = 1$$

Solution 2: Use the property of differentiation

$$\mathscr{L}\{t^n f(t)\} = (-1)^n \frac{\mathrm{d}^n F(s)}{\mathrm{d}s^n}, \quad n = 1, 2, \ldots,$$

with $f(t) = 1$ and $F(s) = \dfrac{1}{s}$. Note that

$$\frac{\mathrm{d}}{\mathrm{d}s}\Big(\frac{1}{s}\Big) = -\frac{1}{s^2}, \qquad \frac{\mathrm{d}^2}{\mathrm{d}s^2}\Big(\frac{1}{s}\Big) = \frac{\mathrm{d}}{\mathrm{d}s}\Big(-\frac{1}{s^2}\Big) = \frac{1\cdot 2}{s^3} = \frac{2!}{s^3},$$

$$\frac{d^3}{ds^3}\left(\frac{1}{s}\right) = \frac{d}{ds}\left(-\frac{2!}{s^3}\right) = -\frac{3\cdot 2!}{s^4} = -\frac{3!}{s^4},$$

$$\cdots\ \cdots$$

$$\frac{d^n}{ds^n}\left(\frac{1}{s}\right) = (-1)^n \frac{n!}{s^{n+1}}.$$

Hence

$$n=1: \quad \mathscr{L}\{t\} = \mathscr{L}\{t\cdot 1\} = -\frac{d}{ds}\left(\frac{1}{s}\right) = \frac{1}{s^2},$$

$$n=2: \quad \mathscr{L}\{t^2\} = \mathscr{L}\{t^2\cdot 1\} = \frac{d^2}{ds^2}\left(\frac{1}{s}\right) = \frac{2!}{s^3},$$

$$\cdots\ \cdots$$

$$n: \quad \mathscr{L}\{t^n\} = \mathscr{L}\{t^n\cdot 1\} = (-1)^n \frac{d^n}{ds^n}\left(\frac{1}{s}\right) = \frac{n!}{s^{n+1}}.$$

Example 6.3

Evaluate $\mathscr{L}\{e^{at}\cos\omega t\}$ and $\mathscr{L}\{e^{at}\sin\omega t\}$.

Note that

$$\mathscr{L}\{\cos\omega t\} = \frac{s}{s^2+\omega^2}, \qquad \mathscr{L}\{\sin\omega t\} = \frac{\omega}{s^2+\omega^2}.$$

Applying the property of shifting $\mathscr{L}\{e^{at}f(t)\} = F(s-a)$, one obtains

$$\mathscr{L}\{e^{at}\cos\omega t\} = \left.\frac{s}{s^2+\omega^2}\right|_{s\to(s-a)} = \frac{s-a}{(s-a)^2+\omega^2},$$

$$\mathscr{L}\{e^{at}\sin\omega t\} = \left.\frac{\omega}{s^2+\omega^2}\right|_{s\to(s-a)} = \frac{\omega}{(s-a)^2+\omega^2}.$$

Example 6.4

Evaluate $\mathscr{L}\{t\sin t\}$ and $\mathscr{L}\{t^2\sin t\}$.

$$\mathscr{L}\{\sin\omega t\} = \frac{\omega}{s^2+\omega^2} \implies \mathscr{L}\{\sin t\} = \frac{1}{s^2+1} = F(s).$$

Apply the property of differentiation $\mathscr{L}\{t^n f(t)\} = (-1)^n F^{(n)}(s)$, $n=1, 2, \ldots$.

$$\frac{dF(s)}{ds} = \frac{d}{ds}\left(\frac{1}{s^2+1}\right) = -\frac{2s}{(s^2+1)^2},$$

$$\frac{d^2F(s)}{ds^2} = \frac{d}{ds}\left\{\frac{dF(s)}{ds}\right\} = \frac{d}{ds}\left\{-\frac{2s}{(s^2+1)^2}\right\} = \frac{6s^2-2}{(s^2+1)^3},$$

$$\therefore\ \mathscr{L}\{t\sin t\} = -\frac{dF(s)}{ds} = \frac{2s}{(s^2+1)^2}, \quad \mathscr{L}\{t^2\sin t\} = \frac{d^2F(s)}{ds^2} = \frac{6s^2-2}{(s^2+1)^3}.$$

Example 6.5

Evaluate $\mathscr{L}\{t\,e^{at}\sin\omega t\}$.

Using the property of shifting $\mathscr{L}\{e^{at}f(t)\} = F(s-a) = \mathscr{L}\{f(t)\}\Big|_{s\to(s-a)}$ leads to

$$\mathscr{L}\{e^{at}\sin\omega t\} = \mathscr{L}\{\sin\omega t\}\Big|_{s\to(s-a)} = \frac{\omega}{s^2+\omega^2}\Big|_{s\to(s-a)} = \frac{\omega}{(s-a)^2+\omega^2}.$$

Applying the property of differentiation $\mathscr{L}\{t\,f(t)\} = -\dfrac{d}{ds}\mathscr{L}\{f(t)\}$

$$\mathscr{L}\{t\cdot e^{at}\sin\omega t\} = -\frac{d}{ds}\mathscr{L}\{e^{at}\sin\omega t\} = -\frac{d}{ds}\left\{\frac{\omega}{(s-a)^2+\omega^2}\right\}$$

$$= \frac{2\omega(s-a)}{\left[(s-a)^2+\omega^2\right]^2}.$$

6.2 The Heaviside Step Function

The Heaviside step function is defined by

$$H(t-a) = \begin{cases} 0, & t<a, \\ 1, & t>a, \end{cases}$$

where a is a real number, and is shown in Figure 6.1.

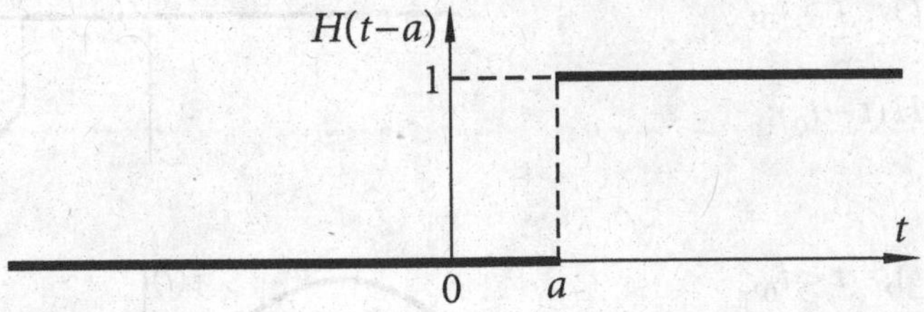

Figure 6.1 Heaviside step function.

The Laplace transform of $H(t-a)$ is given by

$$\mathscr{L}\{H(t-a)\} = \int_0^\infty e^{-st}H(t-a)\,dt$$

$$= \int_0^a e^{-st}H(t-a)\,dt + \int_a^\infty e^{-st}H(t-a)\,dt$$

$$= \int_0^a e^{-st}\cdot 0\,dt + \int_a^\infty e^{-st}\cdot 1\,dt = -\frac{1}{s}e^{-st}\Big|_{t=a}^{\infty} = \frac{1}{s}e^{-as},$$

$$\mathscr{L}\{H(t-a)\} = \frac{1}{s}\mathrm{e}^{-as}, \quad s>0,$$

and, for $s>a\geqslant 0$,

$$\begin{aligned}\mathscr{L}\{f(t-a)H(t-a)\} &= \int_0^\infty \mathrm{e}^{-st} f(t-a)H(t-a)\,\mathrm{d}t \\ &= \int_{-a}^\infty \mathrm{e}^{-s(\tau+a)} f(\tau)H(\tau)\,\mathrm{d}\tau, \qquad t-a=\tau, \\ &= \mathrm{e}^{-as}\int_0^\infty \mathrm{e}^{-s\tau} f(\tau)\,\mathrm{d}\tau = \mathrm{e}^{-as}F(s),\end{aligned}$$

$$\mathscr{L}\{f(t-a)H(t-a)\} = \mathrm{e}^{-as}F(s), \quad s>a\geqslant 0.$$

The Heaviside step function is very useful in dealing with functions with discontinuities or piecewise smooth functions. The following are some examples:

(1)

$$f(t) = \begin{cases} f_1(t), & t<t_0, \\ 0, & t>t_0, \end{cases}$$

$$= f_1(t)\left[1-H(t-t_0)\right];$$

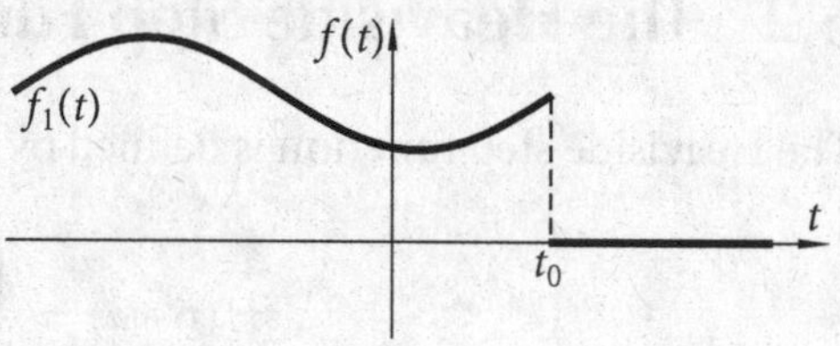

(2)

$$f(t) = \begin{cases} 0, & t<t_0, \\ f_2(t), & t>t_0, \end{cases}$$

$$= f_2(t)H(t-t_0);$$

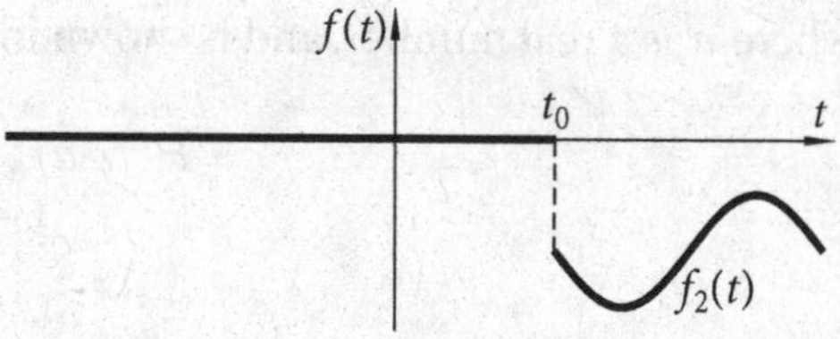

(3)

$$f(t) = \begin{cases} f_1(t), & t<t_0, \\ f_2(t), & t>t_0, \end{cases}$$

$$= \begin{cases} f_1(t)+\left[f_2(t)-f_1(t)\right]\cdot 0, & t<t_0, \\ f_1(t)+\left[f_2(t)-f_1(t)\right]\cdot 1, & t>t_0, \end{cases}$$

$$= f_1(t)+\left[f_2(t)-f_1(t)\right]H(t-t_0)$$

$$= f_1(t)\left[1-H(t-t_0)\right]+f_2(t)H(t-t_0).$$

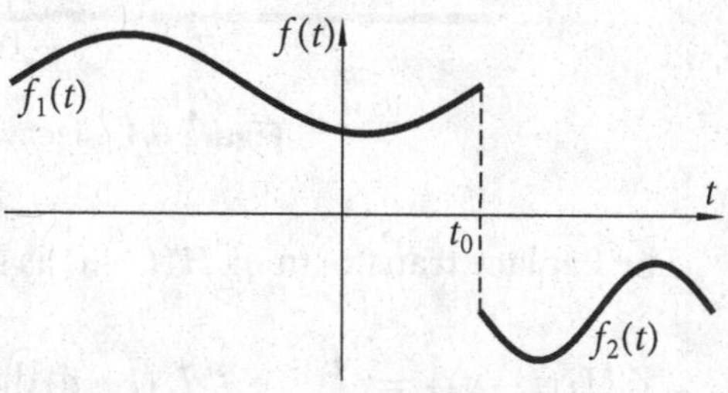

Remarks: Since Case (3) is the combination of Cases (1) and (2), the result obtained reflects this observation.

If function $f(t)$ has nonzero values only in the range of $a<x<b$ as shown in the following figure, than it can be expressed as

$$f(t)=\begin{cases}0, & t<a,\\ g(x), & a<t<b,\\ 0, & t>b,\end{cases}$$

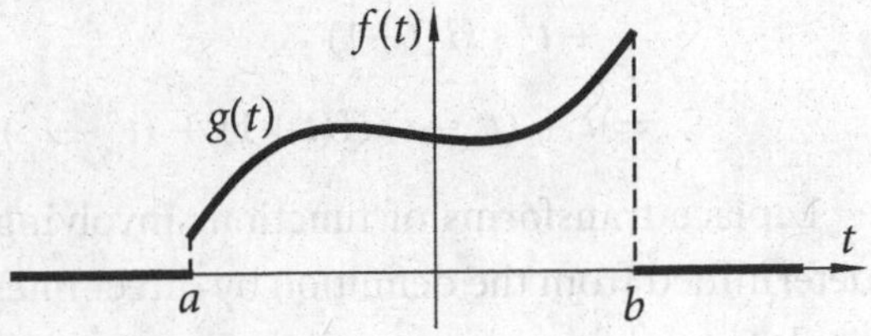

$$= g(x)\left[H(t-a)-H(t-b)\right],$$

since

$$H(t-a)-H(t-b)=\begin{cases}0, & t<a,\\ 1, & a<t<b,\\ 0, & t>b.\end{cases}$$

As a generalization, function $f(t)$ of the following form can be easily written in terms of the Heaviside step function

$$f(t)=\begin{cases}0, & t<t_0,\\ f_1(t), & t_0<t<t_1,\\ f_2(t), & t_1<t<t_2,\\ \dots & \dots\\ f_n(t), & t_{n-1}<t<t_n,\\ 0, & t>t_n,\end{cases}$$

$$= f_1(t)\left[H(t-t_0)-H(t-t_1)\right]+f_2(t)\left[H(t-t_1)-H(t-t_2)\right] + \cdots + f_n(t)\left[H(t-t_{n-1})-H(t-t_n)\right].$$

Example 6.6

Express the following functions in terms of the Heaviside function

1.
$$f(t)=\begin{cases}0, & t<0,\\ 1, & 0<t<1,\\ 2, & 1<t<2,\\ 0, & t>2;\end{cases}$$

2.
$$f(t)=\begin{cases}t, & t<-1,\\ t^2, & -1<t<1,\\ t^3, & t>1.\end{cases}$$

---

1. $f(t)=1\cdot\left[H(t-0)-H(t-1)\right]$ ✍ $f(t)=1,\ 0<t<1$

$\quad +2\cdot\left[H(t-1)-H(t-2)\right]$ ✍ $f(t)=2,\ 1<t<2$

$\quad = H(t)+H(t-1)-2\,H(t-2);$

2. $f(t) = t \cdot [1 - H(t+1)]$ ✍ $f(t)=t,\quad t<-1$

$\quad + t^2 \cdot [H(t+1) - H(t-1)]$ ✍ $f(t)=t^2,\quad -1<t<1$

$\quad + t^3 \cdot H(t-1)$ ✍ $f(t)=t^3,\quad t>1$

$\quad = t + (t^2 - t)\,H(t+1) + (t^3 - t^2)\,H(t-1).$

Laplace transforms of functions involving the Heaviside step function can be determined from the definition by direct integration or using the properties of the Laplace transform.

Example 6.7

Evaluate $\mathscr{L}\{t\,H(t-2)\}$.

Solution 1: Using the definition of Laplace transform

$$\mathscr{L}\{t\,H(t-2)\} = \int_0^\infty e^{-st} \cdot t\,H(t-2)\,dt = \int_2^\infty e^{-st}\,t\,dt$$

$$= -\frac{1}{s}\int_2^\infty t\,d(e^{-st}), \qquad \text{✍ Integration by parts}$$

$$= -\frac{1}{s}\left(t\,e^{-st}\Big|_{t=2}^{\infty} - \int_2^\infty e^{-st}\,dt\right)$$

$$= -\frac{1}{s}\left(-2e^{-2s} + \frac{1}{s}\,e^{-st}\Big|_{t=2}^{\infty}\right) = -\frac{1}{s}\left(-2e^{-2s} - \frac{1}{s}\,e^{-2s}\right)$$

$$= \frac{e^{-2s}}{s}\left(2 + \frac{1}{s}\right).$$

Solution 2: Using the following results

$$\mathscr{L}\{H(t-a)\} = \frac{1}{s}\,e^{-as}, \qquad \mathscr{L}\{t\,f(t)\} = -F'(s)$$

leads to

$$\mathscr{L}\{t\,H(t-2)\} = -\frac{d}{ds}\mathscr{L}\{H(t-2)\} = -\frac{d}{ds}\left(\frac{1}{s}\,e^{-2s}\right)$$

$$= -\left[-\frac{1}{s^2}\,e^{-2s} + \frac{1}{s}\,e^{-2s}(-2)\right] = \frac{e^{-2s}}{s}\left(2 + \frac{1}{s}\right).$$

Solution 3: To use the formula $\mathscr{L}\{f(t-a)\,H(t-a)\} = e^{-as}\,F(s)$ in evaluating $\mathscr{L}\{f(t)\,H(t-a)\}$, one must rewrite $f(t)$ as $f[(t-a)+a] = g(t-a)$.

$$\mathscr{L}\{t\,H(t-2)\} = \mathscr{L}\{[(t-2)+2]\,H(t-2)\}$$

$$= \mathscr{L}\{(t-2)\,H(t-2)\} + 2\,\mathscr{L}\{H(t-2)\}$$

$$= e^{-2s}\,\mathscr{L}\{t\} + 2 \cdot \frac{1}{s}\,e^{-2s} = e^{-2s}\left(\frac{1}{s^2} + \frac{2}{s}\right).$$

Example 6.8

Evaluate $\mathscr{L}\{e^{2t} H(t) - e^{-3t} H(t-4)\}$.

Three methods are used to evaluate $\mathscr{L}\{e^{at} H(t-b)\}$.

Method 1: By definition

$$\mathscr{L}\{e^{at} H(t-b)\} = \int_0^\infty e^{-st} \cdot e^{at} H(t-b)\,dt, \quad b>0$$
$$= \int_b^\infty e^{-(s-a)t}\,dt, \quad ✍ \; H(t-b)=0, \text{ for } t<b$$
$$= -\frac{1}{s-a}\,e^{-(s-a)t}\Big|_{t=b}^{\infty} = \frac{e^{-b(s-a)}}{s-a}, \quad s>a.$$

Method 2: Using the following results

$$\mathscr{L}\{H(t-b)\} = \frac{1}{s}\,e^{-bs}, \qquad \mathscr{L}\{e^{at} f(t)\} = \mathscr{L}\{f(t)\}\Big|_{s\to(s-a)}$$

leads to

$$\mathscr{L}\{e^{at} H(t-b)\} = \mathscr{L}\{H(t-b)\}\Big|_{s\to(s-a)} = \frac{1}{s}\,e^{-bs}\Big|_{s\to(s-a)} = \frac{e^{-b(s-a)}}{s-a}.$$

Method 3: Using the result $\mathscr{L}\{f(t-a)H(t-a)\} = e^{-as}F(s)$ results in

$$\mathscr{L}\{e^{at} H(t-b)\} = \mathscr{L}\{e^{a(t-b)+ab} H(t-b)\} = e^{ab}\,\mathscr{L}\{e^{a(t-b)} H(t-b)\}$$
$$= e^{ab}\cdot e^{-bs}\,\mathscr{L}\{e^{at}\} = e^{-b(s-a)}\,\frac{1}{s-a}.$$

Hence,

$$\mathscr{L}\{e^{2t} H(t) - e^{-3t} H(t-4)\} = \frac{e^{-0\cdot(s-2)}}{s-2} - \frac{e^{-4\cdot(s+3)}}{s+3} = \frac{1}{s-2} - \frac{e^{-4(s+3)}}{s+3}.$$

Example 6.9

Evaluate $F(s) = \mathscr{L}\{\cos 2t\, H(t-\frac{\pi}{8}) + (9t^2+2t-1)\,H(t-2)\}$.

$$F(s) = \mathscr{L}\Big\{\cos\big[2(t-\tfrac{\pi}{8})+\tfrac{\pi}{4}\big]H(t-\tfrac{\pi}{8}) + \big[9(t^2-4t+4)+38t-37\big]H(t-2)\Big\}$$
$$= \mathscr{L}\Big\{\big[\cos 2(t-\tfrac{\pi}{8})\cos\tfrac{\pi}{4} - \sin 2(t-\tfrac{\pi}{8})\sin\tfrac{\pi}{4}\big]H(t-\tfrac{\pi}{8})$$
$$+ \big[9(t-2)^2+38(t-2)+39\big]H(t-2)\Big\}$$
$$= \frac{1}{\sqrt{2}}\,e^{-\frac{\pi}{8}s}\,\mathscr{L}\{\cos 2t - \sin 2t\} + e^{-2s}\,\mathscr{L}\{9t^2+38t+39\}$$

$$= \frac{1}{\sqrt{2}} e^{-\frac{\pi}{8}s} \left(\frac{s}{s^2+2^2} - \frac{2}{s^2+2^2} \right) + e^{-2s} \left(9 \cdot \frac{2!}{s^3} + 38 \cdot \frac{1}{s^2} + 39 \cdot \frac{1}{s} \right)$$

$$= e^{-\frac{\pi}{8}s} \frac{s-2}{\sqrt{2}(s^2+4)} + e^{-2s} \frac{39s^2+38s+18}{s^3}.$$

6.3 Impulse Functions and the Dirac Delta Function

Impulse functions have wide applications in modeling of physical phenomena. For example, consider an elastic ball of mass m moving at velocity v_0 toward a rigid wall as shown in Figure 6.2.

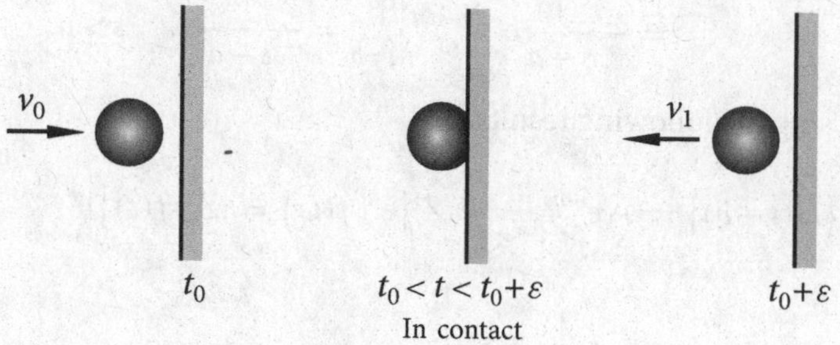

Figure 6.2 An elastic ball colliding with a rigid wall.

At time t_0, the ball collides with the wall; the wall exerts a force $f(t)$ on the ball over a short period of time ε. During this time, the ball is in contact with the wall and the velocity of the ball reduces from v_0 to 0 and then changes its direction, finally leaving the wall with velocity $-v_1$.

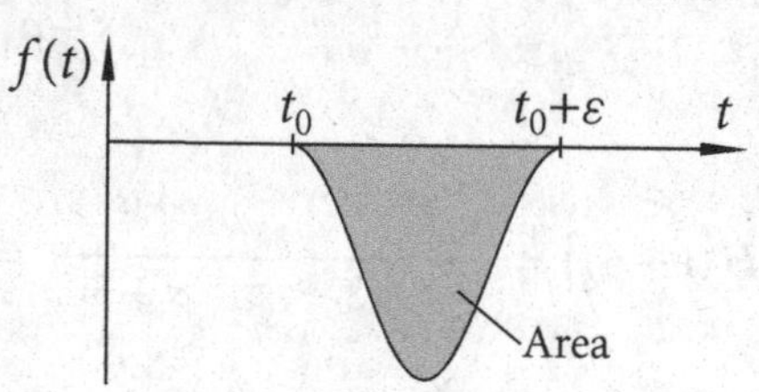

Figure 6.3 Contact force between the elastic ball and rigid wall.

The force $f(t)$ on the ball, shown in Figure 6.3, depends on the contact between the elastic ball and the rigid wall. The force $f(t)$ is negative because it is opposite to the direction of the initial velocity v_0.

The area under the force curve is called the impulse I, i.e.,

$$I = \int_{t_0}^{t_0+\varepsilon} f(t)\,\mathrm{d}t.$$

The *Impulse-Momentum Principle* states that the change in momentum of mass m is equal to the total impulse on m, i.e., $m(-v_1) - mv_0 = I$.

As a mathematical idealization, consider an impulse function $f(t)$ over a time interval $t_0 < t < t_0 + \varepsilon$ with constant amplitude I/ε as shown in Figure 6.4, such that the area under the function $f(t)$, or the impulse, is I.

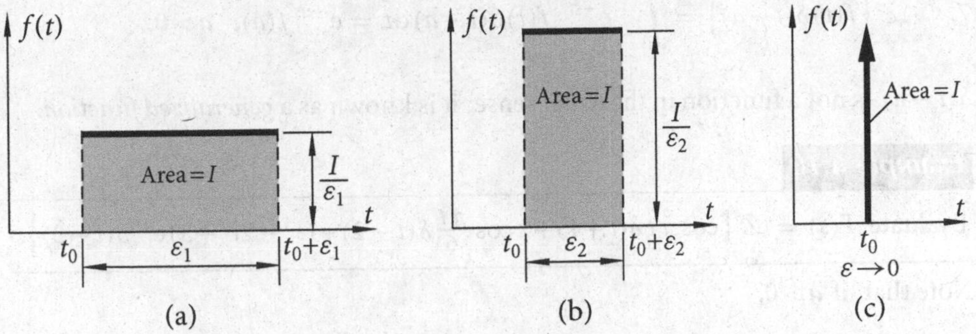

Figure 6.4 The Dirac delta function.

The impulse function $f(t)$ can be expressed in terms of the Heaviside step function

$$f(t) = \frac{I}{\varepsilon}\{H(t-t_0) - H[t-(t_0+\varepsilon)]\}.$$

When ε decreases as depicted in Figure 6.4(a) and Figure 6.4(b), the width of the time interval over which the impulse is defined decreases and the amplitude I/ε of the function increases, while keeping the area under the function constant. For $I = 1$, the limiting function as $\varepsilon \to 0$, i.e.,

$$\lim_{\varepsilon\to 0} f(t) = \lim_{\varepsilon\to 0} \frac{H(t-t_0) - H[t-(t_0+\varepsilon)]}{\varepsilon},$$

is called the unit impulse function or the *Dirac delta function* (Figure 6.4(c)) and is denoted by $\delta(t-t_0)$.

Properties of the Dirac Delta Function

1. $\delta(t-a) = 0$, if $t \neq a$;
2. $\delta(t-a) \to +\infty$, as $t \to a$;
3. $\displaystyle\int_{a-\alpha}^{a+\alpha} \delta(t-a)\,\mathrm{d}t = 1, \quad \alpha > 0;$
4. **Shifting Property** If $g(t)$ is any function,

$$\int_{a-\alpha}^{a+\alpha} g(t)\,\delta(t-a)\,\mathrm{d}t = g(a), \quad \alpha > 0.$$

5. $$\int_{-\infty}^{t} \delta(t-a)\,\mathrm{d}t = H(t-a) = \begin{cases} 0, & t < a, \\ 1, & t > a, \end{cases} \quad \Longrightarrow \quad \frac{\mathrm{d}H(t-a)}{\mathrm{d}t} = \delta(t-a).$$

6. $\mathscr{L}\{\delta(t-a)\} = \int_0^\infty e^{-st}\cdot\delta(t-a)\,dt = e^{-as}, \ a>0.$ ✍ Shifting property

7. $\mathscr{L}\{f(t)\,\delta(t-a)\} = \int_0^\infty e^{-st}\cdot f(t)\,\delta(t-a)\,dt = e^{-as}f(a), \ a>0.$

$\delta(t-a)$ is not a function in the usual sense; it is known as a *generalized function.*

Example 6.10

Evaluate $F(s) = \mathscr{L}\{\cos\pi t\,\delta(t+1)+2\cos\frac{\pi t}{6}\,\delta(t-2)+(t^3-2t^2+5)\,e^{4t}\,\delta(t-3)\}$.

Note that, if $a>0$,

$$\delta(t+a)=0, \ \text{for } t\geqslant 0 \implies \mathscr{L}\{f(t)\,\delta(t+a)\} = \int_0^\infty e^{-st}\cdot f(t)\,\delta(t+a)\,dt = 0,$$

and

$$\mathscr{L}\{f(t)\,\delta(t-a)\} = \int_0^\infty e^{-st}\cdot f(t)\,\delta(t-a)\,dt = e^{-as}f(a).$$

Hence

$$F(s) = 0 + e^{-2s}\cdot 2\cos\left(\frac{\pi}{6}\cdot 2\right) + e^{-3s}\cdot(3^3-2\cdot 3^2+5)\,e^{4\cdot 3} = e^{-2s} + 14e^{12-3s}.$$

Applications of the Dirac Delta Function

Consider a distributed load of intensity $w(x)$ over a length of width ε as shown in Figure 6.5(a).

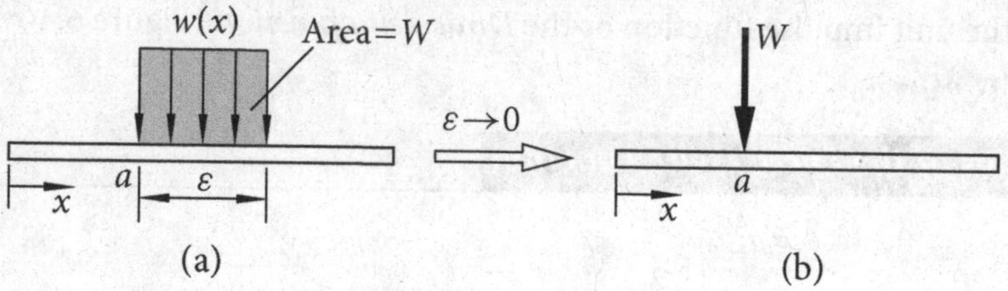

Figure 6.5 Distributed and concentrated loads.

The resultant force of the distributed load is

$$W = \int_{a-\frac{\varepsilon}{2}}^{a+\frac{\varepsilon}{2}} w(x)\,dx.$$

When $\varepsilon\to 0$, the distributed load approaches a concentrated load W, as shown in Figure 6.5(b). In terms of the Dirac delta function, the load can be expressed as $w(x) = W\delta(x-a)$, so that

$$\int_{a-\frac{\varepsilon}{2}}^{a+\frac{\varepsilon}{2}} w(x)\,dx = W\int_{a-\frac{\varepsilon}{2}}^{a+\frac{\varepsilon}{2}} \delta(x-a)\,dx = W.$$

As an example, consider a beam under two concentrated loads W_1 and W_2, applied at $x=a_1$ and $x=a_2$, respectively, and a uniformly distributed load of intensity w_0 over $b_1<x<b_2$ as shown in the following figure.

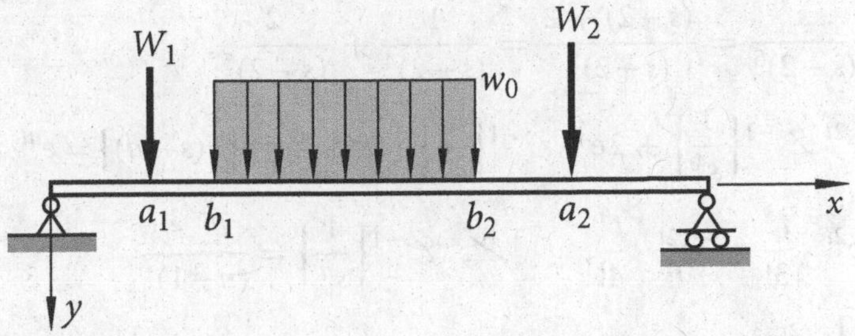

The deflection $y(x)$ of the beam is governed by the differential equation

$$\frac{d^2}{dx^2}\left\{EI(x)\,\frac{d^2y}{dx^2}\right\}=w(x),$$

where $EI(x)$ is the flexural rigidity of the beam, and the loading is given by

$$w(x)=w_0\big[H(x-b_1)-H(x-b_2)\big]+W_1\,\delta(x-a_1)+W_2\,\delta(x-a_2).$$

6.4 The Inverse Laplace Transform

Given the Laplace transform $F(s)$ of function $f(t)$, $F(s)=\mathscr{L}\{f(t)\}$, the inverse Laplace transform is $f(t)=\mathscr{L}^{-1}\{F(s)\}$.

Inverse Laplace transforms of frequently used functions are listed in Table of Inverse Laplace Transforms (Appendix A.5). Hence, to determine the inverse Laplace transform $\mathscr{L}^{-1}\{F(s)\}$, $F(s)$ has to be recast into a combination of the known functions so that the formulas in the Table of Inverse Laplace Transforms can be applied.

Properties of the Inverse Laplace Transform

Corresponding to the properties of the Laplace transform, the following properties of the inverse Laplace transform can be readily obtained.

1. $\mathscr{L}\{\,\cdot\,\}$ is a linear operator $\Longrightarrow$ $\mathscr{L}^{-1}\{\,\cdot\,\}$ **is a linear operator;**

2. $\mathscr{L}\{e^{at}f(t)\}=F(s-a)$ $\Longrightarrow$ $\mathscr{L}^{-1}\{F(s-a)\}=e^{at}\,\mathscr{L}^{-1}\{F(s)\}=e^{at}f(t)$;

3. $\mathscr{L}\{t^n f(t)\}=(-1)^n F^{(n)}(s)$ $\Longrightarrow$ $\mathscr{L}^{-1}\{F^{(n)}(s)\}=(-1)^n t^n f(t)$;

4. $\mathscr{L}\{f(t-a)H(t-a)\}=e^{-as}F(s-a)$ $\Longrightarrow$ $\mathscr{L}^{-1}\{e^{-as}F(s)\}=f(t-a)\,H(t-a)$.

Example 6.11

Evaluate $\mathscr{L}^{-1}\left\{\dfrac{s}{(s-2)^5}\right\}$.

$$\because \quad F(s)=\frac{s}{(s-2)^5}=\frac{(s-2)+2}{(s-2)^5}=\frac{1}{(s-2)^4}+\frac{2}{(s-2)^5},$$

$$\therefore \quad f(t)=\mathrm{e}^{2t}\,\mathscr{L}^{-1}\left\{\frac{1}{s^4}\right\}+2\mathrm{e}^{2t}\,\mathscr{L}^{-1}\left\{\frac{1}{s^5}\right\}, \quad ✍\ \mathscr{L}^{-1}\{F(s-a)\}=\mathrm{e}^{at}\,\mathscr{L}^{-1}\{F(s)\}$$

$$=\mathrm{e}^{2t}\cdot\frac{t^3}{3!}+2\mathrm{e}^{2t}\cdot\frac{t^4}{4!}, \qquad ✍\ \mathscr{L}^{-1}\left\{\frac{1}{s^n}\right\}=\frac{t^{n-1}}{(n-1)!}$$

$$=\frac{1}{12}\mathrm{e}^{2t}t^3(2+t).$$

Example 6.12

Evaluate $\mathscr{L}^{-1}\left\{\dfrac{1+\mathrm{e}^{-3s}}{s^4}\right\}$.

$$F(s)=\frac{1+\mathrm{e}^{-3s}}{s^4}=\frac{1}{s^4}+\mathrm{e}^{-3s}\frac{1}{s^4}.$$

Note that

$$\mathscr{L}^{-1}\left\{\frac{1}{s^n}\right\}=\frac{t^{n-1}}{(n-1)!}, \qquad \mathscr{L}^{-1}\{\mathrm{e}^{-as}F(s)\}=f(t-a)H(t-a).$$

Hence,

$$f(t)=\mathscr{L}^{-1}\{F(s)\}=\frac{t^3}{3!}+\frac{(t-3)^3}{3!}H(t-3)=\frac{1}{6}\left[t^3+(t-3)^3H(t-3)\right].$$

Convolution Integral

Theorem — Convolution Integral

If $\mathscr{L}^{-1}\{F(s)\}=f(t)$ and $\mathscr{L}^{-1}\{G(s)\}=g(t)$, then

$$\mathscr{L}^{-1}\{F(s)\,G(s)\}=\int_0^t f(u)\,g(t-u)\,\mathrm{d}u=\int_0^t g(u)\,f(t-u)\,\mathrm{d}u=\big(f*g\big)(t),$$

in which the integral is known as a *Convolution Integral*.

Example 6.13

Evaluate $\mathscr{L}^{-1}\left\{\dfrac{s}{(s^2+4)^2}\right\}$.

Note that

$$\mathscr{L}^{-1}\left\{\frac{2}{s^2+2^2}\right\}=\sin 2t, \qquad \mathscr{L}^{-1}\left\{\frac{s}{s^2+2^2}\right\}=\cos 2t.$$

Using convolution integral, one has

$$\mathscr{L}^{-1}\left\{\frac{s}{(s^2+4)^2}\right\}=\frac{1}{2}\mathscr{L}^{-1}\left\{\frac{2}{s^2+2^2}\cdot\frac{s}{s^2+2^2}\right\}=\frac{1}{2}\sin 2t * \cos 2t$$

$$=\frac{1}{2}\int_0^t \sin 2u \cos 2(t-u)\,du, \quad ✍ \sin A\cos B=\tfrac{1}{2}\big[\sin(A+B)+\sin(A-B)\big]$$

$$=\frac{1}{4}\int_0^t \big[\sin 2t+\sin(4u-2t)\big]du=\frac{1}{4}\Big[u\sin 2t-\frac{1}{4}\cos(4u-2t)\Big]_{u=0}^{t}$$

$$=\frac{1}{4}\Big[t\sin 2t-\frac{1}{4}\cos 2t+\frac{1}{4}\cos(-2t)\Big]=\frac{1}{4}t\sin 2t.$$

Partial Fractions

Partial fraction decomposition is an essential step in solving ordinary differential equations using the method of Laplace transform. Some important aspects of partial fractions are briefly reviewed.

Consider a fraction $N(x)/D(x)$, where $N(x)$ and $D(x)$ are polynomials in x of degrees n_N and n_D, respectively.

- If $n_N \geqslant n_D$, i.e., the degree of the numerator $N(x)$ is greater than or equal to the degree of the denominator, the fraction can be simplified using long division to yield

$$\frac{N(x)}{D(x)}=P(x)+\frac{N_1(x)}{D(x)},$$

  where $P(x)$ and $N_1(x)$ are both polynomials in x, and the degree of $N_1(x)$ is less than that of $D(x)$. Hence, without loss of generality, the case for $n_N < n_D$ is considered in the following.

- Completely factorize the denominator $D(x)$ into factors of the form

$$(\alpha x+\beta)^m \quad \text{and} \quad (ax^2+bx+c)^n,$$

  where ax^2+bx+c is an unfactorable quadratic.

  For each factor of the form $(\alpha x+\beta)^m$, the partial fraction decomposition includes the following m terms

$$\frac{A_m}{(\alpha x+\beta)^m}+\frac{A_{m-1}}{(\alpha x+\beta)^{m-1}}+\cdots+\frac{A_1}{(\alpha x+\beta)}.$$

For each factor of the form $(ax^2+bx+c)^n$, the partial fraction decomposition includes the following n terms

$$\frac{B_n x+C_n}{(ax^2+bx+c)^n}+\frac{B_{n-1}x+C_{n-1}}{(ax^2+bx+c)^{n-1}}+\cdots+\frac{B_1 x+C_1}{(ax^2+bx+c)}.$$

For example,

$$\frac{5x^3+2x+7}{2x^8+7x^7-10x^5-6x^4-x^3}=\frac{5x^3+2x+7}{x^3(2x-1)(x^2+2x-1)^2}$$
$$=\frac{A_3}{x^3}+\frac{A_2}{x^2}+\frac{A_1}{x}+\frac{B}{2x-1}+\frac{C_2x+D_2}{(x^2+2x-1)^2}+\frac{C_1x+D_1}{x^2+2x-1}.$$

Summing up the right-hand side, the numerator of the resulting fraction is a polynomial of degree 7. Comparing the coefficients of the numerators leads to a system of eight linear algebraic equations for the unknown constants.

The Cover-Up Method

Suppose $D(x)$ has a factor $(x-a)^m$, then the partial fraction decomposition can be written as

$$\frac{N(x)}{D(x)}=\frac{N(x)}{(x-a)^m D_1(x)}=\frac{A_m}{(x-a)^m}+\frac{A_{m-1}}{(x-a)^{m-1}}+\cdots+\frac{A_1}{(x-a)}+\frac{N_1(x)}{D_1(x)},$$

in which $D_1(x)$ does not have $(x-a)$ as a factor. Multiplying both sides of the equation by $(x-a)^m$ yields

$$\frac{N(x)}{D_1(x)}=A_m+A_{m-1}(x-a)+\cdots+A_1(x-a)^{m-1}+\frac{N_1(x)}{D_1(x)}(x-a)^m.$$

Setting $x=a$ gives the value of A_m

$$A_m=\left.\frac{N(x)}{D_1(x)}\right|_{x=a}.$$

This result can be restated as follows: to find A_m, "cover-up" (remove) the term $(x-a)^m$ and set $x=a$:

$$A_m=\left.\frac{N(x)}{\cancel{(x-a)^m}\,D_1(x)}\right|_{x=a}.$$

The cover-up method works only for the **highest power** of repeated **linear factor.**

For example,

$$\frac{4}{(x-1)^2(x+1)}=\frac{A_2}{(x-1)^2}+\frac{A_1}{x-1}+\frac{B}{x+1}.$$

To find A_2, cover-up $(x-1)^2$ and set $x=1$:

$$A_2 = \left.\frac{4}{x+1}\right|_{x=1} = 2.$$

To find B, cover-up $(x+1)$ and set $x=-1$:

$$B = \left.\frac{4}{(x-1)^2}\right|_{x=-1} = 1.$$

A_1 cannot be determined using the cover-up method. But since A_2 and B are known, A_1 can be found by substituting any numerical value (other than 1 and -1) for x in the partial fraction decomposition equation. For instance, setting $x=0$:

$$\frac{4}{(0-1)^2(0+1)} = \frac{2}{(0-1)^2} + \frac{A_1}{0-1} + \frac{1}{0+1} \implies A_1 = -1.$$

Example 6.14

Evaluate $\mathscr{L}^{-1}\left\{\dfrac{8}{(s-1)(s^2+2s+5)}\right\}$.

Solution 1: Using partial fractions

$$F(s) = \frac{8}{(s-1)(s^2+2s+5)} = \frac{A}{s-1} + \frac{Bs+C}{s^2+2s+5}$$

$$= \frac{A(s^2+2s+5)+(Bs+C)(s-1)}{(s-1)(s^2+2s+5)} = \frac{(A+B)s^2+(2A-B+C)s+(5A-C)}{(s-1)(s^2+2s+5)}.$$

To find A, cover-up $(s-1)$ and set $s=1$:

$$A = \left.\frac{8}{s^2+2s+5}\right|_{s=1} = \frac{8}{1+2+5} = 1.$$

Hence, comparing the coefficients of the numerators leads to

$$s^2: \quad A+B=0 \implies B=-A=-1,$$

$$s: \quad 2A-B+C=0 \implies C=B-2A=-1-2\cdot 1=-3,$$

$$1: \quad 5A-C=8.$$ ✍ Use this equation as a check: $5\cdot 1-(-3)=8$.

$$F(s) = \frac{1}{s-1} - \frac{s+3}{s^2+2s+5} = \frac{1}{s-1} - \frac{(s+1)+2}{(s+1)^2+2^2}$$

$$= \frac{1}{s-1} - \frac{(s+1)}{(s+1)^2+2^2} - \frac{2}{(s+1)^2+2^2}.$$

Using the property of shifting $\mathscr{L}^{-1}\{F(s-a)\} = e^{at}f(t)$ along with the results

$$\mathscr{L}^{-1}\left\{\frac{1}{s}\right\} = 1, \quad \mathscr{L}^{-1}\left\{\frac{s}{s^2+\omega^2}\right\} = \cos\omega t, \quad \mathscr{L}^{-1}\left\{\frac{\omega}{s^2+\omega^2}\right\} = \sin\omega t$$

yields

$$f(t) = \mathscr{L}^{-1}\{F(s)\} = e^t - e^{-t}\cos 2t - e^{-t}\sin 2t.$$

Solution 2: Using convolution integral

$$f(t) = \mathscr{L}^{-1}\left\{\frac{8}{(s-1)(s^2+2s+5)}\right\} = 4\mathscr{L}^{-1}\left\{\frac{1}{s-1}\cdot\frac{2}{(s+1)^2+2^2}\right\}$$

$$= 4\,e^t * (e^{-t}\sin 2t) = 4\int_0^t e^{-u}\sin 2u \cdot e^{t-u}\,du$$

$$= 4e^t\int_0^t e^{-2u}\sin 2u\,du = 4e^t\left[\frac{e^{-2u}}{(-2)^2+2^2}(-2\sin 2u - 2\cos 2u)\right]_0^t$$

$$= 4e^t\left[\frac{e^{-2t}}{8}(-2\sin 2t - 2\cos 2t) - \frac{1}{8}(-2)\right]$$

$$= -e^{-t}(\sin 2t + \cos 2t) + e^t.$$

Example 6.15

Evaluate $\mathscr{L}^{-1}\left\{\dfrac{s+1}{(s^2+1)(s^2+9)}\right\}$.

Solution 1: Using partial fractions

$$F(s) = \frac{s+1}{(s^2+1)(s^2+9)} = \frac{As+B}{s^2+1} + \frac{Cs+D}{s^2+9} = \frac{(As+B)(s^2+9)+(Cs+D)(s^2+1)}{(s^2+1)(s^2+9)}$$

$$= \frac{(A+C)s^3 + (B+D)s^2 + (9A+C)s + (9B+D)}{(s^2+1)(s^2+9)}.$$

Hence, comparing the coefficients of the numerators leads to

$$s^3: \quad A + C = 0, \tag{1}$$

$$s^2: \quad B + D = 0, \tag{2}$$

$$s: \quad 9A + C = 1, \tag{3}$$

$$1: \quad 9B + D = 1. \tag{4}$$

$$\therefore \quad \text{Eqn (3)} - \text{Eqn (1)}: \quad 8A = 1 \implies A = \tfrac{1}{8},$$

$$\text{Eqn (4)} - \text{Eqn (2)}: \quad 8B = 1 \implies B = \tfrac{1}{8},$$

$$\text{Eqn (1)}: \quad C = -A = -\tfrac{1}{8},$$

$$\text{Eqn (2)}: \quad D = -B = -\tfrac{1}{8}.$$

$$\therefore\; F(s) = \frac{1}{8}\left(\frac{s+1}{s^2+1} - \frac{s+1}{s^2+9}\right) = \frac{1}{8}\left(\frac{s}{s^2+1^2} + \frac{1}{s^2+1^2} - \frac{s}{s^2+3^2} - \frac{1}{3}\cdot\frac{3}{s^2+3^2}\right),$$

$$f(t) = \mathscr{L}^{-1}\{F(s)\} = \tfrac{1}{8}\big(\cos t + \sin t - \cos 3t - \tfrac{1}{3}\sin 3t\big).$$

Solution 2: Using convolution integral

$$\mathscr{L}^{-1}\left\{\frac{s+1}{(s^2+1)(s^2+9)}\right\} = \mathscr{L}^{-1}\left\{\frac{s}{(s^2+1)(s^2+9)}\right\} + \mathscr{L}^{-1}\left\{\frac{1}{(s^2+1)(s^2+9)}\right\}$$

$$= \frac{1}{3}\mathscr{L}^{-1}\left\{\frac{s}{s^2+1^2}\cdot\frac{3}{s^2+3^2}\right\} + \frac{1}{3}\mathscr{L}^{-1}\left\{\frac{1}{s^2+1^2}\cdot\frac{3}{s^2+3^2}\right\}$$

$$= \frac{1}{3}\big(\cos t * \sin 3t + \sin t * \sin 3t\big)$$

$$= \frac{1}{3}\int_0^t (\cos u + \sin u)\sin 3(t-u)\,du$$

$$= \frac{1}{3}\int_0^t \left\{\frac{1}{2}\big[\sin(3t-2u) - \sin(4u-3t)\big] - \frac{1}{2}\big[\cos(3t-2u) - \cos(4u-3t)\big]\right\} du$$

$$= \tfrac{1}{6}\Big[\tfrac{1}{2}\cos(3t-2u) + \tfrac{1}{4}\cos(4u-3t) + \tfrac{1}{2}\sin(3t-2u) + \tfrac{1}{4}\sin(4u-3t)\Big]_{u=0}^{t}$$

$$= \tfrac{1}{8}\big(\cos t + \sin t - \cos 3t - \tfrac{1}{3}\sin 3t\big).$$

6.5 Solving Differential Equations Using the Laplace Transform

Consider an nth-order linear differential equation with constant coefficients

$$a_n y^{(n)}(t) + a_{n-1} y^{(n-1)}(t) + \cdots + a_1 y'(t) + a_0 y(t) = f(t).$$

Taking the Laplace transform on both sides of the equation and noting that

$$\begin{aligned}
\mathscr{L}\{y(t)\} &= Y(s),\\
\mathscr{L}\{y'(t)\} &= s\,Y(s) - y(0),\\
\mathscr{L}\{y''(t)\} &= s^2 Y(s) - s\,y(0) - y'(0),\\
&\cdots\ \cdots\\
\mathscr{L}\{y^{(n)}(t)\} &= s^n Y(s) - s^{n-1}y(0) - s^{n-2}y'(0) - \cdots - s\,y^{(n-2)}(0) - y^{(n-1)}(0),
\end{aligned}$$

lead to an algebraic equation for $Y(s)$

$$a_n\Big[s^n Y(s) - \sum_{i=1}^{n} s^{n-i}y^{(i-1)}(0)\Big] + a_{n-1}\Big[s^{n-1}Y(s) - \sum_{i=1}^{n-1} s^{n-1-i}y^{(i-1)}(0)\Big] + \cdots$$
$$+ a_1\big[s\,Y(s) - y(0)\big] + a_0 Y(s) = \mathscr{L}\{f(t)\}.$$

Solving for $Y(s)$ yields

$$Y(s) = \frac{\mathscr{L}\{f(t)\} + \sum_{k=1}^{n}\sum_{i=1}^{k} a_k y^{(i-1)}(0)\, s^{k-i}}{\sum_{i=0}^{n} a_i s^i}.$$

Taking the inverse transform $y(t) = \mathscr{L}^{-1}\{Y(s)\}$ results in the solution of the differential equation.

Example 6.16

Solve $y'' - 8y' + 25y = e^{4t}\sin 3t.$

Let $Y(s) = \mathscr{L}\{y(t)\}$. Using the property of shifting, $\mathscr{L}\{e^{at} f(t)\} = F(s-a)$, one has

$$\mathscr{L}\{e^{4t}\sin 3t\} = \mathscr{L}\{\sin 3t\}\Big|_{s\to s-4} = \frac{3}{s^2+3^2}\Big|_{s\to s-4} = \frac{3}{(s-4)^2+3^2}.$$

Taking the Laplace transform of both sides of the differential equation yields

$$\left[s^2 Y(s) - s\,y(0) - y'(0)\right] - 8\left[s\,Y(s) - y(0)\right] + 25\,Y(s) = \frac{3}{(s-4)^2+3^2},$$

or, denoting $y(0) = y_0$, $y'(0) = v_0$,

$$(s^2 - 8s + 25)\,Y(s) = y_0 s + (v_0 - 8y_0) + \frac{3}{(s-4)^2+3^2}.$$

Solving for $Y(s)$ leads to

$$Y(s) = \frac{y_0(s-4)}{(s-4)^2+3^2} + \frac{v_0 - 4y_0}{(s-4)^2+3^2} + \frac{3}{\left[(s-4)^2+3^2\right]^2}.$$

Taking the inverse Laplace transform and using the property of shifting

$$\mathscr{L}^{-1}\{F(s-a)\} = e^{at}\,\mathscr{L}^{-1}\{F(s)\},\quad \mathscr{L}^{-1}\left\{\frac{1}{(s^2+a^2)^2}\right\} = \frac{1}{2a^3}(\sin at - at\cos at),$$

one obtains

$$\begin{aligned} y(t) = \mathscr{L}^{-1}\{Y(s)\} &= e^{4t}\,\mathscr{L}\left\{y_0\cdot\frac{s}{s^2+3^2} + \frac{v_0-4y_0}{3}\cdot\frac{3}{s^2+3^2} + 3\cdot\frac{1}{(s^2+3^2)^2}\right\} \\ &= e^{4t}\left[y_0\cos 3t + \frac{v_0-4y_0}{3}\sin 3t + \frac{3}{2\cdot 3^3}(\sin 3t - 3t\cos 3t)\right] \\ &= e^{4t}\left[y_0\cos 3t + \left(\frac{v_0-4y_0}{3} + \frac{1}{18}\right)\sin 3t - \frac{t}{6}\cos 3t\right]. \end{aligned}$$

Denoting

$$A = y_0, \qquad B = \frac{v_0 - 4y_0}{3} + \frac{1}{18},$$

the general solution of the differential equation can be written as

$$y(t) = e^{4t}\left[A\cos 3t + B\sin 3t - \frac{t}{6}\cos 3t\right].$$

Example 6.17

Solve $y'' + 9y = 18\sin 3t\, H(t-\pi), \quad y(0) = 1, \; y'(0) = 0.$

Let $Y(s) = \mathscr{L}\{y(t)\}$. Taking the Laplace transform of both sides of the differential equation yields

$$\left[s^2 Y(s) - s\,y(0) - y'(0)\right] + 9Y(s) = 18\,\mathscr{L}\{\sin 3t\, H(t-\pi)\},$$

where

$$\mathscr{L}\{\sin 3t\, H(t-\pi)\} = \mathscr{L}\{\sin[3(t-\pi)+3\pi]H(t-\pi)\} \quad ✍ \sin(3\pi+\theta) = -\sin\theta$$

$$= -\mathscr{L}\{\sin 3(t-\pi)H(t-\pi)\} \quad ✍ \mathscr{L}\{f(t-a)H(t-a)\} = e^{-as}\,\mathscr{L}\{f(t)\}$$

$$= -e^{-\pi s}\,\mathscr{L}\{\sin 3t\} = -e^{-\pi s}\frac{3}{s^2+3^2}.$$

Hence, solving for $Y(s)$ leads to

$$Y(s) = \frac{s}{s^2+9} - 54e^{-\pi s}\frac{1}{(s^2+9)^2}.$$

Noting that

$$\mathscr{L}^{-1}\left\{\frac{1}{(s^2+3^2)^2}\right\} = \frac{1}{2\cdot 3^3}(\sin 3t - 3t\cos 3t),$$

and using the property of shifting $\mathscr{L}^{-1}\{e^{as}F(s)\} = f(t-a)H(t-a)$, one has

$$54\,\mathscr{L}^{-1}\left\{e^{-\pi s}\frac{1}{(s^2+3^2)^2}\right\} = \left[\sin 3(t-\pi) - 3(t-\pi)\cos 3(t-\pi)\right]H(t-\pi)$$

$$= \left[-\sin 3t + 3(t-\pi)\cos 3t\right]H(t-\pi).$$

✍ $\sin(\theta - 3\pi) = -\sin(3\pi - \theta) = -\sin\theta, \; \cos(\theta - 3\pi) = \cos(3\pi - \theta) = -\cos\theta$

The solution of the differential equation is

$$y(t) = \mathscr{L}^{-1}\left\{\frac{s}{s^2+3^2}\right\} - 54\,\mathscr{L}^{-1}\left\{e^{-\pi s}\frac{1}{(s^2+3^2)^2}\right\}$$

$$= \cos 3t + \left[\sin 3t - 3(t-\pi)\cos 3t\right]H(t-\pi).$$

Example 6.18

Solve $y'''-y''+4y'-4y=40(t^2+t+1)H(t-2)$, $y(0)=5$, $y'(0)=0$, $y''(0)=10$.

Let $\Upsilon(s)=\mathscr{L}\{y(t)\}$. Taking the Laplace transform of both sides of the differential equation yields

$$\left[s^3\Upsilon(s)-s^2y(0)-sy'(0)-y''(0)\right]-\left[s^2\Upsilon(s)-sy(0)-y'(0)\right]$$
$$+4\left[s\Upsilon(s)-y(0)\right]-4\Upsilon(s)=\mathscr{L}\{40(t^2+t+1)H(t-2)\},$$

where, using $\mathscr{L}\{f(t-a)H(t-a)\}=\mathrm{e}^{-as}\mathscr{L}\{f(t)\}$,

$$\mathscr{L}\{40(t^2+t+1)H(t-2)\}=40\,\mathscr{L}\{\left[(t^2-4t+4)+5t-3\right]H(t-2)\}$$
$$=40\,\mathscr{L}\{\left[(t-2)^2+5(t-2)+7\right]H(t-2)\}$$
$$=40\mathrm{e}^{-2s}\mathscr{L}\{t^2+5t+7\}=40\mathrm{e}^{-2s}\left(\frac{2!}{s^3}+5\cdot\frac{1!}{s^2}+7\cdot\frac{1}{s}\right) \quad ✍\ \mathscr{L}\{t^n\}=\frac{n!}{s^{n+1}}$$
$$=\mathrm{e}^{-2s}\,\frac{40(7s^2+5s+2)}{s^3}.$$

Solving for $\Upsilon(s)$ gives

$$\Upsilon(s)=\frac{5s^2-5s+30}{s^3-s^2+4s-4}+\mathrm{e}^{-2s}\frac{40(7s^2+5s+2)}{s^3(s^3-s^2+4s-4)}.$$

Using partial fractions, one has

$$\frac{5s^2-5s+30}{(s-1)(s^2+4)}=\frac{A}{s-1}+\frac{Bs+C}{s^2+4}=\frac{(A+B)s^2+(-B+C)s+(4A-C)}{(s-1)(s^2+4)}$$

To find A, cover-up $(s-1)$ and set $s=1$

$$A=\left.\frac{5s^2-5s+30}{(s^2+4)}\right|_{s=1}=\frac{5-5+30}{1+4}=6.$$

Comparing the coefficients of the numerators leads to

s^2: $A+B=5 \implies B=5-A=5-6=-1$,

s: $-B+C=-5 \implies C=B-5=-1-5=-6$,

1: $4A-C=30$. ✍ Use this equation as a check: $4\cdot 6-(-6)=30$.

Hence,

$$\mathscr{L}^{-1}\left\{\frac{5s^2-5s+30}{(s-1)(s^2+4)}\right\}=\mathscr{L}^{-1}\left\{\frac{6}{s-1}+\frac{-s-6}{s^2+4}\right\}$$
$$=\mathscr{L}^{-1}\left\{6\cdot\frac{1}{s-1}-\frac{s}{s^2+2^2}-3\cdot\frac{2}{s^2+2^2}\right\}=6\mathrm{e}^t-\cos 2t-3\sin 2t.$$

Similarly, using partial fractions, one has

$$\frac{40(7s^2+5s+2)}{s^3(s-1)(s^2+4)} = \frac{A_3}{s^3} + \frac{A_2}{s^2} + \frac{A_1}{s} + \frac{B}{s-1} + \frac{Cs+D}{s^2+4}.$$

Comparing the coefficients of the numerators leads to

$$\begin{aligned}
s^5:&\quad A_1 + B + C = 0, &(1)\\
s^4:&\quad -A_1 + A_2 - C + D = 0, &(2)\\
s^3:&\quad 4A_1 - A_2 + A_3 + 4B - D = 0, &(3)\\
s^2:&\quad -4A_1 + 4A_2 - A_3 = 280, &(4)\\
s:&\quad -4A_2 + 4A_3 = 200, &(5)\\
1:&\quad -4A_3 = 80, &(6)\\
\text{from Eqn (6)}:&\quad A_3 = -20,\\
\text{from Eqn (5)}:&\quad A_2 = A_3 - 50 = -70,\\
\text{from Eqn (4)}:&\quad A_1 = \tfrac{1}{4}(4A_2 - A_3 - 280) = -135,
\end{aligned}$$

to find B, covering-up $(s-1)$ and setting $s=1$:

$$B = \frac{40(7s^2+5s+2)}{s^3(s^2+4)}\bigg|_{s=1} = \frac{40(7+5+2)}{1\cdot 5} = 112,$$

$$\begin{aligned}
\text{from Eqn (1)}:&\quad C = -A_1 - B = -(-135) - 112 = 23,\\
\text{from Eqn (6)}:&\quad D = A_1 - A_2 + C = -135 - (-70) + 23 = -42.
\end{aligned}$$

Hence,

$$\begin{aligned}
\mathscr{L}^{-1}\left\{\frac{40(7s^2+5s+2)}{s^3(s-1)(s^2+4)}\right\} &= \mathscr{L}^{-1}\left\{-\frac{20}{s^3} - \frac{70}{s^2} - \frac{135}{s} + \frac{112}{s-1} + \frac{23s-42}{s^2+2^2}\right\}\\
&= -135 - 70t - 10t^2 + 112e^t + 23\cos 2t - 21\sin 2t.
\end{aligned}$$

Applying the property of shifting $\mathscr{L}^{-1}\{e^{as}F(s)\} = f(t-a)H(t-a)$, one obtains the solution of the differential equation

$$\begin{aligned}
y(t) &= \mathscr{L}^{-1}\left\{\frac{5s^2-5s+30}{s^3-s^2+4s-4}\right\} + \mathscr{L}^{-1}\left\{e^{-2s}\frac{40(7s^2+5s+2)}{s^3(s^3-s^2+4s-4)}\right\}\\
&= 6e^t - \cos 2t - 3\sin 2t + \big[-135 - 70(t-2) - 10(t-2)^2 + 112e^{(t-2)}\\
&\qquad + 23\cos 2(t-2) - 21\sin 2(t-2)\big]H(t-2)\\
&= 6e^t - \cos 2t - 3\sin 2t + \big[-10t^2 - 30t - 35 + 112e^{t-2}\\
&\qquad + 23\cos(2t-4) - 21\sin(2t-4)\big]H(t-2).
\end{aligned}$$

Remarks:

- Using the method of Laplace transform to solve linear ordinary equations is a general approach, as long as the Laplace transform of the right-hand side function $f(t)$ and the resulting inverse Laplace transform $Y(s)$ can be readily obtained.
- The Laplace transforms of the Heaviside step function and the Dirac delta function are both continuous functions. As a result, **the method of Laplace transform offers great advantage in dealing with differential equations involving the Heaviside step function and the Dirac function.**
- The determination of inverse Laplace transforms is in general a tedious task. As a result, for a linear differential equation with continuous right-hand side function $f(t)$, the methods presented in Chapter 4 are usually more efficient than the method of Laplace transform, especially for higher-order differential equations.

6.6 Applications of the Laplace Transform

6.6.1 *Response of a Single Degree-of-Freedom System*

Consider a single degree-of-freedom system as shown in Figure 6.6 under externally applied force $f(t)$. The initial conditions of the system at time $t=0$ are $x(0)=x_0$ and $\dot{x}(0)=v_0$.

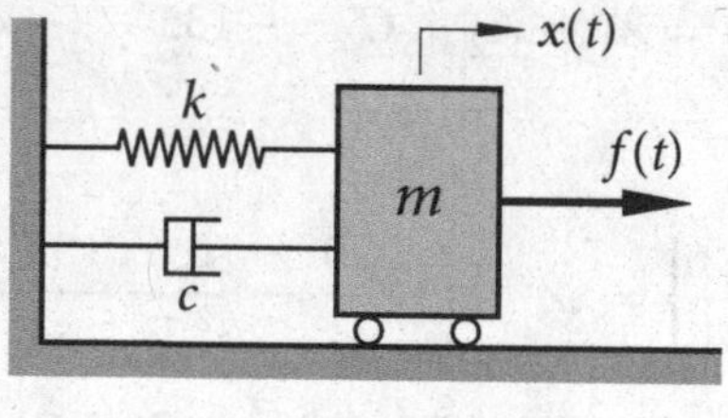

Figure 6.6 A single degree-of-freedom system.

As derived in Section 5.1, the equation of motion of the system is

$$m\ddot{x}+c\dot{x}+kx=f(t), \quad x(0)=x_0, \quad \dot{x}(0)=v_0.$$

Dividing the equation by m and rewriting it in the standard form yield

$$\ddot{x}+2\zeta\omega_0\dot{x}+\omega_0^2 x=\frac{1}{m}f(t), \qquad \omega_0^2=\frac{k}{m}, \quad 2\zeta\omega_0=\frac{c}{m}.$$

Applying the Laplace transform, $\mathscr{L}\{x(t)\}=X(s)$, one obtains

$$\big[s^2X(s)-sx(0)-\dot{x}(0)\big]+2\zeta\omega_0\big[sX(s)-x(0)\big]+\omega_0^2X(s)=\frac{1}{m}\mathscr{L}\{f(t)\}.$$

Solving for $X(s)$ leads to

$$X(s) = \underbrace{\frac{x_0 s + (v_0 + 2\zeta\omega_0 x_0)}{s^2 + 2\zeta\omega_0 s + \omega_0^2}}_{X_{\text{Free}}(s),\ \text{free vibration}} + \underbrace{\frac{1}{m}\cdot\frac{\mathscr{L}\{f(t)\}}{s^2 + 2\zeta\omega_0 s + \omega_0^2}}_{X_{\text{Forced}}(s),\ \text{forced vibration}},$$

in which the first term is the Laplace transform of the response of free vibration $x_{\text{Free}}(t)$, due to the initial conditions $x(0)$ and $\dot{x}(0)$; whereas the second term is the Laplace transform of the response of forced vibration $x_{\text{Forced}}(t)$, due to the externally applied force $f(t)$.

Free Vibration

The response of free vibration is given by

$$x_{\text{Free}}(t) = \mathscr{L}^{-1}\{X_{\text{Free}}(s)\} = \mathscr{L}^{-1}\left\{\frac{x_0 s + (v_0 + 2\zeta\omega_0 x_0)}{s^2 + 2\zeta\omega_0 s + \omega_0^2}\right\}.$$

❧ For $0 \leqslant \zeta < 1$, $\omega_d = \omega_0\sqrt{1-\zeta^2}$,

$$x_{\text{Free}}(t) = \mathscr{L}^{-1}\left\{\frac{x_0(s+\zeta\omega_0) + (v_0+\zeta\omega_0 x_0)}{(s+\zeta\omega_0)^2 + \omega_0^2(1-\zeta^2)}\right\}$$

$$= e^{-\zeta\omega_0 t}\,\mathscr{L}^{-1}\left\{\frac{x_0 s + (v_0+\zeta\omega_0 x_0)}{s^2 + \omega_0^2(1-\zeta^2)}\right\}, \quad ✍\ \mathscr{L}^{-1}\{F(s-a)\} = e^{at}\,\mathscr{L}^{-1}\{F(s)\}$$

$$= e^{-\zeta\omega_0 t}\left[x_0\,\mathscr{L}^{-1}\left\{\frac{s}{s^2+\omega_d^2}\right\} + \frac{v_0+\zeta\omega_0 x_0}{\omega_d}\,\mathscr{L}^{-1}\left\{\frac{\omega_d}{s^2+\omega_d^2}\right\}\right],$$

$$x_{\text{Free}}(t) = e^{-\zeta\omega_0 t}\left(x_0 \cos\omega_d t + \frac{v_0+\zeta\omega_0 x_0}{\omega_d}\sin\omega_d t\right), \quad 0 \leqslant \zeta < 1.$$

❧ For $\zeta = 1$,

$$x_{\text{Free}}(t) = \mathscr{L}^{-1}\left\{\frac{x_0 s + (v_0 + 2\omega_0 x_0)}{s^2 + 2\omega_0 s + \omega_0^2}\right\} = \mathscr{L}^{-1}\left\{\frac{x_0(s+\omega_0) + (v_0+\omega_0 x_0)}{(s+\omega_0)^2}\right\}$$

$$= e^{-\omega_0 t}\,\mathscr{L}^{-1}\left\{\frac{x_0 s + (v_0+\omega_0 x_0)}{s^2}\right\}$$

$$= e^{-\omega_0 t}\left[x_0\,\mathscr{L}^{-1}\left\{\frac{1}{s}\right\} + (v_0+\omega_0 x_0)\,\mathscr{L}^{-1}\left\{\frac{1}{s^2}\right\}\right],$$

$$x_{\text{Free}}(t) = e^{-\omega_0 t}\big[x_0 + (v_0+\omega_0 x_0)\,t\big], \quad \zeta = 1.$$

❧ For $\zeta > 1$,

$$x_{\text{Free}}(t) = \mathscr{L}^{-1}\left\{\frac{x_0(s+\zeta\omega_0) + (v_0+\zeta\omega_0 x_0)}{(s+\zeta\omega_0)^2 - \omega_0^2(\zeta^2-1)}\right\}$$

$$= e^{-\zeta\omega_0 t}\,\mathscr{L}^{-1}\left\{\frac{x_0 s + (v_0+\zeta\omega_0 x_0)}{s^2 - \omega_0^2(\zeta^2-1)}\right\}$$ ✍ Using partial fractions

$$= e^{-\zeta\omega_0 t}\cdot\frac{1}{2}\mathscr{L}^{-1}\left\{\frac{x_0 + \dfrac{v_0+\zeta\omega_0 x_0}{\omega_0\sqrt{\zeta^2-1}}}{s-\omega_0\sqrt{\zeta^2-1}} + \frac{x_0 - \dfrac{v_0+\zeta\omega_0 x_0}{\omega_0\sqrt{\zeta^2-1}}}{s+\omega_0\sqrt{\zeta^2-1}}\right\}$$

$$= e^{-\zeta\omega_0 t}\cdot\frac{1}{2}\left[\left(x_0 + \frac{v_0+\zeta\omega_0 x_0}{\omega_0\sqrt{\zeta^2-1}}\right)e^{\omega_0\sqrt{\zeta^2-1}\,t} + \left(x_0 - \frac{v_0+\zeta\omega_0 x_0}{\omega_0\sqrt{\zeta^2-1}}\right)e^{-\omega_0\sqrt{\zeta^2-1}\,t}\right],$$

$$x_{\text{Free}}(t) = \frac{1}{2\omega_0\sqrt{\zeta^2-1}}\left\{\left[v_0+\omega_0\left(\zeta+\sqrt{\zeta^2-1}\right)x_0\right]e^{-\omega_0(\zeta-\sqrt{\zeta^2-1})t} - \left[v_0+\omega_0\left(\zeta-\sqrt{\zeta^2-1}\right)x_0\right]e^{-\omega_0(\zeta+\sqrt{\zeta^2-1})t}\right\},\quad \zeta>1.$$

Remarks: These results are the same as those obtained in Section 5.1.2.1 and correspond to the complementary solutions.

Forced Vibration

Some examples are studied in the following to illustrate the determination of response of forced vibration using the Laplace transform.

Example 6.19 — Single DOF System under Sinusoidal Excitation

Determine $x_{\text{Forced}}(t)$ of a single degree-of-freedom system with $0 \leqslant \zeta < 1$ under external excitations $f(t) = m\sin\Omega t$ and $f(t) = m\cos\Omega t$.

For $f(t) = m\sin\Omega t$, one has

$$X_{\text{Forced}}(s) = X_{\sin}(s) = \frac{\mathscr{L}\{\sin\Omega t\}}{s^2+2\zeta\omega_0 s+\omega_0^2} = \frac{\Omega}{(s^2+2\zeta\omega_0 s+\omega_0^2)(s^2+\Omega^2)}.$$

Applying partial fractions

$$\frac{\Omega}{(s^2+2\zeta\omega_0 s+\omega_0^2)(s^2+\Omega^2)} = \frac{As+B}{s^2+2\zeta\omega_0 s+\omega_0^2} + \frac{Cs+D}{s^2+\Omega^2}.$$

Summing up the right-hand side and comparing the coefficients of the numerators, one obtains

$$s^3:\quad A + C = 0, \tag{1}$$

$$s^2:\quad B + 2\zeta\omega_0 C + D = 0, \tag{2}$$

$$s:\quad \Omega^2 A + \omega^2 C + 2\zeta\omega_0 D = 0, \tag{3}$$

$$1:\quad \Omega^2 B + \omega_0^2 D = \Omega, \tag{4}$$

$$\text{from Eqn (1)}:\quad A = -C,$$

$$\text{Eqn (2)} \times \Omega^2 - \text{Eqn (4)}: \quad D = \frac{\Omega + 2\zeta\omega_0\Omega^2 C}{\omega_0 - \Omega^2}.$$

Substituting into equation (3) yields

$$C = -\frac{2\zeta\omega_0\Omega}{(\omega_0^2 - \Omega^2)^2 + (2\zeta\omega_0\Omega)^2} = -A \implies D = \frac{\Omega\,(\omega_0^2 - \Omega^2)}{(\omega_0^2 - \Omega^2)^2 + (2\zeta\omega_0\Omega)^2}.$$

From equation (4),

$$B = \frac{1}{\Omega^2}(\Omega - \omega^2 D) = \frac{\Omega\big[(2\zeta\omega_0)^2 + \Omega^2 - \omega_0^2\big]}{(\omega_0^2 - \Omega^2)^2 + (2\zeta\omega_0\Omega)^2}.$$

Hence,

$$X_{\text{sin}}(s) = \frac{A\cdot(s+\zeta\omega_0)}{(s+\zeta\omega_0)^2 + \omega_d^2} + \frac{B - \zeta\omega_0 A}{\omega_d}\cdot\frac{\omega_d}{(s+\zeta\omega_0)^2 + \omega_d^2} + \frac{C\cdot s}{s^2 + \Omega^2} + \frac{D}{\Omega}\cdot\frac{\Omega}{s^2 + \Omega^2}.$$

Taking inverse Laplace transform results in

$$\begin{aligned} x_{\text{sin}}(t) &= \mathscr{L}^{-1}\{X_{\text{sin}}(s)\} \\ &= \mathrm{e}^{-\zeta\omega_0 t}\Big(A\cos\omega_d t + \frac{B - \zeta\omega_0 A}{\sin}\omega_d t\Big) + C\cos\Omega t + \frac{D}{\Omega}\sin\Omega t. \end{aligned}$$

$$\begin{aligned} x_{\text{sin}}(t) &= \mathscr{L}^{-1}\left\{\frac{\mathscr{L}\{\sin\Omega t\}}{s^2 + 2\zeta\omega_0 s + \omega_0^2}\right\} \\ &= \frac{1}{(\omega_0^2 - \Omega^2)^2 + (2\zeta\omega_0\Omega)^2}\bigg\{-2\zeta\omega_0\Omega\cos\Omega t + (\omega_0^2 - \Omega^2)\sin\Omega t \\ &\quad + \mathrm{e}^{-\zeta\omega_0 t}\Big[2\zeta\omega_0\Omega\cos\omega_d t + \frac{\Omega(2\zeta^2\omega_0^2 + \Omega^2 - \omega_0^2)}{\omega_d}\sin\omega_d t\Big]\bigg\}. \end{aligned}$$

Similarly, for $f(t) = m\cos\Omega t$, one has

$$X_{\text{Forced}}(s) = X_{\text{cos}}(s) = \frac{\mathscr{L}\{\cos\Omega t\}}{s^2 + 2\zeta\omega_0 s + \omega_0^2} = \frac{s}{(s^2 + 2\zeta\omega_0 s + \omega_0^2)(s^2 + \Omega^2)},$$

$$\begin{aligned} x_{\text{cos}}(t) &= \mathscr{L}^{-1}\left\{\frac{\mathscr{L}\{\cos\Omega t\}}{s^2 + 2\zeta\omega_0 s + \omega_0^2}\right\} \\ &= \frac{1}{(\omega_0^2 - \Omega^2)^2 + (2\zeta\omega_0\Omega)^2}\bigg\{(\omega_0^2 - \Omega^2)\cos\Omega t + 2\zeta\omega_0\Omega\sin\Omega t \\ &\quad + \mathrm{e}^{-\zeta\omega_0 t}\Big[-(\omega_0^2 - \Omega^2)\cos\omega_d t - \frac{\zeta\omega_0(\omega_0^2 + \Omega^2)}{\omega_d}\sin\omega_d t\Big]\bigg\}. \end{aligned}$$

Remarks: The results $x_{\sin}(t)$ and $x_{\cos}(t)$ are very useful for vibration problems of a single degree-of-freedom system under loads that can be expressed in terms of sinusoidal functions, in which the response of the forced vibration can be expressed in terms of $x_{\sin}(t)$ and $x_{\cos}(t)$. An example using these functions is presented in the following.

Example 6.20 — Vibration of a Vehicle Passing a Speed Bump

In Section 5.3, the vibration of a vehicle passing a speed bump is studied. The equation of motion is solved and the response is obtained separately for two time durations: vehicle on the speed bump and passed the speed bump. Solve this problem again using the Laplace transform.

Since the speed bump occurs for $0 \leqslant x \leqslant b$ or $0 \leqslant t \leqslant T$, $T = b/U$, it can be more easily expressed using the Heaviside step function:

$$y_0(x) = h \sin\Big(\frac{\pi}{b}x\Big)\big[1 - H(x-b)\big], \quad \text{or} \quad y_0(t) = h \sin\Big(\frac{\pi U}{b}t\Big)\big[1 - H(t-T)\big].$$

Referring to Section 5.3, the equation of motion for the relative displacement $z(t) = y(t) - y_0(t)$ becomes, for $t \geqslant 0$,

$$m\ddot{z} + c\dot{z} + kz = mh\Omega^2 \sin\Omega t\big[1 - H(t-T)\big], \qquad \Omega = \frac{\pi U}{b},$$

or, in the standard form as in Section 6.6.1,

$$\ddot{z} + 2\zeta\omega_0\dot{z} + \omega_0^2 z = \frac{1}{m} f(t), \quad 0 < \zeta < 1,$$

where $f(t) = mh\Omega^2 \sin\Omega t\big[1 - H(t-T)\big]$.

Since $z(0) = \dot{z}(0) = 0$, the response of free vibration is $z_{\text{Free}}(t) = 0$.

The Laplace transform $Z_{\text{Forced}}(s)$ of the response of forced vibration is

$$Z_{\text{Forced}}(s) = \frac{1}{m} \cdot \frac{\mathscr{L}\{f(t)\}}{s^2 + 2\zeta\omega_0 s + \omega_0^2},$$

where, using the property $\mathscr{L}\{f(t-a)H(t-a) = \mathrm{e}^{-as}F(s)\}$,

$$\begin{aligned}
&\mathscr{L}\{f(t)\} = mh\Omega^2 \mathscr{L}\big\{\sin\Omega t\big[1 - H(t-T)\big]\big\},\\
&\mathscr{L}\big\{\sin\Omega t\big[1 - H(t-T)\big]\big\} = \mathscr{L}\{\sin\Omega t\} - \mathscr{L}\big\{\sin\big[\Omega(t-T) + \Omega T\big]H(t-T)\big\}\\
&\quad = \mathscr{L}\{\sin\Omega t\} - \mathscr{L}\big\{\big[\sin\Omega(t-T)\cos\Omega T + \cos\Omega(t-T)\sin\Omega T\big]H(t-T)\big\}\\
&\quad = \mathscr{L}\{\sin\Omega t\} - \cos\Omega T \cdot \mathrm{e}^{-Ts}\mathscr{L}\{\sin\Omega t\} - \sin\Omega T \cdot \mathrm{e}^{-Ts}\mathscr{L}\{\cos\Omega t\}\\
&\quad = (1 - \cos\Omega T \cdot \mathrm{e}^{-Ts})\mathscr{L}\{\sin\Omega t\} - \sin\Omega T \cdot \mathrm{e}^{-Ts}\mathscr{L}\{\cos\Omega t\}.
\end{aligned}$$

Referring to Section 6.6.1, in terms of $X_{\sin}(s)$ and $X_{\cos}(s)$, the Laplace transform of the response of forced vibration is

$$Z_{\text{Forced}}(s) = h\Omega^2\left[\left(1-\cos\Omega T\cdot \mathrm{e}^{-Ts}\right)\frac{\mathscr{L}\{\sin\Omega t\}}{s^2+2\zeta\omega_0 s+\omega_0^2} - \sin\Omega T\cdot\mathrm{e}^{-Ts}\frac{\mathscr{L}\{\cos\Omega t\}}{s^2+2\zeta\omega_0 s+\omega_0^2}\right]$$

$$= h\Omega^2\left\{X_{\sin}(s) - \left[\cos\Omega T\cdot\mathrm{e}^{-Ts}X_{\sin}(s) + \sin\Omega T\cdot\mathrm{e}^{-Ts}X_{\cos}(s)\right]\right\}.$$

Taking inverse Laplace transform and using $\mathscr{L}^{-1}\{\mathrm{e}^{-as}F(s)\} = f(t-a)H(t-a)$, the response of forced vibration is, in terms of $x_{\sin}(t)$ and $x_{\cos}(t)$,

$$z_{\text{Forced}} = h\Omega^2\left\{x_{\sin}(t) - \left[\cos\Omega T\cdot x_{\sin}(t-T) + \sin\Omega T\cdot x_{\cos}(t-T)\right]H(t-T)\right\}.$$

Remarks: Using the Heaviside step function, the speed bump and the flat surface can be expressed as a compact analytical equation. Applying the method of Laplace transform, the equation of motion with the loading involving the Heaviside step function can easily be solved. Both the solution procedure and the expression of the response are much simpler than those presented in Section 5.3.

Example 6.21 — Single DOF System under Blast Force

Determine $x_{\text{Forced}}(t)$ with $0<\zeta<1$ and $f(t)$ being a model of a blast force, which can be expressed as $f(t) = \dfrac{f_0}{T}(T-t)\left[1-H(t-T)\right]$.

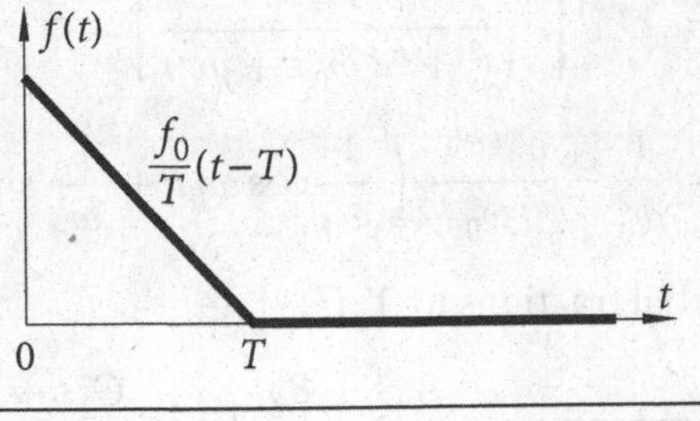

Applying the formula $\mathscr{L}\{f(t-a)H(t-a)\} = \mathrm{e}^{-as}\mathscr{L}\{f(t)\}$, one has

$$\mathscr{L}\{f(t)\} = \frac{f_0}{T}\mathscr{L}\{T-t+(t-T)H(t-T)\} = \frac{f_0}{T}\left(\frac{T}{s}-\frac{1}{s^2}+\frac{1}{s^2}\mathrm{e}^{-Ts}\right),$$

$$\therefore\ X_{\text{Forced}}(s) = \frac{1}{m}\cdot\frac{\mathscr{L}\{f(t)\}}{s^2+2\zeta\omega_0 s+\omega_0^2}$$

$$= \frac{f_0}{mT}\left[T\cdot\underbrace{\frac{1}{s\,(s^2+2\zeta\omega_0 s+\omega_0^2)}}_{Y_1(s)} + \left(-1+\mathrm{e}^{-Ts}\right)\cdot\underbrace{\frac{1}{s^2\,(s^2+2\zeta\omega_0 s+\omega_0^2)}}_{Y_2(s)}\right].$$

Applying partial fractions to $\Upsilon_1(s)$

$$\Upsilon_1(s) = \frac{1}{s\,(s^2+2\zeta\omega_0 s+\omega_0^2)} = \frac{A_1}{s} + \frac{B_1 s + C_1}{s^2+2\zeta\omega_0 s+\omega_0^2}$$
$$= \frac{(A_1+B_1)s^2 + (2\zeta\omega_0 A_1 + C_1)s + \omega_0^2 A_1}{s\,(s^2+2\zeta\omega_0 s+\omega_0^2)},$$

and comparing the coefficients of the numerators, one obtains

$$1:\quad \omega_0^2 A_1 = 1 \implies A_1 = \frac{1}{\omega_0^2},$$
$$s:\quad 2\zeta\omega_0 A_1 + C_1 = 0 \implies C_1 = -2\zeta\omega_0 A_1 = -\frac{2\zeta}{\omega_0},$$
$$s^2:\quad A_1 + B_1 = 0 \implies B_1 = -A_1 = -\frac{1}{\omega_0^2}.$$

Hence,

$$\Upsilon_1(s) = \frac{1}{\omega_0^2}\cdot\frac{1}{s} + \frac{\left[-\frac{1}{\omega_0^2}(s+\zeta\omega_0) + \frac{\zeta}{\omega_0}\right] - \frac{2\zeta}{\omega_0}}{(s+\zeta\omega_0)^2+\omega_d^2}, \qquad \omega_d = \omega_0\sqrt{1-\zeta^2},$$
$$= \frac{1}{\omega_0^2}\cdot\frac{1}{s} - \frac{1}{\omega_0^2}\cdot\frac{(s+\zeta\omega_0)}{(s+\zeta\omega_0)^2+\omega_d^2} - \frac{\zeta}{\omega_0\omega_d}\cdot\frac{\omega_d}{(s+\zeta\omega_0)^2+\omega_d^2},$$

$$\eta_1(t) = \mathscr{L}^{-1}\{\Upsilon_1(s)\} = \frac{1}{\omega_0^2} - \frac{1}{\omega_0^2}\cdot e^{-\zeta\omega_0 t}\cos\omega_d t - \frac{\zeta}{\omega_0\omega_d}\cdot e^{-\zeta\omega_0 t}\sin\omega_d t,$$

$$\eta_1(t) = \mathscr{L}^{-1}\left\{\frac{1}{s\,(s^2+2\zeta\omega_0 s+\omega_0^2)}\right\}$$
$$= \frac{1}{\omega_0^2} - \frac{e^{-\zeta\omega_0 t}}{\omega_0}\left(\frac{1}{\omega_0}\cos\omega_d t + \frac{\zeta}{\omega_d}\sin\omega_d t\right).$$

Similarly, applying partial fractions to $\Upsilon_2(s)$

$$\Upsilon_2(s) = \frac{1}{s^2\,(s^2+2\zeta\omega_0 s+\omega_0^2)} = \frac{A_2 s + B_2}{s^2} + \frac{C_2 s + D_2}{s^2+2\zeta\omega_0 s+\omega_0^2}$$
$$= \frac{(A_2+C_2)s^3 + (2\zeta\omega_0 A_2 + B_2 + D_2)s^2 + (\omega_0^2 A_2 + 2\zeta\omega_0 B_2)s + \omega_0^2 B_2}{s^2\,(s^2+2\zeta\omega_0 s+\omega_0^2)},$$

and comparing the coefficients of the numerators, one obtains

$$1:\quad \omega_0^2 B_2 = 1 \implies B_2 = \frac{1}{\omega_0^2},$$
$$s:\quad \omega_0^2 A_2 + 2\zeta\omega_0 B_2 = 0 \implies A_2 = -\frac{2\zeta}{\omega_0}B_2 = -\frac{2\zeta}{\omega_0^3},$$

$$s^2:\quad 2\zeta\omega_0 A_2 + B_2 + D_2 = 0 \implies D_2 = -2\zeta\omega_0 A_2 - B_2 = \frac{4\zeta^2-1}{\omega_0^2},$$

$$s^3:\quad A_2 + C_2 = 0 \implies C_2 = -A_2 = \frac{2\zeta}{\omega_0^3}.$$

Hence,

$$\begin{aligned}
\Upsilon_2(s) &= -\frac{2\zeta}{\omega_0^3}\cdot\frac{1}{s} + \frac{1}{\omega_0^2}\cdot\frac{1}{s^2} + \frac{\left[\dfrac{2\zeta}{\omega_0^3}(s+\zeta\omega_0) - \dfrac{2\zeta^2}{\omega_0^2}\right] + \dfrac{4\zeta^2-1}{\omega_0^2}}{(s+\zeta\omega_0)^2+\omega_d^2}\\
&= -\frac{2\zeta}{\omega_0^3}\cdot\frac{1}{s} + \frac{1}{\omega_0^2}\cdot\frac{1}{s^2} + \frac{2\zeta}{\omega_0^3}\cdot\frac{(s+\zeta\omega_0)}{(s+\zeta\omega_0)^2+\omega_d^2} + \frac{2\zeta^2-1}{\omega_0^2\omega_d}\cdot\frac{\omega_d}{(s+\zeta\omega_0)^2+\omega_d^2},
\end{aligned}$$

$$\begin{aligned}
\eta_2(t) &= \mathscr{L}^{-1}\{\Upsilon_2(s)\}\\
&= -\frac{2\zeta}{\omega_0^3} + \frac{1}{\omega_0^2}\cdot t + \frac{2\zeta}{\omega_0^3}\cdot e^{-\zeta\omega_0 t}\cos\omega_d t + \frac{2\zeta^2-1}{\omega_0^2\omega_d}\cdot e^{-\zeta\omega_0 t}\sin\omega_d t,
\end{aligned}$$

$$\begin{aligned}
\eta_2(t) &= \mathscr{L}^{-1}\left\{\frac{1}{s^2\,(s^2+2\zeta\omega_0 s+\omega_0^2)}\right\}\\
&= \frac{\omega_0 t - 2\zeta}{\omega_0^3} + \frac{e^{-\zeta\omega_0 t}}{\omega_0^2}\left(\frac{2\zeta}{\omega_0}\cos\omega_d t + \frac{2\zeta^2-1}{\omega_d}\sin\omega_d t\right).
\end{aligned}$$

Noting that $\mathscr{L}^{-1}\{e^{-as}F(s)\} = f(t-a)H(t-a)$, one obtains the response of the forced vibration in terms of functions $\eta_1(t)$ and $\eta_2(t)$

$$x_{\text{Forced}}(t) = \mathscr{L}^{-1}\{X_{\text{Forced}}(s)\} = \frac{f_0}{mT}\left[T\eta_1(t) - \eta_2(t) + \eta_2(t-T)H(t-T)\right].$$

Remarks: The inverse Laplace transforms $\eta_1(t)$ and $\eta_2(t)$ are very useful for vibration problems of a single degree-of-freedom system under piecewise linear loads, in which the response of the forced vibration can be expressed in terms of $\eta_1(t)$ and $\eta_2(t)$.

6.6.2 *Other Applications*

Example 6.22 — Electric Circuit

Consider the circuit shown in the following figure. The switch is moved from position a to b at $t=0$. Derive the differential equation governing $i(t)$ for $t>0$.

The voltage source $V(t)$ gives impulses of amplitude 1 V periodically with period 1 sec as shown. Given that $R_1=R_2=2\,\Omega$, $V_0=4\,\text{V}$, $L=64\,\text{H}$, $\alpha=1$, determine $i(t)$ for the following three cases:

(1) $C=4\,\text{F}$; (2) $C=2\,\text{F}$; (3) $C=\frac{7}{8}\,\text{F}$.

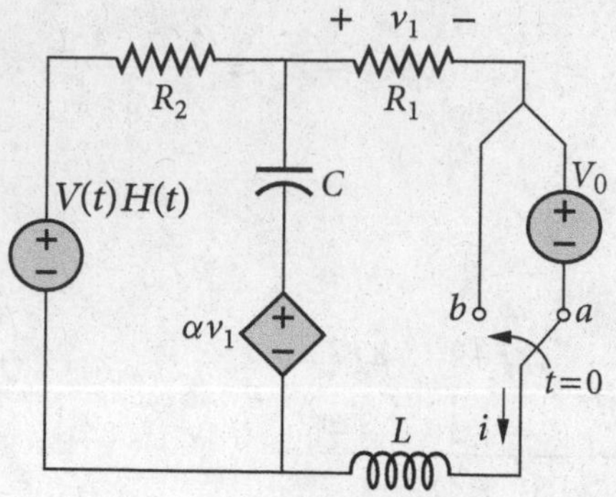

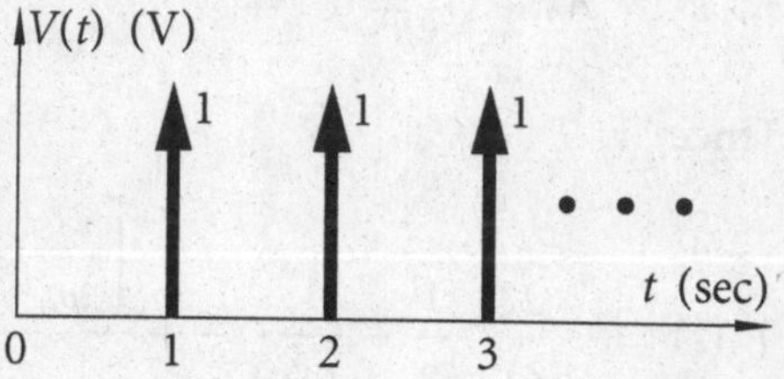

❧ For $t<0$, the switch is at position a and $V(t)H(t)=0$. The inductor behaves as a short circuit and the capacitor behaves as an open circuit.

Applying Kirchhoff's Voltage Law on the bigger mesh gives

$$-v_2(0^-)+v_1(0^-)+V_0+v_L(0^-)=0 \implies -R_2\big[-i(0^-)\big]+R_1 i(0^-)+V_0=0,$$

$$\therefore \quad i(0^-)=-\frac{V_0}{R_1+R_2}, \qquad v_1(0^-)=R_2 i(0^-)=-\frac{R_1}{R_1+R_2}V_0.$$

Applying Kirchhoff's Voltage Law on the right mesh yields

$$-\alpha v_1(0^-)-v_C(0^-)+v_1(0^-)+V_0=0,$$

$$\therefore \quad v_C(0^-)=V_0+(1-\alpha)v_1(0^-)=\frac{\alpha R_1+R_2}{R_1+R_2}V_0.$$

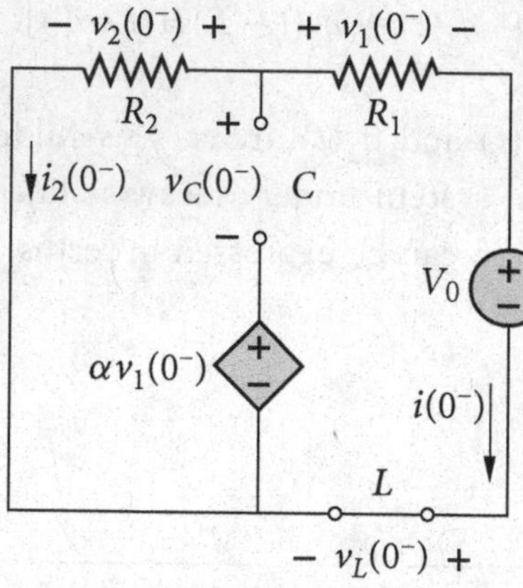

$t\leqslant 0^-$

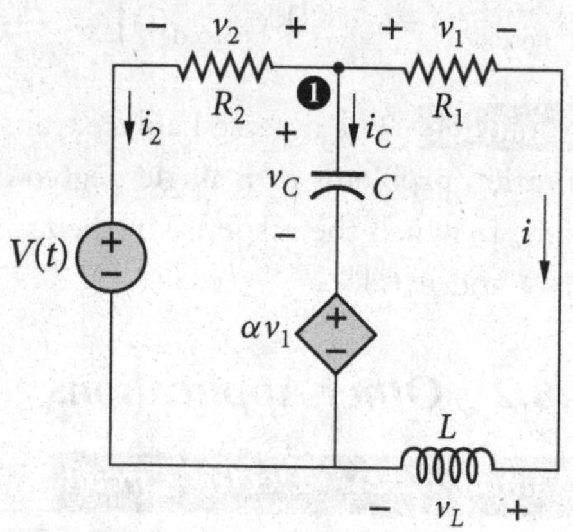

$t\geqslant 0^+$

❧ At $t=0$, the switch is moved from position a to b. Since the current in an inductor cannot change abruptly,

$$i(0^+)=i(0^-)=-\frac{V_0}{R_1+R_2}.$$

Since the voltage across a capacitor cannot change abruptly,

$$v_C(0^+) = v_C(0^-) = \frac{\alpha R_1 + R_2}{R_1 + R_2} V_0.$$

Applying Kirchhoff's Voltage Law on the right mesh yields

$$-\alpha v_1(0^+) - v_C(0^+) + v_1(0^+) + v_L(0^+) = 0.$$

Since $i(0^+) = i(0^-)$, one has $v_1(0^+) = v_1(0^-)$. Hence

$$(1-\alpha)v_1(0^+) - v_C(0^+) = -V_0, \quad v_L(0^+) = L\frac{\mathrm{d}i(0^+)}{\mathrm{d}t} \implies i'(0^+) = \frac{V_0}{L}.$$

❧ For $t > 0$, applying Kirchhoff's Voltage Law on the right mesh

$$-\alpha v_1 - v_C + v_1 + v_L = 0,$$

$$\therefore \quad v_C = (1-\alpha)v_1 + v_L \implies i_C = C\frac{\mathrm{d}v_C}{\mathrm{d}t} = C\Big[(1-\alpha)\frac{\mathrm{d}v_1}{\mathrm{d}t} + \frac{\mathrm{d}v_L}{\mathrm{d}t}\Big].$$

Applying Kirchhoff's Voltage Law on the bigger mesh leads to

$$-V(t) - v_2 + v_1 + v_L = 0 \implies v_2 = -V(t) + v_1 + v_L, \quad i_2 = \frac{v_2}{R_2}.$$

Noting that $v_1 = R_1 i$, $v_L = L\dfrac{\mathrm{d}i}{\mathrm{d}t}$, one has

$$i_C = C\Big[(1-\alpha)R_1\frac{\mathrm{d}i}{\mathrm{d}t} + L\frac{\mathrm{d}^2 i}{\mathrm{d}t^2}\Big], \qquad i_2 = \frac{1}{R_2}\Big[-V(t) + R_1 i + L\frac{\mathrm{d}i}{\mathrm{d}t}\Big].$$

Applying Kirchhoff's Current Law at node 1 yields $i + i_2 + i_C = 0$. Hence

$$i + \frac{1}{R_2}\Big[-V(t) + R_1 i + L\frac{\mathrm{d}i}{\mathrm{d}t}\Big] + C\Big[(1-\alpha)R_1\frac{\mathrm{d}i}{\mathrm{d}t} + L\frac{\mathrm{d}^2 i}{\mathrm{d}t^2}\Big],$$

$$\therefore \quad R_2 CL\frac{\mathrm{d}^2 i}{\mathrm{d}t^2} + \big[L + C(1-\alpha)R_1 R_2\big]\frac{\mathrm{d}i}{\mathrm{d}t} + (R_1 + R_2)i = V(t),$$

with initial conditions

$$i(0^+) = -\frac{V_0}{R_1 + R_2}, \qquad i'(0^+) = \frac{V_0}{L}.$$

The differential equation is of the standard form

$$\frac{\mathrm{d}^2 i}{\mathrm{d}t^2} + 2\zeta\omega_0\frac{\mathrm{d}i}{\mathrm{d}t} + \omega_0^2 i = \frac{1}{m}f(t),$$

where

$$\omega_0 = \sqrt{\frac{R_1 + R_2}{R_2 CL}}, \quad \zeta\omega_0 = \frac{L + C(1-\alpha)R_1 R_2}{2R_2 CL}, \quad m = R_2 CL, \quad f(t) = V(t).$$

Remarks: As discussed in Chapter 5, a second-order circuit is equivalent to a single degree-of-freedom system. Thus, the results obtained for a single degree-of-freedom system can be applied to a second-order circuit.

The solution is $i(t)=i_{\text{Free}}(t)+i_{\text{Forced}}(t)$, where $i_{\text{Free}}(t)$ is the complementary solution or the response of "free" vibration.

Using the Dirac delta function, the forcing function $f(t)=V(t)$ is

$$f(t)=V(t)=1\cdot\delta(t-1)+1\cdot\delta(t-2)+1\cdot\delta(t-3)+\cdots=\sum_{n=1}^{\infty}\delta(t-n).$$

Applying the Laplace transform

$$\mathscr{L}\{\delta(t-a)\}=\mathrm{e}^{-as}\implies\mathscr{L}\{f(t)\}=\sum_{n=1}^{\infty}\mathscr{L}\{\delta(t-n)\}=\sum_{n=1}^{\infty}\mathrm{e}^{-ns}.$$

Case 1. Underdamped System

For $R_1=R_2=2\,\Omega$, $V_0=4\,\text{V}$, $L=64\,\text{H}$, $\alpha=1$, $C=4\,\text{F}$, the differential equation becomes

$$\frac{\mathrm{d}^2 i}{\mathrm{d}t^2}+\frac{1}{8}\frac{\mathrm{d}i}{\mathrm{d}t}+\frac{1}{128}i=\frac{1}{512}V(t),\quad i(0^+)=-1,\quad i'(0^+)=\frac{1}{16},$$

which is of the standard form with $m=512$, $f(t)=V(t)$, and

$$\omega_0=\frac{1}{8\sqrt{2}},\quad 2\zeta\omega_0=\frac{1}{8}\implies\zeta=\frac{1}{\sqrt{2}}<1,\quad \omega_d=\omega_0\sqrt{1-\zeta^2}=\frac{1}{16}.$$

The system is underdamped and

$$i_{\text{Free}}(t)=\mathrm{e}^{-\zeta\omega_0 t}\Big[i(0^+)\cos\omega_d t+\frac{i'(0^+)+\zeta\omega_0 i(0^+)}{\omega_d}\sin\omega_d t\Big]=-\mathrm{e}^{-\frac{t}{16}}\cos\frac{t}{16}.$$

The Laplace transform of the "forced" response $I_{\text{Forced}}(s)=\mathscr{L}\{i_{\text{Forced}}(t)\}$ is

$$\begin{aligned}I_{\text{Forced}}(s)&=\frac{1}{m}\cdot\frac{\mathscr{L}\{f(t)\}}{s^2+2\zeta\omega_0 s+\omega_0^2}=\frac{1}{512}\sum_{n=1}^{\infty}\mathrm{e}^{-ns}\frac{1}{s^2+\frac{1}{8}s+\frac{1}{128}}\\&=\frac{1}{512}\sum_{n=1}^{\infty}\mathrm{e}^{-ns}\frac{1}{\left(\frac{1}{16}\right)}\cdot\frac{\left(\frac{1}{16}\right)}{\left(s+\frac{1}{16}\right)^2+\left(\frac{1}{16}\right)^2}.\end{aligned}$$

Taking the inverse Laplace transform, one has

$$i_{\text{Forced}}(t)=\frac{1}{32}\sum_{n=1}^{\infty}\mathrm{e}^{\frac{n}{16}}\mathscr{L}^{-1}\left\{\mathrm{e}^{-n(s+\frac{1}{16})}\frac{\left(\frac{1}{16}\right)}{\left(s+\frac{1}{16}\right)^2+\left(\frac{1}{16}\right)^2}\right\}\quad ✍\ \begin{aligned}&\mathscr{L}^{-1}\{F(s-a)\}\\&=\mathrm{e}^{at}\mathscr{L}\{F(s)\}\end{aligned}$$

$$= \frac{1}{32}\sum_{n=1}^{\infty} e^{\frac{n}{16}} \cdot e^{-\frac{t}{16}} \mathscr{L}^{-1}\left\{e^{-ns}\frac{\left(\frac{1}{16}\right)}{s^2+\left(\frac{1}{16}\right)^2}\right\} \qquad \mathscr{L}^{-1}\{e^{-as}F(s)\} = f(t-a)H(t-a)$$

$$= \frac{1}{32}\sum_{n=1}^{\infty} e^{-\frac{t-n}{16}} \sin\frac{t-n}{16} H(t-n). \qquad \mathscr{L}^{-1}\left\{\frac{\omega}{s^2+\omega^2}\right\} = \sin\omega t$$

Hence, the current $i(t)$ is

$$i(t) = i_{\text{Free}}(t) + i_{\text{Forced}}(t) = -e^{-\frac{t}{16}}\cos\frac{t}{16} + \frac{1}{32}\sum_{n=1}^{\infty} e^{-\frac{t-n}{16}} \sin\frac{t-n}{16} H(t-n) \text{ (A)}.$$

Case 2. Critically Damped System

For $R_1 = R_2 = 2\,\Omega$, $V_0 = 4\,\text{V}$, $L = 64\,\text{H}$, $\alpha = 1$, $C = 2\,\text{F}$, the differential equation becomes

$$\frac{d^2 i}{dt^2} + \frac{1}{4}\frac{di}{dt} + \frac{1}{64}i = \frac{1}{256}V(t), \quad i(0^+) = -1, \quad i'(0^+) = \frac{1}{16},$$

which is of the standard form with $m = 256$, $f(t) = V(t)$, and

$$\omega_0 = \frac{1}{8}, \quad 2\zeta\omega_0 = \frac{1}{4} \Longrightarrow \zeta = 1.$$

The system is critically damped and

$$i_{\text{Free}}(t) = \left\{i(0^+) + \left[i'(0^+) + \omega_0 i(0^+)\right]t\right\}e^{-\omega_0 t} = -\left(1 + \frac{t}{16}\right)e^{-\frac{t}{8}}.$$

The Laplace transform of the "forced" response $I_{\text{Forced}}(s) = \mathscr{L}\{i_{\text{Forced}}(t)\}$ is

$$I_{\text{Forced}}(s) = \frac{1}{m} \cdot \frac{\mathscr{L}\{f(t)\}}{s^2 + 2\zeta\omega_0 s + \omega_0^2} = \frac{1}{256}\sum_{n=1}^{\infty} e^{-ns}\frac{1}{s^2 + \frac{1}{4}s + \frac{1}{64}}.$$

Take the inverse Laplace transform, one has

$$i_{\text{Forced}}(t) = \frac{1}{256}\sum_{n=1}^{\infty} e^{\frac{n}{8}} \mathscr{L}^{-1}\left\{e^{-n(s+\frac{1}{8})}\frac{1}{\left(s+\frac{1}{8}\right)^2}\right\} \qquad \mathscr{L}^{-1}\{F(s-a)\} = e^{at}\mathscr{L}\{F(s)\}$$

$$= \frac{1}{256}\sum_{n=1}^{\infty} e^{\frac{n}{8}} \cdot e^{-\frac{t}{8}} \mathscr{L}^{-1}\left\{e^{-ns}\frac{1}{s^2}\right\} \qquad \mathscr{L}^{-1}\{e^{-as}F(s)\} = f(t-a)H(t-a)$$

$$= \frac{1}{256}\sum_{n=1}^{\infty} e^{-\frac{t-n}{8}} (t-n)\,H(t-n). \qquad \mathscr{L}^{-1}\left\{\frac{1}{s^2}\right\} = t$$

Hence, the current $i(t)$ is

$$i(t) = i_{\text{Free}}(t) + i_{\text{Forced}}(t) = -\left(1 + \frac{t}{16}\right)e^{-\frac{t}{8}} + \frac{1}{256}\sum_{n=1}^{\infty} e^{-\frac{t-n}{8}}(t-n)\,H(t-n) \text{ (A)}.$$

Case 3. Overdamped System

For $R_1 = R_2 = 2\,\Omega$, $V_0 = 4\,\text{V}$, $L = 64\,\text{H}$, $\alpha = 1$, $C = \frac{7}{8}\,\text{F}$, the differential equation becomes

$$\frac{\mathrm{d}^2 i}{\mathrm{d}t^2} + \frac{4}{7}\frac{\mathrm{d}i}{\mathrm{d}t} + \frac{1}{28}i = \frac{1}{112}V(t), \quad i(0^+) = -1, \quad i'(0^+) = \frac{1}{16},$$

which is of the standard form with $m = 112$, $f(t) = V(t)$, and

$$\omega_0 = \frac{1}{2\sqrt{7}}, \quad 2\zeta\omega_0 = \frac{4}{7} \Longrightarrow \zeta = \frac{4}{\sqrt{7}} > 1, \quad \omega_0\sqrt{\zeta^2-1} = \frac{3}{14}.$$

The system is overdamped and

$$\begin{aligned} i_{\text{Free}}(t) &= \frac{1}{2\omega_0\sqrt{\zeta^2-1}}\Big\{\big[i'(0^+)+\omega_0\big(\zeta+\sqrt{\zeta^2-1}\big)\,i(0^+)\big]\mathrm{e}^{-\omega_0(\zeta-\sqrt{\zeta^2-1})t} \\ &\qquad - \big[i'(0^+)+\omega_0\big(\zeta-\sqrt{\zeta^2-1}\big)\,i(0^+)\big]\mathrm{e}^{-\omega_0(\zeta+\sqrt{\zeta^2-1})t}\Big\} \\ &= \frac{1}{48}\,\mathrm{e}^{-\frac{t}{2}} - \frac{49}{48}\,\mathrm{e}^{-\frac{t}{14}}. \end{aligned}$$

The Laplace transform of the "forced" response $I_{\text{Forced}}(s) = \mathscr{L}\{i_{\text{Forced}}(t)\}$ is

$$I_{\text{Forced}}(s) = \frac{1}{m}\cdot\frac{\mathscr{L}\{f(t)\}}{s^2+2\zeta\omega_0 s+\omega_0^2} = \frac{1}{112}\sum_{n=1}^{\infty}\mathrm{e}^{-ns}\frac{1}{s^2+\frac{4}{7}s+\frac{1}{28}}.$$

Take the inverse Laplace transform, one has

$$\begin{aligned} i_{\text{Forced}}(t) &= \frac{1}{112}\sum_{n=1}^{\infty}\mathrm{e}^{\frac{2n}{7}}\mathscr{L}^{-1}\Big\{\mathrm{e}^{-n(s+\frac{2}{7})}\frac{1}{\big(s+\frac{2}{7}\big)^2-\big(\frac{3}{14}\big)^2}\Big\} && ✍\ \mathscr{L}^{-1}\{F(s-a)\} = \mathrm{e}^{at}\mathscr{L}\{F(s)\} \\ &= \frac{1}{112}\sum_{n=1}^{\infty}\mathrm{e}^{\frac{2n}{7}}\cdot\mathrm{e}^{-\frac{2t}{7}}\mathscr{L}^{-1}\Big\{\mathrm{e}^{-ns}\frac{1}{s^2-\big(\frac{3}{14}\big)^2}\Big\} && ✍\ \mathscr{L}^{-1}\{\mathrm{e}^{-as}F(s)\} = f(t-a)H(t-a) \\ &= \frac{1}{112}\sum_{n=1}^{\infty}\mathrm{e}^{-\frac{2(t-n)}{7}}\frac{1}{\big(\frac{3}{14}\big)}\sinh\frac{3(t-n)}{14}H(t-n) && ✍\ \mathscr{L}^{-1}\Big\{\frac{1}{s^2-a^2}\Big\} = \frac{\sinh at}{a} \\ &= \frac{1}{24}\sum_{n=1}^{\infty}\mathrm{e}^{-\frac{2(t-n)}{7}}\cdot\frac{1}{2}\Big\{\mathrm{e}^{\frac{3(t-n)}{14}}-\mathrm{e}^{-\frac{3(t-n)}{14}}\Big\}H(t-n) && ✍\ \sinh at = \frac{\mathrm{e}^{at}-\mathrm{e}^{-at}}{2} \\ &= \frac{1}{48}\sum_{n=1}^{\infty}\Big(\mathrm{e}^{-\frac{t-n}{14}}-\mathrm{e}^{-\frac{t-n}{2}}\Big)H(t-n). \end{aligned}$$

Hence, the current $i(t)$ is

$$i(t) = i_{\text{Free}}(t)+i_{\text{Forced}}(t) = \frac{1}{48}\mathrm{e}^{-\frac{t}{2}} - \frac{49}{48}\mathrm{e}^{-\frac{t}{14}} + \frac{1}{48}\sum_{n=1}^{\infty}\Big(\mathrm{e}^{-\frac{t-n}{14}}-\mathrm{e}^{-\frac{t-n}{2}}\Big)H(t-n)\ \text{(A)}.$$

Example 6.23 — Beam-Column

Consider the beam-column shown in the following figure. Determine the lateral deflection $y(x)$.

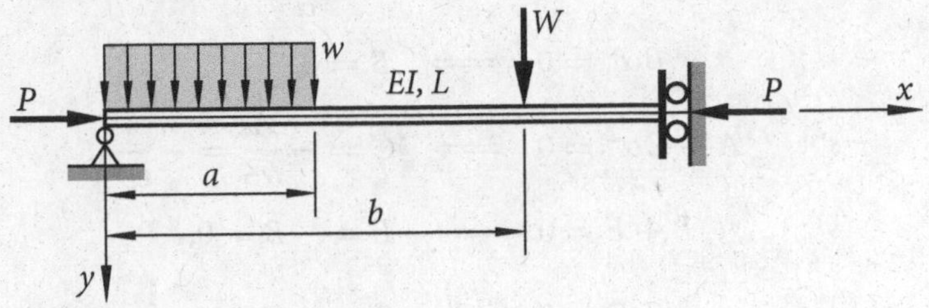

Using the Heaviside step function and the Dirac delta function, the lateral load can be expresses as

$$w(x) = w\big[1 - H(x-a)\big] + W\,\delta(x-b).$$

Following the formulation in Section 5.4, the differential equation becomes

$$\frac{d^4y}{dx^4} + \alpha^2\frac{d^2y}{dx^2} = \hat{w}\big[1 - H(x-a)\big] + \hat{W}\,\delta(x-b), \quad \alpha^2 = \frac{P}{EI}, \quad \hat{w} = \frac{w}{EI}, \quad \hat{W} = \frac{W}{EI}.$$

Since the left end is a hinge support and the right end is a sliding support, the boundary conditions are

$$\begin{aligned} \text{at } x=0: \quad & \text{deflection} = 0 \implies y(0) = 0, \\ & \text{bending moment} = 0 \implies y''(0) = 0, \\ \text{at } x=L: \quad & \text{slope} = 0 \implies y'(L) = 0, \\ & \text{shear force} = 0 \implies V(L) = -EI\,y'''(L) - P\,y'(L) = 0 \\ & \implies y'''(L) = 0. \end{aligned}$$

Applying the Laplace transform $Y(s) = \mathscr{L}\{y(x)\}$, one has

$$\big[s^4 Y(s) - s^3 y(0) - s^2 y'(0) - s\,y''(0) - y'''(0)\big] + \alpha^2\big[s^2 Y(s) - s\,y(0) - y'(0)\big] = \frac{\hat{w}}{s}(1 - e^{-as}) + \hat{W}\,e^{-bs}.$$

Since $y(0) = y''(0) = 0$, solving for $Y(s)$ leads to

$$Y(s) = \frac{y'(0)}{s^2 + \alpha^2} + \frac{\big[y'''(0) + \alpha^2 y'(0)\big] + \hat{W}\,e^{-bs}}{s^2(s^2 + \alpha^2)} + \frac{\hat{w}}{s^3(s^2 + \alpha^2)}(1 - e^{-as}).$$

Applying partial fractions

$$\frac{1}{s^3(s^2 + \alpha^2)} = \frac{A}{s^3} + \frac{B}{s^2} + \frac{C}{s} + \frac{Ds + E}{s^2 + \alpha^2}.$$

Summing up the right-hand side and comparing the coefficients of the numerators, one obtains

$$
\begin{aligned}
1:&\qquad A\alpha^2=1 \implies A=\frac{1}{\alpha^2},\\
s:&\qquad B\alpha^2=0 \implies B=0,\\
s^2:&\qquad A+C\alpha^2=0 \implies C=-\frac{A}{\alpha^2}=-\frac{1}{\alpha^4},\\
s^3:&\qquad B+E=0 \implies E=-B=0,\\
s^4:&\qquad C+D=0 \implies D=-C=-\frac{1}{\alpha^4}.
\end{aligned}
$$

Hence,

$$
\begin{aligned}
\Upsilon(s)=&\ \frac{y'(0)}{\alpha}\frac{\alpha}{s^2+\alpha^2}+\left(\frac{1}{s^2}-\frac{1}{s^2+\alpha^2}\right)\frac{\left[y'''(0)+\alpha^2 y'(0)\right]+\hat{W}\,\mathrm{e}^{-bs}}{\alpha^2}\\
&+\hat{w}\left(\frac{1}{\alpha^2 s^3}-\frac{1}{\alpha^4 s}+\frac{1}{\alpha^4}\frac{s}{s^2+\alpha^2}\right)(1-\mathrm{e}^{-as}).
\end{aligned}
$$

Taking inverse Laplace transform results in

$$
\begin{aligned}
y(x)=\mathscr{L}^{-1}\{\Upsilon(s)\}=&\ y'(0)\cdot x+\left(x-\frac{1}{\alpha}\sin\alpha x\right)\frac{y'''(0)}{\alpha^2}\\
&+\left[(x-b)-\frac{1}{\alpha}\sin\alpha(x-b)\right]\frac{\hat{W}}{\alpha^2}H(x-b)+\hat{w}\left(\frac{x^2}{2\alpha^2}-\frac{1}{\alpha^4}+\frac{1}{\alpha^4}\cos\alpha x\right)\\
&-\hat{w}\left[\frac{(x-a)^2}{2\alpha^2}-\frac{1}{\alpha^4}+\frac{1}{\alpha^4}\cos\alpha(x-a)\right]H(x-a),
\end{aligned}
$$

in which $y'(0)$ and $y'''(0)$ are determined from the boundary conditions at $x=L$. At the right end of the beam-column, $x=L$, i.e., $x>a$, $x>b$, giving $H(x-a)=1$, $H(x-b)=1$. The lateral deflection is simplified as

$$
\begin{aligned}
y(x)=&\ y'(0)\cdot x+\left(x-\frac{1}{\alpha}\sin\alpha x\right)\frac{y'''(0)}{\alpha^2}+\left[(x-b)-\frac{1}{\alpha}\sin\alpha(x-b)\right]\frac{\hat{W}}{\alpha^2}\\
&+\frac{\hat{w}\left[\alpha^2(2ax-a^2)+2\cos\alpha x-2\cos\alpha(x-a)\right]}{2\alpha^4}.
\end{aligned}
$$

Differentiating with respect to x yields

$$
\begin{aligned}
y'(x)=&\ y'(0)+\frac{1-\cos\alpha x}{\alpha^2}y'''(0)+\frac{\hat{W}\left[1-\cos\alpha(x-b)\right]}{\alpha^2}\\
&+\frac{\hat{w}\left[\alpha a-\sin\alpha x+\sin\alpha(x-a)\right]}{\alpha^3},
\end{aligned}
$$

$$
y'''(x)=\cos\alpha x\cdot y'''(0)+\hat{W}\cos\alpha(x-b)+\frac{\hat{w}\left[\sin\alpha x-\sin\alpha(x-a)\right]}{\alpha}.
$$

Denoting

$$W_1 = \frac{1}{\alpha^2}\left\{\hat{W}\left[1-\cos\alpha(L-b)\right] + \frac{\hat{w}}{\alpha}\left[\alpha a - \sin\alpha L + \sin\alpha(L-a)\right]\right\},$$

$$W_2 = \hat{W}\cos\alpha(L-b) + \frac{\hat{w}}{\alpha}\left[\sin\alpha L - \sin\alpha(L-a)\right],$$

and applying the boundary conditions at $x=L$ lead to

$$y'(L) = y'(0) + \frac{1-\cos\alpha x}{\alpha^2}\,y'''(0) + W_1 = 0,$$

$$y'''(L) = \cos\alpha L \cdot y'''(0) + W_2 = 0,$$

which results in

$$y'(0) = -W_1 + \frac{W_2(1-\cos\alpha L)}{\alpha^2\cos\alpha L}, \qquad y'''(0) = -\frac{W_2}{\cos\alpha L}.$$

6.6.3 Beams on Elastic Foundation

Structures that can be modeled as beams placed on elastic foundation are found in many engineering applications, for example, railroad tracks, foundation beams and retaining walls of buildings, and underground infrastructures (Figure 6.7). Networks of beams such as those used in floor systems for ships, buildings, bridges, shells of revolution such as those used in pressure vessels, boilers, containers, and large-span reinforced concrete halls and domes can also be analyzed using the theory of beams on elastic foundation.

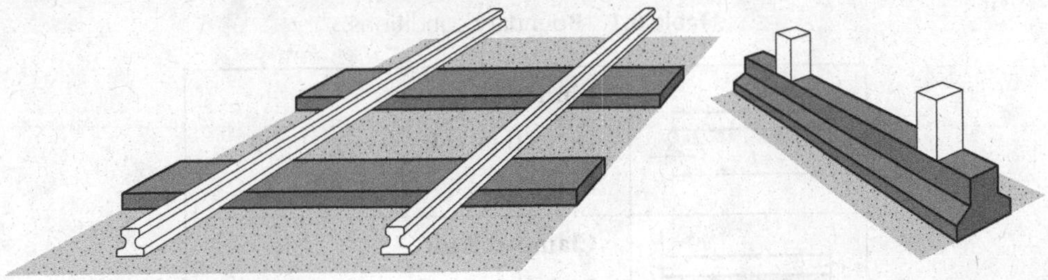

Figure 6.7 Examples of beams on elastic foundation.

Winkler's model of an elastic foundation assumes that the deflection y at any point on the surface of the foundation is proportional to the stress σ at that point, i.e., $\sigma = k_0 y$, where k_0 is called the modulus of the foundation with dimension [force/length^3].

In the study of beams on elastic foundation, let p be the intensity per unit length of the distributed load on the foundation along the length of the beam, i.e., $p=\sigma b$, where b is the width of the beam. Hence, as shown in Figure 6.8(a) and from Winkler's assumption, $p=ky$, where $k=k_0 b$ with dimension [force/length^2].

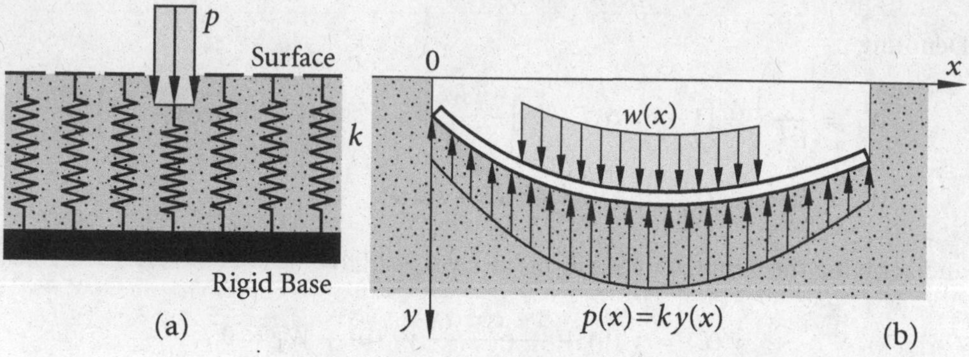

Figure 6.8 Winkler's model of elastic foundation.

For a beam on an elastic foundation under the action of a distributed load $w(x)$ as shown in Figure 6.8(b), the flexural deflection $y(x)$ is governed by the equation

$$EI\,\frac{\mathrm{d}^4 y}{\mathrm{d}x^4} = w(x) - p(x),$$

where EI is the flexural rigidity of the beam. Substituting $p(x) = ky(x)$ into the equation leads to a fourth-order linear ordinary differential equation

$$\frac{\mathrm{d}^4 y}{\mathrm{d}x^4} + 4\beta^4 y = \frac{w(x)}{EI}, \qquad 4\beta^4 = \frac{k}{EI}.$$

The constants in the solution of the differential equation are determined from the boundary conditions of the beam, which are given by the end supports of the beam. Some typical boundary conditions are listed in Table 6.1.

Table 6.1 Boundary conditions.

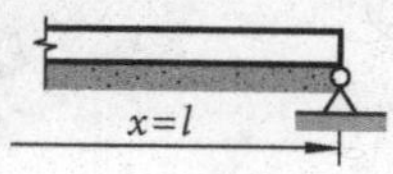	**Pinned End:** Deflection $= 0 \implies y(l) = 0$ Moment $= 0 \implies y''(l) = 0$
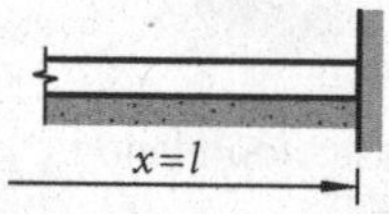	**Clamped End:** Deflection $= 0 \implies y(l) = 0$ Slope $= 0 \implies y'(l) = 0$
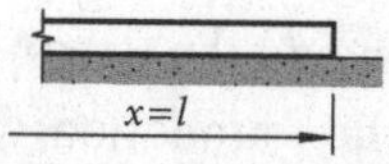	**Free End:** Moment $= 0 \implies y''(l) = 0$ Shear Force $= 0 \implies y'''(l) = 0$
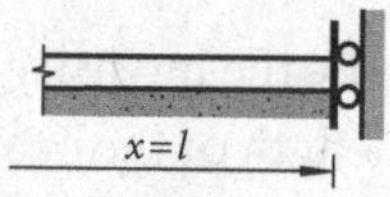	**Sliding End:** Slope $= 0 \implies y'(l) = 0$ Shear Force $= 0 \implies y'''(l) = 0$

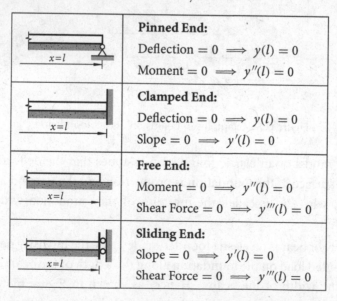

Example 6.24 — Beams on Elastic Foundation

Determine the deflection of a beam free at both ends under a trapezoidally distributed load as shown in the following figure.

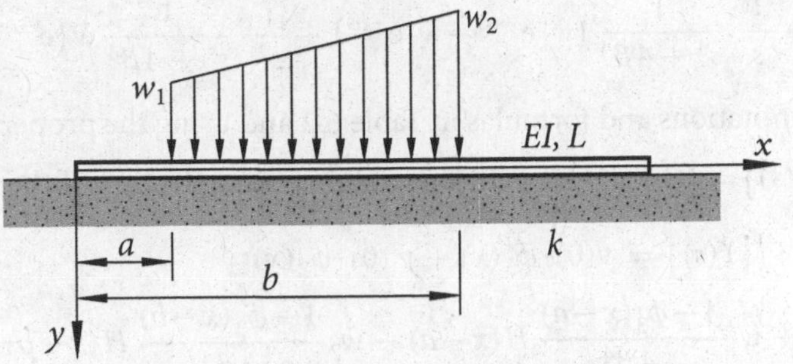

Using the Heaviside step function, the distributed load can be expressed as

$$w(x) = \Big[w_1 + \frac{w_2 - w_1}{b-a}(x-a)\Big]\big[H(x-a) - H(x-b)\big].$$

The differential equation becomes

$$\frac{d^4y}{dx^4} + 4\beta^4 y = \frac{w(x)}{EI} = \big[\hat{w}_1 + \bar{w}\,(x-a)\big]\big[H(x-a) - H(x-b)\big],$$

where

$$\hat{w}_1 = \frac{w_1}{EI}, \qquad \hat{w}_2 = \frac{w_2}{EI}, \qquad \bar{w} = \frac{w_2 - w_1}{(b-a)EI} = \frac{\hat{w}_2 - \hat{w}_1}{b-a},$$

Since both ends are free, the boundary conditions are

$$y''(0) = y'''(0) = y''(L) = y'''(L) = 0.$$

Applying the Laplace transform gives

$$\mathscr{L}\Big\{\big[\hat{w}_1 + \bar{w}\,(x-a)\big]\big[H(x-a) - H(x-b)\big]\Big\}$$

$$= \mathscr{L}\Big\{\big[\hat{w}_1 + \bar{w}\,(x-a)\big]H(x-a)\Big\} - \mathscr{L}\Big\{\big\{\hat{w}_1 + \bar{w}\big[(x-b)+(b-a)\big]\big\}H(x-b)\Big\}$$

$$= \mathscr{L}\Big\{\big[\hat{w}_1 + \bar{w}\,(x-a)\big]H(x-a)\Big\} - \mathscr{L}\Big\{\big[\hat{w}_2 + \bar{w}\,(x-b)\big]H(x-b)\Big\}$$

$$= e^{-as}\,\mathscr{L}\{\hat{w}_1 + \bar{w}x\} - e^{-bs}\,\mathscr{L}\{\hat{w}_2 + \bar{w}x\} \qquad ✍\ \mathscr{L}\{f(x-a)\,H(x-a)\} = e^{-as}F(s)$$

$$= e^{-as}\Big(\frac{\hat{w}_1}{s} + \frac{\bar{w}}{s^2}\Big) - e^{-bs}\Big(\frac{\hat{w}_2}{s} - \frac{\bar{w}}{s^2}\Big), \qquad ✍\ \mathscr{L}\{1\} = \frac{1}{s}, \ \ \mathscr{L}\{x\} = \frac{1}{s^2}$$

and the differential equation becomes

$$\big[s^4 Y(s) - s^3 y(0) - s^2 y'(0) - s\,y''(0) - y'''(0)\big] + 4\beta^4 Y(s)$$

$$= \frac{1}{s}\big(\hat{w}_1 e^{-as} - \hat{w}_2 e^{-bs}\big) + \frac{1}{s^2}\,\bar{w}\big(e^{-as} - e^{-bs}\big).$$

Since $y''(0)=y'''(0)=0$, solving for $\Upsilon(s)$ leads to

$$\Upsilon(s) = y(0)\frac{s^3}{s^4+4\beta^4} + y'(0)\frac{s^2}{s^4+4\beta^4}$$
$$+\frac{1}{s}\cdot\frac{1}{s^4+4\beta^4}\left(\hat{w}_1 e^{-as} - \hat{w}_2 e^{-bs}\right) + \frac{1}{s^2}\cdot\frac{1}{s^4+4\beta^4}\,\bar{w}\left(e^{-as}-e^{-bs}\right).$$

Using the notations and formulas in Table 6.2 and using the property of shifting $\mathscr{L}^{-1}\{e^{-as}F(s)\}=f(t-a)H(t-a)$, one obtains the deflection of the beam

$$y(x) = \mathscr{L}^{-1}\{\Upsilon(s)\} = y(0)\cdot\phi_3(x) + y'(0)\cdot\phi_2(x)$$
$$+\hat{w}_1\,\frac{1-\phi_3(x-a)}{4\beta^4}\,H(x-a) - \hat{w}_2\,\frac{1-\phi_3(x-b)}{4\beta^4}\,H(x-b)$$
$$+\bar{w}\left\{\frac{(x-a)-\phi_2(x-a)}{4\beta^4}\,H(x-a) - \frac{(x-b)-\phi_2(x-b)}{4\beta^4}\,H(x-b)\right\}.$$

For $x>b>a$, $H(x-b)=H(x-a)=1$, and $y(x)$ is simplified as

$$y(x) = y(0)\cdot\phi_3(x) + y'(0)\cdot\phi_2(x) + \frac{1}{4\beta^4}\Big\{\hat{w}_2\,\phi_3(x-b) - \hat{w}_1\,\phi_3(x-a)$$
$$+\bar{w}\left[\phi_2(x-b) - \phi_2(x-a)\right]\Big\}.$$

✍ Noting $\hat{w}_1-\hat{w}_2+\bar{w}(b-a)=0$

Differentiating $y(x)$ with respect to x three times yields

$$y''(x) = y(0)\cdot\phi_3''(x) + y'(0)\cdot\phi_2''(x)$$
$$+\frac{1}{4\beta^4}\Big\{\hat{w}_2\,\phi_3''(x-b) - \hat{w}_1\,\phi_3''(x-a) + \bar{w}\left[\phi_2''(x-b) - \phi_2''(x-a)\right]\Big\}$$
$$= -4\beta^4\left[y(0)\cdot\phi_1(x) + y'(0)\cdot\phi_0(x)\right]$$
$$-\Big\{\hat{w}_2\,\phi_1(x-b) - \hat{w}_1\,\phi_1(x-a) + \bar{w}\left[\phi_0(x-b) - \phi_0(x-a)\right]\Big\},$$

$$y'''(x) = -4\beta^4\left[y(0)\cdot\phi_1'(x) + y'(0)\cdot\phi_0'(x)\right]$$
$$-\Big\{\hat{w}_2\,\phi_1'(x-b) - \hat{w}_1\,\phi_1'(x-a) + \bar{w}\left[\phi_0'(x-b) - \phi_0'(x-a)\right]\Big\}$$
$$= -4\beta^4\left[y(0)\cdot\phi_2(x) + y'(0)\cdot\phi_1(x)\right]$$
$$-\Big\{\hat{w}_2\,\phi_2(x-b) - \hat{w}_1\,\phi_2(x-a) + \bar{w}\left[\phi_1(x-b) - \phi_1(x-a)\right]\Big\}.$$

Using the boundary conditions at $x=L$ leads to

$$y''(L) = -4\beta^4\left[y(0)\cdot\phi_1(L) + y'(0)\cdot\phi_0(L)\right]$$
$$-\Big\{\hat{w}_2\,\phi_1(L-b) - \hat{w}_1\,\phi_1(L-a) + \bar{w}\left[\phi_0(L-b) - \phi_0(L-a)\right]\Big\}$$
$$=0,$$

Table 6.2 Useful formulas of inverse Laplace transforms for beams on elastic foundation.

$$\phi_0(x) = \mathscr{L}^{-1}\left\{\frac{1}{s^4+4\beta^4}\right\} = \frac{1}{4\beta^3}(\sin\beta x\cosh\beta x - \cos\beta x\sinh\beta x),$$

$$\phi_1(x) = \mathscr{L}^{-1}\left\{\frac{s}{s^4+4\beta^4}\right\} = \frac{1}{2\beta^2}\sin\beta x\sinh\beta x,$$

$$\phi_2(x) = \mathscr{L}^{-1}\left\{\frac{s^2}{s^4+4\beta^4}\right\} = \frac{1}{2\beta}(\sin\beta x\cosh\beta x + \cos\beta x\sinh\beta x),$$

$$\phi_3(x) = \mathscr{L}^{-1}\left\{\frac{s^3}{s^4+4\beta^4}\right\} = \cos\beta x\cosh\beta x,$$

$$\mathscr{L}^{-1}\left\{\frac{1}{s}\cdot\frac{1}{s^4+4\beta^4}\right\} = \frac{1-\phi_3(x)}{4\beta^4},$$

$$\mathscr{L}^{-1}\left\{\frac{1}{s^2}\cdot\frac{1}{s^4+4\beta^4}\right\} = \frac{x-\phi_2(x)}{4\beta^4},$$

$$\mathscr{L}^{-1}\left\{\frac{1}{s^3}\cdot\frac{1}{s^4+4\beta^4}\right\} = \frac{x^2-2\phi_1(x)}{8\beta^4},$$

$$\mathscr{L}^{-1}\left\{\frac{1}{s^4}\cdot\frac{1}{s^4+4\beta^4}\right\} = \frac{x^3-6\phi_0(x)}{24\beta^4},$$

$$\phi_0'(x) = \frac{1}{2\beta^2}\sin\beta x\sinh\beta x = \phi_1(x),$$

$$\phi_1'(x) = \frac{1}{2\beta}(\sin\beta x\cosh\beta x + \cos\beta x\sinh\beta x) = \phi_2(x),$$

$$\phi_2'(x) = \cos\beta x\cosh\beta x = \phi_3(x),$$

$$\phi_3'(x) = -\beta(\sin\beta x\cosh\beta x - \cos\beta x\sinh\beta x) = -4\beta^4\phi_0(x),$$

$$\phi_0''(x) = \phi_1'(x) = \phi_2(x),$$

$$\phi_1''(x) = \phi_2'(x) = \phi_3(x),$$

$$\phi_2''(x) = \phi_3'(x) = -4\beta^4\phi_0(x),$$

$$\phi_3''(x) = -4\beta^4\phi_0'(x) = -4\beta^4\phi_1(x),$$

$$\phi_0'''(x) = \phi_2'(x) = \phi_3(x),$$

$$\phi_1'''(x) = \phi_3'(x) = -4\beta^4\phi_0(x),$$

$$\phi_2'''(x) = -4\beta^4\phi_0'(x) = -4\beta^4\phi_1(x),$$

$$\phi_3'''(x) = -4\beta^4\phi_1'(x) = -4\beta^4\phi_2(x).$$

$$\begin{aligned} y'''(L) &= -4\beta^4\big[y(0)\cdot\phi_2(L) + y'(0)\cdot\phi_1(L)\big] \\ &\quad - \Big\{\hat{w}_2\,\phi_2(L-b) - \hat{w}_1\,\phi_2(L-a) + \bar{w}\big[\phi_1(L-b) - \phi_1(L-a)\big]\Big\} \\ &= 0, \end{aligned}$$

which results in two algebraic equations for two unknowns $y(0)$ and $y'(0)$

$$\begin{cases} \phi_1(L)\cdot y(0) + \phi_0(L)\cdot y'(0) = \alpha_2, \\ \phi_2(L)\cdot y(0) + \phi_1(L)\cdot y'(0) = \alpha_3, \end{cases}$$

$$\alpha_2 = -\frac{1}{4\beta^4}\Big\{\hat{w}_2\,\phi_1(L-b) - \hat{w}_1\,\phi_1(L-a) + \bar{w}\big[\phi_0(L-b) - \phi_0(L-a)\big]\Big\},$$

$$\alpha_3 = -\frac{1}{4\beta^4}\Big\{\hat{w}_2\,\phi_2(L-b) - \hat{w}_1\,\phi_2(L-a) + \bar{w}\big[\phi_1(L-b) - \phi_1(L-a)\big]\Big\},$$

and gives

$$y(0) = \frac{\alpha_2\,\phi_1(L) - \alpha_3\,\phi_0(L)}{\phi_1^2(L) - \phi_0(L)\,\phi_2(L)}, \qquad y'(0) = \frac{\alpha_3\,\phi_1(L) - \alpha_2\,\phi_2(L)}{\phi_1^2(L) - \phi_0(L)\,\phi_2(L)}.$$

Example 6.25 — Beams on Elastic Foundation

Determine the deflection of a beam free at both ends under a concentrated load as shown in the following figure.

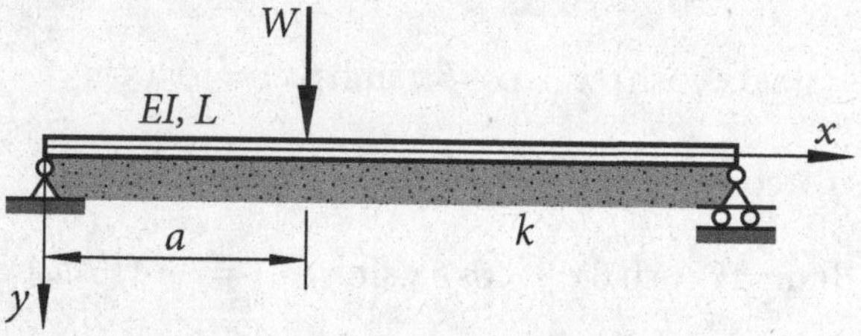

Using the Dirac delta function, the concentrated load can be expressed as

$$w(x) = W\delta(x-a).$$

The differential equation becomes

$$\frac{d^4y}{dx^4} + 4\beta^4 y = \frac{w(x)}{EI} = \bar{W}\,\delta(x-a), \qquad \bar{W} = \frac{W}{EI}.$$

Since both ends of the beam are pinned, the boundary conditions are

$$y(0) = y''(0) = y(L) = y''(L) = 0.$$

Applying the Laplace transform $Y(s) = \mathscr{L}\{y(x)\}$ and $\mathscr{L}\{\delta(x-a)\} = e^{-as}$ gives

$$\big[s^4 Y(s) - s^3 y(0) - s^2 y'(0) - s y''(0) - y'''(0)\big] + 4\beta^4 Y(s) = \bar{W}\,e^{-as}.$$

Since $y(0) = y''(0) = 0$, solving for $\Upsilon(s)$ leads to

$$\Upsilon(s) = y'(0)\,\frac{s^2}{s^4+4\beta^4} + y'''(0)\,\frac{1}{s^4+4\beta^4} + \bar{W}\,\frac{1}{s^4+4\beta^4}\,\mathrm{e}^{-as}.$$

Employing the results and notations as in the previous example, one obtains the deflection of the beam

$$y(x) = \mathscr{L}^{-1}\{\Upsilon(s)\} = y'(0)\cdot\phi_2(x) + y'''(0)\cdot\phi_0(x) + \bar{W}\,\phi_0(x-a)\,H(x-a),$$

in which the unknowns $y'(0)$ and $y'''(0)$ are determined from the boundary conditions $y(L) = y''(L) = 0$. For $x > a$, the deflection $y(x)$ is simplified as

$$y(x) = y'(0)\cdot\phi_2(x) + y'''(0)\cdot\phi_0(x) + \bar{W}\,\phi_0(x-a).$$

Differentiating $y(x)$ with respect to x twice times yields

$$\begin{aligned} y''(x) &= y'(0)\cdot\phi_2''(x) + y'''(0)\cdot\phi_0''(x) + \bar{W}\,\phi_0''(x-a) \\ &= -4\beta^4 y'(0)\cdot\phi_0(x) + y'''(0)\cdot\phi_2(x) + \bar{W}\,\phi_2(x-a). \end{aligned}$$

Applying the boundary conditions at $x = L$ gives

$$\begin{aligned} y(L) &= y'(0)\cdot\phi_2(L) + y'''(0)\cdot\phi_0(L) + \bar{W}\,\phi_0(L-a) = 0, \\ y''(L) &= -4\beta^4 y'(0)\cdot\phi_0(L) + y'''(0)\cdot\phi_2(L) + \bar{W}\,\phi_2(L-a) = 0, \end{aligned}$$

which leads to two algebraic equations for two unknowns $y'(0)$ and $y'''(0)$

$$\begin{aligned} \phi_2(L)\cdot y'(0) + \phi_0(L)\cdot y'''(0) &= \alpha_0, & \alpha_0 &= -\bar{W}\,\phi_0(L-a), \\ -4\beta^4\phi_0(L)\cdot y'(0) + \phi_2(L)\cdot y'''(0) &= \alpha_2, & \alpha_2 &= -\bar{W}\,\phi_2(L-a), \end{aligned}$$

and results in

$$y'(0) = \frac{\alpha_0\phi_2(L) - \alpha_2\phi_0(L)}{\phi_2^2(L) + 4\beta^4\phi_0^2(L)}, \qquad y'''(0) = \frac{\alpha_2\phi_2(L) + 4\beta^4\alpha_0\phi_0(L)}{\phi_2^2(L) + 4\beta^4\phi_0^2(L)}.$$

6.7 Summary

- The Laplace transform $F(s)$ of function $f(t)$ is defined as

$$F(s) = \mathscr{L}\{f(t)\} = \int_0^\infty \mathrm{e}^{-st} f(t)\,\mathrm{d}t, \quad s > 0.$$

Some important properties of Laplace transform and inverse Laplace transform are listed in Table 6.3.

Table 6.3 Properties of Laplace transform and inverse Laplace transform.

$\mathscr{L}\{f(t)\} = F(s)$	$\mathscr{L}^{-1}\{F(s)\} = f(t)$
1. Linear operator	
$\mathscr{L}\{f(t)\}$ is a linear operator	$\mathscr{L}^{-1}\{F(s)\}$ is a linear operator
2. Property of shifting	
$\mathscr{L}\{e^{at} f(t)\} = F(s-a)$	$\mathscr{L}^{-1}\{F(s-a)\} = e^{at} f(t)$
3. Property of differentiation	
$\mathscr{L}\{t^n f(t)\} = (-1)^n \dfrac{d^n F(s)}{ds^n}$	$\mathscr{L}^{-1}\left\{\dfrac{d^n F(s)}{ds^n}\right\} = (-1)^n t^n f(t)$
4. Property of integration	
$\mathscr{L}\left\{\dfrac{f(t)}{t^n}\right\} = \int_s^\infty \cdots \int_s^\infty F(s)(ds)^n$	$\mathscr{L}^{-1}\left\{\int_s^\infty \cdots \int_s^\infty F(s)(ds)^n\right\} = \dfrac{f(t)}{t^n}$
5. Laplace transform of integrals	
$\mathscr{L}\left\{\int_0^t \cdots \int_0^t f(u)(du)^n\right\} = \dfrac{F(s)}{s^n}$	$\mathscr{L}^{-1}\left\{\dfrac{F(s)}{s^n}\right\} = \int_0^t \cdots \int_0^t f(u)(du)^n$
6. Convolution integral	
$\mathscr{L}\{(f*g)(t)\} = F(s)G(s)$ $= \mathscr{L}\left\{\int_0^t f(u)\,g(t-u)\,du\right\}$ $= \mathscr{L}\left\{\int_0^t g(u)\,f(t-u)\,du\right\}$	$\mathscr{L}^{-1}\{F(s)G(s)\} = (f*g)(t)$ $= \int_0^t f(u)\,g(t-u)\,du$ $= \int_0^t g(u)\,f(t-u)\,du$
7. Heaviside function	
$\mathscr{L}\{f(t-a)H(t-a)\} = e^{-as} F(s)$	$\mathscr{L}^{-1}\{e^{-as} F(s)\} = f(t-a)H(t-a)$
8. Dirac delta function	
$\mathscr{L}\{f(t)\delta(t-a)\} = e^{-as} f(a)$	$\mathscr{L}^{-1}\{e^{-as}\} = \delta(t-a)$
9. Laplace transform of derivatives	
$\mathscr{L}\{f'(t)\} = sF(s) - f(0)$ $\mathscr{L}\{f''(t)\} = s^2 F(s) - s f(0) - f'(0)$ $\dots\ \dots$ $\mathscr{L}\{f^{(n)}(t)\} = s^n F(s) - s^{n-1} f(0) - s^{n-2} f'(0) - \cdots - s f^{(n-2)} - f^{(n-1)}(0)$	

- The Heaviside step function defined as

$$H(t-a) = \begin{cases} 0, & t<a, \\ 1, & t>a, \end{cases}$$

is very useful in describing piecewise smooth functions by combining the following results

$$f(t)\left[1-H(t-a)\right] = \begin{cases} f(t), & t<a, \\ 0, & t>a, \end{cases}$$

$$f(t)\left[H(t-a)-H(t-b)\right] = \begin{cases} 0, & t<a, \\ f(t), & a<t<b, \\ 0, & t>b, \end{cases}$$

$$f(t)\,H(t-a) = \begin{cases} 0, & t<a, \\ f(t), & t>a. \end{cases}$$

- The Dirac delta function $\delta(t-a)$ is a mathematical idealization of impulse functions. It is useful in modeling impulse functions, such as concentrated or point loads.
- Applying the Laplace transform to an nth-order linear differential equation with constant coefficients

$$a_n y^{(n)}(t) + a_{n-1} y^{(n-1)}(t) + \cdots + a_1 y'(t) + a_0 y(t) = f(t)$$

converts it into a linear algebraic equation for $Y(s) = \mathscr{L}\{y(t)\}$, which can easily be solved. The solution of the differential equation can be obtained by determining the inverse Laplace transform $y(t) = \mathscr{L}^{-1}\{Y(s)\}$.
- The method of Laplace transform is preferable and advantageous in solving linear ordinary differential equations with the right-hand side functions $f(t)$ involving the Heaviside step function and the Dirac delta function.

Problems

Evaluate the Laplace transform of the following functions.

6.1 $f(t) = 4t^3 - 2t^2 + 5$ ANS $F(s) = \dfrac{24-4s+5s^3}{s^4}$

6.2 $f(t) = 3\sin 2t - 4\cos 5t$ ANS $F(s) = \dfrac{6}{s^2+4} - \dfrac{4s}{s^2+25}$

6.3 $f(t) = e^{-2t}(4\cos 3t + 5\sin 3t)$ ANS $F(s) = \dfrac{4s+23}{s^2+4s+13}$

6.4 $f(t) = 3\cosh 6t + 8\sinh 3t$ ANS $F(s) = \dfrac{3s}{s^2-36} + \dfrac{24}{s^2-9}$

6.5 $f(t) = 3t\cos 2t + t^2 e^t$ ANS $F(s) = \dfrac{3(s^2-4)}{(s^2+4)^2} + \dfrac{2}{(s-1)^3}$

6.6 $f(t) = t\cosh 2t + t^2\sin 5t + t^3$ ANS $F(s) = \dfrac{s^2+4}{(s^2-4)^2} + \dfrac{10(3s^2-25)}{(s^2+25)^3} + \dfrac{6}{s^4}$

6.7 $f(t) = 7e^{-5t}\cos 2t + 9\sinh^2 2t$ ANS $F(s) = \dfrac{7(s+5)}{(s+5)^2+4} + \dfrac{72}{s(s^2-16)}$

6.8 $f(t) = \begin{cases} 0, & t<\pi, \\ \sin t, & t>\pi. \end{cases}$ ANS $F(s) = -\dfrac{e^{-\pi s}}{s^2+1}$

6.9 $f(t) = \begin{cases} 0, & t<1, \\ 4t^2+3t-8, & t>1. \end{cases}$ ANS $F(s) = e^{-s}\left(\dfrac{8}{s^3} + \dfrac{11}{s^2} - \dfrac{1}{s}\right)$

6.10 $f(t) = \begin{cases} 0, & t<1, \\ t^2-1, & 1<t<2, \\ 0, & t>2. \end{cases}$ ANS $F(s) = \dfrac{2(s+1)e^{-s} - (3s^2+4s+2)e^{-2s}}{s^3}$

6.11 $f(t) = \begin{cases} \sin t, & t<\pi, \\ 4\sin 3t, & t>\pi. \end{cases}$ ANS $F(s) = \dfrac{1}{s^2+1} - \dfrac{e^{-\pi s}(11s^2+3)}{(s^2+1)(s^2+9)}$

6.12 $f(t) = \begin{cases} 2t, & 0<t<2, \\ 2+t, & 2<t<4, \\ 10-t, & 4<t<10, \\ 0, & t>10. \end{cases}$ ANS $F(s) = \dfrac{2 - e^{-2s} - 2e^{-4s} + e^{-10s}}{s^2}$

6.13 $f(t) = t^3\,\delta(t-2) + 3\cos 5t\,\delta(t-\pi)$ ANS $F(s) = 8e^{-2s} - 3e^{-\pi s}$

6.14 $f(t) = \sinh 4t\,\delta(t+2) + e^{2t}\,\delta(t-1) + t^2e^{-3t}\,\delta(t-2) + \cos\pi t\,\delta(t-3)$

ANS $F(s) = e^{2-s} + 4e^{-6-2s} - e^{-3s}$

Express the following periodic functions using the Heaviside function or the Dirac function and evaluate the Laplace transform.

6.15

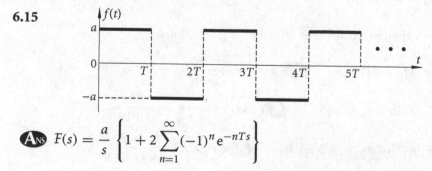

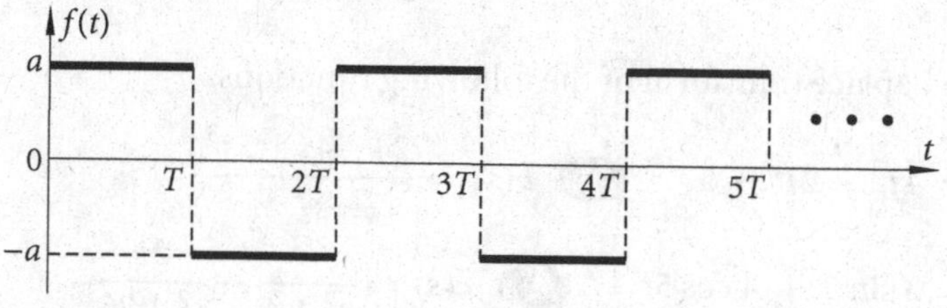

ANS $F(s) = \dfrac{a}{s}\left\{1 + 2\sum_{n=1}^{\infty}(-1)^n e^{-nTs}\right\}$

6.16

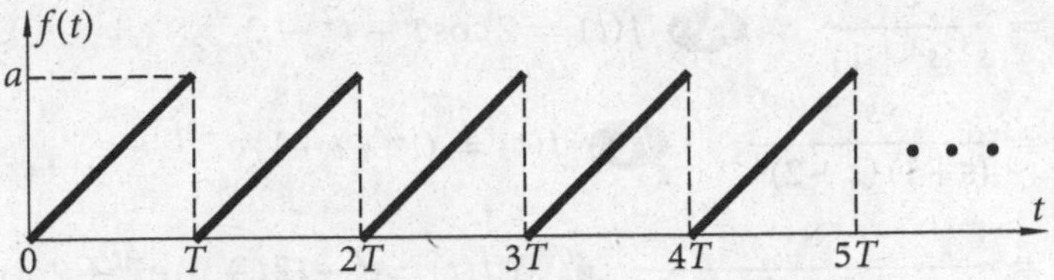

ANS $F(s) = \dfrac{a}{Ts^2} - \dfrac{a}{s}\displaystyle\sum_{n=0}^{\infty} e^{-(n+1)Ts}$

6.17

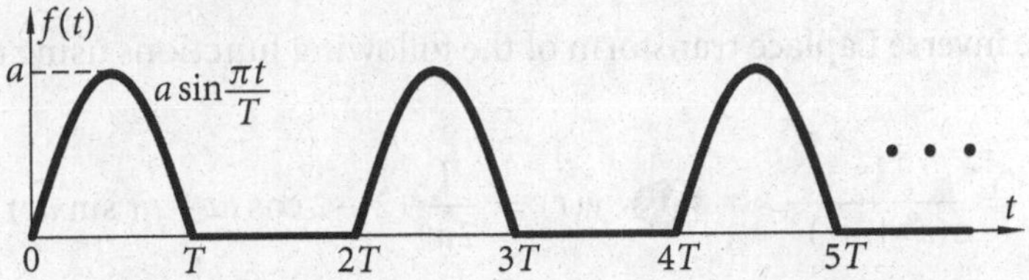

ANS $F(s) = \dfrac{aT\pi}{T^2s^2 + \pi^2}\displaystyle\sum_{n=0}^{\infty} e^{-nTs}$

6.18

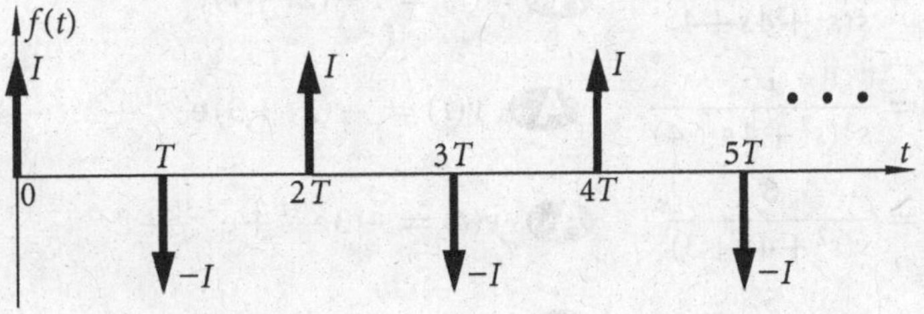

ANS $F(s) = I \cdot \displaystyle\sum_{n=0}^{\infty} (-1)^n e^{-nTs}$

Evaluate the inverse Laplace transform of the following functions.

6.19 $F(s) = \dfrac{s}{(s+1)^3}$ **ANS** $f(t) = \left(t - \frac{1}{2}t^2\right)e^{-t}$

6.20 $F(s) = \dfrac{4(2s+1)}{s^2-2s-3}$ **ANS** $f(t) = 7e^{3t} + e^{-t}$

6.21 $F(s) = \dfrac{3s+2}{s^2+6s+10}$ **ANS** $f(t) = e^{-3t}(3\cos t - 7\sin t)$

6.22 $F(s) = \dfrac{3s^2+2s-1}{s^2-5s+6}$ **ANS** $f(t) = 3\delta(t) + 32e^{3t} - 15e^{2t}$

6.23 $F(s) = \dfrac{30}{(s^2+1)(s^2-9)}$ **ANS** $f(t) = -3\sin t + \sinh 3t$

6.24 $F(s) = \dfrac{13s}{(s^2-4)(s^2+9)}$ **ANS** $f(t) = -\cos 3t + \cosh 2t$

6.25 $F(s) = \dfrac{40s}{(s+1)(s+2)(s^2-9)}$ **ANS** $f(t) = 10e^{-3t} - 16e^{-2t} + 5e^{-t} + e^{3t}$

6.26 $F(s) = \dfrac{2}{s^3(s^2+1)}$ ANS $f(t) = 2\cos t + t^2 - 2$

6.27 $F(s) = \dfrac{s}{(s+1)(s+2)^3}$ ANS $f(t) = (t^2+t+1)\,e^{-2t} - e^{-t}$

6.28 $F(s) = \dfrac{8}{(s-1)(s+1)^2(s^2+1)}$ ANS $f(t) = e^t - (2t+3)\,e^{-t} + 2\cos t - 2\sin t$

6.29 $F(s) = \dfrac{162}{s^3(s^2-9)}$ ANS $f(t) = -9t^2 - 2 + 2\cosh 3t$

Evaluate the inverse Laplace transform of the following functions using *convolution integral.*

6.30 $Y(s) = \dfrac{1}{s(s^2+a^2)^2}$ ANS $y(t) = \dfrac{1}{2a^4}(2 - 2\cos at - at\sin at)$

6.31 $Y(s) = \dfrac{1}{s^2(s^2+a^2)^2}$ ANS $y(t) = \dfrac{1}{2a^5}(2at + at\cos at - 3\sin at)$

6.32 $Y(s) = \dfrac{4}{s(s^2+4s+4)}$ ANS $y(t) = 1 - (2t+1)\,e^{-2t}$

6.33 $Y(s) = \dfrac{16}{s^3(s^2+4s+4)}$ ANS $y(t) = -(2t+3)\,e^{-2t} + 2t^2 - 4t + 3$

6.34 $Y(s) = \dfrac{6}{s(s^2+4s+3)}$ ANS $y(t) = -3\,e^{-t} + e^{-3t} + 2$

6.35 $Y(s) = \dfrac{5}{s(s^2+4s+5)}$ ANS $y(t) = -e^{-2t}(2\sin t + \cos t) + 1$

Solve the following differential equations.

6.36 $y'' + 4y' + 3y = 60\cos 3t, \quad y(0) = 1, \ y'(0) = -1$

ANS $y(t) = 5\,e^{-3t} - 2\,e^{-t} - 2\cos 3t + 4\sin 3t$

6.37 $y'' + y' - 2y = 9e^{-2t}$, $y(0) = 3$, $y'(0) = -6$ ANS $y(t) = e^t - (3t-2)\,e^{-2t}$

6.38 $y'' - y' - 2y = 2t^2 + 1, \quad y(0) = 6, \ y'(0) = 2$

ANS $y(t) = 5\,e^{-t} + 3\,e^{2t} - t^2 + t - 2$

6.39 $y'' + 4y = 8\sin 2t, \quad y(0) = 1, \ y'(0) = 4$

ANS $y(t) = (-2t+1)\cos 2t + 3\sin 2t$

6.40 $y'' - 2y' + y = 4\,e^{-t} + 2\,e^t, \quad y(0) = -1, \ y'(0) = 2$

ANS $y(t) = e^{-t} + (t^2 + 5t - 2)\,e^t$

6.41 $y'' - 2y' + 2y = 8\,e^{-t}\sin t, \quad y(0) = 1, \ y'(0) = -1$

ANS $y(t) = -e^t\sin t + e^{-t}(\cos t + \sin t)$

6.42 $y'' - 2y' + 5y = 8e^t \sin 2t, \quad y(0) = 1, \ y'(0) = -1$

ANS $y(t) = -(2t-1)e^t \cos 2t$

6.43 $y'' + y' - 2y = 54te^{-2t}, \quad y(0) = 6, \ y'(0) = 0$

ANS $y(t) = -(9t^2+6t)e^{-2t} + 6e^t$

6.44 $y'' - y' - 2y = 9e^{2t} H(t-1), \quad y(0) = 6, \ y'(0) = 0$

ANS $y(t) = 4e^{-t} + 2e^{2t} + \big[(3t-4)e^{2t} + e^{3-t}\big]H(t-1)$

6.45 $y'' + 2y' + y = 2\sin t\, H(t-\pi), \quad y(0) = 1, \ y'(0) = 0$

ANS $y(t) = (t+1)e^{-t} - \big[\cos t + (t+1-\pi)e^{\pi-t}\big]H(t-\pi)$

6.46 $y'' + 4y = 8\sin 2t\, H(t-\pi), \quad y(0) = 0, \ y'(0) = 2$

ANS $y(t) = \sin 2t + \big[2(\pi-t)\cos 2t + \sin 2t\big]H(t-\pi)$

6.47 $y'' + 4y = 8(t^2+t-1)H(t-2), \quad y(0) = 1, \ y'(0) = 2$

ANS $y(t) = \sin 2t + \cos 2t + \big[2t^2 + 2t - 3 - 9\cos(2t-4) - 5\sin(2t-4)\big]H(t-2)$

6.48 $y'' - 3y' + 2y = e^t H(t-2), \quad y(0) = 1, \ y'(0) = 2$

ANS $y(t) = e^{2t} + \big[(1-t)e^t + e^{2t-2}\big]H(t-2)$

6.49 $y'' - 5y' + 6y = \delta(t-2), \quad y(0) = -1, \ y'(0) = 1$

ANS $y(t) = -4e^{2t} + 3e^{3t} + \big[e^{3(t-2)} - e^{2(t-2)}\big]H(t-2)$

6.50 $y'' + 4y = 4H(t-\pi) + 2\delta(t-\pi), \quad y(0) = -1, \ y'(0) = 2$

ANS $y(t) = \sin 2t - \cos 2t + (1 + \sin 2t - \cos 2t)H(t-\pi)$

6.51 $y''' - y'' + 4y' - 4y = 10e^{-t}, \quad y(0) = 5, \ y'(0) = -2, \ y''(0) = 0$

ANS $y(t) = -e^{-t} + 5e^t + \cos 2t - 4\sin 2t$

6.52 $y'''' - 5y'' + 4y = 120e^{3t} H(t-1), \ y(0) = 15, \ y'(0) = -6, \ y''(0) = 0,$
$y'''(0) = 0$ ANS $y(t) = 6e^t + 14e^{-t} - 2e^{2t} - 3e^{-2t} + (10e^{t+2} - 5e^{-t+4}$
$- 10e^{2t+1} + 2e^{-2t+5} + 3e^{3t})H(t-1)$

6.53 $y'''' + 3y'' - 4y = 40t^2 H(t-2), \quad y(0) = y'(0) = y''(0) = y'''(0) = 0$

ANS $y(t) = \big[-10t^2 - 15 + 40e^{t-2} + 8e^{2-t} + 7\cos(2t-4) + 4\sin(2t-4)\big]H(t-2)$

6.54 $y'''' + 4y = (2t^2 + t + 1)\delta(t-1), \quad y(0) = 1, \ y'(0) = -2, \ y''(0) = 0,$
$y'''(0) = 0$ ANS $y(t) = e^{-t}\cos t - \sin t \cosh t + \big[\sin(t-1)\cosh(t-1)$
$- \cos(t-1)\sinh(t-1)\big]H(t-1)$

6.55 Determine the lateral deflection $y(x)$ of the beam-column as shown.

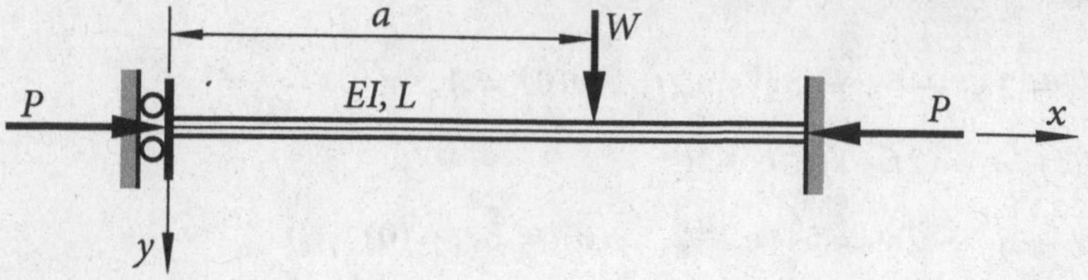

ANS
$$W_0 = \frac{W\big[\alpha(L-a) - \sin\alpha(L-a)\big]}{\alpha^3 EI}, \qquad W_1 = \frac{W\big[1-\cos\alpha(L-a)\big]}{\alpha^2 EI},$$

$$y(0) = \frac{W_1(1-\cos\alpha L)}{\alpha\sin\alpha L} - W_0, \qquad y''(0) = -\frac{W_1\alpha}{\sin\alpha L},$$

$$y(x) = y(0) + \frac{1-\cos\alpha x}{\alpha^2}\,y''(0) + \frac{W\big[\alpha(x-a) - \sin\alpha(x-a)\big]H(x-a)}{\alpha^3 EI}.$$

For the single degree-of-freedom system shown in Figure 6.6, determine the forced vibration response $x_{\text{Forced}}(t)$ due to the externally applied load $f(t)$ shown. The system is assumed to be underdamped, i.e., $0<\zeta<1$.

6.56

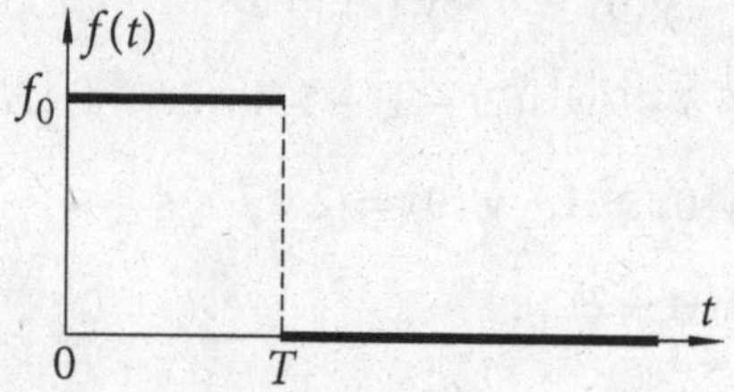

ANS $x_{\text{Forced}}(t) = \dfrac{f_0}{m}\big[\eta_1(t) - \eta_1(t-T)H(t-T)\big]$

6.57

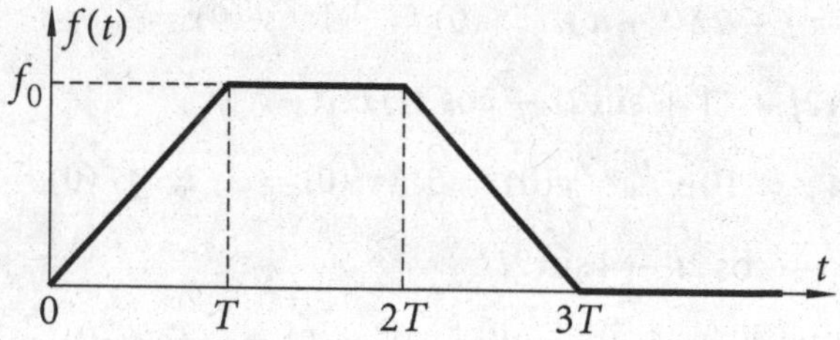

ANS
$$x_{\text{Forced}}(t) = \frac{f_0}{mT}\big[\eta_2(t) - \eta_2(t-T)H(t-T) - \eta_2(t-2T)H(t-2T) + \eta_2(t-3T)H(t-3T)\big]$$

6.58

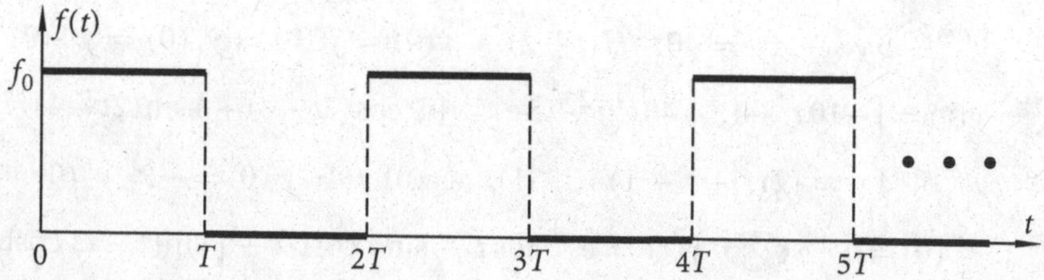

ANS
$$x_{\text{Forced}}(t) = \frac{f_0}{m}\sum_{n=0}^{\infty}\Big[\eta_1(t-2nT)H(t-2nT) - \eta_1\big(t-(2n+1)T\big)H\big(t-(2n+1)T\big)\Big]$$

6.59

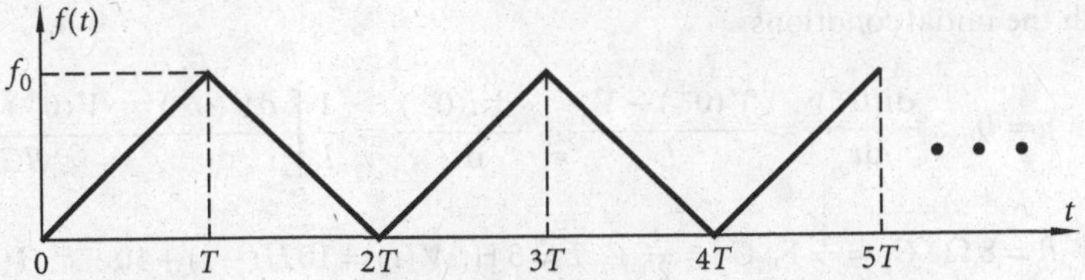

ANS $x_{\text{Forced}}(t)=\dfrac{f_0}{mT}\Bigg\{\eta_2(t)-2\sum_{n=1}^{\infty}\Big[\eta_2\big(t-(2n-1)T\big)H\big(t-(2n-1)T\big)$

$$-\eta_2(t-2nT)H(t-2nT)\Big]\Bigg\}$$

6.60 For the circuit shown in Figure 6.9(a), the current source $I(t)$ is

$$I(t)=I_0\,H(-t)+I_1(t)\,H(t).$$

Show that the differential equation governing $i(t)$ is

$$\frac{\mathrm{d}i}{\mathrm{d}t}+\frac{L+R_1R_2C}{R_1LC}\,i+\frac{R_1+R_2}{R_1LC}\frac{\mathrm{d}i}{\mathrm{d}t}=\frac{I_1(t)}{LC},\qquad i(0^+)=\frac{R_1I_0}{R_1+R_2},\qquad \frac{i(0^+)}{\mathrm{d}t}=0.$$

For $R_1=1\,\Omega$, $R_2=8\,\Omega$, $C=\frac{1}{4}$ F, $L=4$ H, $I(t)=13\sin 2t\,\big[H(t)-H(t-\pi)\big]$ (A), $I_0=0$ A, determine $i(t)$ for $t>0$.

ANS $i(t)=\left(\frac{12}{13}+2t\right)\mathrm{e}^{-3t}-\left[\frac{12}{13}+2(t-\pi)\right]\mathrm{e}^{-3(t-\pi)}H(t-\pi)$

$$+\tfrac{1}{13}(5\sin 2t-12\cos 2t)\big[1-H(t-\pi)\big]\ \text{(A)}$$

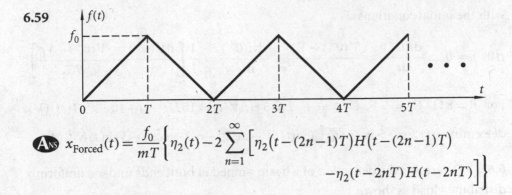

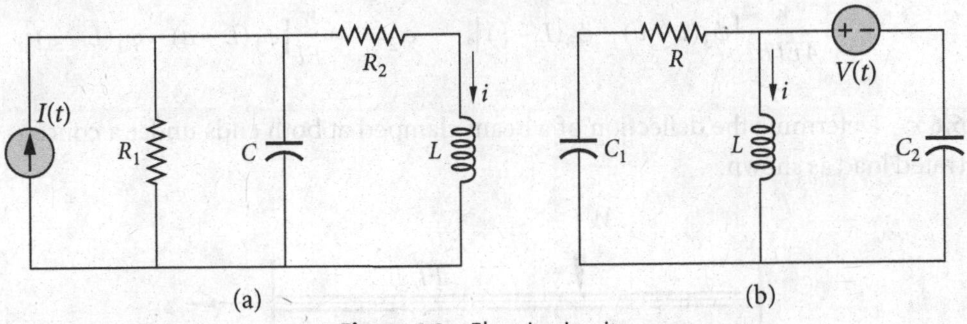

Figure 6.9 Electric circuits.

6.61 For the circuit shown in Figure 6.9(b), the voltage source $V(t)$ is

$$V(t)=V_0\,H(-t)+V_1(t)\,H(t).$$

Show that the differential equation governing $i(t)$ is

$$\frac{\mathrm{d}^3i}{\mathrm{d}t^3}+\frac{C_1+C_2}{RC_1C_2}\frac{\mathrm{d}^2i}{\mathrm{d}t^2}+\frac{1}{LC_2}\frac{\mathrm{d}i}{\mathrm{d}t}+\frac{1}{RLC_1C_2}\,i=\frac{1}{L}\frac{\mathrm{d}^2V_1(t)}{\mathrm{d}t^2}+\frac{1}{RLC_1}\frac{\mathrm{d}V_1(t)}{\mathrm{d}t},$$

with the initial condtions

$$i(0^+) = 0, \quad \frac{di(0^+)}{dt} = \frac{V(0^+) - V_0}{L}, \quad \frac{d^2 i(0^+)}{dt^2} = \frac{1}{L}\left[\frac{dV(0^+)}{dt} - \frac{V(0^+) - V_0}{RC_2}\right].$$

For $R=8\,\Omega$, $C_1 = \frac{1}{4}$ F, $C_2 = \frac{1}{20}$ F, $L=5$ H, $V(t) = 10H(-t) + 10e^{-2t}H(t)$ (V), determine $i(t)$ for $t>0$. ANS $i(t) = -3e^{-2t} + (2 + \cos t - 3\sin t)e^{-t}$ (A)

6.62 Determine the deflection of a beam pinned at both ends under a uniformly distributed load as shown.

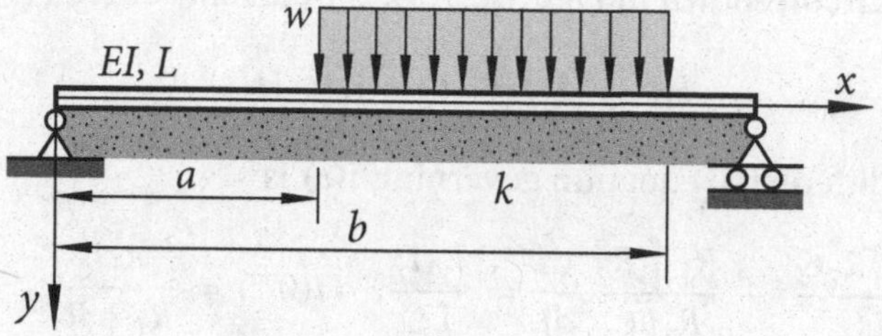

ANS
$$y(x) = y'(0)\cdot\phi_2(x) + y'''(0)\cdot\phi_0(x) + \frac{w}{4EI\beta^4}\Big\{\big[1-\phi_3(x-a)\big]H(x-a) - \big[1-\phi_3(x-b)\big]H(x-b)\Big\},$$

$$y'(0) = \frac{\alpha_0\phi_2(L) - \alpha_2\phi_0(L)}{\phi_2^2(L) + 4\beta^4\phi_0^2(L)}, \qquad y'''(0) = \frac{\alpha_2\phi_2(L) + 4\beta^4\alpha_0\phi_0(L)}{\phi_2^2(L) + 4\beta^4\phi_0^2(L)},$$

$$\alpha_0 = \frac{w}{4EI\beta^4}\big[\phi_3(L-a) - \phi_3(L-b)\big], \qquad \alpha_2 = -\frac{w}{EI}\big[\phi_1(L-a) - \phi_1(L-b)\big].$$

6.63 Determine the deflection of a beam clamped at both ends under a concentrated load as shown.

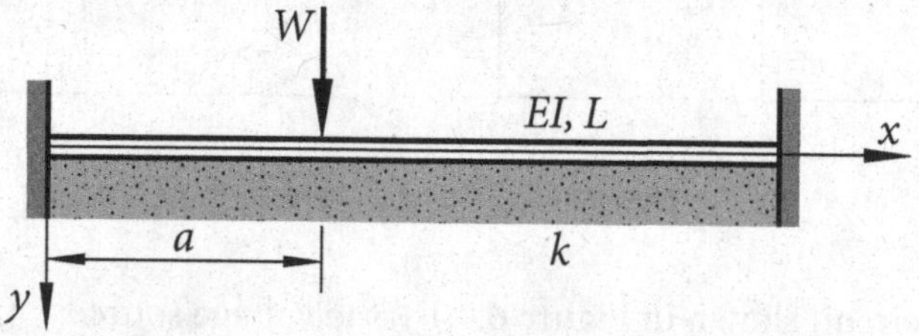

ANS
$$y(x) = y''(0)\cdot\phi_1(x) + y'''(0)\cdot\phi_0(x) + \hat{W}\phi_0(x-a)H(x-a), \qquad \hat{W} = \frac{W}{EI}$$

$$y''(0) = \frac{\alpha_0\phi_1(L) - \alpha_1\phi_0(L)}{\phi_1^2(L) - \phi_0(L)\phi_2(L)}, \qquad y'''(0) = \frac{\alpha_1\phi_1(L) - \alpha_0\phi_2(L)}{\phi_1^2(L) - \phi_0(L)\phi_2(L)},$$

$$\alpha_0 = -\hat{W}\phi_0(L-a), \qquad \alpha_1 = -\hat{W}\phi_1(L-a).$$

6.64 Determine the deflection of a free-clamped beam under a triangularly distributed load as shown.

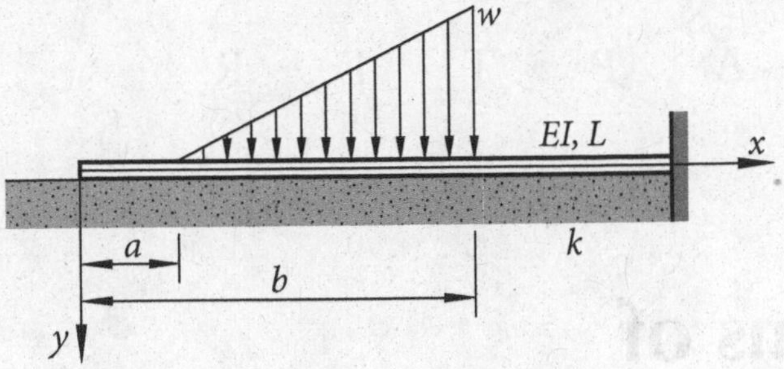

ANS $y(x) = y(0)\cdot\phi_3(x) + y'(0)\cdot\phi_2(x) + \hat{w}\Big\{\big[(x-a)-\phi_2(x-a)\big]H(x-a)$
$-\big[(x-a)-\phi_2(x-b)-(b-a)\phi_3(x-b)\big]H(x-b)\Big\},$

$$\hat{w} = \frac{w}{4(b-a)EI\beta^4},$$

$$y(0) = \frac{\alpha_0\phi_3(L) - \alpha_1\phi_2(L)}{\phi_3^2(L) + 4\beta^4\phi_0(L)\phi_2(L)}, \qquad y'(0) = \frac{\alpha_1\phi_3(L) + 4\beta^4\alpha_0\phi_0(L)}{\phi_3^2(L) + 4\beta^4\phi_0(L)\phi_2(L)},$$

$$\alpha_0 = -\hat{w}\left[\phi_2(L-b) - \phi_2(L-a) + (b-a)\phi_3(L-b)\right],$$
$$\alpha_1 = -\hat{w}\left[\phi_3(L-b) - \phi_3(L-a) - 4\beta^4(b-a)\phi_0(L-b)\right].$$

6.65 Determine the deflection of a sliding-clamped beam under a triangularly distributed load as shown.

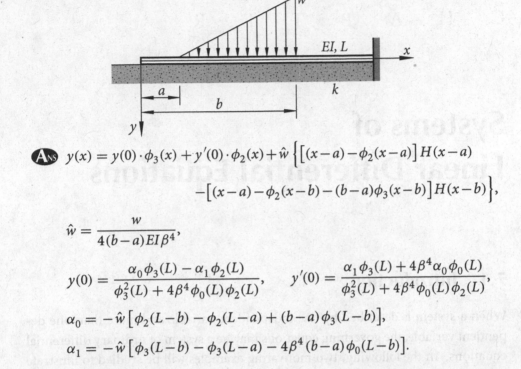

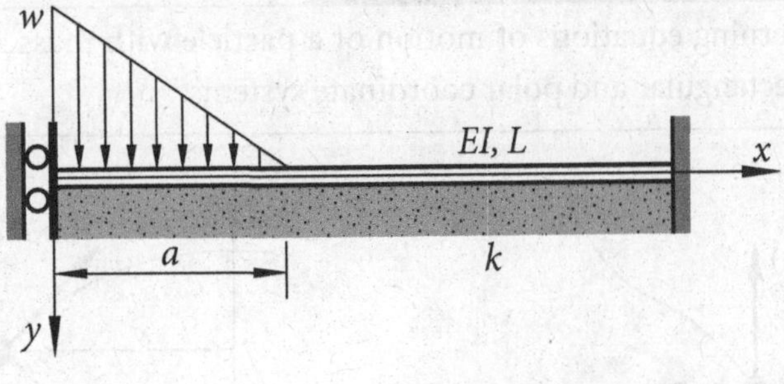

ANS $y(x) = y(0)\cdot\phi_3(x) + y''(0)\cdot\phi_1(x) - \hat{w}\left\{x - a - \phi_2(x) + a\phi_3(x)\right.$
$\left.-\big[(x-a) - \phi_2(x-a)\big]H(x-a)\right\}, \qquad \hat{w} = \dfrac{w}{4aEI\beta^4},$

$$y(0) = \frac{\alpha_0\phi_2(L) - \alpha_1\phi_1(L)}{\phi_1(L)\phi_3(L) + 4\beta^4\phi_0(L)\phi_1(L)}, \quad y''(0) = \frac{\alpha_1\phi_3(L) + 4\beta^4\alpha_0\phi_0(L)}{\phi_1(L)\phi_3(L) + 4\beta^4\phi_0(L)\phi_1(L)},$$

$$\alpha_0 = \hat{w}\left[-\phi_2(L) + a\phi_3(L) + \phi_2(L-a)\right],$$
$$\alpha_1 = \hat{w}\left[-\phi_3(L) - 4a\beta^4\phi_0(L) + \phi_3(L-a)\right].$$

CHAPTER 7

Systems of Linear Differential Equations

7.1 Introduction

When a system is described by one independent variable and more than one dependent variable, the governing equations may be a system of ordinary differential equations. In the following, two motivating examples will be studied to illustrate how systems of differential equations arise in practice.

Example 7.1 — Particle Moving in a Plane

Derive the governing equations of motion of a particle with mass m moving in a plane in both rectangular and polar coordinate systems.

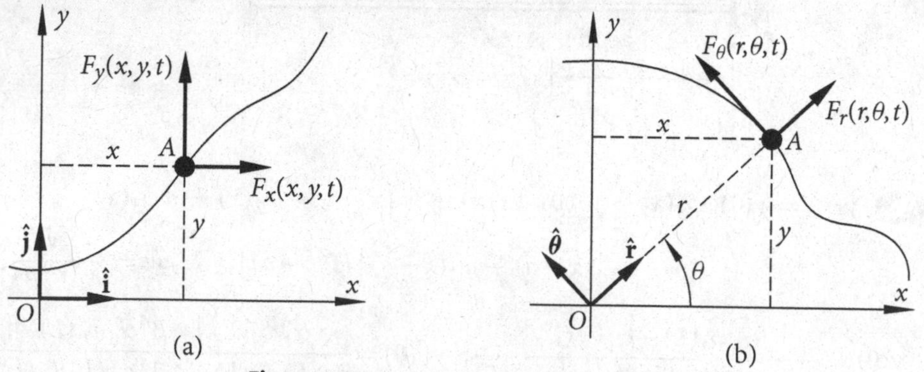

Figure 7.1 A particle moving in a plane.

1. Rectangular Coordinates: The particle moves along a trajectory in the x-y plane as shown in Figure 7.1(a); at time t, its coordinates are (x, y). It is subjected to externally applied forces $F_x(x, y, t)$ in the x-direction and $F_y(x, y, t)$ in the y-direction.

The equations of motion are given by Newton's Second Law

$$\text{in the } x\text{-direction:} \qquad m\ddot{x} = F_x(x, y, t),$$

$$\text{in the } y\text{-direction:} \qquad m\ddot{y} = F_y(x, y, t),$$

which is a system of two second-order ordinary differential equations.

2. Polar Coordinates: At time t, the particle is at point A with polar coordinates (r, θ) as shown in Figure 7.1(b). Since $x = r\cos\theta$, $y = r\sin\theta$, the position vector is given by

$$\mathbf{r} = x\,\hat{\mathbf{i}} + y\,\hat{\mathbf{j}} = r\,(\cos\theta\,\hat{\mathbf{i}} + \sin\theta\,\hat{\mathbf{j}}) = r\,\hat{\mathbf{r}},$$

where $\hat{\mathbf{r}} = \cos\theta\,\hat{\mathbf{i}} + \sin\theta\,\hat{\mathbf{j}}$ is the unit vector in the direction OA or $\mathbf{r}$. The unit vector normal to $\hat{\mathbf{r}}$, denoted as $\hat{\boldsymbol{\theta}}$, is $\hat{\boldsymbol{\theta}} = -\sin\theta\,\hat{\mathbf{i}} + \cos\theta\,\hat{\mathbf{j}}$.

The velocity and the acceleration of the particle are

$$\mathbf{v} = \dot{\mathbf{r}} = \dot{x}\,\hat{\mathbf{i}} + \dot{y}\,\hat{\mathbf{j}}, \qquad \mathbf{a} = \ddot{\mathbf{r}} = \ddot{x}\,\hat{\mathbf{i}} + \ddot{y}\,\hat{\mathbf{j}}.$$

Since, differentiating x and y with respect to time t yields

$$\begin{aligned}
\dot{x} &= \dot{r}\cos\theta - r\sin\theta\cdot\dot{\theta}, \qquad \dot{y} = \dot{r}\sin\theta + r\cos\theta\cdot\dot{\theta},\\
\ddot{x} &= (\ddot{r}\cos\theta - \dot{r}\sin\theta\cdot\dot{\theta}) - (\dot{r}\sin\theta\cdot\dot{\theta} - r\cos\theta\cdot\dot{\theta}^2 - r\sin\theta\cdot\ddot{\theta})\\
&= \ddot{r}\cos\theta - 2\dot{r}\dot{\theta}\sin\theta - r\dot{\theta}^2\cos\theta - r\ddot{\theta}\sin\theta,\\
\ddot{y} &= (\ddot{r}\sin\theta + \dot{r}\cos\theta\cdot\dot{\theta}) + (\dot{r}\cos\theta\cdot\dot{\theta} - r\sin\theta\cdot\dot{\theta}^2 + r\cos\theta\cdot\ddot{\theta})\\
&= \ddot{r}\sin\theta + 2\dot{r}\dot{\theta}\cos\theta - r\dot{\theta}^2\sin\theta + r\ddot{\theta}\cos\theta,
\end{aligned}$$

the acceleration vector becomes

$$\begin{aligned}
\mathbf{a} &= \big(\ddot{r}\cos\theta - 2\dot{r}\dot{\theta}\sin\theta - r\dot{\theta}^2\cos\theta - r\ddot{\theta}\sin\theta\big)\,\hat{\mathbf{i}}\\
&\quad + \big(\ddot{r}\sin\theta + 2\dot{r}\dot{\theta}\cos\theta - r\dot{\theta}^2\sin\theta + r\ddot{\theta}\cos\theta\big)\,\hat{\mathbf{j}}\\
&= (\ddot{r} - r\dot{\theta}^2)\,(\cos\theta\,\hat{\mathbf{i}} + \sin\theta\,\hat{\mathbf{j}}) + (2\dot{r}\dot{\theta} + r\ddot{\theta})\,(-\sin\theta\,\hat{\mathbf{i}} + \cos\theta\,\hat{\mathbf{j}})\\
&= (\ddot{r} - r\dot{\theta}^2)\,\hat{\mathbf{r}} + (2\dot{r}\dot{\theta} + r\ddot{\theta})\,\hat{\boldsymbol{\theta}} = a_r\,\hat{\mathbf{r}} + a_\theta\,\hat{\boldsymbol{\theta}},
\end{aligned}$$

where

$$a_r = \ddot{r} - r\dot{\theta}^2, \qquad a_\theta = 2\dot{r}\dot{\theta} + r\ddot{\theta}$$

are the radial and angular accelerations of the particle in the $\hat{\mathbf{r}}$ and $\hat{\boldsymbol{\theta}}$ directions, respectively. Hence, the equations of motion are, using Newton's Second Law,

$$\text{in the } \hat{\mathbf{r}} \text{ direction:} \quad ma_r = \sum F_r \implies m(\ddot{r} - r\dot{\theta}^2) = F_r(r, \theta, t),$$

$$\text{in the } \hat{\boldsymbol{\theta}} \text{ direction:} \quad ma_\theta = \sum F_\theta \implies m(2\dot{r}\dot{\theta} + r\ddot{\theta}) = F_\theta(r, \theta, t),$$

which is a system of two second-order ordinary differential equations.

Example 7.2 — Vibration of Multiple Story Shear Building

Derive the equations of motion of an n-story shear building as shown in Figure 7.2(a).

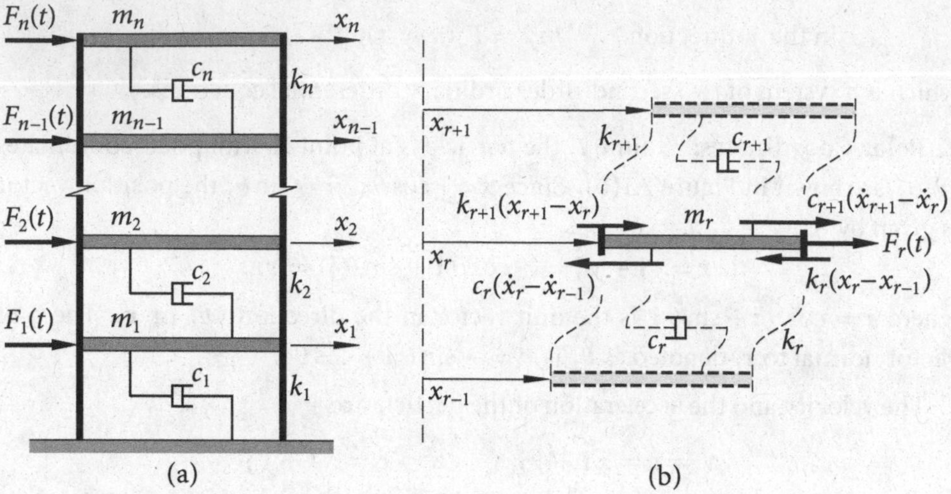

Figure 7.2 An n-story shear building.

The rth floor, $r=1,2,\ldots,n$, is assumed to be rigid with mass m_r and is subjected to externally applied load $F_r(t)$. The combined stiffness of the columns connecting the $(r-1)$th and the rth floors is k_r, and the damping coefficient of the dashpot damper, due to internal friction, between the $(r-1)$th and the rth floors is c_r. The displacement of the rth floor is described by $x_r(t)$.

Consider the motion of the rth floor, whose free-body diagram is shown in Figure 7.2(b). Note that the columns between the $(r-1)$th and the rth floors behave as a spring with stiffness k_r.

Remarks: To determine the shear force applied on the rth floor by the columns between the $(r-1)$th and rth floors, stand on the $(r-1)$th floor to observe the motion of the rth floor. The rth floor is seen to move toward the right with a relative displacement of x_r-x_{r-1}; hence the columns will try to pull the rth floor back to the left, exerting a shear (spring) force of $k_r(x_r-x_{r-1})$ toward the left. Similarly, the damping force is $c_r(\dot{x}_r-\dot{x}_{r-1})$.

On the other hand, to determine the shear force applied on the rth floor by the columns between the rth and $(r+1)$th floors, stand on the $(r+1)$th floor to observe the motion of the rth floor. The rth floor is seen to move toward the left with a relative displacement of $x_{r+1}-x_r$; thus the columns will try to pull the rth floor back to the right, resulting in a shear force of $k_{r+1}(x_{r+1}-x_r)$ toward the right. The damping force is $c_{r+1}(\dot{x}_{r+1}-\dot{x}_r)$.

Applying Newton's Second Law, the equation of motion of the rth floor, $r=1, 2, \ldots, n$, is

$$m_r\ddot{x}_r = F_r(t) + k_{r+1}(x_{r+1}-x_r) + c_{r+1}(\dot{x}_{r+1}-\dot{x}_r) - k_r(x_r-x_{r-1}) - c_r(\dot{x}_r-\dot{x}_{r-1}),$$

or

$$m_r\ddot{x}_r - c_r\dot{x}_{r-1} + (c_r+c_{r+1})\dot{x}_r - c_{r+1}\dot{x}_{r+1} - k_r x_{r-1} + (k_r+k_{r+1})x_r - k_{r+1}x_{r+1} = F_r(t),$$

where $x_0=0$, $c_{n+1}=0$, and $k_{n+1}=0$.

In the matrix form, the equations of motion of the n-story shear building can be written as

$$\mathbf{M}\ddot{\mathbf{x}} + \mathbf{C}\dot{\mathbf{x}} + \mathbf{K}\mathbf{x} = \mathbf{F}(t),$$

where

$$\mathbf{x} = \{x_1, x_2, \ldots, x_n\}^T, \qquad \mathbf{F}(t) = \{F_1(t), F_2(t), \ldots, F_n(t)\}^T$$

are the displacement and load vectors, respectively. **M**, **C**, and **K** are the mass, damping, and stiffness matrices, respectively, given by

$$\mathbf{M} = \operatorname{diag}\{m_1, m_2, \ldots, m_n\},$$

$$\mathbf{C} = \begin{bmatrix} c_1+c_2 & -c_2 & & & \\ -c_2 & c_2+c_3 & -c_3 & & \\ & \ddots & \ddots & \ddots & \\ & & -c_{n-1} & c_{n-1}+c_n & -c_n \\ & & & -c_n & c_n \end{bmatrix},$$

$$\mathbf{K} = \begin{bmatrix} k_1+k_2 & -k_2 & & & \\ -k_2 & k_2+k_3 & -k_3 & & \\ & \ddots & \ddots & \ddots & \\ & & -k_{n-1} & k_{n-1}+k_n & -k_n \\ & & & -k_n & k_n \end{bmatrix}.$$

Hence, the motion of an n-story shear building is governed by a system of n second-order linear ordinary differential equations. The system is equivalent to the mass-spring-damper system shown in Figure 7.3.

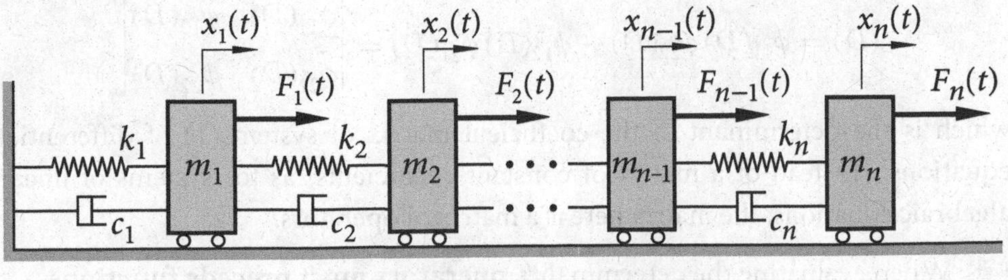

Figure 7.3 An equivalent mass-spring-damper system.

7.2 The Method of Operator

In this section, the method of D-operator is employed to obtain the complementary and particular solutions of systems of linear ordinary differential equations.

7.2.1 Complementary Solutions

Consider a system of two linear ordinary differential equations

$$\phi_{11}(D)\,x_1 + \phi_{12}(D)\,x_2 = 0, \qquad D(\cdot) \equiv \mathrm{d}(\cdot)/\mathrm{d}t, \tag{1a}$$

$$\phi_{21}(D)\,x_1 + \phi_{22}(D)\,x_2 = 0, \tag{1b}$$

where $\phi_{ij}(D)$, $i, j = 1, 2$, are polynomials of D with constant coefficients.

To find x_1, eliminate x_2 as follows:

operate $\phi_{22}(D)$ on (1a): $\quad \phi_{22}(D)\,\phi_{11}(D)\,x_1 + \phi_{22}(D)\,\phi_{12}(D)\,x_2 = 0, \quad$ (2a)

operate $\phi_{12}(D)$ on (1b): $\quad \phi_{12}(D)\,\phi_{21}(D)\,x_1 + \phi_{12}(D)\,\phi_{22}(D)\,x_2 = 0. \quad$ (2b)

Subtracting equation (2b) from (2a) yields

$$\big[\phi_{22}(D)\,\phi_{11}(D) - \phi_{12}(D)\,\phi_{21}(D)\big]\,x_1 = 0. \tag{3}$$

Similarly, to find x_2, eliminate x_1 as follows:

operate $\phi_{21}(D)$ on (1a): $\quad \phi_{21}(D)\,\phi_{11}(D)\,x_1 + \phi_{21}(D)\,\phi_{12}(D)\,x_2 = 0, \quad$ (4a)

operate $\phi_{11}(D)$ on (1b): $\quad \phi_{11}(D)\,\phi_{21}(D)\,x_1 + \phi_{11}(D)\,\phi_{22}(D)\,x_2 = 0. \quad$ (4b)

Subtracting equation (4a) from (4b) yields

$$\big[\phi_{11}(D)\,\phi_{22}(D) - \phi_{21}(D)\,\phi_{12}(D)\big]\,x_2 = 0. \tag{5}$$

It can be seen that equation (3) for x_1 and equation (5) for x_2 have the same form

$$\phi(D)\,x_1 = 0, \qquad \phi(D)\,x_2 = 0,$$

where

$$\phi(D) = \phi_{11}(D)\,\phi_{22}(D) - \phi_{12}(D)\,\phi_{21}(D) = \begin{vmatrix} \phi_{11}(D) & \phi_{12}(D) \\ \phi_{21}(D) & \phi_{22}(D) \end{vmatrix},$$

which is the determinant of the coefficient matrix of system (1) of differential equations. Instead of a matrix of constant coefficients, as for systems of linear algebraic equations, the matrix here is a matrix of operators.

☞ When evaluating the determinants, **operators must precede functions.**

In general, for a system of n linear ordinary differential equations

$$\begin{aligned}
&\phi_{11}(D)\,x_1 + \phi_{12}(D)\,x_2 + \cdots + \phi_{1n}(D)\,x_n = 0,\\
&\phi_{21}(D)\,x_1 + \phi_{22}(D)\,x_2 + \cdots + \phi_{2n}(D)\,x_n = 0,\\
&\cdots\ \cdots\\
&\phi_{n1}(D)\,x_1 + \phi_{n2}(D)\,x_2 + \cdots + \phi_{nn}(D)\,x_n = 0,
\end{aligned} \tag{6}$$

the differential equations for $x_1, x_2, \ldots, x_n$ have the same form

$$\phi(D)\,x_1 = 0, \quad \phi(D)\,x_2 = 0, \quad \ldots, \quad \phi(D)\,x_n = 0,$$

where $\phi(D)$ is the determinant of the coefficient matrix

$$\phi(D) = \begin{vmatrix} \phi_{11}(D) & \phi_{12}(D) & \cdots & \phi_{1n}(D) \\ \phi_{21}(D) & \phi_{22}(D) & \cdots & \phi_{2n}(D) \\ \vdots & \vdots & \cdots & \vdots \\ \phi_{n1}(D) & \phi_{n2}(D) & \cdots & \phi_{nn}(D) \end{vmatrix}.$$

Hence, the unknowns $x_1, x_2, \ldots, x_n$ all have the same characteristic equation $\phi(\lambda)=0$ and, as a result, the same form of complementary solutions.

The complementary solutions of system (6) contain arbitrary constants, the number of which is the degree of polynomial of $\phi(D)$. It is likely that the complementary solutions $x_{1C}, x_{2C}, \ldots, x_{nC}$, written using the roots of the characteristic equation $\phi(\lambda)=0$, will contain more constants. ***The extra constants can be eliminated by substituting the solutions into any one of the original equations*** in system (6).

Example 7.3

Solve $\dfrac{dx}{dt} - 3x - 6y = 0, \quad 3x + \dfrac{dy}{dt} + 3y = 0.$

Using the D-operator, $D(\cdot) \equiv d(\cdot)/dt$, the differential equations become

$$(D-3)\,x - 6y = 0, \tag{1a}$$

$$3x + (D+3)\,y = 0. \tag{1b}$$

The determinant of the coefficient matrix is

$$\phi(D) = \begin{vmatrix} D-3 & -6 \\ 3 & D+3 \end{vmatrix} = (D-3)(D+3) + 18 = D^2 + 9.$$

The characteristic equation is $\phi(\lambda) = \lambda^2 + 9 = 0 \Longrightarrow \lambda = \pm i3$. The complementary solutions of x and y have the same form and are given by

$$x_C = A_1 \cos 3t + B_1 \sin 3t, \qquad y_C = A_2 \cos 3t + B_2 \sin 3t. \tag{2}$$

which contain four arbitrary constants. However, since $\phi(D)$ is a polynomial of degree 2 in D, the complementary solutions should contain only two arbitrary constants.

Substitute solutions (2) into equation (1a) to eliminate the two extra constants

$$
\begin{aligned}
(D-3)x_C - 6y_C &= \big[(-3A_1 \sin 3t + 3B_1 \cos 3t) - 3(A_1 \cos 3t + B_1 \sin 3t)\big] \\
&\quad - 6(A_2 \cos 3t + B_2 \sin 3t) \\
&= (-3A_1 + 3B_1 - 6A_2)\cos 3t + (-3A_1 - 3B_1 - 6B_2)\sin 3t \\
&= 0. \qquad (3)
\end{aligned}
$$

Since $\cos 3t$ and $\sin 3t$ are linearly independent, equation (3) implies that the coefficients of $\cos 3t$ and $\sin 3t$ are zero:

$$
\begin{aligned}
-3A_1 + 3B_1 - 6A_2 = 0 &\implies A_2 = -\tfrac{1}{2}(A_1 - B_1), \\
-3A_1 - 3B_1 - 6B_2 = 0 &\implies B_2 = -\tfrac{1}{2}(A_1 + B_1).
\end{aligned}
$$

Hence, the complementary solutions are

$$x_C = A_1 \cos 3t + B_1 \sin 3t, \quad y_C = -\tfrac{1}{2}(A_1 - B_1)\cos 3t - \tfrac{1}{2}(A_1 + B_1)\sin 3t.$$

Example 7.4

Solve

$$
\begin{aligned}
(D^2 + 3D + 2)x + (D+1)y &= 0, \qquad D(\cdot) \equiv \mathrm{d}(\cdot)/\mathrm{d}t, \qquad (1a) \\
(D+2)x + (D-1)y &= 0. \qquad (1b)
\end{aligned}
$$

The determinant of the coefficient matrix is

$$
\begin{aligned}
\phi(D) &= \begin{vmatrix} D^2+3D+2 & D+1 \\ D+2 & D-1 \end{vmatrix} = \begin{vmatrix} (D+1)\,(D+2) & D+1 \\ D+2 & D-1 \end{vmatrix} \\
&= (D+1)(D+2) \begin{vmatrix} 1 & 1 \\ 1 & D-1 \end{vmatrix} \qquad \text{Take } (D+1) \text{ and } (D+2) \text{ out.} \\
&= (D+1)(D+2)(D-2).
\end{aligned}
$$

Hence, the characteristic equation is

$$\phi(\lambda) = (\lambda+1)(\lambda+2)(\lambda-2) = 0 \implies \lambda = -1, -2, 2.$$

The complementary solutions of x and y have the same form and are given by

$$x_C = C_1 e^{-t} + C_2 e^{-2t} + C_3 e^{2t}, \qquad y_C = D_1 e^{-t} + D_2 e^{-2t} + D_3 e^{2t}. \qquad (2)$$

Since $\phi(D)$ is a polynomial of degree 3 in D, the complementary solutions should contain only three arbitrary constants. The three extra constants in solutions (2) can be eliminated by substituting them into either equation (1a) or (1b). Since equation (1b) is simpler, substituting solutions (2) into equation (1b) yields

$$
\begin{aligned}
&(D+2)x_C + (D-1)y_C \\
&= \big[(-C_1e^{-t} - 2C_2e^{-2t} + 2C_3e^{2t}) + 2(C_1e^{-t} + C_2e^{-2t} + C_3e^{2t})\big] \\
&\quad + \big[(-D_1e^{-t} - 2D_2e^{-2t} + 2D_3e^{2t}) - (D_1e^{-t} + D_2e^{-2t} + D_3e^{2t})\big] \\
&= (C_1 - 2D_1)e^{-t} - 3D_2e^{-2t} + (4C_3 + D_3)e^{2t} = 0. \qquad (3)
\end{aligned}
$$

Since e^{-t}, e^{-2t}, and e^{2t} are linearly independent, each coefficient must be zero:

$$
\begin{aligned}
C_1 - 2D_1 = 0 \quad &\Longrightarrow \quad D_1 = -\tfrac{1}{2}C_1, \\
-3D_2 = 0 \quad &\Longrightarrow \quad D_2 = 0, \\
4C_3 + D_3 = 0 \quad &\Longrightarrow \quad D_3 = -4C_3.
\end{aligned}
$$

Hence, the complementary solutions are

$$
x_C = C_1e^{-t} + C_2e^{-2t} + C_3e^{2t}, \qquad y_C = \tfrac{1}{2}C_1e^{-t} - 4C_3e^{2t}.
$$

7.2.2 *Particular Solutions*

Review of Cramer's Rule

For the following system of n linear algebraic equations

$$
\begin{aligned}
a_{11}x_1 + a_{12}x_2 + \cdots + a_{1n}x_n &= b_1, \\
a_{21}x_1 + a_{22}x_2 + \cdots + a_{2n}x_n &= b_2, \\
\cdots\ \cdots& \\
a_{n1}x_1 + a_{n2}x_2 + \cdots + a_{nn}x_n &= b_n,
\end{aligned}
$$

the solutions are given by

$$
x_i = \frac{\Delta_i}{\Delta}, \qquad i = 1, 2, \ldots, n,
$$

where Δ is the determinant of coefficient matrix, Δ_i is the determinant of the coefficient matrix with the ith column replaced by the right-hand side vector, i.e.,

$$
\Delta = \begin{vmatrix} a_{11} & a_{12} & \cdots & a_{1n} \\ a_{21} & a_{22} & \cdots & a_{2n} \\ \vdots & \vdots & \cdots & \vdots \\ a_{n1} & a_{n2} & \cdots & a_{nn} \end{vmatrix}, \quad
\Delta_i = \begin{vmatrix} a_{11} & \cdots & a_{1,i-1} & b_1 & a_{1,i+1} & \cdots & a_{1n} \\ a_{21} & \cdots & a_{2,i-1} & b_2 & a_{2,i+1} & \cdots & a_{2n} \\ \vdots & \cdots & \vdots & \cdots & \vdots & \cdots & \vdots \\ a_{n1} & \cdots & a_{n,i-1} & \underbrace{b_n}_{i\text{th column}} & a_{n,i+1} & \cdots & a_{nn} \end{vmatrix}.
$$

For a system of linear ordinary differential equations

$$\begin{aligned}
\phi_{11}(D)\,x_1 + \phi_{12}(D)\,x_2 + \cdots + \phi_{1n}(D)\,x_n &= f_1(t),\\
\phi_{21}(D)\,x_1 + \phi_{22}(D)\,x_2 + \cdots + \phi_{2n}(D)\,x_n &= f_2(t),\\
\cdots\cdots\\
\phi_{n1}(D)\,x_1 + \phi_{n2}(D)\,x_2 + \cdots + \phi_{nn}(D)\,x_n &= f_n(t),
\end{aligned}$$

where $D(\cdot) \equiv \mathrm{d}(\cdot)/\mathrm{d}t$, a particular solution is given by, using Cramer's Rule,

$$x_{iP}(t) = \frac{\Delta_i(t)}{\phi(D)}, \qquad i = 1,\ 2,\ \ldots,\ n,$$

where $\phi(D)$ is the determinant of the coefficient matrix as studied in the previous section for complementary solution, $\Delta_i(t)$ is $\phi(D)$ with the ith column being replaced by the right-hand side vector of functions, i.e.,

$$\Delta_i(t) = \begin{vmatrix} \phi_{11}(D) & \cdots & \phi_{1,i-1}(D) & f_1(t) & \phi_{1,i+1}(D) & \cdots & \phi_{1n}(D) \\ \phi_{21}(D) & \cdots & \phi_{2,i-1}(D) & f_2(t) & \phi_{2,i+1}(D) & \cdots & \phi_{2n}(D) \\ \vdots & \cdots & \vdots & \vdots & \vdots & \cdots & \vdots \\ \phi_{n1}(D) & \cdots & \phi_{n,i-1}(D) & f_n(t) & \phi_{n,i+1}(D) & \cdots & \phi_{nn}(D) \end{vmatrix}.$$

☞ It should be emphasized that, **since the elements of the determinant are operators and functions, operators must precede functions when evaluating determinants.** Furthermore, since $\Delta_i(t)$, $i = 1,\ 2,\ \ldots,\ n$, are functions, when determining x_{iP}, $\phi^{-1}(D)$ should precede $\Delta_i(t)$.

Example 7.5

Solve

$$(D-3)\,x - 6y = 0, \qquad D(\cdot) \equiv \mathrm{d}(\cdot)/\mathrm{d}t, \tag{1a}$$

$$3x + (D+3)\,y = 18te^{-3t}. \tag{1b}$$

This system and the system in Example 7.3 have the right-hand sides. Hence, they have the same complementary solutions given by

$$x_C = A_1 \cos 3t + B_1 \sin 3t, \tag{2a}$$

$$y_C = A_2 \cos 3t + B_2 \sin 3t. \tag{2b}$$

A particular solution is given by

$$\Delta_x(t) = \begin{vmatrix} 0 & -6 \\ 18te^{-3t} & D+3 \end{vmatrix} = 108\,te^{-3t},$$

$$\Delta_y(t) = \begin{vmatrix} D-3 & 0 \\ 3 & 18te^{-3t} \end{vmatrix} = 18(e^{-3t} - 3te^{-3t}) - 3\cdot 18te^{-3t} = 18e^{-3t} - 108te^{-3t},$$

$$x_P = \frac{\Delta_x(t)}{\phi(D)} = \frac{108}{D^2+9}(te^{-3t}) = 108e^{-3t}\frac{1}{(D-3)^2+9}t \quad \text{✍ Theorem 2 in Section 4.3.2}$$

$$= 108e^{-3t}\frac{1}{D^2-6D+18}t = 6e^{-3t}\frac{1}{1-(\frac{1}{3}D-\frac{1}{18}D^2)}t$$

$$= 6e^{-3t}\Big[1+(\tfrac{1}{3}D-\tfrac{1}{18}D^2)+\cdots\Big]t \quad \text{✍ Expand the operator in series; stop at } D.$$

$$= 2e^{-3t}(3t+1), \tag{3a}$$

$$y_P = \frac{\Delta_y(t)}{\phi(D)} = \frac{18}{D^2+9}(e^{-3t}) - \frac{108}{D^2+9}(te^{-3t})$$

$$= \frac{18}{(-3)^2+9}e^{-3t} - 2e^{-3t}(3t+1) \quad \text{✍ Apply Theorem 1 in Section 4.3.2 for the first term. Use result of } x_P \text{ for the second term.}$$

$$= -e^{-3t}(6t+1). \tag{3b}$$

The general solutions are

$$x = x_C + x_P = A_1\cos 3t + B_1\sin 3t + 2(3t+1)e^{-3t}, \tag{4a}$$

$$y = y_C + y_P = A_2\cos 3t + B_2\sin 3t - (6t+1)e^{-3t}. \tag{4b}$$

Since $\phi(D)$ is a polynomial of degree 2 in D, the general solutions should contain only two arbitrary constants. The two extra constants can be eliminated by substituting equations (4) into either (1a) or (1b).

Substitute solutions (4) into equation (1b) to eliminate the two extra constants

$$\begin{aligned} &3x + (D+3)y \\ &= 3\big[A_1\cos 3t + B_1\sin 3t + 2(3t+1)e^{-3t}\big] \quad \text{✍ } 3x \\ &\quad + \big[-3A_2\sin 3t + 3B_2\cos 3t - 6e^{-3t} + 3(6t+1)e^{-3t}\big] \quad \text{✍ } Dy \\ &\quad + 3\big[A_2\cos 3t + B_2\sin 3t - (6t+1)e^{-3t}\big] \quad \text{✍ } 3y \\ &= 3(A_1+A_2+B_2)\cos 3t + 3(B_1-A_2+B_2)\sin 3t + 18te^{-3t} \\ &= 18te^{-3t}. \quad \text{✍ Right-hand side of equation (1b)} \end{aligned}$$

Comparing the coefficients of similar terms gives

$$\cos 3t: \quad A_1 + A_2 + B_2 = 0,$$

$$\sin 3t: \quad B_1 - A_2 + B_2 = 0.$$

Since the purpose is to eliminate two arbitrary constants, one can express any two constants in terms of the other two from these equations. Hence

$$A_1 = -(A_2 + B_2), \qquad B_1 = A_2 - B_2,$$

or

$$A_2 = -\tfrac{1}{2}(A_1 - B_1), \qquad B_2 = -\tfrac{1}{2}(A_1 + B_1).$$

The general solutions become

$$x = -(A_2+B_2)\cos 3t + (A_2-B_2)\sin 3t + 2(3t+1)\,\mathrm{e}^{-3t},$$

$$y = A_2\cos 3t + B_2\sin 3t - (6t+1)\,\mathrm{e}^{-3t}.$$

Remarks: In the solution above, the general procedure is followed to illustrate all the steps in solving systems of linear differential equations using the method of operators. However, for this example, it can be solved more easily as follows.

Having obtained the complementary and particular solutions for x, i.e., x_C in (2a) and x_P in (3a), or the general solution x in (4a), the general solution y can be determined from equation (1a)

$$y = \tfrac{1}{6}(D-3)x = \tfrac{1}{6}\Big\{\big[-3A_1\sin 3t + 3B_1\cos 3t + 6\mathrm{e}^{-t} - 3\cdot 2(3t+1)\mathrm{e}^{-3t}\big]$$
$$-3\big[A_1\cos 3t + B_1\sin 3t + 2(3t+1)\mathrm{e}^{-3t}\big]\Big\}$$
$$= -\tfrac{1}{2}(A_1-B_1)\cos 3t - \tfrac{1}{2}(A_1+B_1)\sin 3t - (6t+1)\,\mathrm{e}^{-3t}.$$

Remarks: It is important and efficient to exploit the differential equations to devise an easy way to solve the problem.

Example 7.6

Solve

$$(D-3)x - 6y = 2\cos 3t, \qquad D(\cdot)\equiv \mathrm{d}(\cdot)/\mathrm{d}t, \tag{1a}$$

$$3x + (D+3)y = 2\sin 3t. \tag{1b}$$

This system and the system in Example 7.3 have the right-hand sides. Hence, they have the same complementary solutions given by

$$x_C = A_1\cos 3t + B_1\sin 3t, \qquad y_C = A_2\cos 3t + B_2\sin 3t.$$

A particular solution is given by

$$\Delta_x(t) = \begin{vmatrix} 2\cos 3t & -6 \\ 2\sin 3t & D+3 \end{vmatrix} = 2(D+3)\cos 3t + 12\sin 3t = 6\cos 3t + 6\sin 3t,$$

$$\Delta_y(t) = \begin{vmatrix} D-3 & 2\cos 3t \\ 3 & 2\sin 3t \end{vmatrix} = 2(D-3)\sin 3t - 6\cos 3t = -6\sin 3t,$$

$$x_P = \frac{\Delta_x(t)}{\phi(D)} = \frac{1}{D^2+9}(6\cos 3t + 6\sin 3t),$$

$$y_P = \frac{\Delta_y(t)}{\phi(D)} = \frac{1}{D^2+9}(-6\sin 3t).$$

Theorem 3 fails in Section 4.3.2 fails when evaluating x_P and y_P. Hence, Theorem 4 in Section 4.3.2 must be applied:

$$\because\ \phi(D) = D^2+9, \qquad \phi(\mathrm{i}3) = 0,$$

$$\phi'(D) = 2D, \qquad \phi'(\mathrm{i}3) = \mathrm{i}6,$$

$$\frac{1}{D^2+9}(\mathrm{e}^{\mathrm{i}3t}) = \frac{1}{\phi'(\mathrm{i}3)}t\mathrm{e}^{\mathrm{i}3t} = \frac{1}{\mathrm{i}6}t(\cos 3t + \mathrm{i}\sin 3t) = \frac{t}{6}\sin 3t - \mathrm{i}\frac{t}{6}\cos 3t,$$

$$\therefore\quad x_P = 6\,\mathcal{Re}\Big(\frac{1}{D^2+9}\mathrm{e}^{\mathrm{i}3t}\Big) + 6\,\mathcal{Im}\Big(\frac{1}{D^2+9}\mathrm{e}^{\mathrm{i}3t}\Big)$$

$$= 6\cdot\frac{t}{6}\sin 3t + 6\Big(-\frac{t}{6}\cos 3t\Big) = t(\sin 3t - \cos 3t),$$

$$y_P = -6\,\mathcal{Im}\Big(\frac{1}{D^2+9}\mathrm{e}^{\mathrm{i}3t}\Big) = -6\Big(-\frac{t}{6}\cos 3t\Big) = t\cos 3t.$$

The general solutions are

$$x = x_C + x_P = A_1\cos 3t + B_1\sin 3t + t(\sin 3t - \cos 3t),$$

$$y = y_C + y_P = A_2\cos 3t + B_2\sin 3t + t\cos 3t.$$

Substitute the general solutions into equation (1a) to eliminate the extra constants

$$\begin{aligned}
&(D-3)x - 6y \\
&= \big[-3A_1\sin 3t + 3B_1\cos 3t + (\sin 3t - \cos 3t) + t(3\cos 3t + 3\sin 3t)\big] \\
&\quad - 3\big[A_1\cos 3t + B_1\sin 3t + t(\sin 3t - \cos 3t)\big] \\
&\quad - 6\big(A_2\cos 3t + B_2\sin 3t + t\cos 3t\big) \\
&= (3B_1 - 3A_1 - 6A_2 - 1)\cos 3t + (-3A_1 - 3B_1 - 6B_2 + 1)\sin 3t \\
&= 2\cos 3t. \qquad \text{Right-hand side of equation (1a)}
\end{aligned}$$

Comparing the coefficients of similar terms gives

$$\cos 3t: \quad 3B_1 - 3A_1 - 6A_2 - 1 = 2 \implies A_2 = -\tfrac{1}{2}(A_1 - B_1) - \tfrac{1}{2},$$

$$\sin 3t: \quad -3A_1 - 3B_1 - 6B_2 + 1 = 0 \implies B_2 = -\tfrac{1}{2}(A_1 + B_1) + \tfrac{1}{6}.$$

The general solutions become

$$x = A_1\cos 3t + B_1\sin 3t + t(\sin 3t - \cos 3t),$$

$$y = \left[-\tfrac{1}{2}(A_1 - B_1) - \tfrac{1}{2}\right]\cos 3t + \left[-\tfrac{1}{2}(A_1 + B_1) + \tfrac{1}{6}\right]\sin 3t + t\cos 3t.$$

Remarks: Substitute the **general solutions** into one of the original differential equations to eliminate the extra constants in the complementary solutions. This also serves as a check for the particular solutions obtained.

Example 7.7

Solve
$$(\mathcal{D}^2 - 3\mathcal{D})x - (\mathcal{D} - 2)y = 14t + 7, \qquad \mathcal{D}(\cdot) \equiv \mathrm{d}(\cdot)/\mathrm{d}t, \qquad \text{(1a)}$$
$$(\mathcal{D} - 3)x + \mathcal{D}y = 1. \qquad \text{(1b)}$$

The determinant of the coefficient matrix is

$$\phi(\mathcal{D}) = \begin{vmatrix} \mathcal{D}^2 - 3\mathcal{D} & -(\mathcal{D}-2) \\ \mathcal{D}-3 & \mathcal{D} \end{vmatrix} = \mathcal{D}^2(\mathcal{D}-3) + (\mathcal{D}-2)(\mathcal{D}-3)$$
$$= (\mathcal{D}-3)(\mathcal{D}^2 + \mathcal{D} - 2) = (\mathcal{D}-3)(\mathcal{D}+2)(\mathcal{D}-1) = \mathcal{D}^3 - 2\mathcal{D}^2 - 5\mathcal{D} + 6.$$

The characteristic equation is $\phi(\lambda) = (\lambda+2)(\lambda-1)(\lambda-3) = 0 \implies \lambda = -2, 1, 3$. The complementary solutions are

$$x_C = C_1 \mathrm{e}^{-2t} + C_2 \mathrm{e}^{t} + C_3 \mathrm{e}^{3t}, \qquad y_C = D_1 \mathrm{e}^{-2t} + D_2 \mathrm{e}^{t} + D_3 \mathrm{e}^{3t}.$$

A particular solution is given by

$$\Delta_x(t) = \begin{vmatrix} 14t+7 & -(\mathcal{D}-2) \\ 1 & \mathcal{D} \end{vmatrix} = \mathcal{D}(14t+7) + (\mathcal{D}-2)(1) = 12,$$

$$\Delta_y(t) = \begin{vmatrix} \mathcal{D}^2 - 3\mathcal{D} & 14t+7 \\ \mathcal{D}-3 & 1 \end{vmatrix} = (\mathcal{D}^2 - 3\mathcal{D})(1) - (\mathcal{D}-3)(14t+7) = 42t + 7,$$

$$x_P = \frac{\Delta_x(t)}{\phi(\mathcal{D})} = \frac{1}{6 - 5\mathcal{D} - 2\mathcal{D}^2 + \mathcal{D}^3}(12) = \frac{12}{6} = 2,$$

✍ Special case of polynomial: constant

$$y_P = \frac{\Delta_y(t)}{\phi(\mathcal{D})} = \frac{1}{6 - 5\mathcal{D} - 2\mathcal{D}^2 + \mathcal{D}^3}(42t+7)$$
$$= \frac{1}{6\left[1 - \left(\frac{5}{6}\mathcal{D} + \frac{1}{3}\mathcal{D}^2 - \frac{1}{6}\mathcal{D}^3\right)\right]}(42t+7)$$
$$= \tfrac{1}{6}\left[1 + \left(\tfrac{5}{6}\mathcal{D} + \cdots\right) + \cdots\right](42t+7)$$

✍ Expand the operator in series; stop at $\mathcal{D}$.

$$= \tfrac{1}{6}\left[(42t+7) + \tfrac{5}{6}(42)\right] = 7t + 7.$$

Hence, the general solutions are

$$x = x_C + x_P = C_1 \mathrm{e}^{-2t} + C_2 \mathrm{e}^{t} + C_3 \mathrm{e}^{3t} + 2,$$

$$y = y_C + y_P = D_1 e^{-2t} + D_2 e^{t} + D_3 e^{3t} + 7t + 7.$$

Since $\phi(D)$ is a polynomial of degree 3 in D, the general solutions should contain three arbitrary constants. Substitute the solutions into equation (1b) to eliminate the three extra constants

$$\begin{aligned}(D-3)x + Dy &= (-2C_1 e^{-2t} + C_2 e^{t} + 3C_3 e^{3t}) \\ &\quad - 3(C_1 e^{-2t} + C_2 e^{t} + C_3 e^{3t} + 2) + (-2D_1 e^{-2t} + D_2 e^{t} + 3D_3 e^{3t} + 7) \\ &= (-5C_1 - 2D_1)e^{-2t} + (-2C_2 + D_2)e^{t} + 3D_3 e^{3t} + 1 \\ &= 1. \qquad \text{Right-hand side of equation (1b)}\end{aligned}$$

Comparing the coefficients of similar terms yields

$$\begin{aligned} e^{-2t}: &\qquad -5C_1 - 2D_1 = 0 \implies D_1 = -\tfrac{5}{2}C_1, \\ e^{t}: &\qquad -2C_2 + D_2 = 0 \implies D_2 = 2C_2, \\ e^{3t}: &\qquad 3D_3 = 0 \implies D_3 = 0. \end{aligned}$$

Hence, the general solutions become

$$\begin{aligned} x &= x_C + x_P = C_1 e^{-2t} + C_2 e^{t} + C_3 e^{3t} + 2, \\ y &= y_C + y_P = -\tfrac{5}{2}C_1 e^{-2t} + 2C_2 e^{t} + 7t + 7. \end{aligned}$$

Example 7.8

Solve

$$D y_1 - y_2 = 0, \qquad D(\cdot) \equiv d(\cdot)/dx, \tag{1a}$$

$$D y_2 - y_3 = 0, \tag{1b}$$

$$6y_1 + 11y_2 + (D+6)y_3 = 2e^{-x}. \tag{1c}$$

The determinant of the coefficient matrix is

$$\phi(D) = \det \begin{bmatrix} D & -1 & 0 \\ 0 & D & -1 \\ 6 & 11 & D+6 \end{bmatrix} \begin{matrix} D & -1 \\ 0 & D \\ 6 & 11 \end{matrix} \qquad \text{To evaluate the determinant, add first two columns at right.}$$

$$= D^2(D+6) + 6 + 11D = D^3 + 6D^2 + 11D + 6.$$

The characteristic equation is $\phi(\lambda) = \lambda^3 + 6\lambda^2 + 11\lambda + 6 = 0$. Since

$$(-1)^3 + 6(-1)^2 + 11(-1) + 6 = -1 + 6 - 11 + 6 = 0,$$

$\therefore \lambda = -1$ is a root or $(\lambda+1)$ is a factor. The other factor can be determined using long division and is obtained as $(\lambda+1)(\lambda^2+5\lambda+6) = 0$. Therefore

$$(\lambda+1)(\lambda+2)(\lambda+3) = 0 \implies \lambda = -1,\ -2,\ -3.$$

The complementary solution for y_1 is given by

$$y_{1C} = C_1 e^{-x} + C_2 e^{-2x} + C_3 e^{-3x}.$$

A particular solution for y_1 is given by

$$\Delta_{y1}(x) = \det\begin{bmatrix} 0 & -1 & 0 \\ 0 & D & -1 \\ 2e^{-x} & 11 & D+6 \end{bmatrix} = 2e^{-x}\begin{vmatrix} -1 & 0 \\ D & -1 \end{vmatrix} = 2e^{-x},$$

$$y_{1P} = \frac{\Delta_{y1}(x)}{\phi(D)} = \frac{1}{D^3+6D^2+11D+6}(2e^{-x}).$$

Apply Theorem 4 in Section 4.3.2:

$$\because \quad \phi(D) = D^3+6D^2+11D+6, \quad \phi(-1) = 0,$$

$$\phi'(D) = 3D^2+12D+11, \qquad \phi'(-1) = 2,$$

$$\therefore \quad y_{1P} = \frac{2}{\phi'(-1)} x e^{-x} = x e^{-x}.$$

Hence, the general solution of y_1 is

$$y_1 = y_{1C} + y_{1P} = C_1 e^{-x} + C_2 e^{-2x} + C_3 e^{-3x} + x e^{-x}.$$

Having obtained y_1, the general solutions of y_2 and y_3 can be determined from equations (1a) and (1b), respectively,

$$y_2 = D y_1 = -C_1 e^{-x} - 2C_2 e^{-2x} - 3C_3 e^{-3x} + (1-x)e^{-x},$$

$$y_3 = D y_2 = C_1 e^{-x} + 4C_2 e^{-2x} + 9C_3 e^{-3x} - (2-x)e^{-x}.$$

Method of Variation of Parameters

As for linear ordinary differential equations, when the right-hand sides of a system of linear ordinary differential equations are not of the form

$$e^{\alpha t}\big[(a_0+a_1 t+\cdots+a_k t^k)\cos\beta t + (b_0+b_1 t+\cdots+b_k t^k)\sin\beta t\big],$$

the method of variation of parameters has to be used to obtain particular solutions.

Example 7.9

Solve
$$D x - y = \tan^2 t + 1, \qquad D(\cdot) \equiv d(\cdot)/dt, \tag{1a}$$

$$x + D y = \tan t. \tag{1b}$$

First determine the complementary solutions for the complementary equations

$$D x - y = 0, \qquad x + D y = 0. \tag{2}$$

The characteristic numbers are easily obtained

$$\phi(D) = \begin{vmatrix} D & -1 \\ 1 & D \end{vmatrix} = D^2 + 1 \implies \phi(\lambda) = \lambda^2 + 1 = 0 \implies \lambda = \pm i.$$

The complementary solution of x is

$$x_C = A\cos t + B\sin t. \tag{3a}$$

From the first equation of (2), the complementary solution of y is

$$y_C = Dx_C = -A\sin t + B\cos t. \tag{3b}$$

The method of variation of parameters is then applied to find particular solutions. Vary the parameters, i.e., make constants A and B functions of t, $A \Rightarrow a(t)$, $B \Rightarrow b(t)$, and one has

$$x_P = a(t)\cos t + b(t)\sin t. \qquad y_P = -a(t)\sin t + b(t)\cos t. \tag{4}$$

Substituting equations (4) into (1) yields

$$\begin{aligned} Dx_P - y_P &= \big[a'(t)\cos t - a(t)\sin t + b'(t)\sin t + b(t)\cos t\big] \\ &\quad - \big[-a(t)\sin t + b(t)\cos t\big] \\ &= a'(t)\cos t + b'(t)\sin t = \tan^2 t + 1, \end{aligned} \tag{5a}$$

$$\begin{aligned} x_P + D\,y_P &= \big[a(t)\cos t + b(t)\sin t\big] \\ &\quad + \big[-a'(t)\sin t - a(t)\cos t + b'(t)\cos t - b(t)\sin t\big] \\ &= -a'(t)\sin t + b'(t)\cos t = \tan t. \end{aligned} \tag{5b}$$

Equations (5a) and (5b) give two linear algebraic equations for two unknowns $a'(t)$ and $b'(t)$, which can be solved using Gaussian elimination or Cramer's Rule. To find $a'(t)$, eliminate $b'(t)$ as follows:

$$\text{Eq(5a)} \times \cos t: \qquad a'(t)\cos^2 t + b'(t)\sin t\cos t = (\tan^2 t + 1)\cos t, \tag{6a}$$

$$\text{Eq(5b)} \times \sin t: \qquad -a'(t)\sin^2 t + b'(t)\cos t\sin t = \tan t\sin t, \tag{6b}$$

$$\text{Eq(6a)} - \text{Eq(6b)}: \qquad a'(t) = \tan^2 t\cos t + \cos t - \tan t\sin t = \cos t. \tag{7a}$$

From equation (5b), one obtains

$$b'(t) = \frac{\tan t + a'(t)\sin t}{\cos t} = \frac{\tan t + \cos t\sin t}{\cos t} = \frac{\sin t}{\cos^2 t} + \sin t. \tag{7b}$$

Integrating equations (7) leads to

$$a(t) = \int \cos t \, dt = \sin t,$$

$$b(t) = \int \frac{\sin t}{\cos^2 t} \mathrm{d}t + \int \sin t \,\mathrm{d}t = -\int \frac{1}{\cos^2 t} \mathrm{d}(\cos t) - \cos t = \frac{1}{\cos t} - \cos t.$$

Hence, the general solutions are, using equations (3) and (4),

$$x = x_C + x_P = A\cos t + B\sin t + \sin t \cos t + \left(\frac{1}{\cos t} - \cos t\right) \sin t$$
$$= A\cos t + B\sin t + \tan t,$$
$$y = y_C + y_P = -A\sin t + B\cos t - \sin t \sin t + \left(\frac{1}{\cos t} - \cos t\right) \cos t$$
$$= -A\sin t + B\cos t.$$

Example 7.10

Solve
$$(\mathcal{D}^2+3\mathcal{D}+2)x + (\mathcal{D}+1)y = 0, \qquad \mathcal{D}(\cdot) \equiv \mathrm{d}(\cdot)/\mathrm{d}t, \tag{1a}$$
$$(\mathcal{D}+2)x + (\mathcal{D}-1)y = \frac{8}{\mathrm{e}^{2t}+1}. \tag{1b}$$

First determine the solutions for the complementary equations

$$(\mathcal{D}^2+3\mathcal{D}+2)x + (\mathcal{D}+1)y = 0, \tag{1a}$$
$$(\mathcal{D}+2)x + (\mathcal{D}-1)y = 0. \tag{1b$'$}$$

The determinant of the coefficient matrix is

$$\phi(\mathcal{D}) = \begin{vmatrix} \mathcal{D}^2+3\mathcal{D}+2 & \mathcal{D}+1 \\ \mathcal{D}+2 & \mathcal{D}-1 \end{vmatrix} = \mathcal{D}^3+\mathcal{D}^2-4\mathcal{D}-4 = (\mathcal{D}+1)(\mathcal{D}^2-4).$$

The characteristic equation is $\phi(\lambda) = (\lambda+1)(\lambda^2-4) = 0 \implies \lambda = -1, \pm 2$. The complementary solutions are

$$x_C = C_1\mathrm{e}^{-t} + C_2\mathrm{e}^{2t} + C_3\mathrm{e}^{-2t}, \qquad y_C = D_1\mathrm{e}^{-t} + D_2\mathrm{e}^{2t} + D_3\mathrm{e}^{-2t}.$$

Since $\phi(\mathcal{D})$ is a polynomial of degree 3 in $\mathcal{D}$, the complementary solutions should contain three arbitrary constants. Substitute the complementary solutions into equation (1b$'$) to eliminate the extra constants

$$(\mathcal{D}+2)x_C + (\mathcal{D}-1)y_C = (C_1-2D_1)\mathrm{e}^{-t} + (4C_2+D_2)\mathrm{e}^{2t} - 3D_3\mathrm{e}^{-2t} = 0,$$

which leads to

$$C_1 - 2D_1 = 0 \implies C_1 = 2D_1,$$
$$4C_2 - D_2 = 0 \implies D_2 = -4C_2,$$
$$D_3 = 0.$$

Hence

$$x_C = 2D_1 e^{-t} + C_2 e^{2t} + C_3 e^{-2t}, \tag{2a}$$

$$y_C = D_1 e^{-t} - 4C_2 e^{2t}. \tag{2b}$$

Since equation (1a) is a second-order equation in x, a third equation is needed from differentiating x_C with respect to t

$$x_C' = -2D_1 e^{-t} + 2C_2 e^{2t} - 2C_3 e^{-2t}. \tag{2c}$$

Apply the method of variation of parameters, i.e., make $D_1 \Longrightarrow c_1(t)$, $C_2 \Longrightarrow c_2(t)$, $C_3 \Longrightarrow c_3(t)$ in equations (2) to yield

$$x_P = 2c_1(t)\,e^{-t} + c_2(t)\,e^{2t} + c_3(t)\,e^{-2t}, \tag{3a}$$

$$y_P = c_1(t)\,e^{-t} - 4c_2(t)\,e^{2t}, \tag{3b}$$

$$x_P' = -2c_1(t)\,e^{-t} + 2c_2(t)\,e^{2t} - 2c_3(t)\,e^{-2t}. \tag{3c}$$

Differentiating equation (3a) with respect to t and comparing with (3c) lead to

$$\begin{aligned} x_P' &= 2c_1'(t)\,e^{-t} - 2c_1(t)\,e^{-t} + c_2'(t)\,e^{2t} + 2c_2(t)\,e^{2t} + c_3'(t)\,e^{-2t} - 2c_3(t)\,e^{-2t} \\ &= -2c_1(t)\,e^{-t} + 2c_2(t)\,e^{2t} - 2c_3(t)\,e^{-2t}, \qquad \text{✍ Equation (3c)} \end{aligned}$$

which gives

$$c_1'(t)\,e^{-t} + c_2'(t)\,e^{2t} + c_3'(t)\,e^{-2t} = 0. \tag{4a}$$

Substituting equations (3) into (1a) yields

$$(D^2+3D+2)\,x_P + (D+1)\,y_P = -c_1'(t)\,e^{-t} - 2c_2'(t)\,e^{2t} - 2c_3'(t)\,e^{-2t} = 0, \tag{4b}$$

and into (1b) results in

$$(D+2)\,x_P + (D-1)\,y_P = c_1'(t)\,e^{-t} - 4c_2'(t)\,e^{2t} = \frac{8}{e^{2t}+1}. \tag{4c}$$

Equations (4) provide three linear algebraic equations for three unknowns $c_1'(t)$, $c_2'(t)$, and $c_3'(t)$, which can be solved using Cramer's Rule:

$$\Delta = \begin{vmatrix} e^{-t} & e^{2t} & e^{-2t} \\ -e^{-t} & -2e^{2t} & -2e^{-2t} \\ e^{-t} & -4e^{2t} & 0 \end{vmatrix} = e^{-t}\cdot e^{2t}\cdot e^{-2t} \begin{vmatrix} 1 & 1 & 1 \\ -1 & -2 & -2 \\ 1 & -4 & 0 \end{vmatrix} = -4e^{-t},$$

$$\Delta_1 = \det\begin{pmatrix} 0 & e^{2t} & e^{-2t} \\ 0 & -2e^{2t} & -2e^{-2t} \\ \dfrac{8}{e^{2t}+1} & -4e^{2t} & 0 \end{pmatrix} = \frac{8}{e^{2t}+1}\cdot e^{2t}\cdot e^{-2t} \begin{vmatrix} 1 & 1 \\ -2 & -2 \end{vmatrix} = 0,$$

$$\Delta_2 = \det\begin{bmatrix} e^{-t} & 0 & e^{-2t} \\ -e^{-t} & 0 & -2e^{-2t} \\ e^{-t} & \dfrac{8}{e^{2t}+1} & 0 \end{bmatrix} = -\frac{8}{e^{2t}+1}\cdot e^{-t}\cdot e^{-2t}\begin{vmatrix} 1 & 1 \\ -1 & -2 \end{vmatrix} = \frac{8e^{-3t}}{e^{2t}+1},$$

$$\Delta_3 = \det\begin{bmatrix} e^{-t} & e^{2t} & 0 \\ -e^{-t} & -2e^{2t} & 0 \\ e^{-t} & -4e^{2t} & \dfrac{8}{e^{2t}+1} \end{bmatrix} = \frac{8}{e^{2t}+1}\cdot e^{-t}\cdot e^{2t}\begin{vmatrix} 1 & 1 \\ -1 & -2 \end{vmatrix} = -\frac{8e^{t}}{e^{2t}+1},$$

$$c_1'(t) = \frac{\Delta_1}{\Delta} = 0, \quad c_2'(t) = \frac{\Delta_2}{\Delta} = -\frac{2e^{-2t}}{e^{2t}+1}, \quad c_3'(t) = \frac{\Delta_3}{\Delta} = \frac{2e^{2t}}{e^{2t}+1}.$$

Integrating $c_1'(t)$, $c_2'(t)$, and $c_3'(t)$ result in

$$c_1(t) = 0,$$

$$\begin{aligned} c_2(t) &= -\int \frac{2e^{-2t}}{e^{2t}+1}dt = \int \frac{e^{-2t}}{1+e^{-2t}}d(e^{-2t}) \\ &= \int\Big(1-\frac{1}{1+e^{-2t}}\Big)d(e^{-2t}) = e^{-2t} - \ln(1+e^{-2t}) = e^{-2t} - \ln(e^{2t}+1) + 2t, \end{aligned}$$

$$c_3(t) = \int \frac{2e^{2t}}{e^{2t}+1}dt = \int \frac{1}{e^{2t}+1}d(e^{2t}) = \ln(e^{2t}+1).$$

Hence, the particular solutions are

$$\begin{aligned} x_P &= 2c_1(t)e^{-t} + c_2(t)e^{2t} + c_3(t)e^{-2t} \\ &= \big[e^{-2t} - \ln(e^{2t}+1) + 2t\big]\cdot e^{2t} + \ln(e^{2t}+1)\cdot e^{-2t} \\ &= 1 + (e^{-2t} - e^{2t})\ln(e^{2t}+1) + 2te^{2t}, \\ y_P &= c_1(t)e^{-t} - 4c_2(t)e^{2t} = -4\big[e^{-2t} - \ln(e^{2t}+1) + 2t\big]\cdot e^{2t} \\ &= -4\big[1 - e^{2t}\ln(e^{2t}+1) + 2te^{2t}\big]. \end{aligned}$$

The general solutions are

$$x = x_C + x_P = 2C_1e^{-t} + C_2e^{2t} + C_3e^{-2t} + 1 + (e^{-2t} - e^{2t})\ln(e^{2t}+1) + 2te^{2t},$$

$$y = y_C + y_P = C_1e^{-t} - 4C_2e^{2t} - 4\big[1 - e^{2t}\ln(e^{2t}+1) + 2te^{2t}\big].$$

7.3 The Method of Laplace Transform

The procedure for solving systems of linear ordinary differential equations is very straightforward. Consider a system of n ordinary differential equations for n unknown functions $x_i(t)$, $i = 1, 2, \ldots, n$.

- Take Laplace transform of both sides of the equations, with $X_i(s) = \mathscr{L}\{x_i(t)\}$, $i = 1, 2, \ldots, n$.
- It results in a system of n *algebraic* equations for the n unknown Laplace transforms $X_i(s)$, $i = 1, 2, \ldots, n$, which can be solved using Gaussian elimination or Cramer's Rule.
- The solutions of the system of linear differential equations are obtained by taking inverse Laplace transform $x_i(t) = \mathscr{L}^{-1}\{X_i(s)\}$, $i = 1, 2, \ldots, n$.

Example 7.11

Solve
$$\frac{dx}{dt} - 3x - 6y = 0,$$
$$3x + \frac{dy}{dt} + 3y = 18te^{-3t}, \qquad x(0) = x_0, \;\; y(0) = y_0.$$

Let $X(s) = \mathscr{L}\{x(t)\}$ and $Y(s) = \mathscr{L}\{y(t)\}$. Taking the Laplace transform of both sides of the differential equations yields

$$\big[sX(s) - x(0)\big] - 3X(s) - 6Y(s) = 0,$$
$$3X(s) + \big[sY(s) - y(0)\big] + 3Y(s) = \mathscr{L}\{18te^{-3t}\},$$

where, using $\mathscr{L}\{e^{at}f(t)\} = \mathscr{L}\{f(t)\}\Big|_{s\to s-a}$,

$$\mathscr{L}\{18te^{-3t}\} = 18\,\mathscr{L}\{t\}\Big|_{s\to s+3} = 18 \cdot \frac{1}{s^2}\Big|_{s\to s+3} = \frac{18}{(s+3)^2}.$$

These give two algebraic equations for $X(s)$ and $Y(s)$

$$(s-3)X(s) - 6Y(s) = x_0, \qquad 3X(s) + (s+3)Y(s) = y_0 + \frac{18}{(s+3)^2},$$

which can be solved using Gaussian elimination or Cramer's Rule:

$$\Delta = \begin{vmatrix} s-3 & -6 \\ 3 & s+3 \end{vmatrix} = (s-3)(s+3) + 18 = s^2 + 9,$$

$$\Delta_X = \begin{vmatrix} x_0 & -6 \\ y_0 + \dfrac{18}{(s+3)^2} & s+3 \end{vmatrix} = x_0(s+3) + 6\left[y_0 + \frac{18}{(s+3)^2}\right],$$

$$\Delta_Y = \begin{vmatrix} s-3 & x_0 \\ 3 & y_0 + \dfrac{18}{(s+3)^2} \end{vmatrix} = (s-3)\left[y_0 + \frac{18}{(s+3)^2}\right] - 3x_0,$$

$$\therefore \quad X(s) = \frac{\Delta_X}{\Delta} = \frac{3(x_0 + 2y_0)}{s^2+9} + \frac{x_0\, s}{s^2+9} + \frac{108}{(s+3)^2(s^2+9)},$$

$$\Upsilon(s)=\frac{\Delta_\Upsilon}{\Delta}=\frac{-3(x_0+y_0)}{s^2+9}+\frac{y_0 s}{s^2+9}+\frac{18(s-3)}{(s+3)^2(s^2+9)}.$$

Using partial fractions, one has

$$\frac{108}{(s+3)^2(s^2+9)}=\frac{A_2}{(s+3)^2}+\frac{A_1}{s+3}+\frac{Bs+C}{s^2+9}.$$

To find A_2, cover-up $(s+3)^2$ and set $s=-3$:

$$A_2=\left.\frac{108}{(s^2+9)}\right|_{s=-3}=\frac{108}{(-3)^2+9}=6.$$

Comparing the coefficients of the numerators leads to

$$s^3:\quad A_1+B=0, \tag{1}$$

$$s^2:\quad 3A_1+A_2+6B+C=0, \tag{2}$$

$$s:\quad 9A_1+9B+6C=0, \tag{3}$$

$$1:\quad -27A_1+9A_2+9C=108, \tag{4}$$

$$\text{Eqn (3)}-9\times\text{Eqn (1)}:\quad C=0,$$

$$\text{from Eqn (1)}:\quad B=-A_1,$$

$$\text{from Eqn (2)}:\quad 3A_1+6+6(-A_1)+0=0 \Longrightarrow A_1=2,\quad B=-2.$$

Hence

$$\mathscr{L}^{-1}\left\{\frac{108}{(s+3)^2(s^2+9)}\right\}=\mathscr{L}^{-1}\left\{\frac{6}{(s+3)^2}+\frac{2}{s+3}-\frac{2s}{s^2+3^2}\right\}$$
$$=2\mathrm{e}^{-3t}+6t\mathrm{e}^{-3t}-2\cos 3t.$$

Similarly,

$$\mathscr{L}^{-1}\left\{\frac{18(s-3)}{(s+3)^2(s^2+9)}\right\}=\mathscr{L}^{-1}\left\{-\frac{6}{(s+3)^2}-\frac{1}{s+3}+\frac{s+3}{s^2+3^2}\right\}$$
$$=-\mathrm{e}^{-3t}-6t\mathrm{e}^{-3t}+\cos 3t+\sin 3t.$$

The solutions of the differential equations are

$$x(t)=\mathscr{L}^{-1}\{X(s)\}=\mathscr{L}^{-1}\left\{(x_0+2y_0)\cdot\frac{3}{s^2+3^2}+x_0\cdot\frac{s}{s^2+3^2}+\frac{108}{(s+3)^2(s^2+9)}\right\}$$
$$=(x_0+2y_0)\sin 3t+x_0\cos 3t+2\mathrm{e}^{-3t}+6t\mathrm{e}^{-3t}-2\cos 3t$$
$$=(x_0+2y_0)\sin 3t+(x_0-2)\cos 3t+2(1+3t)\mathrm{e}^{-3t},$$

$$y(t)=\mathscr{L}^{-1}\{\Upsilon(s)\}=\mathscr{L}^{-1}\left\{-(x_0+y_0)\cdot\frac{3}{s^2+3^2}+y_0\cdot\frac{s}{s^2+3^2}+\frac{18(s-3)}{(s+3)^2(s^2+9)}\right\}$$
$$=-(x_0+y_0)\sin 3t+y_0\cos 3t-\mathrm{e}^{-3t}-6t\mathrm{e}^{-3t}+\cos 3t+\sin 3t$$
$$=(1-x_0-y_0)\sin 3t+(1+y_0)\cos 3t-(1+6t)\mathrm{e}^{-3t}.$$

Example 7.12

Solve
$$\frac{dx}{dt}+2x+2\frac{dy}{dt}+5y=0,$$
$$\frac{dx}{dt}+3\frac{dy}{dt}+y=10\sin 2t\,H(t-\pi),\qquad x(0)=x_0,\ \ y(0)=y_0.$$

Let $X(s)=\mathscr{L}\{x(t)\}$ and $Y(s)=\mathscr{L}\{y(t)\}$. Taking the Laplace transform of both sides of the differential equations yields

$$\big[sX(s)-x(0)\big]+2X(s)+2\big[sY(s)-y(0)\big]+5Y(s)=0,$$
$$\big[sX(s)-x(0)\big]+3\big[sY(s)-y(0)\big]+Y(s)=10\,\mathscr{L}\{\sin 2t\,H(t-\pi)\},$$

where

$$\begin{aligned}\mathscr{L}\{\sin 2t\,H(t-\pi)\}&=\mathscr{L}\{\sin 2\big[(t-\pi)+\pi\big]H(t-\pi)\}\\&=\mathscr{L}\{\sin 2(t-\pi)H(t-\pi)\}\quad ✍\ \mathscr{L}\{f(t-a)H(t-a)\}=\mathrm{e}^{-as}\,\mathscr{L}\{f(t)\}\\&=\mathrm{e}^{-\pi s}\,\mathscr{L}\{\sin 2t\}=\mathrm{e}^{-\pi s}\frac{2}{s^2+2^2}.\end{aligned}$$

These two equations lead to two algebraic equations for $X(s)$ and $Y(s)$

$$(s-2)\,X(s)+(2s+5)\,Y(s)=x_0+2y_0,$$
$$sX(s)+(3s+1)\,Y(s)=x_0+3y_0+\mathrm{e}^{-\pi s}\frac{20}{s^2+4},$$

which can be solved using Cramer's Rule:

$$\Delta=\begin{vmatrix}s+2 & 2s+5\\ s & 3s+1\end{vmatrix}=(s+2)(3s+1)-s(2s+5)=s^2+2s+2,$$

$$\Delta_X=\begin{vmatrix}x_0+2y_0 & 2s+5\\ x_0+3y_0+\mathrm{e}^{-\pi s}\dfrac{20}{s^2+4} & 3s+1\end{vmatrix}=x_0s-4x_0-13y_0-\mathrm{e}^{-\pi s}\frac{20\,(2s+5)}{s^2+4},$$

$$\Delta_Y=\begin{vmatrix}s+2 & x_0+2y_0\\ s & x_0+3y_0+\mathrm{e}^{-\pi s}\dfrac{20}{s^2+4}\end{vmatrix}=y_0s+2x_0+6y_0+\mathrm{e}^{-\pi s}\frac{20\,(s+2)}{s^2+4},$$

$$X(s)=\frac{\Delta_X}{\Delta}=\frac{x_0\,(s+1)-5x_0-13y_0}{(s+1)^2+1^2}-\mathrm{e}^{-\pi s}\frac{20\,(2s+5)}{(s^2+4)(s^2+2s+2)},$$

$$Y(s)=\frac{\Delta_Y}{\Delta}=\frac{y_0\,(s+1)+2x_0+5y_0}{(s+1)^2+1^2}+\mathrm{e}^{-\pi s}\frac{20\,(s+2)}{(s^2+4)(s^2+2s+2)}.$$

Using partial fractions, one has

$$-\frac{20(2s+5)}{(s^2+4)(s^2+2s+2)}=\frac{As+B}{s^2+4}+\frac{Cs+D}{s^2+2s+2}.$$

Comparing the coefficients of the numerators leads to

$$\begin{aligned}
s^3:&\quad A+C=0 \implies C=-A, &(1)\\
s^2:&\quad 2A+B+D=0, &(2)\\
s:&\quad 2A+2B+4C=-40, &(3)\\
1:&\quad 2B+4D=-100 \implies B=-2D-50. &(4)
\end{aligned}$$

Substituting equations (1) and (4) into (2) and (3) gives

$$\begin{cases} 2A+(-2D-50)+D=0 \\ 2A+2(-2D-50)+4(-A)=-40 \end{cases} \implies \begin{cases} 2A-D=50, \\ -2A-4D=60, \end{cases}$$

which can easily be solved to yield

$$A=14,\quad D=-22 \implies B=-6,\quad C=-14.$$

Hence

$$\begin{aligned}
\mathscr{L}^{-1}\left\{-\frac{20(2s+5)}{(s^2+4)(s^2+2s+2)}\right\} &= \mathscr{L}^{-1}\left\{\frac{14s-6}{s^2+4}-\frac{14s+22}{s^2+2s+2}\right\}\\
&= \mathscr{L}^{-1}\left\{14\cdot\frac{s}{s^2+2^2}-3\cdot\frac{2}{s^2+2^2}-\frac{14(s+1)+8}{(s+1)^2+1}\right\}\\
&= 14\cos 2t-3\sin 2t-(14\cos t+8\sin t)\,\mathrm{e}^{-t}.
\end{aligned}$$

Similarly,

$$\begin{aligned}
\mathscr{L}^{-1}\left\{\frac{20(s+2)}{(s^2+4)(s^2+2s+2)}\right\} &= \mathscr{L}^{-1}\left\{\frac{-6s+4}{s^2+4}+\frac{6s+8}{s^2+2s+2}\right\}\\
&= -6\cos 2t+2\sin 2t+(6\cos t+2\sin t)\,\mathrm{e}^{-t}.
\end{aligned}$$

The solutions of the differential equations are

$$\begin{aligned}
x(t) &= \mathscr{L}^{-1}\{X(s)\}\\
&= \big[x_0\cos t-(5x_0+13y_0)\sin t\big]\mathrm{e}^{-t}+\Big\{14\cos 2(t-\pi)-3\sin 2(t-\pi)\\
&\qquad -\big[14\cos(t-\pi)+8\sin(t-\pi)\big]\mathrm{e}^{-(t-\pi)}\Big\}H(t-\pi)\\
&= \big[x_0\cos t-(5x_0+13y_0)\sin t\big]\mathrm{e}^{-t}\\
&\qquad +\big[14\cos 2t-3\sin 2t+(14\cos t+8\sin t)\,\mathrm{e}^{-t+\pi}\big]H(t-\pi),
\end{aligned}$$

$$
\begin{aligned}
y(t) &= \mathscr{L}^{-1}\{Y(s)\} \\
&= \big[y_0 \cos t + (2x_0+5y_0)\sin t\big]\mathrm{e}^{-t} + \Big\{-6\cos 2(t-\pi) + 2\sin 2(t-\pi) \\
&\qquad + \big[6\cos(t-\pi) + 2\sin(t-\pi)\big]\mathrm{e}^{-(t-\pi)}\Big\} H(t-\pi) \\
&= \big[y_0 \cos t + (2x_0+5y_0)\sin t\big]\mathrm{e}^{-t} \\
&\qquad + \big[-6\cos 2t + 2\sin 2t - (6\cos t + 2\sin t)\mathrm{e}^{-t+\pi}\big] H(t-\pi).
\end{aligned}
$$

Example 7.13

Solve
$$
\frac{\mathrm{d}^2x}{\mathrm{d}t^2} - 3\frac{\mathrm{d}x}{\mathrm{d}t} - \frac{\mathrm{d}y}{\mathrm{d}t} + 2y = 60\,t\,H(t-1),
$$
$$
\frac{\mathrm{d}x}{\mathrm{d}t} - 3x + \frac{\mathrm{d}y}{\mathrm{d}t} = 0, \qquad x(0)=5,\ y(0)=0,\ x'(0)=15.
$$

Let $X(s)=\mathscr{L}\{x(t)\}$ and $Y(s)=\mathscr{L}\{y(t)\}$. Taking the Laplace transform of both sides of the differential equations yields

$$
\begin{aligned}
&\big[s^2X(s) - s\,x(0) - x'(0)\big] - 3\big[sX(s) - x(0)\big] \\
&\qquad - \big[sY(s) - y(0)\big] + 2Y(s) = 60\,\mathscr{L}\{t\,H(t-1)\}, \\
&\big[sX(s) - x(0)\big] - 3X(s) + \big[sY(s) - y(0)\big] = 0,
\end{aligned}
$$

where

$$
\mathscr{L}\{t\,H(t-1)\} = \mathscr{L}\{\big[(t-1)+1\big]H(t-1)\} = \mathrm{e}^{-s}\,\mathscr{L}\{t+1\} = \mathrm{e}^{-s}\left(\frac{1}{s^2}+\frac{1}{s}\right).
$$

These two equations lead to two algebraic equations for $X(s)$ and $Y(s)$

$$
(s^2-3s)\,X(s) - (s-2)\,Y(s) = 5s + \frac{60\mathrm{e}^{-s}(s+1)}{s^2},
$$
$$
(s-3)\,X(s) + s\,Y(s) = 5,
$$

which can be solved using Cramer's Rule:

$$
\begin{aligned}
\Delta &= \begin{vmatrix} s(s-3) & -(s-2) \\ (s-3) & s \end{vmatrix} = (s-3)\begin{vmatrix} s & -(s-2) \\ 1 & s \end{vmatrix} \\
&= (s-3)(s^2+s-2) = (s-3)(s-1)(s+2),
\end{aligned}
$$

$$
\Delta_X = \begin{vmatrix} 5s + \dfrac{60\mathrm{e}^{-s}(s+1)}{s^2} & -(s-2) \\ 5 & s \end{vmatrix} = 5s^2 + \frac{60\mathrm{e}^{-s}(s+1)}{s} + 5s - 10,
$$

$$
\Delta_Y = \begin{vmatrix} s(s-3) & 5s + \dfrac{60\mathrm{e}^{-s}(s+1)}{s^2} \\ (s-3) & 5 \end{vmatrix} = (s-3)\left\{-\frac{60\mathrm{e}^{-s}(s+1)}{s^2}\right\},
$$

$$\therefore \quad X(s) = \frac{\Delta_X}{\Delta} = \frac{5s^2+5s-10}{(s-3)(s-1)(s+2)} + \frac{60(s+1)}{s(s-3)(s-1)(s+2)}\,e^{-s}$$

$$= \frac{5}{s-3} + \frac{60(s+1)}{s(s-3)(s-1)(s+2)}\,e^{-s},$$

$$Y(s) = \frac{\Delta_Y}{\Delta} = -\frac{60(s+1)}{s^2(s-1)(s+2)}\,e^{-s}.$$

Using partial fractions, one has

$$\frac{60(s+1)}{s(s-3)(s-1)(s+2)} = \frac{A_1}{s} + \frac{B_1}{s-3} + \frac{C_1}{s-1} + \frac{D_1}{s+2}.$$

To find A_1, cover-up s and set $s=0$:

$$A_1 = \left.\frac{60(s+1)}{(s-3)(s-1)(s+2)}\right|_{s=0} = \frac{60(1)}{(-3)(-1)(2)} = 10.$$

Similarly,

$$B_1 = \left.\frac{60(s+1)}{s(s-1)(s+2)}\right|_{s=3} = \frac{60(4)}{(3)(2)(5)} = 8,$$

$$C_1 = \left.\frac{60(s+1)}{s(s-3)(s+2)}\right|_{s=1} = \frac{60(2)}{(1)(-2)(3)} = -20,$$

$$D_1 = \left.\frac{60(s+1)}{s(s-3)(s-1)}\right|_{s=-2} = \frac{60(-1)}{(-2)(-5)(-3)} = 2.$$

Again, using partial fractions, one has

$$\frac{-60(s+1)}{s^2(s-1)(s+2)} = \frac{A_2}{s} + \frac{B_2}{s^2} + \frac{C_2}{s-1} + \frac{D_2}{s+2}.$$

To find B_2, cover-up s^2 and set $s=0$:

$$B_2 = \left.\frac{-60(s+1)}{(s-1)(s+2)}\right|_{s=0} = \frac{-60(1)}{(-1)(2)} = 30.$$

Similarly,

$$C_2 = \left.\frac{-60(s+1)}{s^2(s+2)}\right|_{s=1} = \frac{-60(2)}{(1)(3)} = -40,$$

$$D_2 = \left.\frac{-60(s+1)}{s^2(s-2)}\right|_{s=-2} = \frac{-60(-2)}{(4)(-3)} = -5.$$

To find A_2, set $s=-1$:

$$0 = \frac{A_2}{-1} + \frac{30}{1} + \frac{-40}{-2} + \frac{-5}{1} \implies A_2 = 45.$$

Hence, taking the inverse Laplace transform, the solutions are

$$\begin{aligned}
x(t) &= \mathscr{L}^{-1}\{X(s)\} = \mathscr{L}^{-1}\left\{\frac{5}{s-3} + e^{-s}\left(\frac{10}{s} + \frac{8}{s-3} - \frac{20}{s-1} + \frac{2}{s+2}\right)\right\}\\
&= 5e^{3t} + \Big[10 + 8e^{3t} - 20e^{t} + 2e^{-2t}\Big]_{t\to t-1} H(t-1)\\
&= 5e^{3t} + \big[10 + 8e^{3(t-1)} - 20e^{t-1} + 2e^{-2(t-1)}\big]H(t-1),\\
y(t) &= \mathscr{L}^{-1}\{Y(s)\} = \mathscr{L}^{-1}\left\{e^{-s}\left(\frac{45}{s} + \frac{30}{s^2} - \frac{40}{s-1} - \frac{5}{s+2}\right)\right\}\\
&= \Big[45 + 30t - 40e^{t} - 5e^{-2t}\Big]_{t\to t-1} H(t-1)\\
&= \big[15 + 30t - 40e^{t-1} - 5e^{-2(t-1)}\big]H(t-1).
\end{aligned}$$

7.4 The Matrix Method

Any linear ordinary differential equation or system of linear ordinary differential equations can be written as a system of first-order linear ordinary differential equations. For example, consider the second-order differential equation

$$x'' + 2\zeta\omega_0 x' + \omega_0^2 x = a\sin\Omega t.$$

Denoting $x = x_1$, $x' = x_2$, the differential equation becomes

$$x_2' + 2\zeta\omega_0 x_2 + \omega_0^2 x_1 = a\sin\Omega t.$$

Noting that $x_1' = x_2$, one obtains

$$\begin{Bmatrix} x_1' \\ x_2' \end{Bmatrix} = \begin{Bmatrix} x_2 \\ -2\zeta\omega_0 x_2 - \omega_0^2 x_1 + a\sin\Omega t \end{Bmatrix} = \begin{bmatrix} 0 & 1 \\ -\omega_0^2 & -2\zeta\omega_0 \end{bmatrix}\begin{Bmatrix} x_1 \\ x_2 \end{Bmatrix} + \begin{Bmatrix} 0 \\ a\sin\Omega t \end{Bmatrix},$$

which is a system of two first-order differential equations. Similarly, consider a system of differential equations

$$x''' + 2x' + x - y' = 2\sin 3t, \qquad 4x' + 3x - y'' + 5y' - y = e^{-t}\cos 3t.$$

Letting $x = x_1$, $x' = x_2$, $x'' = x_3$, $y = x_4$, $y = x_5$, the differential equations become

$$x_3' + 2x_2 + x_1 - x_5 = 2\sin 3t, \qquad 4x_2 + 3x_1 - x_5' + 5x_5 - x_4 = e^{-t}\cos 3t,$$

one obtains

$$\begin{Bmatrix} x_1' \\ x_2' \\ x_3' \\ x_4' \\ x_5' \end{Bmatrix} = \begin{Bmatrix} x_2 \\ x_3 \\ -x_1 - 2x_2 + x_5 + 2\sin 3t \\ x_5 \\ 3x_1 + 4x_2 - x_4 + 5x_5 - e^{-t}\cos 3t \end{Bmatrix},$$

or, in the matrix form,

$$\begin{Bmatrix} x_1' \\ x_2' \\ x_3' \\ x_4' \\ x_5' \end{Bmatrix} = \begin{bmatrix} 0 & 1 & 0 & 0 & 0 \\ 0 & 0 & 1 & 0 & 0 \\ -1 & -2 & 0 & 0 & 1 \\ 0 & 0 & 0 & 0 & 1 \\ 3 & 4 & 0 & -1 & 5 \end{bmatrix} \begin{Bmatrix} x_1 \\ x_2 \\ x_3 \\ x_4 \\ x_5 \end{Bmatrix} + \begin{Bmatrix} 0 \\ 0 \\ 2\sin 3t \\ 0 \\ -e^{-t}\cos 3t \end{Bmatrix},$$

which is a system of five first-order differential equations.

Hence, without loss of generality, consider a system of n-dimensional first-order linear ordinary differential equations with constant coefficients of the form

$$\mathbf{x}'(t) = \mathbf{A}\,\mathbf{x}(t) + \mathbf{f}(t),$$

where $\mathbf{x}(t) = \{x_1(t), x_2(t), \ldots, x_n(t)\}^T$, $\mathbf{f}(t) = \{f_1(t), f_2(t), \ldots, f_n(t)\}^T$, and $\mathbf{A}$ is an $n \times n$ matrix with constant entries.

7.4.1 Complementary Solutions

First consider the n-dimensional homogeneous system with $\mathbf{f}(t) = \mathbf{0}$, i.e.,

$$\mathbf{x}'(t) = \mathbf{A}\,\mathbf{x}(t). \tag{1}$$

Seek a solution of the form $\mathbf{x}(t) = e^{\lambda t}\mathbf{v}$, where $\mathbf{v}$ is a constant vector. Substituting into equation (1) yields $\lambda e^{\lambda t}\mathbf{v} = \mathbf{A}\,e^{\lambda t}\mathbf{v}$. Since, $e^{\lambda t} \neq 0$, one obtains

$$(\mathbf{A} - \lambda\,\mathbf{I})\,\mathbf{v} = \mathbf{0}, \tag{2}$$

where $\mathbf{I}$ is the $n \times n$ identity matrix, with 1's on the main diagonal and 0's elsewhere.

Equation (2) is a system of homogeneous linear *algebraic* equations. To have nonzero solutions for $\mathbf{v}$, the determinant of the coefficient matrix must be zero, i.e.,

$$\det(\mathbf{A} - \lambda\,\mathbf{I}) = 0, \tag{3}$$

which leads to the *characteristic equation*, a polynomial equation in λ of degree n.

Distinct Eigenvalues

The n solutions $\lambda_1, \lambda_2, \ldots, \lambda_n$ of the characteristic equation (3) are called the *eigenvalues* of $\mathbf{A}$. Suppose the eigenvalues $\lambda_1, \lambda_2, \ldots, \lambda_n$ are distinct real numbers. A *nonzero* solution $\mathbf{v}_k$ of system (2) with $\lambda = \lambda_k$, i.e.,

$$(\mathbf{A} - \lambda_k\,\mathbf{I})\,\mathbf{v}_k = \mathbf{0}, \qquad k = 1, 2, \ldots, n, \tag{4}$$

is called an *eigenvector* corresponding to eigenvalue λ_k.

From linear algebra it is well known that, if the eigenvalues $\lambda_1, \lambda_2, \ldots, \lambda_n$ are distinct, the corresponding eigenvectors $\mathbf{v}_1, \mathbf{v}_2, \ldots, \mathbf{v}_n$ are linearly independent.

Hence, with n eigenvalue-eigenvector pairs λ_k, $\mathbf{v}_k$, $k=1, 2, \ldots, n$, there are n linearly indenpendent solutions for system (2): $e^{\lambda_1 t}\mathbf{v}_1$, $e^{\lambda_2 t}\mathbf{v}_2$, ..., $e^{\lambda_n t}\mathbf{v}_n$.

Distinct Eigenvalues

Suppose that matrix $\mathbf{A}$ of the homogeneous system $\mathbf{x}'(t) = \mathbf{A}\mathbf{x}(t)$ has distinct eigenvalues $\lambda_1, \lambda_2, \ldots, \lambda_n$ with corresponding eigenvectors $\mathbf{v}_1, \mathbf{v}_2, \ldots, \mathbf{v}_n$. Then the complementary solution of the homogeneous system is

$$\mathbf{x}(t) = C_1 e^{\lambda_1 t}\mathbf{v}_1 + C_2 e^{\lambda_2 t}\mathbf{v}_2 + \cdots + C_n e^{\lambda_n t}\mathbf{v}_n,$$

where $C_1, C_2, \ldots, C_n$ are constants.

The $n \times n$ matrix

$$\mathbf{X}(t) = \left[e^{\lambda_1 t}\mathbf{v}_1,\ e^{\lambda_2 t}\mathbf{v}_2,\ \ldots,\ e^{\lambda_n t}\mathbf{v}_n \right],$$

whose columns are n linearly independent solutions of the homogeneous system, is called a *fundamental matrix* for $\mathbf{x}'(t) = \mathbf{A}\mathbf{x}(t)$.

Using the fundamental matrix, the complementary solution can be written as

$$\mathbf{x}'(t) = \mathbf{A}\mathbf{x}(t) \quad \Longrightarrow \quad \mathbf{x}(t) = \mathbf{X}(t)\,\mathbf{C}, \quad \mathbf{C} = \left\{ C_1, C_2, \ldots, C_n \right\}^T.$$

For the homogeneous system $\mathbf{x}'(t) = \mathbf{A}\mathbf{x}(t)$ with the initial condition $\mathbf{x}(t_0) = \mathbf{x}_0$, one has $\mathbf{x}(t_0) = \mathbf{X}(t_0)\,\mathbf{C} = \mathbf{x}_0 \Rightarrow \mathbf{C} = \mathbf{X}^{-1}(t_0)\,\mathbf{x}_0$,

$$\mathbf{x}'(t) = \mathbf{A}\mathbf{x}(t), \ \ \mathbf{x}(t_0) = \mathbf{x}_0 \quad \Longrightarrow \quad \mathbf{x}(t) = \mathbf{X}(t)\,\mathbf{X}^{-1}(t_0)\,\mathbf{x}_0.$$

Example 7.14

Solve

$$x_1' - x_2' - 6x_2 = 0, \qquad (\cdot)' = \mathrm{d}(\cdot)/\mathrm{d}t, \tag{1}$$

$$x_1' + 2x_2' - 3x_1 = 0. \tag{2}$$

Solve equations (1) and (2) for x_1' and x_2'

$$2 \times \text{Eqn (1)} + \text{Eqn (2)}: \quad 3x_1' - 12x_2 - 3x_1 = 0 \quad \Longrightarrow \quad x_1' = x_1 + 4x_2,$$

$$\text{Eqn (2)} - \text{Eqn (1)}: \quad 3x_2' - 3x_1 + 6x_2 = 0 \quad \Longrightarrow \quad x_2' = x_1 - 2x_2,$$

which can be written in the matrix form as

$$\mathbf{x}'(t) = \mathbf{A}\mathbf{x}(t), \qquad \mathbf{x}(t) = \begin{Bmatrix} x_1 \\ x_2 \end{Bmatrix}, \qquad \mathbf{A} = \begin{bmatrix} 1 & 4 \\ 1 & -2 \end{bmatrix}.$$

The characteristic equation is

$$\det(\mathbf{A}-\lambda\mathbf{I}) = \begin{vmatrix} 1-\lambda & 4 \\ 1 & -2-\lambda \end{vmatrix} = \lambda^2+\lambda-6 = (\lambda+3)(\lambda-2) = 0,$$

and the two eigenvalues are $\lambda_1 = -3$, $\lambda_2 = 2$. The corresponding eigenvectors are obtained as follows.

(1) $\lambda = \lambda_1 = -3$:

$$(\mathbf{A}-\lambda_1\mathbf{I})\,\mathbf{v}_1 = \begin{bmatrix} 4 & 4 \\ 1 & 1 \end{bmatrix} \begin{Bmatrix} v_{11} \\ v_{21} \end{Bmatrix} = \mathbf{0} \implies v_{11}+v_{21} = 0,$$

$$\text{taking } v_{21} = -1, \text{ then } v_{11} = -v_{21} = 1 \implies \mathbf{v}_1 = \begin{Bmatrix} v_{11} \\ v_{21} \end{Bmatrix} = \begin{Bmatrix} 1 \\ -1 \end{Bmatrix}.$$

(2) $\lambda = \lambda_2 = 2$:

$$(\mathbf{A}-\lambda_2\mathbf{I})\,\mathbf{v}_2 = \begin{bmatrix} -1 & 4 \\ 1 & -4 \end{bmatrix} \begin{Bmatrix} v_{12} \\ v_{22} \end{Bmatrix} = \mathbf{0} \implies v_{12}-4v_{22} = 0,$$

$$\text{taking } v_{22} = 1, \text{ then } v_{12} = 4v_{22} = 4 \implies \mathbf{v}_2 = \begin{Bmatrix} v_{12} \\ v_{22} \end{Bmatrix} = \begin{Bmatrix} 4 \\ 1 \end{Bmatrix}.$$

Hence, the complementary solution is

$$\mathbf{x}(t) = C_1\,\mathrm{e}^{\lambda_1 t}\,\mathbf{v}_1 + C_2\,\mathrm{e}^{\lambda_2 t}\,\mathbf{v}_2 = C_1\,\mathrm{e}^{-3t} \begin{Bmatrix} 1 \\ -1 \end{Bmatrix} + C_2\,\mathrm{e}^{2t} \begin{Bmatrix} 4 \\ 1 \end{Bmatrix},$$

or

$$x_1(t) = C_1\,\mathrm{e}^{-3t} + 4C_2\,\mathrm{e}^{2t}, \qquad x_2(t) = -C_1\,\mathrm{e}^{-3t} + C_2\,\mathrm{e}^{2t}.$$

Complex Eigenvalues

Consider the first-order homogeneous system

$$\mathbf{x}'(t) = \mathbf{A}\,\mathbf{x}(t), \tag{1}$$

where $\mathbf{A}$ is a *real* matrix. Suppose $\lambda = \alpha + \mathrm{i}\beta$ is an eigenvalue, i.e., $\det(\mathbf{A}-\lambda\mathbf{I}) = 0$. Since the characteristic equation has real coefficients, then $\bar{\lambda} = \alpha - \mathrm{i}\beta$ is also an eigenvalue.

Let the complex vector $\mathbf{v}$ be an eigenvector corresponding to λ, i.e.,

$$(\mathbf{A}-\lambda\mathbf{I})\,\mathbf{v} = \mathbf{0}. \tag{2}$$

Then $\mathbf{x}_1(t) = \mathrm{e}^{\lambda t}\,\mathbf{v}$ is a solution of the homogeneous system (1).

Taking complex conjugate of equation (2), one has

$$\overline{(\mathbf{A}-\lambda\mathbf{I})\mathbf{v}} = (\bar{\mathbf{A}}-\bar{\lambda}\bar{\mathbf{I}})\bar{\mathbf{v}} = (\mathbf{A}-\bar{\lambda}\mathbf{I})\bar{\mathbf{v}} = \mathbf{0}, \quad ✍ \ \mathbf{A} \text{ and } \mathbf{I} \text{ are real matrices.}$$

implying that $\bar{\mathbf{v}}$ is an eigenvector corresponding to eigenvalue $\bar{\lambda}$. Thus $\mathbf{x}_2(t)=e^{\bar{\lambda}t}\bar{\mathbf{v}}$ is a solution of the homogeneous system (1).

Corresponding to the eigenvalues $\alpha \pm i\beta$, one obtains the complementary solution

$$\mathbf{x}(t) = C_1 e^{\lambda t}\mathbf{v} + C_2 e^{\bar{\lambda}t}\bar{\mathbf{v}}, \qquad \mathbf{v} = \mathbf{v}_R + i\mathbf{v}_I,$$

where C_1 and C_2 are complex constants, and $\mathbf{v}_R$ and $\mathbf{v}_I$ are, respectively, the real and imaginary parts of the eigenvector $\mathbf{v}$. Applying Euler's formula

$$e^{(\alpha \pm i\beta)t} = e^{\alpha t}(\cos\beta t \pm i\sin\beta t)$$

leads to

$$\begin{aligned}\mathbf{x}(t) &= C_1 e^{\alpha t}(\cos\beta t + i\sin\beta t)(\mathbf{v}_R + i\mathbf{v}_I) + C_2 e^{\alpha t}(\cos\beta t - i\sin\beta t)(\mathbf{v}_R - i\mathbf{v}_I)\\ &= e^{\alpha t}\big[(C_1+C_2)(\mathbf{v}_R\cos\beta t - \mathbf{v}_I\sin\beta t) + i(C_1-C_2)(\mathbf{v}_R\sin\beta t + \mathbf{v}_I\cos\beta t)\big].\end{aligned}$$

For the solution $\mathbf{x}(t)$ to be real, one must have $C_1+C_2=A$, $i(C_1-C_2)=B$, where A and B are real constants. This can be accomplished if $\bar{C}_1=C_2$. Hence

$$\begin{aligned}\mathbf{x}(t) &= A\,e^{\alpha t}(\mathbf{v}_R\cos\beta t - \mathbf{v}_I\sin\beta t) + B\,e^{\alpha t}(\mathbf{v}_R\sin\beta t + \mathbf{v}_I\cos\beta t)\\ &= A\,\mathcal{R}e(e^{\lambda t}\mathbf{v}) + B\,\mathcal{I}m(e^{\lambda t}\mathbf{v}).\end{aligned}$$

Complex Eigenvalues

Suppose that matrix $\mathbf{A}$ of the homogeneous system $\mathbf{x}'(t)=\mathbf{A}\mathbf{x}(t)$ is a real matrix. If $\lambda=\alpha+i\beta$ is an eigenvalue with the corresponding eigenvector $\mathbf{v}$, then, corresponding to the eigenvalues $\alpha \pm i\beta$,

$$\begin{aligned}\mathbf{x}_1(t) &= \mathcal{R}e(e^{\lambda t}\mathbf{v}) = e^{\alpha t}\big[\mathcal{R}e(\mathbf{v})\cos\beta t - \mathcal{I}m(\mathbf{v})\sin\beta t\big],\\ \mathbf{x}_2(t) &= \mathcal{I}m(e^{\lambda t}\mathbf{v}) = e^{\alpha t}\big[\mathcal{R}e(\mathbf{v})\sin\beta t + \mathcal{I}m(\mathbf{v})\cos\beta t\big]\end{aligned}$$

are two linearly independent real-valued solutions, or

$$\mathbf{x}(t) = A\,\mathcal{R}e(e^{\lambda t}\mathbf{v}) + B\,\mathcal{I}m(e^{\lambda t}\mathbf{v}).$$

Example 7.15

Solve

$$x_1' + x_1 - 5x_2 = 0, \qquad (\cdot)' = d(\cdot)/dt,$$
$$4x_1 + x_2' + 5x_2 = 0.$$

In the matrix form, the system of differential equations can be written as

$$\mathbf{x}'(t) = \mathbf{A}\,\mathbf{x}(t), \qquad \mathbf{x}(t) = \begin{Bmatrix} x_1 \\ x_2 \end{Bmatrix}, \qquad \mathbf{A} = \begin{bmatrix} -1 & 5 \\ -4 & -5 \end{bmatrix}.$$

The characteristic equation is

$$\det(\mathbf{A} - \lambda\mathbf{I}) = \begin{vmatrix} -1-\lambda & 5 \\ -4 & -5-\lambda \end{vmatrix} = \lambda^2 + 6\lambda + 25 = 0 \implies \lambda = -3 \pm \mathrm{i}4.$$

For eigenvalue $\lambda = -3 + \mathrm{i}4$, the corresponding eigenvector is

$$(\mathbf{A} - \lambda\mathbf{I})\,\mathbf{v} = \begin{bmatrix} 2-\mathrm{i}4 & 5 \\ -4 & -2-\mathrm{i}4 \end{bmatrix} \begin{Bmatrix} v_1 \\ v_2 \end{Bmatrix} = \begin{Bmatrix} (2-\mathrm{i}4)\,v_1 + 5v_2 \\ -4v_1 - (2+\mathrm{i}4)\,v_2 \end{Bmatrix} = \begin{Bmatrix} 0 \\ 0 \end{Bmatrix}.$$

Note that the two equations $(2-\mathrm{i}4)\,v_1 + 5v_2 = 0$ and $4v_1 + (2+\mathrm{i}4)\,v_2 = 0$ are the same. Taking $v_1 = 5$, then $v_2 = -\frac{1}{5}(2-\mathrm{i}4)\,v_1 = -2 + \mathrm{i}4$,

$$\therefore \quad \mathbf{v} = \begin{Bmatrix} v_1 \\ v_2 \end{Bmatrix} = \begin{Bmatrix} 5 \\ -2+\mathrm{i}4 \end{Bmatrix} = \begin{Bmatrix} 5 \\ -2 \end{Bmatrix} + \mathrm{i} \begin{Bmatrix} 0 \\ 4 \end{Bmatrix}.$$

Hence

$$\begin{aligned} \mathrm{e}^{\lambda t}\,\mathbf{v} &= \mathrm{e}^{-3t}\,(\cos 4t + \mathrm{i}\sin 4t) \left(\begin{Bmatrix} 5 \\ -2 \end{Bmatrix} + \mathrm{i} \begin{Bmatrix} 0 \\ 4 \end{Bmatrix} \right) \\ &= \mathrm{e}^{-3t} \left[\left(\begin{Bmatrix} 5 \\ -2 \end{Bmatrix} \cos 4t - \begin{Bmatrix} 0 \\ 4 \end{Bmatrix} \sin 4t \right) + \mathrm{i} \left(\begin{Bmatrix} 5 \\ -2 \end{Bmatrix} \sin 4t + \begin{Bmatrix} 0 \\ 4 \end{Bmatrix} \cos 4t \right) \right]. \end{aligned}$$

Hence, the complementary solution is

$$\begin{aligned} \mathbf{x}(t) = A\,\mathcal{R}e(\mathrm{e}^{\lambda t}\mathbf{v}) + B\,\mathcal{I}m(\mathrm{e}^{\lambda t}\mathbf{v}) &= A\,\mathrm{e}^{-3t} \left(\begin{Bmatrix} 5 \\ -2 \end{Bmatrix} \cos 4t - \begin{Bmatrix} 0 \\ 4 \end{Bmatrix} \sin 4t \right) \\ &\quad + B\,\mathrm{e}^{-3t} \left(\begin{Bmatrix} 5 \\ -2 \end{Bmatrix} \sin 4t + \begin{Bmatrix} 0 \\ 4 \end{Bmatrix} \cos 4t \right), \end{aligned}$$

$$\begin{aligned} \therefore \quad x_1(t) &= 5\mathrm{e}^{-3t}\,(A\cos 4t + B\sin 4t), \\ x_2(t) &= 2\mathrm{e}^{-3t}\big[(-A+2B)\cos 4t - (2A+B)\sin 4t\big]. \end{aligned}$$

Multiple Eigenvalues

For an $n \times n$ matrix $\mathbf{A}$ with constant entries, if its n eigenvalues $\lambda_1, \lambda_2, \ldots, \lambda_n$, either real or complex, are distinct, then the corresponding n eigenvectors $\mathbf{v}_1, \mathbf{v}_2, \ldots, \mathbf{v}_n$ are *linearly independent* and form a complete basis of eigenvectors.

If matrix $\mathbf{A}$ has a repeated eigenvalue with algebraic multiplicity $m>1$ (the number of times the eigenvalue is repeated as a root of the characteristic equation), it is possible that the multiple eigenvalue has m linearly independent eigenvectors. However, it is also possible that there are fewer than m linearly independent eigenvectors; in this case, matrix $\mathbf{A}$ is a *defective* or *deficient* matrix.

In other words, an $n \times n$ matrix is defective if and only if it does not have n linearly independent eigenvectors. A complete basis is formed by augmenting the eigenvectors with *generalized eigenvectors.*

Suppose λ is an eigenvalue of multiplicity m, and there are only $k<m$ linearly independent eigenvectors corresponding to λ. A complete basis of eigenvectors is obtained by including $(m-k)$ *generalized eigenvectors*:

$$(\mathbf{A}-\lambda\mathbf{I})\,\mathbf{v}_i=\mathbf{0} \quad\Longrightarrow\quad \mathbf{v}_i,\ i=1,2,\dots,k,\ \text{linearly independent eigenvectors,}$$

$$\left.\begin{aligned}
(\mathbf{A}-\lambda\mathbf{I})\,\mathbf{v}_{k+1}&=\mathbf{v}_k &&\Longrightarrow (\mathbf{A}-\lambda\mathbf{I})^2\mathbf{v}_{k+1}=\mathbf{0},\\
(\mathbf{A}-\lambda\mathbf{I})\,\mathbf{v}_{k+2}&=\mathbf{v}_{k+1} &&\Longrightarrow (\mathbf{A}-\lambda\mathbf{I})^3\mathbf{v}_{k+2}=\mathbf{0},\\
&\ \ \vdots\\
(\mathbf{A}-\lambda\mathbf{I})\,\mathbf{v}_{m}&=\mathbf{v}_{m-1} &&\Longrightarrow (\mathbf{A}-\lambda\mathbf{I})^{m-k+1}\mathbf{v}_{m}=\mathbf{0}.
\end{aligned}\right\}\ \text{Generalized eigenvectors}$$

Multiple Eigenvalues

Suppose matrix $\mathbf{A}$ of the homogeneous system $\mathbf{x}'(t)=\mathbf{A}\,\mathbf{x}(t)$ has an eigenvalue λ of algebraic multiplicity $m>1$, and a sequence of generalized eigenvectors corresponding to λ is $\mathbf{v}_1, \mathbf{v}_2, \dots, \mathbf{v}_m$. Then, corresponding to the eigenvalues $\lambda, \lambda, \dots, \lambda$ (repeated m times), m *linearly independent* solutions of the homogeneous system are

$$\mathbf{x}_i(t)=e^{\lambda t}\,\mathbf{v}_i, \quad i=1,2,\dots,k, \qquad ✍\ \because\ \mathbf{v}_1,\mathbf{v}_2,\cdots,\mathbf{v}_k \text{ are eigenvectors}$$

$$\mathbf{x}_{k+1}(t)=e^{\lambda t}\left(\mathbf{v}_k t+\mathbf{v}_{k+1}\right),$$

$$\mathbf{x}_{k+2}(t)=e^{\lambda t}\left(\mathbf{v}_k\frac{t^2}{2!}+\mathbf{v}_{k+1}t+\mathbf{v}_{k+2}\right),$$

$$\vdots$$

$$\mathbf{x}_m(t)=e^{\lambda t}\left[\mathbf{v}_k\frac{t^{m-k}}{(m-k)!}+\mathbf{v}_{k+1}\frac{t^{m-k-1}}{(m-k-1)!}+\cdots+\mathbf{v}_{m-2}\frac{t^2}{2!}+\mathbf{v}_{m-1}t+\mathbf{v}_m\right].$$

Example 7.16

Solve

$$\begin{aligned}
x_1'-4x_1+x_2&=0, \qquad (\cdot)'=\mathrm{d}(\cdot)/\mathrm{d}t,\\
3x_1-x_2'+x_2-x_3&=0,\\
x_1-x_3'+x_3&=0.
\end{aligned}$$

In the matrix form, the system of differential equations can be written as

$$\mathbf{x}'(t) = \mathbf{A}\,\mathbf{x}(t), \qquad \mathbf{x}(t) = \begin{Bmatrix} x_1 \\ x_2 \\ x_3 \end{Bmatrix}, \qquad \mathbf{A} = \begin{bmatrix} 4 & -1 & 0 \\ 3 & 1 & -1 \\ 1 & 0 & 1 \end{bmatrix}.$$

The characteristic equation is

$$\det(\mathbf{A} - \lambda\mathbf{I}) = \begin{vmatrix} 4-\lambda & -1 & 0 \\ 3 & 1-\lambda & -1 \\ 1 & 0 & 1-\lambda \end{vmatrix} = -(\lambda^3 - 6\lambda^2 + 12\lambda - 8) = -(\lambda-2)^3 = 0.$$

Hence, $\lambda = 2$ is an eigenvector of multiplicity 3. The eigenvector equation is

$$(\mathbf{A} - \lambda\mathbf{I})\,\mathbf{v}_1 = \begin{bmatrix} 2 & -1 & 0 \\ 3 & -1 & -1 \\ 1 & 0 & -1 \end{bmatrix} \begin{Bmatrix} v_{11} \\ v_{21} \\ v_{31} \end{Bmatrix} = \begin{Bmatrix} 2v_{11} - v_{21} \\ 3v_{11} - v_{21} - v_{31} \\ v_{11} - v_{31} \end{Bmatrix} = \begin{Bmatrix} 0 \\ 0 \\ 0 \end{Bmatrix}.$$

Taking $v_{11} = 1$, then $v_{21} = 2v_{11} = 2$, $v_{31} = v_{11} = 1$,

$$\therefore \quad \mathbf{v}_1 = \begin{Bmatrix} v_{11} \\ v_{21} \\ v_{31} \end{Bmatrix} = \begin{Bmatrix} 1 \\ 2 \\ 1 \end{Bmatrix}.$$

It is not possible to find two more linearly independent eigenvectors. Hence, matrix **A** is defective and a complete basis of eigenvectors is obtained by including two generalized eigenvectors:

$$(\mathbf{A} - \lambda\mathbf{I})\,\mathbf{v}_2 = \mathbf{v}_1 \implies \begin{bmatrix} 2 & -1 & 0 \\ 3 & -1 & -1 \\ 1 & 0 & -1 \end{bmatrix} \begin{Bmatrix} v_{12} \\ v_{22} \\ v_{32} \end{Bmatrix} = \begin{Bmatrix} 2v_{12} - v_{22} \\ 3v_{12} - v_{22} - v_{32} \\ v_{12} - v_{32} \end{Bmatrix} = \begin{Bmatrix} 1 \\ 2 \\ 1 \end{Bmatrix}.$$

Taking $v_{12} = 2$, then $v_{22} = 2v_{12} - 1 = 3$, $v_{32} = v_{12} - 1 = 1$,

$$\therefore \quad \mathbf{v}_2 = \begin{Bmatrix} v_{12} \\ v_{22} \\ v_{32} \end{Bmatrix} = \begin{Bmatrix} 2 \\ 3 \\ 1 \end{Bmatrix}.$$

$$(\mathbf{A} - \lambda\mathbf{I})\,\mathbf{v}_3 = \mathbf{v}_2 \implies \begin{bmatrix} 2 & -1 & 0 \\ 3 & -1 & -1 \\ 1 & 0 & -1 \end{bmatrix} \begin{Bmatrix} v_{13} \\ v_{23} \\ v_{33} \end{Bmatrix} = \begin{Bmatrix} 2v_{13} - v_{23} \\ 3v_{13} - v_{23} - v_{33} \\ v_{13} - v_{33} \end{Bmatrix} = \begin{Bmatrix} 2 \\ 3 \\ 1 \end{Bmatrix}.$$

Taking $v_{13} = 1$, then $v_{23} = 2v_{13} - 2 = 0$, $v_{33} = v_{13} - 1 = 0$,

$$\therefore \quad \mathbf{v}_3 = \begin{Bmatrix} v_{13} \\ v_{23} \\ v_{33} \end{Bmatrix} = \begin{Bmatrix} 1 \\ 0 \\ 0 \end{Bmatrix}.$$

Three linearly independent solutions are

$$\mathbf{x}_1(t) = e^{\lambda t}\mathbf{v}_1 = e^{2t}\begin{Bmatrix}1\\2\\1\end{Bmatrix}, \qquad \mathbf{x}_2(t) = e^{\lambda t}(\mathbf{v}_1 t + \mathbf{v}_2) = e^{2t}\left(\begin{Bmatrix}1\\2\\1\end{Bmatrix} t + \begin{Bmatrix}2\\3\\1\end{Bmatrix}\right),$$

$$\mathbf{x}_3(t) = e^{\lambda t}\left(\mathbf{v}_1\frac{t^2}{2} + \mathbf{v}_2 t + \mathbf{v}_3\right) = e^{2t}\left(\begin{Bmatrix}1\\2\\1\end{Bmatrix}\frac{t^2}{2} + \begin{Bmatrix}2\\3\\1\end{Bmatrix} t + \begin{Bmatrix}1\\0\\0\end{Bmatrix}\right).$$

The complementary solution is

$$\mathbf{x}(t) = C_1\mathbf{x}_1(t) + C_2\mathbf{x}_2(t) + 2C_3\mathbf{x}_3(t)$$

$$= C_1 e^{2t}\begin{Bmatrix}1\\2\\1\end{Bmatrix} + C_2 e^{2t}\left(\begin{Bmatrix}1\\2\\1\end{Bmatrix} t + \begin{Bmatrix}2\\3\\1\end{Bmatrix}\right) + 2C_3 e^{2t}\left(\begin{Bmatrix}1\\2\\1\end{Bmatrix}\frac{t^2}{2} + \begin{Bmatrix}2\\3\\1\end{Bmatrix} t + \begin{Bmatrix}1\\0\\0\end{Bmatrix}\right),$$

$$\therefore \quad x_1(t) = e^{2t}\big[C_3 t^2 + (C_2 + 4C_3)t + (C_1 + 2C_2 + 2C_3)\big],$$

$$x_2(t) = e^{2t}\big[2C_3 t^2 + 2(C_2 + 3C_3)t + (2C_1 + 3C_2)\big],$$

$$x_3(t) = e^{2t}\big[C_3 t^2 + (C_2 + 2C_3)t + (C_1 + C_2)\big].$$

Example 7.17

Solve

$$\mathbf{x}'(t) = \mathbf{A}\mathbf{x}(t), \quad \mathbf{x}(t) = \begin{Bmatrix}x_1\\x_2\\x_3\end{Bmatrix}, \quad \mathbf{A} = \begin{bmatrix}-2 & 1 & -2\\ 1 & -2 & 2\\ 3 & -3 & 5\end{bmatrix}.$$

The characteristic equation is

$$\det(\mathbf{A} - \lambda\mathbf{I}) = \begin{vmatrix}-2-\lambda & 1 & -2\\ 1 & -2-\lambda & 2\\ 3 & -3 & 5-\lambda\end{vmatrix} = -(\lambda^3 - \lambda^2 - 5\lambda - 3)$$

$$= -(\lambda+1)^2(\lambda-3) = 0.$$

Hence, $\lambda_1 = \lambda_2 = -1$ is an eigenvalue of multiplicity 2, and $\lambda_3 = 3$. The eigenvectors are determined as follows.

(1) $\lambda = \lambda_1 = \lambda_2 = -1$:

$$(\mathbf{A} - \lambda\mathbf{I})\mathbf{v} = \mathbf{0} \implies \begin{bmatrix}-1 & 1 & -2\\ 1 & -1 & 2\\ 3 & -3 & 6\end{bmatrix}\begin{Bmatrix}v_1\\v_2\\v_3\end{Bmatrix} = \begin{Bmatrix}-v_1 + v_2 - 2v_3\\ -(-v_1 + v_2 - 2v_3)\\ -3(-v_1 + v_2 - 2v_3)\end{Bmatrix} = \begin{Bmatrix}0\\0\\0\end{Bmatrix},$$

which leads to $v_1 = v_2 - 2v_3$.

Taking $v_2=1,\ v_3=0 \Longrightarrow v_1=1$; taking $v_2=0,\ v_3=1 \Longrightarrow v_1=-2$;

$$\therefore\quad \mathbf{v}_1=\begin{Bmatrix}1\\1\\0\end{Bmatrix},\qquad \mathbf{v}_2=\begin{Bmatrix}-2\\0\\1\end{Bmatrix}.$$

(2) $\lambda=\lambda_3=3$:

$$(\mathbf{A}-\lambda\mathbf{I})\,\mathbf{v}_3=\mathbf{0}\Longrightarrow\begin{bmatrix}-5 & 1 & -2\\ 1 & -5 & 2\\ 3 & -3 & 2\end{bmatrix}\begin{Bmatrix}v_{13}\\v_{23}\\v_{33}\end{Bmatrix}=\begin{Bmatrix}-5v_{13}+v_{23}-2v_{33}\\ v_{13}-5v_{23}+2v_{33}\\ 3v_{13}-3v_{23}+2v_{33}\end{Bmatrix}=\begin{Bmatrix}0\\0\\0\end{Bmatrix}.$$

Taking $v_{33}=3$, then $v_{13}=-1,\ v_{23}=1$,

$$\therefore\quad \mathbf{v}_3=\begin{Bmatrix}v_{13}\\v_{23}\\v_{33}\end{Bmatrix}=\begin{Bmatrix}-1\\1\\3\end{Bmatrix}.$$

The complementary solution is

$$\mathbf{x}(t)=\sum_{k=1}^{3}C_k\,\mathrm{e}^{\lambda_k t}\,\mathbf{v}_k=C_1\mathrm{e}^{-t}\begin{Bmatrix}1\\1\\0\end{Bmatrix}+C_2\mathrm{e}^{-t}\begin{Bmatrix}-2\\0\\1\end{Bmatrix}+C_3\mathrm{e}^{3t}\begin{Bmatrix}-1\\1\\3\end{Bmatrix},$$

$$\therefore\ x_1(t)=(C_1-2C_2)\,\mathrm{e}^{-t}-C_3\,\mathrm{e}^{3t},\ x_2(t)=C_1\,\mathrm{e}^{-t}+C_3\,\mathrm{e}^{3t},\ x_3(t)=C_2\,\mathrm{e}^{-t}+3C_3\,\mathrm{e}^{3t}.$$

Remarks: Although $\lambda=-1$ is an eigenvalue of multiplicity 2, two linearly independent eigenvectors do exist.

7.4.2 Particular Solutions

The method of variation of parameters is applied to find a particular solution of the nonhomogeneous system

$$\mathbf{x}'(t)=\mathbf{A}\mathbf{x}(t)+\mathbf{f}(t).$$

The complementary solution of the homogeneous system $\mathbf{x}'(t)=\mathbf{A}\mathbf{x}(t)$ has been obtained as $\mathbf{x}(t)=\mathbf{X}(t)\,\mathbf{C}$, where $\mathbf{X}(t)$ is a fundamental matrix, whose columns are linearly independent and each is a solution of the homogeneous system, i.e., $\mathbf{X}'(t)=\mathbf{A}\mathbf{X}(t)$, and $\mathbf{C}$ is an n-dimensional constant vector.

Applying the method of variation of parameters, vary the constant vector $\mathbf{C}$ in the complementary solution $\mathbf{x}(t)=\mathbf{X}(t)\,\mathbf{C}$ to make it a vector of functions of t, i.e., $\mathbf{C}\Longrightarrow\mathbf{c}(t)$. Thus a particular solution is assumed to be of the form

$$\mathbf{x}(t)=\mathbf{X}(t)\,\mathbf{c}(t).$$

Differentiating with respect to t yields

$$\mathbf{x}'(t) = \mathbf{X}'(t)\,\mathbf{c}(t) + \mathbf{X}(t)\,\mathbf{c}'(t) = \mathbf{A}\mathbf{x}(t) + \mathbf{f}(t).$$

Substituting $\mathbf{X}'(t) = \mathbf{A}\mathbf{X}(t)$ and $\mathbf{x}(t) = \mathbf{X}(t)\,\mathbf{c}(t)$ yields

$$\mathbf{A}\mathbf{X}(t)\,\mathbf{c}(t) + \mathbf{X}(t)\,\mathbf{c}'(t) = \mathbf{A}\mathbf{X}(t)\,\mathbf{c}(t) + \mathbf{f}(t),$$

$$\therefore \quad \mathbf{X}(t)\,\mathbf{c}'(t) = \mathbf{f}(t) \quad \Longrightarrow \quad \mathbf{c}'(t) = \mathbf{X}^{-1}(t)\,\mathbf{f}(t)..$$

Integrating with respect to t gives

$$\mathbf{c}(t) = \mathbf{C} + \int \mathbf{X}^{-1}(t)\,\mathbf{f}(t)\,dt.$$

Hence, the general solution is given by

$$\mathbf{x}(t) = \mathbf{X}(t)\,\mathbf{c}(t) = \mathbf{X}(t)\left\{\mathbf{C} + \int \mathbf{X}^{-1}(t)\,\mathbf{f}(t)\,dt\right\}.$$

For the nonhomogeneous system $\mathbf{x}'(t) = \mathbf{A}\mathbf{x}(t) + \mathbf{f}(t)$ with the initial condition $\mathbf{x}(t_0) = \mathbf{x}_0$, the general solution can be written as

$$\mathbf{x}(t) = \mathbf{X}(t)\left\{\mathbf{C} + \int_{t_0}^{t} \mathbf{X}^{-1}(t)\,\mathbf{f}(t)\,dt\right\},$$

with

$$\mathbf{x}(t_0) = \mathbf{X}(t_0)\,\mathbf{C} \quad \Longrightarrow \quad \mathbf{C} = \mathbf{X}^{-1}(t_0)\,\mathbf{x}(t_0),$$

which yields

$$\mathbf{x}(t) = \mathbf{X}(t)\left\{\mathbf{X}^{-1}(t_0)\,\mathbf{x}(t_0) + \int_{t_0}^{t} \mathbf{X}^{-1}(t)\,\mathbf{f}(t)\,dt\right\}.$$

To find a particular solution using the method of variation of parameters, one must evaluate the inverse $\mathbf{X}^{-1}(t)$ of a fundamental matrix $\mathbf{X}(t)$. In the following, the Gauss-Jordan method is briefly reviewed.

Gauss-Jordan Method for Finding the Inverse of a Matrix

To find the inverse of an $n \times n$ matrix $\mathbf{A}$, $\det(\mathbf{A}) \neq 0$, augment matrix $\mathbf{A}$ with the $n \times n$ identity matrix $\mathbf{I}$ as

$$\big[\,\mathbf{A}\,\big|\,\mathbf{I}\,\big] = \left[\begin{array}{cccc|cccc} a_{11} & a_{12} & \cdots & a_{1n} & 1 & 0 & \cdots & 0 \\ a_{21} & a_{22} & \cdots & a_{2n} & 0 & 1 & \cdots & 0 \\ \vdots & \vdots & \cdots & \vdots & \vdots & \vdots & \ddots & \vdots \\ a_{n1} & a_{n2} & \cdots & a_{nn} & 0 & 0 & \cdots & 1 \end{array}\right].$$

Apply a series of elementary row operations, such as

- exchange row $\boxed{k}$ and row $\boxed{l}$,
- multiply row $\boxed{k}$ by $\alpha \neq 0$,
- multiply row $\boxed{l}$ by $\beta \neq 0$ and add to row $\boxed{k}$,

to convert the left half of the augmented matrix to the identity matrix. Then the right half of the augmented matrix becomes $\mathbf{A}^{-1}$:

$$\left[\begin{array}{cccc|cccc} 1 & 0 & \cdots & 0 & b_{11} & b_{12} & \cdots & b_{1n} \\ 0 & 1 & \cdots & 0 & b_{21} & b_{22} & \cdots & b_{2n} \\ \vdots & \vdots & \ddots & \vdots & \vdots & \vdots & \cdots & \vdots \\ 0 & 0 & \cdots & 1 & b_{n1} & b_{n2} & \cdots & b_{nn} \end{array}\right] = \left[\, \mathbf{I} \,|\, \mathbf{A}^{-1} \,\right].$$

Use $\boxed{k} \times \alpha + \boxed{l} \times \beta$ to denote the elementary row operations:

- multiply row $\boxed{k}$ by $\alpha \neq 0$;
- multiply row $\boxed{l}$ by $\beta \neq 0$ and add to row $\boxed{k}$.

In particular, for a 2×2 matrix,

$$\mathbf{A} = \begin{bmatrix} a_{11} & a_{12} \\ a_{21} & a_{22} \end{bmatrix} \implies \mathbf{A}^{-1} = \frac{1}{\det(\mathbf{A})} \begin{bmatrix} a_{22} & -a_{12} \\ -a_{21} & a_{11} \end{bmatrix}, \quad \det(\mathbf{A}) = a_{11}a_{22} - a_{12}a_{21}.$$

Example 7.18

Solve
$$x_1' + 3x_1 + 4x_2 = 2\,\mathrm{e}^{-t},$$
$$x_1 - x_2' + x_2 = 0.$$

In the matrix form, the system of differential equations can be written as

$$\mathbf{x}'(t) = \mathbf{A}\,\mathbf{x}(t) + \mathbf{f}(t), \quad \mathbf{x}(t) = \begin{Bmatrix} x_1 \\ x_2 \end{Bmatrix}, \quad \mathbf{A} = \begin{bmatrix} -3 & -4 \\ 1 & 1 \end{bmatrix}, \quad \mathbf{f}(t) = \begin{Bmatrix} 2\,\mathrm{e}^{-t} \\ 0 \end{Bmatrix}.$$

The characteristic equation is

$$\det(\mathbf{A} - \lambda \mathbf{I}) = \begin{vmatrix} -3-\lambda & -4 \\ 1 & 1-\lambda \end{vmatrix} = \lambda^2 + 2\lambda + 1 = 0 \implies \lambda = -1, \ -1.$$

Hence, $\lambda = -1$ is an eigenvalue of multiplicity 2. The eigenvector equation is

$$(\mathbf{A} - \lambda \mathbf{I})\,\mathbf{v}_1 = \begin{bmatrix} -2 & -4 \\ 1 & 2 \end{bmatrix} \begin{Bmatrix} v_{11} \\ v_{21} \end{Bmatrix} = \begin{Bmatrix} 0 \\ 0 \end{Bmatrix} \implies v_{11} + 2v_{21} = 0.$$

Taking $v_{21} = -1$, then $v_{11} = -2v_{21} = 2$,

$$\therefore \quad \mathbf{v}_1 = \begin{Bmatrix} v_{11} \\ v_{21} \end{Bmatrix} = \begin{Bmatrix} 2 \\ -1 \end{Bmatrix}.$$

A second linearly independent eigenvector does not exist. Hence, matrix **A** is defective and a complete basis of eigenvectors is obtained by including a generalized eigenvector:

$$(\mathbf{A}-\lambda\mathbf{I})\,\mathbf{v}_2=\mathbf{v}_1 \implies \begin{bmatrix} -2 & -4 \\ 1 & 2 \end{bmatrix}\begin{Bmatrix} v_{12} \\ v_{22} \end{Bmatrix}=\begin{Bmatrix} 2 \\ -1 \end{Bmatrix} \implies v_{12}+2v_{22}=-1.$$

Taking $v_{22}=-1$, then $v_{12}=-1-2v_{22}=1$,

$$\therefore\quad \mathbf{v}_2=\begin{Bmatrix} v_{12} \\ v_{22} \end{Bmatrix}=\begin{Bmatrix} 1 \\ -1 \end{Bmatrix}.$$

Two linearly independent solutions are

$$\mathbf{x}_1(t)=\mathrm{e}^{\lambda t}\mathbf{v}_1=\mathrm{e}^{-t}\begin{Bmatrix} 2 \\ -1 \end{Bmatrix},\quad \mathbf{x}_2(t)=\mathrm{e}^{\lambda t}(\mathbf{v}_1 t+\mathbf{v}_2)=\mathrm{e}^{-t}\left(\begin{Bmatrix} 2 \\ -1 \end{Bmatrix}t+\begin{Bmatrix} 1 \\ -1 \end{Bmatrix}\right).$$

A fundamental matrix is

$$\mathbf{X}(t)=\big[\mathbf{x}_1(t)\;\;\mathbf{x}_2(t)\big]=\begin{bmatrix} 2\mathrm{e}^{-t} & (2t+1)\mathrm{e}^{-t} \\ -\mathrm{e}^{-t} & -(t+1)\mathrm{e}^{-t} \end{bmatrix},\qquad \det(\mathbf{X})=-\mathrm{e}^{-2t},$$

and its inverse is obtained as

$$\mathbf{X}^{-1}(t)=\frac{1}{-c^{-2t}}\begin{bmatrix} -(t+1)\mathrm{e}^{-t} & -(2t+1)\mathrm{e}^{-t} \\ \mathrm{e}^{-t} & 2\mathrm{e}^{-t} \end{bmatrix}=\begin{bmatrix} (t+1)\mathrm{e}^{t} & (2t+1)\mathrm{e}^{t} \\ -\mathrm{e}^{t} & -2\mathrm{e}^{t} \end{bmatrix}.$$

It is easy to evaluate

$$\int \mathbf{X}^{-1}(t)\,\mathbf{f}(t)\,\mathrm{d}t=\int \begin{bmatrix} (t+1)\mathrm{e}^{t} & (2t+1)\mathrm{e}^{t} \\ -\mathrm{e}^{t} & -2\mathrm{e}^{t} \end{bmatrix}\begin{Bmatrix} 2\mathrm{e}^{-t} \\ 0 \end{Bmatrix}\mathrm{d}t$$

$$=\int \begin{Bmatrix} 2(t+1) \\ -2 \end{Bmatrix}\mathrm{d}t=\begin{Bmatrix} t^2+2t \\ -2t \end{Bmatrix}.$$

The general solution is

$$\mathbf{x}(t)=\mathbf{X}(t)\left\{\mathbf{C}+\int \mathbf{X}^{-1}(t)\,\mathbf{f}(t)\,\mathrm{d}t\right\}=\begin{bmatrix} 2\mathrm{e}^{-t} & (2t+1)\mathrm{e}^{-t} \\ -\mathrm{e}^{-t} & -(t+1)\mathrm{e}^{-t} \end{bmatrix}\begin{Bmatrix} C_1+t^2+2t \\ C_2-2t \end{Bmatrix},$$

$$x_1(t)=\mathrm{e}^{-t}\big[-2t^2+2(C_2+1)t+(2C_1+C_2)\big],\quad x_2(t)=\mathrm{e}^{-t}\big[t^2-C_2t-(C_1+C_2)\big].$$

Example 7.19

Solve

$$x_1'-x_1+x_2=\sec t,$$
$$2x_1-x_2'-x_2=0.$$

In the matrix form, the system of differential equations can be written as

$$\mathbf{x}'(t) = \mathbf{A}\,\mathbf{x}(t), \quad \mathbf{x}(t) = \begin{Bmatrix} x_1 \\ x_2 \end{Bmatrix}, \quad \mathbf{A} = \begin{bmatrix} -1 & 5 \\ -4 & -5 \end{bmatrix}, \quad \mathbf{f}(t) = \begin{Bmatrix} \sec t \\ 0 \end{Bmatrix}.$$

The characteristic equation is

$$\det(\mathbf{A} - \lambda\mathbf{I}) = \begin{vmatrix} 1-\lambda & -1 \\ 2 & -1-\lambda \end{vmatrix} = \lambda^2 + 1 = 0 \implies \lambda = \pm\mathrm{i}.$$

For eigenvalue $\lambda = \mathrm{i}$, the corresponding eigenvector is

$$(\mathbf{A} - \lambda\mathbf{I})\,\mathbf{v} = \begin{bmatrix} 1-\mathrm{i} & -1 \\ 4 & -1-\mathrm{i} \end{bmatrix} \begin{Bmatrix} v_1 \\ v_2 \end{Bmatrix} = \begin{Bmatrix} (1-\mathrm{i})\,v_1 - v_2 \\ 2v_1 - (1+\mathrm{i})\,v_2 \end{Bmatrix} = \begin{Bmatrix} 0 \\ 0 \end{Bmatrix}.$$

Taking $v_1 = 1$, then $v_2 = (1-\mathrm{i})\,v_1 = 1 - \mathrm{i}$,

$$\therefore \quad \mathbf{v} = \begin{Bmatrix} v_1 \\ v_2 \end{Bmatrix} = \begin{Bmatrix} 1 \\ 1-\mathrm{i} \end{Bmatrix} = \begin{Bmatrix} 1 \\ 1 \end{Bmatrix} + \mathrm{i} \begin{Bmatrix} 0 \\ -1 \end{Bmatrix}.$$

Hence, using Euler's formula $\mathrm{e}^{\mathrm{i}\,\theta} = \cos\theta + \mathrm{i}\,\sin\theta$,

$$\begin{aligned} \mathrm{e}^{\lambda t}\,\mathbf{v} &= (\cos t + \mathrm{i}\,\sin t) \left(\begin{Bmatrix} 1 \\ 1 \end{Bmatrix} + \mathrm{i} \begin{Bmatrix} 0 \\ -1 \end{Bmatrix} \right) \\ &= \begin{Bmatrix} 1 \\ 1 \end{Bmatrix} \cos t + \begin{Bmatrix} 0 \\ 1 \end{Bmatrix} \sin t + \mathrm{i} \left(\begin{Bmatrix} 0 \\ -1 \end{Bmatrix} \cos t + \begin{Bmatrix} 1 \\ 1 \end{Bmatrix} \sin t \right). \end{aligned}$$

Two linearly independent real-valued solutions are

$$\mathbf{x}_1(t) = \mathcal{Re}(\mathrm{e}^{\lambda t}\,\mathbf{v}) = \begin{Bmatrix} \cos t \\ \sin t + \cos t \end{Bmatrix}, \qquad \mathbf{x}_2(t) = \mathcal{Im}(\mathrm{e}^{\lambda t}\,\mathbf{v}) = \begin{Bmatrix} \sin t \\ \sin t - \cos t \end{Bmatrix}.$$

A fundamental matrix is

$$\mathbf{X}(t) = \big[\mathbf{x}_1(t) \;\; \mathbf{x}_2(t)\big] = \begin{bmatrix} \cos t & \sin t \\ \sin t + \cos t & \sin t - \cos t \end{bmatrix}, \qquad \det(\mathbf{X}) = -1,$$

and its inverse is obtained as

$$\mathbf{X}^{-1}(t) = \begin{bmatrix} \cos t - \sin t & \sin t \\ \cos t + \sin t & -\cos t \end{bmatrix}.$$

Evaluate the integral

$$\int \mathbf{X}^{-1}(t)\,\mathbf{f}(t)\,\mathrm{d}t = \int \begin{Bmatrix} 1 - \tan t \\ 1 + \tan t \end{Bmatrix} \mathrm{d}t = \begin{Bmatrix} t + \ln|\cos t| \\ t - \ln|\cos t| \end{Bmatrix}.$$

The general solution is

$$\mathbf{x}(t) = \mathbf{X}(t)\left\{\mathbf{C} + \int \mathbf{X}^{-1}(t)\,\mathbf{f}(t)\,\mathrm{d}t\right\}$$

$$= \begin{bmatrix} \cos t & \sin t \\ \sin t + \cos t & \sin t - \cos t \end{bmatrix} \begin{Bmatrix} C_1 + t + \ln|\cos t| \\ C_2 + t - \ln|\cos t| \end{Bmatrix},$$

$$\therefore \quad x_1(t) = (t+C_1)\cos t + (t+C_2)\sin t + (\cos t - \sin t)\ln|\cos t|,$$

$$x_2(t) = (C_1 - C_2)\cos t + (2t + C_1 + C_2)\sin t + 2\cos t\ln|\cos t|.$$

Example 7.20

Solve

$$\mathbf{x}'(t) = \mathbf{A}\mathbf{x}(t) + \mathbf{f}(t), \quad \mathbf{x}(t) = \begin{Bmatrix} x_1 \\ x_2 \\ x_3 \end{Bmatrix}, \quad \mathbf{A} = \begin{bmatrix} 2 & -1 & -1 \\ 2 & -1 & -2 \\ -1 & 1 & 2 \end{bmatrix}, \quad \mathbf{f}(t) = \begin{Bmatrix} 2e^t \\ 4e^{-t} \\ 0 \end{Bmatrix}.$$

The characteristic equation is

$$\det(\mathbf{A} - \lambda\mathbf{I}) = \begin{vmatrix} 2-\lambda & -1 & -1 \\ 2 & -1-\lambda & -2 \\ -1 & 1 & 2-\lambda \end{vmatrix} = -(\lambda^3 - 3\lambda^2 + 3\lambda - 1) = -(\lambda - 1)^3 = 0.$$

Hence, $\lambda = 1$ is an eigenvector of multiplicity 3. The eigenvector equation is

$$(\mathbf{A} - \lambda\mathbf{I})\mathbf{v} = \begin{bmatrix} 1 & -1 & -1 \\ 2 & -2 & -2 \\ -1 & 1 & 1 \end{bmatrix} \begin{Bmatrix} v_1 \\ v_2 \\ v_3 \end{Bmatrix} = \begin{Bmatrix} v_1 - v_2 - v_3 \\ 2(v_1 - v_2 - v_3) \\ -(v_1 - v_2 - v_3) \end{Bmatrix} = \begin{Bmatrix} 0 \\ 0 \\ 0 \end{Bmatrix},$$

which leads to $v_1 = v_2 + v_3$. As a result, there are two linearly independent eigenvectors. Taking $v_{21} = 1$ and $v_{31} = -1$, then $v_{11} = v_{21} + v_{31} = 0$,

$$\therefore \quad \mathbf{v}_1 = \begin{Bmatrix} v_{11} \\ v_{21} \\ v_{31} \end{Bmatrix} = \begin{Bmatrix} 0 \\ 1 \\ -1 \end{Bmatrix}.$$

However, $\mathbf{v}_2$ cannot be chosen arbitrarily; it has to satisfy a condition imposed by $\mathbf{v}_3$, which will be clear in a moment.

A third linearly independent eigenvector does not exist. Hence, matrix $\mathbf{A}$ is defective and a complete basis of eigenvectors is obtained by including one generalized eigenvector:

$$(\mathbf{A} - \lambda\mathbf{I})\mathbf{v}_3 = \mathbf{v}_2 \implies \begin{bmatrix} 1 & -1 & -1 \\ 2 & -2 & -2 \\ -1 & 1 & 1 \end{bmatrix} \begin{Bmatrix} v_{13} \\ v_{23} \\ v_{33} \end{Bmatrix} = \begin{Bmatrix} v_{13} - v_{23} - v_{33} \\ 2(v_{13} - v_{23} - v_{33}) \\ -(v_{13} - v_{23} - v_{33}) \end{Bmatrix} = \begin{Bmatrix} v_{12} \\ v_{22} \\ v_{32} \end{Bmatrix}.$$

If $v_{13}-v_{23}-v_{33}=v_{12}=a$, one must have $v_{22}=2a$, $v_{32}=-a$. Taking $a=1$, then $v_{12}=1$, $v_{22}=2$, $v_{32}=-1$. Taking $v_{13}=v_{23}=0$, then $v_{33}=v_{13}-v_{23}-a=-1$. Hence

$$\mathbf{v}_2=\begin{Bmatrix} v_{12}\\ v_{22}\\ v_{32}\end{Bmatrix}=\begin{Bmatrix} 1\\ 2\\ -1\end{Bmatrix},\qquad \mathbf{v}_3=\begin{Bmatrix} v_{13}\\ v_{23}\\ v_{33}\end{Bmatrix}=\begin{Bmatrix} 0\\ 0\\ -1\end{Bmatrix}.$$

Three linearly independent solutions are

$$\mathbf{x}_1(t)=\mathrm{e}^{\lambda t}\mathbf{v}_1=\mathrm{e}^t\begin{Bmatrix} 0\\ 1\\ -1\end{Bmatrix},\qquad \mathbf{x}_2(t)=\mathrm{e}^{\lambda t}\mathbf{v}_2=\mathrm{e}^t\begin{Bmatrix} 1\\ 2\\ -1\end{Bmatrix},$$

$$\mathbf{x}_3(t)=\mathrm{e}^{\lambda t}\big(\mathbf{v}_2 t+\mathbf{v}_3\big)=\mathrm{e}^t\left(\begin{Bmatrix} 1\\ 2\\ -1\end{Bmatrix}t+\begin{Bmatrix} 0\\ 0\\ -1\end{Bmatrix}\right).$$

A fundamental matrix is

$$\mathbf{X}(t)=\big[\mathbf{x}_1(t)\ \ \mathbf{x}_2(t)\ \ \mathbf{x}_3(t)\big]=\mathrm{e}^t\begin{bmatrix} 0 & 1 & t\\ 1 & 2 & 2t\\ -1 & -1 & -(t+1)\end{bmatrix}.$$

Apply the Gauss-Jordan method to find the inverse of fundamental matrix $\mathbf{X}$:

$$\left[\begin{array}{ccc|ccc} 0 & 1 & t & 1 & 0 & 0\\ 1 & 2 & 2t & 0 & 1 & 0\\ -1 & -1 & -(t+1) & 0 & 0 & 1\end{array}\right]\xrightarrow[\boxed{3}\times(-1)]{\text{Exchange }\boxed{1}\text{ and }\boxed{2}}\left[\begin{array}{ccc|ccc} 1 & 2 & 2t & 0 & 1 & 0\\ 0 & 1 & t & 1 & 0 & 0\\ 1 & 1 & t+1 & 0 & 0 & -1\end{array}\right]$$

$$\xrightarrow{\boxed{3}-\boxed{1}}\left[\begin{array}{ccc|ccc} 1 & 2 & 2t & 0 & 1 & 0\\ 0 & 1 & t & 1 & 0 & 0\\ 0 & -1 & 1-t & 0 & -1 & -1\end{array}\right]\xrightarrow[\boxed{3}+\boxed{2}]{\boxed{1}-\boxed{2}\times 2}\left[\begin{array}{ccc|ccc} 1 & 0 & 0 & -2 & 1 & 0\\ 0 & 1 & t & 1 & 0 & 0\\ 0 & 0 & 1 & 1 & -1 & -1\end{array}\right]$$

$$\xrightarrow{\boxed{2}-\boxed{3}\times t}\left[\begin{array}{ccc|ccc} 1 & 0 & 0 & -2 & 1 & 0\\ 0 & 1 & 0 & 1-t & t & t\\ 0 & 0 & 1 & 1 & -1 & -1\end{array}\right],$$

$$\mathbf{X}^{-1}(t)=\mathrm{e}^{-t}\begin{bmatrix} -2 & 1 & 0\\ 1-t & t & t\\ 1 & -1 & -1\end{bmatrix}.$$

Evaluate the integral

$$\int \mathbf{X}^{-1}(t)\,\mathbf{f}(t)\,\mathrm{d}t=\int\begin{bmatrix} -2\mathrm{e}^{-t} & \mathrm{e}^{-t} & 0\\ (1-t)\mathrm{e}^{-t} & t\mathrm{e}^{-t} & t\mathrm{e}^{-t}\\ \mathrm{e}^{-t} & -\mathrm{e}^{-t} & -\mathrm{e}^{-t}\end{bmatrix}\begin{Bmatrix} 2\mathrm{e}^t\\ 4\mathrm{e}^{-t}\\ 0\end{Bmatrix}\mathrm{d}t$$

$$= \int \begin{Bmatrix} -4+4e^{-2t} \\ 2-2t+4te^{-2t} \\ 2-4e^{-2t} \end{Bmatrix} dt = \begin{Bmatrix} -4t-2e^{-2t} \\ 2t-t^2-(2t+1)e^{-2t} \\ 2t+2e^{-2t} \end{Bmatrix}.$$

The general solution is

$$\mathbf{x}(t) = \mathbf{X}(t)\left\{\mathbf{C} + \int \mathbf{X}^{-1}(t)\,\mathbf{f}(t)\,dt\right\}$$

$$= \begin{bmatrix} 0 & e^t & te^t \\ e^t & 2e^t & 2te^t \\ -e^t & -e^t & -(t+1)e^t \end{bmatrix} \begin{Bmatrix} C_1-4t-2e^{-2t} \\ C_2-t^2+2t-(2t+1)e^{-2t} \\ C_3+2t+2e^{-2t} \end{Bmatrix},$$

$$\therefore \quad x_1(t) = \left[C_2+C_3t+t^2+2t\right]e^t - e^{-t},$$

$$x_2(t) = \left[C_1+2C_2+2C_3t+2t^2\right]e^t - 4e^{-t},$$

$$x_3(t) = -\left[C_1+C_2+C_3(t+1)+t^2\right]e^t + e^{-t}.$$

Example 7.21

Solve

$$x_1' = 3x_1 - 3x_2 + x_3 + 2e^t,$$

$$x_2' = 3x_1 - 2x_2 + 2x_3,$$

$$x_3' = -x_1 + 2x_2, \qquad x_1(0) = 3,\ x_2(0) = 2,\ x_3(0) = 1.$$

In the matrix form, the system of differential equations can be written as

$$\mathbf{x}'(t) = \mathbf{A}\mathbf{x}(t) + \mathbf{f}(t), \quad \mathbf{x}(t) = \begin{Bmatrix} x_1 \\ x_2 \\ x_3 \end{Bmatrix}, \quad \mathbf{A} = \begin{bmatrix} 3 & -3 & 1 \\ 3 & -2 & 2 \\ -1 & 2 & 0 \end{bmatrix}, \quad \mathbf{f}(t) = \begin{Bmatrix} 2e^t \\ 0 \\ 0 \end{Bmatrix}.$$

The characteristic equation is

$$\det(\mathbf{A}-\lambda\mathbf{I}) = \begin{vmatrix} 3-\lambda & -3 & 1 \\ 2 & -2-\lambda & 2 \\ -1 & 2 & -\lambda \end{vmatrix} = -\lambda^3 + \lambda^2 - 2$$

$$= -(\lambda+1)(\lambda^2-2\lambda+2) = 0,$$

which gives the eigenvalues $\lambda = -1,\ 1\pm i$.

(1) $\lambda_1 = -1$:

$$(\mathbf{A}-\lambda_1\mathbf{I})\mathbf{v}_1 = \begin{bmatrix} 4 & -3 & 1 \\ 3 & -1 & 2 \\ -1 & 2 & 1 \end{bmatrix} \begin{Bmatrix} v_{11} \\ v_{21} \\ v_{31} \end{Bmatrix} \implies \begin{cases} 4v_{11}-3v_{21}+v_{31} = 0, & (1) \\ -v_{11}+2v_{21}+v_{31} = 0. & (2) \end{cases}$$

Eqn (1)–Eqn (2): $5v_{11}-5v_{21} = 0$. Taking $v_{21} = 1 \Longrightarrow v_{11} = v_{21} = 1$.

From Eqn (1): $v_{31}=3v_{21}-4v_{11}=-1$. Hence

$$\mathbf{v}_1=\begin{Bmatrix} v_{11}\\ v_{21}\\ v_{31}\end{Bmatrix}=\begin{Bmatrix} 1\\ 1\\ -1\end{Bmatrix}.$$

(2) $\lambda=1+\mathrm{i}$:

$$\begin{bmatrix} 2-\mathrm{i} & -3 & 1\\ 3 & -3-\mathrm{i} & 2\\ -1 & 2 & -1-\mathrm{i}\end{bmatrix}\begin{Bmatrix} v_1\\ v_2\\ v_3\end{Bmatrix}\implies\begin{cases} 3v_1-(3+\mathrm{i})\,v_2+2v_3=0, & (3)\\ -v_1+2v_2-(1+\mathrm{i})\,v_3=0. & (4)\end{cases}$$

Eqn (3) + 3 × Eqn (4): $(3-\mathrm{i})\,v_2-(1+3\mathrm{i})\,v_3=0$.

Taking $v_3=1 \implies v_2=\dfrac{1+3\mathrm{i}}{3-\mathrm{i}}=\dfrac{(1+3\mathrm{i})(3+\mathrm{i})}{(3-\mathrm{i})(3+\mathrm{i})}=\mathrm{i}$.

From Eqn (4): $v_1=2v_2-(1+\mathrm{i})\,v_3=2\mathrm{i}-(1+\mathrm{i})=-1+\mathrm{i}$. Hence

$$\mathbf{v}=\begin{Bmatrix} v_1\\ v_2\\ v_3\end{Bmatrix}=\begin{Bmatrix} -1+\mathrm{i}\\ \mathrm{i}\\ 1\end{Bmatrix},$$

$$\mathrm{e}^{\lambda t}\mathbf{v}=\mathrm{e}^{t}(\cos t+\mathrm{i}\sin t)\left(\begin{Bmatrix} -1\\ 0\\ 1\end{Bmatrix}+\mathrm{i}\begin{Bmatrix} 1\\ 1\\ 0\end{Bmatrix}\right)$$

$$=\mathrm{e}^{t}\left[\begin{Bmatrix} -1\\ 0\\ 1\end{Bmatrix}\cos t-\begin{Bmatrix} 1\\ 1\\ 0\end{Bmatrix}\sin t+\mathrm{i}\left(\begin{Bmatrix} 1\\ 1\\ 0\end{Bmatrix}\cos t+\begin{Bmatrix} -1\\ 0\\ 1\end{Bmatrix}\sin t\right)\right].$$

A fundamental matrix is

$$\mathbf{X}(t)=\left[\mathrm{e}^{\lambda_1 t}\mathbf{v}_1(t)\quad \mathcal{Re}(\mathrm{e}^{\lambda t}\mathbf{v})\quad \mathcal{Im}(\mathrm{e}^{\lambda t}\mathbf{v})\right]$$

$$=\begin{bmatrix} \mathrm{e}^{-t} & -\mathrm{e}^{t}(\cos t+\sin t) & \mathrm{e}^{t}(\cos t-\sin t)\\ \mathrm{e}^{-t} & -\mathrm{e}^{t}\sin t & \mathrm{e}^{t}\cos t\\ -\mathrm{e}^{-t} & \mathrm{e}^{t}\cos t & \mathrm{e}^{t}\sin t\end{bmatrix}.$$

Apply the Gauss-Jordan method to find the inverse of fundamental matrix, $\mathbf{X}^{-1}(t)$:

$$\left[\begin{array}{ccc|ccc} \mathrm{e}^{-t} & -\mathrm{e}^{t}(\cos t+\sin t) & \mathrm{e}^{t}(\cos t-\sin t) & 1 & 0 & 0\\ \mathrm{e}^{-t} & -\mathrm{e}^{t}\sin t & \mathrm{e}^{t}\cos t & 0 & 1 & 0\\ -\mathrm{e}^{-t} & \mathrm{e}^{t}\cos t & \mathrm{e}^{t}\sin t & 0 & 0 & 1\end{array}\right]$$

$$\xrightarrow[\boxed{3}+\boxed{1}]{\boxed{2}-\boxed{1}}\left[\begin{array}{ccc|ccc} \mathrm{e}^{-t} & -\mathrm{e}^{t}(\cos t+\sin t) & \mathrm{e}^{t}(\cos t-\sin t) & 1 & 0 & 0\\ 0 & \mathrm{e}^{t}\cos t & \mathrm{e}^{t}\sin t & -1 & 1 & 0\\ 0 & -\mathrm{e}^{t}\sin t & \mathrm{e}^{t}\cos t & 1 & 0 & 1\end{array}\right].$$

$$\xrightarrow[\boxed{3}\times\cos t+\boxed{2}\times\sin t]{\boxed{1}+\boxed{2}-\boxed{3}}\left[\begin{array}{ccc|ccc} e^{-t} & 0 & 0 & -1 & 1 & -1\\ 0 & e^t\cos t & e^t\sin t & -1 & 1 & 0\\ 0 & 0 & e^t & \cos t-\sin t & \sin t & \cos t\end{array}\right]$$

$$\xrightarrow{\boxed{2}-\boxed{3}\times\sin t}\left[\begin{array}{ccc|ccc} e^{-t} & 0 & 0 & -1 & 1 & -1\\ 0 & e^t\cos t & 0 & -1-\cos t\sin t+\sin^2 t & 1-\sin^2 t & -\sin t\cos t\\ 0 & 0 & e^t & \cos t-\sin t & \sin t & \cos t\end{array}\right]$$

$$\xrightarrow{\substack{\boxed{1}\times e^t\\ \boxed{2}\times e^{-t}/\cos t\\ \boxed{3}\times e^{-t}}}\left[\begin{array}{ccc|ccc} 1 & 0 & 0 & -e^t & e^t & -e^t\\ 0 & 1 & 0 & -e^{-t}(\cos t+\sin t) & e^{-t}\cos t & e^{-t}\sin t\\ 0 & 0 & 1 & -e^{-t}(\cos t-\sin t) & e^{-t}\sin t & e^{-t}\cos t\end{array}\right].$$

Evaluate the integral

$$\int \mathbf{X}^{-1}(t)\,\mathbf{f}(t)\,dt=\int\begin{bmatrix} -e^t & e^t & -e^t\\ -e^{-t}(\cos t+\sin t) & e^{-t}\cos t & e^{-t}\sin t\\ -e^{-t}(\cos t-\sin t) & e^{-t}\sin t & e^{-t}\cos t\end{bmatrix}\begin{Bmatrix}2e^t\\0\\0\end{Bmatrix}dt$$

$$=\int\begin{Bmatrix}-2e^{2t}\\ -2(\cos t+\sin t)\\ 2(\cos t-\sin t)\end{Bmatrix}dt=\begin{Bmatrix}-e^{2t}\\ -2(\sin t-\cos t)\\ 2(\sin t+\cos t)\end{Bmatrix}.$$

Given the initial condition $\mathbf{x}(0)$, the vector $\mathbf{C}$ is given by

$$\mathbf{C}=\mathbf{X}^{-1}(0)\,\mathbf{x}(0)-\left[\int\mathbf{X}^{-1}(t)\,\mathbf{f}(t)\,dt\right]_{t=0}$$

$$=\begin{bmatrix}-1 & 1 & -1\\ -1 & 1 & 0\\ 1 & 0 & 1\end{bmatrix}\begin{Bmatrix}3\\2\\1\end{Bmatrix}-\begin{Bmatrix}-1\\2\\2\end{Bmatrix}=\begin{Bmatrix}-1\\-3\\2\end{Bmatrix}.$$

The solution satisfying the initial condition is

$$\mathbf{x}(t)=\mathbf{X}(t)\left\{\mathbf{C}+\int\mathbf{X}^{-1}(t)\,\mathbf{f}(t)\,dt\right\}$$

$$=\begin{bmatrix} e^{-t} & -e^t(\cos t+\sin t) & e^t(\cos t-\sin t)\\ e^{-t} & -e^t\sin t & e^t\cos t\\ -e^{-t} & e^t\cos t & e^t\sin t\end{bmatrix}\begin{Bmatrix}-1-e^{2t}\\ -3-2(\sin t-\cos t)\\ 2+2(\sin t+\cos t)\end{Bmatrix}$$

$$=\begin{Bmatrix}-e^{-t}+e^t(5\cos t+\sin t-1)\\ -e^{-t}+e^t(2\cos t+3\sin t+1)\\ e^{-t}+e^t(-3\cos t+2\sin t+3)\end{Bmatrix}.$$

7.4.3 Response of Multiple Degrees-of-Freedom Systems

As shown in Section 7.1, the equations of motion of an n degrees-of-freedom system occur naturally as a system of n coupled second-order linear differential equations of the form

$$\mathbf{M}\ddot{\mathbf{x}}(t)+\mathbf{C}\dot{\mathbf{x}}(t)+\mathbf{K}\mathbf{x}(t)=\mathbf{F}(t), \tag{1}$$

where $\mathbf{M}$, $\mathbf{C}$, $\mathbf{K}$ are the mass, damping, and stiffness matrices of dimension $n\times n$, respectively, and $F(t)$ is the load vector of dimension n. Matrices $\mathbf{M}$ and $\mathbf{K}$ are symmetric, i.e., $\mathbf{M}^T=\mathbf{M}$, $\mathbf{K}^T=\mathbf{K}$, and positive definite.

Instead of converting system (1) to a system of $2n$ first-order differential equations, it is more convenient and physically meaningful to study system (1) directly.

Undamped Free Vibration

The equations of motion of undamped free vibration, or complementary solution, are given by

$$\mathbf{M}\ddot{\mathbf{x}}(t)+\mathbf{K}\mathbf{x}(t)=\mathbf{0}. \tag{2}$$

Seeking a solution of the form $\mathbf{x}(t)=\hat{\mathbf{x}}\sin(\omega t+\theta)$ and substituting into equation (2) yield

$$(\mathbf{K}-\omega^2\mathbf{M})\,\hat{\mathbf{x}}\sin(\omega t+\theta)=\mathbf{0}.$$

Since $\sin(\omega t+\theta)$ is not identically zero, one must have

$$(\mathbf{K}-\omega^2\mathbf{M})\,\hat{\mathbf{x}}=\mathbf{0}. \tag{3}$$

Equation (3) is a system of n homogeneous linear *algebraic* equations. To have nonzero solutions for $\hat{\mathbf{x}}$, the determinant of the coefficient matrix must be zero:

$$\det(\mathbf{K}-\omega^2\mathbf{M})=0, \tag{4}$$

which leads to the *characteristic equation*, a polynomial equation in ω^2 of degree n. Equation (4) is also called the *frequency equation*. Since the mass and stiffness matrices $\mathbf{M}$ and $\mathbf{K}$ are symmetric and positive definite, it can be shown that all roots ω^2 of the frequency equation are real and positive. The ith root ω_i, which is called the ith eigenvalue ($\omega_1<\omega_2<\cdots<\omega_n$), is the natural circular frequency of the ith mode of the system or the ith *modal frequency*.

Corresponding to the ith eigenvalue ω_i, a nonzero solution $\hat{\mathbf{x}}_i$ of system (3),

$$(\mathbf{K}-\omega_i^2\mathbf{M})\,\hat{\mathbf{x}}_i=\mathbf{0}, \qquad i=1,2,\ldots,n, \tag{5}$$

is the ith eigenvector or the ith *mode shape*.

The response of the undamped free vibration is then given by

$$\mathbf{x}(t)=a_1\mathbf{x}_1\sin(\omega_1t+\theta_1)+a_2\mathbf{x}_2\sin(\omega_2t+\theta_2)+\cdots+a_n\mathbf{x}_n\sin(\omega_nt+\theta_n),$$

where the $2n$ constants $a_1, a_2, \ldots, a_n, \theta_1, \theta_2, \ldots, \theta_n$ are determined using the initial conditions $\mathbf{x}(0)$ and $\dot{\mathbf{x}}(0)$.

Orthogonality of Mode Shapes

Corresponding to the ith mode, equation (3) becomes

$$\mathbf{K}\hat{\mathbf{x}}_i = \omega_i^2 \mathbf{M}\hat{\mathbf{x}}_i.$$

Multiplying this equation by $\hat{\mathbf{x}}_j^T$ from the left yields

$$\hat{\mathbf{x}}_j^T \mathbf{K}\hat{\mathbf{x}}_i = \omega_i^2 \hat{\mathbf{x}}_j^T \mathbf{M}\hat{\mathbf{x}}_i. \tag{6}$$

Similarly, corresponding to the jth mode, equation (3) becomes

$$\mathbf{K}\hat{\mathbf{x}}_j = \omega_j^2 \mathbf{M}\hat{\mathbf{x}}_j.$$

Multiplying this equation by $\hat{\mathbf{x}}_i^T$ from the left gives

$$\hat{\mathbf{x}}_i^T \mathbf{K}\hat{\mathbf{x}}_j = \omega_j^2 \hat{\mathbf{x}}_i^T \mathbf{M}\hat{\mathbf{x}}_j. \tag{7}$$

Taking transpose of both sides of equation (7) leads to

$$\hat{\mathbf{x}}_j^T \mathbf{K}^T (\hat{\mathbf{x}}_i^T)^T = \omega_j^2 \hat{\mathbf{x}}_j^T \mathbf{M}^T (\hat{\mathbf{x}}_i^T)^T \implies \hat{\mathbf{x}}_j^T \mathbf{K}\hat{\mathbf{x}}_i = \omega_j^2 \hat{\mathbf{x}}_j^T \mathbf{M}\hat{\mathbf{x}}_i. \tag{7'}$$

Subtracting equation (7′) from equation (6) results in

$$(\omega_i^2 - \omega_j^2)\hat{\mathbf{x}}_j^T \mathbf{M}\hat{\mathbf{x}}_i = 0.$$

Since $\omega_i \neq \omega_j$ for different modes $i \neq j$, one has

$$\hat{\mathbf{x}}_j^T \mathbf{M}\hat{\mathbf{x}}_i = 0, \quad i \neq j.$$

From equation (6), one has

$$\hat{\mathbf{x}}_j^T \mathbf{K}\hat{\mathbf{x}}_i = 0, \quad i \neq j.$$

The orthogonality conditions can be written as

$$\hat{\mathbf{x}}_j^T \mathbf{M}\hat{\mathbf{x}}_i = \begin{cases} 0, & j \neq i, \\ m_i, & j = i, \end{cases} \qquad \hat{\mathbf{x}}_j^T \mathbf{K}\hat{\mathbf{x}}_i = \begin{cases} 0, & j \neq i, \\ \omega_i^2 m_i, & j = i. \end{cases} \tag{8}$$

Construct the *modal matrix* from the eigenvectors (mode shapes) $\hat{\mathbf{x}}_1, \hat{\mathbf{x}}_2, \ldots, \hat{\mathbf{x}}_n$ as

$$\mathbf{\Phi} = [\hat{\mathbf{x}}_1, \hat{\mathbf{x}}_2, \ldots, \hat{\mathbf{x}}_n].$$

The orthogonality conditions can then be written as

$$\mathbf{\Phi}^T \mathbf{M} \mathbf{\Phi} = \begin{bmatrix} m_1 & & & \\ & m_2 & & \\ & & \ddots & \\ & & & m_n \end{bmatrix}, \quad \mathbf{\Phi}^T \mathbf{K} \mathbf{\Phi} = \begin{bmatrix} m_1\omega_1^2 & & & \\ & m_2\omega_2^2 & & \\ & & \ddots & \\ & & & m_n\omega_n^2 \end{bmatrix}. \tag{9}$$

Undamped Forced Vibration

The equation of motion of the undamped forced vibration is

$$\mathbf{M}\ddot{\mathbf{x}}(t) + \mathbf{K}\mathbf{x}(t) = \mathbf{F}(t). \tag{10}$$

Letting $\mathbf{x}(t) = \boldsymbol{\Phi}\mathbf{q}(t)$, substituting into equation (10) and multiplying $\boldsymbol{\Phi}^T$ from the left yields

$$\boldsymbol{\Phi}^T\mathbf{M}\boldsymbol{\Phi}\ddot{\mathbf{q}}(t) + \boldsymbol{\Phi}^T\mathbf{K}\boldsymbol{\Phi}\mathbf{q}(t) = \boldsymbol{\Phi}^T\mathbf{F}(t).$$

Using the orthogonality conditions (9), one obtains a set of n uncoupled single degree-of-freedom systems

$$m_i\ddot{q}_i(t) + m_i\omega_i^2 q_i(t) = f_i(t), \qquad i = 1, 2, \ldots, n, \tag{11}$$

where $f_i(t) = \hat{\mathbf{x}}_i^T\mathbf{F}(t)$. Each of equations (11) can be solved using the methods presented in Chapters 5 and 6.

Having obtained $q_i(t)$, the response of the undamped forced vibration is

$$\mathbf{x}(t) = \boldsymbol{\Phi}\mathbf{q}(t) = q_1(t)\,\hat{\mathbf{x}}_1 + q_2(t)\,\hat{\mathbf{x}}_2 + \cdots + q_n(t)\,\hat{\mathbf{x}}_n = \sum_{i=1}^{n} q_i(t)\,\hat{\mathbf{x}}_i. \tag{12}$$

Multiplying equation (12) by $\hat{\mathbf{x}}_j^T\mathbf{M}$ from the left yields

$$\hat{\mathbf{x}}_j^T\mathbf{M}\mathbf{x}(t) = \sum_{i=1}^{n} q_i(t)\,\hat{\mathbf{x}}_j^T\mathbf{M}\hat{\mathbf{x}}_i = m_j\,q_j(t), \qquad j = 1, 2, \ldots, n.$$

Hence, the initial conditions $q_i(0)$ and $\dot{q}_i(0)$ are then given by

$$q_i(0) = \frac{\hat{\mathbf{x}}_i^T\mathbf{M}\mathbf{x}(0)}{m_i}, \qquad \dot{q}_i(0) = \frac{\hat{\mathbf{x}}_i^T\mathbf{M}\dot{\mathbf{x}}(0)}{m_i}. \tag{13}$$

Damped Forced Vibration

The same approach for undamped system can be applied for damped system. Letting $\mathbf{x}(t) = \boldsymbol{\Phi}\mathbf{q}(t)$, substituting into the equation of motion (1) of damped system and multiplying $\boldsymbol{\Phi}^T$ from the left yields

$$\boldsymbol{\Phi}^T\mathbf{M}\boldsymbol{\Phi}\ddot{\mathbf{q}}(t) + \boldsymbol{\Phi}^T\mathbf{C}\boldsymbol{\Phi}\dot{\mathbf{q}}(t) + \boldsymbol{\Phi}^T\mathbf{K}\boldsymbol{\Phi}\mathbf{q}(t) = \boldsymbol{\Phi}^T\mathbf{F}(t).$$

Assuming that the orthogonality condition applies to the damping matrix

$$\hat{\mathbf{x}}_j^T\mathbf{C}\,\hat{\mathbf{x}}_i = \begin{cases} 0, & j \neq i, \\ c_i, & j = i, \end{cases}$$

the equations of motion are then decoupled

$$m_i\ddot{q}_i(t) + c_i\dot{q}_i(t) + m_i\omega_i^2 q_i(t) = f_i(t), \qquad i = 1, 2, \ldots, n,$$

or, in the standard form,

$$\ddot{q}_i(t) + 2\zeta_i\omega_i\dot{q}_i(t) + \omega_i^2 q_i(t) = \frac{f_i(t)}{m_i}, \qquad i = 1, 2, \ldots, n, \tag{14}$$

where $\zeta_i = c_i/(2\omega_i m_i)$ is the ith *modal damping coefficient* (ratio).

Remarks: In practice, it is often not practical to set up the damping matrix **C** by evaluating its elements. It is generally more convenient and physically reasonable to define the damping of a multiple degrees-of-freedom system by specifying the modal damping coefficients ζ_i, $i = 1, 2, \ldots, n$, because the modal damping coefficient ζ_i can be determined experimentally or estimated with adequate precision in many engineering applications.

A more detailed discussion of damping in a multiple degrees-of-freedom system is beyond the scope of this book, and can be found in standard textbooks on structural dynamics.

Except for the damping terms, the procedure of analysis for damped forced vibration is the same as that for undamped forced vibration.

An example of vibration of a two-story shear building is present in Section 8.4.

7.5 Summary

In this chapter, three methods, i.e., the method of operator, the method of Laplace transform, and the matrix method, are introduced for solving systems of linear ordinary differential equations.

7.5.1 The Method of Operator

The method of operator is an extension of the approach presented in Chapter 4 for nth-order linear ordinary differential equations with constant coefficients.

Consider a system of linear ordinary differential equations with independent variable t and n dependent variables $x_1, x_2, \ldots, x_n$,

$$\phi_{11}(D)\,x_1 + \phi_{12}(D)\,x_2 + \cdots + \phi_{1n}(D)\,x_n = f_1(t), \qquad D(\cdot) \equiv \mathrm{d}(\cdot)/\mathrm{d}t,$$

$$\phi_{21}(D)\,x_1 + \phi_{22}(D)\,x_2 + \cdots + \phi_{2n}(D)\,x_n = f_2(t),$$

$$\cdots\cdots$$

$$\phi_{n1}(D)\,x_1 + \phi_{n2}(D)\,x_2 + \cdots + \phi_{nn}(D)\,x_n = f_n(t).$$

The determinant of the coefficient matrix is

$$\phi(D) = \left|\phi_{ij}(D)\right|.$$

Suppose the order of operator $\phi(D)$ is N.

Complementary Solutions

The characteristic equation is $\phi(\lambda)=0$, which is a polynomial equation of degree N and gives N characteristic numbers $\lambda_1, \lambda_2, \ldots, \lambda_N$. The complementary solutions for $x_1, x_2, \ldots, x_n$ all have the same form and can easily be obtained from the characteristic numbers, denoted as $x_{iC}(t;\, C_{i1}, C_{i2}, \ldots, C_{iN})$, $i=1, 2, \ldots, n$.

Particular Solutions

When the right-hand side functions $f_i(t)$, $i=1, 2, \ldots, N$, are of the form

$$e^{\alpha t}\big[(a_0 + a_1 t + \cdots + a_k t^k)\cos\beta t + (b_0 + b_1 t + \cdots + b_k t^k)\sin\beta t\big],$$

it is advantageous to obtain particular solutions using the method of D-operator

$$x_{iP}(t) = \frac{\Delta_i(t)}{\phi(D)}, \qquad i = 1, 2, \ldots, n,$$

where $\Delta_i(t)$ is obtained by replacing the ith column of $\phi(D)$ with the right-hand side vector.

Otherwise, particular solutions have to be obtained using the method of variation of parameters.

General Solutions

The general solutions are

$$x_i(t) = x_{iC}(t;\, C_{i1}, C_{i2}, \ldots, C_{iN}) + x_{iP}(t), \qquad i = 1, 2, \ldots, n.$$

Since the order of $\phi(D)$ is N, there should only be N arbitrary constants in the general solutions. The extra constants can be eliminated by substituting the general solutions into one of the original differential equations.

7.5.2 The Method of Laplace Transform

The procedure of the method of Laplace transform for solving systems of linear ordinary differential equations is the most straightforward.

Applying the Laplace transform to a system of linear differential equations converts it to a system of linear algebraic equations for the Laplace transforms, which can easily be solved using Gaussian elimination or Cramer's Rule. The solutions of the system of linear ordinary differential equations can then be obtained by finding the inverse Laplace transforms.

The method of Laplace transform is advantageous in solving linear differential equations with the right-hand side functions involving the Heaviside step function and the Dirac delta function. The restriction of this method is that the Laplace transform of the right-hand sides should be easily obtained.

7.5.3 The Matrix Method

The matrix method is the most general and systematic approach, especially in dealing with systems of higher dimensions. However, this method is the most difficult to master because of the challenging concepts in eigenvalues and eigenvectors, particularly when multiple eigenvalues are involved.

Rewrite a system of linear ordinary differential equations in the standard form of a system of n first-order linear ordinary differential equations

$$\mathbf{x}'(t)=\mathbf{A}\mathbf{x}(t)+\mathbf{f}(t),\quad \mathbf{x}(t)=\begin{Bmatrix} x_1 \\ x_2 \\ \vdots \\ x_n \end{Bmatrix},\quad \mathbf{A}=\begin{bmatrix} a_{11} & a_{12} & \cdots & a_{1n} \\ a_{21} & a_{22} & \cdots & a_{2n} \\ \vdots & \vdots & \cdots & \vdots \\ a_{n1} & a_{n2} & \cdots & a_{nn} \end{bmatrix},\quad \mathbf{f}(t)=\begin{Bmatrix} f_1(t) \\ f_2(t) \\ \vdots \\ f_n(t) \end{Bmatrix}.$$

Complementary solutions are the solutions of the homogeneous differential equations with $\mathbf{f}(t)=\mathbf{0}$ being set to zero: $\mathbf{x}'(t)=\mathbf{A}\mathbf{x}(t)$. The characteristic equation of matrix $\mathbf{A}$ is

$$\det(\mathbf{A}-\lambda\mathbf{I})=0,$$

which is a polynomial equation in λ of degree n. A solution of the characteristic equation λ_k is called an eigenvalue and the corresponding eigenvector $\mathbf{v}_k$ is given by

$$(\mathbf{A}-\lambda_k\mathbf{I})\,\mathbf{v}_k=\mathbf{0}.$$

Case 1. Distinct Eigenvalues

When the n solutions $\lambda_1, \lambda_2, \ldots, \lambda_n$ of the characteristic equations are distinct, the n corresponding eigenvectors $\mathbf{v}_1, \mathbf{v}_2, \ldots, \mathbf{v}_n$ will be linearly independent. n linearly independent solutions for the homogeneous system are

$$\mathbf{x}_1(t)=e^{\lambda_1 t}\mathbf{v}_1,\quad \mathbf{x}_2(t)=e^{\lambda_2 t}\mathbf{v}_2,\quad \ldots,\quad \mathbf{x}_n(t)=e^{\lambda_n t}\mathbf{v}_n.$$

Case 2. Complex Eigenvalues

Suppose $\lambda=\alpha+i\beta$ is an eigenvalue with the corresponding eigenvector $\mathbf{v}$. Then $\bar{\lambda}=\alpha-i\beta$ is also an eigenvalue. Corresponding to the eigenvalues $\alpha\pm i\beta$, two linearly independent real-valued solutions of the homogeneous system are

$$\mathbf{x}_1(t)=\mathcal{Re}(e^{\lambda t}\mathbf{v})=e^{\alpha t}\big[\mathcal{Re}(\mathbf{v})\cos\beta t-\mathcal{Im}(\mathbf{v})\sin\beta t\big],$$

$$\mathbf{x}_2(t)=\mathcal{Im}(e^{\lambda t}\mathbf{v})=e^{\alpha t}\big[\mathcal{Re}(\mathbf{v})\sin\beta t+\mathcal{Im}(\mathbf{v})\cos\beta t\big].$$

Case 3. Multiple Eigenvalues

Suppose λ is an eigenvalue of multiplicity m, and there are k linearly independent eigenvectors corresponding to λ. If $k<m$, then matrix $\mathbf{A}$ is defective and a complete basis of eigenvectors is obtained by including $m-k$ generalized eigenvectors

obtained as follows:

$$
\begin{aligned}
&(\mathbf{A}-\lambda\mathbf{I})\,\mathbf{v}_i=\mathbf{0} &&\Longrightarrow \mathbf{v}_i,\ i=1,2,\ldots,k, \text{ linearly independent eigenvectors,}\\
&(\mathbf{A}-\lambda\mathbf{I})\,\mathbf{v}_{k+1}=\mathbf{v}_k &&\Longrightarrow (\mathbf{A}-\lambda\mathbf{I})^2\mathbf{v}_{k+1}=\mathbf{0},\\
&(\mathbf{A}-\lambda\mathbf{I})\,\mathbf{v}_{k+2}=\mathbf{v}_{k+1} &&\Longrightarrow (\mathbf{A}-\lambda\mathbf{I})^3\mathbf{v}_{k+2}=\mathbf{0}, \quad \text{Generalized eigenvectors}\\
&\quad\vdots\\
&(\mathbf{A}-\lambda\mathbf{I})\,\mathbf{v}_m=\mathbf{v}_{m-1} &&\Longrightarrow (\mathbf{A}-\lambda\mathbf{I})^{m-k+1}\mathbf{v}_m=\mathbf{0}.
\end{aligned}
$$

m linearly independent solutions of the homogeneous system are

$$
\begin{aligned}
\mathbf{x}_i(t)&=\mathrm{e}^{\lambda t}\,\mathbf{v}_i,\quad i=1,2,\ldots,k,\\
\mathbf{x}_{k+1}(t)&=\mathrm{e}^{\lambda t}\left(\mathbf{v}_k t+\mathbf{v}_{k+1}\right),\\
\mathbf{x}_{k+2}(t)&=\mathrm{e}^{\lambda t}\left(\mathbf{v}_k\frac{t^2}{2!}+\mathbf{v}_{k+1}t+\mathbf{v}_{k+2}\right),\\
&\ \ \vdots\\
\mathbf{x}_m(t)&=\mathrm{e}^{\lambda t}\left[\mathbf{v}_k\frac{t^{m-k}}{(m-k)!}+\mathbf{v}_{k+1}\frac{t^{m-k-1}}{(m-k-1)!}+\cdots+\mathbf{v}_{m-2}\frac{t^2}{2!}+\mathbf{v}_{m-1}t+\mathbf{v}_m\right].
\end{aligned}
$$

Complementary Solutions

In general, if $\mathbf{x}_1(t), \mathbf{x}_2(t), \ldots, \mathbf{x}_n(t)$ are n linearly independent solutions of the homogeneous system, a fundamental matrix is

$$\mathbf{X}(t)=\left[\mathbf{x}_1(t)\quad \mathbf{x}_2(t)\quad \cdots\quad \mathbf{x}_n(t)\right].$$

The complementary solution is given by

$$\mathbf{x}_C(t)=\mathbf{X}(t)\,\mathbf{C}=C_1\,\mathbf{x}_1(t)+C_2\,\mathbf{x}_2(t)+\cdots+C_n\,\mathbf{x}_n(t).$$

General Solutions

A particular solution is $\mathbf{x}_P(t)=\mathbf{X}(t)\displaystyle\int\mathbf{X}^{-1}(t)\,\mathbf{f}(t)\,\mathrm{d}t$.

The general solution of the nonhomogeneous system is given by

$$\mathbf{x}(t)=\mathbf{x}_C(t)+\mathbf{x}_P(t)=\mathbf{X}(t)\left\{\mathbf{C}+\int\mathbf{X}^{-1}(t)\,\mathbf{f}(t)\,\mathrm{d}t\right\}.$$

If the nonhomogeneous system satisfies the initial condition $\mathbf{x}(t_0)=\mathbf{x}_0$, then

$$\mathbf{x}(t)=\mathbf{X}(t)\left\{\mathbf{X}^{-1}(t_0)\,\mathbf{x}_0+\int_{t_0}^{t}\mathbf{X}^{-1}(t)\,\mathbf{f}(t)\,\mathrm{d}t\right\}.$$

Problems

The Method of Operator

Solve the following systems of differential equations using the method of operator.

7.1 $(D+2)x-y=0,\quad x+(D-2)y=0,\quad D(\cdot)\equiv d(\cdot)/dt$

ANS $x=C_1e^{\sqrt{3}t}+C_2e^{-\sqrt{3}t},\ y=(2+\sqrt{3})C_1e^{\sqrt{3}t}+(2-\sqrt{3})C_2e^{-\sqrt{3}t}$

7.2 $(2D+1)x-(5D+4)y=0,\ (3D-2)x-(4D-1)y=0,\ D(\cdot)\equiv d(\cdot)/dt$

ANS $x=3C_1e^{t}+C_2e^{-t},\ \ y=C_1e^{t}+C_2e^{-t}$

7.3 $(D-1)x+3y=0,\quad 3x-(D-1)y=0,\quad D(\cdot)\equiv d(\cdot)/dt$

ANS $x=e^{t}(A\cos 3t+B\sin 3t),\ \ y=e^{t}(A\sin 3t-B\cos 3t)$

7.4 $(D^2+D)x+(D-2)y=0,\quad (D+1)x-Dy=0,\quad D(\cdot)\equiv d(\cdot)/dt$

ANS $x=C_1e^{t}+C_2e^{-t}+C_3e^{-2t},\quad y=2C_1e^{t}+\frac{1}{2}C_3e^{-2t}$

7.5 $(D^2-3)x-4y=0,\quad x+(D^2+1)y=0,\quad D(\cdot)\equiv d(\cdot)/dt$

ANS $x=-2e^{t}(C_1+C_2+C_2t)-2e^{-t}(C_3-C_4+C_4t)$

$y=e^{t}(C_1+C_2t)+e^{-t}(C_3+C_4t)$

7.6 $\dfrac{dy_1}{dx}-y_2=0,\ 4y_1+\dfrac{dy_2}{dx}-4y_2-2y_3=0,\ -2y_1+y_2+\dfrac{dy_3}{dx}+y_3=0$

ANS $y_1=C_1+C_2e^{x}+C_3e^{2x},\ y_2=C_2e^{x}+2C_3e^{2x},\ y_3=2C_1+\frac{1}{2}C_2e^{x}$

7.7 $(D-2)y_1+3y_2-3y_3=0,\quad -4y_1+(D+5)y_2-3y_3=0,$

$-4y_1+4y_2+(D-2)y_3=0,\quad D(\cdot)\equiv d(\cdot)/dx$

ANS $y_1=C_1e^{-x}+C_3e^{2x},\ \ y_2=C_1e^{-x}+C_2e^{-2x}+C_3e^{2x}$

$y_3=C_2e^{-2x}+C_3e^{2x}$

7.8 $(D+1)x+2y=8,\quad 2x+(D-2)y=2e^{-t}-8$

ANS $x=C_1e^{-2t}+C_2e^{3t}+e^{-t},\quad y=\frac{1}{2}C_1e^{-2t}-2C_2e^{3t}+4$

7.9 $\dfrac{dx}{dt}=2x-3y+te^{-t},\quad \dfrac{dy}{dt}=2x-3y+e^{-t}$

ANS $x=C_1+C_2e^{-t}-t^2e^{-t},\quad y=\frac{2}{3}C_1+C_2e^{-t}-(t^2-t)e^{-t}$

7.10 $(D-1)x-2y=e^{t},\quad -4x+(D-3)y=1$

ANS $x=C_1e^{-t}+C_2e^{5t}+\frac{1}{4}e^{t}-\frac{2}{5},\ y=-C_1e^{-t}+2C_2e^{5t}-\frac{1}{2}e^{t}+\frac{1}{5}$

7.11 $(D-4)x+3y=\sin t, \quad -2x+(D+1)y=-2\cos t$

ANS $x=C_1e^t+C_2e^{2t}+\cos t-2\sin t, \quad y=C_1e^t+\frac{2}{3}C_2e^{2t}+2\cos t-2\sin t$

7.12 $\dfrac{dx}{dt}-y=0, \quad -x+\dfrac{dy}{dt}=e^t+e^{-t}$

ANS $x=C_1e^t+C_2e^{-t}+\frac{1}{2}te^t-\frac{1}{2}te^{-t}$

$y=\left(C_1+\frac{1}{2}\right)e^t+\left(-C_2-\frac{1}{2}\right)e^{-t}+\frac{1}{2}te^t+\frac{1}{2}te^{-t}$

7.13 $(D+2)x+5y=0, \quad -x+(D-2)y=\sin 2t$

ANS $x=A\cos t+B\sin t+\frac{5}{3}\sin 2t$

$y=-\frac{1}{5}(2A+B)\cos t+\frac{1}{5}(A-2B)\sin t-\frac{2}{3}(\cos 2t+\sin 2t)$

7.14 $(D-2)x+2Dy=-4e^{2t}, \quad (2D-3)x+(3D-1)y=0$

ANS $x=C_1e^{-2t}+C_2e^t+5e^{2t}, \quad y=-C_1e^{-2t}+\frac{1}{2}C_2e^t-e^{2t}$

7.15 $(3D+2)x+(D-6)y=5e^t, \quad (4D+2)x+(D-8)y=5e^t+2t-3$

ANS $x=A\cos 2t+B\sin 2t+2e^t-3t+5, \quad y=B\cos 2t-A\sin 2t+e^t-t$

7.16 $(D-5)x+3y=2e^{3t}, \quad -x+(D-1)y=5e^{-t}$

ANS $x=C_1e^{2t}+3C_2e^{4t}-e^{-t}-4e^{3t}, \; y=C_1e^{2t}+C_2e^{4t}-2e^{-t}-2e^{3t}$

7.17 $(D-2)x+y=0, \quad x+(D-2)y=-5e^t\sin t$

ANS $x=C_1e^t+C_2e^{3t}+e^t(2\cos t-\sin t), \quad y=C_1e^t-C_2e^{3t}+e^t(3\cos t+\sin t)$

7.18 $(D+4)x+2y=\dfrac{2}{e^t-1}, \quad 6x-(D-3)y=\dfrac{3}{e^t-1}$

ANS $x=C_1+2C_2e^{-t}+2e^{-t}\ln\left|e^t-1\right|, \quad y=-2C_1-3C_2e^{-t}-3e^{-t}\ln\left|e^t-1\right|$

7.19 $(D-1)x+y=\sec t, \quad -2x+(D+1)y=0$

ANS $x=C_1\cos t+C_2\sin t+t(\cos t+\sin t)+(\cos t-\sin t)\ln\left|\cos t\right|$

$y=(C_1-C_2)\cos t+(C_1+C_2)\sin t+2t\sin t+2\cos t\ln\left|\cos t\right|$

The Method of Laplace Transform

Solve the following differential equations using the method of Laplace transform.

7.20 $\dfrac{dx}{dt}-x-2y=16te^t, \quad 2x-\dfrac{dy}{dt}-2y=0, \quad x(0)=4, \; y(0)=0$

ANS $x=-e^t(12t+13)+e^{-3t}+16e^{2t}, \quad y=-2e^t(4t+3)-2e^{-3t}+8e^{2t}$

7.21 $\dfrac{dx}{dt}-2x+y=5e^t\cos t, \; x+\dfrac{dy}{dt}-2y=10e^t\sin t, \; x(0)=y(0)=0$

ANS $x=5e^t(1-\cos t+\sin t), \quad y=5e^t(1-\cos t)$

7.22 $\dfrac{dx}{dt} - 4x + 3y = \sin t, \quad 2x - \dfrac{dy}{dt} - y = 2\cos t, \quad x(0) = x_0,\ y(0) = y_0$

ANS $x = (-2x_0+3y_0-4)\,e^t + 3(x_0-y_0+1)\,e^{2t} + \cos t - 2\sin t$

$y = (-2x_0+3y_0-4)\,e^t + 2(x_0-y_0+1)\,e^{2t} + 2\cos t - 2\sin t$

7.23 $\dfrac{dx}{dt} - 2x - y = 2e^t, \quad x - \dfrac{dy}{dt} + 2y = 3e^{4t}, \quad x(0) = x_0,\ y(0) = y_0$

ANS $x = \frac{1}{2}(x_0+y_0+4)\,e^{3t} + \frac{1}{2}(x_0-y_0+2t-2)\,e^t - e^{4t}$

$y = \frac{1}{2}(x_0+y_0+4)\,e^{3t} - \frac{1}{2}(x_0-y_0+2t)\,e^t - 2e^{4t}$

7.24 $\dfrac{d^2x}{dt^2} + \dfrac{dx}{dt} + \dfrac{dy}{dt} - 2y = 40e^{3t}, \quad \dfrac{dx}{dt} + x - \dfrac{dy}{dt} = 36e^t$

$x(0) = 1,\ y(0) = 3,\ x'(0) = 1$

ANS $x = 22e^{-2t} - 33e^{-t} - 3e^t(2t-3) + 3e^{3t}, \quad y = 11e^{-2t} - 12e^t(t+1) + 4e^{3t}$

7.25 $\dfrac{dx}{dt} - 2x - y = 2e^t, \quad \dfrac{dy}{dt} - 2y - 4z = 4e^{2t}, \quad x - \dfrac{dz}{dt} - z = 0$

$x(0) = 9,\ y(0) = 3,\ z(0) = 1$

ANS $x = 3t + 2 + 2e^t - 3e^{2t} + 8e^{3t}, \quad y = -6t - 1 - 4e^t + 8e^{3t}$

$z = 3t - 1 + e^t - e^{2t} + 2e^{3t}$

7.26 $\dfrac{d^2x}{dt^2} + 2x - 2\dfrac{dy}{dt} = 0, \quad 3\dfrac{dx}{dt} + \dfrac{d^2y}{dt^2} - 8y = 240e^t$

$x(0) = y(0) = x'(0) = y'(0) = 0$

ANS $x = 12\cos 2t - 24\sin 2t - 10e^{-2t} + 30e^{2t} - 32e^t$

$y = -12\cos 2t - 6\sin 2t + 15e^{-2t} + 45e^{2t} - 48e^t$

7.27 $\dfrac{dx}{dt} - x - 2y = 0, \quad x - \dfrac{dy}{dt} = 15\cos t\, H(t-\pi), \quad x(0) = x_0,\ y(0) = y_0$

ANS $x = \frac{2}{3}(x_0+y_0)\,e^{2t} + \frac{1}{3}(x_0-2y_0)\,e^{-t} + \big[4e^{2(t-\pi)} + 5e^{-(t-\pi)}$

$\qquad + 9\cos t + 3\sin t\big]H(t-\pi)$

$y = \frac{1}{3}(x_0+y_0)\,e^{2t} - \frac{1}{3}(x_0-2y_0)\,e^{-t} + \big[2e^{2(t-\pi)} - 5e^{-(t-\pi)}$

$\qquad - 3\cos t - 6\sin t\big]H(t-\pi)$

7.28 $\dfrac{dx}{dt} - x + y = 2\sin t\,[1 - H(t-\pi)], \quad 2x - \dfrac{dy}{dt} - y = 0, \quad x(0) = y(0) = 0$

ANS $x = (t+1)\sin t - t\cos t + \big[-(t-\pi+1)\sin t + (t-\pi)\cos t\big]H(t-\pi)$

$y = 2(\sin t - t\cos t) + 2\big[-\sin t + (t-\pi)\cos t\big]H(t-\pi)$

7.29 $2\dfrac{dx}{dt}+x-5\dfrac{dy}{dt}-4y=28e^{t}H(t-2),\quad 3\dfrac{dx}{dt}-2x-4\dfrac{dy}{dt}+y=0$

$x(0)=2,\ y(0)=0$

ANS $x=-e^{-t}+3e^{t}+\left[5e^{4-t}-(6t-7)e^{t}\right]H(t-2)$

$y=-e^{-t}+e^{t}+\left[5e^{4-t}-(2t+1)e^{t}\right]H(t-2)$

The Matrix Method

Solve the following differential equations using the matrix method, in which $(\cdot)'=d(\cdot)/dt$.

7.30 $x_1'=x_1-x_2,\quad x_2'=-4x_1+x_2$

ANS $x_1=C_1e^{-t}+C_2e^{3t},\quad x_2=2C_1e^{-t}-2C_2e^{3t}$

7.31 $x_1'=x_1-3x_2,\quad x_2'=3x_1+x_2$

ANS $x_1=e^{t}(A\cos 3t+B\sin 3t),\quad x_2=e^{t}(A\sin 3t-B\cos 3t)$

7.32 $x_1'=5x_1+3x_2,\ x_2'=-3x_1-x_2,\ x_1(0)=1,\ x_2(0)=-2$

ANS $x_1=(-3t+1)e^{2t},\quad x_2=(3t-2)e^{2t}$

7.33 $x_1'=2x_1-x_2+x_3,\quad x_2'=x_1+2x_2-x_3,\quad x_3'=x_1-x_2+2x_3$

ANS $x_1=C_2e^{2t}+C_3e^{3t},\quad x_2=C_1e^{t}+C_2e^{2t},\quad x_3=C_1e^{t}+C_2e^{2t}+C_3e^{3t}$

7.34 $x_1'=3x_1-x_2+x_3,\quad x_2'=x_1+x_2+x_3,\quad x_3'=4x_1-x_2+4x_3$

ANS $x_1=C_1e^{t}+C_2e^{2t}+C_3e^{5t},\quad x_2=C_1e^{t}-2C_2e^{2t}+C_3e^{5t}$

$x_3=-C_1e^{t}-3C_2e^{2t}+3C_3e^{5t}$

7.35 $x_1'=2x_1+x_2,\quad x_2'=x_1+3x_2-x_3,\quad x_3'=-x_1+2x_2+3x_3$

ANS $x_1=Ce^{2t}+e^{3t}(A\cos t+B\sin t),\ x_2=e^{3t}\left[(A+B)\cos t+(B-A)\sin t\right]$

$x_3=Ce^{2t}+e^{3t}\left[(2A-B)\cos t+(2B+A)\sin t\right]$

7.36 $x_1'=3x_1-2x_2-x_3,\quad x_2'=3x_1-4x_2-3x_3,\quad x_3'=2x_1-4x_2$

ANS $x_1=(C_1+2C_2)e^{2t}+C_3e^{-5t},\ x_2=C_2e^{2t}+3C_3e^{-5t},\ x_3=C_1e^{2t}+2C_3e^{-5t}$

7.37 $x_1'=x_1-x_2+x_3,\quad x_2'=x_1+x_2-x_3,\quad x_3'=-x_2+2x_3$

$x_1(0)=1,\quad x_2(0)=-2,\quad x_3(0)=0$

ANS $x_1=te^{t}+e^{2t},\quad x_2=(t-2)e^{t},\quad x_3=(t-1)e^{t}+e^{2t}$

7.38 $x_1'=-x_1+x_2-2x_3,\quad x_2'=4x_1+x_2,\quad x_3'=2x_1+x_2-x_3$

ANS $x_1=\left[C_1+C_2(t-1)\right]e^{-t},\quad x_2=-\left[2C_1+C_2(2t-1)\right]e^{-t}+2C_3e^{t}$

$x_3=-(C_1+C_2t)e^{-t}+C_3e^{t}$

7.39 $x_1' = 2x_1 + x_2 + 26\sin t, \quad x_2' = 3x_1 + 4x_2$

ANS $x_1 = C_1 e^t + C_2 e^{5t} - 10\cos t - 11\sin t, \; x_2 = -C_1 e^t + 3C_2 e^{5t} + 9\cos t + 6\sin t$

7.40 $x_1' = -x_1 + 8x_2 + 9t, \quad x_2' = x_1 + x_2 + 3e^{-t}$

ANS $x_1 = 2C_1 e^{3t} + 4C_2 e^{-3t} - 3e^{-t} + t - 1, \quad x_2 = C_1 e^{3t} - C_2 e^{-3t} - t$

7.41 $x_1' = -x_1 + 2x_2, \quad x_2' = -3x_1 + 4x_2 + \dfrac{e^{3t}}{e^{2t}+1}$

ANS $x_1 = C_1 e^t + 2C_2 e^{2t} - e^t \ln(e^{2t}+1) + 2e^{2t} \tan^{-1} e^t$

$x_2 = C_1 e^t + 3C_2 e^{2t} - e^t \ln(e^{2t}+1) + 3e^{2t} \tan^{-1} e^t$

7.42 $x_1' = -4x_1 - 2x_2 + \dfrac{2}{e^t - 1}, \quad x_2' = 6x_1 + 3x_2 - \dfrac{3}{e^t - 1}$

ANS $x_1 = 2\left(C_1 + \ln\left|e^t - 1\right|\right) e^{-t} + C_2, \; x_2 = -3\left(C_1 + \ln\left|e^t - 1\right|\right) e^{-t} - 2C_2$

7.43 $x_1' = x_1 + x_2 + e^{2t}, \quad x_2' = -2x_1 + 3x_2$

ANS $x_1 = e^{2t}(C_1 \cos t + C_2 \sin t - 1), \; x_2 = e^{2t}\left[(C_1 + C_2)\cos t + (C_2 - C_1)\sin t - 2\right]$

7.44 $x_1' = -x_1 - 5x_2, \quad x_2' = x_1 + x_2 + \dfrac{4}{\sin 2t}$

ANS $x_1 = (C_1 - 2C_2 + 10t)\cos 2t + (2C_1 + C_2)\sin 2t - 5\sin 2t \ln\left|\sin 2t\right|$

$x_2 = -(C_1 + 2t)\cos 2t - (C_2 - 4t)\sin 2t + (2\cos 2t + \sin 2t)\ln\left|\sin 2t\right|$

7.45 $x_1' = 2x_1 + x_2 + 27t, \quad x_2' = -x_1 + 4x_2$

ANS $x_1 = (C_1 + C_2 t) e^{3t} - 12t - 5, \quad x_2 = \left[C_1 + C_2(t+1)\right] e^{3t} - 3t - 2$

7.46 $x_1' = 3x_1 - x_2 + e^t, \quad x_2' = 4x_1 - x_2$

ANS $x_1 = e^t\left[C_1 + C_2(t+1) + t^2 + t\right], \quad x_2 = e^t\left[2C_1 + C_2(2t+1) + 2t^2\right]$

7.47 $x_1' = 3x_1 - 2x_2, \quad x_2' = 2x_1 - x_2 + 35e^t t^{3/2}$

ANS $x_1 = e^t\left[C_1 + C_2(2t+1) - 8t^{7/2}\right], \; x_2 = e^t(C_1 + 2C_2 t - 8t^{7/2} + 14t^{5/2})$

7.48 $x_1' = x_1 - x_2 + x_3, \quad x_2' = x_1 + x_2 - x_3 + 6e^{-t}, \quad x_3' = 2x_1 - x_2$

ANS $x_1 = -(C_1 + 1)e^{-t} + C_2 e^t + C_3 e^{2t}, \quad x_2 = 3(C_1 - 1)e^{-t} + C_2 e^t$

$x_3 = (5C_1 - 1)e^{-t} + C_2 e^t + C_3 e^{2t}$

7.49 $x_1' = x_1 - 2x_2 - x_3, \quad x_2' = -x_1 + x_2 + x_3 + 12t, \quad x_3' = x_1 - x_3$

ANS $x_1 = 3C_1 e^{2t} + 6t^2 + 6t + (C_3 + 3), \quad x_2 = -2C_1 e^{2t} + C_2 e^{-t} - 6$

$x_3 = C_1 e^{2t} - 2C_2 e^{-t} + 6t^2 - 6t + (C_3 + 9)$

7.50 $x_1' = -3x_1 + 4x_2 - 2x_3 + e^t, \quad x_2' = x_1 + x_3, \quad x_3' = 6x_1 - 6x_2 + 5x_3$

ANS $x_1 = C_1 e^{-t} + (-t + C_2 + 1)e^t, \quad x_2 = (-t + C_2 - 1)e^t + C_3 e^{2t}$

$x_3 = -C_1 e^{-t} - 3e^t + 2C_3 e^{2t}$

7.51 $x_1' = x_1 - x_2 - x_3 + 4e^t, \quad x_2' = x_1 + x_2, \quad x_3' = 3x_1 + x_3$

ANS $x_1 = 2e^t(-C_2 \sin 2t + C_3 \cos 2t), \quad x_2 = e^t(C_1 + 1 + C_2 \cos 2t + C_3 \sin 2t)$

$x_3 = e^t(-C_1 + 3 + 3C_2 \cos 2t + 3C_3 \sin 2t)$

7.52 $x_1' = 2x_1 - x_2 + 2x_3, \quad x_2' = x_1 + 2x_3, \quad x_3' = -2x_1 + x_2 - x_3 + 4\sin t$

ANS $x_1 = -2(C_2 + 2t)\cos t - 2(C_3 - 1)\sin t$

$x_2 = 2C_1 e^t - 2(C_2 + 2t)\cos t - 2(C_3 - 1)\sin t$

$x_3 = C_1 e^t + (C_2 - C_3 + 2t - 1)\cos t + (C_2 + C_3 + 2t - 1)\sin t$

7.53 $x_1' = 4x_1 - x_2 - x_3 + e^{3t}, \quad x_2' = x_1 + 2x_2 - x_3, \quad x_3' = x_1 + x_2 + 2x_3$

$x_1(0) = 1, \; x_2(0) = 2, \; x_3(0) = 3$

ANS $x_1 = 5e^{2t} + 2(t - 2)e^{3t}, \quad x_2 = 5e^{2t} + (t - 3)e^{3t}, \quad x_3 = 5e^{2t} + (t - 2)e^{3t}$

7.54 $x_1' = 2x_1 - x_2 - x_3 + 2e^{2t}, \; x_2' = 3x_1 - 2x_2 - 3x_3, \; x_3' = -x_1 + x_2 + 2x_3$

ANS $x_1 = C_1 + (C_2 + C_3)e^t + 3e^{2t}, \quad x_2 = 3C_1 + C_2 e^t + 3e^{2t}$

$x_3 = -C_1 + C_3 e^t - e^{2t}$

7.55 $x_1' = 2x_1 - x_3 + 24t, \; x_2' = x_1 - x_2, \; x_3' = 3x_1 - x_2 - x_3$

ANS $x_1 = t^4 + 8t^3 + (C_3 + 12)t^2 + (C_2 + 2C_3)t + (C_1 + C_2 + 2C_3)$

$x_2 = t^4 + 4t^3 + C_3 t^2 + C_2 t + (C_1 + 2C_3)$

$x_3 = 2t^4 + 12t^3 + 2C_3 t^2 + 2(C_2 + C_3)t + (2C_1 + C_2 + 2C_3)$

CHAPTER 8

Applications of Systems of Linear Differential Equations

In this chapter, examples are presented to illustrate engineering applications of systems of linear differential equations.

8.1 Mathematical Modeling of Mechanical Vibrations

In many engineering applications, such as vibration of mechanical systems, the systems are usually complex and have to be modeled as multiple degrees-of-freedom systems, resulting in systems of linear ordinary differential equations. Modeling a complex engineering system as an appropriate, mathematically tractable problem and establishing the governing differential equations are often the first challenging step. In this section, a free vibration problem of a simple two degrees-of-freedom system is first considered to illustrated the basic procedure. The equations of motion of a more complex problem, i.e., the vibration of an automobile, which is modeled as a four degrees-of-freedom system, are then established.

Example 8.1 — Two Degrees-of-Freedom System

Two uniform rods AB and CD of mass density per unit length ρ are hinged at A and C. Rotational springs of stiffnesses κ_1 and κ_2 provide resistance to rotations of end A and end C, respectively, as shown. The lengths of AB and CD are L_1 and L_2, respectively. Rod AB carries a concentrated mass M at end B. The rods are connected by a spring of stiffness k. When the rods are hanging freely, they are vertical and there is no force in the spring.

The motion of the system is described by two angles θ_1 and θ_2. Consider only small oscillations, i.e., $|\theta_1| \ll 1$ and $|\theta_2| \ll 1$, and neglect the effect of gravity.

1. Set up the differential equations governing the angles of rotation θ_1 and θ_2.
2. For the special case when $L_1 = L$, $L_2 = 2L$, $M = \rho L$, $\kappa_1 = \kappa_2 = 0$, and initial conditions $\theta_1(0) = \theta_{10}$, $\theta_2(0) = \theta_{20}$, $\dot{\theta}_1(0) = \dot{\theta}_2(0) = 0$, determine the responses of the free vibration.
3. For the special case when $L_1 = L_2 = L$, $M = 0$, $\kappa_1 = \kappa_2 = kL^2$, discuss the responses of the free vibration.

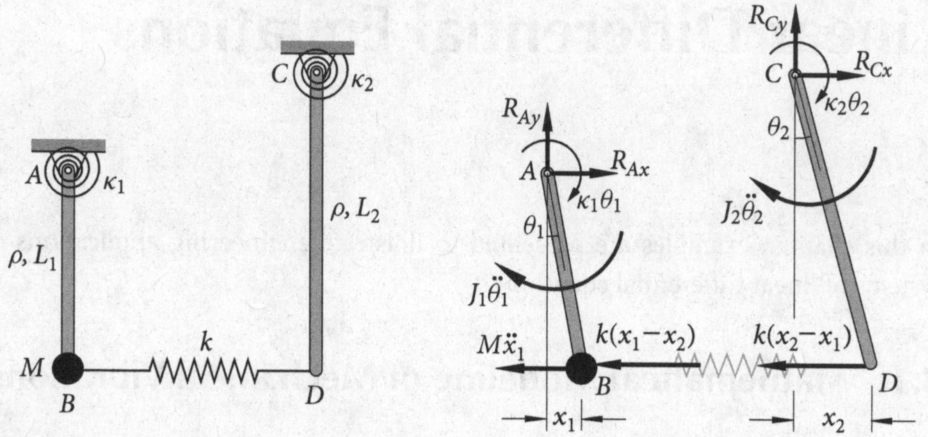

1. When the rotation of rod AB about point A is θ_1, the horizontal displacement of point B is $x_1 = L_1 \sin\theta_1 \approx L_1\theta_1$. The mass of rod AB is $m_1 = \rho L_1$ and its moment of inertia about point A is

$$J_1 = \frac{1}{3} m_1 L_1^2 = \frac{1}{3} \rho L_1^3.$$

When the rotation of rod CD about point C is θ_2, the horizontal displacement of point D is $x_2 = L_2 \sin\theta_2 \approx L_2\theta_2$. The mass of rod CD is $m_2 = \rho L_2$ and its moment of inertia about point C is

$$J_2 = \frac{1}{3} m_2 L_2^2 = \frac{1}{3} \rho L_2^3.$$

To establish the differential equations governing the angles θ_1 and θ_2, consider the free-body diagrams of rods AB and CD, respectively. D'Alembert's Principle is applied to set up the equation of motion.

❧ **Rod *AB*.** To find the spring force applied at end B, observe the spring at end D: end B moves toward the right by x_1 so that the spring is compressed and the spring force pushes end B toward the left. Since end D also moves toward the right by x_2, the net compression of the spring is $x_1 - x_2$; hence, the spring force is $k(x_1 - x_2)$.

Since the acceleration of point B is $\ddot{x}_1 = L_1\ddot{\theta}_1$ toward the right, the inertia force of mass M is $ML_1\ddot{\theta}_1$ toward the left.

Since the angular acceleration of rod AB about point A is $\ddot{\theta}_1$ counterclockwise, the inertia moment of rod AB is $J_1\ddot{\theta}_1$ clockwise.

Replace the hinge at A by two reaction force components R_{Ax} and R_{Ay}. The rotational spring at A provides a clockwise resisting moment $\kappa_1\theta_1$.

Applying D'Alembert's Principle and summing up moments about point A yield

$$\curvearrowright \sum M_A = 0: \quad J_1\ddot{\theta}_1 + ML_1\ddot{\theta}_1 \cdot L_1 + k(x_1 - x_2)\cdot L_1 + \kappa_1\theta_1 = 0,$$

$$\therefore \quad \Big(\frac{1}{3}\rho L_1^2 + ML_1\Big)\ddot{\theta}_1 + k(L_1\theta_1 - L_2\theta_2) + \frac{\kappa_1}{L_1}\theta_1 = 0.$$

❧ **Rod CD.** To find the spring force applied at end D, observe the spring at end B: end D moves toward the right by x_2 so that the spring is extended and the spring force pulls end D toward the left. Since end B also moves toward the right by x_1, the net extension of the spring is $x_2 - x_1$; hence, the spring force is $k(x_2 - x_1)$.

Since the angular acceleration of rod CD about point C is $\ddot{\theta}_2$ counterclockwise, the inertia moment of rod CD is $J_2\ddot{\theta}_2$ clockwise.

Replace the hinge at C by two reaction force components R_{Cx} and R_{Cy}. The rotational spring at C provides a clockwise resisting moment $\kappa_2\theta_2$.

Applying D'Alembert's Principle and summing up moments about point C give

$$\curvearrowright \sum M_C = 0: \quad J_2\ddot{\theta}_2 + k(x_2 - x_1)\cdot L_2 + \kappa_2\theta_2 = 0,$$

$$\therefore \quad \frac{1}{3}\rho L_2^2\ddot{\theta}_2 + k(L_2\theta_2 - L_1\theta_1) + \frac{\kappa_2}{L_2}\theta_2 = 0.$$

2. When $L_1 = L$, $L_2 = 2L$, $M = \rho L$, $\kappa_1 = \kappa_2 = 0$, the equations of motion become

$$\Big[\frac{1}{3}\rho L^2 + (\rho L)L\Big]\ddot{\theta}_1 + k\big[L\theta_1 - (2L)\theta_2\big] = 0,$$

$$\frac{1}{3}\rho(2L)^2\ddot{\theta}_2 + k\big[(2L)\theta_2 - L\theta_1\big] = 0,$$

$$\therefore \quad \ddot{\theta}_1 + \frac{\omega_0^2}{3}(\theta_1 - 2\theta_2) = 0, \qquad \omega_0^2 = \frac{9k}{4\rho L}.$$

$$\ddot{\theta}_2 + \frac{\omega_0^2}{3}(2\theta_2 - \theta_1) = 0,$$

Using the D-operator, $D \equiv \mathrm{d}/\mathrm{d}t$, the equations of motion can be written as

$$\Big(D^2 + \frac{\omega_0^2}{3}\Big)\theta_1 - \frac{2\omega_0^2}{3}\theta_2 = 0,$$

$$-\frac{\omega_0^2}{3}\theta_1 + \Big(D^2 + \frac{2\omega_0^2}{3}\Big)\theta_2 = 0.$$

The determinant of the coefficient matrix is

$$\phi(D) = \begin{vmatrix} D^2 + \dfrac{\omega_0^2}{3} & -\dfrac{2\omega_0^2}{3} \\ -\dfrac{\omega_0^2}{3} & D^2 + \dfrac{2\omega_0^2}{3} \end{vmatrix} = D^4 + \omega_0^2 D^2.$$

The characteristic equation is then given by $\phi(\lambda) = 0$, which gives

$$\lambda^4 + \omega_0^2 \lambda^2 = 0 \implies \lambda^2(\lambda^2 + \omega_0^2) = 0 \implies \lambda = 0, 0, \pm i\omega_0.$$

Hence, the complementary solution for θ_2 is given by

$$\theta_2(t) = C_0 + C_1 t + A\cos\omega_0 t + B\sin\omega_0 t.$$

The complementary solution of θ_1 can be obtained as

$$\begin{aligned} \theta_1(t) &= \frac{3}{\omega_0^2}\Big(D^2 + \frac{2\omega_0^2}{3}\Big)\theta_2(t) \\ &= \frac{3}{\omega_0^2}(-A\omega_0^2\cos\omega_0 t - B\omega_0^2\sin\omega_0 t) + 2(C_0 + C_1 t + A\cos\omega_0 t + B\sin\omega_0 t) \\ &= 2(C_0 + C_1 t) - A\cos\omega_0 t - B\sin\omega_0 t. \end{aligned}$$

Hence, the complementary solutions or the responses of free vibration of the system are given by

$$\begin{aligned} \theta_1(t) &= 2(C_0 + C_1 t) - (A\cos\omega_0 t + B\sin\omega_0 t), \\ \theta_2(t) &= \underbrace{(C_0 + C_1 t)}_{\text{Nonoscillatory motion}} + \underbrace{(A\cos\omega_0 t + B\sin\omega_0 t)}_{\text{Harmonic oscillation at } \omega_0}, \end{aligned}$$

in which the first two terms correspond to nonoscillatory motion with $\theta_1 = 2\theta_2$ (the length of the spring is not changed); whereas the last two terms correspond to harmonic oscillation with $\theta_1 = -\theta_2$ at circular frequency ω_0. These two motions are shown schematically in the following figure.

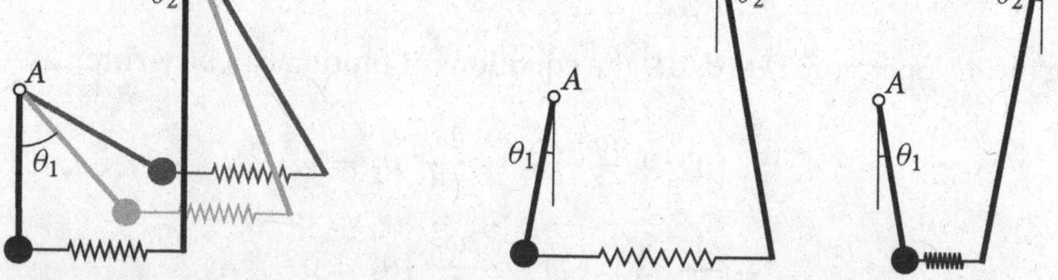

Nonoscillatory motion   Harmonic oscillation at frequency ω_0

For the initial conditions $\theta_1(0)=\theta_{10}$, $\theta_2(0)=\theta_{20}$, $\dot{\theta}_1(0)=\dot{\theta}_2(0)=0$, and noting

$$\dot{\theta}_1(t)=2C_1+A\omega_0 \sin\omega_0 t-B\omega_0 \cos\omega_0 t,\quad \dot{\theta}_2(t)=C_1-A\omega_0 \sin\omega_0 t+B\omega_0 \cos\omega_0 t,$$

one has

$$\theta_1(0)=2C_0-A=\theta_{10},\qquad \theta_2(0)=C_0+A=\theta_{20},$$
$$\dot{\theta}_1(0)=2C_1-B\omega_0=0,\qquad \dot{\theta}_2(0)=C_1+B\omega_0=0,$$

which gives

$$C_0=\frac{1}{3}(\theta_{10}+\theta_{20}),\quad C_1=0,\quad A=\frac{1}{3}(2\theta_{20}-\theta_{10}),\quad B=0.$$

The responses of the free vibration are

$$\theta_1(t)=\frac{2}{3}(\theta_{10}+\theta_{20})-\frac{1}{3}(2\theta_{20}-\theta_{10})\cos\omega_0 t,$$
$$\theta_2(t)=\frac{1}{3}(\theta_{10}+\theta_{20})+\frac{1}{3}(2\theta_{20}-\theta_{10})\cos\omega_0 t.$$

3. When $L_1=L$, $L_2=L$, $M=0$, $\kappa_1=\kappa_2=kL^2$, the equations of motion become

$$\frac{1}{3}\rho L^2\ddot{\theta}_1+k(L\theta_1-L\theta_2)+\frac{kL^2}{L}\theta_1=0,$$
$$\frac{1}{3}\rho L^2\ddot{\theta}_2+k(L\theta_2-L\theta_1)+\frac{kL^2}{L}\theta_2=0,$$

which yield

$$\ddot{\theta}_1+\omega^2(2\theta_1-\theta_2)=0,\qquad \omega^2=\frac{3k}{\rho L}.$$
$$\ddot{\theta}_2+\omega^2(2\theta_2-\theta_1)=0,$$

Using the D-operator, $D\equiv \mathrm{d}/\mathrm{d}t$, the equations of motion can be written as

$$(D^2+2\omega^2)\theta_1-\omega^2\theta_2=0,$$
$$-\omega^2\theta_1+(D^2+2\omega^2)\theta_2=0.$$

The determinant of the coefficient matrix is

$$\phi(D)=\begin{vmatrix} D^2+2\omega^2 & -\omega^2 \\ -\omega^2 & D^2+2\omega^2 \end{vmatrix}=D^4+4\omega^2D^2+3\omega^4.$$

The characteristic equation is then given by $\phi(\lambda)=0$, which gives

$$\lambda^4+4\omega^2\lambda^2+3\omega^4=0\implies(\lambda^2+\omega^2)(\lambda^2+3\omega^2)=0\implies\lambda=\pm\mathrm{i}\omega,\ \pm\mathrm{i}\sqrt{3}\omega.$$

Hence, the complementary solution for θ_1 is given by

$$\theta_1(t)=A_1\cos\omega t+B_1\sin\omega t+A_2\cos\sqrt{3}\omega t+B_2\sin\sqrt{3}\omega t.$$

The complementary solution of θ_2 can be obtained as

$$\begin{aligned}\theta_2(t) &= \frac{1}{\omega^2}(D^2+2\omega^2)\theta_1(t)\\ &= \frac{1}{\omega^2}\big[(-A\omega^2\cos\omega t - B\omega^2\sin\omega t - A_2\cdot 3\omega^2\cos\sqrt{3}\omega t - B_2\cdot 3\omega^2\sin\sqrt{3}\omega t)\\ &\qquad + 2\omega^2(A_1\cos\omega t + B_1\sin\omega t + A_2\cos\sqrt{3}\omega t + B_2\sin\sqrt{3}\omega t)\big]\\ &= A_1\cos\omega t + B_1\sin\omega t - A_2\cos\sqrt{3}\omega t - B_2\sin\sqrt{3}\omega t.\end{aligned}$$

The complementary solutions or the responses of free vibration of the system are

$$\theta_1(t) = (A_1\cos\omega t + B_1\sin\omega t) + (A_2\cos\sqrt{3}\omega t + B_2\sin\sqrt{3}\omega t),$$

$$\theta_2(t) = \underbrace{(A_1\cos\omega t + B_1\sin\omega t)}_{\text{First Mode Oscillation at }\omega} - \underbrace{(A_2\cos\sqrt{3}\omega t + B_2\sin\sqrt{3}\omega t)}_{\text{Second Mode Oscillation at }\sqrt{3}\omega}.$$

The four constants A_1, B_1, A_2, and B_2 are determined from the initial conditions $\theta_1(0)$, $\theta_2(0)$, $\dot{\theta}_1(0)$, and $\dot{\theta}_1(0)$. The first two terms in the solutions correspond to the first mode of harmonic oscillation at circular frequency $\omega_1 = \omega$ with $\theta_1 = \theta_2$; whereas the last two terms correspond to the second mode of oscillation at circular frequency $\omega_2 = \sqrt{3}\omega$ with $\theta_1 = -\theta_2$. Note that natural frequencies are always ordered in ascending order, i.e., $\omega_1 < \omega_2$. These two motions are shown schematically in the following figure.

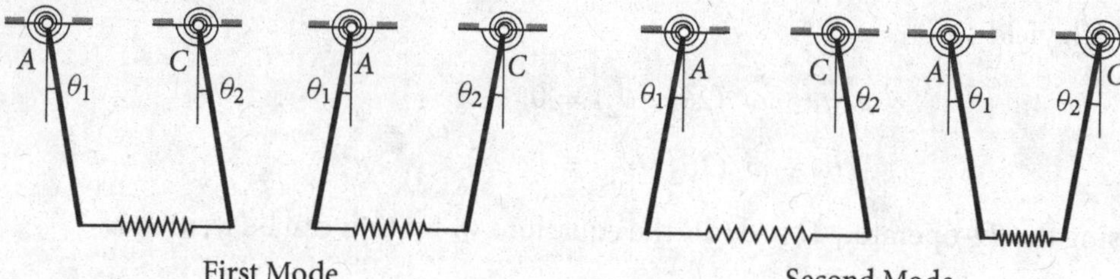

First Mode
Oscillation at frequency ω, $\theta_1 = \theta_2$

Second Mode
Oscillation at frequency $\sqrt{3}\omega$, $\theta_1 = -\theta_2$

Example 8.2 — Vibration of an Automobile

To study the dynamic response of an automobile on a wavy road, a simplified mechanical model shown in Figures 8.1(a) and (b) is used.

- The body of the vehicle is modeled as a rigid body of mass m and moment of inertia J about its center of gravity C.
- The front shock absorbers are modeled as a spring of stiffness k_f and a damper of damping coefficient c_f. The rear shock absorbers are modeled as a spring of stiffness k_r and a damper of damping coefficient c_r.
- The front axle is modeled as a point mass m_f. The rear axle is modeled as a point mass m_r.

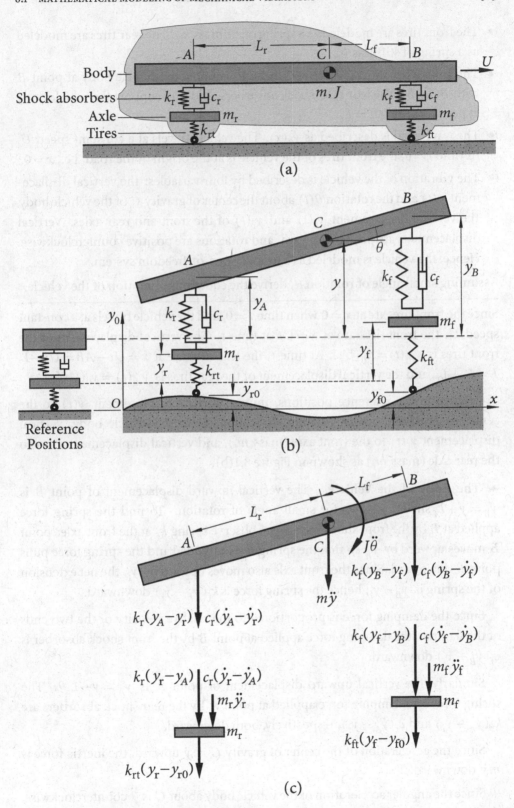

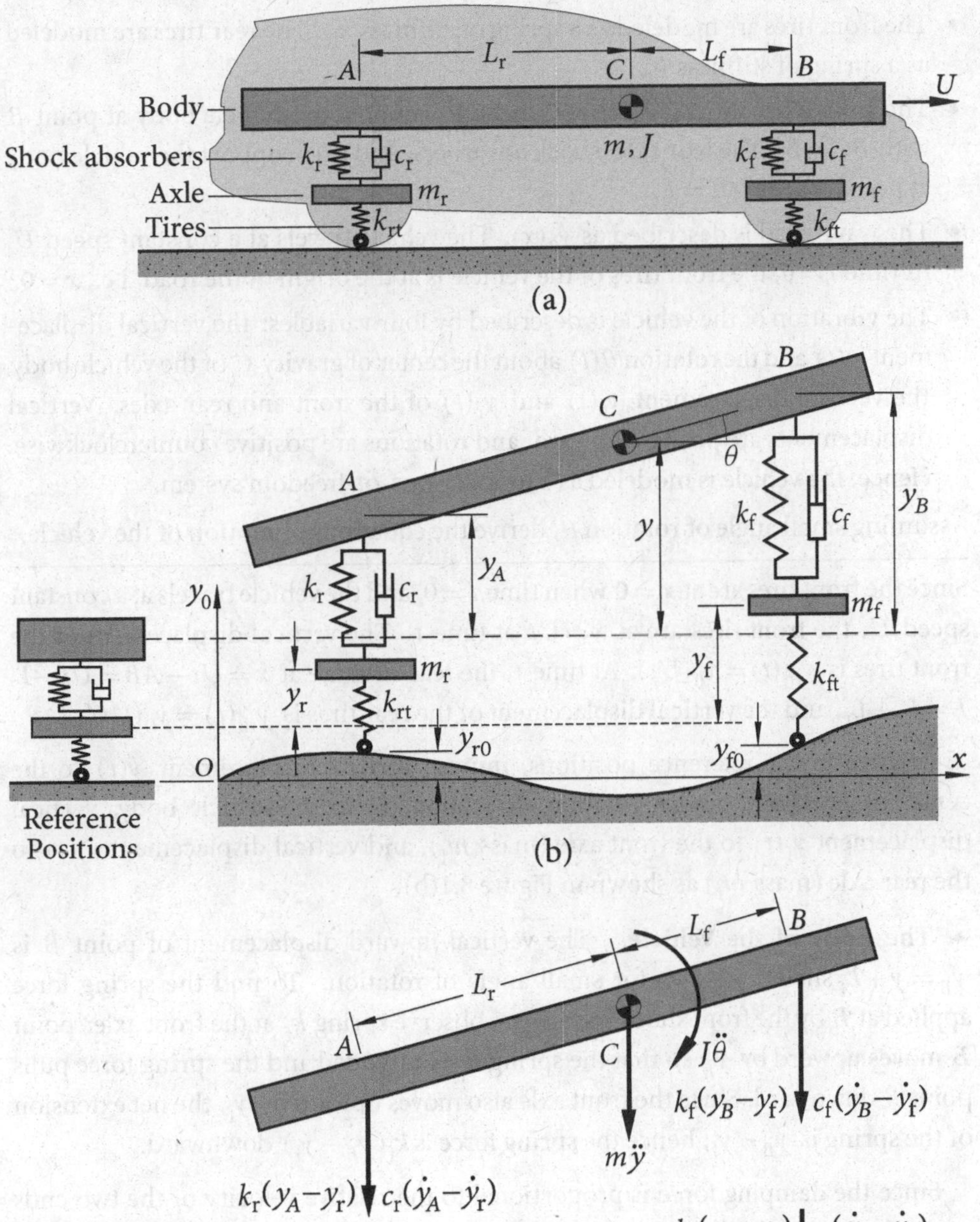

Figure 8.1 An automobile moving on a wavy road.

- The front tires are modeled as a spring of stiffness $k_{\rm ft}$. The rear tires are modeled as a spring of stiffness $k_{\rm rt}$.
- The front axle, shock absorbers, and tires support the vehicle body at point B with $BC = L_{\rm f}$. The rear axle, shock absorbers, and tires support the vehicle body at point A with $AC = L_{\rm r}$.
- The wavy road is described as $y_0(x)$. The vehicle travels at a constant speed U. At time $t = 0$, the front tires of the vehicle is at the origin of the road, i.e., $x = 0$.
- The vibration of the vehicle is described by four variables: the vertical displacement $y(t)$ and the rotation $\theta(t)$ about the center of gravity C of the vehicle body, the vertical displacement $y_{\rm f}(t)$ and $y_{\rm r}(t)$ of the front and rear axles. Vertical displacements are positive upward, and rotations are positive counterclockwise. Hence, the vehicle is modeled as a four degrees-of-freedom system.

Assuming small angle of rotation θ, derive the equations of motion of the vehicle.

Since the front tires are at $x = 0$ when time $t = 0$, and the vehicle travels at a constant speed U, the front tires are at $x = Ut$ at time t. The vertical displacement of the front tires is $y_{\rm f0}(t) = y_0(Ut)$. At time t, the rear tires are at $x = Ut - AB = Ut - L$, $L = L_{\rm f} + L_{\rm r}$, and the vertical displacement of the rear tires is $y_{\rm r0}(t) = y_0(Ut - L)$.

Relative to the reference positions, impose vertical displacement $y(t)$ to the center of gravity C and a rotation $\theta(t)$ about C to the vehicle body, vertical displacement $y_{\rm f}(t)$ to the front axle (mass $m_{\rm f}$), and vertical displacement $y_{\rm r}(t)$ to the rear axle (mass $m_{\rm r}$) as shown in Figure 8.1(b).

- **The Body of the Vehicle.** The vertical upward displacement of point B is $y_B = y + L_{\rm f} \sin\theta \approx y + L_{\rm f}\theta$ for small angle of rotation. To find the spring force applied at B by the front shock absorbers, observe spring $k_{\rm f}$ at the front axle: point B moves upward by y_B so that the spring $k_{\rm f}$ is extended and the spring force pulls point B downward. Since the front axle also moves upward by $y_{\rm f}$, the net extension of the spring is $y_B - y_{\rm f}$; hence the spring force is $k_{\rm f}(y_B - y_{\rm f})$ downward.

Since the damping force is proportional to the relative velocity of the two ends of the damper, the damping force applied at point B by the front shock absorber is $c_{\rm f}(\dot{y}_B - \dot{y}_{\rm f})$ downward.

Similarly, the vertical upward displacement of point A is $y_A = y - L_{\rm r}\theta$. The spring force and damping force applied at point A by the rear shock absorbers are $k_{\rm r}(y_A - y_{\rm r})$ and $c_{\rm r}(\dot{y}_A - \dot{y}_{\rm r})$, respectively, both downward.

Since the acceleration of the center of gravity C is $\ddot{y}$ upward, the inertia force is $m\ddot{y}$ downward.

Since the angular acceleration of the vehicle body about C is $\ddot{\theta}$ counterclockwise, the inertia moment is $J\ddot{\theta}$ clockwise.

Applying D'Alembert's Principle, the free-body of the vehicle body as shown in Figure 8.1(c) is in dynamic equilibrium. Hence

$$\downarrow \sum F_y = 0: \quad m\ddot{y} + c_f(\dot{y}_B - \dot{y}_f) + k_f(y_B - y_f) + c_r(\dot{y}_A - \dot{y}_r) + k_r(y_A - y_r) = 0,$$

$$\curvearrowright \sum M_C = 0: \quad J\ddot{\theta} + \big[c_f(\dot{y}_B - \dot{y}_f) + k_f(y_B - y_f)\big] L_f - \big[c_r(\dot{y}_A - \dot{y}_r) + k_r(y_A - y_r)\big] L_r = 0.$$

❧ **The Front Axle.** To determine the spring force applied on the front axle m_f by the front shock absorbers, observe spring k_f at point B: the front axle m_f moves upward by y_f so that the spring k_f is compressed and the spring force pushes the front axle m_f downward. Since point B also moves upward by y_B, the net compression of the spring is $y_f - y_B$; hence the spring force is $k_f(y_f - y_B)$ downward. Similarly, the damping force applied on the front axle m_f by the front shock absorber is $c_f(\dot{y}_f - \dot{y}_B)$ downward.

To determine the spring force applied on the front axle m_f by the front tires, observe spring k_{ft} at the bottom of the front tires: the front axle m_f moves upward by y_f so that the spring k_{ft} is extended and the spring force pulls the front axle m_f downward. Since the bottom of the front tires also moves upward by y_{f0}, the net extension of the spring is $y_f - y_{f0}$; hence the spring force is $k_{ft}(y_f - y_{f0})$ downward.

Since the acceleration of the front axle m_f is $\ddot{y}_f$ upward, the inertia force is $m_f\ddot{y}_f$ downward.

Applying D'Alembert's Principle, the free-body of the front axle as shown in Figure 8.1(c) is in dynamic equilibrium. Hence

$$\downarrow \sum F_y = 0: \quad m_f\ddot{y}_f + c_f(\dot{y}_f - \dot{y}_B) + k_f(y_f - y_B) + k_{ft}(y_f - y_{f0}) = 0.$$

❧ **The Rear Axle.** Similarly, the equation of motion of the rear axle is

$$\downarrow \sum F_y = 0: \quad m_r\ddot{y}_r + c_r(\dot{y}_r - \dot{y}_A) + k_r(y_r - y_A) + k_{rt}(y_r - y_{r0}) = 0.$$

Noting $y_B = y + L_f\theta$, $y_A = y - L_r\theta$, $y_{f0}(t) = y_0(Ut)$, $y_{r0}(t) = y_0(Ut - L)$, the equations of motion of the automobile are four second-order differential equations

$$m\ddot{y} + (c_f + c_r)\dot{y} + (c_f L_f - c_r L_r)\dot{\theta} - c_f\dot{y}_f - c_r\dot{y}_r + (k_f + k_r)y + (k_f L_f - k_r L_r)\theta - k_f y_f - k_r y_r = 0,$$

$$J\ddot{\theta} + (c_f L_f - c_r L_r)\dot{y} + (c_f L_f^2 + c_r L_r^2)\dot{\theta} - c_f L_f \dot{y}_f + c_r L_r \dot{y}_r + (k_f L_f - k_r L_r)y + (k_f L_f^2 + k_r L_r^2)\theta - k_f L_f y_f + k_r L_r y_r = 0,$$

$$m_f\ddot{y}_f - c_f\dot{y} - c_f L_f\dot{\theta} + c_f\dot{y}_f - k_f y - k_f L_f\theta + (k_f + k_{ft})y_f = k_{ft}\, y_0(Ut),$$

$$m_r\ddot{y}_r - c_r\dot{y} + c_r L_r\dot{\theta} + c_r\dot{y}_r - k_r y + k_r L_r\theta + (k_r + k_{rt})y_r = k_{rt}\, y_0(Ut - L).$$

8.2 Vibration Absorbers or Tuned Mass Dampers

In engineering applications, many systems can be modeled as single degree-of-freedom systems. For example, a machine mounted on a structure can be modeled using a mass-spring-damper system, in which the machine is considered to be rigid with mass m and the supporting structure is equivalent to a spring k and a damper c, as shown in Figure 8.2. The machine is subjected to a sinusoidal force $F_0 \sin \Omega t$, which can be an externally applied load or due to imbalance in the machine.

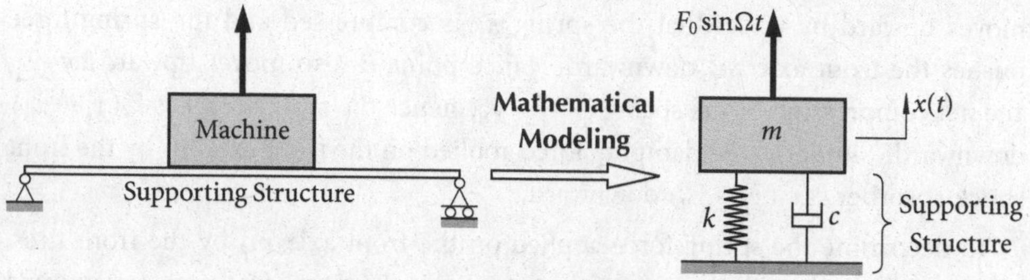

Figure 8.2 A machine mounted on a structure.

From Chapter 5 on the response of a single degree-of-freedom system, it is well known that when the excitation frequency Ω is close to the natural frequency of the system $\omega_0 = \sqrt{k/m}$, vibration of large amplitude occurs. In particular, when the system is undamped, i.e., $c=0$, resonance occurs when $\Omega = \omega_0$, in which the amplitude of the response grows linearly with time.

To reduce the vibration of the system, a vibration absorber or a tuned mass damper (TMD), which is an auxiliary mass-spring-damper system, is mounted on the main system as shown in Figure 8.3(a). The mass, spring stiffness, and damping coefficient of the viscous damper are m_a, k_a, and c_a, respectively, where the subscript "a" stands for "auxiliary."

To derive the equation of motion of the main mass m, consider its free-body diagram as shown in Figure 8.3(b). Since mass m moves upward, spring k is extended and spring k_a is compressed.

- Because of the displacement x of mass m, the extension of spring k is x. Hence the spring k exerts a downward force kx and the damper c exerts a downward force $c\dot{x}$ on mass m.
- Because the mass m_a also moves upward a distance x_a, the net compression in spring k_a is $x - x_a$. Hence the spring k_a and damper c_a exert downward forces $k_a(x - x_a)$ and $c_a(\dot{x} - \dot{x}_a)$, respectively, on mass m.

Newton's Second Law requires

$$\uparrow m\ddot{x} = \sum F: \quad m\ddot{x} = -kx - c\dot{x} - k_a(x - x_a) - c_a(\dot{x} - \dot{x}_a) + F_0 \sin \Omega t,$$

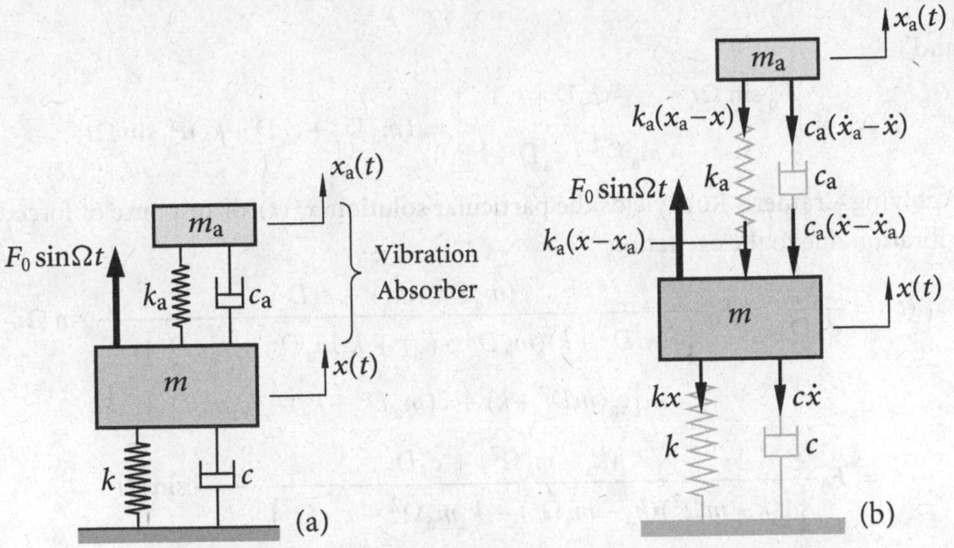

Figure 8.3 A vibration absorber mounted on the main system.

or

$$m\ddot{x} + (c + c_a)\dot{x} + (k + k_a)x - c_a\dot{x}_a - k_a x_a = F_0 \sin \Omega t.$$

Similarly, consider the free-body diagram of mass m_a. Since mass m_a moves upward a distance $x_a(t)$, spring k_a is extended. The net extension of spring k_a is $x_a - x$. Hence, the spring k_a and damper c_a exert downward forces $k_a(x_a - x)$ and $c_a(\dot{x}_a - \dot{x})$, respectively. Applying Newton's Second Law gives

$$\uparrow m_a\ddot{x}_a = \sum F: \quad m_a\ddot{x}_a = -k_a(x_a - x) - c_a(\dot{x}_a - \dot{x}),$$

$$\therefore \quad m_a\ddot{x}_a + c_a\dot{x}_a + k_a x_a - c_a\dot{x} - k_a x = 0.$$

The equations of motion can be written using the D-operator as

$$\left[mD^2 + (c + c_a)D + (k + k_a)\right]x - (c_a D + k_a)x_a = F_0 \sin \Omega t,$$

$$-(c_a D + k_a)x + (m_a D^2 + c_a D + k_a)x_a = 0.$$

Because of the existence of damping, the responses of free vibration (complementary solutions) decay exponentially and approach zero as time increases. Hence, it is practically more important and useful to study responses of *forced vibration* (*particular solutions*). The determinant of the coefficient matrix is

$$\phi(D) = \begin{vmatrix} mD^2 + (c + c_a)D + (k + k_a) & -(c_a D + k_a) \\ -(c_a D + k_a) & m_a D^2 + c_a D + k_a \end{vmatrix}$$

$$= \left[mD^2 + (c + c_a)D + (k + k_a)\right](m_a D^2 + c_a D + k_a) - (c_a D + k_a)^2$$

$$= \left[(mD^2 + k)(m_a D^2 + k_a) + k_a m_a D^2 + c_a c D^2\right]$$
$$+ \left[c_a(mD^2 + k) + c(m_a D^2 + k_a) + c_a m_a D^2\right]D,$$

and

$$\Delta_1 = \begin{vmatrix} F_0 \sin \Omega t & -(c_a D + k_a) \\ 0 & m_a D^2 + c_a D + k_a \end{vmatrix} = (m_a D^2 + c_a D + k_a) F_0 \sin \Omega t.$$

Applying Cramer's Rule yields the particular solution $x_P(t)$ or response of forced vibration due to the excitation

$$x_P(t) = \frac{\Delta_1}{\phi(D)} = F_0 \frac{(m_a D^2 + k_a) + c_a D}{\left\{ \left[(mD^2+k)(m_a D^2 + k_a) + k_a m_a D^2 + c_a c D^2 \right] + \left[c_a (mD^2 + k) + c(m_a D^2 + k_a) + c_a m_a D^2 \right] D \right\}} \sin \Omega t$$

$$= F_0 \frac{(k_a - m_a \Omega^2) + c_a D}{\left\{ \left[(k - m\Omega^2)(k_a - m_a \Omega^2) - k_a m_a \Omega^2 - c_a c \Omega^2 \right] + \left[c_a (k - m\Omega^2) + c(k_a - m_a \Omega^2) - c_a m_a \Omega^2 \right] D \right\}} \sin \Omega t$$

✍ Theorem 3 of Chapter 4: replace D^2 by $-\Omega^2$.

$$= F_0 \frac{(k_a - m_a \Omega^2) + c_a D}{A + BD} \sin \Omega t,$$

where

$$A = (k - m\Omega^2)(k_a - m_a \Omega^2) - k_a m_a \Omega^2 - c_a c \Omega^2,$$

$$B = c_a (k - m\Omega^2) + c(k_a - m_a \Omega^2) - c_a m_a \Omega^2.$$

Hence

$$x_P(t) = F_0 \frac{\left[(k_a - m_a \Omega^2) + c_a D \right](A - BD)}{(A + BD)(A - BD)} \sin \Omega t$$

$$= F_0 \frac{(k_a - m_a \Omega^2) A - c_a B D^2 + \left[-(k_a - m_a \Omega^2) B + c_a A \right] D}{A^2 - B^2 D^2} \sin \Omega t$$

$$= F_0 \frac{\left[(k_a - m_a \Omega^2) A + c_a B \Omega^2 \right] \sin \Omega t + \left[-(k_a - m_a \Omega^2) B + c_a A \right] \Omega \cos \Omega t}{A^2 + B^2 \Omega^2}$$

✍ Replace D^2 by $-\Omega^2$ and then evaluate using $D \sin \Omega t = \Omega \cos \Omega t$.

$$= F_0 \frac{\sqrt{\left[(k_a - m_a \Omega^2) A + c_a B \Omega^2 \right]^2 + \left[-(k_a - m_a \Omega^2) B + c_a A \right]^2 \Omega^2}}{A^2 + B^2 \Omega^2} \sin(\Omega t + \varphi)$$

✍ Use $a \sin \theta + b \cos \theta = \sqrt{a^2 + b^2} \sin(\theta + \varphi), \quad \varphi = \tan^{-1}(b/a)$.

$$= F_0 \frac{\sqrt{\left[(k_a - m_a \Omega^2)^2 + c_a^2 \Omega^2 \right](A^2 + B^2 \Omega^2)}}{A^2 + B^2 \Omega^2} \sin(\Omega t + \varphi)$$

$$= F_0 \sqrt{\frac{(k_a - m_a \Omega^2)^2 + c_a^2 \Omega^2}{A^2 + B^2 \Omega^2}} \sin(\Omega t + \varphi).$$

The Dynamic Magnification Factor (DMF) for mass m is

$$\text{DMF} = \frac{\left|x_P(t)\right|_{\max}}{x_{\text{static}}} = \frac{F_0\sqrt{\dfrac{(k_a - m_a\Omega^2)^2 + c_a^2\Omega^2}{A^2 + B^2\Omega^2}}}{\dfrac{F_0}{k}} = k\sqrt{\frac{(k_a - m_a\Omega^2)^2 + c_a^2\Omega^2}{A^2 + B^2\Omega^2}}.$$

Adopting the following notations

$$\omega_0^2 = \frac{k}{m}, \quad c = 2m\zeta\omega_0, \quad r = \frac{\Omega}{\omega_0}, \quad \mu = \frac{m_a}{m}, \quad \omega_a^2 = \frac{k_a}{m_a}, \quad c_a = 2m_a\zeta_a\omega_0, \quad r_a = \frac{\omega_a}{\omega_0},$$

one has

$$\begin{aligned}
(k_a - m_a\Omega^2)^2 + c_a^2\Omega^2 &= m_a\left(\frac{k_a}{m_a} - \Omega^2\right)^2 + (2m_a\zeta_a\omega_0)^2\Omega^2 \\
&= m_a^2\omega_0^4\left[\left(\frac{\omega_a^2 - \Omega^2}{\omega_0^2}\right)^2 + \left(\frac{2\zeta_a\omega_0\Omega}{\omega_0^2}\right)^2\right] = m_a^2\omega_0^4[(r_a^2 - r^2)^2 + (2\zeta_a r)^2],
\end{aligned}$$

$$\begin{aligned}
A^2 + B^2\Omega^2 &= \left[(k - m\Omega^2)(k_a - m_a\Omega^2) - k_a m_a\Omega^2 - c_a c\Omega^2\right]^2 \\
&\quad + \left[c_a(k - m\Omega^2) + c(k_a - m_a\Omega^2) - c_a m_a\Omega^2\right]^2\Omega^2 \\
&= \left[m_a m\left(\frac{k}{m} - \Omega^2\right)\left(\frac{k_a}{m_a} - \Omega^2\right) - m_a^2\frac{k_a}{m_a}\Omega^2 - (2m_a\zeta_a\omega_0)(2m\zeta\omega_0)\Omega^2\right]^2 \\
&\quad + \left[(2m_a\zeta_a\omega_0)m\left(\frac{k}{m} - \Omega^2\right) + (2m\zeta\omega_0)m_a\left(\frac{k_a}{m_a} - \Omega^2\right) - (2m_a\zeta_a\omega_0)m_a\Omega^2\right]^2\Omega^2 \\
&= \left[m_a m\omega_0^4\left(\frac{\omega_0^2 - \Omega^2}{\omega_0^2}\right)\left(\frac{\omega_a^2 - \Omega^2}{\omega_0^2}\right) - m_a^2\omega_0^4\frac{\omega_a^2}{\omega_0^2}\frac{\Omega^2}{\omega_0^2} - m_a m\omega_0^4\cdot 4\zeta_a\zeta\frac{\Omega^2}{\omega_0^2}\right]^2 \\
&\quad + \left[m_a m\omega_0^3\cdot 2\zeta_a\left(\frac{\omega_0^2 - \Omega^2}{\omega_0^2}\right) + m_a m\omega_0^3\cdot 2\zeta\left(\frac{\omega_a^2 - \Omega^2}{\omega_0^2}\right) - m_a^2\omega_0^3\cdot 2\zeta_a\frac{\Omega^2}{\omega_0^2}\right]^2\Omega^2 \\
&= m_a^2 m^2\omega_0^8\Big\{\left[(1 - r^2)(r_a^2 - r^2) - \mu r_a^2 r^2 - 4\zeta_a\zeta r^2\right]^2 \\
&\qquad + 4r^2\left[\zeta_a(1 - r^2) + \zeta(r_a^2 - r^2) - \mu\zeta_a r^2\right]^2\Big\}.
\end{aligned}$$

The Dynamic Magnification Factor becomes

$$\text{DMF} = \left\{\frac{(r_a^2 - r^2)^2 + (2\zeta_a r)^2}{\left[(1 - r^2)(r_a^2 - r^2) - \mu r_a^2 r^2 - 4\zeta_a\zeta r^2\right]^2 + 4r^2\left[\zeta_a(1 - r^2 - \mu r^2) + \zeta(r_a^2 - r^2)\right]^2}\right\}^{\frac{1}{2}}.$$

For the special case when $\mu = 0$, $r_a = 0$, $\zeta_a = 0$, the Dynamic Magnification Factor reduces to

$$\text{DMF} = \frac{1}{\sqrt{(1 - r^2)^2 + (2\zeta r)^2}},$$

which recovers the DMF of a single degree-of-freedom system, i.e., the main system without the auxiliary vibration absorber or TMD.

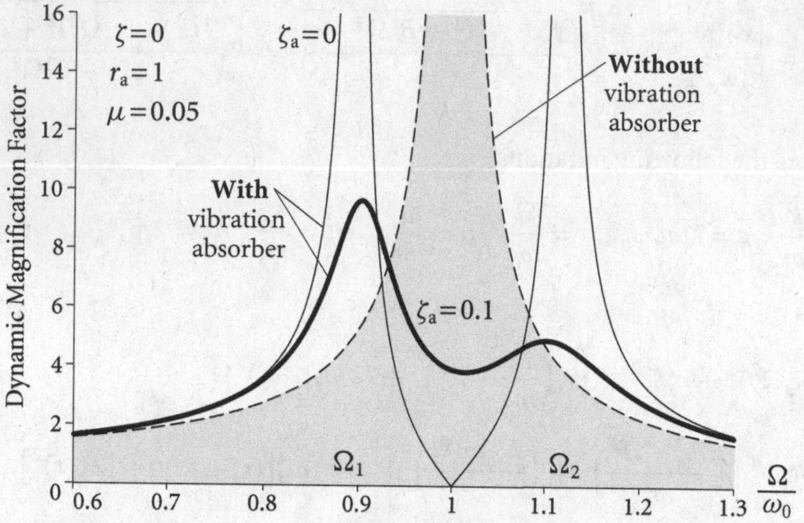

Figure 8.4 Dynamic Magnification Factor for $\zeta = 0$.

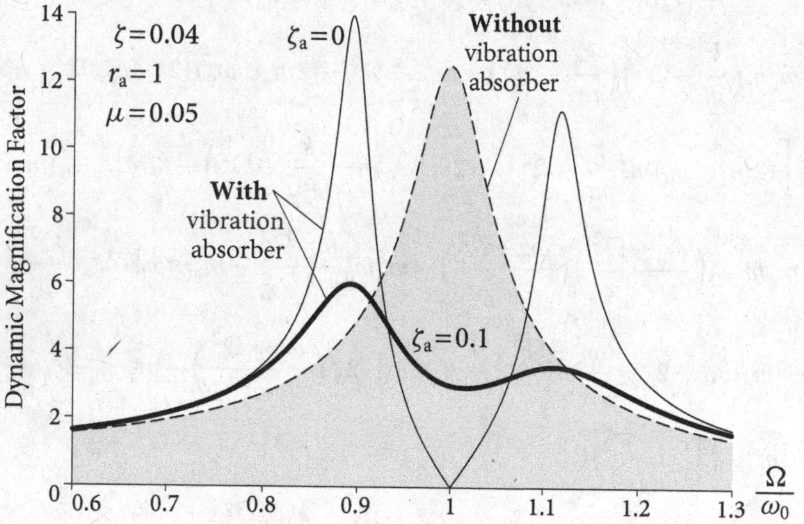

Figure 8.5 Dynamic Magnification Factor for $\zeta = 0.04$.

The Dynamic Magnification Factors for an undamped main system, i.e., $\zeta = 0$, are shown in Figure 8.4. Without the vibration absorber or TMD, the single degree-of-freedom system is in resonance when $r = 1$ or $\Omega = \omega_0$, where the amplitude of the response grows linearly with time or DMF approaches infinite.

In order to reduce the vibration of the main system at resonance, a vibration absorber or TMD is attached to the main mass m. The vibration absorber is usually tuned so that $\omega_a = \omega_0$ or $r_a = 1$, hence the name tuned mass damper. In practice, the mass of the vibration absorber or TMD is normally much smaller than that of the main mass, i.e., $m_a \ll m$ or $\mu \ll 1$; in Figure 8.4, μ is taken as $1/20 = 0.05$.

If the vibration absorber or TMD is undamped, i.e., $\zeta_a = 0$, then DMF$=0$ when $\Omega = \omega_0$, meaning that the vibration absorber eliminates vibration of the main mass m at the resonant frequency $\Omega = \omega_0$. However, it is seen that the vibration absorber or TMD introduces two resonant frequencies Ω_1 and Ω_2, at which the amplitude of vibration of the main mass m is infinite. In practice, the excitation frequency Ω must be kept away from the frequencies Ω_1 and Ω_2.

In order not to introduce extra resonant frequencies, vibration absorbers or TMD are usually damped. A typical result of DMF is shown in Figure 8.4 for $\zeta_a = 0.1$. It is seen that the vibration of the main mass m is effectively suppressed for all excitation frequencies. By varying the value of ζ_a, an optimal vibration absorber can be designed.

When the main system is also damped, typical results of DMF are shown in Figure 8.5. Similar conclusions can also be drawn.

As an application, Figure 8.6 shows a schematic illustration of the tuned mass damper (TMD) in Taipei 101. Taipei 101 is a landmark skyscraper located in Taipei, Taiwan. The building has 101 floors above ground and 5 floors underground; the height of the top of its spire or Pinnacle is 509.2 m. It has been extolled as one of the Seven New Wonders of the World and Seven Wonders of Engineering.

Taiwan lies in one of the most earthquake-prone regions of the world. A catastrophic earthquake on September 21, 1999, measured 7.6 on the Richter scale,

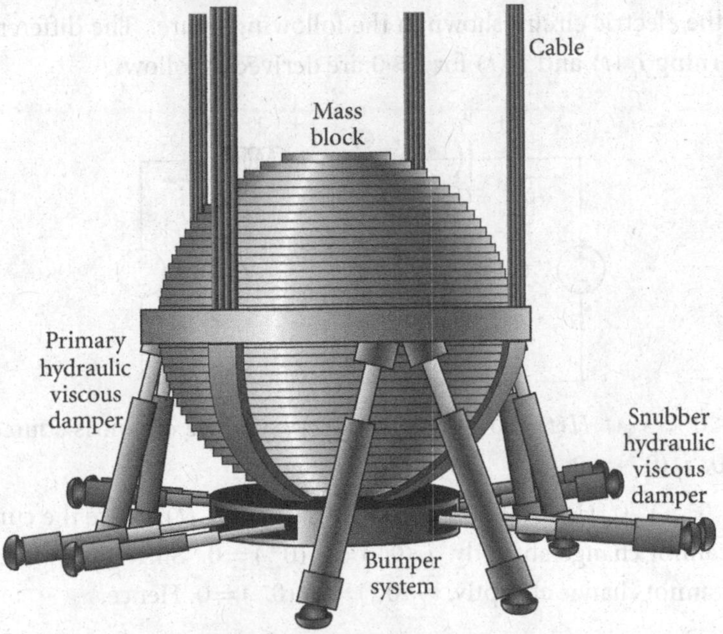

Figure 8.6 Tuned mass damper (TMD) in Taipei 101.

killed over 2000 people, and seriously injured over 11,000 people in northern and central Taiwan. From summer to fall, Taiwan is also affected by typhoons.

Taipei 101 is designed to withstand the strongest earthquakes likely to occur in a 2500-year cycle and gale winds of 60 m/sec.

A tuned mass damper (TMD), the largest and heaviest of its type in the world, is installed between the 87th and 91st floors. It consists of a sphere of 5.5 m in diameter constructed from 41 steel plates with a total weight of 660 metric tons (equivalent to 0.26 percent of building weight). The sphere is suspended from four groups of steel cables (each group has four 9-cm diameter cables) and supported by eight primary hydraulic viscous dampers as shown in Figure 8.6. A bumper system of eight snubber hydraulic viscous dampers placed at the bottom of the sphere absorbs vibration impacts, particularly in major typhoons or earthquakes where the movement of the TMD exceeds 1.5 m. The period of the TMD is tuned to be the same as that of the building, approximately 7 sec. The TMD helps to stabilize the tower to withstand earthquakes measuring above the magnitude of 7.0 and reduces the building's vibration by as much as 45% in strong winds.

Another two tuned mass dampers, each weighing 6 metric tons, are installed at the Pinnacle to reduce wind-induced fatigue of its steel structure, which vibrates approximately 180,000 cycles every year. The two Pinnacle-TMD's reduce the vibration of the Pinnacle by 40%.

8.3 An Electric Circuit

Consider the electric circuit shown in the following figure. The differential equations governing $i_C(t)$ and $i_L(t)$ for $t>0$ are derived as follows.

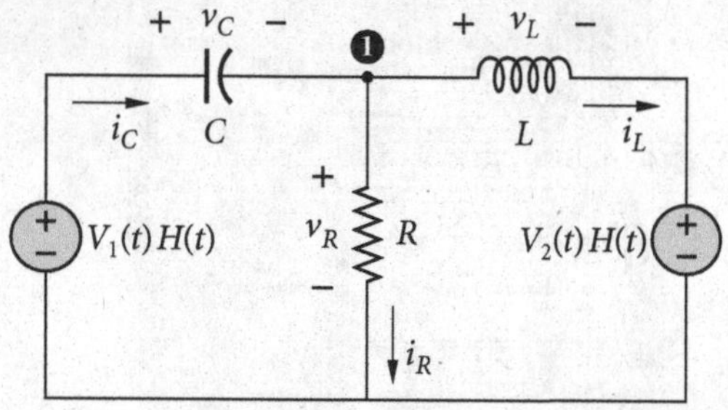

❧ For $t \leqslant 0^-$, $V_1(t)H(t)=0$ and $V_2(t)H(t)=0$. The circuit is source free and $v_C(0^-)=0$, $i_L(0^-)=0$.

❧ For $t \geqslant 0^+$, $V_1(t)H(t)=V_1(t)$ and $V_2(t)H(t)=V_2(t)$. Since the current in an inductor cannot change abruptly, $i_L(0^+)=i_L(0^-)=0$. Since the voltage across a capacitor cannot change abruptly, $v_C(0^+)=v_C(0^-)=0$. Hence,

$$v_R(0^+) = v_1(0^+) \implies i_C(0^+) = i_R(0^+) = \frac{V_1(0^+)}{R}.$$

Applying Kirchhoff's Current Law at node 1 yields $i_R = i_C - i_L$.
Applying Kirchhoff's Voltage Law on the left mesh gives

$$-V_1(t) + v_C + v_R = 0 \implies \frac{1}{C}\int_{-\infty}^{t} i_C \,\mathrm{d}t + R i_R = V_1(t).$$

Differentiating with respect to t leads to

$$\frac{1}{C} i_C + R\frac{\mathrm{d}(i_C - i_L)}{\mathrm{d}t} = \frac{\mathrm{d}V_1(t)}{\mathrm{d}t}.$$

Applying Kirchhoff's Voltage Law on the right mesh results in

$$-v_R + v_L + V_2(t) = 0 \implies R(i_C - i_L) - L\frac{\mathrm{d}i_L}{\mathrm{d}t} = V_2(t).$$

Hence, the system of differential equations governing $i_C(t)$ and $i_L(t)$ are

$$R\frac{\mathrm{d}i_C}{\mathrm{d}t} + \frac{1}{C} i_C - R\frac{\mathrm{d}i_L}{\mathrm{d}t} = \frac{\mathrm{d}V_1(t)}{\mathrm{d}t},$$

$$R i_C - L\frac{\mathrm{d}i_L}{\mathrm{d}t} - R i_L = V_2(t), \qquad i_C(0^+) = \frac{V_1(0^+)}{R}, \quad i_L(0^+) = 0.$$

Suppose that $R = 2\,\Omega$, $C = \frac{5}{8}$ F, $L = 1\,H$, $V_1(t) = 6\,$V, and $V_2(t)$ is as shown in the following figure, which can be expressed as $V_2(t) = 24t - 24(t-1)H(t-1)$.

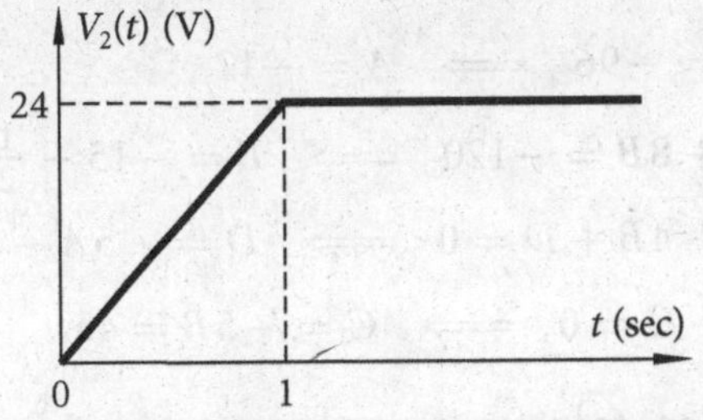

The system of differential equations becomes

$$5\frac{\mathrm{d}i_C}{\mathrm{d}t} + 4 i_C - 5\frac{\mathrm{d}i_L}{\mathrm{d}t} = 0,$$

$$2 i_C - \frac{\mathrm{d}i_L}{\mathrm{d}t} - 2 i_L = 24t - 24(t-1)H(t-1), \qquad i_C(0^+) = 3, \quad i_L(0^+) = 0.$$

The method of Laplace transform and the matrix method will be applied to solve the differential equations for $i_C(t)$ and $i_L(t)$.

Method of Laplace Transform

Let $I_C(s) = \mathscr{L}\{i_C(t)\}$ and $I_L(s) = \mathscr{L}\{i_L(t)\}$. Taking the Laplace transform of both sides of the differential equations gives

$$5\big[s I_C(s) - i_C(0^+)\big] + 4 I_C(s) - 5\big[s I_L(s) - i_L(0^+)\big] = 0,$$

$$2I_C(s) - \big[sI_L(s) - i_L(0^+)\big] - 2I_L(s) = \frac{24(1-e^{-s})}{s^2}.$$

These equations lead to two algebraic equations for $I_C(s)$ and $I_L(s)$

$$(5s+4)I_C(s) - 5sI_L(s) = 15, \quad 2I_C(s) - (s+2)I_L(s) = \frac{24(1-e^{-s})}{s^2},$$

which can be solved using Cramer's Rule

$$\Delta = \begin{vmatrix} 5s+4 & -5s \\ 2 & -(s+2) \end{vmatrix} = -(5s^2+4s+8),$$

$$\Delta_L = \begin{vmatrix} 5s+4 & 15 \\ 2 & \dfrac{24(1-e^{-s})}{s^2} \end{vmatrix} = \frac{120s+96}{s^2}(1-e^{-s}) - 30,$$

$$\therefore \quad I_L(s) = \frac{\Delta_L}{\Delta} = \frac{-120s-96}{s^2(5s^2+4s+8)}(1-e^{-s}) + \frac{30}{5s^2+4s+8}.$$

Using partial fractions, one has

$$\frac{-120s-96}{s^2(5s^2+4s+8)} = \frac{A}{s^2} + \frac{B}{s} + \frac{Cs+D}{5s^2+4s+8}.$$

Comparing the coefficients of the numerators leads to

$$\begin{aligned} 1:&\quad 8A = -96 \implies A = -12, \\ s:&\quad 4A + 8B = -120 \implies B = -15 - \tfrac{1}{2}A = -9, \\ s^2:&\quad 5A + 4B + D = 0 \implies D = -5A - 4B = 96, \\ s^3:&\quad 5B + C = 0 \implies C = -5B = 45. \end{aligned}$$

Taking the inverse Laplace transform, one has

$$\begin{aligned} i_L(t) &= \mathscr{L}^{-1}\left\{\left(\frac{-12}{s^2} + \frac{-9}{s} + \frac{45s+96}{5s^2+4s+8}\right)(1-e^{-s}) + \frac{30}{5s^2+4s+8}\right\} \\ &= \mathscr{L}^{-1}\left\{\left(-\frac{12}{s^2} - \frac{9}{s}\right)(1-e^{-s}) + \frac{9\left(s+\frac{2}{5}\right) + 13\left(\frac{6}{5}\right)}{\left(s+\frac{2}{5}\right)^2 + \left(\frac{6}{5}\right)^2}\left[1 - e^{\frac{2}{5}}\cdot e^{-(s+\frac{2}{5})}\right] \right. \\ &\qquad \left. + \frac{5\left(\frac{6}{5}\right)}{\left(s+\frac{2}{5}\right)^2 + \left(\frac{6}{5}\right)^2}\right\} \qquad ✍ \ \mathscr{L}^{-1}\{F(s-a)\} = e^{at}\,\mathscr{L}\{F(s)\} \\ &= -12t - 9 - \big[-12(t-1)H(t-1) - 9H(t-1)\big] \quad ✍ \ \mathscr{L}^{-1}\{e^{-as}F(s)\} \\ &\qquad\qquad\qquad\qquad\qquad\qquad\qquad\qquad = f(t-a)H(t-a) \\ &\quad + e^{-\frac{2t}{5}}\,\mathscr{L}^{-1}\left\{\frac{9s + 13\left(\frac{6}{5}\right)}{s^2 + \left(\frac{6}{5}\right)^2}\left(1 - e^{\frac{2}{5}}\cdot e^{-s}\right) + \frac{5\left(\frac{6}{5}\right)}{s^2 + \left(\frac{6}{5}\right)^2}\right\} \end{aligned}$$

$$= -12t - 9 + (12t-3)H(t-1) + e^{-\frac{2t}{5}}\left\{\left(9\cos\frac{6t}{5} + 13\sin\frac{6t}{5}\right)\right.$$

$$\left. - e^{\frac{2}{5}}\left[9\cos\frac{6(t-1)}{5} + 13\sin\frac{6(t-1)}{5}\right]H(t-1) + 5\sin\frac{6t}{5}\right\}$$

$$= -12t - 9 + (12t-3)H(t-1) + e^{-\frac{2t}{5}}\left(9\cos\frac{6t}{5} + 18\sin\frac{6t}{5}\right)$$

$$- e^{-\frac{2(t-1)}{5}}\left[9\cos\frac{6(t-1)}{5} + 13\sin\frac{6(t-1)}{5}\right]H(t-1).$$

The current $i_C(t)$ can also be obtained in the same way as i_L. However, since $i_C(t)$ can be expressed in terms of $i_L(t)$ from the second differential equation of the system, one can obtain i_C using

$$i_C(t) = 12t - 12(t-1)H(t-1) + i_L(t) + \frac{1}{2}\frac{di_L(t)}{dt}.$$

Noting that $\dfrac{dH(t-a)}{dt} = \delta(t-a)$, $f(t)\,\delta(t-a) = f(a)\,\delta(t-a)$, and using the product rule, one has

$$\frac{di_L}{dt} = -12 + 12H(t-1) + (12t-3)\,\delta(t-1)$$

$$-\frac{2}{5}e^{-\frac{2t}{5}}\left(9\cos\frac{6t}{5} + 18\sin\frac{6t}{5}\right) + e^{-\frac{2t}{5}}\left(-\frac{54}{5}\sin\frac{6t}{5} + \frac{108}{5}\cos\frac{6t}{5}\right)$$

$$+\frac{2}{5}e^{-\frac{2(t-1)}{5}}\left[9\cos\frac{6(t-1)}{5} + 13\sin\frac{6(t-1)}{5}\right]H(t-1)$$

$$- e^{-\frac{2(t-1)}{5}}\left[-\frac{54}{5}\sin\frac{6(t-1)}{5} + \frac{78}{5}\cos\frac{6(t-1)}{5}\right]H(t-1)$$

$$- e^{-\frac{2(t-1)}{5}}\left[9\cos\frac{6(t-1)}{5} + 13\sin\frac{6(t-1)}{5}\right]\delta(t-1)$$

$$= -12 + 12H(t-1) + 18e^{-\frac{2t}{5}}\left(\cos\frac{6t}{5} - \sin\frac{6t}{5}\right)$$

$$+ e^{-\frac{2(t-1)}{5}}\left[-12\cos\frac{6(t-1)}{5} + 16\sin\frac{6(t-1)}{5}\right]H(t-1),$$

$$\therefore\ i_C(t) = -15 + 15H(t-1) + 9e^{-\frac{2t}{5}}\left(2\cos\frac{6t}{5} + \sin\frac{6t}{5}\right)$$

$$- 5e^{-\frac{2(t-1)}{5}}\left[3\cos\frac{6(t-1)}{5} + \sin\frac{6(t-1)}{5}\right]H(t-1).$$

Matrix Method

From the second differential equation, one has

$$\frac{di_L}{dt} = 2i_C - 2i_L - 24t + 24(t-1)H(t-1).$$

Substituting into the first differential equation yields

$$\frac{di_L}{dt} = -\frac{4}{5}i_C + \frac{di_C}{dt} = \frac{6}{5}i_C - 2i_L - 24t + 24(t-1)H(t-1).$$

In the matrix form, the system of differential equations can be written as

$$\frac{d\mathbf{i}(t)}{dt} = \mathbf{A}\mathbf{i}(t), \quad \mathbf{i}(t) = \begin{Bmatrix} i_C \\ i_L \end{Bmatrix}, \quad \mathbf{A} = \begin{bmatrix} \frac{6}{5} & -2 \\ 2 & -2 \end{bmatrix},$$

$$\mathbf{f}(t) = -24\big[t+(t-1)H(t-1)\big]\begin{Bmatrix} 1 \\ 1 \end{Bmatrix}, \quad \mathbf{i}(0) = \begin{Bmatrix} i_C(0^+) \\ i_L(0^+) \end{Bmatrix} = \begin{Bmatrix} 3 \\ 0 \end{Bmatrix}.$$

The characteristic equation is

$$\det(\mathbf{A}-\lambda\mathbf{I}) = \begin{vmatrix} \frac{6}{5}-\lambda & -2 \\ 2 & -2-\lambda \end{vmatrix} = \tfrac{1}{5}(5\lambda^2+4\lambda+8) = 0 \implies \lambda = -\tfrac{2}{5} \pm \mathrm{i}\tfrac{6}{5}.$$

For eigenvalue $\lambda = -\frac{2}{5}+\mathrm{i}\frac{6}{5}$, the corresponding eigenvector is

$$(\mathbf{A}-\lambda\mathbf{I})\mathbf{v} = \begin{bmatrix} \frac{8}{5}-\mathrm{i}\frac{6}{5} & -2 \\ 2 & -\frac{8}{5}-\mathrm{i}\frac{6}{5} \end{bmatrix}\begin{Bmatrix} v_1 \\ v_2 \end{Bmatrix} = \begin{Bmatrix} (\frac{8}{5}-\mathrm{i}\frac{6}{5})v_1 - 2v_2 \\ 2v_1 + (-\frac{8}{5}-\mathrm{i}\frac{6}{5})v_2 \end{Bmatrix} = \begin{Bmatrix} 0 \\ 0 \end{Bmatrix}.$$

Taking $v_1 = 5$, then $v_2 = \frac{1}{2}(\frac{8}{5}-\mathrm{i}\frac{6}{5})v_1 = 4-\mathrm{i}3$,

$$\therefore \quad \mathbf{v} = \begin{Bmatrix} v_1 \\ v_2 \end{Bmatrix} = \begin{Bmatrix} 5 \\ 4-\mathrm{i}3 \end{Bmatrix} = \begin{Bmatrix} 5 \\ 4 \end{Bmatrix} + \mathrm{i}\begin{Bmatrix} 0 \\ -3 \end{Bmatrix}.$$

Hence, using Euler's formula $e^{\mathrm{i}\theta} = \cos\theta + \mathrm{i}\sin\theta$,

$$e^{\lambda t}\mathbf{v} = e^{-\frac{2t}{5}}\Big(\cos\frac{6t}{5} + \mathrm{i}\sin\frac{6t}{5}\Big)\left(\begin{Bmatrix} 5 \\ 4 \end{Bmatrix} + \mathrm{i}\begin{Bmatrix} 0 \\ -3 \end{Bmatrix}\right)$$

$$= e^{-\frac{2t}{5}}\left[\begin{Bmatrix} 5 \\ 4 \end{Bmatrix}\cos\frac{6t}{5} - \begin{Bmatrix} 0 \\ -3 \end{Bmatrix}\sin\frac{6t}{5} + \mathrm{i}\left(\begin{Bmatrix} 0 \\ -3 \end{Bmatrix}\cos\frac{6t}{5} + \begin{Bmatrix} 5 \\ 4 \end{Bmatrix}\sin\frac{6t}{5}\right)\right].$$

A fundamental matrix is

$$\mathbf{I}(t) = \big[\mathcal{Re}(e^{\lambda t}\mathbf{v}),\ \mathcal{Im}(e^{\lambda t}\mathbf{v})\big] = e^{-\frac{2t}{5}}\begin{bmatrix} 5\cos\frac{6t}{5} & 5\sin\frac{6t}{5} \\ 3\sin\frac{6t}{5} + 4\cos\frac{6t}{5} & 4\sin\frac{6t}{5} - 3\cos\frac{6t}{5} \end{bmatrix},$$

and its inverse is obtained as

$$\mathbf{I}^{-1}(t) = -\frac{e^{\frac{2t}{5}}}{15}\begin{bmatrix} 4\sin\frac{6t}{5} - 3\cos\frac{6t}{5} & -5\sin\frac{6t}{5} \\ -3\sin\frac{6t}{5} - 4\cos\frac{6t}{5} & 5\cos\frac{6t}{5} \end{bmatrix}.$$

Evaluate the quantity

$$\mathbf{I}^{-1}(0)\,\mathbf{i}(0)+\int_0^t \mathbf{I}^{-1}(t)\,\mathbf{f}(t)\,\mathrm{d}t=-\frac{1}{15}\begin{bmatrix}-3 & 0\\ -4 & 5\end{bmatrix}\begin{Bmatrix}3\\ 0\end{Bmatrix}$$

$$+\int_0^t -\frac{e^{\frac{2t}{5}}}{15}\begin{bmatrix}4\sin\frac{6t}{5}-3\cos\frac{6t}{5} & -5\sin\frac{6t}{5}\\ -3\sin\frac{6t}{5}-4\cos\frac{6t}{5} & 5\cos\frac{6t}{5}\end{bmatrix}\begin{Bmatrix}1\\ 1\end{Bmatrix}\Big\{-24\big[t+(t-1)H(t-1)\big]\Big\}\mathrm{d}t$$

$$=\begin{Bmatrix}\frac{3}{5}\\ \frac{4}{5}\end{Bmatrix}-\frac{8}{5}\int_0^t\begin{Bmatrix}e^{\frac{2t}{5}}\Big(\sin\frac{6t}{5}+3\cos\frac{6t}{5}\Big)t\\ e^{\frac{2t}{5}}\Big(3\sin\frac{6t}{5}-\cos\frac{6t}{5}\Big)t\end{Bmatrix}\mathrm{d}t$$

$$+\frac{8}{5}\int_1^t\begin{Bmatrix}e^{\frac{2t}{5}}\Big(\sin\frac{6t}{5}+3\cos\frac{6t}{5}\Big)(t-1)\\ e^{\frac{2t}{5}}\Big(3\sin\frac{6t}{5}-\cos\frac{6t}{5}\Big)(t-1)\end{Bmatrix}\mathrm{d}t\cdot H(t-1)$$

$$=\left\{\begin{matrix}\frac{18}{5}+e^{\frac{2t}{5}}\Big(-4t\sin\frac{6t}{5}-3\cos\frac{6t}{5}+\sin\frac{6t}{5}\Big)\\ \quad+\Big[e^{\frac{2t}{5}}\Big(4t\sin\frac{6t}{5}+3\cos\frac{6t}{5}-5\sin\frac{6t}{5}\Big)-e^{\frac{2}{5}}\Big(3\cos\frac{6}{5}-\sin\frac{6}{5}\Big)\Big]H(t-1)\\ \frac{9}{5}+e^{\frac{2t}{5}}\Big(4t\cos\frac{6t}{5}-\cos\frac{6t}{5}-3\sin\frac{6t}{5}\Big)\\ \quad-\Big[e^{\frac{2t}{5}}\Big(4t\cos\frac{6t}{5}-5\cos\frac{6t}{5}-3\sin\frac{6t}{5}\Big)+e^{\frac{2}{5}}\Big(\cos\frac{6}{5}+3\sin\frac{6}{5}\Big)\Big]H(t-1)\end{matrix}\right\}.$$

Substituting into the solution of the system of differential equations

$$\mathbf{i}(t)=\mathbf{I}(t)\left\{\mathbf{I}^{-1}(0)\,\mathbf{i}(0)+\int_0^t \mathbf{I}^{-1}(t)\,\mathbf{f}(t)\,\mathrm{d}t\right\},$$

expanding, and simplifying by combining terms using trigonometric identities, one obtains the same results as obtained using the method of Laplace transform.

Remarks: The method of Laplace transform, while applicable, is easier than the matrix method; whereas the matrix method is the most general approach.

8.4 Vibration of a Two-Story Shear Building

Consider the vibration of an undamped two-story shear building under harmonic excitations with $m_1=m_2=m$, $k_1=k_2=k$, $F_1(t)=F_2(t)=ma\cos\Omega t$, as shown in Figure 8.7. A two-story shear building is a two degrees-of-freedom system, i.e., it needs two variables x_1 and x_2 to describe its motion.

From Example 7.2 in Chapter 7, equations of motion are given by

$$m\ddot{x}_1+2kx_1-kx_2=ma\cos\Omega t,$$

$$m\ddot{x}_2-kx_1+kx_2=ma\cos\Omega t,$$

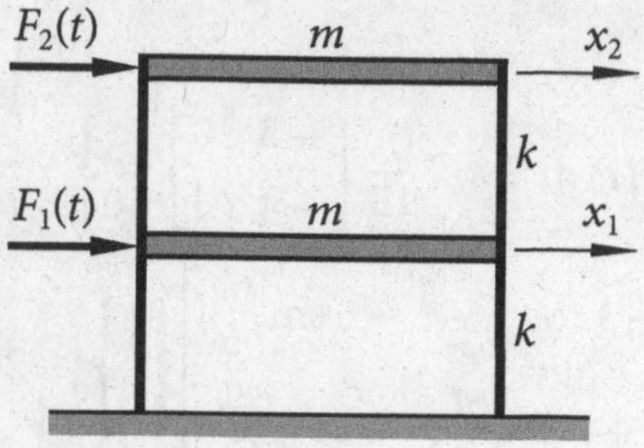

Figure 8.7 A two-story shear building.

or, using the notations $\omega_0 = \sqrt{k/m}$,

$$\ddot{x}_1 + 2\omega_0^2 x_1 - \omega_0^2 x_2 = a\cos\Omega t,$$
$$\ddot{x}_2 - \omega_0^2 x_1 + \omega_0^2 x_2 = a\cos\Omega t.$$

In the matrix form, the equations of motion are

$$\mathbf{I}\ddot{\mathbf{x}} + \mathbf{K}\mathbf{x} = \mathbf{F}(t),$$

where

$$\mathbf{K} = \begin{bmatrix} 2\omega_0^2 & -\omega_0^2 \\ -\omega_0^2 & \omega_0^2 \end{bmatrix}, \qquad \mathbf{F}(t) = \begin{Bmatrix} a\cos\Omega t \\ a\cos\Omega t \end{Bmatrix}.$$

The analysis of free and forced vibration follows the procedure presented in Section 7.4.3.

8.4.1 Free Vibration—Complementary Solutions

The frequency equation is given by

$$\left|\mathbf{K} - \omega^2\mathbf{I}\right| = \begin{vmatrix} 2\omega_0^2 - \omega^2 & -\omega_0^2 \\ -\omega_0^2 & \omega_0^2 - \omega^2 \end{vmatrix} = 0,$$

$$\omega^4 - 3\omega_0^2\omega^2 + \omega_0^4 = 0 \implies \omega^2 = \frac{3\pm\sqrt{5}}{2}\omega_0^2,$$

$$\omega = \sqrt{\frac{3\pm\sqrt{5}}{2}}\,\omega_0 = \frac{\sqrt{6\pm2\sqrt{5}}}{2}\,\omega_0 = \frac{\sqrt{(\sqrt{5})^2\pm2\sqrt{5}+1}}{2}\,\omega_0 = \frac{\sqrt{5}\pm1}{2}\,\omega_0.$$

Hence, the eigenvalues or modal frequencies are

$$\omega_1 = \frac{\sqrt{5}-1}{2}\,\omega_0, \qquad \omega_2 = \frac{\sqrt{5}+1}{2}\,\omega_0.$$

☞ Natural frequencies are always ordered in ascending order, i.e., $\omega_1 < \omega_2 < \dots$.

Corresponding to ω_1, the first mode shape $\hat{\mathbf{x}}_1$ is the eigenvector given by

$$\begin{bmatrix} 2\omega_0^2-\omega_1^2 & -\omega_0^2 \\ -\omega_0^2 & \omega_0^2-\omega_1^2 \end{bmatrix} \begin{Bmatrix} \hat{x}_{11} \\ \hat{x}_{21} \end{Bmatrix} = \mathbf{0}.$$

Taking $\hat{x}_{11}=1$, then

$$\hat{x}_{21} = \frac{2\omega_0^2-\omega_1^2}{\omega_0^2} = 2 - \frac{3-\sqrt{5}}{2} = \frac{1+\sqrt{5}}{2}.$$

Corresponding to ω_2, the second mode shape $\hat{\mathbf{x}}_2$ is the eigenvector given by

$$\begin{bmatrix} 2\omega_0^2-\omega_2^2 & -\omega_0^2 \\ -\omega_0^2 & \omega_0^2-\omega_2^2 \end{bmatrix} \begin{Bmatrix} \hat{x}_{12} \\ \hat{x}_{22} \end{Bmatrix} = \mathbf{0}.$$

Taking $\hat{x}_{12}=1$, then

$$\hat{x}_{22} = \frac{2\omega_0^2-\omega_2^2}{\omega_0^2} = 2 - \frac{3+\sqrt{5}}{2} = \frac{1-\sqrt{5}}{2}.$$

Hence, the two mode shapes are given by

$$\hat{\mathbf{x}}_1 = \begin{Bmatrix} \hat{x}_{11} \\ \hat{x}_{21} \end{Bmatrix} = \begin{Bmatrix} 1 \\ \frac{1+\sqrt{5}}{2} \end{Bmatrix}, \qquad \hat{\mathbf{x}}_2 = \begin{Bmatrix} \hat{x}_{12} \\ \hat{x}_{22} \end{Bmatrix} = \begin{Bmatrix} 1 \\ \frac{1-\sqrt{5}}{2} \end{Bmatrix}.$$

The responses of free vibration (or complementary solution) are of the form

$$\mathbf{x}_C(t) = a_1\,\hat{\mathbf{x}}_1 \sin(\omega_1 t+\theta_1) + a_2\,\hat{\mathbf{x}}_2 \sin(\omega_2 t+\theta_2),$$

where the constants a_1, θ_1, a_2, θ_2 are determined from the initial conditions $\mathbf{x}(0)$ and $\dot{\mathbf{x}}(0)$.

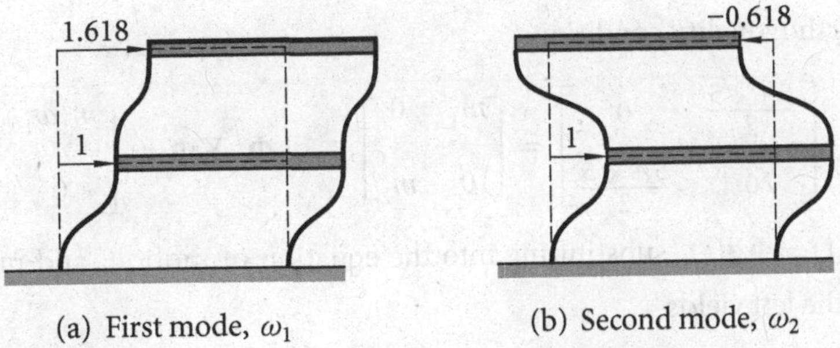

Figure 8.8 Mode shapes of a two-story shear building.

If the initial conditions are such that $a_2=0$, $\theta_2=0$, the system is vibrating in the first mode with natural circular frequency ω_1. The responses of free vibration are of the form

$$\mathbf{x}_C(t) = \begin{Bmatrix} x_{1C}(t) \\ x_{2C}(t) \end{Bmatrix} = a_1 \begin{Bmatrix} 1 \\ \frac{1+\sqrt{5}}{2} \end{Bmatrix} \sin(\omega_1 t+\theta_1).$$

The ratio of the amplitudes of responses of the second and first floors is

$$\frac{x_{2C}(t)\big|_{\text{amplitude}}}{x_{1C}(t)\big|_{\text{amplitude}}} = \frac{\frac{1+\sqrt{5}}{2}}{1} = \frac{1.618}{1}.$$

If the initial conditions are such that $a_1=0$, $\theta_1=0$, the system is vibrating in the second mode with natural circular frequency ω_2. The responses of free vibration are of the form

$$\mathbf{x}_C(t) = \begin{Bmatrix} x_{1C}(t) \\ x_{2C}(t) \end{Bmatrix} = a_2 \begin{Bmatrix} 1 \\ \frac{1-\sqrt{5}}{2} \end{Bmatrix} \sin(\omega_2 t+\theta_2).$$

The ratio of the amplitudes of responses of the second and first floors is

$$\frac{x_{2C}(t)\big|_{\text{amplitude}}}{x_{1C}(t)\big|_{\text{amplitude}}} = \frac{\frac{1-\sqrt{5}}{2}}{1} = \frac{-0.618}{1}.$$

The mode shapes $\hat{\mathbf{x}}_1$ and $\hat{\mathbf{x}}_2$ giving the ratios of the amplitudes of vibration of x_{2C} and x_{1C} are shown in Figure 8.8. In general, the responses of the system are linear combinations of the first and second modes.

8.4.2 Forced Vibration—General Solutions

The modal matrix is

$$\mathbf{\Phi} = [\hat{\mathbf{x}}_1, \hat{\mathbf{x}}_2] = \begin{bmatrix} 1 & 1 \\ \frac{1+\sqrt{5}}{2} & \frac{1-\sqrt{5}}{2} \end{bmatrix},$$

with the orthogonality conditions

$$\mathbf{\Phi}^T \mathbf{I} \mathbf{\Phi} = \begin{bmatrix} \frac{5+\sqrt{5}}{2} & 0 \\ 0 & \frac{5-\sqrt{5}}{2} \end{bmatrix} = \begin{bmatrix} m_1 & 0 \\ 0 & m_2 \end{bmatrix}, \qquad \mathbf{\Phi}^T \mathbf{K} \mathbf{\Phi} = \begin{bmatrix} m_1\omega_1^2 & 0 \\ 0 & m_2\omega_2^2 \end{bmatrix}.$$

Letting $\mathbf{x}(t)=\mathbf{\Phi}\mathbf{q}(t)$, substituting into the equation of motion, and multiplying $\mathbf{\Phi}^T$ from the left yields

$$\mathbf{\Phi}^T \mathbf{\Phi}\ddot{\mathbf{q}}(t) + \mathbf{\Phi}^T \mathbf{K}\mathbf{\Phi}\mathbf{q}(t) = \mathbf{\Phi}^T \mathbf{F}(t),$$

i.e.,

$$m_1\ddot{q}_1(t) + m_1\omega_1^2 q_1(t) = f_1(t), \qquad m_2\ddot{q}_2(t) + m_2\omega_2^2 q_2(t) = f_2(t),$$

where

$$\begin{Bmatrix} f_1(t) \\ f_2(t) \end{Bmatrix} = \mathbf{\Phi}^T \mathbf{F}(t) = \begin{bmatrix} 1 & \frac{1+\sqrt{5}}{2} \\ 1 & \frac{1-\sqrt{5}}{2} \end{bmatrix} \begin{Bmatrix} a\cos\Omega t \\ a\cos\Omega t \end{Bmatrix} = \begin{Bmatrix} \frac{3+\sqrt{5}}{2} \\ \frac{3-\sqrt{5}}{2} \end{Bmatrix} a\cos\Omega t,$$

or

$$\ddot{q}_1(t) + \omega_1^2 q_1(t) = \bar{f}_1(t), \qquad \ddot{q}_2(t) + \omega_2^2 q_2(t) = \bar{f}_2(t),$$

where

$$\bar{f}_1(t) = \frac{f_1(t)}{m_1} = \frac{5+\sqrt{5}}{10}\, a\cos\Omega t, \quad \bar{f}_2(t) = \frac{f_2(t)}{m_2} = \frac{5-\sqrt{5}}{10}\, a\cos\Omega t.$$

Using the method of operators, particular solutions, or responses of the system due to forcing, are given by

$$q_{1P}(t) = \frac{5+\sqrt{5}}{10}\,\frac{a}{D^2+\omega_1^2}\cos\Omega t, \quad q_{2P}(t) = \frac{5-\sqrt{5}}{10}\,\frac{a}{D^2+\omega_2^2}\cos\Omega t.$$

Hence,

$$\mathbf{x}_P(t) = \mathbf{\Phi}\,\mathbf{q}_P(t) = \begin{bmatrix} 1 & 1 \\ \frac{1+\sqrt{5}}{2} & \frac{1-\sqrt{5}}{2} \end{bmatrix} \begin{Bmatrix} \frac{5+\sqrt{5}}{10}\,\frac{a}{D^2+\omega_1^2}\cos\Omega t \\ \frac{5-\sqrt{5}}{10}\,\frac{a}{D^2+\omega_2^2}\cos\Omega t \end{Bmatrix}$$

$$= a \begin{Bmatrix} \frac{5+\sqrt{5}}{10}\,\frac{1}{D^2+\omega_1^2}\cos\Omega t + \frac{5-\sqrt{5}}{10}\,\frac{1}{D^2+\omega_2^2}\cos\Omega t \\ \frac{5+3\sqrt{5}}{10}\,\frac{1}{D^2+\omega_1^2}\cos\Omega t + \frac{5-3\sqrt{5}}{10}\,\frac{1}{D^2+\omega_2^2}\cos\Omega t \end{Bmatrix}.$$

The general solutions, or the responses of the system, are given by

$$\mathbf{x}(t) = \mathbf{x}_C(t) + \mathbf{x}_P(t),$$

where, as obtained earlier,

$$\mathbf{x}_C(t) = a_1\,\hat{\mathbf{x}}_1 \sin(\omega_1 t+\theta_1) + a_2\,\hat{\mathbf{x}}_2 \sin(\omega_2 t+\theta_2),$$

in which the constants $a_1, \theta_1, a_2, \theta_2$ are determined from the initial conditions $\mathbf{x}(0)$ and $\dot{\mathbf{x}}(0)$.

Depending on the values of the excitation frequency Ω, the particular solutions will have different forms.

1. $\Omega \neq \omega_1$ or ω_2, i.e., No Resonance

The following operators can be evaluated using Theorem 3 of Chapter 4 as

$$\frac{1}{D^2+\omega_i^2}\cos\Omega t = \frac{1}{\omega_i^2-\Omega^2}\cos\Omega t, \quad i=1,2. \qquad \text{✍ Replace } D^2 \text{ by } -\Omega^2$$

Hence,

$$\mathbf{x}_P(t) = \begin{Bmatrix} \dfrac{5+\sqrt{5}}{10}\dfrac{1}{\omega_1^2-\Omega^2} + \dfrac{5-\sqrt{5}}{10}\dfrac{1}{\omega_2^2-\Omega^2} \\ \dfrac{5+3\sqrt{5}}{10}\dfrac{1}{\omega_1^2-\Omega^2} + \dfrac{5-3\sqrt{5}}{10}\dfrac{1}{\omega_2^2-\Omega^2} \end{Bmatrix} a\cos\Omega t.$$

2. $\Omega = \omega_1$, i.e., Resonance in the First Mode

A particular solution $q_{1P}(t)$ can be evaluated using Theorem 4 of Chapter 4

$$\frac{1}{D^2+\omega_1^2}\cos\omega_1 t = \mathcal{Re}\Big(\frac{1}{D^2+\omega_1^2}\mathrm{e}^{\mathrm{i}\omega_1 t}\Big),$$

$$\phi(D) = D^2+\omega_1^2, \quad \phi(\mathrm{i}\omega_1) = (\mathrm{i}\omega_1)^2+\omega_1^2 = 0,$$

$$\phi'(D) = 2D, \quad \phi'(\mathrm{i}\omega_1) = 2\mathrm{i}\omega_1,$$

$$\mathcal{Re}\Big(\frac{1}{D^2+\omega_1^2}\mathrm{e}^{\mathrm{i}\omega_1 t}\Big) = \mathcal{Re}\Big[\frac{1}{\phi'(\mathrm{i}\omega_1)}t\,\mathrm{e}^{\mathrm{i}\omega_1 t}\Big] = \mathcal{Re}\Big[\frac{t}{2\mathrm{i}\omega_1}(\cos\omega_1 t + \mathrm{i}\sin\omega_1 t)\Big]$$

$$= \frac{\sqrt{5}+1}{4\omega_0}t\sin\omega_1 t.$$

A particular solution $q_{2P}(t)$ can be evaluated using Theorem 3 of Chapter 4

$$\frac{1}{D^2+\omega_2^2}\cos\omega_1 t = \frac{1}{\omega_2^2-\omega_1^2}\cos\omega_1 t = \frac{\sqrt{5}}{5\omega_0^2}\cos\omega_1 t.$$ ✍ Replace D^2 by $-\omega_1^2$

Hence,

$$\mathbf{x}_P(t) = a\begin{Bmatrix} \underbrace{\dfrac{5+3\sqrt{5}}{20}}_{\gamma_1}\cdot\dfrac{1}{\omega_0}\,t\sin\omega_1 t + \dfrac{\sqrt{5}-1}{10}\cdot\dfrac{1}{\omega_0^2}\cos\omega_1 t \\ \underbrace{\dfrac{5+2\sqrt{5}}{10}}_{\gamma_2}\cdot\dfrac{1}{\omega_0}\,t\sin\omega_1 t + \dfrac{\sqrt{5}-3}{10}\cdot\dfrac{1}{\omega_0^2}\cos\omega_1 t \end{Bmatrix}.$$

Note that, when $\Omega=\omega_1$, the terms involving $(t\sin\omega_1 t)$ have amplitudes growing linearly in time t and are dominant in the responses of the system. The ratio of the amplitudes of these two terms is

$$\frac{\gamma_2}{\gamma_1} = \frac{5+2\sqrt{5}}{10}\times\frac{20}{5+3\sqrt{5}} = \frac{1+\sqrt{5}}{2} = \frac{1.618}{1},$$

which conforms to the mode shape of the first mode as depicted in Figure 8.8(a), indicating that the system is in resonance in the first mode.

3. $\Omega = \omega_2$, i.e., Resonance in the Second Mode

A particular solution $q_{1P}(t)$ can be evaluated using Theorem 3 of Chapter 4

$$\frac{1}{D^2+\omega_1^2}\cos\omega_2 t = \frac{1}{\omega_1^2-\omega_2^2}\cos\omega_2 t = -\frac{\sqrt{5}}{5\omega_0^2}\cos\omega_2 t. \quad \text{Replace } D^2 \text{ by } -\omega_2^2$$

A particular solution $q_{2P}(t)$ can be evaluated using Theorem 4 of Chapter 4

$$\frac{1}{D^2+\omega_2^2}\cos\omega_2 t = \mathcal{Re}\Big(\frac{1}{D^2+\omega_2^2}\,e^{i\omega_2 t}\Big),$$

$$\phi(D) = D^2+\omega_2^2, \quad \phi(i\omega_2) = (i\omega_2)^2+\omega_2^2 = 0,$$

$$\phi'(D) = 2D, \quad \phi'(i\omega_2) = 2i\omega_2,$$

$$\mathcal{Re}\Big(\frac{1}{D^2+\omega_2^2}\,e^{i\omega_2 t}\Big) = \mathcal{Re}\Big[\frac{1}{\phi'(i\omega_2)}\,t\,e^{i\omega_2 t}\Big] = \mathcal{Re}\Big[\frac{t}{2i\omega_2}(\cos\omega_2 t + i\sin\omega_2 t)\Big]$$

$$= \frac{\sqrt{5}-1}{4\omega_0}\,t\sin\omega_2 t.$$

Hence,

$$\mathbf{x}_P(t) = a\left\{\begin{array}{l} -\dfrac{\sqrt{5}+1}{10}\cdot\dfrac{1}{\omega_0^2}\cos\omega_2 t + \underbrace{\dfrac{3\sqrt{5}-5}{20}}_{\gamma_1}\cdot\dfrac{1}{\omega_0}\,t\sin\omega_2 t \\ -\dfrac{\sqrt{5}+3}{10}\cdot\dfrac{1}{\omega_0^2}\cos\omega_2 t + \underbrace{\dfrac{2\sqrt{5}-5}{10}}_{\gamma_2}\cdot\dfrac{1}{\omega_0}\,t\sin\omega_2 t \end{array}\right\}.$$

Note that, when $\Omega=\omega_2$, the terms involving $(t\sin\omega_2 t)$ have amplitudes growing linearly in time t and are dominant in the responses of the system. The ratio of the amplitudes of these two terms is

$$\frac{\gamma_2}{\gamma_1} = \frac{2\sqrt{5}-5}{10}\times\frac{20}{3\sqrt{5}-5} = \frac{1-\sqrt{5}}{2} = \frac{-0.618}{1},$$

which conforms to the mode shape of the second mode as depicted in Figure 8.8(b), indicating that the system is in resonance in the second mode.

Problems

8.1 An object is launched with initial velocity v_0 at an angle of θ_0 as shown in the following figure. Neglect the air resistance.

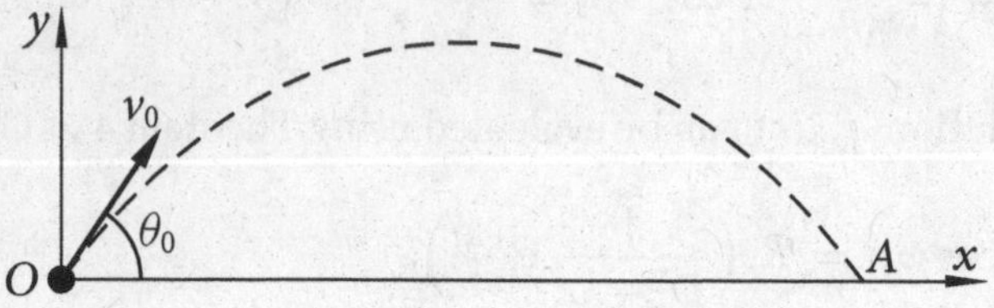

1. Show that the equation of the projectile is given by

$$y = x \tan\theta_0 - \frac{g x^2}{2 v_0^2 \cos^2\theta_0}.$$

2. Determine the maximum horizontal distance $L = OA$ and the time T it takes to reach point A.

ANS $x_{\text{max}} = \dfrac{v_0^2 \sin 2\theta_0}{g}, \quad T = \dfrac{2 v_0 \sin\theta_0}{g}$

8.2 A light rod AB is supported at each end by two similar springs with spring stiffness k, and carries two objects of concentrated mass m, one at end B and the other at the center C, as shown in the following figure. Neglect the effect of gravity.

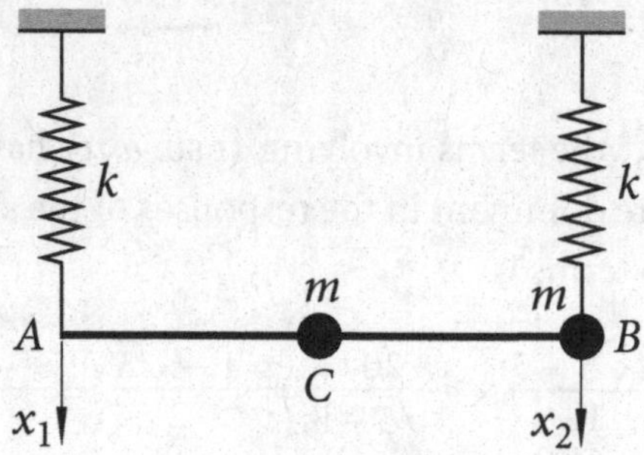

1. Show that the equations of motion are given by

$$m\ddot{x}_1 + 3m\ddot{x}_2 + 2kx_1 + 2kx_2 = 0,$$
$$m\ddot{x}_2 - kx_1 + kx_2 = 0.$$

2. Show that the two natural circular frequencies of vibration of the system are

$$\omega_1 = \sqrt{\frac{(3-\sqrt{5})\,k}{m}}, \qquad \omega_2 = \sqrt{\frac{(3+\sqrt{5})\,k}{m}}.$$

8.3 Two light rods AB and CD, each of length $2L$, are hinged at A and C and carry concentrated masses of m each at B and D. The rods are supported by three springs, each of stiffness k, as shown in the following figure. The motion of the system is described by two angles θ_1 and θ_2. Consider only small oscillations, i.e., $|\theta_1| \ll 1$ and $|\theta_2| \ll 1$, and neglect the effect of gravity.

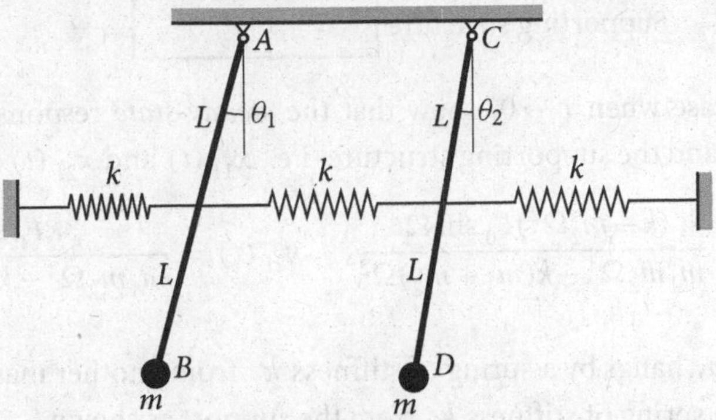

1. Show that the equations of motion are given by

$$\ddot{\theta}_1 + \frac{k}{4m}(2\theta_1 - \theta_2) = 0, \qquad \ddot{\theta}_2 + \frac{k}{4m}(-\theta_1 + 2\theta_2) = 0.$$

2. Determine the two natural circular frequencies ω_1, ω_2, and the ratio of the amplitudes θ_1/θ_2 for the two modes of vibration.

ANS First mode: $\omega_1 = \dfrac{1}{2}\sqrt{\dfrac{k}{m}}, \qquad \dfrac{\theta_1}{\theta_2} = 1$

Second mode: $\omega_2 = \dfrac{1}{2}\sqrt{\dfrac{3k}{m}}, \qquad \dfrac{\theta_1}{\theta_2} = -1$

8.4 In many engineering applications, the structures or foundations vibrate when the machines mounted on them operate. For example, the engines of an airplane mounted on wings excite the wings when they operate. To reduce the vibration of the supporting structures, vibration isolators are used to connect between the machines and the supporting structures.

The following figure shows the model of a machine of mass m_1 supported on a structure of mass m_2. The vibration isolator is modeled by a spring of stiffness k and a dashpot damper of damping coefficient c. The machine is subjected to a harmonic load $F(t) = F_0 \sin \Omega t$.

1. Show that the equations of motion are given by

$$m_1\ddot{x}_1 + c\dot{x}_1 - c\dot{x}_2 + kx_1 - kx_2 = F_0 \sin \Omega t,$$
$$m_2\ddot{x}_2 + c\dot{x}_2 - c\dot{x}_1 + kx_2 - kx_1 = 0.$$

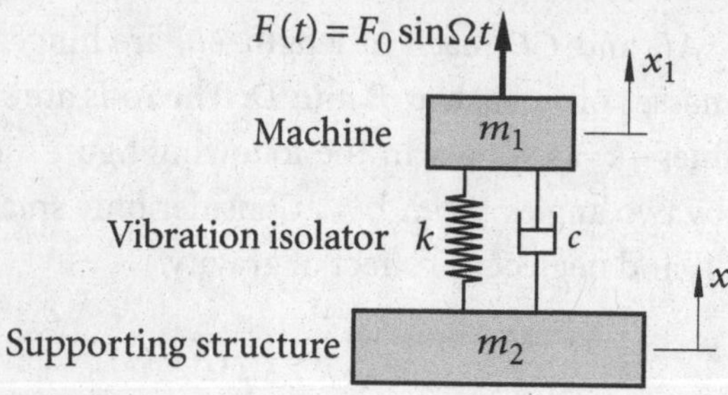

2. For the case when $c \to 0$, show that the *steady-state* responses of both the machine and the supporting structure, i.e., $x_{1P}(t)$ and $x_{2P}(t)$ are given by

$$x_{1P}(t) = \frac{(k - m_2\Omega^2)F_0 \sin \Omega t}{m_1 m_2 \Omega^4 - k(m_1 + m_2)\Omega^2}, \qquad x_{2P}(t) = \frac{kF_0 \sin \Omega t}{m_1 m_2 \Omega^4 - k(m_1 + m_2)\Omega^2}.$$

8.5 A mass m_1 hangs by a spring of stiffness k_1 from another mass m_2 which in turn hangs by a spring of stiffness k_2 from the support as shown.

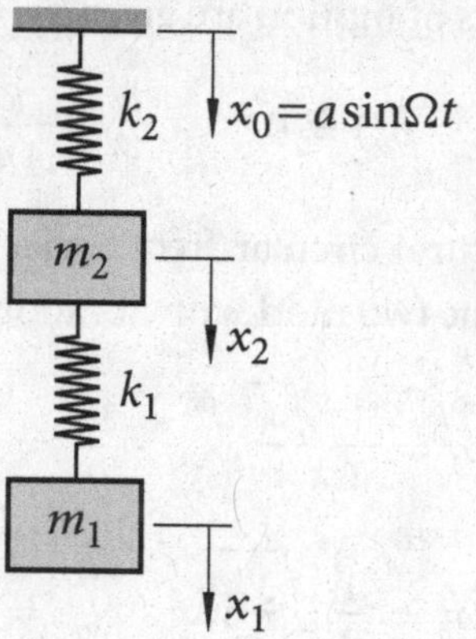

1. Show that the two natural circular frequencies of vibration are given by the equation

$$m_1 m_2 \omega^4 - \left[m_1(k_1 + k_2) + m_2 k_1\right]\omega^2 + k_1 k_2 = 0.$$

2. If the support vibrates with $x_0(t) = a \cos \Omega t$, show that the amplitudes of the forced vibration are

$$a_1 = \frac{k_1 k_2 a}{m_1 m_2 \Omega^4 - \left[m_1(k_1 + k_2) + m_2 k_1\right]\Omega^2 + k_1 k_2},$$

$$a_2 = \frac{(k_1 - m_1\Omega^2)k_2 a}{m_1 m_2 \Omega^4 - \left[m_1(k_1 + k_2) + m_2 k_1\right]\Omega^2 + k_1 k_2}.$$

8.6 A mass m, supported by an elastic structure which may be modeled as a spring with stiffness k, is subjected to a simple harmonic disturbing force of maximum value F_0 and frequency $\Omega = 2\pi f$, i.e., $F(t) = F_0 \sin \Omega t$. To reduce the amplitude of vibration of this system, a vibration absorber is attached to the mass. The absorber consists of a mass of m_a on a spring of stiffness k_a.

1. Show that response of forced vibration of mass m is

$$x_P(t) = \frac{(k_a - m_a\Omega^2)F_0 \sin \Omega t}{mm_a\Omega^4 - \left[mk_a + m_a(k+k_a)\right]\Omega^2 + kk_a}.$$

2. If $m = 250$ kg, $k = 1$ kN/m, $F_0 = 100$ N, $f = 35$ Hz, $m_a = 0.5$ kg, and the amplitude of forced vibration of mass m is to be reduced to zero, determine the value of k_a. ANS $k_a = 24.18$ kN/m

8.7 For the system shown in the following figure, show that the equations of motion are given by

$$m_1\ddot{x}_1 + (k_1+k_2)x_1 - k_2x_2 = F_1(t),$$
$$m_2\ddot{x}_2 + (k_2+k_3)x_2 - k_2x_1 = F_2(t).$$

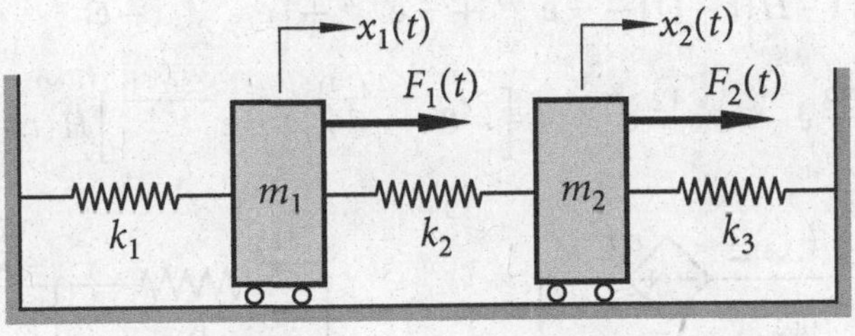

Consider the special case $m_1 = m_2 = 1$, $k_1 = k_2 = k_3 = 1$, and the initial conditions $x_1(0) = x_2(0) = 0$, $\dot{x}_1(0) = \dot{x}_2(0) = 0$. Using the matrix method, determine the responses of the system for the following excitations:

1. $F_1(t) = 3\sin 2t$, $F_2(t) = 0$;

ANS $x_1 = \sin t + \sqrt{3}\sin\sqrt{3}t - 2\sin 2t$, $x_2 = \sin t - \sqrt{3}\sin\sqrt{3}t + \sin 2t$

2. $F_1(t) = 4\sin t$, $F_2(t) = 0$;

ANS $x_1 = 2\sin t - t\cos t - \dfrac{1}{\sqrt{3}}\sin\sqrt{3}t$, $x_2 = -t\cos t + \dfrac{1}{\sqrt{3}}\sin\sqrt{3}t$

3. $F_1(t) = 12\sin\sqrt{3}t$, $F_2(t) = 0$;

ANS $x_1 = 3\sqrt{3}\sin t - 2\sin\sqrt{3}t - \sqrt{3}t\cos\sqrt{3}t$

$x_2 = 3\sqrt{3}\sin t - 4\sin\sqrt{3}t + \sqrt{3}t\cos\sqrt{3}t$

4. $F_1(t) = 12\sin t$, $F_2(t) = 12\sin\sqrt{3}t$.

ANS $x_1 = (6+3\sqrt{3})\sin t - 3t\cos t - (4+\sqrt{3})\sin\sqrt{3}t + \sqrt{3}t\cos\sqrt{3}t$

$x_2 = 3\sqrt{3}\sin t - 3t\cos t + (-2+\sqrt{3})\sin\sqrt{3}t - \sqrt{3}t\cos\sqrt{3}t$

8.8 For the circuit shown in Figure 8.9(a), the voltage source is given by

$$V(t) = V_0 H(-t) + V(t)H(t).$$

Show that the differential equations governing $i_1(t)$ and $i_2(t)$ for $t > 0$ are

$$L\frac{di_1}{dt} + Ri_1 - L\frac{di_2}{dt} = V(t), \qquad (R-\alpha)\frac{di_1}{dt} + \alpha\frac{di_2}{dt} + \frac{1}{C}i_2 = \frac{dV(t)}{dt},$$

with the initial conditions

$$i_1(0^+) = \frac{V(0^+)}{R}, \quad i_2(0^+) = \frac{V(0^+) - V_0}{R}.$$

For $R = 6\,\Omega$, $C = 1$ F, $L = 1$ H, $\alpha = 2$, and

$$V(t) = \begin{cases} 6\text{ V}, & t \leqslant 0^-, \\ 30\text{ V}, & 0^+ \leqslant t < 1, \\ 0, & t > 1, \end{cases}$$

determine $i_1(t)$ and $i_2(t)$ for $t > 0$.

ANS $i_1(t) = 5\big[1 - H(t-1)\big] - \frac{4}{5}e^{-\frac{2t}{3}} + \frac{4}{5}e^{-\frac{3t}{2}} + \Big[e^{-\frac{2(t-1)}{3}} - e^{-\frac{3(t-1)}{2}}\Big]H(t-1)$ (A)

$i_1(t) = \frac{32}{5}e^{-\frac{2t}{3}} - \frac{12}{5}e^{-\frac{3t}{2}} + \Big[-8e^{-\frac{2(t-1)}{3}} + 3e^{-\frac{3(t-1)}{2}}\Big]H(t-1)$ (A)

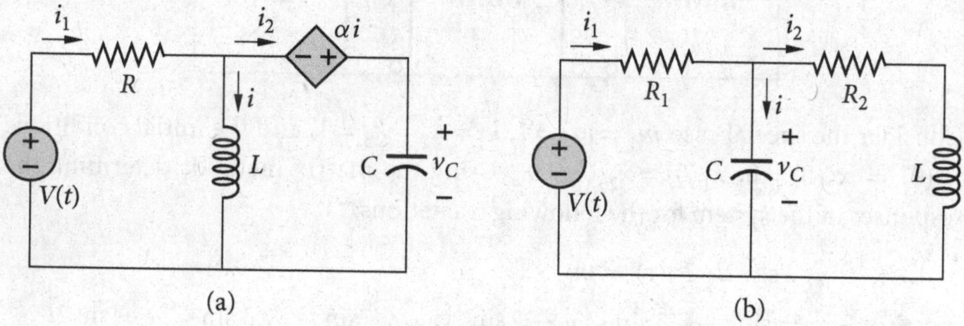

Figure 8.9 Electric circuits.

8.9 For the circuit shown in Figure 8.9(b), the voltage source is given by

$$V(t) = V_0 H(-t) + V(t)H(t).$$

Show that the differential equations governing $i_1(t)$ and $i_2(t)$ for $t > 0$ are

$$R_1 i_1 + R_2 i_2 + L\frac{di_2}{dt} = V(t), \qquad R_1\frac{di_1}{dt} + \frac{1}{C}i_1 - \frac{1}{C}i_2 = \frac{dV(t)}{dt},$$

with the initial conditions

$$i_1(0^+) = \frac{1}{R_1}\Big[V(0^+) - \frac{R_2}{R_1 + R_2}V_0\Big], \quad i_2(0^+) = \frac{V_0}{R_1 + R_2}.$$

For $R_1 = 1\,\Omega$, $R_2 = 3\,\Omega$, $C = 1$ F, $L = 1$ H, and

$$V(t) = \begin{cases} 4\text{ V}, & t \leqslant 0^-, \\ 8\text{ V}, & 0^+ \leqslant t < 2, \\ 12\text{ V}, & t > 2, \end{cases}$$

determine $i_1(t)$ and $i_2(t)$ for $t > 0$.

ANS $i_1(t) = 2 + (3+2t)\,e^{-2t} + \left[1 + (-1+2t)\,e^{-2(t-2)}\right] H(t-2)$ (A)

$i_2(t) = 2 - (1+2t)\,e^{-2t} + \left[1 + (3-2t)\,e^{-2(t-2)}\right] H(t-2)$ (A)

8.10 For the circuit shown in the following figure, the voltage source is given by $V(t) = V_0 H(-t) + V(t) H(t)$. Show that the differential equations governing $i_1(t)$ and $i_2(t)$ for $t > 0$ are

$$\frac{i_2 - i_1}{C} + L_2 \frac{d^2 i_2}{dt} + R_2 \frac{di_2}{dt} = 0, \qquad R_1 i_1 + L_1 \frac{di_1}{dt} + L_2 \frac{di_2}{dt} + R_2 i_2 = V(t),$$

with the initial conditions

$$i_1(0^+) = \frac{V_0}{R_1 + R_2}, \quad i_2(0^+) = \frac{V_0}{R_1 + R_2}, \quad \frac{di_2(0^+)}{dt} = 0.$$

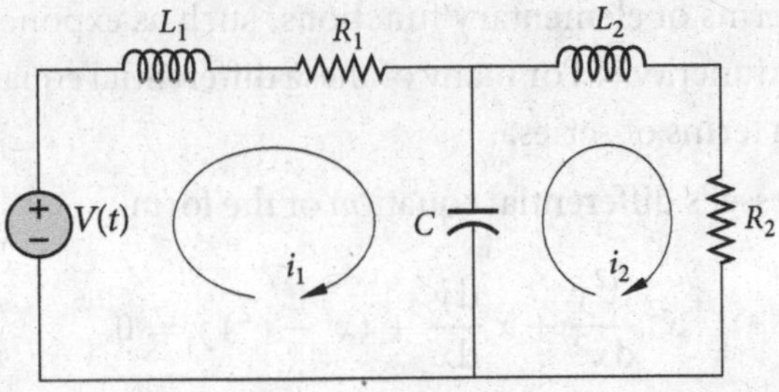

For $R_1 = R_2 = 1\,\Omega$, $C = 8$ F, $L_1 = L_2 = 1$ H, and $V(t) = 8H(-t) + 8e^{-\frac{t}{2}} H(t)$ (V), find $i_1(t)$ and $i_2(t)$ for $t > 0$.

ANS $i_1(t) = (8 - t^2)\,e^{-\frac{t}{2}} - 4e^{-t}$ (A), $\quad i_2(t) = (8 + t^2)\,e^{-\frac{t}{2}} - 4e^{-t}$ (A)

CHAPTER 9

Series Solutions of Differential Equations

Various analytical methods have been presented in previous chapters for solving ordinary differential equations to obtain exact solutions. However, in applied mathematics, science, and engineering applications, there are a large number of differential equations, especially those with variable coefficients, that cannot be solved exactly in terms of elementary functions, such as exponential, logarithmic, and trigonometric functions. For many of these differential equations, it is possible to find solutions in terms of series.

For example, Bessel's differential equation of the form

$$x^2\frac{d^2y}{dx^2}+x\frac{dy}{dx}+(x^2-\nu^2)\,y=0,$$

where ν is an arbitrary real or complex number, finds many applications in engineering disciplines. Some examples include heat conduction in a cylindrical object, vibration of a thin circular or annular membrane, and electromagnetic waves in a cylindrical waveguide. Bessel's equation cannot be solved exactly in terms of elementary functions; it can be solved using series, which were first defined by Daniel Bernoulli and then generalized by Friedrich Bessel and are known as Bessel functions.

The objective of this chapter is to present the essential techniques for solving such ordinary differential equations, in particular second-order linear ordinary differential equations with variable coefficients.

Before explaining how series can be used to solve ordinary differential equations, some relevant results on power series are briefly reviewed in the following section.

9.1 Review of Power Series

Definition — Power Series

A *power series* is an infinite series of the form

$$\sum_{n=0}^{\infty} a_n (x-x_0)^n = a_0 + a_1 (x-x_0) + a_2 (x-x_0)^2 + a_3 (x-x_0)^3 + \cdots, \quad (1)$$

where $a_0, a_1, a_2, \ldots$ are constants, and x_0 is a fixed number.

This series usually arises as the Taylor series of some function $f(x)$. If $x_0 = 0$, the power series becomes

$$\sum_{n=0}^{\infty} a_n x^n = a_0 + a_1 x + a_2 x^2 + a_3 x^3 + \cdots.$$

Convergence of a Power Series

Power series (1) is *convergent* at x if the limit

$$\lim_{N\to\infty} \sum_{n=0}^{N} a_n (x-x_0)^n$$

exists and is finite. Otherwise, the power series is *divergent*. A power series will converge for some values of x and may diverge for other values. Series (1) is always convergent at $x = x_0$.

If power series (1) is convergent for all x in the interval $|x-x_0| < r$ and is divergent whenever $|x-x_0| > r$, where $0 \leqslant r \leqslant \infty$, then r is called the *radius of convergence* of the power series.

The radius of convergence r is given by

$$r = \lim_{n\to\infty} \left| \frac{a_n}{a_{n+1}} \right|$$

if this limit exists.

Four very important power series are

$$\frac{1}{1-x} = 1 + x + x^2 + x^3 + \cdots = \sum_{n=0}^{\infty} x^n, \quad -1 < x < 1,$$

$$e^x = 1 + x + \frac{x^2}{2!} + \frac{x^3}{3!} + \cdots = \sum_{n=0}^{\infty} \frac{x^n}{n!}, \quad -\infty < x < \infty,$$

$$\sin x = x - \frac{x^3}{3!} + \frac{x^5}{5!} - \frac{x^7}{7!} + \cdots = \sum_{n=0}^{\infty} (-1)^n \frac{x^{2n+1}}{(2n+1)!}, \quad -\infty < x < \infty,$$

$$\cos x = 1 - \frac{x^2}{2!} + \frac{x^4}{4!} - \frac{x^6}{6!} + \cdots = \sum_{n=0}^{\infty} (-1)^n \frac{x^{2n}}{(2n)!}, \quad -\infty < x < \infty.$$

Operations of Power Series

Suppose functions $f(x)$ and $g(x)$ can be expanded into power series as

$$f(x) = \sum_{n=0}^{\infty} a_n (x-x_0)^n, \qquad \text{for } |x-x_0| < r_1,$$

$$g(x) = \sum_{n=0}^{\infty} b_n (x-x_0)^n, \qquad \text{for } |x-x_0| < r_2.$$

Then, for $|x-x_0| < r,\ r = \min(r_1, r_2)$,

$$f(x) \pm g(x) = \sum_{n=0}^{\infty} (a_n \pm b_n)(x-x_0)^n,$$

i.e., the power series of the sum or difference of the functions can be obtained by termwise addition and substraction. For multiplication,

$$f(x)\,g(x) = \left[\sum_{m=0}^{\infty} a_m (x-x_0)^m\right]\left[\sum_{n=0}^{\infty} b_n (x-x_0)^n\right] = \sum_{m=0}^{\infty}\sum_{n=0}^{\infty} a_m b_n (x-x_0)^{m+n}$$

$$= \sum_{n=0}^{\infty}\left(\sum_{m=0}^{n} a_m b_{n-m}\right)(x-x_0)^n,$$

and for division,

$$\frac{f(x)}{g(x)} = \frac{\sum_{n=0}^{\infty} a_n (x-x_0)^n}{\sum_{n=0}^{\infty} b_n (x-x_0)^n} = \sum_{n=0}^{\infty} c_n (x-x_0)^n$$

$$\Longrightarrow \sum_{n=0}^{\infty} a_n (x-x_0)^n = \left[\sum_{n=0}^{\infty} b_n (x-x_0)^n\right]\left[\sum_{n=0}^{\infty} c_n (x-x_0)^n\right],$$

in which c_n can be obtained by expanding the right-hand side and comparing coefficients of $(x-x_0)^n,\ n = 0, 1, 2, \ldots$.

If the power series of $f(x)$ is convergent in the interval $|x-x_0| < r_1$, then $f(x)$ is continuous and has continuous derivatives of all orders in this interval. The derivatives can be obtained by differentiating the power series termwise

$$f'(x) = \sum_{n=1}^{\infty} a_n n (x-x_0)^{n-1}, \qquad \text{for } |x-x_0| < r_1.$$

The integral of $f(x)$ can be obtained by integrating the power series termwise

$$\int f(x)\,\mathrm{d}x = \sum_{n=0}^{\infty} \frac{a_n (x-x_0)^{n+1}}{n+1} + C, \qquad \text{for } |x-x_0| < r_1,$$

$$\xrightarrow{n+1=m} \sum_{m=1}^{\infty} \frac{a_{m-1} (x-x_0)^m}{m} + C.$$

Change the index of summation.

Definition — Analytic Function

A function $f(x)$ defined in the interval I containing x_0 is said to be *analytic* at x_0 if $f(x)$ can be expressed as a power (Taylor) series $f(x)=\sum_{n=0}^{\infty} a_n(x-x_0)^n$, which has a positive radius of convergence.

Example 9.1

Determine the radius of convergence for

(1) $\displaystyle\sum_{n=0}^{\infty} \frac{1}{2^n n}(x-1)^n$ (2) $\displaystyle\sum_{n=0}^{\infty} \frac{(n!)^3}{(3n)!}x^n$

(1) $a_n = \dfrac{1}{2^n n}$, $a_{n+1} = \dfrac{1}{2^{n+1}(n+1)}$,

$$r = \lim_{n\to\infty}\left|\frac{a_n}{a_{n+1}}\right| = \lim_{n\to\infty} \frac{\dfrac{1}{2^n n}}{\dfrac{1}{2^{n+1}(n+1)}} = \lim_{n\to\infty} \frac{2^{n+1}(n+1)}{2^n n}$$

$$= \lim_{n\to\infty} 2\left(1+\frac{1}{n}\right) = 2.$$

(2) $a_n = \dfrac{(n!)^3}{(3n)!}$, $a_{n+1} = \dfrac{[(n+1)!]^3}{[3(n+1)]!}$,

$$r = \lim_{n\to\infty}\left|\frac{a_n}{a_{n+1}}\right| = \lim_{n\to\infty} \frac{\dfrac{(n!)^3}{(3n)!}}{\dfrac{[(n+1)\cdot n!]^3}{(3n+3)!}}$$

$$= \lim_{n\to\infty}\left[\frac{(n!)^3}{(3n)!}\,\frac{(3n+3)(3n+2)(3n+1)(3n)!}{(n+1)^3(n!)^3}\right]$$

$$= \lim_{n\to\infty}\frac{(3n+3)(3n+2)(3n+1)}{(n+1)^3} = \lim_{n\to\infty}\frac{27(n+1)\left(n+\frac{2}{3}\right)\left(n+\frac{1}{3}\right)}{(n+1)^3} = 27.$$

Example 9.2

Expand $\dfrac{1}{x(x+1)}$ as a power series in $x-1$.

Letting $t=x-1$ yields

$$\frac{1}{x(x+1)} = \frac{1}{(t+1)(t+2)} = \frac{1}{1+t} - \frac{1}{2+t} = \frac{1}{1+t} - \frac{1}{2}\,\frac{1}{1+\dfrac{t}{2}}$$

$$= \sum_{n=0}^{\infty}(-t)^n - \frac{1}{2}\sum_{n=0}^{\infty}\left(-\frac{t}{2}\right)^n = \sum_{n=0}^{\infty}(-1)^n\left(1-\frac{1}{2^{n+1}}\right)t^n$$

$$= \sum_{n=0}^{\infty} (-1)^n \left(1 - \frac{1}{2^{n+1}}\right)(x-1)^n.$$

Since the interval of convergence of the power series of $\dfrac{1}{1+t}$ is $-1<t<1$ and that of $\dfrac{1}{1+t/2}$ is $-2<t<2$, hence the region of convergence of the power series of $\dfrac{1}{x(x+1)}$ is $-1<t<1$ or $0<(x=t+1)<2$.

Example 9.3

Expand $\dfrac{1}{(1-x)^3}$ as a power series in x.

$$\frac{1}{(1-x)^3} = \frac{1}{2}\left(\frac{1}{1-x}\right)'' = \frac{1}{2}\left(\sum_{n=0}^{\infty} x^n\right)'', \quad -1<x<1,$$

$$= \frac{1}{2}\sum_{n=2}^{\infty} n(n-1)x^{n-2}, \quad -1<x<1,$$

$$\xrightarrow{n-2=m} \frac{1}{2}\sum_{m=0}^{\infty}(m+2)(m+1)x^m.$$

✍ Change the index of summation.

Example 9.4

Expand $\ln(1+x)$ as a power series in x.

$$\ln(1+x) = \int \frac{1}{1+x}\,dx = \int \left[\sum_{n=0}^{\infty}(-1)^n x^n\right] dx, \quad -1<x<1,$$

$$= \sum_{n=0}^{\infty}(-1)^n \frac{x^{n+1}}{n+1} \xrightarrow{n+1=m} \sum_{m=1}^{\infty}(-1)^{m-1}\frac{x^m}{m}, \quad -1<x<1.$$

9.2 Series Solution about an Ordinary Point

Two simple ordinary differential equations with closed-form solutions are considered first as motivating examples.

Motivating Example 1

Consider the first-order ordinary differential equation

$$y' - y = 0.$$

Let the solution of the equation be in the form of a power series

$$y(x) = \sum_{n=0}^{\infty} a_n x^n, \quad |x|<r,$$

for some $r>0$, where a_n are constants to be determined. Differentiating $y(x)$ with respect to x yields

$$y'(x) = \sum_{n=1}^{\infty} a_n n x^{n-1} \xrightarrow{n-1=m} \sum_{m=0}^{\infty} a_{m+1}(m+1)x^m.$$

Change the index of summation.

Substituting into the differential equation leads to

$$\sum_{n=0}^{\infty} a_{n+1}(n+1)x^n - \sum_{n=0}^{\infty} a_n x^n = 0 \implies \sum_{n=0}^{\infty}\big[(n+1)a_{n+1} - a_n\big]x^n = 0.$$

For this equation to be true, the coefficient of x^n, $n=0, 1, \ldots$, must be zero:

$$\begin{aligned}
x^0:&\quad a_1 - a_0 = 0 \implies a_1 = a_0,\\
x^1:&\quad 2a_2 - a_1 = 0 \implies a_2 = \frac{1}{2}a_1 = \frac{1}{2!}a_0,\\
x^2:&\quad 3a_3 - a_2 = 0 \implies a_3 = \frac{1}{3}a_2 = \frac{1}{3}\cdot\frac{1}{2!}a_0 = \frac{1}{3!}a_0,\\
&\vdots\\
x^n:&\quad (n+1)a_{n+1} - a_n = 0 \implies a_{n+1} = \frac{1}{n+1}a_n = \frac{1}{n+1}\cdot\frac{1}{n!}a_0 = \frac{1}{(n+1)!}a_0.
\end{aligned}$$

Hence, the solution is

$$\begin{aligned}
y(x) &= a_0 + a_0 x + \frac{1}{2!}a_0 x^2 + \frac{1}{3!}a_0 x^3 + \cdots + \frac{1}{n!}a_0 x^2 + \cdots\\
&= a_0\Big(1 + x + \frac{1}{2!}x^2 + \frac{1}{3!}x^3 + \cdots + \frac{1}{n!}x^n + \cdots\Big)\\
&= a_0 e^x, \qquad a_0 \text{ is an arbitrary constant,}
\end{aligned}$$

which recovers the general solution of $y' - y = 0$.

Motivating Example 2

Consider the second-order ordinary differential equation

$$y'' + y = 0.$$

Suppose that the solution is in the form of a power series $y(x) = \sum_{n=0}^{\infty} a_n x^n$, $|x|<r$, for some $r>0$, where a_n are constants to be determined. Differentiating $y(x)$ with respect to x twice yields

$$\begin{aligned}
y'(x) &= \sum_{n=1}^{\infty} a_n n x^{n-1},\\
y''(x) &= \sum_{n=2}^{\infty} a_n n(n-1)x^{n-2} \xrightarrow{n-2=m} \sum_{m=0}^{\infty} a_{m+2}(m+2)(m+1)x^m.
\end{aligned}$$

Substituting into the differential equation leads to

$$\sum_{n=0}^{\infty} a_{n+2}\,(n+2)(n+1)x^n + \sum_{n=0}^{\infty} a_n x^n = 0 \implies \sum_{n=0}^{\infty}\big[(n+2)(n+1)\,a_{n+1} + a_n\big]x^n = 0.$$

For this equation to be true, the coefficient of x^n, $n=0,1,\ldots,$ must be zero:

$$\begin{aligned}
x^0:&\quad 2\cdot 1\,a_2 + a_0 = 0 \implies a_2 = -\frac{1}{2!}a_0,\\
x^1:&\quad 3\cdot 2\,a_3 + a_1 = 0 \implies a_3 = -\frac{1}{3!}a_1,\\
x^2:&\quad 4\cdot 3\,a_4 + a_2 = 0 \implies a_4 = -\frac{1}{4\cdot 3}a_2 = \frac{1}{4!}a_0,\\
x^3:&\quad 5\cdot 4\,a_5 + a_3 = 0 \implies a_5 = -\frac{1}{5\cdot 4}a_3 = \frac{1}{5!}a_1,\\
x^4:&\quad 6\cdot 5\,a_6 + a_4 = 0 \implies a_6 = -\frac{1}{6\cdot 5}a_4 = -\frac{1}{6!}a_0,\\
x^5:&\quad 7\cdot 6\,a_7 + a_5 = 0 \implies a_7 = -\frac{1}{7\cdot 6}a_5 = -\frac{1}{7!}a_1,\\
&\ \vdots
\end{aligned}$$

In general, for $k=1,2,3,\ldots,$

$$a_{2k} = (-1)^k\frac{1}{(2k)!}a_0, \qquad a_{2k+1} = (-1)^k\frac{1}{(2k+1)!}a_1.$$

Hence, the solution is

$$\begin{aligned}
y(x) &= \sum_{n=0}^{\infty} a_n x^n = a_0\sum_{k=0}^{\infty}(-1)^k\frac{x^{2k}}{(2k)!} + a_1\sum_{k=0}^{\infty}(-1)^k\frac{x^{2k+1}}{(2k+1)!}\\
&= a_0\cos x + a_1\sin x, \qquad a_0 \text{ and } a_1 \text{ are arbitrary constants,}
\end{aligned}$$

which recovers the general solution of $y''+y=0$.

Remarks: These two examples show that it is possible to solve an ordinary differential equation using power series.

Definition — Ordinary Point

Consider the nth-order linear ordinary differential equation

$$y^{(n)}(x) + p_{n-1}(x)\,y^{(n-1)}(x) + p_{n-2}(x)\,y^{(n-2)}(x) + \cdots + p_0(x)\,y(x) = f(x).$$

A point x_0 is called an *ordinary point* of the given differential equation if each of the coefficients $p_0(x)$, $p_1(x), \ldots, p_{n-1}(x)$ and $f(x)$ is analytic at $x=x_0$, i.e., $p_i(x)$, for $i=0,1,\ldots,n-1$, and $f(x)$ can be expressed as power series about x_0 that are convergent for $|x-x_0|<r$, $r>0$,

$$p_i(x) = \sum_{n=0}^{\infty} p_{i,n}\,(x-x_0)^n, \qquad f(x) = \sum_{n=0}^{\infty} f_n\,(x-x_0)^n.$$

Theorem — Series Solution about an Ordinary Point

Suppose that x_0 is an ordinary point of the nth-order linear ordinary differential equation

$$y^{(n)} + p_{n-1}(x)\,y^{(n-1)} + p_{n-2}(x)\,y^{(n-2)} + \cdots + p_0(x)\,y = f(x),$$

i.e., the coefficients $p_0(x)$, $p_1(x), \ldots, p_{n-1}(x)$ and $f(x)$ are all analytic at $x = x_0$ and each can be expressed as a power series about x_0 convergent for $|x - x_0| < r$, $r > 0$. Then every solution of this differential equation can be expanded in one and only one way as a power series in $(x - x_0)$

$$y(x) = \sum_{n=0}^{\infty} a_n (x - x_0)^n, \quad |x - x_0| < R,$$

where the radius of convergence $R \geqslant r$.

Example 9.5 — Legendre Equation

Find the power series solution in x of the Legendre equation

$$(1 - x^2)y'' - 2xy' + p(p+1)y = 0, \quad p > 0.$$

The differential equation can be written as

$$y'' + p_1(x)y' + p_0(x)y = 0, \qquad p_1(x) = -\frac{2x}{1-x^2}, \quad p_0(x) = \frac{p(p+1)}{1-x^2}.$$

Both $p_1(x)$ and $p_0(x)$ can be expanded in power series as

$$p_1(x) = -2x \cdot \frac{1}{1-x^2} = -2x \sum_{n=0}^{\infty} (x^2)^n = -2 \sum_{n=0}^{\infty} x^{2n+1}, \quad |x| < 1,$$

$$p_0(x) = p(p+1) \cdot \frac{1}{1-x^2} = p(p+1) \sum_{n=0}^{\infty} (x^2)^n = p(p+1) \sum_{n=0}^{\infty} x^{2n}, \quad |x| < 1.$$

Hence, $x = 0$ is an ordinary point and a unique power series solution exists

$$y(x) = \sum_{n=0}^{\infty} a_n x^n, \quad |x| < 1,$$

where a_n, $n = 0, 1, \ldots,$ are constants to be determined. Differentiating $y(x)$ with respect to x yields

$$y'(x) = \sum_{n=1}^{\infty} n\,a_n x^{n-1}, \qquad y''(x) = \sum_{n=2}^{\infty} n(n-1)\,a_n x^{n-2}.$$

Substituting y, y', and y'' into the differential equation yields

$$(1 - x^2) \sum_{n=2}^{\infty} n(n-1)\,a_n x^{n-2} - 2x \sum_{n=1}^{\infty} n\,a_n x^{n-1} + p(p+1) \sum_{n=0}^{\infty} a_n x^n = 0,$$

or, noting that

$$\sum_{n=2}^{\infty} n(n-1)\,a_n x^{n-2} \xrightarrow{n-2=m} \sum_{m=0}^{\infty} (m+2)(m+1)\,a_{m+2} x^m,$$

Change the index of summation.

one has

$$\sum_{n=0}^{\infty} (n+2)(n+1)\,a_{n+2} x^n - \sum_{n=2}^{\infty} n(n-1)\,a_n x^n - \sum_{n=1}^{\infty} 2n a_n x^n + \sum_{n=0}^{\infty} p(p+1)\,a_n x^n = 0.$$

For this equation to be true, the coefficient of x^n, $n = 0, 1, \ldots$, must be zero:

$$x^0: \quad 2\cdot 1 a_2 + p(p+1)\,a_0 = 0 \implies a_2 = -\frac{p(p+1)}{2!}\,a_0,$$

$$x^1: \quad 3\cdot 2 a_3 - 2a_1 + p(p+1)\,a_1 = 0 \implies a_3 = -\frac{(p-1)(p+2)}{3!}\,a_1.$$

For $n \geqslant 2$, the coefficient of x^n gives

$$(n+2)(n+1)\,a_{n+2} - n(n-1)\,a_n - 2n a_n + p(p+1)\,a_n = 0$$

$$\implies a_{n+2} = -\frac{(p-n)[p+(n+1)]}{(n+2)(n+1)}\,a_n.$$

Hence,

$$x^2: \quad a_4 = -\frac{(p-2)(p+3)}{4\cdot 3}\,a_2 = -\frac{(p-2)(p+3)}{4\cdot 3}\left[-\frac{p(p+1)}{2!}\,a_0\right]$$

$$= (-1)^2\,\frac{p(p+1)(p-2)(p+3)}{4!}\,a_0,$$

$$x^3: \quad a_5 = -\frac{(p-3)(p+4)}{5\cdot 4}\,a_3 = -\frac{(p-3)(p+4)}{5\cdot 4}\left[-\frac{(p-1)(p+2)}{3!}\,a_1\right]$$

$$= (-1)^2\,\frac{(p-1)(p+2)(p-3)(p+4)}{5!}\,a_1,$$

$$x^4: \quad a_6 = -\frac{(p-4)(p+5)}{6\cdot 5}\,a_4$$

$$= -\frac{(p-4)(p+5)}{6\cdot 5}\left[(-1)^2\,\frac{p(p+1)(p-2)(p+3)}{4!}\,a_0\right]$$

$$= (-1)^3\,\frac{p(p+1)(p-2)(p+3)(p-4)(p+5)}{6!}\,a_0,$$

$$x^5: \quad a_7 = -\frac{(p-5)(p+6)}{7\cdot 6}\,a_5$$

$$= -\frac{(p-5)(p+6)}{7\cdot 6}\left[(-1)^2\,\frac{(p-1)(p+2)(p-3)(p+4)}{5!}\,a_1\right]$$

$$= (-1)^3\,\frac{(p-1)(p+2)(p-3)(p+4)(p-5)(p+6)}{7!}\,a_1,$$

$\cdots\ \cdots$

In general,

$$a_{2k} = (-1)^k \frac{p(p+1)(p-2)(p+3)\cdots(p-2k+2)(p+2k-1)}{(2k)!} a_0$$
$$= \frac{(-1)^k}{(2k)!} \prod_{i=1}^{k} \left[(p-2i+2)(p+2i-1)\right] a_0,$$
$$a_{2k+1} = (-1)^k \frac{(p-1)(p+2)(p-3)(p+4)\cdots(p-2k+1)(p+2k)}{(2k+1)!} a_1$$
$$= \frac{(-1)^k}{(2k+1)!} \prod_{i=1}^{k} \left[(p-2i+1)(p+2i)\right] a_1.$$

Thus, the power series solution of Legendre equation is

$$y(x) = a_0 \sum_{k=0}^{\infty} \frac{(-1)^k}{(2k)!} \prod_{i=1}^{k} \left[(p-2i+2)(p+2i-1)\right] x^{2k}$$
$$+ a_1 \sum_{k=0}^{\infty} \frac{(-1)^k}{(2k+1)!} \prod_{i=1}^{k} \left[(p-2i+1)(p+2i)\right] x^{2k+1}, \quad |x| < 1,$$

where a_0 and a_1 are arbitrary constants.

Example 9.6

Find the power series solution in x of the equation $xy'' + y\ln(1-x) = 0$, $|x| < 1$.

The differential equation can be written as

$$y'' + \frac{\ln(1-x)}{x} y = 0, \quad |x| < 1,$$

which is of the form

$$y'' + p_1(x)y' + p_0(x)y = 0, \qquad p_1(x) = 0, \quad p_0(x) = \frac{\ln(1-x)}{x}.$$

Since

$$\frac{1}{1-x} = \sum_{n=0}^{\infty} x^n, \quad |x| < 1,$$

integrating both sides of the equation with respect to x yields

$$\ln(1-x) = -\int \frac{1}{1-x}\,\mathrm{d}x = -\int \sum_{n=0}^{\infty} x^n\,\mathrm{d}x = -\sum_{n=0}^{\infty} \frac{x^{n+1}}{n+1}, \quad |x| < 1,$$
$$\therefore \quad p_0(x) = \frac{\ln(1-x)}{x} = -\sum_{n=0}^{\infty} \frac{x^n}{n+1}, \quad |x| < 1.$$

Hence, both $p_0(x)$ and $p_1(x)$ can be expanded in power series, leading to $x=0$ being an ordinary point. The solution of the differential equation can be expressed in a power series

$$y(x)=\sum_{n=0}^{\infty}a_n x^n,\qquad |x|<1,$$

where a_n, $n=0,1,\ldots$, are constants to be determined. Differentiating $y(x)$ with respect to x gives

$$y'(x)=\sum_{n=1}^{\infty}n\,a_n x^{n-1},\qquad y''(x)=\sum_{n=2}^{\infty}n\,(n-1)\,a_n x^{n-2},\qquad |x|<1.$$

Substituting into the differential equation results in

$$\sum_{n=2}^{\infty}n\,(n-1)\,a_n x^{n-2}-\sum_{n=0}^{\infty}\frac{x^n}{n+1}\cdot\sum_{n=0}^{\infty}a_n x^n=0.$$

Noting that

$$\sum_{n=2}^{\infty}n\,(n-1)\,a_n x^{n-2}\xrightarrow{n-2=m}\sum_{m=0}^{\infty}(m+2)(m+1)\,a_{m+2}x^m,$$

$$\sum_{n=0}^{\infty}\frac{x^n}{n+1}\cdot\sum_{n=0}^{\infty}a_n x^n=\sum_{n=0}^{\infty}\sum_{m=0}^{n}\left(\frac{x^m}{m+1}\cdot a_{n-m}x^{n-m}\right)=\sum_{n=0}^{\infty}\left(\sum_{m=0}^{n}\frac{a_{n-m}}{m+1}\right)x^n,$$

one obtains

$$\sum_{n=0}^{\infty}\left[(n+2)(n+1)\,a_{n+2}-\sum_{m=0}^{n}\frac{a_{n-m}}{m+1}\right]x^n=0.$$

For this equation to be true, the coefficient of x^n, $n=0,1,\ldots$, must be zero:

$$a_{n+2}=\frac{1}{(n+2)(n+1)}\sum_{m=0}^{n}\frac{a_{n-m}}{m+1}.$$

Hence,

$$n=0:\ a_2=\frac{1}{2\cdot1}\,a_0=\frac{a_0}{2},$$

$$n=1:\ a_3=\frac{1}{3\cdot2}\left(a_1+\frac{a_0}{2}\right)=\frac{a_0}{12}+\frac{a_1}{6},$$

$$n=2:\ a_4=\frac{1}{4\cdot3}\left(a_2+\frac{a_1}{2}+\frac{a_0}{3}\right)=\frac{1}{12}\left(\frac{a_0}{2}+\frac{a_1}{2}+\frac{a_0}{3}\right)=\frac{5a_0}{72}+\frac{a_1}{24},$$

$$n=3:\ a_5=\frac{1}{5\cdot4}\left(a_3+\frac{a_2}{2}+\frac{a_1}{3}+\frac{a_0}{4}\right)=\frac{1}{20}\left[\left(\frac{a_0}{12}+\frac{a_1}{6}\right)+\frac{1}{2}\left(\frac{a_0}{2}\right)+\frac{a_1}{3}+\frac{a_0}{4}\right]$$
$$=\frac{7a_0}{240}+\frac{a_1}{40},$$

$$n=4:\ a_6=\frac{1}{6\cdot5}\left(a_4+\frac{a_3}{2}+\frac{a_2}{3}+\frac{a_1}{4}+\frac{a_0}{5}\right)$$
$$=\frac{1}{30}\left[\left(\frac{5a_0}{72}+\frac{a_1}{24}\right)+\frac{1}{2}\left(\frac{a_0}{12}+\frac{a_1}{6}\right)+\frac{1}{3}\left(\frac{a_0}{2}\right)+\frac{a_1}{4}+\frac{a_0}{5}\right]=\frac{43a_0}{2700}+\frac{a_1}{80}.$$

It is difficult to obtain the general expression for a_n. Stopping at x^6, the series solution is given by

$$y(x) = \sum_{n=0}^{\infty} a_n x^n = a_0 + a_1 x + \frac{a_0}{2}x^2 + \left(\frac{a_0}{12}+\frac{a_1}{6}\right)x^3 + \left(\frac{5a_0}{72}+\frac{a_1}{24}\right)x^4$$
$$+\left(\frac{7a_0}{240}+\frac{a_1}{40}\right)x^5 + \left(\frac{43a_0}{2700}+\frac{a_1}{80}\right)x^6 + \cdots$$
$$= a_0\left(1+\frac{x^2}{2}+\frac{x^3}{12}+\frac{5x^4}{72}+\frac{7x^5}{240}+\frac{43x^6}{2700}+\cdots\right)$$
$$+a_1\left(x+\frac{x^3}{6}+\frac{x^4}{24}+\frac{x^5}{40}+\frac{x^6}{80}+\cdots\right),$$

where a_0 and a_1 are arbitrary constants.

Example 9.7

Find the power series solution in x of the equation $y''' - xy'' + (x-2)y' + y = 0$.

The differential equation is of the form

$$y''' + p_2(x)y'' + p_1(x)y' + p_0(x)y = 0, \quad p_2(x) = -x, \quad p_1(x) = x-2, \quad p_0(x) = 1.$$

Each of $p_0(x)$, $p_1(x)$ and $p_1(x)$ can be expressed in power series. Hence, $x=0$ is an ordinary point and there exists a unique power series solution

$$y(x) = \sum_{n=0}^{\infty} a_n x^n, \quad -\infty < x < \infty,$$

where a_n, $n = 0, 1, \ldots$, are constants to be determined. Differentiating with respect to x yields, for $-\infty < x < \infty$,

$$y' = \sum_{n=1}^{\infty} n a_n x^{n-1}, \quad y'' = \sum_{n=2}^{\infty} n(n-1)a_n x^{n-2}, \quad y''' = \sum_{n=3}^{\infty} n(n-1)(n-2)a_n x^{n-3}.$$

Substituting into the differential equation results in

$$\sum_{n=3}^{\infty} n(n-1)(n-2)a_n x^{n-3} - \sum_{n=2}^{\infty} n(n-1)a_n x^{n-1}$$
$$+\sum_{n=1}^{\infty} n a_n x^n - 2\sum_{n=1}^{\infty} n a_n x^{n-1} + \sum_{n=0}^{\infty} a_n x^n = 0.$$

Changing the indices of summations

$$\sum_{n=3}^{\infty} n(n-1)(n-2)a_n x^{n-3} \xrightarrow{n-3=m} \sum_{m=0}^{\infty} (m+3)(m+2)(m+1)a_{m+3}x^m,$$

$$\sum_{n=2}^{\infty} n(n-1)a_n x^{n-1} \xrightarrow{n-1=m} \sum_{m=1}^{\infty} (m+1)m a_{m+1}x^m,$$

$$\sum_{n=1}^{\infty} n a_n x^{n-1} \xrightarrow{n-1=m} \sum_{m=0}^{\infty} (m+1) a_{m+1} x^m,$$

one obtains

$$\sum_{n=0}^{\infty} (n+3)(n+2)(n+1) a_{n+3} x^n - \sum_{n=1}^{\infty} (n+1) n a_{n+1} x^n$$
$$+ \sum_{n=1}^{\infty} n a_n x^n - 2\sum_{n=0}^{\infty} (n+1) a_{n+1} x^n + \sum_{n=0}^{\infty} a_n x^n = 0,$$

$$\therefore \quad \sum_{n=0}^{\infty} \left[(n+3)(n+2)(n+1) a_{n+3} - 2(n+1) a_{n+1} + a_n\right] x^n$$
$$+ \sum_{n=1}^{\infty} \left[-(n+1) n a_{n+1} + n a_n\right] x^n = 0.$$

For this equation to be true, the coefficient of x^n, $n=0, 1, \ldots$, must be zero. When $n=0$, one has

$$3\cdot 2\cdot 1 a_3 - 2\cdot 1 a_1 + a_0 = 0 \Longrightarrow a_3 = -\frac{a_0}{6} + \frac{a_1}{3}.$$

For $n \geqslant 1$, one obtains

$$\left[(n+3)(n+2)(n+1) a_{n+3} - 2(n+1) a_{n+1} + a_n\right] + \left[-(n+1) n a_{n+1} + n a_n\right] = 0,$$

$$\therefore \quad a_{n+3} = -\frac{a_n}{(n+3)(n+2)} + \frac{a_{n+1}}{n+3}.$$

Hence,

$$n=1: \quad a_4 = -\frac{a_1}{4\cdot 3} + \frac{a_2}{4} = -\frac{a_1}{12} + \frac{a_2}{4},$$

$$n=2: \quad a_5 = -\frac{a_2}{5\cdot 4} + \frac{a_3}{5} = -\frac{a_2}{20} + \frac{1}{5}\left(-\frac{a_0}{6} + \frac{a_1}{3}\right) = -\frac{a_0}{30} + \frac{a_1}{15} - \frac{a_2}{20},$$

$$n=3: \quad a_6 = -\frac{a_3}{6\cdot 5} + \frac{a_4}{6} = -\frac{1}{30}\left(-\frac{a_0}{6} + \frac{a_1}{3}\right) + \frac{1}{6}\left(-\frac{a_1}{12} + \frac{a_2}{4}\right)$$
$$= \frac{a_0}{180} - \frac{a_1}{40} + \frac{a_2}{24}.$$

It is difficult to obtain the general expression for a_n. Stopping at x^6, the series solution is given by

$$y(x) = \sum_{n=0}^{\infty} a_n x^n = a_0 + a_1 x + a_2 x^2 + \left(-\frac{a_0}{6} + \frac{a_1}{3}\right) x^3 + \left(-\frac{a_1}{12} + \frac{a_2}{4}\right) x^4$$
$$+ \left(-\frac{a_0}{30} + \frac{a_0}{15} - \frac{a_2}{20}\right) x^5 + \left(\frac{a_0}{180} - \frac{a_1}{40} + \frac{a_2}{24}\right) x^6 + \cdots$$
$$= a_0\left(1 - \frac{x^3}{6} - \frac{x^5}{30} + \frac{x^6}{180} + \cdots\right) + a_1\left(x + \frac{x^3}{3} - \frac{x^4}{12} + \frac{x^5}{15} - \frac{x^6}{40} + \cdots\right)$$
$$+ a_2\left(x^2 + \frac{x^4}{4} - \frac{x^5}{20} + \frac{x^6}{24} + \cdots\right),$$

where a_0, a_1, and a_2 are arbitrary constants.

9.3 Series Solution about a Regular Singular Point

Definition — Singular Point

Consider the nth-order linear homogeneous ordinary differential equation

$$y^{(n)} + p_{n-1}(x)\,y^{(n-1)} + p_{n-2}(x)\,y^{(n-2)} + \cdots + p_0(x)\,y = 0.$$

- A point x_0 is called a *singular point* of the given differential equation if it is not an ordinary point, i.e., not all of the coefficients $p_0(x)$, $p_1(x)$, …, $p_{n-1}(x)$ are analytic at $x = x_0$.
- A point x_0 is a *regular singular point* of the given differential equation if it is not an ordinary point, i.e., not all of the coefficients $p_k(x)$ are analytic, but all of $(x-x_0)^{n-k}\,p_k(x)$ are analytic for $k = 0, 1, \ldots, n-1$.
- A point x_0 is an *irregular singular point* of the given differential equation if it is neither an ordinary point nor a regular singular point.

Consider the second-order linear homogeneous ordinary differential equation

$$y'' + P(x)y' + Q(x)y = 0.$$

If $x = 0$ is a regular singular point, then $xP(x)$ and $x^2Q(x)$ can be expanded as power series

$$xP(x) = \sum_{n=0}^{\infty} P_n x^n, \quad x^2Q(x) = \sum_{n=0}^{\infty} Q_n x^n, \quad |x| < r,$$

which leads to

$$P(x) = \sum_{n=0}^{\infty} P_n x^{n-1}, \quad Q(x) = \sum_{n=0}^{\infty} Q_n x^{n-2}, \quad |x| < r, \ x \neq 0.$$

Seek the power series solution of the differential equation of the form

$$y(x) = x^{\alpha} \cdot \sum_{n=0}^{\infty} a_n x^n = \sum_{n=0}^{\infty} a_n x^{n+\alpha}, \quad 0 < x < r,$$

which is called a *Frobenius* series solution. Differentiating with respect to x yields

$$y'(x) = \sum_{n=0}^{\infty} (n+\alpha)\,a_n x^{n+\alpha-1}, \quad y''(x) = \sum_{n=0}^{\infty} (n+\alpha)(n+\alpha-1)\,a_n x^{n+\alpha-2}.$$

Substituting into the differential equation results in

$$\sum_{n=0}^{\infty} (n+\alpha)(n+\alpha-1)\,a_n x^{n+\alpha-2} + \sum_{n=0}^{\infty} P_n x^{n-1} \cdot \sum_{n=0}^{\infty} (n+\alpha)\,a_n x^{n+\alpha-1}$$
$$+ \sum_{n=0}^{\infty} Q_n x^{n-2} \cdot \sum_{n=0}^{\infty} a_n x^{n+\alpha} = 0.$$

Noting that

$$\sum_{n=0}^{\infty} P_n x^{n-1} \cdot \sum_{n=0}^{\infty} (n+\alpha) a_n x^{n+\alpha-1} = \sum_{n=0}^{\infty} \sum_{m=0}^{n} P_{n-m} x^{n-m-1} \cdot (m+\alpha) a_m x^{m+\alpha-1}$$

$$= \sum_{n=0}^{\infty} \left[\sum_{m=0}^{n} (m+\alpha) P_{n-m} a_m \right] x^{n+\alpha-2},$$

$$\sum_{n=0}^{\infty} Q_n x^{n-2} \cdot \sum_{n=0}^{\infty} a_n x^{n+\alpha} = \sum_{n=0}^{\infty} \left(\sum_{m=0}^{n} Q_{n-m} a_m \right) x^{n+\alpha-2},$$

one obtains

$$\sum_{n=0}^{\infty} \left\{ (n+\alpha)(n+\alpha-1) a_n + \sum_{m=0}^{n} \left[(m+\alpha) P_{n-m} + Q_{n-m} \right] a_m \right\} x^{n+\alpha-2} = 0.$$

For this equation to be true, the coefficient of $x^{n+\alpha-2}$, $n=0, 1, \ldots,$ must be zero. For $n=0$, one has

$$\left[\alpha(\alpha-1) + \alpha P_0 + Q_0 \right] a_0 = 0,$$

which implies either $a_0 = 0$ or $\alpha(\alpha-1) + \alpha P_0 + Q_0 = 0$. For $n \geqslant 1$, one obtains

$$(n+\alpha)(n+\alpha-1) a_n + \sum_{m=0}^{n} \left[(m+\alpha) P_{n-m} + Q_{n-m} \right] a_m = 0.$$

$$\therefore \quad a_n = -\frac{1}{(n+\alpha)(n+\alpha-1) + (n+\alpha) P_0 + Q_0} \sum_{m=0}^{n-1} \left[(m+\alpha) P_{n-m} + Q_{n-m} \right] a_m.$$

Case 1. If $a_0 = 0$, then $a_1 = a_2 = \cdots = 0$, resulting in the zero solution $y(x) = 0$.

Case 2. If $a_0 \neq 0$, then

$$\alpha(\alpha-1) + \alpha P_0 + Q_0 = 0,$$

which is called the *indicial equation*. Solving this quadratic equation for α, one obtains two roots α_1 and α_2.

Hence, in order to have a nonzero solution, it is required that $a_0 \neq 0$ and α is a root of the indicial equation.

Remarks: If a series solution about a point $x = x_0 \neq 0$ is to be determined, one can change the independent variable to $t = x - x_0$ and then solve the resulting differential equation about $t = 0$. If a solution valid for $x < 0$ is to be determined, let $t = -x$ and then solve the resulting differential equation.

Fuchs' Theorem — Series Solution about a Regular Singular Point

For the second-order linear homogeneous ordinary differential equation

$$y''(x) + P(x)\,y'(x) + Q(x)\,y(x) = 0,$$

if $x = 0$ is a regular singular point, then

$$x\,P(x) = \sum_{n=0}^{\infty} P_n x^n, \quad x^2 Q(x) = \sum_{n=0}^{\infty} Q_n x^n, \quad |x| < r.$$

Suppose that the indicial equation

$$\alpha(\alpha - 1) + \alpha P_0 + Q_0 = 0$$

has two real roots α_1 and α_2, $\alpha_1 \geqslant \alpha_2$. Then the differential equation has at least one *Frobenius* series solution given by

$$y_1(x) = x^{\alpha_1} \sum_{n=0}^{\infty} a_n x^n, \quad a_0 \neq 0, \quad 0 < x < r,$$

where the coefficients a_n can be determined by substituting $y_1(x)$ into the differential equation. A second linearly independent solution is obtained as follows:

1. If $\alpha_1 - \alpha_2$ is not equal to an integer, then a second Frobenius series solution is given by

$$y_2(x) = x^{\alpha_2} \sum_{n=0}^{\infty} b_n x^n, \quad 0 < x < r,$$

   in which the coefficients b_n can be determined by substituting $y_2(x)$ into the differential equation.

2. If $\alpha_1 = \alpha_2 = \alpha$, then

$$y_2(x) = y_1(x) \ln x + x^{\alpha} \sum_{n=0}^{\infty} b_n x^n, \quad 0 < x < r,$$

   in which the coefficients b_n can be determined by substituting $y_2(x)$ into the differential equation, once $y_1(x)$ is known. In this case, the second solution $y_2(x)$ is not a Frobenius series solution.

3. If $\alpha_1 - \alpha_2$ is a positive integer, then

$$y_2(x) = a\,y_1(x) \ln x + x^{\alpha_2} \sum_{n=0}^{\infty} b_n x^n, \quad 0 < x < r,$$

   where the coefficients b_n and a can be determined by substituting y_2 into the differential equation, once y_1 is known. The parameter a may be zero, in which case the second solution $y_2(x)$ is also a Frobenius series solution.

The general solution of the differential equation is then given by

$$y(x) = C_1\,y_1(x) + C_2\,y_2(x).$$

Example 9.8

Obtain series solution about $x=0$ of the equation

$$2x^2y''+x(2x+1)y'-y=0.$$

The differential equation is of the form

$$y''+P(x)y'+Q(x)y=0,\quad P(x)=\frac{2x+1}{2x},\quad Q(x)=-\frac{1}{2x^2}.$$

Obviously, $x=0$ is a singular point. Note that

$$xP(x)=\frac{2x+1}{2}=\frac{1}{2}+x+0\cdot x^2+0\cdot x^3+\cdots\implies P_0=\frac{1}{2},$$
$$x^2Q(x)=-\frac{1}{2}=-\frac{1}{2}+0\cdot x+0\cdot x^2+0\cdot x^3+\cdots\implies Q_0=-\frac{1}{2}.$$

Both $xP(x)$ and $x^2Q(x)$ are analytic at $x=0$ and can be expanded as power series that are convergent for $|x|<\infty$. Hence, $x=0$ is a regular singular point.

The indicial equation is $\alpha(\alpha-1)+\alpha P_0+Q_0=0$:

$$\alpha(\alpha-1)+\alpha\cdot\frac{1}{2}-\frac{1}{2}=0\implies(\alpha+\tfrac{1}{2})(\alpha-1)=0\implies\alpha_1=1,\ \alpha_2=-\frac{1}{2}.$$

Thus the equation has a Frobenius series solution of the form

$$y_1(x)=x^{\alpha_1}\sum_{n=0}^{\infty}a_nx^n=\sum_{n=0}^{\infty}a_nx^{n+1},\quad a_0\neq0,\quad 0<x<\infty,$$

where a_n, $n=0,1,\ldots$, are constants to be determined. Differentiating with respect to x yields

$$y_1'(x)=\sum_{n=0}^{\infty}(n+1)a_nx^n,\quad y_1''(x)=\sum_{n=1}^{\infty}(n+1)na_nx^{n-1}.$$

Substituting y_1, y_1', and y_1'' into the differential equation results in

$$\sum_{n=1}^{\infty}2(n+1)na_nx^{n+1}+\sum_{n=0}^{\infty}2(n+1)a_nx^{n+2}+\sum_{n=0}^{\infty}(n+1)a_nx^{n+1}-\sum_{n=0}^{\infty}a_nx^{n+1}=0.$$

Changing the indices of the summations

$$\sum_{n=1}^{\infty}2(n+1)na_nx^{n+1}\xrightarrow{n+1=m}\sum_{m=2}^{\infty}2m(m-1)a_{m-1}x^m,$$
$$\sum_{n=0}^{\infty}2(n+1)a_nx^{n+2}\xrightarrow{n+2=m}\sum_{m=2}^{\infty}2(m-1)a_{m-2}x^m,$$
$$\sum_{n=0}^{\infty}na_nx^{n+1}\xrightarrow{n+1=m}\sum_{m=1}^{\infty}(m-1)a_{m-1}x^m,$$

one obtains

$$\sum_{n=2}^{\infty}\left[2n(n-1)\,a_{n-1}+2(n-1)\,a_{n-2}\right]x^n+\sum_{n=1}^{\infty}(n-1)\,a_{n-1}x^n=0.$$

For this equation to be true, the coefficient of x^n, $n=1, 2, \dots,$ must be zero. For $n=1$, one has

$$0\cdot a_0=0 \implies a_0\neq 0 \text{ is arbitrary; take } a_0=1.$$

For $n\geqslant 2$, one has

$$2n(n-1)\,a_{n-1}+2(n-1)\,a_{n-2}+(n-1)\,a_{n-1}=0 \implies a_{n-1}=-\frac{2a_{n-2}}{2n+1}.$$

Hence,

$$n=2:\quad a_1=-\frac{2a_0}{2\cdot 2+1}=-\frac{2}{5},$$

$$n=3:\quad a_2=-\frac{2a_1}{2\cdot 3+1}=(-1)^2\frac{2^2}{7\cdot 5},$$

$$\vdots$$

$$n+1:\quad a_n=-\frac{2a_{n-1}}{2(n+1)+1}=(-1)^n\frac{2^n}{(2n+3)(2n+1)\cdots 5}=(-1)^n\frac{3\cdot 2^n}{(2n+3)!!},$$

where $(2n+3)!!=(2n+3)(2n+1)\cdots 5\cdot 3\cdot 1$ is the double factorial. The first Frobenius series solution is

$$y_1(x)=\sum_{n=0}^{\infty}a_n x^{n+1}=\sum_{n=0}^{\infty}(-1)^n\frac{3\cdot 2^n}{(2n+3)!!}x^{n+1},\qquad 0<x<\infty.$$

Since $\alpha_1-\alpha_2=\frac{3}{2}$, according to Fuchs' Theorem, a second linearly independent solution is also a Frobenius series given by

$$y_2(x)=x^{\alpha_2}\sum_{n=0}^{\infty}b_n x^n=\sum_{n=0}^{\infty}b_n x^{n-\frac{1}{2}},\quad b_0\neq 0,\quad 0<x<\infty,$$

$$y_2'(x)=\sum_{n=0}^{\infty}\left(n-\tfrac{1}{2}\right)b_n x^{n-\frac{3}{2}},\quad y_2''(x)=\sum_{n=0}^{\infty}\left(n-\tfrac{1}{2}\right)\left(n-\tfrac{3}{2}\right)b_n x^{n-\frac{5}{2}}.$$

Substituting y_2, y_2', and y_2'' into the differential equation leads to

$$2x^2\sum_{n=0}^{\infty}\left(n-\tfrac{1}{2}\right)\left(n-\tfrac{3}{2}\right)b_n x^{n-\frac{5}{2}}+(2x^2+x)\sum_{n=0}^{\infty}\left(n-\tfrac{1}{2}\right)b_n x^{n-\frac{3}{2}}-\sum_{n=0}^{\infty}b_n x^{n-\frac{1}{2}}=0,$$

$$\sum_{n=0}^{\infty}\left\{\left[2\left(n-\tfrac{1}{2}\right)\left(n-\tfrac{3}{2}\right)+\left(n-\tfrac{1}{2}\right)-1\right]b_n x^{n-\frac{1}{2}}+2\left(n-\tfrac{1}{2}\right)b_n x^{n+\frac{1}{2}}\right\}=0.$$

Multiplying this equation by $x^{\frac{1}{2}}$ yields

$$\sum_{n=0}^{\infty}\Big[n(2n-3)\,b_n x^n + (2n-1)\,b_n x^{n+1}\Big] = 0.$$

Changing the index of the summation

$$\sum_{n=0}^{\infty}(2n-1)\,b_n x^{n+1} \xrightarrow{\;n+1=m\;} \sum_{m=1}^{\infty}(2m-3)\,b_{m-1}x^m,$$

one obtains

$$\sum_{n=0}^{\infty}n(2n-3)\,b_n x^n + \sum_{n=1}^{\infty}(2n-3)\,b_{n-1}x^n = 0.$$

For this equation to be true, the coefficient of x^n, $n=0, 1, \ldots,$ must be zero. For $n=0$, one has

$$0\cdot(-3)\,b_0 = 0 \implies b_0 \neq 0 \text{ is arbitrary; take } b_0 = 1.$$

For $n \geqslant 1$, one has

$$n(2n-3)\,b_n + (2n-3)\,b_{n-1} = 0 \implies b_n = -\frac{b_{n-1}}{n}.$$

Hence,

$$b_1 = -\frac{b_0}{1} = -\frac{1}{1}, \quad b_2 = -\frac{b_1}{2} = (-1)^2\frac{1}{2!}, \quad b_3 = -\frac{b_2}{3} = (-1)^3\frac{1}{3!}, \quad \ldots$$

$$\therefore \quad b_n = -\frac{b_{n-1}}{n} = (-1)^n\frac{1}{n!}.$$

Thus, a second linearly independent solution is

$$y_2(x) = x^{-\frac{1}{2}}\sum_{n=0}^{\infty}b_n x^n = x^{-\frac{1}{2}}\sum_{n=0}^{\infty}(-1)^n\frac{x^n}{n!} = x^{-\frac{1}{2}}\mathrm{e}^{-x}.$$

The general solution of the differential equation is

$$y(x) = C_1 y_1(x) + C_2 y_2(x) = C_1\sum_{n=0}^{\infty}(-1)^n\frac{3\cdot 2^n}{(2n+3)!!}x^{n+1} + C_2 x^{-\frac{1}{2}}\mathrm{e}^{-x}.$$

9.3.1 Bessel's Equation and Its Applications

9.3.1.1 Solutions of Bessel's Equation

Bessel's equation of the form

$$x^2y'' + xy' + (x^2 - \nu^2)\,y = 0, \quad x > 0,$$

where $\nu \geqslant 0$ is a constant, is of great importance in applied mathematics and has numerous applications in engineering and science. Furthermore, in solving Bessel's equation using series, it exhibits all possibilities in Fuchs' Theorem. As a result, it is an excellent example to illustrate the procedure and nuances for solving a second-order differential equation using series about a regular singular point.

Bessel's equation is of the form

$$y'' + P(x)y' + Q(x)y = 0, \quad P(x) = \frac{1}{x}, \quad Q(x) = \frac{x^2 - \nu^2}{x^2}.$$

It is obvious that $x = 0$ is a singular point. Since

$$xP(x) = 1 = 1 + 0\cdot x + 0\cdot x^2 + \cdots \implies P_0 = 1,$$
$$x^2 Q(x) = x^2 - \nu^2 = -\nu^2 + 0\cdot x + x^2 + 0\cdot x^3 + 0\cdot x^4 + \cdots \implies Q_0 = -\nu^2,$$

both $xP(x)$ and $x^2Q(x)$ are analytic at $x = 0$ and can be expanded as power series convergent for $|x| < \infty$. Hence, $x = 0$ is a regular singular point.

The indicial equation is $\alpha(\alpha - 1) + \alpha P_0 + Q_0 = 0$:

$$\alpha(\alpha - 1) + \alpha \cdot 1 - \nu^2 = 0 \implies \alpha - \nu^2 = 0 \implies \alpha_1 = \nu, \ \alpha_2 = -\nu.$$

Bessel's equation has a Frobenius series solution of the form

$$y_1(x) = x^{\nu} \sum_{n=0}^{\infty} a_n x^n = \sum_{n=0}^{\infty} a_n x^{n+\nu}, \quad a_0 \neq 0, \quad 0 < x < \infty.$$

Differentiating with respect to x yields

$$y_1'(x) = \sum_{n=0}^{\infty} (n+\nu) a_n x^{n+\nu-1}, \quad y_1''(x) = \sum_{n=0}^{\infty} (n+\nu)(n+\nu-1) a_n x^{n+\nu-2}.$$

Substituting y_1, y_1', and y_1'' into Bessel's equation results in

$$x^2 \sum_{n=0}^{\infty} (n+\nu)(n+\nu-1) a_n x^{n+\nu-2} + x \sum_{n=0}^{\infty} (n+\nu) a_n x^{n+\nu-1} + (x^2 - \nu^2) \sum_{n=0}^{\infty} a_n x^{n+\nu} = 0.$$

Changing the index of the summation

$$\sum_{n=0}^{\infty} a_n x^{n+\nu+2} \xrightarrow{n+2=m} \sum_{m=2}^{\infty} a_{m-2} x^{m+\nu} = \sum_{n=2}^{\infty} a_{n-2} x^{n+\nu},$$

one obtains

$$x^{\nu} \left\{ \sum_{n=0}^{\infty} \left[(n+\nu)(n+\nu-1) + (n+\nu) - \nu^2 \right] a_n x^n + \sum_{n=2}^{\infty} a_{n-2} x^n \right\} = 0,$$

$$x^{\nu} \neq 0 \implies \sum_{n=0}^{\infty} n(n+2\nu) a_n x^n + \sum_{n=2}^{\infty} a_{n-2} x^n = 0.$$

For this equation to be true, the coefficient of x^n, $n = 0, 1, \ldots$, must be zero:

$$x^0: \quad 0\cdot(0+2\nu)\, a_0 = 0 \implies a_0 \neq 0 \text{ is arbitrary},$$

$$x^1: \quad 1\cdot(1+2\nu)\,a_1 = 0 \implies a_1 = 0.$$

For $n \geqslant 2$, one obtains

$$x^n: \quad n(n+2\nu)\,a_n + a_{n-2} = 0 \implies a_n = -\frac{a_{n-2}}{n(n+2\nu)}.$$

Hence, $a_{2n+1} = 0$, for $n = 0, 1, \ldots$, and

$$a_2 = -\frac{a_0}{2(2+2\nu)} = -\frac{a_0}{2^2\cdot 1(1+\nu)},$$

$$a_4 = -\frac{a_2}{4(4+2\nu)} = -\frac{a_2}{2^2\cdot 2(2+\nu)} = (-1)^2\frac{a_0}{2^4\cdot 2!(1+\nu)(2+\nu)},$$

$$\cdots \quad \cdots$$

$$a_{2n} = (-1)^n\frac{a_0}{2^{2n}\cdot n!(1+\nu)(2+\nu)\cdots(n+\nu)},$$

and

$$y_1(x) = a_0 x^\nu \sum_{n=0}^{\infty} (-1)^n \frac{1}{n!(1+\nu)(2+\nu)\cdots(n+\nu)}\left(\frac{x}{2}\right)^{2n}, \quad 0 < x < \infty.$$

To simplify the solution, use the *Gamma function* defined by

$$\Gamma(\nu+1) = \int_0^\infty t^\nu e^{-t}\,dt, \quad \nu > 0.$$

Using integration by parts, it is easy to show that

$$\Gamma(\nu+1) = -\int_0^\infty t^\nu\,d(e^{-t}) = -t^\nu e^{-t}\Big|_{t=0}^{\infty} + \int_0^\infty e^{-t}\cdot\nu t^{\nu-1}\,dt$$

$$= \nu\int_0^\infty t^{\nu-1}e^{-t}\,dt = \nu\,\Gamma(\nu),$$

$$\therefore \quad \Gamma(n+\nu+1) = (n+\nu)\,\Gamma(n+\nu) = (n+\nu)(n+\nu-1)\,\Gamma(n+\nu-1) = \cdots$$

$$= (n+\nu)(n+\nu-1)\cdots(1+\nu)\,\Gamma(1+\nu).$$

When $\nu = k$ is an integer, one obtains

$$\Gamma(1) = \int_0^\infty e^{-t}\,dt = -e^{-t}\Big|_{t=0}^{\infty} = 1,$$

$$\Gamma(2) = 1\cdot\Gamma(1) = 1, \quad \Gamma(3) = 2\cdot\Gamma(2) = 2\cdot 1 = 2!, \quad \cdots$$

$$\therefore \quad \Gamma(k+1) = k\cdot\Gamma(k) = k!.$$

Hence, letting $a_0 = \left[2^\nu\,\Gamma(1+\nu)\right]^{-1}$, the first Frobenius series solution is

$$y_1(x) = J_\nu(x),$$

where $J_\nu(x)$ is called the *Bessel function of the first kind of order* ν given by

$$J_\nu(x) = \frac{1}{2^\nu\,\Gamma(1+\nu)}\,x^\nu \sum_{n=0}^{\infty}(-1)^n \frac{1}{n!\,(1+\nu)(2+\nu)\cdots(n+\nu)}\left(\frac{x}{2}\right)^{2n},$$

$$\therefore\quad J_\nu(x) = \sum_{n=0}^{\infty}(-1)^n \frac{1}{n!\,\Gamma(n+\nu+1)}\left(\frac{x}{2}\right)^{2n+\nu}, \qquad 0<x<\infty.$$

According to Fuchs' Theorem, the form of a second linearly independent solution depends on whether the difference of the roots of the indicial equation, i.e., $\alpha_1-\alpha_2=2\nu$, is noninteger, zero, or a positive integer.

Case 1. 2ν is not an integer

A second Frobenius series solution is

$$y_2(x) = x^{-\nu}\sum_{n=0}^{\infty} b_n x^n, \qquad 0<x<\infty.$$

Following the same procedure, it is easy to show that, for $n=1,2,\ldots,$

$$a_{2n-1}=0, \qquad a_{2n} = (-1)^n \frac{b_0}{2^{2n}\cdot n!\,(1-\nu)(2-\nu)\cdots(n-\nu)},$$

$$y_2(x) = b_0 x^{-\nu}\sum_{n=0}^{\infty}(-1)^n \frac{1}{n!\,(1-\nu)(2-\nu)\cdots(n-\nu)}\left(\frac{x}{2}\right)^{2n}, \qquad 0<x<\infty.$$

Letting $b_0=\left[2^{-\nu}\Gamma(1-\nu)\right]^{-1}$, one obtains

$$y_2(x) = \sum_{n=0}^{\infty}(-1)^n \frac{1}{n!\,\Gamma(n-\nu+1)}\left(\frac{x}{2}\right)^{2n-\nu} = J_{-\nu}(x), \qquad 0<x<\infty.$$

The general solution is

$$y(x) = C_1 J_\nu(x) + C_2 J_{-\nu}(x).$$

The general solution can also be written as

$$y(x) = D_1 J_\nu(x) + D_2 Y_\nu(x),$$

where

$$Y_\nu(x) = \frac{J_\nu(x)\cos\nu\pi - J_{-\nu}(x)}{\sin\nu\pi}$$

is the *Bessel function of the second kind of order* ν.

Case 2. $\nu=0$, then $\alpha_1=\alpha_2=0$

The first Frobenius series solution is simplified as

$$y_1(x) = J_0(x) = \sum_{n=0}^{\infty}(-1)^n \frac{1}{(n!)^2}\left(\frac{x}{2}\right)^{2n}, \qquad 0<x<\infty.$$

A second linearly independent solution is

$$y_2(x) = y_1(x) \ln x + \sum_{n=0}^{\infty} b_n x^n, \quad 0 < x < \infty.$$

Differentiating with respect to x yields

$$y_2'(x) = y_1' \ln x + \frac{y_1}{x} + \sum_{n=1}^{\infty} n b_n x^{n-1},$$

$$y_2''(x) = y_1'' \ln x + \frac{2y_1'}{x} - \frac{y_1}{x^2} + \sum_{n=2}^{\infty} n(n-1) b_n x^{n-2}.$$

Substituting y_2, y_2', and y_2'' into Bessel's equation results in, with $\nu = 0$,

$$(x^2 y_1'' + x y_1' + x^2 y_1) \ln x + 2x y_1' + \sum_{n=2}^{\infty} n(n-1) b_n x^n + \sum_{n=1}^{\infty} n b_n x^n + \sum_{n=0}^{\infty} b_n x^{n+2} = 0.$$

Since $y_1(x)$ is a solution of Bessel's equation, $x^2 y_1'' + x y_1' + x^2 y_1 = 0$, and noting

$$2x y_1' = 2x \sum_{n=1}^{\infty} (-1)^n \frac{1}{(n!)^2} \frac{2n \cdot x^{2n-1}}{2^{2n}} = \sum_{n=1}^{\infty} (-1)^n \frac{4n}{(n!)^2} \left(\frac{x}{2}\right)^{2n},$$

one obtains

$$\sum_{n=1}^{\infty} (-1)^n \frac{4n}{(n!)^2} \left(\frac{x}{2}\right)^{2n} + \sum_{n=2}^{\infty} n(n-1) b_n x^n + \sum_{n=1}^{\infty} n b_n x^n + \sum_{n=2}^{\infty} b_{n-2} x^n = 0.$$

For this equation to be true, the coefficient of x^n, $n = 0, 1, \ldots$, must be zero.

From the coefficient of x^1, one has $1 \cdot b_1 = 0 \implies b_1 = 0$.

From the coefficient of x^n, $n \geqslant 1$, one obtains $b_{2n+1} = 0$, and

$$(-1)^n \frac{4n}{(n!)^2} \left(\frac{1}{2}\right)^{2n} + \left[2n(2n-1) + 2n\right] b_{2n} + b_{2n-2} = 0,$$

$$\therefore \quad b_{2n} = (-1)^{n+1} \frac{1}{n\,(n!)^2} \left(\frac{1}{2}\right)^{2n} - \frac{b_{2n-2}}{(2n)^2}.$$

For simplicity, take $b_0 = 0$. Using mathematical induction, it can be shown that

$$b_{2n} = (-1)^{n+1} \frac{1 + \frac{1}{2} + \frac{1}{3} + \cdots + \frac{1}{n}}{(n!)^2} \left(\frac{1}{2}\right)^{2n}.$$

Hence, a second linearly independent solution is

$$y_2(x) = J_0(x) \ln x + \sum_{n=1}^{\infty} (-1)^{n+1} \frac{1 + \frac{1}{2} + \frac{1}{3} + \cdots + \frac{1}{n}}{(n!)^2} \left(\frac{x}{2}\right)^{2n}, \quad 0 < x < \infty,$$

or, in terms of the *Bessel function of the second kind of order* 0, $Y_0(x)$,

$$y_2(x) = \frac{\pi}{2}\, Y_0(x) + (\ln 2 - \gamma)\, J_0(x), \quad 0 < x < \infty.$$

The Bessel function of the second kind of order 0 is defined as, for $0 < x < \infty$,

$$Y_0(x) = \frac{2}{\pi}\left\{\left(\ln\frac{x}{2} + \gamma\right) J_0(x) + \sum_{n=1}^{\infty} (-1)^{n+1} \frac{1 + \frac{1}{2} + \frac{1}{3} + \cdots + \frac{1}{n}}{(n!)^2} \left(\frac{x}{2}\right)^{2n}\right\},$$

in which

$$\gamma = 0.57721566490153\cdots = \lim_{n\to\infty}\left(\sum_{k=1}^{n} \frac{1}{k} - \ln n\right)$$

is the Euler constant. The general solution is given by

$$y(x) = C_1\, J_0(x) + C_2\, Y_0(x).$$

Case 3. ν is a positive integer

The first Frobenius series solution is simplified as

$$y_1(x) = J_\nu(x) = \sum_{n=0}^{\infty} (-1)^n \frac{1}{n!(n+\nu)!}\left(\frac{x}{2}\right)^{2n+\nu}, \quad 0 < x < \infty.$$

A second linearly independent solution is

$$y_2(x) = a\, y_1(x) \ln x + x^{-\nu} \sum_{n=0}^{\infty} b_n x^n, \quad 0 < x < \infty,$$

$$y_2'(x) = a\left(y_1' \ln x + \frac{y_1}{x}\right) + \sum_{n=0}^{\infty} (n-\nu) b_n x^{n-\nu-1},$$

$$y_2''(x) = a\left(y_1'' \ln x + \frac{2y_1'}{x} - \frac{y_1}{x^2}\right) + \sum_{n=0}^{\infty} (n-\nu)(n-\nu-1) b_n x^{n-\nu-2}.$$

Substituting into Bessel's equation results in

$$a\left[x^2 y_1'' + x y_1' + (x^2 - \nu^2) y_1\right] \ln x + 2ax y_1' + \sum_{n=0}^{\infty} (n-\nu)(n-\nu-1) b_n x^{n-\nu}$$

$$+ \sum_{n=0}^{\infty} (n-\nu) b_n x^{n-\nu} + \sum_{n=0}^{\infty} b_n x^{n-\nu+2} - \sum_{n=0}^{\infty} \nu^2 b_n x^{n-\nu} = 0.$$

Since $y_1(x)$ is a solution of Bessel's equation, $x^2 y_1'' + x y_1' + (x^2 - \nu^2) y_1 = 0$; noting

$$2ax y_1' = 2ax \sum_{n=0}^{\infty} (-1)^n \frac{(2n+\nu)\cdot x^{2n+\nu-1}}{n!(n+\nu)!\, 2^{2n+\nu}} = \sum_{n=0}^{\infty} (-1)^n \frac{2a(2n+\nu)}{n!(n+\nu)!}\left(\frac{x}{2}\right)^{2n+\nu},$$

and multiplying the equation by x^ν yield

$$\sum_{n=0}^{\infty} (-1)^n \frac{2^{\nu+1} a(2n+\nu)}{n!(n+\nu)!}\left(\frac{x}{2}\right)^{2(n+\nu)} + \sum_{n=0}^{\infty} n(n-2\nu) b_n x^n + \sum_{n=2}^{\infty} b_{n-2} x^n = 0.$$

For this equation to be true, the coefficient of x^n, $n = 0, 1, \ldots,$ must be zero.

❧ From the coefficient of x^n, $0 \leqslant n < 2\nu$, the first summation in the above equation has no contribution,

$$x^0: \quad 0\cdot(0-2\nu)\,b_0 = 0 \implies b_0 \text{ is arbitrary; take } b_0 = 1,$$

$$x^1: \quad 1\cdot(1-2\nu)\,b_1 = 0 \implies b_1 = 0.$$

From the coefficient of x^n, $2 \leqslant n < 2\nu$, one has

$$n\cdot(n-2\nu)\,b_n + b_{n-2} = 0 \implies b_n = \frac{b_{n-2}}{n(2\nu-n)},$$

$$n=2: \quad b_2 = \frac{b_0}{2\,(2\nu-2)} = \frac{1}{2^2\cdot 1\,(\nu-1)},$$

$$n=4: \quad b_4 = \frac{b_2}{4\,(2\nu-4)} = \frac{1}{2^4\cdot 2!\,(\nu-1)(\nu-2)},$$

$$\vdots$$

$$n=2k: \quad b_{2k} = \frac{b_{2k-2}}{2k\,(2\nu-2k)} = \frac{1}{2^{2k}\cdot k!\,(\nu-1)(\nu-2)\cdots(\nu-k)} = \frac{(\nu-k-1)!}{2^{2k}k!\,(\nu-1)!}.$$

It is easy to see that $b^{2n+1} = 0$, for all $n = 0, 1, 2, \ldots$.

❧ From the coefficient of $x^{2\nu}$, one obtains

$$\frac{2^{\nu+1}a\nu}{\nu!}\left(\frac{x}{2}\right)^{2\nu} + b_{2\nu-2} = 0 \implies a = -2^{\nu-1}(\nu-1)!\,b_{2(\nu-1)} = -\frac{1}{2^{\nu-1}(\nu-1)!}.$$

The value of $b_{2\nu}$ is arbitrary; for simplicity, take $b_{2\nu} = 0$.

❧ From the coefficient of $x^{2(n+\nu)}$, $n \geqslant 1$, one obtains

$$(-1)^n\frac{2^{\nu+1}a(2n+\nu)}{n!(n+\nu)!}\left(\frac{1}{2}\right)^{2(n+\nu)} + (2n+2\nu)(2n)\,b_{2(n+\nu)} + b_{2(n-1+\nu)} = 0,$$

$$\therefore \quad b_{2(n+\nu)} = (-1)^{n+1}\frac{2^{\nu-1}a(2n+\nu)}{n(n+\nu)\,n!(n+\nu)!}\left(\frac{1}{2}\right)^{2(n+\nu)} - \frac{b_{2(n-1+\nu)}}{2^2 n(n+\nu)}.$$

Using mathematical induction, it can be shown that

$$b_{2(n+\nu)} = (-1)^{n+1}\frac{2^{\nu-1}aA_n}{n!(n+\nu)!}\left(\frac{1}{2}\right)^{2(n+\nu)},$$

where

$$A_n = \left(\frac{1}{1} + \frac{1}{2} + \cdots + \frac{1}{n}\right) + \left(\frac{1}{1+\nu} + \frac{1}{2+\nu} + \cdots + \frac{1}{n+\nu}\right).$$

Hence, a second linearly independent solution is

$$y_2(x) = a\,J_\nu(x)\,\ln x + x^{-\nu}\left\{\sum_{n=0}^{\nu-1}\frac{(\nu-n-1)!}{n!(\nu-1)!}\left(\frac{x}{2}\right)^{2n}\right.$$

$$\left. + a\sum_{n=1}^{\infty}(-1)^{n+1}\frac{2^{\nu-1}A_n}{n!(n+\nu)!}\left(\frac{x}{2}\right)^{2(n+\nu)}\right\}, \qquad 0 < x < \infty.$$

Using the notation

$$1+\frac{1}{2}+\cdots+\frac{1}{n}=\psi(n+1)+\gamma, \qquad \psi(1)=-\gamma,$$

where $\psi(n)=\Gamma'(n)/\Gamma(n)$ is the psi function,

$y_2(x)$ can be expressed in terms of the *Bessel function of the second kind of order* ν, $Y_\nu(x)$, as

$$y_2(x)=a\left\{J_\nu(x)\ln x+\frac{x^{-\nu}}{a}\sum_{n=0}^{\nu-1}\frac{(\nu-n-1)!}{n!(\nu-1)!}\left(\frac{x}{2}\right)^{2n}\right.$$
$$\left.+\frac{1}{2}\sum_{n=0}^{\infty}(-1)^{n+1}\frac{\psi(n+1)+\psi(n+\nu+1)-\psi(\nu+1)+\gamma}{n!(n+\nu)!}\left(\frac{x}{2}\right)^{2n+\nu}\right\}$$
$$=a\left\{\left[J_\nu(x)\ln\frac{x}{2}-\frac{1}{2}\sum_{n=0}^{\nu-1}\frac{(\nu-n-1)!}{n!}\left(\frac{x}{2}\right)^{2n-\nu}\right.\right.$$
$$\left.-\frac{1}{2}\sum_{n=0}^{\infty}(-1)^n\frac{\psi(n+1)+\psi(n+\nu+1)}{n!(n+\nu)!}\left(\frac{x}{2}\right)^{2n+\nu}\right]$$
$$\left.-\frac{1}{2}\left[\gamma-\psi(\nu+1)-2\ln 2\right]J_\nu(x)\right\}$$
$$=a\left\{\frac{\pi}{2}Y_\nu(x)-\frac{1}{2}\left[\gamma-\psi(\nu+1)-2\ln 2\right]J_\nu(x)\right\},$$

where, for $0<x<\infty$,

$$Y_\nu(x)=\frac{2}{\pi}J_\nu(x)\ln\frac{x}{2}-\frac{1}{\pi}\sum_{n=0}^{\nu-1}\frac{(\nu-n-1)!}{n!}\left(\frac{x}{2}\right)^{2n-\nu}$$
$$-\frac{1}{\pi}\sum_{n=0}^{\infty}(-1)^n\frac{\psi(n+1)+\psi(n+\nu+1)}{n!(n+\nu)!}\left(\frac{x}{2}\right)^{2n+\nu}.$$

Using the Bessel functions of the first and second kinds, the general solution is

$$y(x)=C_1J_\nu(x)+C_2Y_\nu(x).$$

Remarks: This is the case that, when $\alpha_1-\alpha_2$ is a positive integer, the second solution contains the logarithmic term $\ln x$.

Case 4. $\nu=k+\frac{1}{2}$, $k=0,1,\ldots,$ *and* $\alpha_1-\alpha_2=2k+1$ *is a positive integer*

The first Frobenius series solution becomes

$$y_1(x)=J_{k+\frac{1}{2}}(x)=\sum_{n=0}^{\infty}(-1)^n\frac{1}{n!\,\Gamma\left(n+k+\frac{3}{2}\right)}\left(\frac{x}{2}\right)^{2n+k+\frac{1}{2}}, \qquad 0<x<\infty,$$

$$y_1'(x)=\sum_{n=0}^{\infty}(-1)^n\frac{2n+k+\frac{1}{2}}{2n!\,\Gamma\left(n+k+\frac{3}{2}\right)}\left(\frac{x}{2}\right)^{2n+k-\frac{1}{2}}.$$

A second linearly independent solution is

$$y_2(x) = a\, y_1(x) \ln x + x^{-(k+\frac{1}{2})} \sum_{n=0}^{\infty} b_n x^n, \quad 0 < x < \infty,$$

$$y_2'(x) = a\left(y_1' \ln x + \frac{y_1}{x}\right) + \sum_{n=0}^{\infty} \left(n-k-\tfrac{1}{2}\right) b_n x^{n-k-\frac{3}{2}},$$

$$y_2''(x) = a\left(y_1'' \ln x + \frac{2y_1'}{x} - \frac{y_1}{x^2}\right) + \sum_{n=0}^{\infty} \left(n-k-\tfrac{1}{2}\right)\left(n-k-\tfrac{3}{2}\right) b_n x^{n-k-\frac{5}{2}}.$$

Substituting into Bessel's equation results in

$$a\left[x^2 y_1'' + x y_1' + (x^2 - \nu^2) y_1\right] \ln x + 2ax y_1' + \sum_{n=0}^{\infty} n(n-2k-1) b_n x^{n-k-\frac{1}{2}} + \sum_{n=0}^{\infty} b_n x^{n-k+\frac{3}{2}} = 0.$$

Noting that $x^2 y_1'' + x y_1' + (x^2 - \nu^2) y_1 = 0$ and multiplying the equation by $x^{k+\frac{1}{2}}$ lead to

$$a \sum_{n=0}^{\infty} (-1)^n \frac{2^{k+\frac{3}{2}}\left(2n+k+\frac{1}{2}\right)}{n!\, \Gamma\left(n+k+\frac{3}{2}\right)} \left(\frac{x}{2}\right)^{2n+2k+1} + \sum_{n=0}^{\infty} n(n-2k-1) b_n x^n + \sum_{n=0}^{\infty} b_n x^{n+2} = 0.$$

For this equation to be true, the coefficient of x^n, $n = 0, 1, \ldots$, must be zero.

For the coefficient of x^n, $0 \leqslant n < 2k+1$, the first summation in the above equation has no contribution,

$$x^0: \quad 0 \cdot (0-2k-1)\, b_0 = 0 \implies b_0 \text{ is arbitrary},$$

$$x^1: \quad 1 \cdot (1-2k-1)\, b_1 = 0 \implies b_1 = 0.$$

It is easy to see that

$$b_{2m+1} = -\frac{b_{2m-1}}{(2m+1)(2m-2k)} = 0, \quad 1 < m < k.$$

From the coefficient of x^{2k+1}, one has

$$a \cdot \frac{k+\frac{1}{2}}{2^{k-\frac{1}{2}}\, \Gamma\left(k+\frac{3}{2}\right)} + (2k+1) \cdot 0 \cdot b_{2k+1} + b_{2k-1} = 0 \implies a = 0.$$

With $a = 0$, the coefficient of x^n gives

$$n(n-2\nu)\, b_n + b_{n-2} = 0 \implies b_{2m-1} = 0, \;\; b_{2m} = -\frac{b_{2m-2}}{2^2 m(m-\nu)}, \quad m = 1, 2, \ldots.$$

It can be easily shown that

$$b_{2m} = (-1)^m \frac{b_0}{2^{2m} m!\, (1-\nu)(2-\nu) \cdots (m-\nu)}.$$

Taking $b_0=[2^{-\nu}\Gamma(1-\nu)]^{-1}$, one obtains

$$b_{2m}=\frac{(-1)^m}{m!\,\Gamma(m-\nu+1)}\left(\frac{1}{2}\right)^{2m-\nu},$$

$$y_2(x)=x^{-\nu}\sum_{n=0}^{\infty}b_{2n}x^{2n}=\sum_{n=0}^{\infty}\frac{(-1)^n}{n!\,\Gamma(n-\nu+1)}\left(\frac{x}{2}\right)^{2n-\nu}=J_{-\nu}(x),\qquad 0<x<\infty.$$

The general solution is

$$y(x)=C_1\,J_\nu(x)+C_2\,J_{-\nu}(x),$$

which is the same as Case 1 when 2ν is not an integer.

Remarks: This is the case that, when $\alpha_1-\alpha_2$ is a positive integer, the second solution does not contain the logarithmic term $\ln x$ and is a Frobenius series.

In practice, Bessel's equation rarely appears in its standard form. A second-order linear ordinary differential equation of the form

$$\frac{d^2y}{dx^2}+\frac{1-2\alpha}{x}\frac{dy}{dx}+\left[(\beta\rho x^{\rho-1})^2+\frac{\alpha^2-\nu^2\rho^2}{x^2}\right]y=0,\quad x>0, \tag{1}$$

where α, β, ν, ρ are constants, can be transformed to Bessel's equation

$$\xi^2\frac{d^2\eta}{d\xi^2}+\xi\frac{d\eta}{d\xi}+(\xi^2-\nu^2)\eta=0,\quad \xi>0. \tag{2}$$

This result is established as follows.

Changing the variable $\xi=\beta x^\rho$, one has

$$\xi\frac{dy}{d\xi}=\xi\frac{dy/dx}{d\xi/dx}=\beta x^\rho\frac{1}{\beta\rho x^{\rho-1}}\frac{dy}{dx}=\frac{x}{\rho}\frac{dy}{dx}\implies\xi\frac{d(\cdot)}{d\xi}=\frac{x}{\rho}\frac{d(\cdot)}{dx}.$$

Equation (2) becomes

$$\xi\frac{d}{d\xi}\left(\xi\frac{d\eta}{d\xi}\right)+(\xi^2-\nu^2)\eta=0\implies\frac{x}{\rho}\frac{d}{dx}\left(\frac{x}{\rho}\frac{d\eta}{dx}\right)+(\beta^2x^{2\rho}-\nu^2)\eta=0.$$

Making the transformation $\eta=x^{-\alpha}y$, one has

$$x\frac{d\eta}{dx}=x^{1-\alpha}\frac{dy}{dx}-\alpha x^{-\alpha}y,$$

$$x\frac{d}{dx}\left(x\frac{d\eta}{dx}\right)=x^{2-\alpha}\frac{d^2y}{dx^2}+(1-2\alpha)x^{1-\alpha}\frac{dy}{dx}+\alpha^2x^{-\alpha}y.$$

Hence, the differential equation becomes

$$x^{2-\alpha}\frac{d^2y}{dx^2}+(1-2\alpha)x^{1-\alpha}\frac{dy}{dx}+\alpha^2x^{-\alpha}y+(\beta^2x^{2\rho}-\nu^2)\rho^2x^{-\alpha}y=0,$$

which leads to equation (1).

If the solution of Bessel's equation (2) is denoted as $\eta = \mathcal{B}_\nu(\xi)$, then the solution of equation (1) is $y = x^\alpha \mathcal{B}_\nu(\beta x^\rho)$.

Solutions of Bessel's Equation

- Denote the solution of Bessel's equation

$$x^2 y'' + xy' + (x^2 - \nu^2)y = 0, \quad x > 0,$$

where $\nu \geqslant 0$ is a constant, as $y(x) = \mathcal{B}_\nu(x)$, where

$$\mathcal{B}_\nu(x) = C_1 J_\nu(x) + C_2 Y_\nu(x),$$

in which $J_\nu(x)$ and $Y_\nu(x)$ are the Bessel functions of the first and second kinds, respectively, of order ν. When $\nu \neq 0, 1, 2, \ldots$, the solution can also be written as

$$\mathcal{B}_\nu(x) = C_1 J_\nu(x) + C_2 J_{-\nu}(x), \quad \nu \neq 0, 1, 2, \ldots.$$

- The solution of the differential equation

$$\frac{d^2y}{dx^2} + \frac{1-2\alpha}{x}\frac{dy}{dx} + \left[(\beta\rho x^{\rho-1})^2 + \frac{\alpha^2 - \nu^2\rho^2}{x^2}\right] y = 0, \quad x > 0,$$

where α, β, ν, ρ are constants, is given by

$$y(x) = x^\alpha \mathcal{B}_\nu(\beta x^\rho).$$

Some useful formulas of Bessel functions are

$$J_{\nu-1}(x) + J_{\nu+1}(x) = \frac{2\nu}{x} J_\nu(x),$$

$$J'_\nu(x) = J_{\nu-1}(x) - \frac{\nu}{x} J_\nu(x) = -J_{\nu+1}(x) + \frac{\nu}{x} J_\nu(x) = \tfrac{1}{2}\left[J_{\nu-1}(x) - J_{\nu+1}(x)\right],$$

$$\left(\frac{d}{x\,dx}\right)^m \left[x^\nu J_\nu(x)\right] = x^{\nu-m} J_{\nu-m}(x),$$

$$\left(\frac{d}{x\,dx}\right)^m \left[x^{-\nu} J_\nu(x)\right] = (-1)^m x^{-\nu-m} J_{\nu+m}(x).$$

Bessel functions of the second kind $Y_\nu(x)$ satisfy the same recurrence relations as Bessel functions of the first kind $J_\nu(x)$.

9.3.2 Applications of Bessel's Equation

Example 9.9 — Buckling of a Tapered Column

Consider the stability of a tapered column of length L fixed at the base $x=0$ and free at the top $x=L$. The column is subjected to an axial compressive load P at the top. The cross-section of the column is of circular shape, with radii r_0 at the base and $r_1 < r_0$ at the top, respectively, varying linearly along the length x. The modulus of elasticity for the column material is E. Determine the buckling load P_{cr} when the column loses its stability.

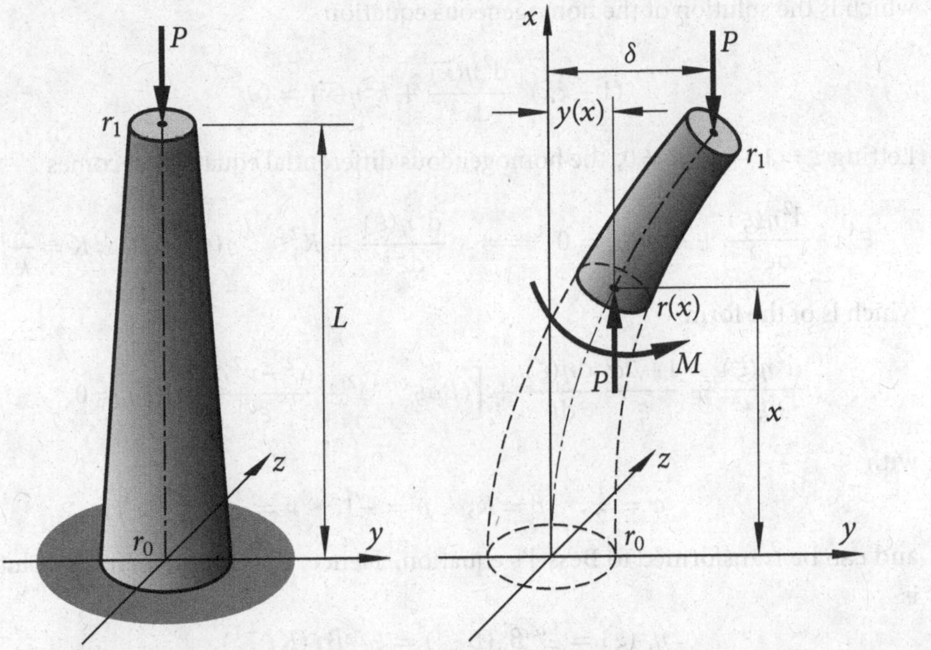

Figure 9.1 Buckling of a fixed-free tapered column.

The deflected shape of the column is shown in Figure 9.1. Consider the equilibrium of a segment between x and L. The lateral deflection of the column at x is $y(x)$. The bending moment at x is $M(x) = P[\delta - y(x)]$, where δ is the deflection at the free end of the column. The moment-curvature relation requires that

$$EI(x)\,y''(x) = M(x) = P[\delta - y(x)],$$

where $I(x)$ is the moment of inertia of the circular cross-section at x given by

$$I(x) = \frac{\pi}{4} r^4(x) = \frac{\pi}{4}\left[r_0\left(1 - \frac{r_0 - r_1}{r_0}\cdot\frac{x}{L}\right)\right]^4 = I_0\,(1-\kappa\bar{x})^4,$$

in which

$$I_0 = \frac{\pi r_0^4}{4}, \qquad \kappa = \frac{r_0 - r_1}{r_0}, \qquad \bar{x} = \frac{x}{L}.$$

Letting $\eta = y/L$ and $\bar{\delta} = \delta/L$, the moment-curvature relation leads to a second-order linear differential equation of the form

$$EI_0\,(1-\kappa\bar{x})^4\cdot\frac{1}{L}\,\frac{\mathrm{d}^2\eta(\bar{x})}{\mathrm{d}\bar{x}^2} + P[L\eta(\bar{x})] = P\,(L\bar{\delta})$$

$$\therefore\quad (1-\kappa\bar{x})^4\,\frac{\mathrm{d}^2\eta(\bar{x})}{\mathrm{d}\bar{x}^2} + k^2\,\eta(\bar{x}) = k^2\,\bar{\delta}, \qquad k^2 = \frac{PL^2}{EI_0}.$$

The general solution is

$$\eta(\bar{x}) = \eta_C(\bar{x}) + \eta_P(\bar{x}),$$

where $\eta_P(\bar{x}) = \bar{\delta}$ is a particular solution and $\eta_C(\bar{x})$ is the complementary solution, which is the solution of the homogeneous equation

$$(1-\kappa\bar{x})^4 \frac{d^2\eta(\bar{x})}{d\bar{x}^2} + k^2\eta(\bar{x}) = 0.$$

Letting $\xi = 1-\kappa\bar{x}$, $\kappa \neq 0$, the homogeneous differential equation becomes

$$\xi^4\kappa^2 \frac{d^2\eta(\xi)}{d\xi^2} + k^2\eta(\xi) = 0 \implies \frac{d^2\eta(\xi)}{d\xi^2} + K^2\xi^{-4}\eta(\xi) = 0, \quad K = \frac{k}{\kappa},$$

which is of the form

$$\frac{d^2\eta(\xi)}{d\xi^2} + \frac{1-2\alpha}{\xi}\frac{d\eta(\xi)}{d\xi} + \left[(\beta\rho\xi^{\rho-1})^2 + \frac{\alpha^2-\nu^2\rho^2}{\xi^2}\right]\eta(\xi) = 0,$$

with

$$\alpha = \frac{1}{2}, \quad \beta = K, \quad \rho = -1, \quad \nu = \frac{1}{2},$$

and can be transformed to Bessel's equation. Hence, the complementary solution is

$$\eta_C(\xi) = \xi^\alpha \mathcal{B}_\nu(\beta\xi^\rho) = \xi^{\frac{1}{2}} \mathcal{B}_{\frac{1}{2}}(K\xi^{-1}).$$

The deflection of the column is then given by

$$\eta(\xi) = \xi^{\frac{1}{2}}\left[C_1 J_{\frac{1}{2}}(K\xi^{-1}) + C_2 J_{-\frac{1}{2}}(K\xi^{-1})\right] + \bar{\delta},$$

where C_1, C_2, and $\bar{\delta}$ are constants to be determined using the boundary conditions

$$\text{at } x=0 \text{ or } \xi=1: \quad \eta(\xi)=0, \quad \eta'(\xi)=0,$$

$$\text{at } x=L \text{ or } \xi=1-\kappa: \quad \eta(\xi)=\bar{\delta}.$$

Note that

$$J'_{\frac{1}{2}}(x) = J_{\frac{1}{2}-1}(x) - \frac{\frac{1}{2}}{x} J_{\frac{1}{2}}(x) = J_{-\frac{1}{2}}(x) - \frac{1}{2x} J_{\frac{1}{2}}(x),$$

$$J'_{-\frac{1}{2}}(x) = -J_{-\frac{1}{2}+1}(x) + \frac{-\frac{1}{2}}{x} J_{-\frac{1}{2}}(x) = -J_{\frac{1}{2}}(x) - \frac{1}{2x} J_{-\frac{1}{2}}(x).$$

Differentiating $\eta'(\xi)$ with respect to ξ yields

$$\eta'(\xi) = \frac{1}{2\sqrt{\xi}}\left[C_1 J_{\frac{1}{2}}\Big(\frac{K}{\xi}\Big) + C_2 J_{-\frac{1}{2}}\Big(\frac{K}{\xi}\Big)\right] + \sqrt{\xi}\left\{C_1\left[J_{-\frac{1}{2}}\Big(\frac{K}{\xi}\Big) - \frac{\xi}{2K} J_{\frac{1}{2}}\Big(\frac{K}{\xi}\Big)\right]\right.$$
$$\left. + C_2\left[-J_{\frac{1}{2}}\Big(\frac{K}{\xi}\Big) - \frac{\xi}{2K} J_{-\frac{1}{2}}\Big(\frac{K}{\xi}\Big)\right]\right\} \cdot \Big(-\frac{K}{\xi^2}\Big),$$

$$\therefore \quad \eta'(1) = \left[J_{\frac{1}{2}}(K) - K\,J_{-\frac{1}{2}}(K)\right] C_1 + \left[J_{-\frac{1}{2}}(K) + K\,J_{\frac{1}{2}}(K)\right] C_2 = 0.$$

At $x = L$ or $\xi = 1 - \kappa$:

$$\eta(1-\kappa) = \sqrt{1-\kappa}\left[C_1\,J_{\frac{1}{2}}(K_L) + C_2\,J_{-\frac{1}{2}}(K_L)\right] + \bar{\delta} = \bar{\delta}, \quad K_L = \frac{K}{1-\kappa},$$

$$\therefore \quad J_{\frac{1}{2}}(K_L)\,C_1 + J_{-\frac{1}{2}}(K_L)\,C_2 = 0.$$

These give two linear homogeneous algebraic equations for C_1 and C_2. To have nontrivial solutions, the determinant of the coefficient matrix must be zero:

$$\begin{vmatrix} J_{\frac{1}{2}}(K) - K\,J_{-\frac{1}{2}}(K) & J_{-\frac{1}{2}}(K) + K\,J_{\frac{1}{2}}(K) \\ J_{\frac{1}{2}}(K_L) & J_{-\frac{1}{2}}(K_L) \end{vmatrix} = 0,$$

$$\therefore \quad J_{-\frac{1}{2}}(K_L)\left[J_{\frac{1}{2}}(K) - K\,J_{-\frac{1}{2}}(K)\right] - J_{\frac{1}{2}}(K_L)\left[J_{-\frac{1}{2}}(K) + K\,J_{\frac{1}{2}}(K)\right] = 0,$$

which is called the *buckling equation*.

For a given value of $\kappa = (r_0 - r_1)/r_0$, the roots of this algebraic equation K_n, $n = 1, 2, \ldots$, can be determined, from which the nth buckling load can be found

$$K = \frac{k}{\kappa}, \quad k^2 = \frac{PL^2}{EI_0} \quad \Longrightarrow \quad P_n = (p_n\pi)^2\frac{EI_0}{L^2}, \quad p_n = \frac{\kappa K_n}{\pi}, \qquad n = 1, 2, 3, \ldots.$$

A numerical method must be used to solve the nonlinear buckling equation to obtain the roots K_n. Because of its remarkable ability in handling special functions, symbolic computation software, such as *Maple*, is well-suited for solving the buckling equation, as illustrated in Section 12.2.

Some numerical results are shown in the following table for the first three buckling loads; the last row gives the results for the prismatic column with $r_1 = r_0$:

$$P_n = \left(\frac{2n-1}{2}\cdot\pi\right)^2\frac{EI_0}{L^2}, \quad p_n = \frac{2n-1}{2}, \quad n = 1, 2, 3, \ldots.$$

Buckling loads for a fixed-free tapered column with circular cross-section.

		$n=1$		$n=2$		$n=3$	
r_1/r_0	κ	K_1	p_1	K_2	p_2	K_3	p_3
0.5	0.5	2.0288	0.3229	4.9132	0.7820	7.9787	1.2698
0.6	0.4	2.8606	0.3642	7.2735	0.9261	11.9067	1.5160
0.7	0.3	4.2094	0.4020	11.2033	1.0698	18.4523	1.7621
0.8	0.2	6.8620	0.4369	19.0592	1.2133	31.5427	2.0081
0.9	0.1	14.7465	0.4694	42.6226	1.3567	70.8129	2.2540
0.99	0.01	156.1429	0.4970	466.7386	1.4857	777.6715	2.4754
1			**0.5**		**1.5**		**2.5**

Example 9.10 — Ascending Motion of a Rocket

Consider the ascending motion of a rocket of initial mass m_0 (including shell and fuel). The fuel is consumed at a constant rate $q = -\mathrm{d}m/\mathrm{d}t$ and is expelled at a constant speed u relative to the rocket. At time t, the mass of the rocket is $m(t) = m_0 - qt$. If the velocity of the rocket is $v = v_0$ at $t = t_0$, determine the velocity $v(t)$.

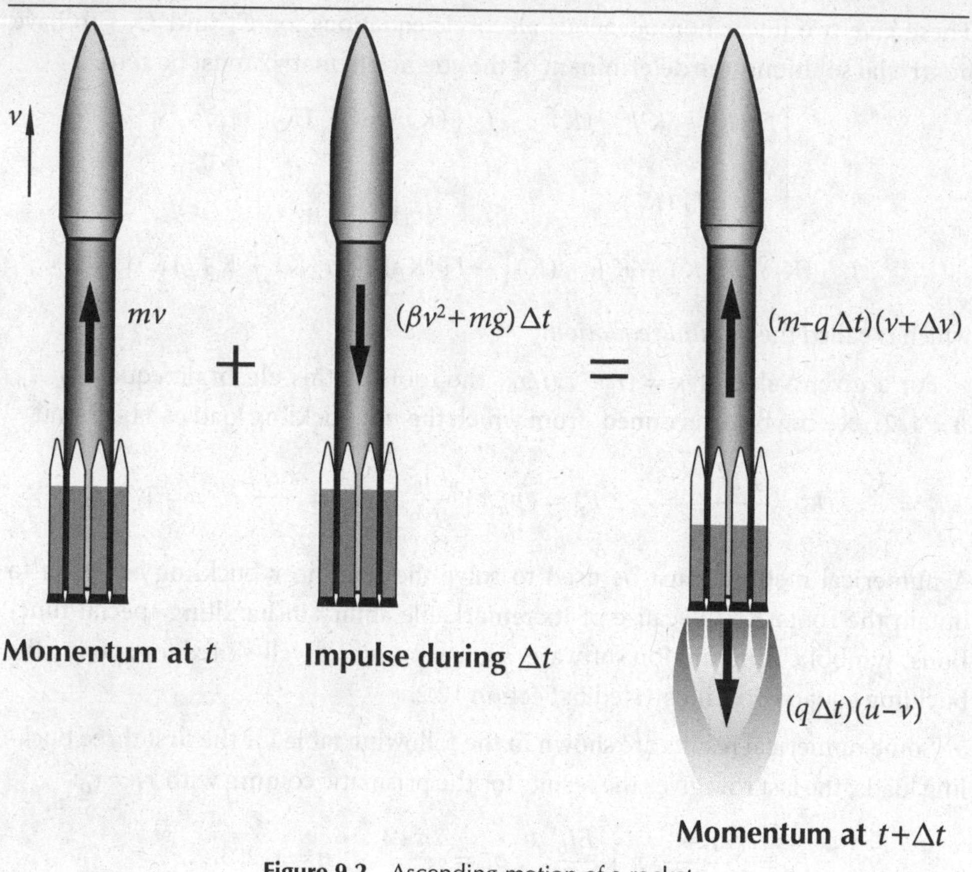

Figure 9.2 Ascending motion of a rocket.

The aerodynamic drag force F_d, which is opposed to the direction of motion, is

$$F_d = \frac{1}{2}\rho v^2 C_d A,$$

where A is the the frontal (or projected) area of the rocket, ρ is the density of the air, and C_d is the dimensionless drag coefficient, a number used to model the complex dependencies of drag on shape inclination and some flow conditions. Note that the drag force F_d depends not on the velocity but on the velocity squared. If the fluid properties are considered constant, the drag force can be written as

$$F_d = \beta v^2,$$

where β is the damping coefficient.

To set up the equation of motion of the rocket, apply the *Impulse-Momentum Principle* between time t and time $t+\Delta t$:

(Momentum at time t) + (Impulse during Δt) = (Momentum at time $t+\Delta t$),

where, as shown in Figure 9.2,

$$\begin{aligned}
\text{Momentum at time } t &= m(t)v(t),\\
\text{Impulse during } \Delta t &= -\big[\beta v^2(t)+m(t)g\big]\Delta t,\\
\text{Momentum at time } t+\Delta t &= m(t+\Delta t)\,v(t+\Delta t) - (q\Delta t)\big[u-v(t)\big]\\
&= \big[m(t)-q\Delta t\big]\big[v(t)+\Delta v\big] - (q\Delta t)\big[u-v(t)\big].
\end{aligned}$$

Hence

$$m(t)v(t) - \big[\beta v^2(t)+m(t)g\big]\Delta t = \big[m(t)-q\Delta t\big]\big[v(t)+\Delta v\big] - (q\Delta t)\big[u-v(t)\big],$$

and, by taking the limit $\Delta t \to 0$, one obtains the equation of motion of a rocket moving upward at high speed during the propelled phase

$$\underbrace{m(t)\frac{\mathrm{d}v(t)}{\mathrm{d}t}}_{\substack{\text{Inertia}\\ \text{force}}} + \underbrace{\beta v^2(t)}_{\substack{\text{Drag}\\ \text{force}}} + \underbrace{m(t)g}_{\text{Gravity}} - \underbrace{qu}_{\substack{\text{Thrust}\\ \text{force}}} = 0,$$

which is a first-order nonlinear differential equation with variable coefficients. The equation can be reduced to Bessel's equation by the following change of variables.

Letting the velocity be

$$v(t) = \frac{m(t)\,\dot{V}(t)}{\beta\,V(t)},$$

where $V(t)$ is the new transformed "velocity," one has

$$\frac{\mathrm{d}v}{\mathrm{d}t} = \frac{m(t)\,\ddot{V}(t)}{\beta\,V(t)} - \frac{m(t)\,\dot{V}^2(t)}{\beta\,V^2(t)} + \frac{\dot{m}(t)\,\dot{V}(t)}{\beta\,V(t)}, \qquad \dot{m}(t) = -q.$$

Substituting into the equation of motion yields

$$m^2(t)\frac{\mathrm{d}^2V(t)}{\mathrm{d}t^2} - q\,m(t)\frac{\mathrm{d}V(t)}{\mathrm{d}t} + \big[\beta g\,m(t) - \beta qu\big]V(t) = 0.$$

Now changing the time t to the dimensionless variable τ

$$\tau = \frac{2}{q}\sqrt{\beta g\,m(t)} \implies \sqrt{m(t)} = \frac{q}{2\sqrt{\beta g}}\,\tau,$$

one has

$$\frac{\mathrm{d}V}{\mathrm{d}t} = \frac{\mathrm{d}V}{\mathrm{d}\tau}\frac{\mathrm{d}\tau}{\mathrm{d}t} = -\frac{2\beta g}{q\tau}\frac{\mathrm{d}V}{\mathrm{d}\tau},$$

$$\frac{d^2V}{dt^2} = \frac{d}{d\tau}\left(-\frac{2\beta g}{q\tau}\frac{dV}{d\tau}\right)\frac{d\tau}{dt} = \frac{4\beta^2 q^2}{q^2\tau^2}\left(\frac{d^2V}{d\tau^2} - \frac{1}{\tau}\frac{dV}{d\tau}\right).$$

The equation of motion becomes Bessel's equation

$$\tau^2\frac{d^2V(\tau)}{d\tau^2} + \tau\frac{dV(\tau)}{d\tau} + (\tau^2 - \nu^2)\,V(\tau) = 0, \quad \nu = 2\sqrt{\frac{\beta u}{q}}.$$

The solution of Bessel's equation is

$$V(\tau) = C_1 J_\nu(\tau) + C_2 Y_\nu(\tau),$$
$$\frac{dV(\tau)}{d\tau} = C_1\left[\frac{\nu}{\tau}J_\nu(\tau) - J_{\nu+1}(\tau)\right] + C_2\left[\frac{\nu}{\tau}Y_\nu(\tau) - Y_{\nu+1}(\tau)\right].$$

The velocity of the rocket is

$$v(\tau) = \frac{m(t)\,\dot{V}(t)}{\beta V(t)} = \frac{m(\tau)\cdot\left[-\dfrac{2\beta g}{q\tau}\dfrac{dV(\tau)}{d\tau}\right]}{\beta V(\tau)}$$
$$= \frac{\dfrac{q^2\tau^2}{4\beta g}\cdot\left(-\dfrac{2\beta g}{q\tau}\right)\left\{\dfrac{\nu}{\tau}[C_1 J_\nu(\tau) + C_2 Y_\nu(\tau)] - [C_1 J_{\nu+1}(\tau) + C_2 Y_{\nu+1}(\tau)]\right\}}{\beta[C_1 J_\nu(\tau) + C_2 Y_\nu(\tau)]}$$
$$= \frac{q}{2\beta}\left[\tau\cdot\frac{C J_{\nu+1}(\tau) + Y_{\nu+1}(\tau)}{C J_\nu(\tau) + Y_\nu(\tau)} - \nu\right], \quad C = \frac{C_1}{C_2}.$$

The constant C is determined by the initial condition $v = v_0$ when $t = t_0$ or $\tau = \tau_0$, and is given by

$$C = -\frac{A\,Y_\nu(\tau_0) - Y_{\nu+1}(\tau_0)}{A\,J_\nu(\tau_0) - J_{\nu+1}(\tau_0)}, \quad A = \frac{1}{\tau_0}\left(\frac{2\beta v_0}{q} + \nu\right), \quad \tau_0 = \frac{2}{q}\sqrt{\beta g\, m(t_0)}.$$

9.4 Summary

- A point x_0 is an *ordinary point* of the linear ordinary differential equation

$$y^{(n)}(x) + p_{n-1}(x)\,y^{(n-1)}(x) + p_{n-2}(x)\,y^{(n-2)}(x) + \cdots + p_0(x)\,y(x) = f(x),$$

if each of the coefficients $p_0(x), p_1(x), \ldots, p_{n-1}(x)$ and $f(x)$ is analytic at $x = x_0$, i.e., each of them can be expressed as a power series about x_0 that is convergent for $|x - x_0| < r$, $r > 0$. Every solution of this differential equation can be expanded in one and only one way as a power series in $(x - x_0)$

$$y(x) = \sum_{n=0}^{\infty} a_n (x - x_0)^n, \quad |x - x_0| < R, \quad R \geqslant r.$$

❧ Consider the second-order linear homogeneous ordinary differential equation

$$y''(x) + P(x)\,y'(x) + Q(x)\,y(x) = 0.$$

If $x=0$ is a regular singular point, then

- it is not an ordinary point, i.e., not all $P(x)$ and $Q(x)$ are analytic at $x=0$;
- $x\,P(x)$ and $x^2 Q(x)$ are analytic at $x=0$, i.e.,

$$x\,P(x) = \sum_{n=0}^{\infty} P_n x^n, \quad x^2 Q(x) = \sum_{n=0}^{\infty} Q_n x^n, \quad |x| < r.$$

Let $\alpha_1 \geqslant \alpha_2$ be the two real roots of the indicial equation $\alpha(\alpha-1)+\alpha P_0 + Q_0 = 0$. The differential equation has at least one *Frobenius* series solution given by

$$y_1(x) = x^{\alpha_1} \sum_{n=0}^{\infty} a_n x^n, \quad a_0 \neq 0, \quad 0 < x < r.$$

A second linearly independent solution is given as follows:

- If $\alpha_1 - \alpha_2$ is not an integer, then a second Frobenius solution is

$$y_2(x) = x^{\alpha_2} \sum_{n=0}^{\infty} b_n x^n, \quad 0 < x < r.$$

- If $\alpha_1 = \alpha_2 = \alpha$, then

$$y_2(x) = y_1(x)\ln x + x^{\alpha} \sum_{n=0}^{\infty} b_n x^n, \quad 0 < x < r,$$

  which is not a Frobenius series solution.
- If $\alpha_1 - \alpha_2$ is a positive integer, then

$$y_2(x) = a\,y_1(x)\ln x + x^{\alpha_2} \sum_{n=0}^{\infty} b_n x^n, \quad 0 < x < r,$$

  which is a Frobenius series solution if $a=0$.

The general solution of the differential equation is then given by

$$y(x) = C_1\,y_1(x) + C_2\,y_2(x).$$

❧ The solution of Bessel's equation

$$x^2 y'' + x y' + (x^2 - \nu^2)\,y = 0, \quad x > 0,$$

where $\nu \geqslant 0$ is a constant, is

$$y(x) = \mathcal{B}_\nu(x) = C_1\,J_\nu(x) + C_2\,Y_\nu(x),$$

in which $J_\nu(x)$ and $Y_\nu(x)$ are the Bessel functions of the first and second kinds, respectively, of order ν. When $\nu \neq 0, 1, 2, \ldots$, the solution can also be written as

$$y(x) = \mathcal{B}_\nu(x) = C_1\,J_\nu(x) + C_2\,J_{-\nu}(x), \quad \nu \neq 0, 1, 2, \ldots.$$

The differential equation

$$\frac{d^2y}{dx^2}+\frac{1-2\alpha}{x}\frac{dy}{dx}+\left[(\beta\rho x^{\rho-1})^2+\frac{\alpha^2-\nu^2\rho^2}{x^2}\right]y=0,\quad x>0,$$

where α, β, ν, ρ are constants, can be transformed to Bessel's equation and the solution is given by

$$y(x)=x^{\alpha}\,\mathcal{B}_\nu(\beta x^{\rho}).$$

Problems

9.1 Show that the general solution of the Airy equation

$$y''-xy=0,$$

is, for $|x|<\infty$,

$$y(x)=a_0\left\{1+\sum_{n=1}^{\infty}\frac{\prod_{k=1}^{n}(3k-2)}{(3n)!}x^{3n}\right\}+a_1\left\{x+\sum_{n=1}^{\infty}\frac{\prod_{k=1}^{n}(3k-1)}{(3n+1)!}x^{3n+1}\right\}.$$

9.2 Show that the general solution of the equation

$$(1+x^2)y''+4xy'+2y=0,$$

is, for $|x|<1$,

$$y(x)=\sum_{n=0}^{\infty}(-1)^n\left(a_0x^{2n}+a_1x^{2n+1}\right)=\frac{a_0+a_1x}{1+x^2}.$$

Determine the general solution of the following differential equations in terms of power series about $x=0$.

9.3 $y'''+xy=0$.

ANS $$y(x)=a_0\left\{1+\sum_{n=1}^{\infty}(-1)^n\frac{\prod_{k=1}^{n}(4k-3)}{(4n)!}x^{4n}\right\}$$

$$+a_1\left\{x+\sum_{n=1}^{\infty}(-1)^n\frac{\prod_{k=1}^{n}(4k-2)}{(4n+1)!}x^{4n+1}\right\}+a_2\left\{x^2+2\cdot\sum_{n=1}^{\infty}(-1)^n\frac{\prod_{k=1}^{n}(4k-1)}{(4n+2)!}x^{4n+2}\right\}$$

9.4 $(1-x^2)y''+y=0$.

ANS $$y(x)=a_0\left\{1-\frac{x^2}{2}-\sum_{n=2}^{\infty}\frac{\prod_{k=1}^{n-1}(4k^2-2k-1)}{(2n)!}x^{2n}\right\}$$

$$+a_1\left\{x-\frac{x^3}{6}-\sum_{n=2}^{\infty}\frac{\prod_{k=1}^{n-1}(4k^2+2k-1)}{(2n+1)!}x^{2n+1}\right\}$$

9.5 $y'' - 2x^2 y = 0.$

ANS $$y(x) = a_0\left\{1 + \sum_{n=1}^{\infty} \frac{2^n}{\prod_{k=1}^{n}(4k)(4k-1)} x^{4n}\right\} + a_1\left\{x + \sum_{n=1}^{\infty} \frac{2^n}{\prod_{k=1}^{n}(4k+1)(4k)} x^{4n+1}\right\}$$

9.6 $y'' - 2x^2 y' + xy = 0.$ ANS $$y(x) = a_0\left\{1 - \frac{x^3}{6} - \sum_{n=2}^{\infty} \frac{\prod_{k=1}^{n-1}(6k-1)(3k+1)}{(3n)!} x^{3n}\right\}$$
$$+ a_1\left\{x + \sum_{n=1}^{\infty} \frac{\prod_{k=0}^{n-1}(6k+1)(3k+2)}{(3n+1)!} x^{3n+1}\right\}$$

9.7 $(x^2-1)y'' + (4x-1)y' + 2y = 0.$ ANS $$y(x) = a_0\left(1 + x^2 - \frac{x^3}{3} + \frac{13x^4}{12} - \frac{11x^5}{20} + \cdots\right) + a_1\left(x - \frac{x^2}{2} + \frac{7x^3}{6} - \frac{19x^4}{24} + \frac{53x^5}{40} - \cdots\right)$$

9.8 $y'' + (1+\cos x)y = 0.$ ANS $$y(x) = a_0\left(1 - x^2 + \frac{5x^4}{24} - \frac{23x^6}{720} + \frac{19x^8}{5040} - \frac{271x^{10}}{725760} + \cdots\right) + a_1\left(x - \frac{x^3}{3} + \frac{7x^5}{120} - \frac{13x^7}{1680} + \frac{151x^9}{181440} - \cdots\right)$$

9.9 $y'' + y' \sin x + y\cos x = 0.$ ANS $$y(x) = a_0\left(1 - \frac{x^2}{2} + \frac{x^4}{6} - \frac{31x^6}{720} + \frac{379x^8}{40320} - \frac{1639x^{10}}{907200} + \cdots\right) + a_1\left(x - \frac{x^3}{3} + \frac{x^5}{10} - \frac{59x^7}{2520} + \frac{31x^9}{6480} - \cdots\right)$$

9.10 Show that the general solution of the modified Bessel's equation of order 0

$$xy'' + y' - xy = 0, \quad x > 0,$$

is, for $0 < x < \infty$,

$$y(x) = (C_1 + C_2 \ln x)\sum_{n=0}^{\infty} \frac{x^{2n}}{[(2n)!!]^2} - C_2 \sum_{n=1}^{\infty} \frac{1 + \frac{1}{2} + \frac{1}{3} + \cdots + \frac{1}{n}}{[(2n)!!]^2} x^{2n}.$$

9.11 Show that the first Frobenius series solution of the Laguerre equation

$$xy'' + (1-x)y' + ky = 0, \quad k = \text{nonnegative integer},$$

is given by

$$y_1(x) = \sum_{n=0}^{k} (-1)^n \frac{k!}{(k-n)!\,(n!)^2} x^n.$$

For the case of $k=3$, show that a second linearly independent solution is given by

$$y_2(x) = \left(1 - 3x + \frac{3x^2}{2} - \frac{x^3}{6}\right)\ln x + 7x - \frac{23x^2}{4} + \frac{11x^3}{12} - \sum_{n=4}^{\infty} \frac{6\cdot(n-4)!}{(n!)^2} x^n.$$

Determine two linearly independent solutions of the following equations.

9.12 $x^2y'' + (x-2x^2)y' - xy = 0.$

ANS $$y_1(x) = 1+x+\frac{3x^2}{4}+\frac{5x^3}{12}+\frac{35x^4}{192}+\frac{21x^5}{320}+\cdots$$

$$y_2(x) = y_1(x)\ln x + \left(-\frac{x^2}{4}-\frac{x^3}{4}-\frac{19x^4}{128}-\frac{25x^5}{384}-\cdots\right)$$

9.13 $x^2y'' - (2x+x^2)y' + 2y = 0.$

ANS $$y_1(x) = x^2\left(1+x+\frac{x^2}{2}+\frac{x^3}{6}+\frac{x^4}{24}+\frac{x^5}{120}+\cdots\right) = x^2\sum_{n=0}^{\infty}\frac{x^n}{n!} = x^2\,\mathrm{e}^x$$

$$y_2(x) = y_1(x)\ln x + x\left(1+x-\frac{x^3}{4}-\frac{5x^4}{36}-\frac{13x^5}{288}-\cdots\right)$$

9.14 $x^2y'' + \left(\frac{1}{2}x+x^2\right)y' + xy = 0.$

ANS $$y_1(x) = \sqrt{x}\left(1-x+\frac{x^2}{2}-\frac{x^3}{6}+\frac{x^4}{24}-\frac{x^5}{120}+\cdots\right) = \sqrt{x}\sum_{n=0}^{\infty}\frac{(-x)^n}{n!} = \sqrt{x}\,\mathrm{e}^{-x}$$

$$y_2(x) = 1-2x+\frac{4x^2}{3}-\frac{8x^3}{15}+\frac{16x^4}{105}-\frac{32x^5}{945}+\cdots$$

9.15 $x^2y'' + (x-x^2)y' - (x+1)y = 0.$

ANS $$y_1(x) = x\left(1+\frac{2x}{3}+\frac{x^2}{4}+\frac{x^3}{15}+\frac{x^4}{72}+\frac{x^5}{420}+\cdots\right)$$

$$y_2(x) = x^{-1}\left(1+x^2+\frac{2x^3}{3}+\frac{x^4}{4}+\frac{x^5}{15}+\cdots\right)$$

9.16 $x^2y'' + 2xy' - (x^2+2)y = 0.$

ANS $$y_1(x) = x\left(1+\frac{x^2}{10}+\frac{x^4}{280}+\frac{x^6}{15120}+\frac{x^8}{1330560}+\frac{x^{10}}{172972800}+\cdots\right)$$

$$y_2(x) = x^{-2}\left(1-\frac{x^2}{2}-\frac{x^4}{8}-\frac{x^6}{144}-\frac{x^8}{5760}-\frac{x^{10}}{403200}-\cdots\right)$$

9.17 $xy'' - 2xy' - y = 0.$

ANS $$y_1(x) = x\left(1+\frac{3x}{2}+\frac{5x^2}{4}+\frac{35x^3}{48}+\frac{21x^4}{64}+\frac{77x^5}{640}+\cdots\right)$$

$$y_2(x) = y_1(x)\ln x + \left(1+x+\frac{x^2}{4}-\frac{x^3}{3}-\frac{79x^4}{192}-\frac{499x^5}{1920}+\cdots\right)$$

9.18 $xy'' + 2y' - xy = 0.$

ANS $$y_1(x) = 1+\frac{x^2}{6}+\frac{x^4}{120}+\frac{x^6}{5040}+\frac{x^8}{362880}+\frac{x^{10}}{39916800}+\cdots$$

$$y_2(x) = x^{-1}\left(1+\frac{x^2}{2}+\frac{x^4}{24}+\frac{x^6}{720}+\frac{x^8}{40320}+\frac{x^{10}}{3628800}+\cdots\right)$$

9.19 $x^2y'' - x^2y' + 2(x-1)y = 0.$

ANS $y_1(x) = x^2,\ y_2(x) = x^2 \ln x + x^{-1}\left(-2-3x-3x^2+\dfrac{x^4}{4}+\dfrac{x^5}{40}+\cdots\right)$

9.20 $xy'' + y' - xy = 0.$

ANS $y_1(x) = 1+\dfrac{x^2}{4}+\dfrac{x^4}{64}+\dfrac{x^6}{2304}+\dfrac{x^8}{147456}+\dfrac{x^{10}}{14745600}+\cdots$

$$y_2(x) = y_1(x)\ln x + \left(-\frac{x^2}{4}-\frac{3x^4}{128}-\frac{11x^6}{13824}-\frac{25x^8}{1769472}-\frac{137x^{10}}{884736000}-\cdots\right)$$

9.21 Consider the stability of a simply supported tapered column of length L. The column is subjected to an axial compressive load P. The cross-section of the column is of rectangular shape. The thickness of the column is constant h, and the width changes from b_0 at the bottom to b_1 ($b_1 < b_0$) at the top linearly along the length x as shown in Figure 9.3. The modulus of elasticity for the column material is E. Determine the first three buckling loads when the column loses its stability in the x-y plane for $b_1/b_0 = 0.5, 0.6, 0.7, 0.8, 0.9, 0.99$.

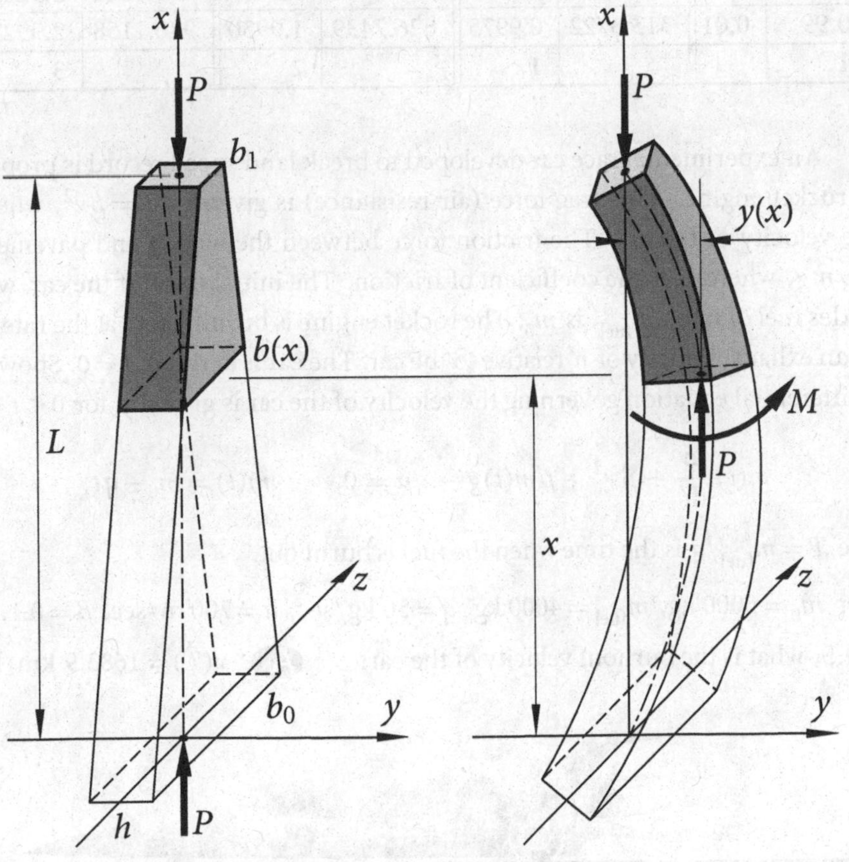

Figure 9.3 Buckling of a simply supported tapered column.

ANS Moment-curvature relation: $\dfrac{d^2\eta(\xi)}{d\xi^2} + K^2\xi^{-1}\eta(\xi) = 0$

$$K = \frac{k}{\kappa}, \quad k^2 = \frac{PL^2}{EI_0}, \quad \xi = 1 - \kappa\bar{x}, \quad \kappa = 1 - \frac{b_1}{b_0}, \quad \bar{x} = \frac{x}{L}, \quad I_0 = \frac{b_0 h^3}{12}$$

Buckling equation: $J_1(2K)\,Y_1(2K_L) - J_1(2K_L)\,Y_1(2K) = 0, \quad K_L = K\sqrt{1-\kappa}$

Buckling loads: $P_n = (p_n\pi)^2 \dfrac{EI_0}{L^2}, \quad p_n = \dfrac{\kappa K_n}{\pi}, \quad n = 1, 2, 3, \ldots,$

For a prismatic beam with $b_1 = b_0$, $p_n = n$, $n = 1, 2, 3, \ldots$

Buckling loads for a simply supported tapered column with rectangular cross-section.

		$n=1$		$n=2$		$n=3$	
b_1/b_0	κ	K_1	p_1	K_2	p_2	K_3	p_3
0.5	0.5	5.3873	0.8574	10.7384	1.7091	16.0973	2.5620
0.6	0.4	6.9860	0.8895	13.9463	1.7757	20.9123	2.6626
0.7	0.3	9.6283	0.9194	19.2393	1.8372	28.8541	2.7554
0.8	0.2	14.8858	0.9477	29.7611	1.8947	44.6387	2.8418
0.9	0.1	30.6131	0.9744	61.2213	1.9487	91.8306	2.9231
0.99	0.01	313.3722	0.9975	626.7439	1.9950	940.1158	2.9925
1			**1**		**2**		**3**

9.22 An experimental race car developed to break land speed record is propelled by a rocket engine. The drag force (air resistance) is given by $R = \beta v^2$, where v is the velocity of the car. The friction force between the wheels and pavement is $F = \mu mg$, where μ is the coefficient of friction. The initial mass of the car, which includes fuel of mass m_{fuel}, is m_0. The rocket engine is burning fuel at the rate of q with an exhaust velocity of u relative to the car. The car is at rest at $t = 0$. Show that the differential equation governing the velocity of the car is given by, for $0 \leqslant t \leqslant T$,

$$m(t)\frac{dv}{dt} + \beta v^2 + \mu m(t) g - qu = 0, \qquad m(t) = m_0 - qt,$$

where $T = m_{\text{fuel}}/q$ is the time when the fuel is burnt out.

For $m_0 = 5000$ kg, $m_{\text{fuel}} = 4000$ kg, $q = 50$ kg/sec, $u = 700$ m/sec, $\beta = 0.1$, and $\mu = 0.5$, what is the burnout velocity of the car? ANS $v(T) = 1683.9$ km/hr

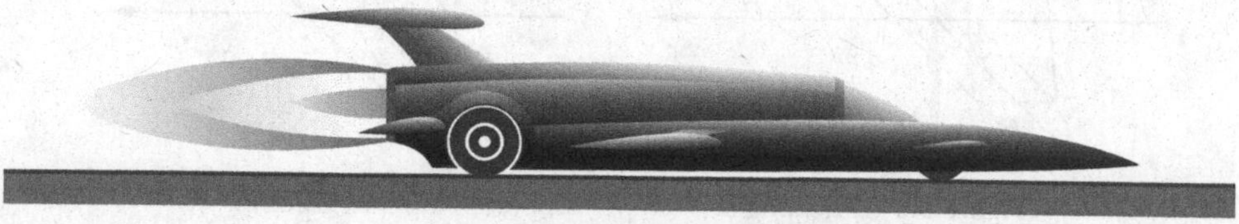

CHAPTER 10

Numerical Solutions of Differential Equations

In previous chapters, various analytical methods are introduced to solve first-order and simple higher-order differential equations (Chapter 2), linear differential equations with constant coefficients (Chapters 4 and 6), systems of linear differential equations with constant coefficients (Chapter 7), and linear differential equations with variable coefficients (Chapter 9).

However, in practical applications, there are many equations, especially nonlinear differential equations and differential equations with variable coefficients, which cannot be solved analytically. In these situations, numerical approaches have to be applied to obtain numerical solutions.

In this chapter, a number of classical numerical methods are presented, through which the concepts of error and stability are introduced.

10.1 Numerical Solutions of First-Order Initial Value Problems

Consider the first-order differential equation

$$\frac{\mathrm{d}y}{\mathrm{d}x} = f(x,y), \quad y(x_0) = y_0.$$

The solution $y(x)$ is required to satisfy the initial condition, i.e., $y = y_0$ at $x = x_0$. The differential equation, along with the initial condition, is therefore called an *initial value problem*, as discussed in Chapter 1. Discretize the independent variable x at points $x_0, x_1, x_2, \ldots$, in which $x_{i+1} = x_i + h$, $i = 0, 1, 2, \ldots$, and h is called the *stepsize*.

Knowing the initial point (x_0, y_0), solutions of the differential equation at the discrete points $x_1, x_2, \ldots$ are required. Adopt the following notations:

- $y(x_i)$, $i=1,2,\ldots$, is the exact value of the solution of the differential equation at $x=x_i$;
- y_i, $i=1,2,\ldots$, is a numerical approximation of the solution of the differential equation at $x=x_i$.

Expand $y(x+\alpha h)$ in Taylor series about x as

$$y(x+\alpha h) = y(x) + \frac{\alpha h}{1!}y'(x) + \frac{(\alpha h)^2}{2!}y''(x) + \frac{(\alpha h)^3}{3!}y'''(x) + \cdots.$$

By selecting different values of α and truncating the Taylor series at different orders of h, various approximate schemes can be developed.

10.1.1 The Euler Method or Constant Slope Method

Taking $\alpha=1$, $x=x_i$, and keeping only the first two terms in the Taylor series give

$$y(x_i+h) \approx y(x_i) + hy'(x_i),$$

or, noting $y'(x_i)=f(x_i, y_i)$,

$$y_{i+1} = y_i + h\, f(x_i, y_i), \quad i = 0, 1, 2, \ldots.$$

This formula forwards the approximate solution y_i at x_i a step h to y_{i+1} at x_{i+1}, and is called the (*forward*) *Euler method*.

Note that, unless the slope $y'=f(x, y)=\text{constant}$ for $x_i \leqslant x \leqslant x_{i+1}$, $i=0, 1, \ldots$, y_{i+1} may not be the true value of y at x_{i+1}. Since the Euler method assumes the slope y' for $x_i \leqslant x \leqslant x_{i+1}$ as a constant $f(x_i, y_i)$ using the value at the left end x_i, it is also called the *constant slope method*. A schematic diagram for the Euler method is shown in Figure 10.1.

Example 10.1

For the initial value problem $y'=x\,y^2-y$, $y(0)=0.5$, determine $y(1.0)$ using the Euler method with $h=0.5$.

The differential equation is a Bernoulli equation and can be written as

$$\frac{1}{y^2}y' + \frac{1}{y} = x.$$

Letting $u=\dfrac{1}{y}$, one has $\dfrac{du}{dx} = -\dfrac{1}{y^2}\dfrac{dy}{dx}$, which leads to $-\dfrac{du}{dx}+u=x$, a first-order linear equation of the form

$$\frac{du}{dx} + P(x)\,u = Q(x), \qquad \text{with} \quad P(x)=-1, \quad Q(x)=-x.$$

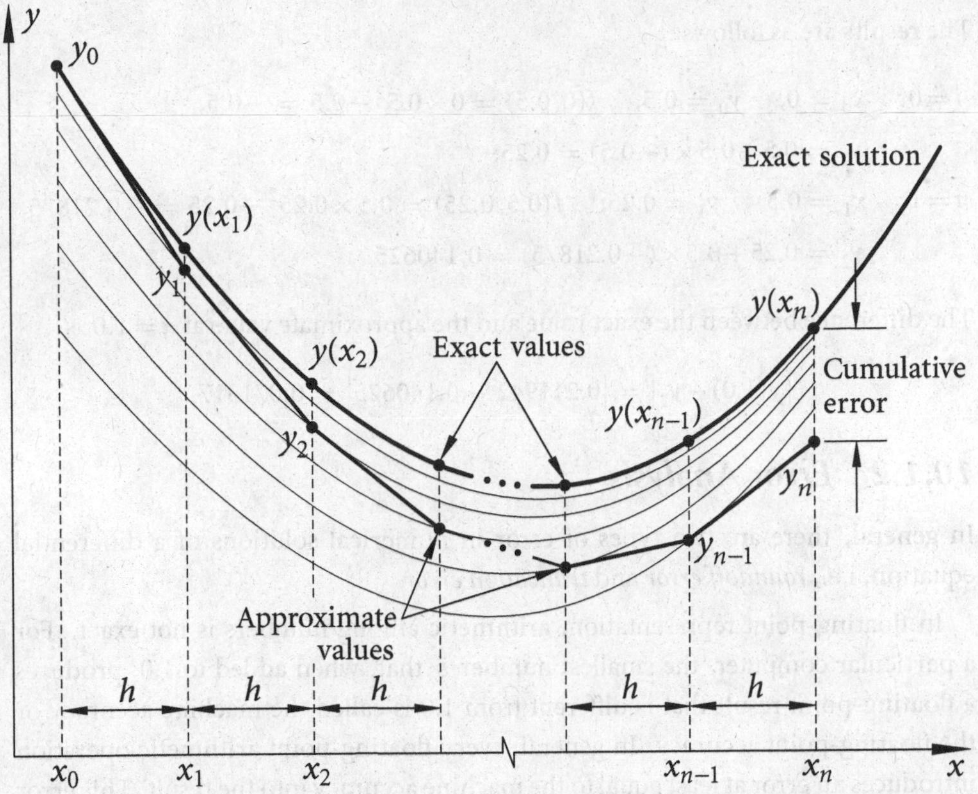

Figure 10.1 The Euler method.

It is easy to evaluate

$$\int P(x)\,\mathrm{d}x = -x, \quad \mathrm{e}^{\int P(x)\,\mathrm{d}x} = \mathrm{e}^{-x}, \quad \mathrm{e}^{-\int P(x)\,\mathrm{d}x} = \mathrm{e}^{x},$$

$$\int Q(x)\,\mathrm{e}^{\int P(x)\,\mathrm{d}x}\,\mathrm{d}x = \int -x\mathrm{e}^{-x} = \mathrm{e}^{-x}(x+1).$$

The general solution is

$$u = \frac{1}{y} = \mathrm{e}^{-\int P(x)\,\mathrm{d}x}\left[\int Q(x)\,\mathrm{e}^{\int P(x)\,\mathrm{d}x}\,\mathrm{d}x + C\right] = \mathrm{e}^{x}\left[\mathrm{e}^{-x}(x+1)+C\right] = 1+x+C\mathrm{e}^{x}.$$

Using the initial condition $y(0)=0.5$,

$$\frac{1}{0.5} = 1+0+C\cdot\mathrm{e}^{0} = 1+C \implies C=1.$$

Hence the exact solution of the initial value problem is

$$y(x) = \frac{1}{1+x+\mathrm{e}^{x}} \implies y(0.5) = 0.317589, \quad y(1.0) = 0.211942.$$

The Euler method is, with $h=0.5$,

$$f(x_i, y_i) = x_i\, y_i^2 - y_i, \qquad y_{i+1} = y_i + h f(x_i, y_i).$$

The results are as follows

$$\begin{aligned}
i=0:\quad & x_0 = 0, \quad y_0 = 0.5, \quad f(0,0.5) = 0\times 0.5^2 - 0.5 = -0.5,\\
& y_1 = 0.5+0.5\times(-0.5) = 0.25;\\
i=1:\quad & x_1 = 0.5, \quad y_1 = 0.25, \quad f(0.5,0.25) = 0.5\times 0.25^2 - 0.25 = -0.21875,\\
& y_2 = 0.25+0.5\times(-0.21875) = 0.140625.
\end{aligned}$$

The difference between the exact value and the approximate value at $x=1.0$ is

$$|y(1.0)-y_2| = |0.211942 - 0.140625| = 0.071317.$$

10.1.2 Error Analysis

In general, there are two types of error in numerical solutions of a differential equation, i.e., *roundoff error* and *truncation error.*

In floating-point representation, arithmetic among numbers is not exact. For a particular computer, the smallest number ε that, when added to 1.0, produces a floating-point result that is different from 1.0 is called the machine accuracy or the floating-point accuracy. In general, every floating-point arithmetic operation introduces an error at least equal to the machine accuracy into the result. This error is known as *roundoff error* and is a characteristic of computer hardware. Roundoff errors are cumulative. Depending on the algorithm used, a calculation involving n arithmetic operations might have a total roundoff error between $\sqrt{n}\varepsilon$ and $n\varepsilon$.

The discrepancy between the exact result and the result obtained through a numerical algorithm is called the *truncation error.* Truncation error is introduced in the process of numerically approximating a continuous solution by evaluating it at a finite number of discrete points. It is the error that a numerical algorithm would have if it were run on an infinite-precision computer. Truncation error can be reduced, at the cost of computation effort, by developing and selecting algorithms that are of higher orders of the stepsizes.

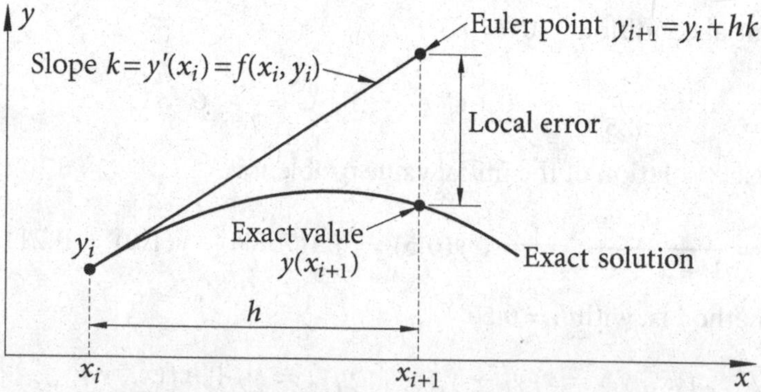

Figure 10.2 Local error of the Euler method.

In most cases, truncation error is independent of the roundoff error. A numerical result thus includes the accumulation of truncation errors due to the approximation in the numerical algorithm and the roundoff errors associated with the machine accuracy and the number of operations performed.

Truncation Error of the Euler Method

To study the truncation error in the $(i+1)$th step, refer to Figure 10.2, in which $y(x_{i+1})$ is the exact solution satisfying $y=y_i$ at $x=x_i$.

Consider the Taylor series

$$\begin{aligned} y(x_{i+1}) = y(x_i+h) &= y(x_i) + \frac{h}{1!}y'(x_i) + \frac{h^2}{2!}y''(x_i) + \cdots \\ &= y(x_i) + h\,y'(x_i) + \frac{1}{2}h^2\,y''(r_i), \quad x_i \leqslant r_i \leqslant x_{i+1}. \end{aligned}$$

The *local error* or *truncation error* at the $(i+1)$th step is

$$\left|y(x_{i+1}) - y_{i+1}\right| = \left|\frac{h^2}{2!}y''(r_i)\right| \leqslant m_i h^2, \qquad \left|y''(r_i)\right| \leqslant 2m_i, \quad x_i \leqslant r_i \leqslant x_{i+1}.$$

Starting from the initial point $y(x_0)=y_0$, the *cumulative error* at $x=x_n$ after following the Euler method for n steps is (Figure 10.1)

$$\begin{aligned} \left|y(x_n) - y_n\right| &\leqslant \sum_{i=0}^{n-1} m_i h^2 \leqslant Mnh^2, && M = \max\{m_1, m_2, \ldots, m_n\}, \\ &= M(x_n - x_0)\,h, && nh = x_n - x_0, \\ &= Ch, && M(x_n - x_0) = C. \end{aligned}$$

The cumulative error is of the first order of h, i.e., $\mathcal{O}(h)$ or $\left|y(x_n) - y_n\right| \leqslant Ch$.

Table 10.1 Comparison of solutions and cumulative errors for different stepsizes.

		$h=0.5$		$h=0.2$		$h=0.1$	
x_i	$y(x_i)$	y_i	$\lvert y(x_i)-y_i\rvert$	y_i	$\lvert y(x_i)-y_i\rvert$	y_i	$\lvert y(x_i)-y_i\rvert$
0.0	0.5	0.5	0	0.5	0	0.5	0
0.1	0.453480					0.45	0.003480
0.2	0.412984			0.4	0.012984	0.407025	0.005959
0.3	0.377379					0.369636	0.007743
0.4	0.345802			0.3264	0.019402	0.336771	0.009031
0.5	0.317589	0.25	0.067589			0.307631	0.009959
0.6	0.292217			0.269642	0.022574	0.281599	0.010617
0.7	0.269269					0.258197	0.011072
0.8	0.248414			0.224439	0.023975	0.237044	0.011370
0.9	0.229379					0.217835	0.011544
1.0	0.211942	0.140625	0.071317	0.187611	0.024330	0.200322	0.011619

As a numerical example, Table 10.1 shows the solutions and the cumulative errors for the initial value problem in Example 10.1 with different stepsizes. Since the cumulative error is of the order $\mathcal{O}(h)$, the accuracy of numerical approximation can be increased by using a smaller stepsize.

However, there are two caveats with increasing numerical accuracy by decreasing stepsize:

- It is not efficient. It is more preferable to use higher order algorithms that are more accurate and efficient.
- The use of arbitrarily small stepsizes in numerical computation is prevented by roundoff error. When the stepsize is small enough, the local truncation error may be smaller than the roundoff error. Furthermore, to determine the solution of a differential equation at some point x, a smaller stepsize h means a larger number of steps x/h; the accuracy may actually decrease as the roundoff errors accumulate while the computation steps along.

The Euler method is the simplest algorithm for numerically solving an ordinary differential equation. However, it is not recommended for any practical use because it is not very accurate when compared to other higher order methods, such as the Runge-Kutta methods, which will be introduced in Section 10.1.5. It is presented here for understanding the essential procedure of solving an ordinary differential numerically and the concept of error analysis.

10.1.3 The Backward Euler Method

Taking $\alpha = -1$, $x = x_{i+1}$, and keeping only the first two terms in the Taylor series give

$$y(x_{i+1} - h) \approx y(x_{i+1}) - h y'(x_{i+1}),$$

or, noting $y'(x_{i+1}) = f(x_{i+1}, y_{i+1})$,

$$y_i = y_{i+1} - h\, f(x_{i+1}, y_{i+1}),$$

or

$$y_{i+1} = y_i + h\, f(x_{i+1}, y_{i+1}), \quad i = 0, 1, 2, \ldots .$$

This formula is called the *backward Euler method*, since y_i is derived in terms of y_{i+1}.

If function $f(x, y)$ is linear in y, then $f(x, y) = a(x) y + b(x)$ and

$$y_i = y_{i+1} - h\big[a(x_{i+1})\, y_{i+1} + b(x_{i+1})\big].$$

y_{i+1} can be expressed explicitly in terms of y_i

$$y_{i+1} = \frac{y_i + h \cdot b(x_{i+1})}{1 - h \cdot a(x_{i+1})}.$$

Otherwise, the backward Euler method gives an *implicit* equation for y_{i+1}. A root-finding algorithm has to be applied to solve the nonlinear algebraic equation for y_{i+1}.

The backward Euler method is one of a class of numerical techniques known as *implicit methods*. The main disadvantage of implicit methods is that nonlinear algebraic equations must be solved at each step. However, the most important advantage of implicit methods over explicit methods is that they give better stability, which will be illustrated in Section 10.3.

Example 10.2

For the initial value problem $y' = x\,y^2 - y$, $y(0) = 0.5$, determine $y(1.0)$ using the backward Euler method with $h = 0.5$.

The backward Euler method is, with $f(x, y) = x\,y^2 - y$ and $h = 0.5$,

$$y_{i+1} = y_i + h\,f(x_{i+1}, y_{i+1}).$$

The results are as follows

$i = 0$: $x_0 = 0$, $y_0 = 0.5$,

$$y_1 = 0.5 + 0.5 \times (0.5 \times y_1^2 - y_1) \implies 0.25\,y_1^2 - 1.5\,y_1 + 0.5 = 0,$$

$$y_1 = 0.354249, \quad \cancel{5.6645751};$$

✍ Since the difference between y_{n+1} and y_n, $|y_{n+1} - y_n|$, is of the order $O(h^2)$, the root 0.354249 is selected. The root 5.6645751 is not reasonable and should be discarded.

$i = 1$: $x_1 = 0.5$, $y_1 = 0.354249$,

$$y_2 = 0.354249 + 0.5 \times (1.0 \times y_2^2 - y_2) \implies 0.5\,y_2^2 - 1.5\,y_2 + 0.354249 = 0,$$

$$y_2 = 0.258427, \quad \cancel{2.741573};$$

Cumulative error: $|y(1.0) - y_2| = |0.211942 - 0.258427| = 0.046486.$

✍ Similarly, the root 0.258427 is selected and the root 2.741573 is discarded.

Remarks: For nonlinear algebraic equations, there are usually more than one real root; care must be taken to pick the correct root, especially when the roots are closely spaced.

10.1.4 Improved Euler Method—Average Slope Method

The Euler method $y_{i+1} = y_i + h\,f(x_i, y_i)$ is not symmetric; it uses the slope $f(x_i, y_i)$ at the left end point of the interval $[x_i, x_{i+1}]$ as the actual slope of the solution $y'(x)$ over the entire interval.

On the other hand, the backward Euler method $y_{i+1} = y_i + hf(x_{i+1}, y_{i+1})$ uses the slope $f(x_{i+1}, y_{i+1})$ at the right end point of the interval $[x_i, x_{i+1}]$ as the actual slope of the solution $y'(x)$ over the entire interval.

The improved Euler method combines the Euler method and the backward Euler method. It uses the average of the slopes $f(x_i, y_i)$ at the left end point and $f(x_{i+1}, y_{i+1})$ at the right end point of the interval $[x_i, x_{i+1}]$ as the actual slope of the solution $y'(x)$ over the entire interval, i.e.,

$$y_{i+1} = y_i + \frac{h}{2}\big[f(x_i, y_i) + f(x_{i+1}, y_{i+1})\big], \quad i = 0, 1, 2, \ldots.$$

The *improved Euler method* or the *average slope method* is also called the *trapezoidal rule method* because of its close relation to the trapezoidal rule for integration

$$y(x_{i+1}) - y(x_i) = \int_{x_i}^{x_{i+1}} y'(x)\,\mathrm{d}x \approx \frac{h}{2}\big[y'(x_i) + y'(x_{i+1})\big].$$

Expanding $y'(x_{i+1})$ in Taylor series yields

$$y'(x_{i+1}) = y'(x_i + h) = y'(x_i) + \frac{h}{1!}y''(x_i) + O(h^2).$$

Noting that

$$y_{i+1} = y_i + \tfrac{1}{2}h\big[f(x_i, y_i) + f(x_{i+1}, y_{i+1})\big] = y_i + \tfrac{1}{2}h\big[y'(x_i) + y'(x_{i+1})\big],$$

the truncation error of the improved Euler method at the $(i+1)$th step is

$$\begin{aligned}\big|y(x_{i+1}) - y_{i+1}\big| &= \Big|\big[y(x_i) + hy'(x_i) + \tfrac{1}{2}h^2 y''(x_i) + O(h^3)\big] \\ &\qquad - \Big\{y_i + \tfrac{1}{2}h\big\{y'(x_i) + \big[y'(x_i) + hy''(x_i) + O(h^2)\big]\big\}\Big\}\Big| \\ &= O(h^3).\end{aligned}$$

Hence, the truncation error of the improved Euler method in each step is of the order $O(h^3)$, and the cumulative error is of the second order of h, i.e., $O(h^2)$ or $|y(x_n) - y_n| \leqslant Ch^2$.

The improved Euler method is obviously implicit and shares the advantage of implicit methods in terms of stability, which is particularly important in solving stiff equations (see Section 10.3 for a brief discussion).

On the other hand, it also shares the disadvantage of having to solve a nonlinear algebraic equation at each step with the backward Euler method. To overcome this problem, a numerical technique known as *predictor-corrector* can be applied, in which the value y_{i+1} on the right-hand side of the equation is approximated using an explicit method, such as the Euler method. Unfortunately, while the improved Euler method is rendered explicit by applying the predictor-corrector technique, it also loses its stability characteristic as an implicit method.

A schematic diagram is shown in Figure 10.3 to illustrate the procedure of the improved Euler predictor-corrector method.

Improved Euler Predictor-Corrector Method

At the $(i+1)$th step, $i=0, 1, 2, \dots,$

(1)	$k_1 = f(x_i, y_i)$	Slope at the left end point x_i
(2) **Predictor**	$y^{\mathrm{P}}_{i+1} = y_i + hk_1$	Predict y at x_{i+1} using the Euler method
(3)	$k_2 = f(x_{i+1}, y^{\mathrm{P}}_{i+1})$	Predicted slope at the right end point x_{i+1}
(4)	$k = \dfrac{k_1 + k_2}{2}$	The averaged slope is used on $[x_i, x_{i+1}]$
(5) **Corrector**	$y_{i+1} = y_i + hk$	Improved Euler point

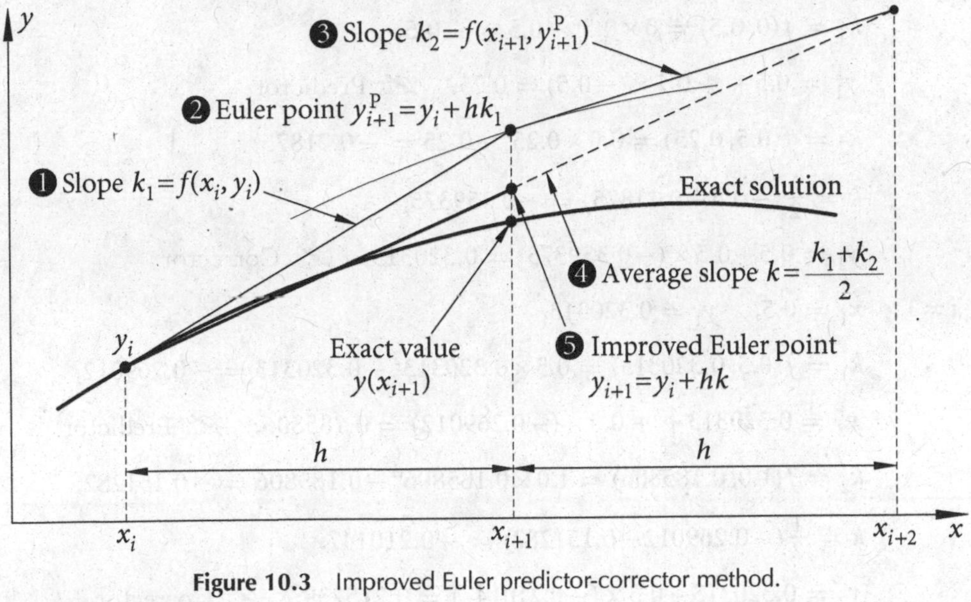

Figure 10.3 Improved Euler predictor-corrector method.

Example 10.3

For the initial value problem $y'=xy^2-y$, $y(0)=0.5$, determine $y(1.0)$ using the improved Euler method and the improved Euler predictor-corrector method with $h=0.5$.

(1) The improved Euler method is, with $f(x, y)=xy^2-y$ and $h=0.5$,

$$y_{i+1} = y_i + \tfrac{1}{2}h\big[f(x_i, y_i) + f(x_{i+1}, y_{i+1})\big].$$

The results are as follows

$i=0$: $x_0=0$, $y_0=0.5$,

$$y_1 = 0.5 + \tfrac{1}{2} \times 0.5 \times \big[(0 \times 0.5^2 - 0.5) + (0.5 \times y_1^2 - y_1)\big]$$

$$\Longrightarrow 0.125\, y_1^2 - 1.25\, y_1 + 0.375 = 0 \implies y_1 = 0.309584, \ \text{~~9.690416~~};$$

$i=1:\ x_1 = 0.5, \quad y_1 = 0.309584,$

$$y_2 = 0.309584 + \tfrac{1}{2} \times 0.5 \times \big[(0.5 \times 0.309584^2 - 0.309584) + (1.0 \times y_2^2 - y_2)\big]$$

$$\Longrightarrow 0.25\, y_2^2 - 1.25\, y_2 + 0.281373 = 0 \implies y_2 = 0.236262, \ \text{~~4.763738~~};$$

Cumulative error: $\big|y(1.0) - y_2\big| = \big|0.211942 - 0.236262\big| = 0.024321.$

(2) The improved Euler predictor-corrector method is

$$k_1 = f(x_i, y_i), \quad y_{i+1}^{\rm P} = y_i + h k_1, \quad k_2 = f(x_{i+1}, y_{i+1}^{\rm P}), \quad k = \frac{k_1 + k_2}{2}, \quad y_{i+1} = y_i + hk.$$

The results are as follows

$i=0:\ x_0 = 0, \quad y_0 = 0.5,$

$k_1 = f(0, 0.5) = 0 \times 0.5^2 - 0.5 = -0.5,$

$y_1^{\rm P} = 0.5 + +0.5 \times (-0.5) = 0.25,$ ✍ Predictor

$k_2 = f(0.5, 0.25) = 0.5 \times 0.25^2 - 0.25 = -0.21875,$

$k = \tfrac{1}{2}(-0.5 - 0.21875) = -0.359375,$

$y_1 = 0.5 + 0.5 \times (-0.359375) = 0.320313;$ ✍ Corrector

$i=1:\ x_1 = 0.5, \quad y_1 = 0.320313,$

$k_1 = f(0.5, 0.320313) = 0.5 \times 0.320313^2 - 0.320313 = -0.269012,$

$y_2^{\rm P} = 0.320313 + +0.5 \times (-0.269012) = 0.185806,$ ✍ Predictor

$k_2 = f(1.0, 0.185806) = 1.0 \times 0.185806^2 - 0.185806 = -0.151282,$

$k = \tfrac{1}{2}(-0.269012 - 0.151282) = -0.210147,$

$y_2 = 0.320313 + 0.5 \times (-0.210147) = 0.215239;$ ✍ Corrector

Cumulative error: $\big|y(1.0) - y_2\big| = \big|0.211942 - 0.215239\big| = 0.003297.$

10.1.5 The Runge-Kutta Methods

Taking $\alpha = 1$, $x = x_i$, and keeping only the first three terms in the Taylor series give

$$y(x_i + h) \approx y(x_i) + h\, y'(x_i) + \frac{h^2}{2} y''(x_i). \tag{1}$$

Since $y'(x) = f(x, y)$, differentiating with respect to x yields

$$y''(x) = \frac{\partial f}{\partial x} + \frac{\partial f}{\partial y}\frac{{\rm d}y}{{\rm d}x} = \frac{\partial f}{\partial x} + f\frac{\partial f}{\partial y}. \tag{2}$$

On the other hand, applying the Taylor series for function $F(x_1, x_2)$ of two variables

$$F(x_1, x_2) = F(a_1, a_2) + \left[(x_1-a_1)\frac{\partial F}{\partial x_1} + (x_2-a_2)\frac{\partial F}{\partial x_2}\right] + \frac{1}{2!}\left[(x_1-a_1)^2\frac{\partial^2 F}{\partial x_1^2} + 2(x_1-a_1)(x_2-a_2)\frac{\partial^2 F}{\partial x_1 \partial x_2} + (x_2-a_2)^2\frac{\partial^2 F}{\partial x_2^2}\right] + \cdots$$

to function $f(x+\frac{1}{2}h, y+\frac{1}{2}hf)$ yields, keeping only the first two terms,

$$f(x+\tfrac{1}{2}h, y+\tfrac{1}{2}hf) \approx f(x,y) + \left[(\tfrac{1}{2}h)\frac{\partial f}{\partial x} + (\tfrac{1}{2}hf)\frac{\partial f}{\partial y}\right]$$

$$= f(x,y) + \tfrac{1}{2}h\,y''(x). \quad \text{Using equation (2)}$$

Hence, an approximation of $y''(x)$ is given by

$$y''(x) \approx \frac{2}{h}\left[f(x+\tfrac{1}{2}h, y+\tfrac{1}{2}hf) - f(x,y)\right]. \tag{3}$$

Substituting equation (3) into equation (1) results in

$$y(x_i+h) \approx y(x_i) + hf(x_i, y_i) + h\left[f\left(x_i+\tfrac{1}{2}h, y_i+\tfrac{1}{2}hf(x_i, y_i)\right) - f(x_i, y_i)\right],$$

or

$$y_{i+1} = y_i + hf\left(x_i+\tfrac{1}{2}h, y_i+\tfrac{1}{2}hf(x_i, y_i)\right).$$

From this equation, the *second-order Runge-Kutta method* or *midpoint method* can be developed.

A schematic diagram is shown in Figure 10.4 to illustrate the procedure of the second-order Runge-Kutta method. The cumulative error is of the second order of h, i.e., $O(h^2)$ or $|y(x_n)-y_n| \leqslant Ch^2$.

Example 10.4

For the initial value problem $y' = xy^2 - y$, $y(0) = 0.5$, determine $y(1.0)$ using the second-order Runge-Kutta method with $h = 0.5$.

The second-order Runge-Kutta method is, with $f(x,y) = xy^2 - y$ and $h = 0.5$,

$$k_1 = f(x_i, y_i), \quad y_{i+\frac{1}{2}} = y_i + \tfrac{1}{2}hk_1, \quad k_2 = f(x_i+\tfrac{1}{2}h, y_{i+\frac{1}{2}}), \quad y_{i+1} = y_i + hk_2.$$

The results are as follows

$i=0$: $x_0 = 0$, $y_0 = 0.5$,

$$k_1 = f(0, 0.5) = 0\times 0.5^2 - 0.5 = -0.5,$$

Second-Order Runge-Kutta Method

At the $(i+1)$th step, $i=0, 1, 2, \ldots,$

(1) $k_1 = f(x_i, y_i)$	Slope at the left end point x_i
(2) $y_{i+\frac{1}{2}} = y_i + \frac{1}{2}hk_1$	Predicted y at the midpoint $x_i+\frac{1}{2}h$ using the Euler method
(3) $k_2 = f(x_i+\frac{1}{2}h, y_{i+\frac{1}{2}})$	Predicted slope at the midpoint $x_i+\frac{1}{2}h$
(4) $k = k_2$	Slope on $[x_i, x_{i+1}]$ is taken as the midpoint slope
(5) $y_{i+1} = y_i + hk$	Second-order Runge-Kutta method

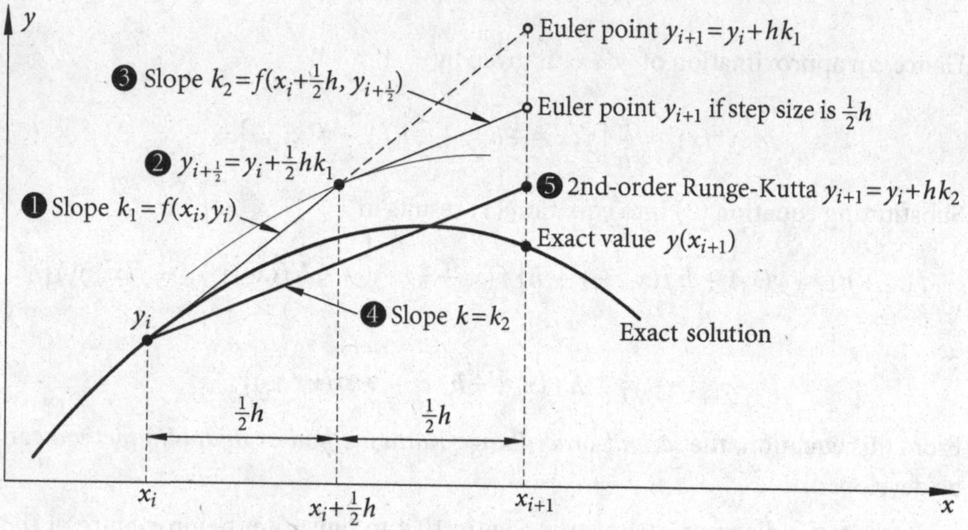

Figure 10.4 Second-order Runge-Kutta method.

$$y_{0+\frac{1}{2}} = 0.5+\frac{1}{2}\times 0.5\times(-0.5) = 0.375,$$

$$k_2 = f(0.25, 0.375) = 0.25\times 0.375^2-0.375 = -0.339844,$$

$$y_1 = 0.5+0.5\times(-0.339844) = 0.330078;$$

$i=1$: $x_1 = 0.5, \quad y_1 = 0.330078,$

$$k_1 = f(0.5, 0.330078) = 0.5\times 0.330078^2-0.330078 = -0.275602,$$

$$y_{1+\frac{1}{2}} = 0.330078+\frac{1}{2}\times 0.5\times(-0.275602) = 0.261178,$$

$$k_2 = f(0.75, 0.261178) = 0.75\times 0.261178^2-0.261178 = -0.210017,$$

$$y_2 = 0.330078+0.5\times(-0.210017) = 0.225069;$$

Cumulative error: $\left|y(1.0)-y_2\right| = \left|0.211942 - 0.225069\right| = 0.013128.$

Higher-order Runge-Kutta methods can also be developed. For most practical applications, the *fourth-order Runge-Kutta method* is the most popular method.

A schematic diagram is shown in Figure 10.5 to illustrate the procedure of the fourth-order Runge-Kutta method. The cumulative error is of the fourth order of h, i.e., $\mathcal{O}(h^4)$ or $|y(x_n)-y_n| \leqslant Ch^4$.

Example 10.5

For the initial value problem $y'=xy^2-y$, $y(0)=0.5$, determine $y(1.0)$ using the fourth-order Runge-Kutta method with $h=0.5$.

The fourth-order Runge-Kutta method is, with $f(x,y)=xy^2-y$ and $h=0.5$,

$$k_1=f(x_i,y_i),\quad y_{i+\frac{1}{2}}=y_i+\frac{1}{2}hk_1,\quad k_2=f(x_i+\frac{1}{2}h,y_{i+\frac{1}{2}}),\quad \tilde{y}_{i+\frac{1}{2}}=y_i+\frac{1}{2}hk_2,$$

$$k_3=f(x_i+\frac{1}{2}h,\tilde{y}_{i+\frac{1}{2}}),\quad \bar{y}_{i+1}=y_i+hk_3,\quad k_4=f(x_i+h,\bar{y}_{i+1}),$$

$$k=\frac{1}{6}(k_1+2k_2+2k_3+k_4),\quad y_{i+1}=y_i+hk.$$

The results are as follows

$i=0$: $x_0=0$, $y_0=0.5$,

$$k_1=f(0,0.5)=0\times 0.5^2-0.5=-0.5,$$

$$y_{0+\frac{1}{2}}=0.5+\frac{1}{2}\times 0.5\times(-0.5)=0.375,$$

$$k_2=f(0.25,0.375)=0.25\times 0.375^2-0.375=-0.339844,$$

$$\tilde{y}_{0+\frac{1}{2}}=0.5+\frac{1}{2}\times 0.5\times(-0.339844)=0.415039,$$

$$k_3=f(0.25,0.415039)=0.25\times 0.415039^2-0.415039=-0.371975,$$

$$\bar{y}_1=0.5+0.5\times(-0.371975)=0.314012,$$

$$k_4=f(0.5,0.314012)=0.5\times 0.314012^2-0.314012=-0.264711,$$

$$k=\frac{1}{6}(-0.5-2\times 0.339844-2\times 0.371975-0.264711)=-0.364725,$$

$$y_1=0.5+0.5\times(-0.364725)=0.317638;$$

$i=1$: $x_1=0.5$, $y_1=0.317638$,

$$k_1=f(0.5,0.317638)=0.5\times 0.317638^2-0.317638=-0.267191,$$

$$y_{1+\frac{1}{2}}=0.317638+\frac{1}{2}\times 0.5\times(-0.267191)=0.250840,$$

$$k_2=f(0.75,0.250840)=0.75\times 0.250840^2-0.250840=-0.203649,$$

$$\tilde{y}_{1+\frac{1}{2}}=0.317638+\frac{1}{2}\times 0.5\times(-0.203649)=0.266725,$$

Fourth-Order Runge-Kutta Method

At the $(i+1)$th step, $i=0, 1, 2, \ldots,$

(1) $k_1 = f(x_i, y_i)$	Slope at the left end point x_i
(2) $y_{i+\frac{1}{2}} = y_i + \frac{1}{2}hk_1$	Predicted y at the midpoint $x_i+\frac{1}{2}h$ using the Euler method
(3) $k_2 = f(x_i+\frac{1}{2}h, y_{i+\frac{1}{2}})$	Predicted slope at the midpoint $x_i+\frac{1}{2}h$
(4) $\tilde{k} = k_2$	Slope on $\left[x_i, x_i+\frac{1}{2}h\right]$ is taken as the midpoint slope
(5) $\tilde{y}_{i+\frac{1}{2}} = y_i + \frac{1}{2}h\tilde{k}$	Improved predicted y at the midpoint $x_i+\frac{1}{2}h$ (similar to second-order Runge-Kutta method)
(6) $k_3 = f(x_i+\frac{1}{2}h, \tilde{y}_{i+\frac{1}{2}})$	Improved predicted slope at the midpoint $x_i+\frac{1}{2}h$
(7) $\bar{k} = k_3$	Slope on $\left[x_i, x_i+\frac{1}{2}h\right]$ is taken as the improved midpoint slope
(8) $\bar{y}_{i+1} = y_i + h\bar{k}$	Predicted y at the right end point x_{i+1}
(9) $k_4 = f(x_i+h, \bar{y}_{i+1})$	Predicted slope at the right end point x_{i+1}
(10) $k = \dfrac{k_1+2k_2+2k_3+k_4}{6}$	Slope on $\left[x_i, x_i+h\right]$ is taken as the weighted average slope
(11) $y_{i+1} = y_i + hk$	Fourth-order Runge-Kutta method

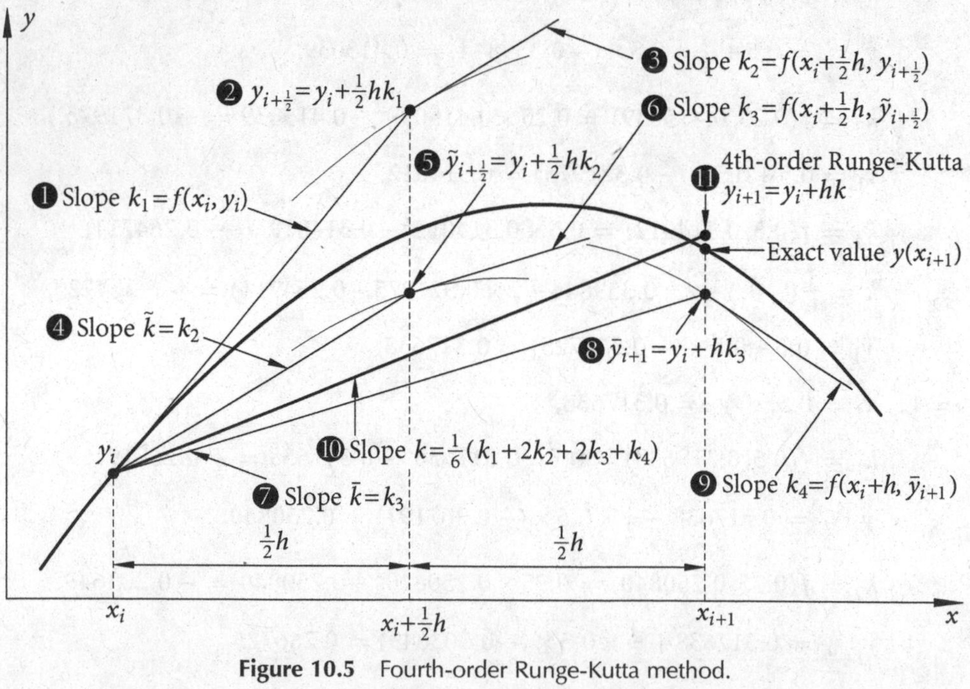

Figure 10.5 Fourth-order Runge-Kutta method.

$$k_3 = f(0.75, 0.266725) = 0.75\times 0.266725^2 - 0.266725 = -0.213369,$$

$$\bar{y}_2 = 0.317638 + 0.5\times(-0.213369) = 0.210953,$$

$$k_4 = f(1.0, 0.210953) = 1.0\times 0.210953^2 - 0.210953 = -0.166452,$$

$$k = \frac{1}{6}(-0.267191 - 2\times 0.203649 - 2\times 0.213369 - 0.166452) = -0.211280,$$

$$y_2 = 0.317638 + 0.5\times(-0.211280) = 0.211998;$$

Cumulative error: $\big|y(1.0) - y_2\big| = \big|0.211942 - 0.211998\big| = 0.000056.$

10.2 Numerical Solutions of Systems of Differential Equations

As discussed in Section 7.4, any ordinary differential equation or system of differential equations can be expressed as a system of first-order differential equations. For example, consider the following equations of motion of a two-dimensional nonlinear system

$$\ddot{q}_1 + \beta_1\dot{q}_1 + \omega_1^2(1 + \gamma_{11}q_1^2 + \gamma_{12}q_2^2)\,q_1 = a_1\cos\Omega t,$$
$$\ddot{q}_2 + \beta_2\dot{q}_2 + \omega_2^2(1 + \gamma_{21}q_1^2 + \gamma_{22}q_2^2)\,q_2 = a_2\cos\Omega t.$$

Letting $x=t$, $y_1=q_1$, $y_2=q_2$, $y_3=\dot{q}_1$, and $y_4=\dot{q}_2$ changes the equations to

$$\dot{y}_3 + \beta_1 y_3 + \omega_1^2(1 + \gamma_{11}y_1^2 + \gamma_{12}y_2^2)\,y_1 = a_1\cos\Omega x,$$
$$\dot{y}_4 + \beta_2 y_4 + \omega_2^2(1 + \gamma_{21}y_1^2 + \gamma_{22}y_2^2)\,y_2 = a_2\cos\Omega x.$$

In the matrix form, one has

$$\begin{Bmatrix} \dot{y}_1 \\ \dot{y}_2 \\ \dot{y}_3 \\ \dot{y}_4 \end{Bmatrix} = \begin{Bmatrix} y_3 \\ y_4 \\ a_1\cos\Omega x - \omega_1^2(1 + \gamma_{11}y_1^2 + \gamma_{12}y_2^2)\,y_1 - \beta_1 y_3 \\ a_2\cos\Omega x - \omega_2^2(1 + \gamma_{21}y_1^2 + \gamma_{22}y_2^2)\,y_2 - \beta_2 y_4 \end{Bmatrix}.$$

Hence, without loss of generality, consider the initial value problem of a system of m first-order differential equations

$$\mathbf{y}' = \mathbf{f}(x, \mathbf{y}), \qquad \mathbf{y}(x_0) = \mathbf{y}_0,$$

where x is the independent variable and $\mathbf{y} = \{y_1, y_2, \ldots, y_m\}^T$ is an m-dimensional vector of dependent variables, and $\mathbf{f}(x, \mathbf{y}) = \{f_1, f_2, \ldots, f_m\}^T$ is an m-dimensional vector of functions of x, y_1, y_2, $\ldots$, y_m.

The numerical methods presented in Section 10.1 for a first-order ordinary differential equation can be readily extended to a system of first-order ordinary differential equations.

1. The (Forward) Euler Method or Constant Slope Method

$$\mathbf{y}_{i+1} = \mathbf{y}_i + h\,\mathbf{f}(x_i, \mathbf{y}_i), \quad i = 0, 1, 2, \ldots.$$

2. The Backward Euler Method

$$\mathbf{y}_{i+1} = \mathbf{y}_i + h\,\mathbf{f}(x_{i+1}, \mathbf{y}_{i+1}), \quad i = 0, 1, 2, \ldots.$$

3. Improved Euler Method or Average Slope Method

$$\mathbf{y}_{i+1} = \mathbf{y}_i + \tfrac{1}{2}h\big[\mathbf{f}(x_i, \mathbf{y}_i) + \mathbf{f}(x_{i+1}, \mathbf{y}_{i+1})\big], \quad i = 0, 1, 2, \ldots.$$

4. Improved Euler Predictor-Corrector Method

At the $(i+1)$th step, $i = 0, 1, 2, \ldots,$

(1)	$\mathbf{k}_1 = \mathbf{f}(x_i, \mathbf{y}_i)$	Vector of slopes at the left end point x_i
(2) Predictor	$\mathbf{y}^{P}_{i+1} = \mathbf{y}_i + h\mathbf{k}_1$	Predict $\mathbf{y}$ at x_{i+1} using the Euler method
(3)	$\mathbf{k}_2 = \mathbf{f}(x_{i+1}, \mathbf{y}^{P}_{i+1})$	Predicted slopes at the right end point x_{i+1}
(4)	$\mathbf{k} = \dfrac{\mathbf{k}_1 + \mathbf{k}_2}{2}$	Vector of averaged slopes is used on $[x_i, x_{i+1}]$
(5) Corrector	$\mathbf{y}_{i+1} = \mathbf{y}_i + h\mathbf{k}$	Improved Euler point

5. Second-Order Runge-Kutta Method

At the $(i+1)$th step, $i = 0, 1, 2, \ldots,$

(1) $\mathbf{k}_1 = \mathbf{f}(x_i, \mathbf{y}_i)$	Vector of slopes at the left end point x_i
(2) $\mathbf{y}_{i+\frac{1}{2}} = \mathbf{y}_i + \frac{1}{2}h\mathbf{k}_1$	Predicted $\mathbf{y}$ at the midpoint $x_i + \frac{1}{2}h$ using the Euler method
(3) $\mathbf{k}_2 = \mathbf{f}(x_i + \frac{1}{2}h, \mathbf{y}_{i+\frac{1}{2}})$	Vector of predicted slopes at the midpoint $x_i + \frac{1}{2}h$
(4) $\mathbf{k} = \mathbf{k}_2$	Slopes on $[x_i, x_{i+1}]$ are taken as the midpoint slopes
(5) $\mathbf{y}_{i+1} = \mathbf{y}_i + h\mathbf{k}$	Second-order Runge-Kutta method

6. Fourth-Order Runge-Kutta Method

At the $(i+1)$th step, $i = 0, 1, 2, \ldots,$

(1) $\mathbf{k}_1 = \mathbf{f}(x_i, \mathbf{y}_i)$	Vector of slopes at the left end point x_i
(2) $\mathbf{y}_{i+\frac{1}{2}} = \mathbf{y}_i + \frac{1}{2}h\mathbf{k}_1$	Vector of predicted $\mathbf{y}$ at the midpoint $x_i + \frac{1}{2}h$ using the Euler method
(3) $\mathbf{k}_2 = \mathbf{f}(x_i + \frac{1}{2}h, \mathbf{y}_{i+\frac{1}{2}})$	Vector of predicted slopes at the midpoint $x_i + \frac{1}{2}h$

(4) $\tilde{\mathbf{k}} = \mathbf{k}_2$	Slopes on $[x_i,\ x_i+\frac{1}{2}h]$ are taken as the midpoint slopes
(5) $\tilde{\mathbf{y}}_{i+\frac{1}{2}} = \mathbf{y}_i + \frac{1}{2}h\tilde{\mathbf{k}}$	Improved predicted $\mathbf{y}$ at the midpoint $x_i+\frac{1}{2}h$ (similar to second-order Runge-Kutta method)
(6) $\mathbf{k}_3 = \mathbf{f}(x_i+\frac{1}{2}h,\ \tilde{\mathbf{y}}_{i+\frac{1}{2}})$	Improved predicted slopes at the midpoint $x_i+\frac{1}{2}h$
(7) $\bar{\mathbf{k}} = \mathbf{k}_3$	Slopes on $[x_i,\ x_i+\frac{1}{2}h]$ are taken as the improved midpoint slope
(8) $\bar{\mathbf{y}}_{i+1} = \mathbf{y}_i + h\bar{\mathbf{k}}$	Predicted $\mathbf{y}$ at the right end point x_{i+1}
(9) $\mathbf{k}_4 = \mathbf{f}(x_i+h,\ \bar{\mathbf{y}}_{i+1})$	Vector of predicted slopes at the right end x_{i+1}
(10) $\mathbf{k} = \dfrac{\mathbf{k}_1+2\mathbf{k}_2+2\mathbf{k}_3+\mathbf{k}_4}{6}$	Slopes on $[x_i,\ x_i+h]$ are taken as the weighted average slopes
(11) $\mathbf{y}_{i+1} = \mathbf{y}_i + h\mathbf{k}$	Fourth-order Runge-Kutta method

Example 10.6

Consider the initial value problem $2y'' - y'^3 \sin 2x = 0$, $y(0) = 0$, $y'(0) = 1$.

1. Determine the exact solution $y(x)$ of the initial value problem.
2. Determine $y(0.5)$ using the fourth-order Runge-Kutta method with $h = 0.5$.

(1) This second-order equation is of the type of y absent. Letting $u = y'$ and $u' = y''$ give $2u' = u^3 \sin 2x$. The equation is now variable separable

$$\int 2\frac{du}{u^3} = \int \sin 2x\, dx + C \implies \frac{1}{u^2} = \frac{1}{2}\cos 2x - C.$$

Using the initial condition $u = y' = 1$ at $x = 0$ yields

$$\frac{1}{1^2} = \frac{1}{2}\cos 0 - C \implies C = -\frac{1}{2}.$$

Hence,

$$\frac{1}{u^2} = \frac{1+\cos 2x}{2} = \cos^2 x \implies u = y' = \sec x.$$

✍ Since $u = y' = 1 > 0$ at $x = 0$, the + sign is taken.

Integrating with respect to x results in

$$y = \int \sec x\, dx + D = \ln\left|\sec x + \tan x\right| + D.$$

Using the initial condition $y = 0$ at $x = 0$ gives $0 = \ln\left|\sec 0 + \tan 0\right| + D$, which leads to $D = 0$ and

$$y = \ln\left|\sec x + \tan x\right|.$$

(2) Letting $y_1 = y$, $y_2 = y'$, one has

$$2y_2' - y_2^3 \sin 2x = 0 \implies y_2' = \frac{1}{2} y_2^3 \sin 2x,$$

or, in the matrix form,

$$\begin{Bmatrix} y_1' \\ y_2' \end{Bmatrix} = \begin{Bmatrix} y_2 \\ \frac{1}{2} y_2^3 \sin 2x \end{Bmatrix} = \begin{Bmatrix} f_1(x, \mathbf{y}) \\ f_2(x, \mathbf{y}) \end{Bmatrix} = \begin{Bmatrix} f_1(x; y_1, y_2) \\ f_2(x; y_1, y_2) \end{Bmatrix}, \quad \mathbf{y} = \begin{Bmatrix} y_1 \\ y_2 \end{Bmatrix}.$$

The fourth-order Runge-Kutta method is

$$\mathbf{k}_1 = \mathbf{f}(x_i, \mathbf{y}_i), \quad \mathbf{y}_{i+\frac{1}{2}} = \mathbf{y}_i + \tfrac{1}{2} h \mathbf{k}_1, \quad \mathbf{k}_2 = \mathbf{f}(x_i + \tfrac{1}{2}h, \mathbf{y}_{i+\frac{1}{2}}), \quad \tilde{\mathbf{y}}_{i+\frac{1}{2}} = \mathbf{y}_i + \tfrac{1}{2} h \mathbf{k}_2,$$

$$\mathbf{k}_3 = \mathbf{f}(x_i + \tfrac{1}{2}h, \tilde{\mathbf{y}}_{i+\frac{1}{2}}), \quad \bar{\mathbf{y}}_{i+1} = \mathbf{y}_i + h\mathbf{k}_3, \quad \mathbf{k}_4 = \mathbf{f}(x_i + h, \bar{\mathbf{y}}_{i+1}),$$

$$\mathbf{k} = \tfrac{1}{6}(\mathbf{k}_1 + 2\mathbf{k}_2 + 2\mathbf{k}_3 + \mathbf{k}_4), \quad \mathbf{y}_{i+1} = \mathbf{y}_i + h\mathbf{k}.$$

The results are as follows

$i=0$: $x_0 = 0$, $y_{0,1} = y(0) = 0$, $y_{0,2} = y'(0) = 1$, $h = 0.5$,

$$\mathbf{k}_1 = \begin{Bmatrix} f_1(0; 0, 1) \\ f_2(0; 0, 1) \end{Bmatrix} = \begin{Bmatrix} 1 \\ \frac{1}{2} \times 1^3 \times \sin 0 \end{Bmatrix} = \begin{Bmatrix} 1 \\ 0 \end{Bmatrix},$$

$$\mathbf{y}_{0+\frac{1}{2}} = \begin{Bmatrix} 0 \\ 1 \end{Bmatrix} + \tfrac{1}{2} \times 0.5 \times \begin{Bmatrix} 1 \\ 0 \end{Bmatrix} = \begin{Bmatrix} 0.25 \\ 1 \end{Bmatrix},$$

$$\mathbf{k}_2 = \begin{Bmatrix} f_1(0.25; 0.25, 1) \\ f_2(0.25; 0.25, 1) \end{Bmatrix} = \begin{Bmatrix} 1 \\ \frac{1}{2} \times 1^3 \times \sin 0.5 \end{Bmatrix} = \begin{Bmatrix} 1 \\ 0.239713 \end{Bmatrix},$$

$$\tilde{\mathbf{y}}_{0+\frac{1}{2}} = \begin{Bmatrix} 0 \\ 1 \end{Bmatrix} + \tfrac{1}{2} \times 0.5 \times \begin{Bmatrix} 1 \\ 0.239713 \end{Bmatrix} = \begin{Bmatrix} 0.25 \\ 1.059928 \end{Bmatrix},$$

$$\mathbf{k}_3 = \begin{Bmatrix} f_1(0.25; 0.25, 1.059928) \\ f_2(0.25; 0.25, 1.059928) \end{Bmatrix} = \begin{Bmatrix} 1.059928 \\ \frac{1}{2} \times 1.059928^3 \times \sin 0.5 \end{Bmatrix} = \begin{Bmatrix} 1.059928 \\ 0.285444 \end{Bmatrix},$$

$$\bar{\mathbf{y}}_1 = \begin{Bmatrix} 0 \\ 1 \end{Bmatrix} + \tfrac{1}{2} \times 0.5 \times \begin{Bmatrix} 1.059928 \\ 0.285444 \end{Bmatrix} = \begin{Bmatrix} 0.529964 \\ 1.142722 \end{Bmatrix},$$

$$\mathbf{k}_4 = \begin{Bmatrix} f_1(0.25; 0.529964, 1.142722) \\ f_2(0.25; 0.529964, 1.142722) \end{Bmatrix} = \begin{Bmatrix} 1.142722 \\ \dfrac{1.142722^3 \times \sin 1.0}{2} \end{Bmatrix} = \begin{Bmatrix} 1.142722 \\ 0.627814 \end{Bmatrix},$$

$$\mathbf{k} = \frac{1}{2}\left(\begin{Bmatrix} 1 \\ 0 \end{Bmatrix} + 2 \times \begin{Bmatrix} 1 \\ 0.239713 \end{Bmatrix} + 2 \times \begin{Bmatrix} 1.059928 \\ 0.285444 \end{Bmatrix} + \begin{Bmatrix} 1.142722 \\ 0.627814 \end{Bmatrix} \right) = \begin{Bmatrix} 1.043763 \\ 0.279688 \end{Bmatrix},$$

$$\mathbf{y}_1 = \begin{Bmatrix} y_{1,1} \\ y_{1,2} \end{Bmatrix} = \begin{Bmatrix} y_1(0.5) \\ y_2(0.5) \end{Bmatrix} = \begin{Bmatrix} 0 \\ 1 \end{Bmatrix} + 0.5 \times \begin{Bmatrix} 1.043763 \\ 0.279688 \end{Bmatrix} = \begin{Bmatrix} 0.521882 \\ 1.139844 \end{Bmatrix}.$$

The exact solution at $x=0.5$ is

$$\mathbf{y}(0.5) = \begin{Bmatrix} \ln|\sec x + \tan x| \\ \sec x \end{Bmatrix}\Bigg|_{x=0.5} = \begin{Bmatrix} \ln|\sec 0.5 + \tan 0.5| \\ \sec 0.5 \end{Bmatrix} = \begin{Bmatrix} 0.522238 \\ 1.139494 \end{Bmatrix},$$

and the error is

$$\mathbf{y}(0.5) - \mathbf{y}_1 = \begin{Bmatrix} 0.522238 \\ 1.139494 \end{Bmatrix} - \begin{Bmatrix} 0.521882 \\ 1.139844 \end{Bmatrix} = \begin{Bmatrix} 0.000356 \\ -0.000350 \end{Bmatrix}.$$

10.3 Stiff Differential Equations

Consider the simple first-order ordinary differential equation

$$y'(x) = -ay, \qquad y(0) = y_0, \quad a>0.$$

The solution of this initial value problem is $y(x) = y_0 \mathrm{e}^{-x}$, which approaches zero as $x \to \infty$.

Applying the Euler method gives

$$y_{i+1} = y_i + h(-ay_i) \implies y_{i+1} = (1-ah)\,y_i, \quad i = 0, 1, 2, \ldots.$$

Noting that $y_0 = y(0) = 1$, one has

$$y_n = (1-ah)^n y_0, \quad n = 1, 2, \ldots.$$

Since $a>0$, for $0<h<2/a$, $|1-ah|<1$ and $|y_n| \to 0$ when $n \to \infty$. However, if $h>2/a$, $|1-ah|>1$ and $|y_n| \to \infty$ when $n \to \infty$. Hence, the Euler method is *unstable* for this initial value problem when $h>2/a$. The Euler method is therefore *conditionally stable.*

Using the backward Euler method yields

$$y_{i+1} = y_i + h(-ay_{i+1}) \implies y_{i+1} = \frac{y_i}{1+ah}, \quad i = 0, 1, 2, \ldots,$$

or

$$y_n = \frac{1}{(1+ah)^n}\, y_0.$$

It is clear that for any stepsize $h>0$, $|y_n| \to 0$ when $n \to \infty$. Hence, the backward Euler method is *unconditionally stable* for all $h>0$.

Similarly, using the improved Euler method leads to

$$y_{i+1} = y_i + \tfrac{1}{2}h\big[(-ay_i) + (-ay_{i+1})\big] \implies y_{i+1} = \frac{1-\frac{1}{2}ah}{1+\frac{1}{2}ah}\, y_i.$$

For any stepsize $h>0$, $|y_n| \to 0$ when $n \to \infty$, and the improved Euler method is *unconditionally stable.*

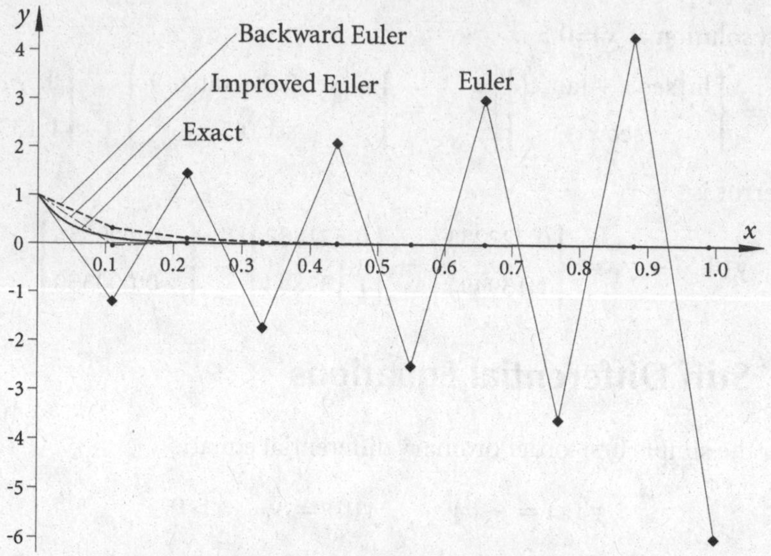

Figure 10.6 Illustration of unstable Euler method.

Figure 10.6 shows the numerical results obtained using the Euler method, the backward Euler method, and the improved Euler method for $a=20$, $h=0.11>2/a$. It can be clearly seen that the Euler method is unstable.

In this example, $y(x)$ varies rapidly—decaying exponentially at the rate of e^{-20x}. Small stepsizes must be taken to ensure the stability of explicit methods, such as the Euler method.

In general, implicit methods give better stability. When large stepsizes are used, accuracies in numerical solutions decrease; however, stability is still maintained.

An ordinary differential equation problem is *stiff* if its solution has two components, one varying slowly and the other varying rapidly, so that a numerical method must take small stepsizes to obtain satisfactory results.

A practical example of a stiff equation is the equation of motion of an overdamped single degree-of-freedom system

$$y''(x) + 2\zeta\omega_0 y'(x) + \omega_0^2 y(x) = a \sin \Omega x, \qquad \zeta > 1,$$

where x is the time parameter. If the system is at rest when $x=0$, then $y(0)=0$ and $y'(0)=0$.

For $\zeta=2$, $\omega_0=\Omega=10$, $a=400$, the response of the system can be solved analytically using the approaches presented in Chapter 4 to yield

$$y(x) = \underbrace{\frac{3-2\sqrt{3}}{6}\mathrm{e}^{-10(2+\sqrt{3})x} + \frac{3+2\sqrt{3}}{6}\mathrm{e}^{-10(2-\sqrt{3})x}}_{\text{Transient response}} \underbrace{- \cos 10x}_{\substack{\text{Steady-state}\\ \text{response}}}.$$

The first two terms decay exponentially as $x \to \infty$. After some time, the contribution of these two terms in the response becomes negligible; these two exponentially decaying terms correspond to the *transient response.* As a result, the last sinusoidal term is the *steady-state response,* which is the term that survives or dominates for large time x.

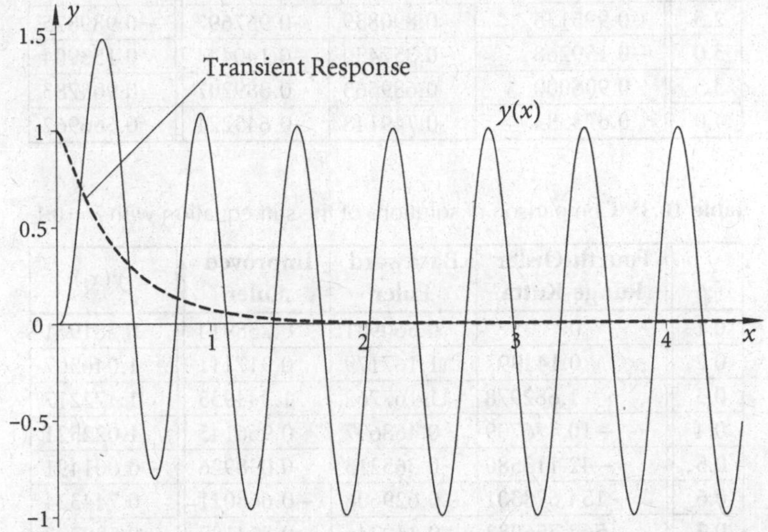

Figure 10.7 Response of an overdamped single degree-of-freedom system.

The total response $y(x)$ and the transient response are shown in Figure 10.7. It is obvious that, after some time x, only the steady-state response is dominant and is important in application.

However, in numerically solving the equation of motion for response $y(x)$, the situation is very different. Since the transient response varies rapidly and the steady-state response changes slowly, the differential equation is stiff. The selection of the stepsizes for the stiff equation is dictated by the transient response. To guarantee the stability of numerical algorithms, small stepsizes must be chosen.

Table 10.2 shows the solutions of the overdamped single degree-of-freedom system solved with $h=0.05$, using the fourth-order Runge-Kutta method, the backward Euler method, and the improved Euler method. Both the fourth-order Runge-Kutta method and the improved Euler method yield satisfactory results, with cumulative errors at $x=4.0$ being 0.99% and 3.26%, respectively.

When the stepsize is increased to $h=0.1$ (Table 10.3), the fourth-order Runge-Kutta method is unstable, with the solution growing exponentially. The backward Euler method and the improved Euler method, both of which are implicit methods, are still stable, although the accuracies are reduced due to the increase in stepsize.

Table 10.2 Comparison of solutions of the stiff equation with $h = 0.05$.

x_i	Fourth-Order Runge-Kutta	Backward Euler	Improved Euler	$y(x_i)$
0.5	−0.007319	−0.197616	0.008708	−0.001491
1.0	0.914348	0.682814	0.899324	0.912976
1.5	0.785655	0.829904	0.755871	0.779044
2.0	−0.400632	−0.142329	−0.404011	−0.403012
2.5	−0.995135	−0.890839	−0.967697	−0.989875
3.0	−0.159268	−0.357430	−0.140444	−0.153904
3.5	0.906000	0.689663	0.889207	0.903783
4.0	0.673584	0.749148	0.645224	0.666962

Table 10.3 Comparison of solutions of the stiff equation with $h = 0.1$.

x_i	Fourth-Order Runge-Kutta	Backward Euler	Improved Euler	$y(x_i)$
0.1	0	0.560981	0.258914	0.281961
0.2	0.142997	1.167179	0.917111	1.046507
0.3	−1.682978	1.167762	1.344935	1.472217
0.4	−10.476709	0.468697	0.966145	1.022521
0.5	−42.113580	−0.365213	0.038926	−0.001491
0.6	−154.678301	−0.629606	−0.668021	−0.744324
0.7	−562.756082	−0.130746	−0.564187	−0.588792
0.8	−2052.495719	0.633761	0.208191	0.271801
0.9	−7494.944646	0.930297	0.903682	1.007744
1.0	−27372.057913	0.461990	0.855766	0.912976

Consideration of stiff equations is mainly on the efficiency of the numerical algorithms. Even for conditionally stable algorithms, as long as the stepsizes are small enough, they can solve stiff problems, although they may take a long time to do it with small stepsizes.

10.4 Summary

In this chapter, some classical methods for numerically solving ordinary differential equations are introduced.

Explicit Methods

- The (forward) Euler method. The cumulative error is of the order $\mathcal{O}(h)$.
- The improved Euler predictor-corrector method. The cumulative error is of the order $\mathcal{O}(h^2)$.
- The second-order and fourth-order Runge-Kutta methods, with the cumulative errors being of the orders $\mathcal{O}(h^2)$ and $\mathcal{O}(h^4)$, respectively.

Implicit Methods

- The backward Euler method. The cumulative error is of the order $\mathcal{O}(h)$.
- The improved Euler method. The cumulative error is of the order $\mathcal{O}(h^2)$.

There are two types of errors in numerical solutions of ordinary differential equations. The roundoff error is due to the finite lengths of floating-point representations and is a characteristic of computer hardware. Roundoff errors are cumulative. Truncation error arises when numerically approximating a continuous solution by evaluating at a finite number of discrete points. Accuracy of numerical approximation can be improved by selecting algorithms that are of higher orders of stepsize h and using small stepsizes, as long as they are not so small that the local truncation error may be smaller than the roundoff error.

A differential equation is stiff if its solution has two components with vastly different rates of change. Selection of algorithms and stepsizes is often determined by the stability of the method, affected mainly by the component of solution with the fastest rate of change. Implicit methods usually give better stability. Small stepsizes must be used to obtain results with satisfactory numerical accuracy.

In solving practical application problems, it is important for an engineer to be able to describe a physical problem using a mathematical model, to establish the governing differential equations following physical laws, to select suitable methods for solving the differential equations, and to interpret the results obtained.

The purpose of this chapter is, through the introduction of some simple classical methods, to present the concept of numerical solutions of differential equations, the errors involved in the approximation, and caveats in selecting numerical algorithms. It is not the objective of this chapter nor is it essential for an engineer to master and implement numerical algorithms that are of high accuracy and efficiency.

With the rapid and constant development of computer software, sophisticated, high-performance numerical algorithms have been and are still being developed and implemented in numerical libraries such as

- the *Numerical Recipes* series in various programming languages including C, C++, and Fortran (authored by W.H. Press, B.P. Flannery, S.A. Teukolsky, and W.T. Vetterling, and published by the Cambridge University Press);
- the free GNU Scientific Library for C and C++ programmers;
- the commercial IMSL and NAG Libraries.

Symbolic and numerical computation software, such as *Maple*, *Mathematica*, and *Matlab*, all have built-in high-level commands for numerical solutions of ordinary differential equations.

Application examples solved using *Maple* will be presented in Chapter 12.

Problems

10.1 Consider the initial value problem $y' = -xy^3 + 5y,\ y(0) = 1$.

1. Show that the exact solution is $y = \dfrac{10}{\sqrt{20x - 2 + 102\,e^{-10x}}}$.
2. Determine $y(0.4)$ using the *(forward) Euler method* (FE) and the *backward Euler method* (BE) with $h = 0.2$, and determine the cumulative errors by comparing with the exact solution.

   ANS $y_{FE}(0.4) = 3.6800;\quad y_{BE}(0.4) = 3.3187$

10.2 Consider the initial value problem $y' = -e^{2x}y,\ y(0) = 1$.

1. Show that the exact solution is $y = e^{(1-e^{2x})/2}$.
2. Determine $y(2.0)$ using the *(forward) Euler method* (FE) and the *backward Euler method* (BE) with $h = 0.5$.

   ANS $y_{FE}(2.0) = -4.3754;\quad y_{BE}(2.0) = 0.00028894$
3. Determine the cumulative errors by comparing with the exact solution and discuss the results.

10.3 Consider the initial value problem $y' = (\sin^2 x - y)\cos x,\ y(0) = 2$.

1. Show that the exact solution is $y = \sin^2 x - 2\sin x + 2$.
2. Determine $y(1.0)$ using the *improved Euler method* (IE) and the *improved Euler predictor-corrector method* (IEPC) with $h = 0.5$, and determine the cumulative errors by comparing with the exact solution.

   ANS $y_{IE}(1.0) = 1.0031;\quad y_{IEPC}(1.0) = 1.0706$

10.4 Consider the initial value problem $y' = -y^5 + 20y,\ y(0) = 1$.

1. Show that the exact solution is $y = \left(\dfrac{20}{1 + 19e^{-80x}}\right)^{\frac{1}{4}}$.
2. Determine $y(0.2)$ using the *improved Euler method* (IE) and the *improved Euler predictor-corrector method* (IEPC) with $h = 0.1$.

   ANS $y_{IE}(0.2) = 2.1337;\quad y_{IEPC}(0.2) = -8.7568 \times 10^{11}$
3. Determine the cumulative errors by comparing with the exact solution and discuss the results.

10.5 Consider the initial value problem $y' = -3y + 3,\ y(1) = 0$.

1. Show that the exact solution is $y = 1 - e^{-3(x-1)}$.

2. Determine $y(3.0)$ using the *improved Euler method* (IE) and the *improved Euler predictor-corrector method* (IEPC) with $h=0.5$ and $h=1.0$, respectively.

ANS $h=0.5$: $y_{\text{IE}}(3.0)=0.99958$; $y_{\text{IEPC}}(3.0)=0.84741$

$h=1.0$: $y_{\text{IE}}(3.0)=0.96$; $y_{\text{IEPC}}(3.0)=-5.25$

3. Determine the cumulative errors by comparing with the exact solution and discuss the results.

10.6 Consider the initial value problem $y' = \cos^2 \dfrac{y}{x} + \dfrac{y}{x}$, $y(1)=0$.

1. Show that the exact solution is $y=x\tan^{-1}(\ln x)$.
2. Determine $y(2.0)$ using the *(forward) Euler method* (FE), the *improved Euler predictor-corrector method* (IEPC), the *second-order Runge-Kutta method* (RK2), and the *fourth-order Runge-Kutta method* (RK4) with $h=0.5$, and determine the cumulative errors by comparing with the exact solution.

ANS $y_{\text{FE}}(2.0)=1.1131$; $y_{\text{IEPC}}(2.0)=1.1865$

$y_{\text{RK2}}(2.0)=1.2161$; $y_{\text{RK4}}(2.0)=1.2115$

10.7 Consider the initial value problem $y' = \sin^2(x-y)$, $y(0)=0$.

1. Show that the exact solution is $y=x-\tan^{-1}x$.
2. Determine $y(1.0)$ using the *second-order Runge-Kutta method* (RK2) and the *fourth-order Runge-Kutta method* (RK4) with $h=0.5$, and determine the cumulative errors by comparing with the exact solution.

ANS $y_{\text{RK2}}(1.0)=0.22256$; $y_{\text{RK4}}(1.0)=0.21493$

10.8 Consider the initial value problem $y' = \dfrac{y}{x}(x^2+3\ln y)$, $y(1)=1$.

1. Show that the exact solution is $y=e^{x^3-x^2}$.
2. Determine $y(1.4)$ using the *second-order Runge-Kutta method* (RK2) and the *fourth-order Runge-Kutta method* (RK4) with $h=0.2$, and determine the cumulative errors by comparing with the exact solution.

ANS $y_{\text{RK2}}(1.4)=2.0066$; $y_{\text{RK4}}(1.4)=2.1844$

10.9 Consider the initial value problem

$$\begin{Bmatrix} y_1' \\ y_2' \end{Bmatrix} = \begin{bmatrix} 8 & 9 \\ -18 & -19 \end{bmatrix} \begin{Bmatrix} y_1 \\ y_2 \end{Bmatrix}, \qquad \begin{Bmatrix} y_1(0) \\ y_2(0) \end{Bmatrix} = \begin{Bmatrix} 2 \\ -3 \end{Bmatrix}.$$

1. Show that the exact solution is

$$\begin{Bmatrix} y_1 \\ y_2 \end{Bmatrix} = \begin{Bmatrix} 1 \\ -1 \end{Bmatrix} e^{-x} + \begin{Bmatrix} 1 \\ -2 \end{Bmatrix} e^{-10x}.$$

2. Determine $\mathbf{y}(0.4)$ using the *(forward) Euler method* (FE) and the *backward Euler method* (BE) with $h=0.1$ and $h=0.2$, respectively.

ANS $h=0.1$: $\mathbf{y}_{\text{FE}}(0.4)=\{0.6561, -0.6561\}^T$; $\mathbf{y}_{\text{BE}}(0.4)=\{0.74551, -0.80801\}^T$

$h=0.2$: $\mathbf{y}_{\text{FE}}(0.4)=\{1.64, -2.64\}^T$; $\mathbf{y}_{\text{BE}}(0.4)=\{0.80556, -0.91667\}^T$

3. Determine the cumulative errors by comparing with the exact solution and discuss the results.

10.10 Consider the initial value problem $1+y'^2=2yy''$, $y(1)=2$, $y'(1)=1$.

1. Show that the exact solution is $y=\frac{1}{4}x^2+\frac{1}{2}x+\frac{5}{4}$.
2. Convert the second-order equation into a system of two first-order ordinary differential equations.
3. Determine $y(2.0)$ using the *improved Euler predictor-corrector method* (IEPC) with $h=0.5$, and determine the cumulative error by comparing with the exact solution. ANS $y_{\text{IEPC}}(2.0)=3.2518$

10.11 Consider the initial value problem $y''=e^x y'^2$, $y(0)=0$, $y'(0)=-\frac{1}{2}$.

1. Show that the exact solution is $y=\ln(1+e^{-x})-\ln 2$.
2. Convert the second-order equation into a system of two first-order ordinary differential equations.
3. Determine $y(1.0)$ using the *improved Euler predictor-corrector method* (IEPC) with $h=0.5$, and determine the cumulative error by comparing with the exact solution. ANS $y_{\text{IEPC}}(1.0)=-0.37883$

10.12 Consider the initial value problem $x^3y''-x^2y'=3-x^2$, $y(1)=3$, $y'(1)=2$.

1. Show that the exact solution is $y=\frac{1}{x}+x+x^2$.
2. Convert the second-order equation into a system of two first-order ordinary differential equations.
3. Determine $y(2.0)$ using the *fourth-order Runge-Kutta method* (RK4) with $h=0.5$, and determine the cumulative error by comparing with the exact solution. ANS $y_{\text{RK4}}(2.0)=6.5083$

10.13 Show that the local truncation error of the *second-order Runge-Kutta method* in each step is of the order $O(h^3)$, and the cumulative error is of the order $O(h^2)$, or $|y(x_n)-y_n|\leqslant Ch^2$.

C H A P T E R 11

Partial Differential Equations

Partial differential equations found wide applications in various engineering disciplines. The study of partial differential equations is a vast and complex subject. In this chapter, the method of separation of variables is introduced, which, when applicable, converts a partial differential equation into a set of ordinary differential equations. As applications, the method of separation of variables is applied to study the flexural vibration of continuous beams and heat conduction.

11.1 Simple Partial Differential Equations

Some simple partial differential equations can be solved by direct integration with respect to the independent variables.

Example 11.1

Solve $\quad \dfrac{\partial^2 u}{\partial x \partial y} = 12xy^2 + 8x^3 e^{2y}, \quad u_x(x,0) = 4x, \quad u(0,y) = 3.$

The partial differential equation has two boundary conditions $u_x(x,0) = 4x$ and $u(0,y) = 3$, which are required to be satisfied by the solution.

Because of the boundary condition $u_x(x,0) = 4x$, it is easier to obtain u_x first so that this boundary condition can be applied. Integrating the differential equation with respect to y, while keeping x fixed, yields

$$\frac{\partial u}{\partial x} = 4xy^3 + 4x^3 e^{2y} + f(x).$$

Applying the boundary condition $u_x(x,0) = 4x$ leads to

$$u_x(x,0) = 4x \cdot 0^3 + 4x^3 e^{2 \cdot 0} + f(x) = 4x^3 + f(x) = 4x \implies f(x) = 4x - 4x^3.$$

Hence

$$\frac{\partial u}{\partial x} = 4xy^3 + 4x^3 e^{2y} + 4x - 4x^3.$$

Integrating with respect to x, while keeping y fixed, gives

$$u = 2x^2y^3 + x^4e^{2y} + 2x^2 - x^4 + C.$$

The constant of integration C can be determined from the boundary condition $u(0, y) = 3$:

$$u(0, y) = 2 \cdot 0^2 \cdot y^3 + 0^4 \cdot e^{2y} + 2 \cdot 0^2 - 0^4 + C = C = 3 \implies C = 3.$$

The solution of the partial differential equation is

$$u(x, y) = 2x^2y^3 + x^4e^{2y} + 2x^2 - x^4 + 3.$$

11.2 Method of Separation of Variables

The method of separation of variables is illustrated through a few examples.

Example 11.2

Solve $\quad x\dfrac{\partial u}{\partial x} - y\dfrac{\partial u}{\partial y} + 2u = 0, \quad u(1,1) = 3, \quad u(2,2) = 48.$

The solution of the partial differential equation $u(x, y)$ is a function of x and y. Apply the method of separation of variables by assuming

$$u(x, y) = X(x) \cdot Y(y),$$

i.e., consider $u(x, y)$ as the product of a function of x and a function of y, hence the name "separation of variables." Substituting into the differential equation yields

$$x\left(\frac{dX}{dx}Y\right) - y\left(X\frac{dY}{dy}\right) + 2XY = 0.$$

Dividing the equation by XY leads to

$$x\frac{\dfrac{dX}{dx}Y}{XY} - y\frac{X\dfrac{dY}{dy}}{XY} + 2\frac{XY}{XY} = 0 \implies \underbrace{\frac{x}{X}\frac{dX}{dx} + 2}_{\text{A function of } x \text{ only}} = \underbrace{\frac{y}{Y}\frac{dY}{dy}}_{\text{A function of } y \text{ only}}.$$

For a function of x only to be equal to a function of y only, they must be equal to the same constant k, i.e.,

$$\frac{x}{X}\frac{dX}{dx} + 2 = \frac{y}{Y}\frac{dY}{dy} = k.$$

The X-equation gives

$$\frac{x}{X}\frac{\mathrm{d}X}{\mathrm{d}x} = k-2. \qquad \text{✍ First-order ODE, variable separable}$$

The solution is given by

$$\int \frac{1}{X}\,\mathrm{d}X = \int \frac{k-2}{x}\,\mathrm{d}x + C \implies \ln|X| = (k-2)\ln|x| + \ln C \implies X(x) = Cx^{k-2}.$$

The Y-equation yields

$$\frac{y}{Y}\frac{\mathrm{d}Y}{\mathrm{d}y} = k. \qquad \text{✍ First-order ODE, variable separable}$$

The solution is easily obtained as

$$\int \frac{1}{Y}\,\mathrm{d}Y = \int \frac{k}{y}\,\mathrm{d}y + D \implies \ln|Y| = k\ln|y| + \ln D \implies Y(y) = Dy^k.$$

The solution of the differential equation is given by

$$u(x,y) = X(x)\cdot Y(y) = Cx^{k-2}\cdot Dy^k = Ax^{k-2}y^k, \quad A = CD.$$

The constants A and k can be determined from the boundary condition $u(1,1)=3$ and $u(2,2)=48$:

$$u(1,1) = A\cdot 1^{k-2}\cdot 1^k = A = 3,$$
$$u(2,2) = A\cdot 2^{k-2}\cdot 2^k = 3\cdot 2^{2k-2} = 48 \implies 2^{2k-2} = 16 \implies k = 3.$$

Hence, the solution of the partial differential equation is

$$u(x,y) = 3xy^3.$$

Example 11.3

Find the solution of the heat conduction problem

$$\frac{\partial^2 u}{\partial x^2} = 9\frac{\partial u}{\partial t}, \tag{1}$$

with the initial condition

$$u(x,0) = 2\sin\frac{3\pi x}{L}, \qquad \text{for } 0 \leqslant x \leqslant L; \tag{2}$$

and the boundary condition

$$u(0,t) = 0, \quad u(L,t) = 0, \qquad \text{for } t > 0. \tag{3}$$

The solution $u(x,t)$ is a function of x and t. Apply the method of separation of variables and let

$$u(x,t) = X(x)\cdot T(t),$$

i.e., the product of a function of x and a function of t. Substituting into the differential equation (1) yields

$$\frac{d^2X}{dx^2}T = 9\left(X\frac{dT}{dt}\right).$$

Dividing by XT leads to

$$\frac{\dfrac{d^2X}{dx^2}T}{XT} = 9\frac{X\dfrac{dT}{dt}}{XT} \implies \underbrace{\frac{1}{X}\frac{d^2X}{dx^2}}_{\text{A function of } x \text{ only}} = \underbrace{\frac{9}{T}\frac{dT}{dt}}_{\text{A function of } t \text{ only}}.$$

For a function of x only to be equal to a function of t only, both of them should be equal to the same constant k, i.e.,

$$\frac{1}{X}\frac{d^2X}{dx^2} = \frac{9}{T}\frac{dT}{dt} = k.$$

Because the nonzero initial condition (2) involves the independent variable x, the X-equation is solved first

$$\frac{d^2X}{dx^2} - kX = 0. \qquad \text{Second-order linear ODE}$$

The characteristic equation is $\lambda^2 - k = 0$.

Remarks: Recall the results for linear ordinary differential equations (Chapter 4). Depending on the value of k, the roots of the characteristic equation $\lambda^2 - k = 0$ and the solution of the differential equation have different forms

1. $\lambda = \pm\beta \implies X(x) = C_1 e^{-\beta x} + C_2 e^{\beta x} = A\cosh\beta x + B\sinh\beta x$
2. $\lambda = 0, 0 \implies X(x) = C_0 + C_1 x$
3. $\lambda = \pm i\beta \implies X(x) = A\cos\beta x + B\sin\beta x$

The values of k depend on the boundary condtions.

Because the initial condition (2) is a sinusoidal function, the roots of the characteristic equation must be a pair of imaginary numbers in order for the complementary solution $X(x)$ to have sinusoidal functions. As a result, it is required that $k = -\beta^2$. Hence $\lambda = \pm i\beta$, and the solution is

$$X(x) = A\cos\beta x + B\sin\beta x.$$

Comparing the solution with the initial condition (2), it is obvious that

$$\beta = \frac{3\pi}{L} \implies k = -\beta^2 = -\frac{9\pi^2}{L^2}.$$

The T-equation gives

$$\frac{9}{T}\frac{\mathrm{d}T}{\mathrm{d}t} = -\frac{9\pi^2}{L^2}, \qquad \text{First-order ODE, variable separable}$$

and the solution is

$$\int \frac{1}{T}\,\mathrm{d}T = \int -\frac{\pi^2}{L^2}\,\mathrm{d}t + C \implies \ln|T| = -\frac{\pi^2}{L^2}t + \ln|C|,$$

$$\therefore \quad T(t) = C\exp\left(-\frac{\pi^2}{L^2}t\right).$$

Hence the solution of the partial differential equation (1) is

$$u(x,t) = X(x)\,T(t) = \left(A\cos\frac{3\pi x}{L} + B\sin\frac{3\pi x}{L}\right)\cdot C\exp\left(-\frac{\pi^2}{L^2}t\right).$$

The constant C can be absorbed into A and B as follows

$$u(x,t) = \left(AC\cos\frac{3\pi x}{L} + BC\sin\frac{3\pi x}{L}\right)\exp\left(-\frac{\pi^2}{L^2}t\right),$$

where the constants AC and BC can be renamed as A and B to yield

$$u(x,t) = \left(A\cos\frac{3\pi x}{L} + B\sin\frac{3\pi x}{L}\right)\exp\left(-\frac{\pi^2}{L^2}t\right).$$

Using the initial condition (2) results in

$$u(x,0) = \left(A\cos\frac{3\pi x}{L} + B\sin\frac{3\pi x}{L}\right)\exp\left(-\frac{\pi^2}{L^2}\cdot 0\right)$$
$$= A\cos\frac{3\pi x}{L} + B\sin\frac{3\pi x}{L} = 2\sin\frac{3\pi x}{L}.$$

Comparing the coefficients of sine and cosine terms gives $A=0$ and $B=2$; hence

$$u(x,t) = 2\sin\frac{3\pi x}{L}\exp\left(-\frac{\pi^2}{L^2}t\right).$$

Applying the boundary conditions (3) yields

$$u(0,t) = 2\sin\frac{3\pi\cdot 0}{L}\exp\left(-\frac{\pi^2}{L^2}t\right) = 0, \qquad \text{BC satisfied}$$

$$u(L,t) = 2\sin\frac{3\pi\cdot L}{L}\exp\left(-\frac{\pi^2}{L^2}t\right) = 0; \qquad \text{BC satisfied}$$

hence the boundary conditions (3) are automatically satisfied.

Therefore, the solution of the partial differential equation (1) is

$$u(x,t) = 2\sin\frac{3\pi x}{L}\exp\left(-\frac{\pi^2}{L^2}t\right).$$

Example 11.4

Find the solution of the boundary value problem

$$\frac{\partial^2 u}{\partial x^2} = 9\frac{\partial^2 u}{\partial t^2}, \tag{1}$$

with the boundary conditions

$$u(0,t) = 0, \qquad u(10,t) = 0, \tag{2}$$

and the initial conditions

$$u_t(x,0) = \sin\frac{\pi x}{2} + 2\sin 2\pi x, \tag{3}$$

$$u(x,0) = 6\sin\frac{\pi x}{2} + 12\sin 2\pi x. \tag{4}$$

The solution $u(x,t)$ is a function of x and t. Separate the variables x and t, and let $u(x,t)=X(x)\,T(t)$. Substituting into the differential equation (1) yields

$$\frac{d^2X}{dx^2}T = 9X\frac{d^2T}{dt^2}.$$

Dividing the equation by XT leads to

$$\frac{\dfrac{d^2X}{dx^2}T}{XT} = 9\frac{X\dfrac{d^2T}{dt^2}}{XT} \qquad \Longrightarrow \qquad \underbrace{\frac{1}{X}\frac{d^2X}{dx^2}}_{\text{A function of } x \text{ only}} = \underbrace{\frac{9}{T}\frac{d^2T}{dt^2}}_{\text{A function of } t \text{ only}}.$$

For a function of x only to be equal to a function of t only, they should both be equal to the same constant k, i.e.,

$$\frac{1}{X}\frac{d^2X}{dx^2} = \frac{9}{T}\frac{d^2T}{dt^2} = k.$$

Since the nonzero initial conditions (3) and (4) involve the independent variable x, the X-equation is solved first

$$\frac{d^2X}{dx^2} - kX = 0. \qquad \text{Second-order linear ODE}$$

The characteristic equation is $\lambda^2 - k = 0$. Since the initial conditions (3) and (4) contain sinusoidal functions in x, the complementary solution of $X(x)$ must be a sinusoidal function. As a result, the roots of the characteristic equation must be a pair of imaginary numbers. Hence $k = -\beta^2$, which leads to $\lambda = \pm i\beta$. The solution is

$$X(x) = A\cos\beta x + B\sin\beta x.$$

Comparing the solution with the initial conditions (3) and (4), it is seen that β takes two values and the corresponding solutions are

$$\beta_1 = \frac{\pi}{2} \implies X_1(x) = A_1 \cos\frac{\pi x}{2} + B_1 \sin\frac{\pi x}{2},$$

$$\beta_2 = 2\pi \implies X_2(x) = A_2 \cos 2\pi x + B_2 \sin 2\pi x.$$

The T-equation is

$$\frac{d^2T}{dt^2} - \frac{k}{9}T = 0. \qquad \text{Second-order linear ODE}$$

The characteristic equation is $\lambda^2 - \frac{k}{9} = 0 \implies \lambda = \pm i\frac{\beta}{3}$, and the solutions are

$$\beta_1 = \frac{\pi}{2} \implies \lambda_1 = \pm i\frac{\pi}{6} \implies T_1(t) = C_1 \cos\frac{\pi t}{6} + D_1 \sin\frac{\pi t}{6},$$

$$\beta_2 = 2\pi \implies \lambda_2 = \pm i\frac{2\pi}{3} \implies T_2(t) = C_2 \cos\frac{2\pi t}{3} + D_2 \sin\frac{2\pi t}{3}.$$

The solutions of the differential equation (1) are

$$\begin{aligned}\beta_1 = \frac{\pi}{2}:\quad u_1(x,t) &= X_1(x)\,T_1(t)\\ &= \left(A_1 \cos\frac{\pi x}{2} + B_1 \sin\frac{\pi x}{2}\right)\left(C_1 \cos\frac{\pi t}{6} + D_1 \sin\frac{\pi t}{6}\right),\\ \beta_2 = 2\pi:\quad u_2(x,t) &= X_2(x)\,T_2(t)\\ &= \left(A_2 \cos 2\pi x + B_2 \sin 2\pi x\right)\left(C_2 \cos\frac{2\pi t}{3} + D_2 \sin\frac{2\pi t}{3}\right).\end{aligned}$$

Since differential equation (1) is linear, the sum of solutions is also a solution; hence

$$\begin{aligned}u(x,t) &= u_1(x,t) + u_2(x,t)\\ &= \left(A_1 \cos\frac{\pi x}{2} + B_1 \sin\frac{\pi x}{2}\right)\left(C_1 \cos\frac{\pi t}{6} + D_1 \sin\frac{\pi t}{6}\right)\\ &\quad + \left(A_2 \cos 2\pi x + B_2 \sin 2\pi x\right)\left(C_2 \cos\frac{2\pi t}{3} + D_2 \sin\frac{2\pi t}{3}\right).\end{aligned}$$

Constants $A_1, A_2, \ldots, D_2$ are obtained using the boundary and initial conditions.

Using initial condition (4)

$$\begin{aligned}u(x,0) &= \left(A_1 \cos\frac{\pi x}{2} + B_1 \sin\frac{\pi x}{2}\right)C_1 + \left(A_2 \cos 2\pi x + B_2 \sin 2\pi x\right)C_2\\ &= 6 \sin\frac{\pi x}{2} + 12 \sin 2\pi x,\end{aligned}$$

which leads to $A_1 = A_2 = 0$. Hence, $u(x,t)$ becomes

$$\begin{aligned}u(x,t) &= B_1 \sin\frac{\pi x}{2}\left(C_1 \cos\frac{\pi t}{6} + D_1 \sin\frac{\pi t}{6}\right)\\ &\quad + B_2 \sin 2\pi x\left(C_2 \cos\frac{2\pi t}{3} + D_2 \sin\frac{2\pi t}{3}\right).\end{aligned}$$

Constant B_1 can be absorbed into C_1 and D_1, and constant B_2 can be absorbed into C_2 and D_2 as follows

$$u(x,t) = \sin\frac{\pi x}{2}\left(B_1C_1 \cos\frac{\pi t}{6} + B_1D_1 \sin\frac{\pi t}{6}\right)$$
$$+ \sin 2\pi x\left(B_2C_2 \cos\frac{2\pi t}{3} + B_2D_2 \sin\frac{2\pi t}{3}\right),$$

where, by renaming the constants,

$$B_1C_1 \Rightarrow C_1, \quad B_1D_1 \Rightarrow D_1, \quad B_2C_2 \Rightarrow C_2, \quad B_2D_2 \Rightarrow D_2,$$

one obtains

$$u(x,t) = \sin\frac{\pi x}{2}\left(C_1 \cos\frac{\pi t}{6} + D_1 \sin\frac{\pi t}{6}\right) + \sin 2\pi x\left(C_2 \cos\frac{2\pi t}{3} + D_2 \sin\frac{2\pi t}{3}\right).$$

Using initial condition (4) again gives

$$u(x,0) = \sin\frac{\pi x}{2}\cdot C_1 + \sin 2\pi x \cdot C_2 = 6\sin\frac{\pi x}{2} + 12\sin 2\pi x,$$

which yields $C_1 = 6$ and $C_2 = 12$.

Since

$$u_t(x,t) = \sin\frac{\pi x}{2}\left(-C_1\cdot\frac{\pi}{6}\sin\frac{\pi t}{6} + D_1\cdot\frac{\pi}{6}\cos\frac{\pi t}{6}\right)$$
$$+ \sin 2\pi x\left(-C_2\cdot\frac{2\pi}{3}\sin\frac{2\pi t}{3} + D_2\cdot\frac{2\pi}{3}\cos\frac{2\pi t}{3}\right),$$

applying initial condition (3) yields

$$u_t(x,0) = \sin\frac{\pi x}{2}\cdot D_1\frac{\pi}{6} + \sin 2\pi x\cdot D_2\frac{2\pi}{3} = \sin\frac{\pi x}{2} + 2\sin 2\pi x,$$

which leads to $D_1 = 6/\pi$ and $D_2 = 3/\pi$.

Applying boundary conditions (2) gives

$$u(0,t) = \sin\frac{\pi\cdot 0}{2}\left(C_1\cos\frac{\pi t}{6} + D_1\sin\frac{\pi t}{6}\right)$$
$$+ \sin(2\pi\cdot 0)\left(C_2\cos\frac{2\pi t}{3} + D_2\sin\frac{2\pi t}{3}\right) = 0,$$
$$u(10,t) = \sin\frac{\pi\cdot 10}{2}\left(C_1\cos\frac{\pi t}{6} + D_1\sin\frac{\pi t}{6}\right)$$
$$+ \sin(2\pi\cdot 10)\left(C_2\cos\frac{2\pi t}{3} + D_2\sin\frac{2\pi t}{3}\right) = 0;$$

hence boundary conditions (2) are automatically satisfied.

The solution of the partial differential equation (1) is

$$u(x,t) = \sin\frac{\pi x}{2}\left(6\cos\frac{\pi t}{6} + \frac{6}{\pi}\sin\frac{\pi t}{6}\right) + \sin 2\pi x\left(12\cos\frac{2\pi t}{3} + \frac{3}{\pi}\sin\frac{2\pi t}{3}\right).$$

11.3 Application—Flexural Motion of Beams

11.3.1 Formulation—Equation of Motion

Consider the flexural vibration of a uniform elastic beam as shown in Figure 11.1. The beam is subjected to a transverse dynamic load $w(x,t)$. Let ρ be the mass density per unit volume of the beam, L the length (span) of the beam, A the cross-sectional area, I the moment of inertia about the neutral axis, and $v(x,t)$ the transverse displacement of the central axis.

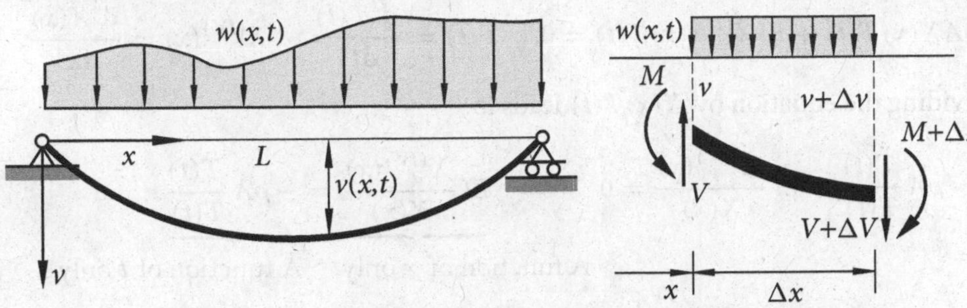

Figure 11.1 Flexural motion of a uniform elastic beam.

To set up the equation of motion, study the free-body of an infinitesimal segment of length Δx shown in Figure 11.1. The beam segment is subjected to the portion of the dynamic load $w(x,t)$, which is considered to be constant over the small segment Δx, the shear force V and the bending moment M on the left end, and the shear force $V+\Delta V$ and the bending moment $M+\Delta M$ on the right end.

Summing up the moments about the midpoint of the beam segment gives

$$\curvearrowleft \sum M = 0: \qquad (M+\Delta M) - M + V\cdot\frac{\Delta x}{2} + (V+\Delta V)\cdot\frac{\Delta x}{2} = 0,$$

which, after neglecting higher-order terms and taking the limit $\Delta x \to 0$, leads to

$$\Delta M + V\cdot\Delta x = 0 \implies V = -\frac{\partial M}{\partial x}.$$

Applying Newton's Second Law in the vertical direction

$$\downarrow ma = \sum F: \qquad (\rho A \Delta x)\frac{\partial^2 v}{\partial t^2} = (V+\Delta V) - V + w(x,t)\Delta x,$$

$$\rho A\frac{\partial^2 v}{\partial t^2} = \frac{\partial V}{\partial x} + w(x,t) \implies \rho A\frac{\partial^2 v}{\partial t^2} = -\frac{\partial^2 M}{\partial x^2} + w(x,t).$$

Using the moment-curvature relationship of an elastic element $EI\dfrac{\partial^2 v}{\partial x^2} = M$, the equation of motion becomes

$$\rho A\frac{\partial^2 v}{\partial t^2} + EI\frac{\partial^4 v}{\partial x^4} = w(x,t).$$

11.3.2 Free Vibration

For free vibration, the externally applied dynamic load $w(x,t)=0$, and the equation of motion becomes

$$\rho A \frac{\partial^2 v}{\partial t^2} + EI \frac{\partial^4 v}{\partial x^4} = 0.$$

Apply the method of separation of variables and let $v(x,t)=X(x)\cdot T(t)$. Substituting into equation of motion yields

$$\rho A X(x)\,\ddot{T}(t) + EI\,X^{(IV)}(x)\,T(t) = 0, \quad \ddot{T}(t) = \frac{\mathrm{d}^2 T(t)}{\mathrm{d}t^2}, \quad X^{(IV)}(x) = \frac{\mathrm{d}^4 X(x)}{\mathrm{d}x^4}.$$

Dividing the equation by $X(x)\,T(t)$ leads to

$$\rho A \frac{\ddot{T}(t)}{T(t)} + EI \frac{X^{(IV)}(x)}{X(x)} = 0 \implies \underbrace{EI \frac{X^{(IV)}(x)}{X(x)}}_{\text{A function of } x \text{ only}} = \underbrace{-\rho A \frac{\ddot{T}(t)}{T(t)}}_{\text{A function of } t \text{ only}}.$$

For a function of x only to be equal to a function of t only, each of them must be equal to the same constant k, i.e.,

$$EI \frac{X^{(IV)}(x)}{X(x)} = -\rho A \frac{\ddot{T}(t)}{T(t)} = k.$$

The T-equation gives

$$\ddot{T}(t) + \frac{k}{\rho A} T(t) = 0,$$ ✍ Second-order linear ODE

and the characteristic equation is $\lambda^2 + k/(\rho A) = 0$. In order to have oscillatory solution in time t, i.e., the solution of $T(t)$ is a sinusoidal function, the roots of the characteristic equation must be a pair of complex (imaginary) numbers. Hence, $k/(\rho A) = \omega^2$ and the characteristic numbers are $\lambda = \pm\mathrm{i}\omega$. The solution is

$$T(t) = A_1 \cos \omega t + B_1 \sin \omega t = a \cos(\omega t - \varphi),$$

where A_1, B_1 or a, φ are real constants, $\omega > 0$ is undetermined at this point.

The X-equation becomes

$$X^{(IV)}(x) - \frac{k}{EI} X(x) = 0 \implies X^{(IV)}(x) - \beta^4 X(x) = 0, \quad \beta = \left(\frac{\rho A}{EI}\omega^2\right)^{\frac{1}{4}}.$$

This is a fourth-order linear ordinary differential equation. The characteristic equation is $\lambda^4 - \beta^4 = 0 \implies \lambda^2 = \beta^2, -\beta^2 \implies \lambda = \pm\beta, \pm\mathrm{i}\beta$. The solution is

$$X(x) = A \cos \beta x + B \sin \beta x + C \cosh \beta x + D \sinh \beta x,$$

where A, B, C, D are real constants.

The response is

$$v(x,t) = X(x)\,T(t)$$
$$= (A\cos\beta x + B\sin\beta x + C\cosh\beta x + D\sinh\beta x)\cdot a\cos(\omega t - \varphi),$$

in which the constant a can be absorbed into A, B, C, D to yield

$$v(x,t) = (A\cos\beta x + B\sin\beta x + C\cosh\beta x + D\sinh\beta x)\cos(\omega t - \varphi).$$

For a simply supported beam, both ends are pinned or hinged and the boundary conditions are, at $x=0$ and L,

$$v(x,t) = 0,$$
$$M(x,t) = 0 \implies EI\,\frac{\partial^2 v(x,t)}{\partial x^2} = 0 \implies \frac{\partial^2 v(x,t)}{\partial x^2} = 0.$$

Since

$$\frac{\partial^2 v(x,t)}{\partial x^2} = (-A\cos\beta x - B\sin\beta x + C\cosh\beta x + D\sinh\beta x)\,\beta^2\cos(\omega t - \varphi),$$

applying the boundary conditions results in

$$v(0,t) = 0: \quad (A + C)\ \cos(\omega t - \varphi) = 0,$$
$$v(L,t) = 0: \quad (A\cos\beta L + B\sin\beta L + C\cosh\beta L + D\sinh\beta L)\cdot\cos(\omega t - \varphi) = 0,$$
$$\left.\frac{\partial^2 v(x,t)}{\partial x^2}\right|_{x=0} = 0: \quad (-A + C)\ \beta^2\cos(\omega t - \varphi) = 0,$$
$$\left.\frac{\partial^2 v(x,t)}{\partial x^2}\right|_{x=L} = 0: \quad (-A\cos\beta L - B\sin\beta L + C\cosh\beta L + D\sinh\beta L)\ \beta^2\cdot\cos(\omega t - \varphi) = 0.$$

Since $\cos(\omega t - \varphi)$ is not identically equal to zero t, the shaded terms in the above boundary conditions must be zero.

The first and third boundary conditions give

$$A + C = 0, \quad -A + C = 0 \implies A = C = 0.$$

The second and fourth boundary conditions are simplified as

$$B\sin\beta L + D\sinh\beta L = 0,$$
$$-B\sin\beta L + D\sinh\beta L = 0.$$

Adding these two equations yields

$$2D\sinh\beta L = 0.$$

Since $\beta > 0$ and $\sinh \beta L \neq 0$, one must have $D = 0$, and the fourth boundary condition becomes

$$B \sin \beta L = 0.$$

This equation implies

1. $B = 0$, which leads to $v(x, t) \equiv 0$, i.e., there is no vibration. This case is not interesting.
2. $B \neq 0$, $\sin \beta L = 0$, which yields $\beta L = \pi, 2\pi, 3\pi, \ldots$, i.e.,

$$\beta_n = \frac{n\pi}{L}, \quad \omega_n = \sqrt{\frac{EI}{\rho A}}\, \beta_n^2 = \left(\frac{n\pi}{L}\right)^2 \sqrt{\frac{EI}{\rho A}}, \quad n = 1, 2, \ldots.$$

The solutions of the equation of motion for a simply supported beam are

$$v_n(x, t) = B_n \sin \frac{n\pi x}{L} \cos(\omega_n t - \varphi_n), \quad n = 1, 2, \ldots,$$

where ω_n is the natural circular frequency of the nth mode of vibration of the beam, and

$$X_n(x) = \sin \frac{n\pi x}{L}$$

is the nth mode shape. The first three mode shapes of a simply supported beam are shown in Figure 11.2.

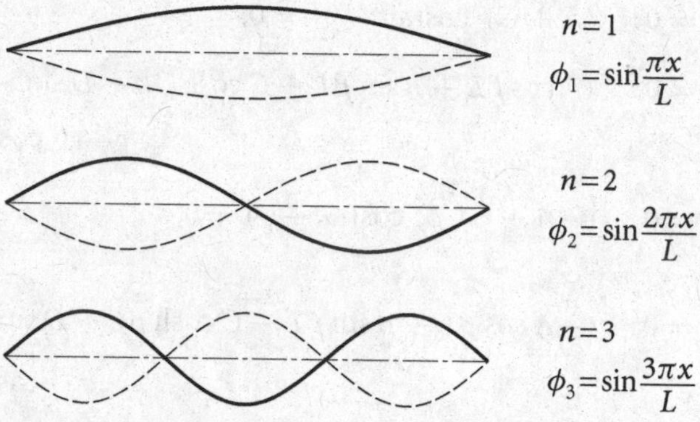

Figure 11.2 First three mode shapes of a simply supported beam.

Note that, since n can be any positive integer, a continuous beam is a system with infinitely many degrees-of-freedom.

Since the equation of motion is linear, the sum of any two solutions is also a solution; hence, the response of free vibration is given by

$$v(x, t) = \sum_{n=1}^{\infty} v_n(x, t) = \sum_{n=1}^{\infty} B_n \sin \frac{n\pi x}{L} \cos(\omega_n t - \varphi_n)$$
$$= \sum_{n=1}^{\infty} \sin \frac{n\pi x}{L} (a_n \cos \omega_n t + b_n \sin \omega_n t),$$

where B_n, φ_n or a_n, b_n are determined by the initial conditions at $t = 0$.

Example 11.5 — Free Flexural Vibration of a Simply Supported Beam

A uniform beam of length L is lifted from its right support and then dropped as shown in the following figure. The beam rotates about its left end. At time $t=0$, the right end of the beam hits the right support with velocity V_0, and the right end is held in contact with the right support for $t>0$, i.e., the beam is simply supported. Determine the response of the beam for time $t>0$.

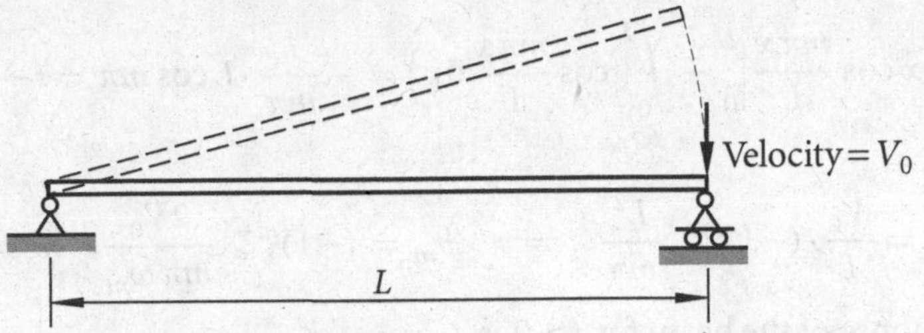

Suppose the beam rotates as a rigid body, at time $t=0$, the velocity of the beam is

$$\left.\frac{\partial v(x,t)}{\partial t}\right|_{t=0} = \frac{x}{L}V_0,$$

and the displacement is $v(x,t)\big|_{t=0}=0$. The response of the beam is given by

$$v(x,t) = \sum_{n=1}^{\infty} \sin\frac{n\pi x}{L}\left(a_n\cos\omega_n t + b_n\sin\omega_n t\right), \qquad \omega_n = \left(\frac{n\pi}{L}\right)^2\sqrt{\frac{EI}{\rho A}}.$$

Applying the initial conditions

$$v(x,0) = \sum_{n=1}^{\infty} \sin\frac{n\pi x}{L}\cdot a_n \quad\Longrightarrow\quad a_n = 0,$$

and the response is reduced to

$$v(x,t) = \sum_{n=1}^{\infty} b_n\sin\frac{n\pi x}{L}\,\sin\omega_n t.$$

Using the second boundary condition gives

$$\left.\frac{\partial v(x,t)}{\partial t}\right|_{t=0} = \left.\sum_{n=1}^{\infty} b_n\sin\frac{n\pi x}{L}\cdot\omega_n\cos\omega_n t\right|_{t=0} = \sum_{n=1}^{\infty} b_n\omega_n\sin\frac{n\pi x}{L} = \frac{x}{L}V_0.$$

Note that this initial condition is expressed in Fourier sine series in x. Some relevant results of Fourier series are summarized on pages 470–471.

To determine the coefficients b_n, multiplying the equation by $\sin\dfrac{m\pi x}{L}$, $m=1, 2, \ldots$, and integrating with respect to x from 0 to L yield

$$\sum_{n=1}^{\infty} b_n\omega_n\int_0^L \sin\frac{n\pi x}{L}\sin\frac{m\pi x}{L}\,\mathrm{d}x = \frac{V_0}{L}\int_0^L x\sin\frac{m\pi x}{L}\,\mathrm{d}x.$$

Note the orthogonality property of the sine functions

$$\int_0^L \sin\frac{n\pi x}{L}\sin\frac{m\pi x}{L}\,\mathrm{d}x = \begin{cases} 0, & m \neq n, \\ \frac{1}{2}L, & m = n, \end{cases}$$

and

$$\int_0^L x\sin\frac{m\pi x}{L}\,\mathrm{d}x = -\frac{L}{m\pi}\int_0^L x\,\mathrm{d}\left(\cos\frac{m\pi x}{L}\right)$$ ✍ Integration by parts

$$= -\frac{L}{m\pi}\left(x\cos\frac{m\pi x}{L}\Big|_0^L - \int_0^L \cos\frac{m\pi x}{L}\,\mathrm{d}x\right) = -\frac{L}{m\pi}\cdot L\cos m\pi = (-1)^{m+1}\frac{L^2}{m\pi}.$$

Hence

$$b_m\omega_m\cdot\frac{L}{2} = \frac{V_0}{L}\cdot(-1)^{m+1}\frac{L^2}{m\pi} \implies b_m = (-1)^{m+1}\frac{2V_0}{m\pi\,\omega_m}, \quad m = 1, 2, \ldots,$$

and the response of the beam for $t > 0$ is

$$v(x,t) = \frac{2V_0}{\pi}\sum_{n=1}^{\infty}(-1)^{n+1}\frac{1}{n\omega_n}\sin\frac{n\pi x}{L}\sin\omega_n t.$$

Fourier Series

A periodic function $f(x)$ of period $2T$ can be expressed in the *Fourier series*

$$f(x) = \frac{a_0}{2} + \sum_{n=1}^{\infty}\left(a_n\cos\frac{n\pi x}{T} + b_n\sin\frac{n\pi x}{T}\right).$$

The coefficients a_n and b_n can be obtained using the orthogonality properties of the sine and cosine functions

$$\int_{-T}^{T}\cos\frac{n\pi x}{T}\cos\frac{m\pi x}{T}\,\mathrm{d}x = \int_{-T}^{T}\sin\frac{n\pi x}{T}\sin\frac{m\pi x}{T}\,\mathrm{d}x = \begin{cases} 0, & m \neq n, \\ T, & m = n, \end{cases}$$

$$\int_{-T}^{T}\cos\frac{n\pi x}{T}\sin\frac{m\pi x}{T}\,\mathrm{d}x = 0.$$

To find a_n, multiplying the Fourier series by $\cos\frac{m\pi x}{T}$, $m = 0, 1, 2, \ldots$, and integrating with respect to x from $-T$ to T yield

$$\int_{-T}^{T} f(x)\cos\frac{m\pi x}{T}\,\mathrm{d}x = \frac{a_0}{2}\int_{-T}^{T}\cos\frac{m\pi x}{T}\,\mathrm{d}x$$

$$+\sum_{n=1}^{\infty}a_n\int_{-T}^{T}\cos\frac{n\pi x}{T}\cos\frac{m\pi x}{T}\,\mathrm{d}x + \sum_{n=1}^{\infty}b_n\int_{-T}^{T}\sin\frac{n\pi x}{T}\cos\frac{m\pi x}{T}\,\mathrm{d}x,$$

which gives, after applying the orthogonality properties,

$$a_n = \frac{1}{T}\int_{-T}^{T} f(x)\cos\frac{n\pi x}{T}\,\mathrm{d}x, \quad n = 0, 1, 2, \ldots.$$

Similarly, to find b_n, multiplying the Fourier series by $\sin\dfrac{m\pi x}{T}$, $m=1,2,\ldots$, and integrating with respect to x from $-T$ to T yield

$$b_n = \frac{1}{T}\int_{-T}^{T} f(x)\sin\frac{n\pi x}{T}\,dx, \quad n=1,2,\ldots.$$

The Fourier series converges to function $\bar{f}(x)$, which equals $f(x)$ at points of continuity or the average of the two limits of $f(x)$ at points of discontinuity.

❧ If $f(x)$ is an even function, i.e., $f(-x)=f(x)$, then $b_n=0$, and the Fourier series reduces to the *Fourier cosine series*

$$f(x) = \frac{a_0}{2} + \sum_{n=1}^{\infty} a_n \cos\frac{n\pi x}{T}, \quad a_n = \frac{2}{T}\int_0^T f(x)\cos\frac{n\pi x}{T}\,dx, \quad n=0,1,2,\ldots.$$

❧ If $f(x)$ is an odd function, i.e., $f(-x)=-f(x)$, then $a_n=0$, and the Fourier series reduces to the *Fourier sine series*

$$f(x) = \sum_{n=1}^{\infty} b_n \sin\frac{n\pi x}{T}, \quad b_n = \frac{2}{T}\int_0^T f(x)\sin\frac{n\pi x}{T}\,dx, \quad n=1,2,\ldots.$$

11.3.3 Forced Vibration

Consider the case of a simply supported beam subjected to a dynamic concentrated load $P\sin\Omega t$ applied at $x=a$ as shown in Figure 11.3.

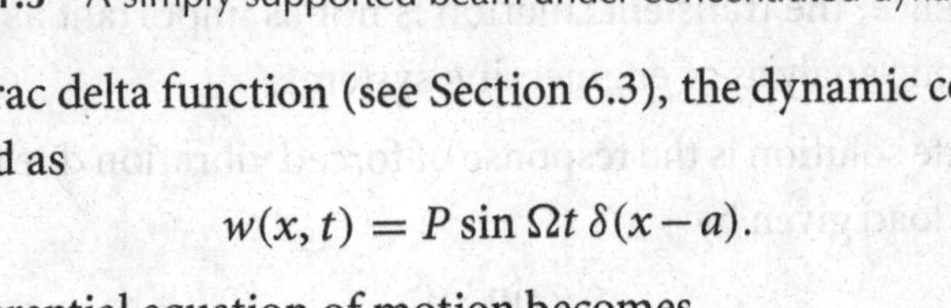

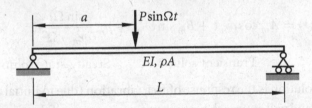

Figure 11.3 A simply supported beam under concentrated dynamic load.

Using the Dirac delta function (see Section 6.3), the dynamic concentrated load can be expressed as

$$w(x,t) = P\sin\Omega t\,\delta(x-a).$$

The partial differential equation of motion becomes

$$\rho A\frac{\partial^2 v}{\partial t^2} + EI\frac{\partial^4 v}{\partial x^4} = P\sin\Omega t\,\delta(x-a).$$

Applying the method of separation of variables and using the results obtained for free vibration, let

$$v(x,t) = \sum_{n=1}^{\infty} q_n(t)\sin\frac{n\pi x}{L}.$$

Substituting into the equation of motion yields

$$\rho A \sum_{n=1}^{\infty} \ddot{q}_n(t) \sin\frac{n\pi x}{L} + EI \sum_{n=1}^{\infty} q_n(t) \cdot \left(\frac{n\pi}{L}\right)^4 \sin\frac{n\pi x}{L} = P \sin\Omega t \, \delta(x-a).$$

Multiplying the equation by $\sin\dfrac{m\pi x}{L}$, $m=1,2,\ldots$, and integrating with respect to x from 0 to L give

$$\sum_{n=1}^{\infty} \left[\rho A \ddot{q}_n(t) + EI\left(\frac{n\pi}{L}\right)^4 q_n(t)\right] \int_0^L \sin\frac{n\pi x}{L} \sin\frac{m\pi x}{L} \, dx$$
$$= P \sin\Omega t \int_0^L \delta(x-a) \sin\frac{m\pi x}{L} \, dx.$$

Using the orthogonality of sine function and the property of Dirac delta function $\int_{a-\varepsilon}^{a+\varepsilon} f(x)\delta(x-a)\,dx = f(a)$, one obtains, for $m=1,2,\ldots$,

$$\left[\rho A \ddot{q}_m(t) + EI\left(\frac{m\pi}{L}\right)^4 q_m(t)\right] \cdot \frac{L}{2} = P \sin\Omega t \sin\frac{m\pi a}{L},$$

i.e.,

$$\ddot{q}_m(t) + \omega_m^2 q_m(t) = p_m \sin\Omega t, \quad \omega_m = \left(\frac{m\pi}{L}\right)^2 \sqrt{\frac{EI}{\rho A}}, \quad p_m = \frac{2P}{\rho AL} \sin\frac{m\pi a}{L}.$$

This is a system of infinitely many *uncoupled* second-order linear ordinary differential equations. Each equation can be solved separately using the methods presented in Chapter 4, and the solutions are given by

$$q_m(t) = \underbrace{A_m \cos\omega_m t + B_m \sin\omega_m t}_{\text{Transient solution}} + \underbrace{p_m \frac{\sin\Omega t}{\omega_m^2 - \Omega^2}}_{\text{Steady-state solution}}.$$

The transient solution is the response of free vibration (due to initial displacements and velocities) and will approach zero when time increases if there is some damping in the system. Hence, the transient solution is not as important as the steady-state solution in dynamic analysis of engineering systems.

The steady-state solution is the response of forced vibration due to the externally applied dynamic load given by

$$q_{m,\,\text{steady-state}}(t) = p_m \frac{\sin\Omega t}{\omega_m^2 - \Omega^2}, \qquad m = 1, 2, \ldots.$$

Therefore, the steady-state response of the beam is

$$v(x,t) = \sum_{n=1}^{\infty} q_n(t) \sin\frac{n\pi x}{L} = \sum_{n=1}^{\infty} \frac{p_n}{\omega_n^2 - \Omega^2} \sin\frac{n\pi x}{L} \sin\Omega t.$$

It is seen that, when $\Omega=\omega_n$, the beam is in resonance in the nth mode, leading to large amplitude of vibration in the nth mode.

11.4 Application—Heat Conduction

11.4.1 Formulation—Heat Equation

Heat conduction is the transfer of heat from warm areas to cooler areas. Consider two parallel planes, a distance Δx apart, in a solid as shown in Figure 11.4(a). Suppose the temperatures at the two planes are T and $T+\Delta T$, respectively. In time Δt, the quantity of heat entering an area A of the plane at x is $Q(x)=Q_x$ and the quantity of heat leaving an area A of the plane at $x+\Delta x$ is $Q(x+\Delta x)=Q_x+\Delta Q_x$. The quantity of heat flowing between the planes through area A in time Δt is

$$(Q_x+\Delta Q_x)-Q_x = kA\frac{T-(T+\Delta T)}{\Delta x}\Delta t \implies \Delta Q_x = -kA\frac{\Delta T}{\Delta x}\Delta t,$$

where k is the *coefficient of thermal conductivity*, and the negative sign indicates that heat flows in the direction of falling temperature.

Fourier's Law of Heat Conduction

The rate of heat conduction, dQ_x/dt, is proportional to the area A measured normal to the direction of heat flow, and to the temperature gradient dT/dt in the direction of the heat flow, i.e.,

$$\frac{dQ_x}{dt} = -kA\frac{dT}{dx},$$

where k is the coefficient of thermal conductivity.

Equation of Heat Conduction

Consider the volume element, shown in Figure 11.4(b), in a homogeneous (uniform constituency) and isotropic (same properties in all directions) solid with constant coefficient of thermal conductivity k.

From Fourier's Law of Heat Conduction, the total quantity of heat entering the face $dy\,dz$ at x in time dt is given by

$$dQ_x = -k\,(dy\,dz)\frac{\partial T}{\partial x}dt.$$

Denote the heat leaving the face $dy\,dz$ at $x+dx$ as dQ_{x+dx}. Since

$$\frac{dQ_{x+dx}-dQ_x}{dx} = \frac{\partial(dQ_x)}{\partial x},$$

the heat flowing in the volume element in the x-direction in time dt is

$$dQ_{x+dx}-dQ_x = \frac{\partial(dQ_x)}{\partial x}dx = -k\,(dx\,dy\,dz)\frac{\partial^2 T}{\partial x^2}dt.$$

Similar analyses in the other two directions give the corresponding equations.

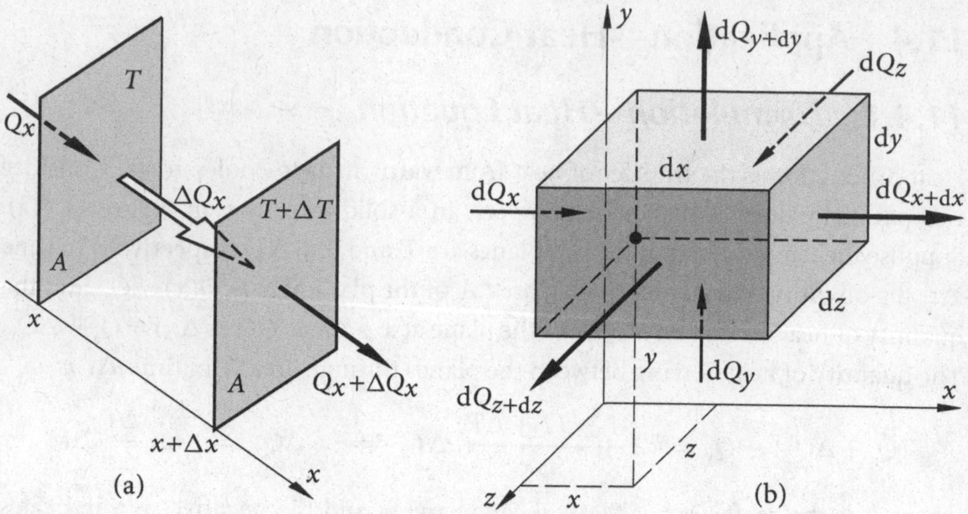

Figure 11.4 Heat conduction.

Suppose that the solid generates heat at a rate $q_g(x, y, z)$ per unit time. The total internal heat generated in time dt in the volume element is

$$dQ_g = q_g(x, y, z)\,(dx\,dy\,dz)\,dt.$$

In time dt, the internal energy of the volume element is increased by

$$dU = \rho C\,(dx\,dy\,dz)\,\frac{\partial T}{\partial t}\,dt,$$

where ρ is the density and C is the *specific heat* of the solid.

Applying the Principle of Conservation of Energy of the volume element gives

$$(dQ_{x+dx} - dQ_x) + (dQ_{y+dy} - dQ_y) + (dQ_{z+dz} - dQ_z) = dQ_g - dU,$$

which results in

$$\frac{\partial^2 T}{\partial x^2} + \frac{\partial^2 T}{\partial y^2} + \frac{\partial^2 T}{\partial z^2} = -\frac{q_g(x, y, z)}{k} + \frac{1}{\alpha}\frac{\partial T}{\partial t},$$

where $\alpha = k/(\rho C)$ is the *thermal diffusivity* of the solid. Some important special cases (without internal heat generation) can be reduced from this general equation:

One-Dimensional Transient Heat Conduction

$$\frac{\partial^2 T}{\partial x^2} = \frac{1}{\alpha}\frac{\partial T}{\partial t}.$$

✍ Fourier's equation in one-dimension

Two-Dimensional Steady-State Heat Conduction

$$\frac{\partial^2 T}{\partial x^2} + \frac{\partial^2 T}{\partial y^2} = 0.$$

✍ Laplace's equation in two-dimensions

Three-Dimensional Steady-State Heat Conduction

$$\frac{\partial^2 T}{\partial x^2} + \frac{\partial^2 T}{\partial y^2} + \frac{\partial^2 T}{\partial z^2} = 0.$$

✍ Laplace's equation in three-dimensions

Boundary Conditions

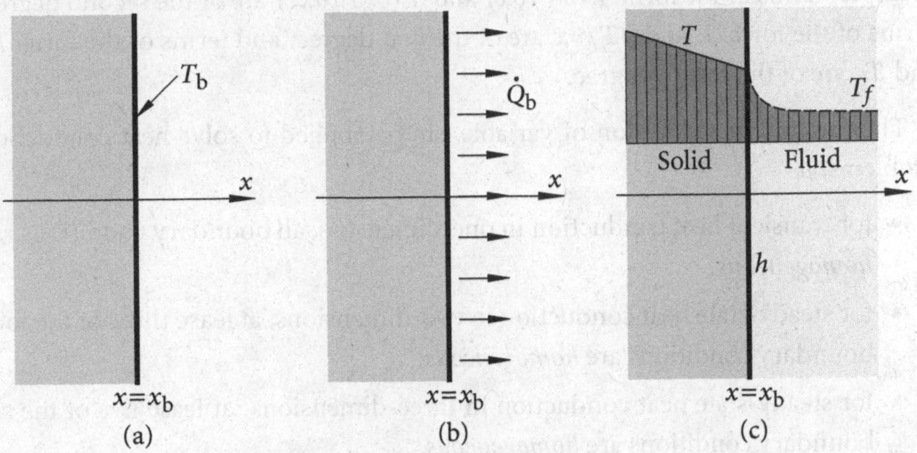

Figure 11.5 Boundary conditions.

Boundary Condition of the First Kind

In this case, the temperature at a boundary is specified as T_b as shown in Figure 11.5(a). Then

$$T(x_b, y, z, t) = T_b.$$

Boundary Condition of the Second Kind

In this case, the heat flux (the rate of heat transfer per unit area of the solid) at a boundary is specified as $\dot{Q}_b$ as shown in Figure 11.5(b). Then

$$-k\left(\frac{\partial T}{\partial x}\right)_{x=x_b} = \dot{Q}_b.$$

In the case when the boundary is *insulated*, the heat flux is zero, i.e., $\dot{Q}_b = 0$, which leads to

$$\left.\frac{\partial T}{\partial x}\right|_{x=x_b} = 0.$$

Boundary Condition of the Third Kind

In this case, the body is in contact with a convecting fluid as shown in Figure 11.5(c). At the boundary, the heat conduction in the solid equals the heat convected by the fluid, i.e.,

$$-k\left(\frac{\partial T}{\partial x}\right)_{x=x_b} = h\left[T|_{x=x_b} - T_f\right],$$

where T_f is the temperature of the convecting fluid far away from and unaffected by the boundary, and h is the *heat transfer coefficient*, which includes the combined effects of conduction and convection in the fluid.

A boundary condition is *homogeneous* if all of its terms, other than zero, are of the *first degree* in the unknown function (temperature) and its derivatives. For example, terms of the form $T(\partial T/\partial x)$ and $T(\partial^2 T/\partial x^2)$ are of the second degree; terms of the form T and $\partial T/\partial x$ are of the first degree; and terms of the form T_b and T_f are of the zeroth degree.

The method of separation of variable can be applied to solve heat conduction problems, if

- for transient heat conduction in one-dimension, all boundary conditions are *homogeneous.*
- for steady-state heat conduction in two-dimensions, at lease three of the four boundary conditions are *homogeneous;*
- for steady-state heat conduction in three-dimensions, at lease five of the six boundary conditions are *homogeneous.*

Four heat conduction problems are considered in the following subsections.

11.4.2 Two-Dimensional Steady-State Heat Conduction

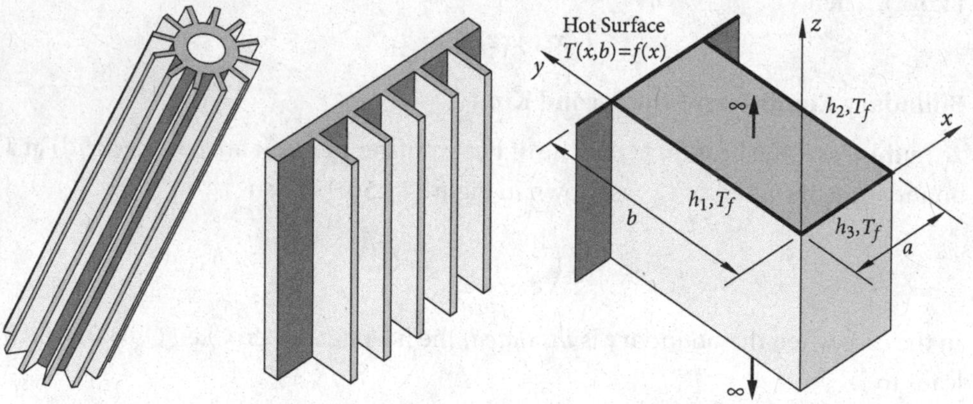

Figure 11.6 Finned surface and two-dimensional heat conduction.

The rate of convective heat transfer from a primary surface is directly proportional to the surface exposed to the fluid. It is a common practice in engineering to increase this area by attaching thin strips of metal or fins to the primary surface. Two typical examples of finned surfaces are shown in Figure 11.6.

A long straight fin is modeled as a long (in the z-direction) conductive bar of rectangular cross-section ($a \times b$) attached to a hot surface with one of its sides as shown in Figure 11.6. The coefficient of thermal conductivity is k. A cold fluid of temperature T_f flows around the three remaining sides of the bar. The bar augments the heat transfer from the hot surface to the fluid.

The temperature changes in the z-direction are assumed to be negligible. As a result, the heat conduction in the fin is two-dimensional, i.e., in the x- and y-dimensions. This assumption is justified from the fact that the length of the fin in the z-direction is very large compared to both a and b, making end effects unimportant.

The heat transfer coefficients along the three sides of the fin are constant and are denoted by h_1, h_2, and h_3. The temperature at the boundary $y=b$ is known and is given by $f(x)$.

Of interest is the steady-state temperature distribution within the fin, from which the heat flux (the rate of heat transfer per unit area) through the base of the fin can also be determined.

The heat equation for the two-dimensional steady-state problem is

$$\frac{\partial^2 T}{\partial x^2}+\frac{\partial^2 T}{\partial y^2}=0, \quad 0\leqslant x\leqslant a, \; 0\leqslant y\leqslant b,$$

Two-dimensional Laplace's equation

with the boundary conditions

$$\text{at } x=0:\; k\frac{\partial T}{\partial x}=h_1(T-T_f), \qquad \text{at } x=a:\; -k\frac{\partial T}{\partial x}=h_2(T-T_f),$$

$$\text{at } y=0:\; k\frac{\partial T}{\partial y}=h_3(T-T_f), \qquad \text{at } y=b:\quad T=f(x).$$

Consider the special case $h_1\to\infty$, $h_2\to\infty$, $h_3\to\infty$, $f(x)=T_0$. Very large heat transfer coefficient h implies "perfect" thermal contact, allowing one to set the surface temperature of the solid equal to the fluid temperature T_f in contact with the solid. Hence, the boundary conditions become

$$T(0,y)=T_f, \quad T(a,y)=T_f, \quad T(x,0)=T_f, \quad T(x,b)=T_0,$$

in which all four of the boundary conditions are nonhomogenous.

To apply the method of separation of variables to solve this two-dimensional steady-state heat conduction problem, at least three of the four boundary conditions must be homogeneous. Hence, by defining a new variable $\hat{T}(x,y)=T(x,y)-T_f$, the two-dimensional steady-state problem becomes

$$\frac{\partial^2 \hat{T}}{\partial x^2}+\frac{\partial^2 \hat{T}}{\partial y^2}=0,$$

with the boundary conditions

$$\hat{T}(0,y)=0, \quad \hat{T}(a,y)=0, \quad \hat{T}(x,0)=0, \quad \hat{T}(x,b)=T_0-T_f.$$

Applying the method of separation of variables, letting

$$\hat{T}(x, y) = X(x) \cdot Y(y),$$

and substituting in the two-dimensional Laplace's equation yield

$$\frac{\mathrm{d}^2 X(x)}{\mathrm{d}x^2} Y(y) + X(x) \frac{\mathrm{d}^2 Y(y)}{\mathrm{d}y^2} = 0.$$

Dividing the equation by $X(x)\,Y(y)$ leads to

$$\underbrace{\frac{1}{X(x)} \frac{\mathrm{d}^2 X(x)}{\mathrm{d}x^2}}_{\text{A function of } x \text{ only}} = \underbrace{-\frac{1}{Y(y)} \frac{\mathrm{d}^2 Y(y)}{\mathrm{d}y^2}}_{\text{A function of } y \text{ only}} = -\omega^2,$$

in which, for a function of x only to be equal to a function of y only, each of them must be equal to the same constant $-\omega^2$.

The X-equation gives

$$\frac{\mathrm{d}^2 X}{\mathrm{d}x^2} + \omega^2 X = 0, \quad \text{✍ Second-order linear ODE}$$

and the characteristic equation is $\lambda^2 + \omega^2 = 0 \Longrightarrow \lambda = \pm \mathrm{i}\omega$. The solution is

$$X(x) = A \cos \omega x + B \sin \omega x.$$

The Y-equation gives

$$\frac{\mathrm{d}^2 Y}{\mathrm{d}y^2} - \omega^2 Y = 0, \quad \text{✍ Second-order linear ODE}$$

and the characteristic equation is $\lambda^2 - \omega^2 = 0 \Longrightarrow \lambda = \pm \omega$. The solution is

$$Y(y) = C \cosh \omega y + D \sinh \omega y.$$

The solution of the two-dimensional Laplace's equation is

$$\hat{T}(x, y) = X(x)\,Y(y) = (A \cos \omega x + B \sin \omega x)(C \cosh \omega y + D \sinh \omega y),$$

in which the constants A, B, C, D, and $\omega > 0$ are constants to be determined using the boundary conditions.

Apply the boundary conditions

$$\hat{T}(0, y) = A \cdot (C \cosh \omega y + D \sinh \omega y) = 0 \Longrightarrow A = 0,$$

$$\hat{T}(x, 0) = (B \sin \omega x) \cdot C = 0 \Longrightarrow C = 0.$$

✍ B cannot be zero; otherwise, it will lead to zero solution.

The solution becomes

$$\hat{T}(x, y) = B \sin \omega x \cdot D \sinh \omega y = F \sin \omega x \sinh \omega y. \qquad \text{Rename } F = B \cdot D$$

Use the boundary condition

$$\hat{T}(a, y) = F \sin \omega a \sinh \omega y = 0 \implies \sin \omega a = 0 \implies \omega a = n\pi, \; n = 1, 2, \ldots,$$

$$\therefore \quad \omega_n = \frac{n\pi}{a}, \quad n = 1, 2, \ldots \implies \hat{T}_n(x, y) = F_n \sin \omega_n x \sinh \omega_n y.$$

Since the heat conduction problem is linear, any linear combination of solutions is also a solution. Hence, the general solution is

$$\hat{T}(x, y) = \sum_{n=1}^{\infty} \hat{T}_n(x, y) = \sum_{n=1}^{\infty} F_n \sin \frac{n\pi x}{a} \sinh \frac{n\pi y}{a}.$$

Apply the boundary condition

$$\hat{T}(x, b) = \sum_{n=1}^{\infty} F_n \sinh \frac{n\pi b}{a} \cdot \sin \frac{n\pi x}{a} = T_0 - T_f,$$

which is expressed in Fourier sine series in x. Noting that

$$\int_0^a \sin \frac{m\pi x}{a} \, dx = \frac{a}{m\pi}[1 - (-1)^m], \qquad \int_0^a \sin \frac{n\pi x}{a} \sin \frac{m\pi x}{a} \, dx = \begin{cases} 0, & n \neq m, \\ \frac{1}{2}a, & n = m, \end{cases}$$

multiplying the equation by $\sin \frac{m\pi x}{a}$, $m = 1, 2, \ldots$, integrating with respect to x from 0 to a yield

$$\sum_{n=1}^{\infty} F_n \sinh \frac{n\pi b}{a} \int_0^a \sin \frac{n\pi x}{a} \sin \frac{m\pi x}{a} \, dx = (T_0 - T_f) \int_0^a \sin \frac{m\pi x}{a} \, dx,$$

$$F_m \sinh \frac{m\pi b}{a} \cdot \frac{a}{2} = (T_0 - T_f) \frac{a}{m\pi} [1 - (-1)^m],$$

$$\therefore \quad F_m = \begin{cases} 0, & m = 2n, \quad n = 1, 2, \ldots \\ \dfrac{4(T_0 - T_f)}{m\pi \sinh \dfrac{m\pi b}{a}}, & m = 2n - 1, \quad n = 1, 2, \ldots \end{cases}$$

Hence, the distribution of temperature of the two-dimensional steady-state heat conduction is

$$\begin{aligned} T(x, y) &= T_f + \hat{T}(x, y) \\ &= T_f + \frac{4(T_0 - T_f)}{\pi} \sum_{n=1}^{\infty} \frac{\sin \dfrac{(2n-1)\pi x}{a} \sinh \dfrac{(2n-1)\pi y}{a}}{(2n-1) \sinh \dfrac{(2n-1)\pi b}{a}}. \end{aligned}$$

11.4.3 One-Dimensional Transient Heat Conduction

Consider a wall or plate of infinite size and of thickness L, as shown in Figure 11.7, which is suddenly exposed to fluids in motion on both of its surfaces. The coefficient of thermal conductivity of the wall or plate is k. Suppose the wall has an initial temperature distribution $T(x, 0) = f(x)$. The temperatures of the fluids and the heat transfer coefficients on the left-hand and right-hand sides of the wall are T_{f1}, h_1 and T_{f2}, h_2, respectively.

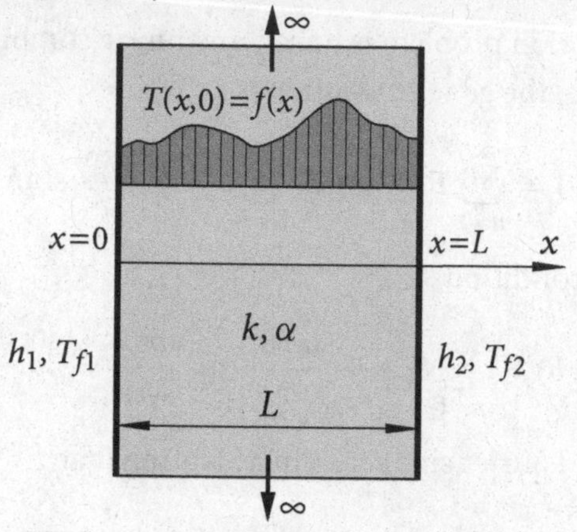

Figure 11.7 An infinite wall.

Because the wall or plate is infinitely large, the heat transfer process is simplified as one-dimensional (in the x-dimension).

The differential equation (Fourier's equation in one-dimension), the initial condition, and the boundary conditions of this one-dimensional transient heat conduction problem are

$$\frac{\partial T}{\partial t} = \alpha \frac{\partial^2 T}{\partial x^2}, \qquad 0 \leqslant x \leqslant L, \quad t \geqslant 0,$$

Initial Condition (IC): $T = f(x)$, at $t = 0$,

Boundary Conditions (BCs): $k \dfrac{\partial T}{\partial x} = h_1 (T - T_{f1})$, at $x = 0$,

$$-k \frac{\partial T}{\partial x} = h_2 (T - T_{f2}), \qquad \text{at } x = L.$$

This mathematical model has many engineering applications.

- The infinite wall is a model of a flat wall of a heat exchanger, which is initially isothermal at $T = T_0$. The operation of the heat exchanger is initiated at $t = 0$; two different fluids of temperatures T_{f1} and T_{f2}, respectively, are flowing along the sides of the wall.
- The infinite wall is a model of a wall in a building or a furnace. One side of the wall is suddenly exposed to a higher temperature T_{f1} due to fire occurring in a room or the ignition of flames in the furnace.

❧ The infinite plate is a model of a large plate, whose thickness L is much smaller than the other two dimensions. Immediately after it is manufactured, the hot plate is immersed in a cold liquid bath of temperature $T_{f1}=T_{f2}=T_f$. This heat treatment process is known as the quenching process.

Suppose $T(x,0)=T_0$ and the heat transfer coefficients $h_1=h_2=h\to\infty$. The initial and boundary conditions become

$$T(x,0)=T_0, \qquad T(0,t)=T_{f1}, \qquad T(L,t)=T_{f2},$$

in which both of the boundary conditions are not homogeneous. To apply the method of separation of variables, both boundary conditions must be converted to homogeneous boundary conditions. Let

$$u(x)=T_{f1}+\frac{T_{f2}-T_{f1}}{L}\,x \implies u(0)=T_{f1}, \quad u(L)=T_{f2}.$$

Defining a new variable $\hat{T}(x,t)=T(x,t)-u(x)$, the differential equation, initial condition, and boundary conditions become

$$\frac{1}{\alpha}\frac{\partial \hat{T}}{\partial t}=\frac{\partial^2 \hat{T}}{\partial x^2},$$

$$\text{IC:} \quad \hat{T}(x,0)=T(x,0)-u(x)=(T_0-T_{f1})-\frac{T_{f2}-T_{f1}}{L}x,$$

$$\text{BCs:} \quad \hat{T}(0,t)=T(0,t)-u(0)=0,$$

$$\hat{T}(L,t)=T(L,t)-u(L)=0.$$

Applying the method of separation of variables, letting $\hat{T}(x,t)=X(x)\cdot V(t)$, and substituting in the differential equation yield

$$\frac{1}{\alpha}X(x)\frac{\mathrm{d}V(t)}{\mathrm{d}t}=\frac{\mathrm{d}^2X(x)}{\mathrm{d}x^2}V(t).$$

Dividing the equation by $X(x)\,V(t)$ leads to

$$\underbrace{\frac{1}{\alpha}\frac{1}{V(t)}\frac{\mathrm{d}V(t)}{\mathrm{d}t}}_{\text{A function of } t \text{ only}}=\underbrace{\frac{1}{X(x)}\frac{\mathrm{d}^2X(x)}{\mathrm{d}x^2}}_{\text{A function of } x \text{ only}}=-\omega^2,$$

in which, for a function of t only to be equal to a function of x only, each of them must be equal to the same constant $-\omega^2$.

The X-equation gives

$$\frac{\mathrm{d}^2X}{\mathrm{d}x^2}+\omega^2X=0, \qquad \text{✍ Second-order linear ODE}$$

and the characteristic equation is $\lambda^2+\omega^2=0 \Longrightarrow \lambda=\pm i\omega$. The solution is

$$X(x) = A\cos\omega x + B\sin\omega x.$$

The V-equation gives

$$\frac{1}{V}\,dV = -\alpha\omega^2\,dt, \quad \text{✍ First-order ODE, variable separable}$$

and, by integrating both sides, the solution is

$$\ln|V| = -\alpha\omega^2 t + C \Longrightarrow V(t) = Ce^{-\alpha\omega^2 t}.$$

The solution of the partial differential equation is

$$\hat{T}(x,t) = X(x)\,V(t) = e^{-\alpha\omega^2 t}(A\cos\omega x + B\sin\omega x),$$

in which the constants A, B, and $\omega>0$ are constants to be determined using the initial and boundary conditions.

Apply the boundary conditions

$$\hat{T}(0,t) = e^{-\alpha\omega^2 t}\cdot A = 0 \Longrightarrow A=0,$$

$$\hat{T}(L,t) = e^{-\alpha\omega^2 t}\cdot B\sin\omega L = 0 \Longrightarrow \sin\omega L=0 \Longrightarrow \omega L=n\pi, \; n=1,2,\ldots,$$

✍ B cannot be zero; otherwise, it will lead to zero solution.

$$\therefore \quad \omega_n = \frac{n\pi}{L}, \; n=1,2,\ldots \Longrightarrow \hat{T}_n(x,t) = B_n e^{-\alpha\omega_n^2 t}\sin\omega_n x.$$

Since the heat conduction problem is linear, any linear combination of solutions is also a solution. Hence, the general solution is

$$\hat{T}(x,t) = \sum_{n=1}^{\infty}\hat{T}_n(x,t) = \sum_{n=1}^{\infty}B_n e^{-\alpha\omega_n^2 t}\sin\frac{n\pi x}{L}.$$

Apply the initial condition

$$\hat{T}(x,0) = \sum_{n=1}^{\infty}B_n\sin\frac{n\pi x}{L} = (T_0-T_{f1}) - \frac{T_{f2}-T_{f1}}{L}x,$$

which is expressed in Fourier sine series in x.

Multiplying the equation by $\sin\dfrac{m\pi x}{L}$, $m=1,2,\ldots$, integrating with respect to x from 0 to L, and using the orthogonality condition of the sine function yield

$$\sum_{n=1}^{\infty}B_n\int_0^L \sin\frac{n\pi x}{L}\sin\frac{m\pi x}{L}\,dx = \int_0^L\left[(T_0-T_{f1}) - \frac{T_{f2}-T_{f1}}{L}x\right]\sin\frac{m\pi x}{L}\,dx,$$

$$B_m\cdot\frac{L}{2} = \frac{L}{m\pi}\big[(T_0-T_{f1}) + (-1)^{m+1}(T_0-T_{f2})\big],$$

$$\therefore \quad B_m = \frac{2}{m\pi}\left[(T_0 - T_{f1}) + (-1)^{m+1}(T_0 - T_{f2})\right], \quad m = 1, 2, \ldots.$$

Hence, the temperature of the one-dimensional transient heat conduction is

$$T(x,t) = u(x) + \hat{T}(x,t)$$
$$= T_{f1} + \frac{T_{f2} - T_{f1}}{L}x + \sum_{n=1}^{\infty} \frac{2}{n\pi}\left[(T_0 - T_{f1}) + (-1)^{n+1}(T_0 - T_{f2})\right] \times \exp\left\{-\alpha\left(\frac{n\pi}{L}\right)^2 t\right\} \sin\frac{n\pi x}{L}.$$

11.4.4 *One-Dimensional Transient Heat Conduction on a Semi-Infinite Interval*

In areas where the atmospheric temperature remains below 0°C for prolonged periods of time, the freezing of water in underground pipes is a major concern. Fortunately, the soil remains relatively warm during those periods, and it takes weeks for the subfreezing temperatures to reach the water mains in the ground. Thus, the soil effectively serves as an insulation to protect the water from the freezing atmospheric temperature.

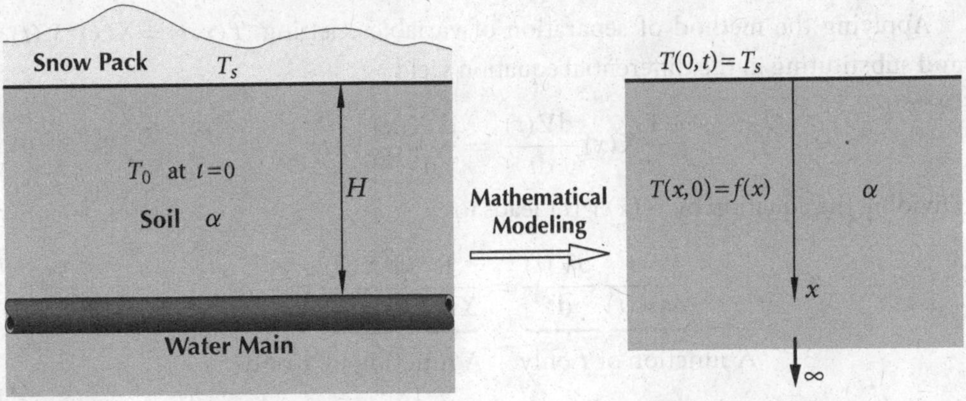

Figure 11.8 One-dimensional transient heat conduction on a semi-infinite interval.

Figure 11.8 shows the ground at a particular location covered with snow pack at temperature T_s. The average thermal diffusivity of the soil is α at that location. If the ground has an initial uniform temperature of T_0, it is of interest to know the change of temperature with time so that the minimum burial depth H of the water mains can be determined to prevent the water from freezing.

The problem can be modeled as heat conduction on a semi-infinite body as shown in Figure 11.8, which is similar to the heat conduction problem of an infinite plate considered in Section 11.4.3 except that the thickness of the plate $L \to \infty$. The temperature changes with time and the depth x, and is therefore a one-dimensional heat conduction problem.

The differential equation (Fourier's equation in one-dimension), the initial condition, and the boundary condition of this one-dimensional transient heat conduction problem are

$$\frac{\partial T}{\partial t} = \alpha \frac{\partial^2 T}{\partial x^2}, \qquad 0 \leqslant x < \infty, \quad t \geqslant 0,$$

$$\text{Initial Condition (IC)}: \qquad T(x,0) = T_0,$$

$$\text{Boundary Condition (BC)}: \qquad T(0,t) = T_s,$$

in which the boundary condition is not homogeneous. To apply the method of separation of variables, the boundary condition must be converted to homogeneous boundary condition.

Defining a new variable $\hat{T}(x,t) = T(x,t) - T_s$, the differential equation, initial condition, and boundary conditions become

$$\frac{1}{\alpha}\frac{\partial \hat{T}}{\partial t} = \frac{\partial^2 \hat{T}}{\partial x^2}, \qquad 0 \leqslant x < \infty, \quad t > 0,$$

$$\text{IC}: \quad \hat{T}(x,0) = T(x,0) - T_s = T_0 - T_s,$$

$$\text{BC}: \quad \hat{T}(0,t) = 0.$$

Applying the method of separation of variables, letting $\hat{T}(x,t) = X(x) \cdot V(t)$, and substituting in the differential equation yield

$$\frac{1}{\alpha} X(x) \frac{\mathrm{d}V(t)}{\mathrm{d}t} = \frac{\mathrm{d}^2 X(x)}{\mathrm{d}x^2} V(t).$$

Dividing the equation by $X(x)\,V(t)$ leads to

$$\underbrace{\frac{1}{\alpha}\frac{1}{V(t)}\frac{\mathrm{d}V(t)}{\mathrm{d}t}}_{\text{A function of } t \text{ only}} = \underbrace{\frac{1}{X(x)}\frac{\mathrm{d}^2 X(x)}{\mathrm{d}x^2}}_{\text{A function of } x \text{ only}} = -\omega^2,$$

in which, for a function of t only to be equal to a function of x only, each of them must be equal to the same constant $-\omega^2$.

The X-equation gives

$$\frac{\mathrm{d}^2 X}{\mathrm{d}x^2} + \omega^2 X = 0, \qquad \text{✍ Second-order linear ODE}$$

and the characteristic equation is $\lambda^2 + \omega^2 = 0 \implies \lambda = \pm \mathrm{i}\,\omega$. The solution is

$$X(x) = A \cos \omega x + B \sin \omega x.$$

The boundary condition $\hat{T}(0,t) = X(0)\,V(t) = 0$ leads to $X(0) = 0$. Hence,

$$X(0) = A = 0 \implies X(x) = B \sin \omega x.$$

The V-equation gives

$$\frac{1}{V}\,\mathrm{d}V = -\alpha\omega^2\,\mathrm{d}t,$$ ✍ First-order ODE, variable separable

and, by integrating both sides, the solution is

$$\ln|V| = -\alpha\omega^2 t + C \implies V(t) = C\mathrm{e}^{-\alpha\omega^2 t}.$$

The solution of the partial differential equation is

$$\hat{T}(x,t) = X(x)\,V(t) = B\sin\omega x\cdot C\mathrm{e}^{-\alpha\omega^2 t} \implies \hat{T}(x,t) = B(\omega)\,\mathrm{e}^{-\alpha\omega^2 t}\sin\omega x,$$

in which the coefficient B depends on ω. Since this solution is valid for all values of $\omega > 0$, the general solution is obtained by integrating over $0 \leqslant \omega < \infty$:

$$\hat{T}(x,t) = \int_0^\infty B(\omega)\,\mathrm{e}^{-\alpha\omega^2 t}\sin\omega x\,\mathrm{d}\omega.$$

Remarks: For one-dimensional heat conduction in a plate of finite thickness L (Section 11.4.3), there are countably infinitely many values of ω, i.e., $\omega = n\pi/L$, $n = 1, 2, \ldots$. The solution of the heat conduction equation is expressed as a Fourier series in x. For one-dimensional heat conduction on a semi-infinite interval, the thickness of the plate $L \to \infty$. There are uncountably infinitely many values of ω. The summation in the Fourier series becomes integration, and the Fourier series in x becomes Fourier integral in x.

Fourier Integral

For a nonperiodic continuous function $f(x)$, $-\infty < x < +\infty$, the corresponding *Fourier integral* can be written as

$$f(x) = \int_0^\infty \big[a(\omega)\cos\omega x + b(\omega)\sin\omega x\big]\,\mathrm{d}\omega,$$

where

$$a(\omega) = \frac{1}{\pi}\int_{-\infty}^{+\infty} f(x)\cos\omega x\,\mathrm{d}x, \qquad b(\omega) = \frac{1}{\pi}\int_{-\infty}^{+\infty} f(x)\sin\omega x\,\mathrm{d}x.$$

- If $f(x)$ is an even function in $-\infty < x < +\infty$, i.e., $f(-x) = f(x)$, then $b(\omega) = 0$

$$f(x) = \int_0^\infty a(\omega)\cos\omega x\,\mathrm{d}\omega, \qquad a(\omega) = \frac{2}{\pi}\int_0^{+\infty} f(x)\cos\omega x\,\mathrm{d}x,$$

which is called the *Fourier cosine integral*.

- If $f(x)$ is an odd function in $-\infty < x < +\infty$, i.e., $f(-x) = -f(x)$, then $a(\omega) = 0$

$$f(x) = \int_0^\infty b(\omega)\sin\omega x\,\mathrm{d}\omega, \qquad b(\omega) = \frac{2}{\pi}\int_0^{+\infty} f(x)\sin\omega x\,\mathrm{d}x,$$

which is called the *Fourier sine integral*.

Apply the initial condition

$$\hat{T}(x,0) = \int_0^\infty B(\omega)\sin\omega x\,\mathrm{d}\omega = T_0 - T_s,$$

which is a Fourier sine integral, and

$$B(\omega) = \frac{2(T_0 - T_s)}{\pi}\int_0^\infty \sin\omega\xi\,\mathrm{d}\xi.$$

Hence, the solution is

$$\hat{T}(x,t) = \int_{\omega=0}^\infty \left[\frac{2(T_0-T_s)}{\pi}\int_{\xi=0}^\infty \sin\omega\xi\,\mathrm{d}\xi\right]\mathrm{e}^{-\alpha\omega^2 t}\sin\omega x\,\mathrm{d}\omega$$

$$= \frac{2(T_0-T_s)}{\pi}\int_{\xi=0}^\infty\left[\int_{\omega=0}^\infty \sin\omega\xi\,\sin\omega x\cdot\mathrm{e}^{-\alpha\omega^2 t}\,\mathrm{d}\omega\right]\mathrm{d}\xi.$$

Noting that

$$\sin\omega x\sin\omega\xi = \tfrac{1}{2}\left[\cos\omega(x-\xi) - \cos\omega(x+\xi)\right],$$

$$\int_0^\infty \mathrm{e}^{-a\omega^2}\cos\omega x\,\mathrm{d}\omega = \sqrt{\frac{\pi}{4a}}\exp\left(-\frac{x^2}{4a}\right),$$

one has

$$\int_0^\infty \sin\omega\xi\,\sin\omega x\cdot\mathrm{e}^{-\alpha\omega^2 t}\,\mathrm{d}\omega$$

$$= \frac{1}{2}\left[\int_0^\infty \cos\omega(x-\xi)\,\mathrm{e}^{-\alpha\omega^2 t}\,\mathrm{d}\omega - \int_0^\infty \cos\omega(x+\xi)\,\mathrm{e}^{-\alpha\omega^2 t}\,\mathrm{d}\omega\right]$$

$$= \frac{1}{2}\sqrt{\frac{\pi}{4\alpha t}}\left[\exp\left\{-\frac{(x-\xi)^2}{4\alpha t}\right\} - \exp\left\{-\frac{(x+\xi)^2}{4\alpha t}\right\}\right].$$

Hence, the solution becomes

$$\hat{T}(x,t) = \frac{T_0 - T_s}{\sqrt{4\pi\alpha t}}\int_0^\infty\left[\exp\left\{-\frac{(x-\xi)^2}{4\alpha t}\right\} - \exp\left\{-\frac{(x+\xi)^2}{4\alpha t}\right\}\right]\mathrm{d}\xi.$$

Using the *error function* and the *complementary error function* defined as

$$\mathrm{erf}(x) = \frac{2}{\sqrt{\pi}}\int_0^x \mathrm{e}^{-t^2}\mathrm{d}t, \qquad \mathrm{erfc}(x) = 1 - \mathrm{erf}(x) = \frac{2}{\sqrt{\pi}}\int_x^\infty \mathrm{e}^{-t^2}\mathrm{d}t,$$

one has

$$\int_{\xi=0}^\infty \exp\left\{-\frac{(x-\xi)^2}{4\alpha t}\right\}\mathrm{d}\xi = \int_{\eta=-\frac{x}{\sqrt{4\alpha t}}}^\infty \mathrm{e}^{-\eta^2}\sqrt{4\alpha t}\,\mathrm{d}\eta, \qquad \text{✍ Letting } \eta = \frac{\xi - x}{\sqrt{4\alpha t}}$$

$$= \sqrt{\pi\alpha t}\cdot\frac{2}{\sqrt{\pi}}\int_{\eta=-\frac{x}{\sqrt{4\alpha t}}}^\infty \mathrm{e}^{-\eta^2}\mathrm{d}\eta = \sqrt{\pi\alpha t}\left[1 - \mathrm{erf}\left(-\frac{x}{\sqrt{4\alpha t}}\right)\right].$$

The solution can then be simplified as

$$\hat{T}(x,t) = \frac{T_0 - T_s}{\sqrt{4\pi\alpha t}} \cdot \sqrt{\pi\alpha t} \left\{ \left[1 - \text{erf}\left(-\frac{x}{\sqrt{4\alpha t}} \right) \right] - \left[1 - \text{erf}\left(\frac{x}{\sqrt{4\alpha t}} \right) \right] \right\}$$

$$= \frac{T_0 - T_s}{2} \left\{ \text{erf}\left(\frac{x}{\sqrt{4\alpha t}} \right) - \text{erf}\left(-\frac{x}{\sqrt{4\alpha t}} \right) \right\}, \quad \text{erf}(-x) = -\text{erf}(x)$$

$$\therefore \quad T(x,t) = T_s + \hat{T}(x,t) = T_s + (T_0 - T_s)\,\text{erf}\left(\frac{x}{\sqrt{4\alpha t}} \right).$$

Example 11.6

The ground at a particular location is covered with snow pack at $T_s = -10°\text{C}$ for a continuous period of three months (90 days). The average thermal diffusivity of the soil is $\alpha = 0.15 \times 10^{-6}\ \text{m}^2/\text{sec}$ at that location. Assuming an initial uniform temperature of $T_0 = 15°\text{C}$ for the ground, determine the minimum burial depth H of the water mains to prevent the water from freezing.

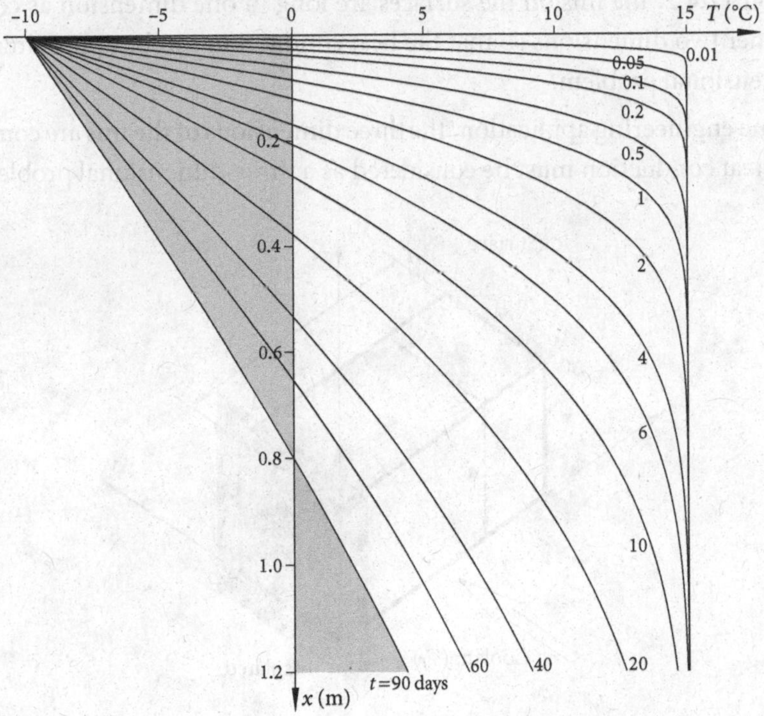

Figure 11.9 Distribution of temperature in the soil.

Given that $T_s = -10°\text{C}$ and $T_0 = 15°\text{C}$, it is required that $T(H,t) > 0$ for $t \leqslant 90$ days, i.e., $t \leqslant 90 \times 24 \times 3600 = 0.7776 \times 10^7$ sec. Since

$$T(x,t) = T_s + (T_0 - T_s)\,\text{erf}\left(\frac{x}{\sqrt{4\alpha t}} \right),$$

one has

$$T(H,t) = T_s + (T_0 - T_s)\,\text{erf}\Big(\frac{H}{\sqrt{4\alpha t}}\Big) > 0,$$

$$\text{erf}\left(\frac{H}{\sqrt{4\alpha t}}\right) > -\frac{T_s}{T_0 - T_s} = -\frac{-10}{15-(-10)} = 0.4.$$

Using the Table of Error Functions or a mathematical software, such as *Maple* (see Chapter 12), it can be found that

$$\frac{H}{\sqrt{4\alpha t}} > 0.3708 \implies H > 0.3708\sqrt{4\cdot(0.15\times10^{-6})\cdot(0.7776\times10^{7})} = 0.80\ \text{m}.$$

Hence, the minimum burial depth of the water mains is 0.80 m to prevent the water from freezing. Changes of the soil temperature T with the depth x are shown in Figure 11.9 for various values of time t.

11.4.5 Three-Dimensional Steady-State Heat Conduction

In Section 11.4.2, the fins on the surfaces are long in one dimension as compared to the other two dimensions so that the heat conduction can be approximated as a two-dimensional problem.

In some engineering application, the three dimensions of the fins are comparable and the heat conduction must be considered as a three-dimensional problem.

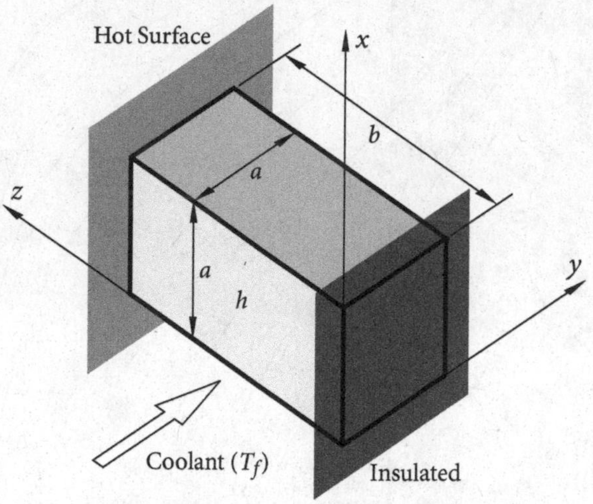

Figure 11.10 Finned surface and three-dimensional heat conduction.

In Figure 11.10, a fin of length b with square cross-section $a \times a$ is mounted on a hot surface, while the other end of the fin is *insulated*. The coefficient of thermal conductivity of the fin is k. The temperature distribution at the hot surface is given by $T(x, y, b) = f(x, y)$. A fluid coolant of temperature T_f flows past the fin. The heat transfer coefficient between the fin surface and the coolant is h. Of interest is the steady-state temperature distribution within the fin.

The heat equation for the three-dimensional steady-state problem is given by the three-dimensional Laplace's equation

$$\frac{\partial^2 T}{\partial x^2}+\frac{\partial^2 T}{\partial y^2}+\frac{\partial^2 T}{\partial z^2}=0,\quad 0\leqslant x\leqslant a,\ 0\leqslant y\leqslant a,\ 0\leqslant z\leqslant b,$$

with the boundary conditions

$$\text{at } x=0:\quad k\frac{\partial T}{\partial x}=h(T-T_f),\qquad \text{at } x=a:\quad -k\frac{\partial T}{\partial x}=h(T-T_f),$$

$$\text{at } y=0:\quad k\frac{\partial T}{\partial y}=h(T-T_f),\qquad \text{at } y=a:\quad -k\frac{\partial T}{\partial y}=h(T-T_f),$$

$$\text{at } z=0:\quad \frac{\partial T}{\partial z}=0,\qquad \text{at } z=b:\quad T=f(x,y).$$

Consider the special case $f(x,y)=T_0$ and $h\to\infty$. The boundary conditions become

$$T(0,y,z)=T_f,\qquad T(a,y,z)=T_f,$$
$$T(x,0,z)=T_f,\qquad T(x,a,z)=T_f,$$
$$\frac{T(x,y,0)}{\partial z}=0,\qquad T(x,y,b)=T_0.$$

in which all six of the boundary conditions are nonhomogenous.

To apply the method of separation of variables to solve this three-dimensional steady-state heat conduction problem, five of the six boundary conditions must be homogeneous. Hence, by defining a new variable $\hat{T}(x,y,z)=T(x,y,z)-T_f$, the three-dimensional steady-state problem becomes

$$\frac{\partial^2 \hat{T}}{\partial x^2}+\frac{\partial^2 \hat{T}}{\partial y^2}+\frac{\partial^2 \hat{T}}{\partial z^2}=0,$$

with the boundary conditions

$$\hat{T}(0,y,z)=0,\qquad \hat{T}(a,y,z)=0,$$
$$\hat{T}(x,0,z)=0,\qquad \hat{T}(x,a,z)=0,$$
$$\frac{\hat{T}(x,y,0)}{\partial z}=0,\qquad \hat{T}(x,y,b)=T_0-T_f.$$

Applying the method of separation of variables, letting

$$\hat{T}(x,y,z)=X(x)\cdot Y(y)\cdot Z(z),$$

and substituting in the three-dimensional Laplace's equation yield

$$\frac{\mathrm{d}^2X(x)}{\mathrm{d}x^2}Y(y)Z(z)+X(x)\frac{\mathrm{d}^2Y(y)}{\mathrm{d}y^2}Z(z)+X(x)Y(y)\frac{\mathrm{d}^2Z(z)}{\mathrm{d}z^2}=0.$$

Dividing the equation by $X(x)Y(y)Z(z)$ leads to

$$\underbrace{\frac{1}{X(x)}\frac{d^2X(x)}{dx^2}}_{\text{A function of } x \text{ only}} = \underbrace{-\left[\frac{1}{Y(y)}\frac{d^2Y(y)}{dy^2} + \frac{1}{Z(z)}\frac{d^2Z(z)}{dz^2}\right]}_{\text{A function of } y \text{ and } z \text{ only}} = -\omega^2,$$

in which, for a function of x only to be equal to a function of y and z only, each of them must be equal to the same constant $-\omega^2$. Furthermore,

$$\underbrace{\frac{1}{Y(y)}\frac{d^2Y(y)}{dy^2}}_{\text{A function of } y \text{ only}} = \underbrace{\omega^2 - \frac{1}{Z(z)}\frac{d^2Z(z)}{dz^2}}_{\text{A function of } z \text{ only}} = -\Omega^2,$$

in which, for a function of y only to be equal to a function of z only, each of them must be equal to the same constant $-\Omega^2$.

The X-equation gives

$$\frac{d^2X}{dx^2} + \omega^2 X = 0, \quad \text{✍ Second-order linear ODE}$$

and the characteristic equation is $\lambda^2 + \omega^2 = 0 \Longrightarrow \lambda = \pm i\,\omega$. The solution is

$$X(x) = A_1 \cos \omega x + B_1 \sin \omega x.$$

The Y-equation gives

$$\frac{d^2Y}{dy^2} + \Omega^2 Y = 0, \quad \text{✍ Second-order linear ODE}$$

and the characteristic equation is $\lambda^2 + \Omega^2 = 0 \Longrightarrow \lambda = \pm i\,\Omega$. The solution is

$$Y(y) = A_2 \cos \Omega y + B_2 \sin \Omega y.$$

The Z-equation gives

$$\frac{d^2Z}{dz^2} - (\omega^2 + \Omega^2) Z = 0, \quad \text{✍ Second-order linear ODE}$$

and the characteristic equation is $\lambda^2 - (\omega^2 + \Omega^2) = 0 \Longrightarrow \lambda = \pm \nu, \ \nu = \sqrt{\omega^2 + \Omega^2}$. The solution is

$$Z(z) = A_3 \cosh \nu z + B_3 \sinh \nu z.$$

The solution of the three-dimensional Laplace's equation is

$$\hat{T}(x, y, z) = X(x)\,Y(y)\,Z(z) = (A_1 \cos \omega x + B_1 \sin \omega x)\,(A_2 \cos \Omega y + B_2 \sin \Omega y) \times (A_3 \cosh \nu z + B_3 \sinh \nu z),$$

in which the constants $A_1, B_1, A_2, B_2, A_3, B_3, \omega > 0$, and $\Omega > 0$ are constants to be determined using the boundary conditions.

Apply the boundary conditions

$$\begin{aligned}
\hat{T}(0, y, z) &= A_1 \cdot (A_2 \cos \Omega y + B_2 \sin \Omega y)(A_3 \cosh \nu z + B_3 \sinh \nu z) \\
&= 0 \implies A_1 = 0, \\
\hat{T}(x, 0, z) &= (B_1 \sin \omega x) \cdot A_2 \cdot (A_3 \cosh \nu z + B_3 \sinh \nu z) = 0 \implies A_2 = 0, \\
\left.\frac{\partial \hat{T}}{\partial z}\right|_{z=0} &= (B_1 \sin \omega x)(B_2 \sin \Omega y) \cdot \nu (A_3 \sinh \nu z + B_3 \cosh \nu z)\Big|_{z=0} \\
&= (B_1 \sin \omega x)(B_2 \sin \Omega y) \cdot \nu B_3 = 0 \implies B_3 = 0.
\end{aligned}$$

The solution becomes

$$\hat{T}(x, y, z) = B_1 \sin \omega x \cdot B_2 \sin \Omega y \cdot A_3 \cosh \nu z = F \sin \omega x \cdot \sin \Omega y \cdot \cosh \nu z.$$

Use the boundary conditions

$$\begin{aligned}
&\hat{T}(a, y, z) = F \sin \omega a \cdot \sin \Omega y \cdot \cosh \nu z = 0 \implies \sin \omega a = 0, \\
&\therefore \quad \omega a = m\pi, \quad m = 1, 2, \ldots \implies \omega_m = \frac{m\pi}{a}, \quad m = 1, 2, \ldots, \\
&\hat{T}(x, a, z) = F \sin \omega x \cdot \sin \Omega a \cdot \cosh \nu z = 0 \implies \sin \Omega a = 0, \\
&\therefore \quad \Omega a = n\pi, \quad n = 1, 2, \ldots \implies \Omega_n = \frac{n\pi}{a}, \quad n = 1, 2, \ldots,
\end{aligned}$$

which gives, for $m = 1, 2, \ldots$, and $n = 1, 2, \ldots$,

$$\hat{T}_{mn}(x, y, z) = F_{mn} \sin \frac{m\pi x}{a} \cdot \sin \frac{n\pi y}{a} \cdot \cosh \frac{\sqrt{m^2 + n^2}\,\pi z}{a}.$$

Since the heat conduction problem is linear, any linear combination of the solutions is also a solution. Hence, the general solution is

$$\hat{T}(x, y, z) = \sum_{m=1}^{\infty} \sum_{n=1}^{\infty} F_{mn} \sin \frac{m\pi x}{a} \cdot \sin \frac{n\pi y}{a} \cdot \cosh \frac{\sqrt{m^2 + n^2}\,\pi z}{a}.$$

Apply the boundary condition

$$\begin{aligned}
\hat{T}(x, y, b) &= \sum_{m=1}^{\infty} \sum_{n=1}^{\infty} F_{mn} \sin \frac{m\pi x}{a} \cdot \sin \frac{n\pi y}{a} \cdot \cosh \frac{\sqrt{m^2 + n^2}\,\pi b}{a} \\
&= \sum_{m=1}^{\infty} \underbrace{\left(\sum_{n=1}^{\infty} F_{mn} \cosh \frac{\sqrt{m^2 + n^2}\,\pi b}{a} \cdot \sin \frac{n\pi y}{a} \right)}_{G_m} \sin \frac{m\pi x}{a} = T_0 - T_f,
\end{aligned}$$

which is expressed in Fourier sine series in x with G_m, $m=1,2,\ldots$, being the coefficients. Multiplying the equation by $\sin\dfrac{i\pi x}{a}$, $i=1,2,\ldots$, and integrating with respect to x from 0 to a yield

$$G_i\cdot\frac{a}{2}=(T_0-T_f)\cdot\frac{a}{i\pi}\big[1-(-1)^i\big] \implies G_i=\frac{2(T_0-T_f)}{i\pi}\big[1-(-1)^i\big],$$

$$\therefore \quad \sum_{n=1}^{\infty}\left(F_{in}\cosh\frac{\sqrt{i^2+n^2}\,\pi b}{a}\right)\sin\frac{n\pi y}{a}=G_i,$$

which is expressed in Fourier sine series in y. Multiplying the equation by $\sin\dfrac{j\pi y}{a}$, $j=1,2,\ldots$, and integrating with respect to y from 0 to a yield

$$F_{ij}\cosh\frac{\sqrt{i^2+j^2}\,\pi b}{a}\cdot\frac{a}{2}=G_i\cdot\frac{a}{j\pi}\big[1-(-1)^j\big],$$

$$\therefore \quad F_{ij}=\frac{4(T_0-T_f)}{ij\pi^2\cosh\dfrac{\sqrt{i^2+j^2}\,\pi b}{a}}\big[1-(-1)^i\big]\big[1-(-1)^j\big].$$

Hence, the temperature distribution of the steady-state heat conduction is

$$T(x,y,z)=T_f+\hat{T}(x,y,z)$$

$$=T_f+\frac{4(T_0-T_f)}{\pi^2}\sum_{i=1}^{\infty}\sum_{j=1}^{\infty}\frac{\big[1-(-1)^i\big]\big[1-(-1)^j\big]}{ij\cosh\dfrac{\sqrt{i^2+j^2}\,\pi b}{a}}\sin\frac{i\pi x}{a}\sin\frac{j\pi y}{a}\cosh\frac{\sqrt{i^2+j^2}\,\pi z}{a}$$

$$=T_f+\frac{16(T_0-T_f)}{\pi^2}\sum_{m=1}^{\infty}\sum_{n=1}^{\infty}\frac{1}{(2m-1)(2n-1)\cosh\dfrac{\sqrt{(2m-1)^2+(2n-1)^2}\,\pi b}{a}}$$

$$\times\sin\frac{(2m-1)\pi x}{a}\cdot\sin\frac{(2n-1)\pi y}{a}\cdot\cosh\frac{\sqrt{(2m-1)^2+(2n-1)^2}\,\pi z}{a}.$$

11.5 Summary

The method of separation of variables may be applied to solve certain partial differential equations, which have many important engineering applications. When the method is applicable, it converts a partial differential equation into a set of ordinary differential equations.

Consider a partial differential equation governing $u(x,y)$, the procedure of the method of separation of variables is as follows:

1. Separate the function $u(x,y)$, which is a function of x and y, to a product of $X(x)$ and $Y(y)$, i.e., $u(x,y)=X(x)\cdot Y(y)$.

2. Substitute into the partial differential equation. Manipulate the resulting equation so that one side of the equation is a function of x only, and the other side of the equation is a function of y only. For this to be true, both of them must be equal to the same constant k:

$$\underbrace{f\left(x, X, \frac{dX}{dx}, \frac{d^2X}{dx^2}, \ldots\right)}_{\text{A function of } x \text{ only}} = \underbrace{g\left(y, Y, \frac{dY}{dy}, \frac{d^2Y}{dy^2}, \ldots\right)}_{\text{A function of } y \text{ only}} = k.$$

3. One then obtains two *ordinary* differential equations:

$$f\left(x, X, \frac{dX}{dx}, \frac{d^2X}{dx^2}, \ldots\right) = k, \quad \text{and} \quad g\left(y, Y, \frac{dY}{dy}, \frac{d^2Y}{dy^2}, \ldots\right) = k.$$

The values of the constant k and the constant coefficients in the solutions are determined by the boundary and initial conditions. For linear problems, any linear combination of solutions is also a solution.

Problems

11.1 Find the solution to the following partial differential equation:

$$x\frac{\partial u}{\partial x} = u + y\frac{\partial u}{\partial y}; \quad u(1,1) = 2, \quad u(1,2) = 8.$$

ANS $u(x,y) = 2x^3y^2$

11.2 Find the solution of the heat conduction problem:

$$\frac{1}{4}\frac{\partial^2 u}{\partial x^2} = \frac{\partial u}{\partial t};$$

$$u(x,0) = 3\sin\frac{4\pi x}{L}, \quad \text{for } 0 \leqslant x \leqslant L; \quad u(0,t) = 0, \quad u(L,t) = 0, \quad \text{for } t > 0.$$

ANS $u(x,t) = 3\sin\dfrac{4\pi x}{L}\exp\left(\dfrac{-4\pi^2 t}{L^2}\right)$

11.3 Solve the following boundary value problem:

$$2\frac{\partial u}{\partial t} = \frac{\partial^2 u}{\partial x^2}; \quad u(0,t) = 0, \quad u(\pi,t) = 0, \quad u(x,0) = 2\sin 3x - 5\sin 4x.$$

ANS $u(x,t) = 2e^{-\frac{9}{2}t}\sin 3x - 5e^{-8t}\sin 4x$

11.4 Find the solution of the heat conduction problem:

$$\frac{\partial^2 u}{\partial x^2} = \frac{\partial u}{\partial t}; \quad u(0,t) = 0, \quad u(1,t) = 0, \quad u(x,0) = \sin 2\pi x - \sin 5\pi x.$$

ANS $u(x,t) = e^{-4\pi^2 t}\sin 2\pi x - e^{-25\pi^2 t}\sin 5\pi x$

11.5 Solve the following boundary value problem:

$$\frac{\partial^2 u}{\partial t^2} = \frac{\partial^2 u}{\partial x^2};$$

$$u(0,t)=0,\ u(L,t)=0,\ u(x,0)=0,\ u_t(x,0)=5\sin\frac{\pi x}{L}-3\sin\frac{3\pi x}{L}.$$

ANS $u(x,t) = \dfrac{5L}{\pi}\sin\dfrac{\pi x}{L}\sin\dfrac{\pi t}{L} - \dfrac{L}{\pi}\sin\dfrac{3\pi x}{L}\sin\dfrac{3\pi t}{L}$

11.6 Find the solution of the boundary value problem:

$$9\,\frac{\partial^2 u}{\partial t^2} = \frac{\partial^2 u}{\partial x^2};$$

$$u(0,t)=0,\ u(\pi,t)=0,\ u_t(x,0)=2\sin x-3\sin 2x,\ u(x,0)=0.$$

ANS $u(x,t) = 6\sin x\sin\dfrac{t}{3} - \dfrac{9}{2}\sin 2x\sin\dfrac{2t}{3}$

11.7 A cable is under constant tension F as shown in the following figure.

1. Show that the transverse vibration $v(x,t)$ is governed by the equation

$$\frac{\partial^2 v}{\partial t^2} = \frac{F}{m}\frac{\partial^2 v}{\partial x^2},$$

where m is the mass of the cable per unit length.

2. The cable is fixed at both ends and is imposed on the following initial conditions

$$v(x,0) = a\sin\frac{\pi x}{L},\qquad \left.\frac{\partial v(x,t)}{\partial t}\right|_{t=0} = 0.$$

Determine the response of the cable.

ANS $v(x,t) = a\sin\dfrac{\pi x}{L}\cos\left(\dfrac{\pi}{L}\sqrt{\dfrac{F}{m}}\,t\right)$

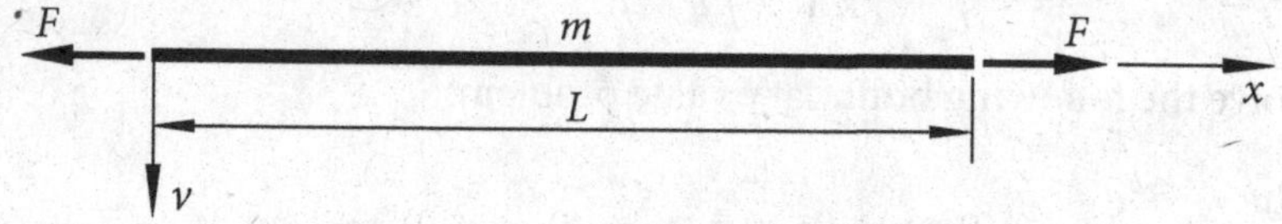

11.8 The simply supported beam is subjected to a constant moving concentrated load P as shown in the following figure. The load P is at the left end of the beam when $t=0$ and moves at constant speed U toward the right. The beam is at rest at time $t=0$. Adopt the following notations:

$$\Omega_m = \frac{m\pi U}{L},\qquad \omega_m = \left(\frac{m\pi}{L}\right)^2\sqrt{\frac{EI}{\rho A}},\qquad m=1,2,\ldots.$$

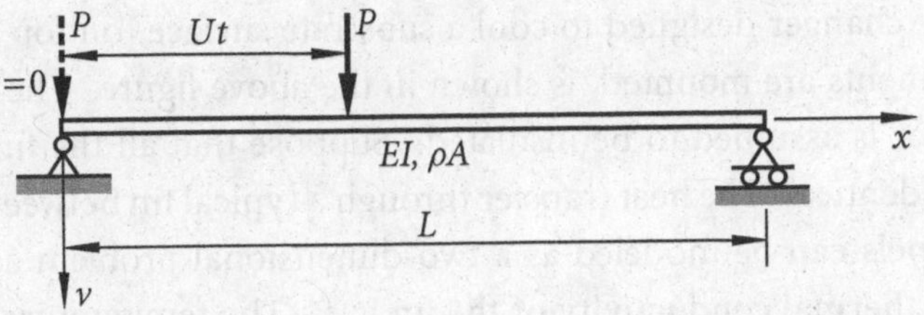

1. For the nonresonant case, i.e., $\Omega_m \neq \omega_m \implies U \neq \frac{m\pi}{L}\sqrt{\frac{EI}{\rho A}}$, $m = 1, 2, \ldots$, show that the response of the beam is, for $0 \leqslant t \leqslant \frac{L}{U}$, i.e., when the load is on the beam,

$$v(x,t) = \frac{2P}{\rho AL}\sum_{m=1}^{\infty}\frac{1}{\omega_m^2 - \Omega_m^2}\left(-\frac{\Omega_m}{\omega_m}\sin\omega_m t + \sin\Omega_m t\right)\sin\frac{m\pi x}{L}.$$

2. If the structure is resonant in the rth mode, i.e., $\omega_r = \Omega_r$, or $U = \frac{r\pi}{L}\sqrt{\frac{EI}{\rho A}}$, show that the response of the beam is, for $0 \leqslant t \leqslant \frac{L}{U}$,

$$v(x,t) = \frac{2P}{\rho AL}\left\{\sum_{\substack{m=1\\ m\neq r}}^{\infty}\frac{1}{\omega_m^2 - \Omega_m^2}\left(-\frac{\Omega_m}{\omega_m}\sin\omega_m t + \sin\Omega_m t\right)\sin\frac{m\pi x}{L} + \frac{1}{2\omega_r^2}(\sin\omega_r t - \omega_r t\cos\omega_r t)\sin\frac{r\pi x}{L}\right\}.$$

11.9 Electronic components mounted on substrate surfaces may result in high temperatures affecting the operation and reliability of electronic equipment. To ensure the reliable operation of the equipment, efficient removal of heat from the electronic components is desired.

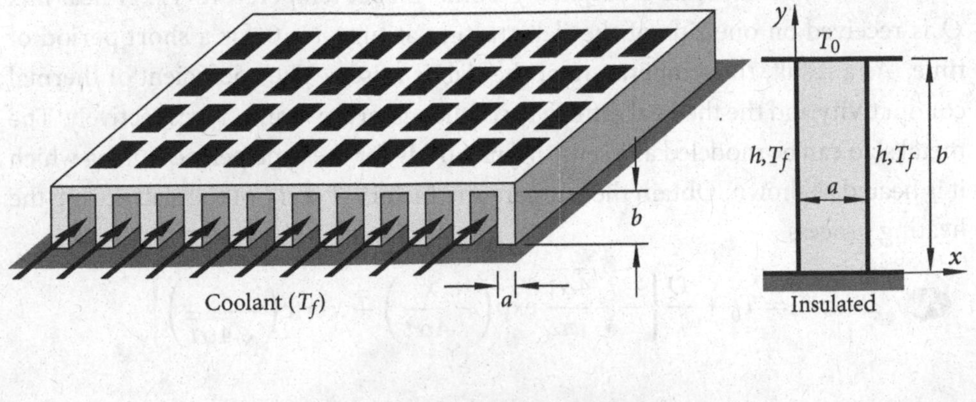

The heat exchanger designed to cool a substrate surface, on top of which electronic components are mounted, is shown in the above figure. The bottom of the heat exchanger is assumed to be insulated. Suppose that all the fins and coolant channels are identical. The heat transfer through a typical fin between two adjacent coolant channels can be modeled as a two-dimensional problem as shown. The coefficient of thermal conductivity of the fin is k. The temperature at the base of the fin (and the substrate surface) is T_0. The temperature of the coolant is T_f and the heat transfer coefficient $h \to \infty$.

Determine the steady-state temperature distribution in a typical fin.

ANS $$T(x,y) = T_f + \frac{4(T_0 - T_f)}{\pi} \sum_{n=1}^{\infty} \frac{\sin\dfrac{(2n-1)\pi x}{a} \cosh\dfrac{(2n-1)\pi y}{a}}{(2n-1)\cosh\dfrac{(2n-1)\pi b}{a}}$$

11.10 The infinite wall as shown in the following figure is a model of a wall in a furnace. Suppose the wall is initially isothermal at temperature T_0. The left-hand side of the wall is suddenly exposed to a high temperature T_f due to the ignition of flames in the furnace. The heat transfer coefficient on the left-hand side of the wall is $h \to \infty$. The right-hand side of the wall is insulated. Determine the temperature history $T(x,t)$.

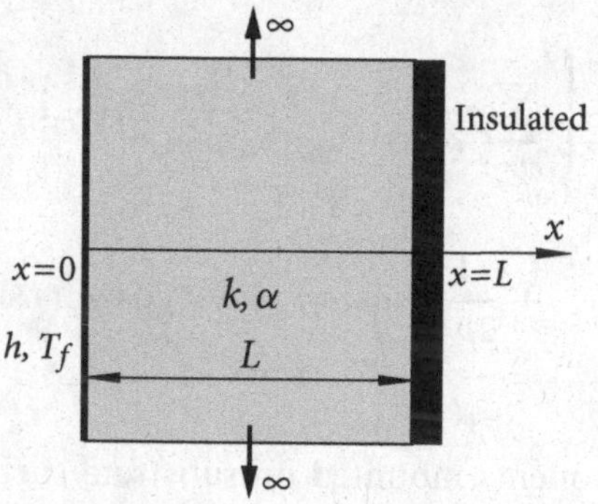

ANS $$T(x,t) = T_f + \sum_{n=1}^{\infty} \frac{4(T_0 - T_f)}{(2n-1)\pi} \exp\left\{-\alpha\left[\frac{(2n-1)\pi}{2L}\right]^2 t\right\} \sin\frac{(2n-1)\pi x}{2L}$$

11.11 A *thick* metal slab is originally isothermal at temperature T_0. A heat flux $\dot{Q}$ is received on one side of the slab starting at time $t=0$ for a short period of time. As a result, the temperature of the slab is raised. The coefficient of thermal conductivity and the thermal diffusivity of the metal are k and α, respectively. The metal slab can be modeled as a semi-infinite body for the time period during which it is heated as shown. Obtain the temperature history $T(x,t)$ of the slab during the heating process.

ANS $$T(x,t) = T_0 + \frac{\dot{Q}}{k}\left\{-\sqrt{\frac{4\alpha t}{\pi}}\exp\left(-\frac{x^2}{4\alpha t}\right) + x\,\mathrm{erfc}\left(\frac{x}{\sqrt{4\alpha t}}\right)\right\}$$

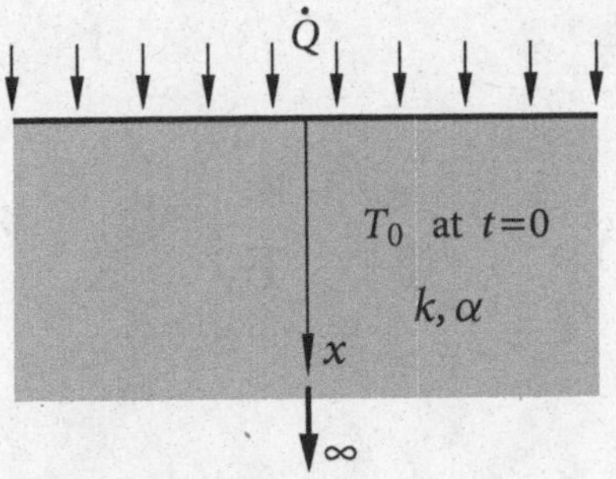

Hint: Use the integrals

$$\int_{\xi=0}^{\infty}\left[\exp\left\{-\frac{(x-\xi)^2}{4\alpha t}\right\}+\exp\left\{-\frac{(x+\xi)^2}{4\alpha t}\right\}\right]d\xi = 2\sqrt{\pi\alpha t}$$

$$\int_{\xi=0}^{\infty}\xi\left[\exp\left\{-\frac{(x-\xi)^2}{4\alpha t}\right\}+\exp\left\{-\frac{(x+\xi)^2}{4\alpha t}\right\}\right]d\xi$$
$$= 4\alpha t\exp\left(-\frac{x^2}{4\alpha t}\right)+x\sqrt{4\pi\alpha t}\,\mathrm{erf}\left(\frac{x}{\sqrt{4\alpha t}}\right)$$

11.12 A fin of length b with square cross-section $a\times a$ is mounted on a hot surface as shown in the following figure. The coefficient of thermal conductivity of the fin is k. The temperature at the hot surface is T_0. A fluid coolant of temperature T_f flows past the fin. The heat transfer coefficient between the fin surface and the coolant is $h\to\infty$.

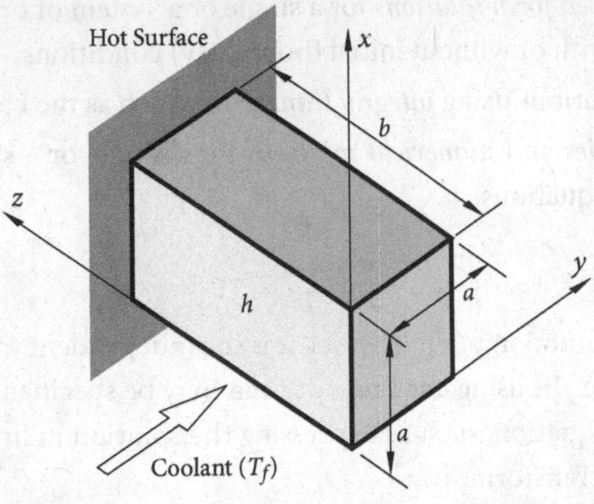

Show that the steady-state temperature distribution $T(x,y,z)$ in the fin is

$$T_f+\frac{16(T_0-T_f)}{\pi^2}\sum_{m=1}^{\infty}\sum_{n=1}^{\infty}\frac{1}{(2m-1)(2n-1)\sinh\dfrac{\sqrt{(2m-1)^2+(2n-1)^2}\,\pi b}{a}}$$
$$\times\sin\frac{(2m-1)\pi x}{a}\cdot\sin\frac{(2n-1)\pi y}{a}\cdot\sinh\frac{\sqrt{(2m-1)^2+(2n-1)^2}\,\pi z}{a}.$$

CHAPTER 12

Solving Ordinary Differential Equations Using *Maple*

Computer software packages for symbolic computations, such as *Maple* and *Mathematica*, are very useful in mathematical analysis. In this chapter, *Maple* commands commonly used in solving ordinary differential equations are introduced.

`dsolve` is a general ordinary differential equation solver, which can handle various types of ordinary differential equation problems, including

- compute *closed form solutions* for a single or a system of ordinary differential equations, with or without initial (boundary) conditions;
- compute solutions using *integral transforms*, such as the Laplace transform;
- compute *series* and *numerical solutions* for a single or a system of ordinary differential equations.

Calling Sequence

✍ Solve `ODE` for function `y(x)`, in which `x` is the independent variable and `y` is the dependent variable. In using `dsolve`, `options` may be specified to instruct *Maple* how to solve the equation, such as expressing the solution in implicit equation or using the Laplace transform.

```
dsolve(ODE,y(x),options)
```

✍ Solve `ODE` together with initial conditions `IC` for function `y(x)`.

```
dsolve({ODE,IC},y(x),options)
```

In the following, various examples are solved using `dsolve` to illustrate the procedures. The output from *Maple* is formatted somewhat for better presentation.

12.1 Closed-Form Solutions of Differential Equations

12.1.1 *Simple Ordinary Differential Equations*

First-order and simple higher-order ordinary differential equations studied in Chapter 2 can be easily solved using *Maple*.

Example 12.1 (Example 2.9)

Solve $y' = (x+y)^2$.

```
>ODE:=diff(y(x),x)=(x+y(x))^2;
```
✍ Define the ODE.

$$\text{ODE} := \frac{\mathrm{d}}{\mathrm{d}x}y(x) = (x+y(x))^2$$

```
>sol:=dsolve(ODE,y(x));
```
✍ Solve the ODE using `dsolve`.

$$\text{sol} := y(x) = -x - \tan(-x + \_C1)$$

✍ _C1 is an arbitrary constant.

Example 12.2 (Example 2.8)

Solve $\dfrac{\mathrm{d}y}{\mathrm{d}x} = \dfrac{x-y+5}{2x-2y-2}$.

```
>ODE:=diff(y(x),x)=(x-y(x)+5)/(2*x-2*y(x)-2):
```
✍ Define the ODE.

```
>sol:=dsolve(ODE,y(x));
```
✍ Solve the ODE using `dsolve` without options.

$$\text{sol} := y(x) = x - 6\,\text{LambertW}\left(-\tfrac{1}{6}\,\mathrm{e}^{x/12}\,\_C1\,\mathrm{e}^{-7/6}\right) - 7$$

✍ The result is in terms of a special function, because *Maple* tries to solve a nonlinear equation to obtain an explicit solution.

✍ To overcome this problem, use the `implicit` option to force the solution in the form of an implicit equation.

```
>sol_implicit:=dsolve(ODE,y(x),implicit);
```

$$\text{sol\_implicit} := -x + 2\,y(x) - 12\ln((y(x) - x + 7) - \_C1 = 0$$

Example 12.3 (Example 2.45)

Solve $yy'' = y'^2(1 - y'\sin y - yy'\cos y)$.

✍ In *Maple*, the *n*th-order derivative of $y(x)$ with respect to x is `diff(y(x),x$n)`.

```
>ODE:=y(x)*diff(y(x),x$2)=diff(y(x),x)^2*(1-diff(y(x),x)*sin(y(x))
 -y(x)*diff(y(x),x)*cos(y(x))):
```
✍ Define the ODE.

```
>sol:=dsolve(ODE,y(x));
```
✍ Solve the ODE using `dsolve` without options.

$$\text{sol} := y(x) = \_C1, \quad -\cos(y(x)) + \_C1\ln(y(x)) - x - \_C2 = 0$$

✍ An explicit or implicit solution can be tested using odetest, which checks the validity of the solution by substituting it into the ODE. If the solution is valid, odetest returns 0. When there are many solutions, use map to apply odetest to each element of sol.

```
>map(odetest,[sol],ODE);
```

$$[0, 0]$$

✍ The first 0 indicates that the first solution sol[1] is valid; the second 0 indicates that the second solution sol[2] is valid.

Example 12.4 (Example 2.33)

Solve $2xyy' = y^2 - 2x^3, \quad y(1) = 2.$

```
>ODE:=2*x*y(x)*diff(y(x),x)=y(x)^2-2*x^3:
```
✍ Define the ODE.

```
>IC:=y(1)=2;
```
✍ Define the initial condition IC.

$$IC := y(1) = 2$$

✍ Solve the ODE and IC using dsolve without options.

```
>sol:=dsolve({ODE,IC},y(x));
```

$$sol := y(x) = \sqrt{-x^3 + 5x}$$ ✍ Explicit solution.

✍ Use odetest to check if the result satisfies the ODE and IC.

```
>odetest(sol, [ODE, IC]);
```

$$[0, 0]$$

✍ The first 0 indicates that the solution satisfies the ODE; the second 0 indicates that the solution satisfies the IC.

✍ Solve the ODE and IC using dsolve with implicit option.

```
>sol_implicit:=dsolve({ODE,IC},y(x),implicit);
```

$$sol\_implicit := y(x)^2 + x^3 - 5x = 0$$ ✍ Implicit solution.

```
>odetest(sol_implicit, [ODE, IC]);
```

$$[0, 0]$$

Example 12.5 (Example 2.1)

Solve $\dfrac{dy}{dx} + \dfrac{1}{y}\,e^{y^2+3x} = 0.$

```
>ODE:=diff(y(x),x)+1/y(x)*exp(y(x)^2+3*x)=0:
```
✍ Define the ODE.

```
>sol:=dsolve(ODE,y(x));
```
✍ Solve the ODE using dsolve without options.

$$sol := y(x) = \sqrt{\ln\left(\frac{3}{2}\,\frac{1}{e^{3x}+3\_C1}\right)}, \quad y(x) = -\sqrt{\ln\left(\frac{3}{2}\,\frac{1}{e^{3x}+3\_C1}\right)}$$

✍ Solve the ODE using dsolve with implicit option.

```
>sol_implicit:=dsolve(ODE,y(x),implicit);
```

$$sol\_implicit := \frac{1}{3}\,e^{3x} - \frac{1}{2}\,e^{-y(x)^2} + \_C1 = 0$$

Example 12.6 (Example 2.24)

Solve $y^2\,\mathrm{d}x + (xy + y^2 - 1)\,\mathrm{d}y = 0.$

```
>ODE:=y(x)^2+(x*y(x)+y(x)^2-1)*diff(y(x),x)=0:
```
✍ Define the ODE.

```
>sol:=dsolve(ODE,y(x));
```
✍ Solve the ODE using dsolve without options.

$$\mathrm{sol} := y(x) = \mathrm{e}^{\mathrm{RootOf}(2\_Z - 2x\mathrm{e}^{-\_Z} - (\mathrm{e}^{-\_Z})^2 + 2\_C1)}$$

✍ The solution is in terms of RootOf(···), because *Maple* tries to solve a nonlinear equation to obtain an explicit expression for $y(x)$.

✍ Solve the ODE using dsolve with implicit option.

```
>sol_implicit:=dsolve(ODE,y(x),implicit);
```

$$\mathrm{sol\_implicit} := x - \frac{-\frac{1}{2}y(x)^2 + \ln(y(x)) + \_C1}{y(x)} = 0$$

✍ Simplify the result selecting the left-hand side of the equation using lhs and extracting the numerator using numer.

```
>sol_implicit:=numer(lhs(sol_implicit))=0;
```

$$\mathrm{sol\_implicit} := 2xy(x) + y(x)^2 - 2\ln(y(x)) - 2\_C1 = 0$$

Example 12.7 (Problem 2.92)

Solve $2xy' - y = \ln y'$.

```
>ODE:=2*x*diff(y(x),x)-y(x)=ln(diff(y(x),x)):
```
✍ Define the ODE.

✍ Solve the ODE using dsolve with implicit option.

```
>sol_implicit:=dsolve(ODE,y(x),implicit);
```

$$\mathrm{sol\_implicit} := \left[x(\_T) = \frac{\_T + \_C1}{\_T^2},\ y(\_T) = \frac{2(\_T + \_C1)}{\_T} - \ln(\_T)\right]$$

✍ This ODE is of the type solvable for variable x or y. The solution is in the form of parametric equations with $\_T$ being the parameter.

Example 12.8 (Example 2.37)

Solve $y = x\left\{\frac{\mathrm{d}y}{\mathrm{d}x} + \sqrt{1 + \left(\frac{\mathrm{d}y}{\mathrm{d}x}\right)^2}\right\}.$

```
>ODE:=y(x)=x*(diff(y(x),x)+sqrt(1+diff(y(x),x)^2)):
```
✍ Define the ODE.

```
>sol:=dsolve(ODE,y(x));
```
✍ Solve the ODE using dsolve without options.

$$\mathrm{sol} := \frac{\_C1}{\sqrt{\frac{(x^2+y(x)^2)^2}{y(x)^2x^2}}\left(-\frac{1}{2}\frac{1-y(x)^2+x^2}{x\,y(x)} + \frac{1}{2}\sqrt{\frac{(x^2+y(x)^2)^2}{y(x)^2x^2}}\right)} + x = 0$$

✍ To simplify the result, use `simplify` with option `sqrt` and assume that both x and $y(x)$ are positive.

```
>sol:=simplify(sol,sqrt) assuming x::positive, y(x)::positive;
```

$$\frac{x\left(\_C1\,x+x^2+y(x)^2\right)}{x^2+y(x)^2}=0$$

✍ Simplify the result further: take the left-hand side of the equation using `lhs`; extract the numerator using `numer`; and divide by x.

```
>sol:=numer(lhs(sol))/x=0;
```

$$\_C1\,x+x^2+y(x)^2=0$$

✍ Solve the ODE using `dsolve` with `implicit` option.

```
>sol_implicit:=dsolve(ODE,y(x),implicit);
```

$$\text{sol\_implicit} := \left[x(T)=\frac{\_C1}{\sqrt{1+T^2}\,e^{\operatorname{arcsinh}(T)}},\quad y(T)=\frac{\_C1\left(T+\sqrt{1+T^2}\right)}{\sqrt{1+T^2}\,e^{\operatorname{arcsinh}(T)}}\right]$$

```
>simplify(sol_implicit);
```

✍ Use `simplify` to simplify the result.

$$\left[x(T)=\frac{\_C1}{\sqrt{1+T^2}\left(T+\sqrt{1+T^2}\right)},\quad y(T)=\frac{\_C1}{\sqrt{1+T^2}}\right]$$

✍ For this ODE, `dsolve` with option `implicit` yields solution in the parametric form.

Example 12.9 (Example 2.16)

Solve $\left(\frac{1}{y}\sin\frac{x}{y}-\frac{y}{x^2}\cos\frac{y}{x}+1\right)\mathrm{d}x+\left(\frac{1}{x}\cos\frac{y}{x}-\frac{x}{y^2}\sin\frac{x}{y}+\frac{1}{y^2}\right)\mathrm{d}y=0.$

```
>ODE:=1/y(x)*sin(x/y(x))-y(x)/x^2*cos(y(x)/x)+1+(1/x*cos(y(x)/x)
 -x/y(x)^2*sin(x/y(x))+1/y(x)^2)*diff(y(x),x)=0:
```

✍ Define the ODE.

✍ Solve the ODE using `dsolve` and simplify the result using `simplify`.

```
>sol:=simplify(dsolve(ODE,y(x)));
```

$$\text{sol} := -\frac{1+2y(x)\cos^2\dfrac{x}{2y(x)}-2y(x)-\_C1\,y(x)-2y(x)\sin\dfrac{y(x)}{2x}\cos\dfrac{y(x)}{2x}-x\,y(x)}{y(x)}=0$$

✍ Simplify the result further: take the left-hand side of the equation using `lhs`; extract the numerator using `numer`; simplify the trigonometric functions using `combine`; and collect terms of $y(x)$ using `collect`.

```
>sol:=collect(combine(numer(lhs(sol))),y(x))=0;
```

$$\text{sol} := -1+\left(\sin\frac{y(x)}{x}-\cos\frac{x}{y(x)}+x+1+\_C1\right)y(x)=0$$

✍ The "1" before $\_C1$ can be absorbed in the constant $\_C1$ using `algsubs`.

```
>sol:=algsubs(1+_C1=C[1],sol);
```

$$\text{sol} := -1 + \left(\sin\frac{y(x)}{x} - \cos\frac{x}{y(x)} + x + C_1\right) y(x) = 0$$

Example 12.10 (Example 2.4)

Solve $\dfrac{dy}{dx} + \dfrac{x}{y} + 2 = 0, \quad y(0) = 1.$

`>ODE:=diff(y(x),x)+x/y(x)+2=0:` ✍ Define the ODE.

`>IC:=y(0)=1:` ✍ Define the IC.

`>sol:=dsolve({ODE,IC},y(x));` ✍ Solve the ODE and IC using `dsolve`.

$$\text{sol} := y(x) = -\frac{x\,(1 + \text{LambertW}(-x))}{\text{LambertW}(-x)}$$

✍ Without the `implicit` option, the solution is in terms of a special function.

✍ Now solve the ODE using `dsolve` with the `implicit` option.

```
>sol_implicit:=dsolve({ODE,IC},y(x),implicit);
```

sol_implicit :=

✍ With the `implicit` option, *Maple* does not render a solution if the initial condition is included.

✍ Solve the ODE using the `implicit` option but without the initial condition.

```
>sol_implicit:=dsolve(ODE,y(x),implicit);
```

$$\text{sol\_implicit} := -\_C1 + \frac{\ln\left(\frac{y(x)+x}{x}\right) y(x) + \ln\left(\frac{y(x)+x}{x}\right) x + x}{y(x)+x} + \ln(x) = 0$$

✍ The solution contains terms with x appearing in the denominator; hence, the initial condition at $x=0$ cannot be applied directly. This is the reason that *Maple* does not give a solution when both the `implicit` option and the initial condition `IC` are imposed.

✍ To simplify the implicit solution, take the left-hand side of the solution using `lhs`, and extract the numerator using `numer`.

```
>SOL1:=numer(lhs(sol_implicit));
```

$$\text{SOL1} := -\_C1\, y(x) - \_C1\, x + \ln\left(\frac{y(x)+x}{x}\right) y(x) + \ln\left(\frac{y(x)+x}{x}\right) x + x$$
$$+ \ln(x)\, y(x) + \ln(x)\, x$$

✍ The solution can be further simplified by combining the logarithmic terms using `combine` and assuming that x is positive.

```
>SOL2:=combine(SOL1) assuming x::positive;
```

$$SOL2 := -\_C1\, y(x) - \_C1\, x + y(x)\ln(y(x)+x) + x\ln(y(x)+x) + x$$

✍ The initial condition $y(0) = 1$ can now be applied using subs.

```
>eqn:=subs({x=0,y(x)=1},SOL2);
```

$$eqn := -\_C1 + \ln(1)$$

✍ The constant $\_C1$ is solved from the resulting algebraic equation using `solve`.

```
>_C1:=solve(eqn,_C1);
```

$$\_C1 := 0$$

✍ The particular solution in the implicit form is obtained.

```
>sol_implicit:=collect(SOL2,ln);
```

$$sol\_implicit := (y(x)+x)\ln(y(x)+x) + x$$

✍ Verify that the solution obtained satisfies both the ODE and IC using `odetest`.

```
>odetest(sol_implicit,[ODE,IC]);
```

$$[0, 0]$$

✍ To plot an implicit equation, use `implicitplot` in the `plots` package.

```
>with(plots):
```
✍ Load the `plots` package.

```
>implicitplot(sol_implicit,x=-5..5,y(x)=-0..10,view=[-5..1,0..9],
 numpoints=100000,thickness=3,labels=["x","y"],
 tickmarks=[[-5,-4,-3,-2,-1,0,1],[0,1,2,3,4,5,6,7,8,9]]);
```

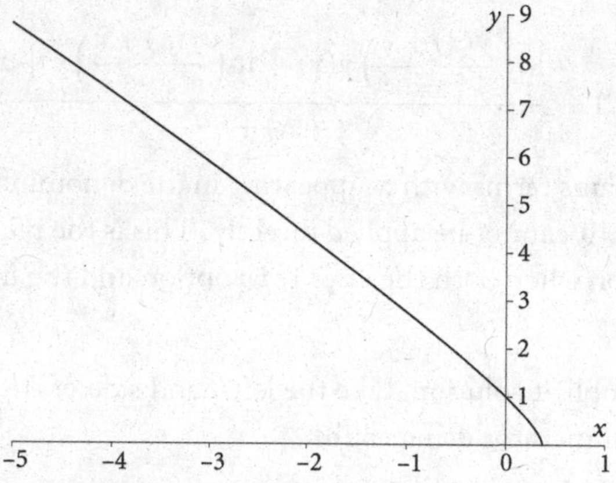

Example 12.11 (Example 2.35)

Solve $x = \dfrac{dy}{dx} + \left(\dfrac{dy}{dx}\right)^4$.

```
>ODE:=x=diff(y(x),x)+diff(y(x),x)^4:
```
✍ Define the ODE.

```
>sol:=dsolve(ODE,y(x));
```
✍ Solve the ODE using dsolve.

$$sol := y(x) = \frac{4}{5}\,x\,\mathrm{RootOf}(-x+\_Z+\_Z^4) - \frac{3}{10}\,\mathrm{RootOf}(-x+\_Z+\_Z^4)^2 + \_C1$$

✍ The solution is expressed in terms of a root of a polynomial equation of degree four, i.e., $\_Z^4+\_Z-x=0$, in which $\_Z$ is the unknown. Using the option `implicit` in `dsolve` does not help.

✍ Extract the value of $y(x)$ from the solution `sol` using `eval`.

```
>eval(y(x),sol);
```

$$\frac{4}{5}\,x\,\mathrm{RootOf}(-x+\_Z+\_Z^4) - \frac{3}{10}\,\mathrm{RootOf}(-x+\_Z+\_Z^4)^2 + \_C1$$

✍ This result is then assigned to y, defined as a function of x, using `unapply`. The ditto operator, `%`, is a shorthand notation for the previous result.

```
>y:=unapply(%,x);
```

$$y := x \rightarrow \frac{4}{5}\,x\,\mathrm{RootOf}(-x+\_Z+\_Z^4) - \frac{3}{10}\,\mathrm{RootOf}(-x+\_Z+\_Z^4)^2 + \_C1$$

✍ These two lines can be combined as

```
>y:=unapply(eval(y(x),sol),x);
```

✍ This task can also be accomplished by using

```
>y:=unapply(rhs(sol),x);
```

✍ The ODE is of the type solvable for variable x; the solution can be expressed in the form of parametric equations. Rewrite the fourth-degree polynomial equation and solve for x in terms of the parameter $\_Z$.

```
>eqn:=RootOf(-x+_Z+_Z^4,_Z)=_Z:
>xp:=solve(eqn,x);
```

$$xp := \_Z + \_Z^4$$

✍ Variable y is also expressed in terms of the parameter $\_Z$.

```
>yp:=simplify(eval(subs(x=xp,subs(eqn,y(x)))));
```

$$yp := \frac{1}{2}\,\_Z^2 + \frac{4}{5}\,\_Z^5 + \_C1$$

✍ For a better presentation, the parameter $\_Z$ is replaced by p.

```
>xp:=subs(_Z=p,xp); yp:=subs(_Z=p,yp);
```

$$xp := p + p^4, \qquad yp := \frac{1}{2}\,p^2 + \frac{4}{5}\,p^5 + \_C1$$

Remarks: Most of the first-order and simple higher-order ODEs studied in Chapter 2 are nonlinear equations with closed-form solutions in the form of nonlinear equations or parametric equations; `implicit` is one of the most important options in obtaining a solution in terms of elementary functions (exponential, logarithmic, and trigonometric functions). As illustrated in the examples, in many cases, one has to direct *Maple* interactively in order to obtain an useful result.

12.1.2 Linear Ordinary Differential Equations

Linear ordinary differential equations studied in Chapter 4 can be easily solved using *Maple* as shown in the following examples.

Example 12.12

Solve $y''' - y'' + y' - y = 8x\sin x + 5e^{2x} + x^2$, $y(0)=2$, $y'(0)=9$, $y''(0)=5$.

✍ In *Maple*, $y'(a)=b \Longrightarrow$ `D(y)(a)=b`, and $y^{(n)}(a)=b \Longrightarrow$ `(D@@n)(y)(a)=b`.

✍ Define the ODE and initial conditions ICs.

```
>ODE:=diff(y(x),x$3)-diff(y(x),x$2)+diff(y(x),x)-y(x)=8*x*sin(x)
 +5*exp(2*x)+x^2:
>ICs:=y(0)=2, D(y)(0)=9, (D@@2)(y)(0)=5:
```

✍ Solve the ODE with ICs using `dsolve`. Sort the result by collecting the sine terms and then the cosine terms.

```
>sol:=collect(collect(dsolve({ODE,ICs},y(x)),sin),cos);
```

$$\text{sol} := y(x) = (-x^2-3x+6)\sin x + (x^2-x-3)\cos x + 4e^x + e^{2x} - x^2 - 2x$$

Example 12.13 (Example 4.29)

Solve $y''' - y' = \dfrac{e^x}{1+e^x}$.

```
>ODE:=diff(y(x),x$3)-diff(y(x),x)=exp(x)/(1+exp(x)):
```
✍ Define the ODE.

✍ Solve the ODE using `dsolve`. Sort the result by collecting exponential terms.

```
>sol:=collect(dsolve(ODE,y(x)),exp);
```

$$\text{sol} := y(x) = \left(\_C1 - \tfrac{1}{2}\ln(1+e^x) + \tfrac{1}{2}\ln(e^x)\right)e^x - \left(\_C2 + \tfrac{1}{2}\ln(1+e^x)\right)e^{-x} + \_C3 + \tfrac{1}{2} - \ln(1+e^x)$$

✍ The $\frac{1}{2}$ after $\_C3$ can be absorbed in the arbitrary constant $\_C3$.

Example 12.14

Solve $x^2y'' - xy' + 2y = x(\ln x)^4 + 4x\sin(\ln x)$.

✍ Define the ODE. This is an Euler differential equation.

```
>ODE:=x^2*diff(y(x),x$2)-x*diff(y(x),x)+2*y(x)=x*(ln(x))^4
 +4*x*sin(ln(x)):
>sol:=dsolve(ODE,y(x));
```
✍ Solve the ODE using `dsolve`.

$$\text{sol} := y(x) = \_C1\,x\cos(\ln x) + \_C2\,x\sin(\ln x) + x\left((\ln x)^4 - 12(\ln x)^2 + 24 + \sin(\ln x) - 2\ln x\cos(\ln x)\right)$$

12.1.3 The Laplace Transform

Integral transforms, such as the Laplace transform and the Fourier transform, are available by loading the `inttrans` package using `with(inttrans)`. In *Maple*, the Heaviside step function $H(t-a)$ is `Heaviside(t-a)`, and the Dirac delta function $\delta(t-a)$ is `Dirac(t-a)`.

`>with(inttrans):` ✍ Load the `inttrans` package.

✍ Given a function $f(t)$.

```
>f:=t*cosh(2*t)+t^2*sin(5*t)+t^3+sin(t)*Heaviside(t-Pi);
```

$$f := t\cosh 2t + t^2 \sin 5t + t^3 + \sin t \,\mathrm{Heaviside}(t-\pi)$$

`>F:=laplace(f,t,s);` ✍ Evaluate the Laplace transform using `laplace`.

$$F := \frac{1}{2(s-2)^2} + \frac{1}{2(s+2)^2} + \frac{10(3s^2-25)}{(s^2+25)^3} + \frac{6}{s^4} + \frac{e^{-\pi s}}{s^2+1}$$

✍ Given the Laplace transform of a function $G(s)$.

```
>G:=(s-3)/(s^2-6*s+25)+2/(s+2)^3+exp(-2*s)*(2+1/(s^2+1));
```

$$G := \frac{s-3}{s^2-6s+25} + \frac{2}{(s+2)^3} + e^{-2s}\left(2+\frac{1}{s^2+1}\right)$$

✍ Evaluate inverse Laplace transform using `invlaplace`.

```
>g:=invlaplace(G,s,t);
```

$$g := e^{3t}\cos 4t + t^2 e^{-2t} + 2\,\mathrm{Dirac}(t-2) + \mathrm{Heaviside}(t-2)\sin(t-2)$$

✍ When evaluating the inverse Laplace transform by hand, one frequently needs to perform partial fraction decomposition, which can be easily done using *Maple*.

`>F:=8*(s+2)/(s-1)/(s+1)^2/(s^2+1)/(s^2+9);` ✍ Define a fraction.

$$F := \frac{8(s+2)}{(s-1)(s+1)^2(s^2+1)(s^2+9)}$$

✍ Perform partial fraction decomposition using `convert` with option `parfrac`, in which s is the variable.

```
>convert(F,parfrac,s);
```

$$\frac{3}{10(s-1)} - \frac{1}{5(s+1)^2} - \frac{27}{50(s+1)} + \frac{s-3}{4(s^2+1)} + \frac{s-11}{100(s^2+9)}$$

When an ODE is solved using `dsolve` with the option `method=laplace`, *Maple* forces the equation to be solved by the method of Laplace transform.

Example 12.15 (Example 6.18)

Solve $y''' - y'' + 4y' - 4y = 40(t^2 + t + 1)H(t-2)$, $y(0) = 5$, $y'(0) = 0$, $y''(0) = 10$.

```
>ODE:=diff(y(t),t$3)-diff(y(t),t$2)+4*diff(y(t),t)-4*y(t)
 =40*(t^2+t+1)*Heaviside(t-2):
```
✍ Define the ODE.

```
>ICs:=y(0)=5, D(y)(0)=0, (D@@2)(y)(0)=10:
```
✍ Define the ICs.

✍ Solve the ODE and ICs using `dsolve` with the option `method=laplace`. The trigonometric terms are simplified using `combine`. The result is simplified by first collecting terms with `Heaviside(t-2)` and then exponential terms using `collect`.

```
>sol:=collect(collect(combine(dsolve({ODE,ICs},y(t),method=laplace)),
 Heaviside(t-2)),exp);
```

$$\text{sol} := y(t) = 6\,e^t + 112\big(1 - \text{Heaviside}(2-t)\big)\,e^{t-2} + \big(23\cos(2t-4) \\ -21\sin(2t-4) - 10t^2 - 30t - 35\big)\,\text{Heaviside}(t-2) - \cos 2t - 3\sin 2t$$

Example 12.16

Solve $y''' - y'' + 4y' - 4y = 10e^{-t}$.

✍ Define the ODE.

```
>ODE:=diff(y(t),t$3)-diff(y(t),t$2)+4*diff(y(t),t)-4*y(t)=10*exp(-t):
```

✍ Solve the ODE using `dsolve` with the option `method=laplace`. Simplified the result by collecting sine, cosine, and then exponential terms using `collect`.

```
>collect(collect(collect(dsolve(ODE,y(t),method=laplace),sin),cos),exp);
```

$$y(t) = -e^{-t} + \big(1 + \tfrac{4}{5}\,y(0) + \tfrac{1}{5}\,y''(0)\big)\,e^t + \tfrac{1}{5}\big(y(0) - y''(0)\big)\cos 2t \\ + \big(-1 - \tfrac{2}{5}\,y(0) + \tfrac{1}{2}\,y'(0) - \tfrac{1}{10}\,y''(0)\big)\sin 2t$$

```
>dsolve(ODE,y(t));
```
✍ Solve the ODE using `dsolve` without any option.

$$y(t) = -e^{-t} + \_C1\,e^t + \_C2\cos 2t + \_C3\sin 2t$$

Remarks: Using the option `method=laplace` to solve an ODE by the method of Laplace transform is done "behind-the-scene." When the initial conditions are specified, there is no difference between the particular solutions obtained with and without the `method=laplace` option.

However, when the initial conditions are not specified, the general solution obtained with the `method=laplace` option is expressed in terms of the unknown initial conditions $y(0)$, $y'(0)$, $y''(0)$, ...; whereas the general solution obtained without the `method=laplace` option is given in terms of arbitrary constants $\_C1$, $\_C2$, ...

12.1.4 Systems of Ordinary Differential Equations

Example 12.17

Solve $y_1' - y_2 = 0, \quad y_2' - y_3 = 2H(x-1), \quad 6y_1 + 11y_2 + y_3' + 6y_3 = \mathrm{e}^{-x}.$

```
>ODE[1]:=diff(y[1](x),x)-y[2](x)=0:
```
Define the ODEs.

```
>ODE[2]:=diff(y[2](x),x)-y[3](x)=2*Heaviside(x-1):
>ODE[3]:=6*y[1](x)+11*y[2](x)+diff(y[3](x),x)+6*y[3](x)=exp(-x):
```

Solve the ODEs using `dsolve`. The result is simplified by collecting first the exponential terms and then terms involving `Heaviside(x-1)` using `collect`.

```
>sol:=collect(collect(dsolve({ODE[1],ODE[2],ODE[3]},{y[1](x),y[2](x),
 y[3](x)}),exp),Heaviside(x-1));
```

$$\begin{aligned}
\mathrm{sol} := \Big\{ y_1(x) &= \left(-\mathrm{e}^{-3(x-1)} + 4\mathrm{e}^{-2(x-1)} - 5\mathrm{e}^{-(x-1)} + 2\right)\mathrm{Heaviside}(x-1) \\
&+ \left(_C1 + \frac{3_C2}{2} + \frac{_C3}{2}\right)\mathrm{e}^{-3x} - (3_C1 + 4_C2 + _C3)\,\mathrm{e}^{-2x} \\
&+ \left(3_C1 + \frac{5_C2}{2} + \frac{_C3}{2} + \frac{x}{2} - \frac{3}{4}\right)\mathrm{e}^{-x}, \\
y_2(x) &= \left(3\mathrm{e}^{-3(x-1)} - 8\mathrm{e}^{-2(x-1)} + 5\mathrm{e}^{-(x-1)}\right)\mathrm{Heaviside}(x-1) \\
&+ \left(-3_C1 - \frac{9_C2}{2} - \frac{3_C3}{2}\right)\mathrm{e}^{-3x} + (6_C1 + 2_C2 + 8_C3)\,\mathrm{e}^{-2x} \\
&+ \left(-3_C1 - \frac{5_C2}{2} - \frac{_C3}{2} - \frac{x}{2} + \frac{5}{4}\right)\mathrm{e}^{-x}, \\
y_3(x) &= \left(-9\mathrm{e}^{-3(x-1)} + 16\mathrm{e}^{-2(x-1)} - 5\mathrm{e}^{-(x-1)-2}\right)\mathrm{Heaviside}(x-1) \\
&+ \left(9_C1 + \frac{27_C2}{2} + \frac{9_C3}{2}\right)\mathrm{e}^{-3x} - (12_C1 + 16_C2 + 4_C3)\,\mathrm{e}^{-2x} \\
&+ \left(3_C1 + \frac{5_C2}{2} + \frac{_C3}{2} - \frac{7}{4}\right)\mathrm{e}^{-x} \Big\}
\end{aligned}$$

Example 12.18

Solve $x'' + 3x' + 2x + y' + y = 4t^2 + 2 + 40t\cos 2t,$

$x' + 2x + y' - y = 192t\,\mathrm{e}^{2t} + 5\sin 2t, \quad x(0) = 0,\ x'(0) = 0,\ y(0) = 0.$

```
>ODE[1]:=diff(x(t),t$2)+3*diff(x(t),t)+2*x(t)+diff(y(t),t)+y(t)
 =4*t^2+2+40*t*cos(2*t):
```
Define the ODEs.

```
>ODE[2]:=diff(x(t),t)+2*x(t)+diff(y(t),t)-y(t)=192*t*exp(2*t)
 +5*sin(2*t):
>ICs:=x(0)=0,D(x)(0)=0,y(0)=0:
```
Define the initial conditions ICs.

✍ Solve the ODEs and ICs using `dsolve`. The result is simplified by collecting the exponential terms, the cosine terms, and the sine terms using `collect`.

```
>sol:=collect(collect(collect(dsolve({ODE[1],ODE[2],ICs},{x(t),y(t)}),
 exp),cos),sin);
```

$$\text{sol} := \Big\{x(t) = \big(4t+\tfrac{21}{40}\big)\sin 2t + \big(-3t+\tfrac{16}{5}\big)\cos 2t + \tfrac{77}{16}\,\mathrm{e}^{-2t} - \tfrac{148}{15}\,\mathrm{e}^{-t} + \big(-24t^2+12t-\tfrac{151}{48}\big)\mathrm{e}^{2t} + t^2 - 4t + 5,$$
$$y(t) = \big(2t-\tfrac{101}{20}\big)\sin 2t + \big(6t+\tfrac{7}{20}\big)\cos 2t - \tfrac{74}{15}\,\mathrm{e}^{-t} + \big(96t^2+\tfrac{7}{12}\big)\mathrm{e}^{2t} + 2t^2 - 2t + 4\Big\}$$

✍ Use `odetest` to check if the solution satisfies the ODEs and ICs.

```
>odetest(sol,[ODE[1],ODE[2],ICs]);
```

[0, 0, 0, 0, 0] ✍ All of the two ODEs and three ICs are satisfied.

Eigenvalues and Eigenvectors of a Matrix

In solving systems of linear ordinary differential equations using the matrix method, as studied in Chapter 7, one needs to find the eigenvalues and the corresponding eigenvectors (sometimes the generalized eigenvectors if there are multiple eigenvalues) of the system matrix **A**. This task can be easily accomplished using *Maple*.

`>with(LinearAlgebra):` ✍ Load the `LinearAlgebra` package.

`>A:=Matrix([[1,-1],[2,-1]]);` ✍ Define matrix **A**.

$$\mathbf{A} := \begin{bmatrix} 1 & -1 \\ 2 & -1 \end{bmatrix}$$

✍ Evaluate the eigenvalues and eigenvectors using `Eigenvectors`, stored in $\boldsymbol{\lambda}$ and $\mathbf{u}$, respectively. In *Maple*, i is denoted as *I*.

```
>(lambda,v):=Eigenvectors(A);
```

$$\boldsymbol{\lambda}, \mathbf{v} := \begin{bmatrix} I \\ -I \end{bmatrix}, \begin{bmatrix} \frac{1}{2}+\frac{1}{2}I & \frac{1}{2}-\frac{1}{2}I \\ 1 & -1 \end{bmatrix}$$

As studied in Chapter 7, if matrix **A** has a multiple eigenvalue λ with multiplicity $m>1$ and if there are fewer than m linearly independent eigenvectors corresponding to λ, then matrix **A** is defective. In this case, a complete basis of eigenvectors is obtained by including generalized eigenvectors.

In matrix theory, if matrix **A** of dimension $n\times n$ has n linearly independent eigenvectors, it can be reduced to a diagonal matrix **D**, i.e., $\mathbf{Q}^{-1}\mathbf{A}\mathbf{Q}=\mathbf{D}$, in which the diagonal elements of **D** are the eigenvalues of **A** and the columns of **Q** are the corresponding eigenvectors.

However, if matrix **A** is defective, it can be reduced to the Jordan form **J**, i.e., $\mathbf{Q}^{-1}\mathbf{A}\mathbf{Q}=\mathbf{J}$, in which the diagonal elements of the Jordan form **J** are the eigenvalues of **A** and the columns of **Q** are the eigenvectors and generalized eigenvectors of **A**.

```
>with(LinearAlgebra):
```
✍ Load the LinearAlgebra package.

```
>A:=Matrix([[1,0,1],[0,1,-1],[0,0,2]]);
```
✍ Define matrix **A**.

$$\mathbf{A} := \begin{bmatrix} 1 & 0 & 1 \\ 0 & 1 & -1 \\ 0 & 0 & 2 \end{bmatrix}$$

```
>(lambda,v):=Eigenvectors(A);
```
✍ Evaluate the eigenvalues and eigenvectors.

$$\boldsymbol{\lambda}, \mathbf{v} := \begin{bmatrix} 1 \\ 1 \\ 2 \end{bmatrix}, \begin{bmatrix} 0 & 1 & 1 \\ 1 & 0 & -1 \\ 0 & 0 & 1 \end{bmatrix}$$

✍ Although matrix **A** has an eigenvalue $\lambda=1$ of multiplicity 2, it does have two linearly independent eigenvectors.

```
>A:=Matrix([[4,-1,0],[3,1,-1],[1,0,1]]);
```
✍ Define matrix **A**.

$$\mathbf{A} := \begin{bmatrix} 4 & -1 & 0 \\ 3 & 1 & -1 \\ 1 & 0 & 1 \end{bmatrix}$$

```
>(lambda,v):=Eigenvectors(A);
```
✍ Evaluate the eigenvalues and eigenvectors.

$$\boldsymbol{\lambda}, \mathbf{v} := \begin{bmatrix} 2 \\ 2 \\ 2 \end{bmatrix}, \begin{bmatrix} 1 & 0 & 0 \\ 2 & 0 & 0 \\ 1 & 0 & 0 \end{bmatrix}$$

✍ Matrix **A** has an eigenvalue $\lambda=2$ of multiplicity 3; but it has only one eigenvector, because the last two columns of **v** are zero vector.

✍ Use JordanForm to determine the Jordan form **J** and matrix **Q**.

```
>(J,Q):=JordanForm(A,output=['J','Q']);
```

$$\mathbf{J}, \mathbf{Q} := \begin{bmatrix} 2 & 1 & 0 \\ 0 & 2 & 1 \\ 0 & 0 & 2 \end{bmatrix}, \begin{bmatrix} 1 & 2 & 1 \\ 2 & 3 & 0 \\ 1 & 1 & 0 \end{bmatrix}$$

✍ In matrix **Q**, the first column is the eigenvector and the last two columns are the generalized eigenvectors.

12.2 Series Solutions of Differential Equations

Special Functions

✍ *Maple* can be used to evaluate special functions effectively.

`>convert(GAMMA(n+1),factorial);` ✍ Convert Gamma function to factorial.

$$n! \qquad ✍\ \Gamma(n+1) = n!$$

`>GAMMA(1/2);` ✍ Evaluate $\Gamma\left(\frac{1}{2}\right)$.

$$\sqrt{\pi} \qquad ✍\ \Gamma\left(\tfrac{1}{2}\right) = \sqrt{\pi}$$

`>evalf(GAMMA(1/3));` ✍ Evaluate $\Gamma\left(\frac{1}{3}\right)$ using floating-point arithmetic.

$$2.678938537 \qquad ✍\ \Gamma\left(\tfrac{1}{3}\right) = 2.678938537$$

`>evalf(BesselJ(1/3,1));` ✍ Evaluate $J_{\frac{1}{3}}(1)$ using floating-point arithmetic.

$$0.7308764022 \qquad ✍\ J_{\frac{1}{3}}(1) = 0.7308764022$$

✍ Expand $J_0(x)$ in series about $x=0$ to the order $\mathcal{O}(x^{12})$.

`>J[0]:=series(BesselJ(0,x),x,12);`

$$J_0 := 1 - \frac{x^2}{4} + \frac{x^4}{64} - \frac{x^6}{2304} + \frac{x^8}{147456} - \frac{x^{10}}{14745600} + \mathcal{O}(x^{12})$$

✍ Convert J_0 to polynomial by dropping the order term $\mathcal{O}(x^{12})$.

`>J[0]:=convert(J[0],polynom);`

$$J_0 := 1 - \frac{x^2}{4} + \frac{x^4}{64} - \frac{x^6}{2304} + \frac{x^8}{147456} - \frac{x^{10}}{14745600}$$

✍ Expand $Y_0(x)$ in series about $x=0$ to the order $\mathcal{O}(x^{12})$ and then convert to polynomial. Unfortunately, the result is not presented in a very clear format.

`>Y[0]:=convert(series(BesselY(0,x),x,12),polynom);`

$$Y_0 := \frac{2(-\ln 2 + \ln x)}{\pi} + \frac{2\gamma}{\pi} + \left(-\frac{1}{2}\frac{-\ln 2 + \ln x}{\pi} - \frac{-\frac{1}{2} + \frac{\gamma}{2}}{\pi}\right)x^2 + \cdots$$

✍ To factor out $2/\pi$, divide Y_0 by $2/\pi$ first (and then multiply by $2/\pi$ later).

✍ Since *Maple* does not allow collecting in terms of $(\ln x - \ln 2 + \gamma)$, replace $(\ln x - \ln 2 + \gamma)$ by T and then `collect` in terms of T.

`>Y[0]:=collect(simplify(subs(ln(x)=T+ln(2)-gamma,Y[0]/(2/Pi))),T);`

$$Y_0 := \left(1 - \frac{x^2}{4} + \frac{x^4}{64} - \frac{x^6}{2304} + \frac{x^8}{147456} - \frac{x^{10}}{14745600}\right)T$$
$$+ \frac{x^2}{4} - \frac{3x^4}{128} + \frac{11x^6}{13824} - \frac{25x^8}{1769472} + \frac{137x^{10}}{884736000}$$

✍ Replace T with $\ln(x/2)+\gamma$ and multiply by $2/\pi$ to obtain final result.

```
>Y[0]:=2/Pi*subs(T=ln(x/2)+gamma,Y[0]);
```

$$Y_0 := \frac{2}{\pi}\left[\left(1-\frac{x^2}{4}+\frac{x^4}{64}-\frac{x^6}{2304}+\frac{x^8}{147456}-\frac{x^{10}}{14745600}\right)\left(\ln\frac{x}{2}+\gamma\right) + \frac{x^2}{4}-\frac{3x^4}{128}+\frac{11x^6}{13824}-\frac{25x^8}{1769472}+\frac{137x^{10}}{884736000}\right]$$

✍ Plot $J_0(x)$ and $Y_0(x)$.

```
>plot({BesselJ(0,x),BesselY(0,x)},x=0..20,y=-1..1,numpoints=1000);
```

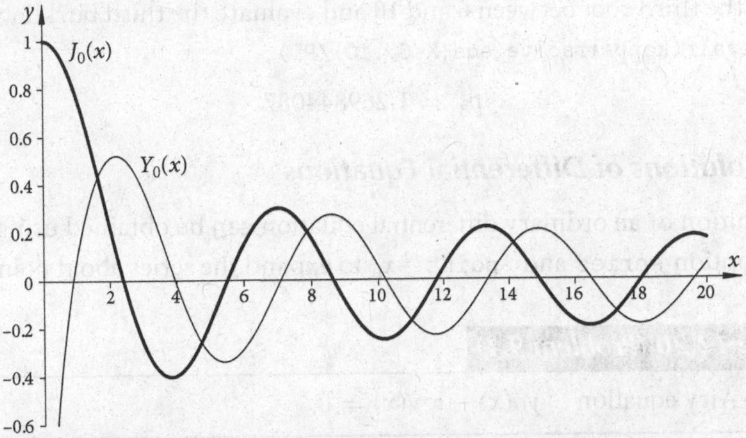

Example 12.19 — Buckling of a Tapered Column (Example 9.9)

Evaluate the first three buckling parameters p_n, $n=1, 2, 3$, of the fixed-free tapered column ($r_1=0.5r_0$) with circular cross-section considered in Section 9.3.2.

✍ Define the buckling equation.

```
>eqn:=BesselJ(-1/2,K/(1-kappa))*(BesselJ(1/2,K)-K*BesselJ(-1/2,K))
  -BesselJ(1/2,K/(1-kappa))*(BesselJ(-1/2,K)+K*BesselJ(1/2,K)):
>kappa:=(r[0]-r[1])/r[0]:
```

✍ $\kappa=(r_0-r_1)/r_0$

```
>r[1]:=0.5*r[0]:
```

✍ $r_1=0.5r_0$

```
>plot(eqn,K=0..15);
```

✍ Plot the buckling equation for $0\leqslant K\leqslant 15$.

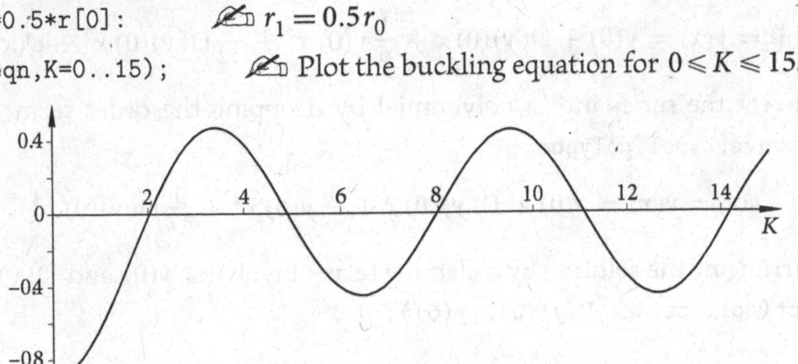

```
>K[1]:=fsolve(eqn,K=1..3);
```
✍ Find the first root K_1 between 1 and 3.

$$K_1 := 2.028757838$$

```
>p[1]:=evalf(kappa*K[1]/Pi);
```
✍ The first buckling load p_1.

$$p_1 := 0.3228868382$$

```
>K[2]:=fsolve(eqn,K=4..6);
```
✍ Find the second root K_2 between 4 and 6.

$$K_2 := 4.913180439$$

```
>p[2]:=evalf(kappa*K[2]/Pi);
```
✍ The second buckling load p_2.

$$p_2 := 0.7819569531$$

✍ Find the third root between 6 and 10 and evaluate the third buckling load p_3.

```
>p[3]:=evalf(kappa*fsolve(eqn,K=6..10)/Pi);
```

$$p_3 := 1.269844087$$

Series Solutions of Differential Equations

Series solution of an ordinary differential equation can be obtained using `dsolve` with the option `series` and `'point'=`x_0 to expand the series about point $x = x_0$.

Example 12.20 (Problem 9.1)

Solve the Airy equation $y''(x) - x\,y(x) = 0.$

```
>ODE:=diff(y(x),x$2)-x*y(x)=0:
```
✍ Define the ODE.

✍ Solve the ODE without any option. The solution is given in terms of the Airy functions.

```
>dsolve(ODE,y(x));
```

$$y(x) = \_C1\,\mathrm{AiryAi}(x) + \_C2\,\mathrm{AiryBi}(x)$$

✍ Solve the ODE with the `series` option. The series is expanded about `'point'` $x = 0$. The default order of the series expansion is 6, i.e., $O(x^6)$.

```
>sol:=dsolve(ODE,y(x),series,'point'=0);
```

$$sol := y(x) = y(0) + \mathrm{D}(y)(0)\,x + \frac{1}{6}\,y(0)\,x^3 + \frac{1}{12}\mathrm{D}(y)(0)\,x^4 + O(x^6)$$

✍ Convert the series into a polynomial by dropping the order term $O(x^6)$.

```
>sol:=convert(sol,polynom);
```

$$sol := y(x) = y(0) + \mathrm{D}(y)(0)\,x + \frac{1}{6}\,y(0)\,x^3 + \frac{1}{12}\mathrm{D}(y)(0)\,x^4$$

✍ Rearranging the solution by collecting terms involving $y(0)$ and $\mathrm{D}(y)(0)$.

```
>collect(collect(sol,D(y)(0)),y(0));
```

$$y(x) = \left(1 + \frac{x^3}{6}\right) y(0) + \left(x + \frac{x^4}{12}\right) \mathrm{D}(y)(0)$$

✍ Expand the series about $x=1$ using 'point'=1. Convert to polynomial and then rearrange the solution by collecting terms involving $y(1)$ and $D(y)(1)$.

```
>sol:=dsolve(ODE,y(x),series,'point'=1):
>sol:=convert(sol,polynom):
>collect(collect(sol,D(y)(1)),y(1));
```

$$y(x) = \left[1+\frac{(x-1)^2}{2}+\frac{(x-1)^3}{6}+\frac{(x-1)^4}{24}+\frac{(x-1)^5}{30}\right]y(1) + \left[(x-1)+\frac{(x-1)^3}{6}+\frac{(x-1)^4}{12}+\frac{(x-1)^5}{120}\right]D(y)(1)$$

✍ The order of the series expansion can be changed using Order.

```
>Order:=11:
```
✍ Change the order of series expansion to $O(x^{11})$.

```
>sol:=dsolve(ODE,y(x),series,'point'=0):
>sol:=convert(sol,polynom):
>collect(collect(sol,D(y)(0)),y(0));
```

$$y(x) = \left(1+\frac{x^3}{6}+\frac{x^6}{180}+\frac{x^9}{12960}\right)y(0) + \left(x+\frac{x^4}{12}+\frac{x^7}{504}+\frac{x^{10}}{45360}\right)D(y)(0)$$

✍ When the ODE is a homogeneous linear ordinary differential equation with polynomial coefficients, dsolve with option 'formal_solution' gives a set of formal solutions with the specified coefficients at the given 'point'. In the following, the result is formatted for better presentation.

```
>Formal_Solution:=dsolve(ODE,y(x),'formal_solution','point'=0);
```

$$\text{Formal\_Solution} := y(x) = C_1\,\Gamma\left(\tfrac{2}{3}\right)\sum_{n=0}^{\infty}\frac{9^{-n}x^{3n}}{\Gamma(n+1)\,\Gamma\left(n+\frac{2}{3}\right)} + C_2\,\frac{2\pi}{\Gamma\left(\frac{2}{3}\right)}\sum_{n=0}^{\infty}\frac{3^{-\frac{3}{2}-2n}x^{3n+1}}{\Gamma(n+1)\,\Gamma\left(n+\frac{4}{3}\right)}$$

Example 12.21

Solve the Riccati-Bessel equation $x^2y''(x)+\left[x^2-k(k+1)\right]y(x)=0$, for $k=-\frac{1}{2}$.

✍ x is a regular singular point. The roots of the indicial equation are $\alpha_1=\alpha_2=\frac{1}{2}$.

```
>ODE:=x^2*diff(y(x),x$2)+(x^2-k*(k+1))*y(x)=0:
```
✍ Define the ODE.

```
>k:=-1/2:
```
✍ Set the value of k to $-\frac{1}{2}$.

✍ Solve the ODE without any option. The solution is given in terms of the Bessel functions $J_0(x)$ and $Y_0(x)$.

```
>dsolve(ODE,y(x));
```

$$y(x) = \_C1\,\sqrt{x}\,\text{BesselJ}(0,x) + \_C2\,\sqrt{x}\,\text{BesselY}(0,x)$$

✍ Obtain formal solutions using `dsolve` with option `'formal_solution'`.

```
>Formal_Solution:=dsolve(ODE,y(x),'formal_solution','point'=0);
```

$$\text{Formal\_Solution} := y(x) = C_1\sqrt{x}\sum_{n=0}^{\infty}\frac{(-1)^n 4^{-n} x^{2n}}{\Gamma(n+1)^2}$$

✍ For this ODE, only one formal solution is given. It seems that, in many cases, *Maple* is not able to give all the formal solutions.

`>Order:=11:` ✍ Set the order of series expansion to $O(x^{11})$.

`>sol:=dsolve(ODE,y(x),series,'point'=0):` ✍ Obtain series solution.

```
>sol:=convert(sol,polynom);
```

$$\begin{aligned}\text{sol} := y(x) = {}&\_C1\sqrt{x}\left(1-\frac{x^2}{4}+\frac{x^4}{64}-\frac{x^6}{2304}+\frac{x^8}{147456}-\frac{x^{10}}{14745600}\right)\\ &+\_C2\sqrt{x}\left[\left(1-\frac{x^2}{4}+\frac{x^4}{64}-\frac{x^6}{2304}+\frac{x^8}{147456}-\frac{x^{10}}{14745600}\right)\ln x\right.\\ &\left.+\left(\frac{x^2}{4}-\frac{3x^4}{128}+\frac{11x^6}{13824}-\frac{25x^8}{1769472}+\frac{137x^{10}}{884736000}\right)\right]\end{aligned}$$

Example 12.22 (Problem 9.11)

Solve the Laguerre equation $xy''(x)+(1-x)y'(x)+ky(x)=0$, for $k=4$.

✍ x is a regular singular point. The roots of the indicial equation are $\alpha_1=\alpha_2=0$.

`>ODE:=x*diff(y(x),x$2)+(1-x)*diff(y(x),x)+k*y(x)=0:` ✍ Define the ODE.

`>k:=4:` ✍ Set the value of k to 4.

✍ Solve the ODE without any option. The solution is given in terms of the exponential integral Ei$(1,-x)$.

```
>dsolve(ODE,y(x));
```

$$\begin{aligned}y(x) = {}&\_C1\,(24-96x+72x^2-16x^3+x^4)+\frac{\_C2}{576}\big[(24-96x+72x^2-16x^3\\ &+x^4)\,\mathrm{Ei}(1,-x)+e^x(-50+58x-15x^2+x^3)\big]\end{aligned}$$

✍ Obtain formal solutions using `dsolve` with option `'formal_solution'`.

```
>Formal_Solution:=dsolve(ODE,y(x),'formal_solution','point'=0);
```

$$\begin{aligned}\text{Formal\_Solution} := y(x) = {}&\_C1\left(1-4x+3x^2-\frac{2x^3}{3}+\frac{x^4}{24}\right)+\_C2\Big[(600\\ &-2400x+1800x^2-400x^3+25x^4)\ln x+5400x-6450x^2+1900x^3-\frac{625x^4}{4}\\ &+14400\sum_{n=5}^{\infty}\frac{x^n}{\Gamma(n+1)\;n(n-1)(n-2)(n-3)(n-4)}\Big]\end{aligned}$$

✍ For this ODE, *Maple* does give two formal solutions. The second solution can be simplified by converting the Gamma function to factorial, and absorbing the constant 600 into _C2 by using _C2=_C3/600. This makes the polynomial in front of $\ln x$ the same as the first solution, consistent with Fuchs' Theorem.

```
>collect(collect(collect(simplify(subs(_C2=_C3/600,
 convert(Formal_Solution,factorial))),ln(x)),_C3),_C1);
```

$$y(x) = \_C1\left(1-4x+3x^2-\frac{2x^3}{3}+\frac{x^4}{24}\right)+\_C3\left[\left(1-4x+3x^2-\frac{2x^3}{3}+\frac{x^4}{24}\right)\ln x + 9x-\frac{43x^2}{4}+\frac{19x^3}{6}-\frac{25x^4}{96}+24\sum_{n=5}^{\infty}\frac{x^n}{n!\,n(n-1)(n-2)(n-3)(n-4)}\right]$$

```
>Order:=10:
```
✍ Set the order of series expansion to $O(x^{10})$.

```
>sol:=dsolve(ODE,y(x),series,'point'=0):
```
✍ Obtain series solution.

```
>convert(sol,polynom);
```

$$y(x) = \_C1\left(1-4x+3x^2-\frac{2x^3}{3}+\frac{x^4}{24}\right)+\_C2\left[\left(1-4x+3x^2-\frac{2x^3}{3}+\frac{x^4}{24}\right)\ln x +9x-\frac{43x^2}{4}+\frac{19x^3}{6}-\frac{25x^4}{96}+\frac{x^5}{600}-\frac{x^6}{21600}+\frac{x^7}{529200}-\frac{x^8}{11289600}+\frac{x^9}{228614400}\right]$$

12.3 Numerical Solutions of Differential Equations

Numerical solution of an ordinary differential equation is accomplished by `dsolve` with the option `numeric` or `type=numeric`. The default method for initial value problems is a Runge-Kutta-Fehlberg method that produces a fifth-order accurate solution. If the option `stiff=true` is specified, then the differential equation is regarded as a stiff equation, and the default method is a Rosenbrock method.

For most initial value problems, the default approach is sufficient. However, if an equation is known to be stiff, it is more efficient to specify the `stiff=true` option as illustrated in the following example.

```
>with(LinearAlgebra):
```
✍ Load the `LinearAlgebra` package.

```
>A:=Matrix([[99998,99999],[-199998,-199999]]);
```
✍ Define matrix **A**.

$$\mathbf{A} := \begin{bmatrix} 99998 & 99999 \\ -199998 & -199999 \end{bmatrix}$$

```
>(lambda,v):=Eigenvectors(A);
```
✍ Evaluate the eigenvalues and eigenvectors.

$$\lambda, \mathbf{v} := \begin{bmatrix} -100000 \\ -1 \end{bmatrix}, \begin{bmatrix} -1/2 & -1 \\ 1 & 1 \end{bmatrix}$$

✍ Define the ODEs as $\mathbf{z}'(x) = \mathbf{A}\mathbf{z}(x)$ with ICs $z_1(0) = 2$, $z_2(0) = -3$.

```
>ODE[1]:=diff(z[1](x),x)=A[1,1]*z[1](x)+A[1,2]*z[2](x):
>ODE[2]:=diff(z[2](x),x)=A[2,1]*z[1](x)+A[2,2]*z[2](x):
>ICs := z[1](0)=2, z[2](0)=-3:
```

✍ Solve the system of ODEs with ICs using `dsolve`.

```
>sol:=dsolve({ODE[1],ODE[2],ICs},{z[1](x),z[2](x)});
```

$$\mathrm{sol} := \left\{z_1(x) = \mathrm{e}^{-100000x} + \mathrm{e}^{-x},\ z_2(x) = -2\mathrm{e}^{-100000x} - \mathrm{e}^{-x}\right\}$$

✍ $z_1(x)$ is assigned to y_1 as a function of x.

```
>y[1]:=unapply(eval(z[1](x),sol[1]),x);
```

$$y_1 := x \to \mathrm{e}^{-100000x} + \mathrm{e}^{-x}$$

✍ Solve the system of ODEs with ICs numerically using `dsolve` with the option `numeric`. The system is solved using the default Runge-Kutta-Fehlberg method. The maximum number of evaluating the right-hand side functions is set to `maxfun=1000000` with the default being 30000.

```
>yn_nonstiff:=dsolve({ODE[1],ODE[2],ICs},{z[1](x),z[2](x)},numeric,
 maxfun=1000000);
```

yn_nonstiff := proc(x_rkf45) ... end proc

✍ Solve the system numerically again using `dsolve` with the option `numeric`. The option `stiffness=true` is specified so that the system is regarded as stiff and solved using the default Rosenbrock method.

```
>yn_stiff:=dsolve({ODE[1],ODE[2],ICs},{z[1](x),z[2](x)},numeric,
 stiff=true);
```

yn_stiff := proc(x_rosenbrock) ... end proc

✍ Evaluate the value of $z_1(x) = y_1(x)$ at $x = 10.0$ using three approaches.

`>y[1](10.0);` ✍ Numerical value evaluated from the analytical expression

0.00004539992976

✍ Numerical value for which the system is not regarded as stiff.

```
>yn_nonstiff(10.0);
```

Error, (in yn_nonstiff) cannot evaluate the solution further right of 6.1284837, maxfun limit exceeded (see ?dsolve,maxfun for details)

✍ Due to the extremely small stepsizes that the Runge-Kutta-Fehlberg method has to take, the solution cannot go beyond $x = 6.1284837$ when `maxfun` $= 10^6$. When `maxfun` $= 10^7$, the solution cannot go beyond $x = 61.294082$. The Runge-Kutta-Fehlberg method is not efficient for this stiff equation.

`>yn_stiff(10.0);` ✍ Numerical value for which the system is regarded as stiff.

$$[x = 10.0,\ z_1(x) = 0.000045480505899753,\ z_2(x) = -0.000045480505899753]$$

Example 12.23 — Dynamical Response of Parametrically Excited System

Consider the parametrically excited nonlinear system given by

$$\ddot{x} + \beta\dot{x} - (1 + \mu\cos\Omega t)\,x + \alpha x^3 = 0.$$

Examples of this equation are found in many applications of mechanics, especially in problems of dynamic stability of elastic systems. In particular, the transverse vibration of a buckled column under the excitation of a periodic end displacement is described by this equation. The system is called parametrically excited because the forcing term $\mu\cos\Omega t$ appears in the coefficient (parameter) of the equation.

✍ It is a good practice to put `restart` at the beginning of each program so that *Maple* can start fresh if the program has to be rerun.

```
>restart:
>with(plots):
```
✍ Load the `plots` package.

```
>ODE:=diff(x(t),t$2)+beta*diff(x(t),t)-(1+mu*cos(Omega*t))*x(t)
  +alpha*x(t)^3=0:
```
✍ Define the ODE.

```
>ICs:=x(0)=0,D(x)(0)=0.1:
```
✍ Define the ICs: $x(0)=0$, $\dot{x}(0)=0.1$.

Periodic Motion ($\mu = 0.3$)

```
>alpha:=1.0: beta:=0.2: Omega:=1.0: mu:=0.3:
```
✍ Assign the parameters.

✍ Solve the system numerically using `dsolve` with option `numeric`.

```
>sol:=dsolve({ODE,ICs},x(t),numeric,maxfun=1000000):
```

✍ Plot the time series $x(t)$ versus t, $(x_1 = x)$. ✍ Figure 12.1(a)

```
>odeplot(sol,[t,x(t)],t=0..500,numpoints=10000,labels=["t","x1"],
  tickmarks=[[0,100,200,300,400,500],[-1.5,-1,-0.5,0,0.5,1,1.5]]);
```

✍ Plot the time series $\dot{x}(t)$ versus t, $(x_2 = \dot{x})$. ✍ Figure 12.1(b)

```
>odeplot(sol,[t,D(x)(t)],t=0..500,numpoints=10000,labels=["t","x2"],
  tickmarks=[[0,100,200,300,400,500],[-0.8,-0.6,-0.4,-0.2,0,0.2,0.4,
  0.6,0.8]]);
```

✍ Plot the phase portrait $\dot{x}(t)$ versus $x(t)$. ✍ Figure 12.1(c)

```
>odeplot(sol,[x(t),D(x)(t)],t=0..500,numpoints=10000,view=[-1.8..1.8,
  -1.0..1.0], tickmarks=[[-1.8,-1.2,-0.6,0.6,1.2,1.8],[-1,-0.75,-0.5,
  -0.25,0.25,0.5,0.75,1]],axes=normal,labels=["x1","x2"]);
```

✍ When $\mu = 0.3$, after some transient part, the response of the system will settle down to periodic motion.

Chaotic Motion ($\mu = 0.4$)

```
>alpha:=1.0: beta:=0.2: Omega:=1.0: mu:=0.4:
```
✍ Assign the parameters.

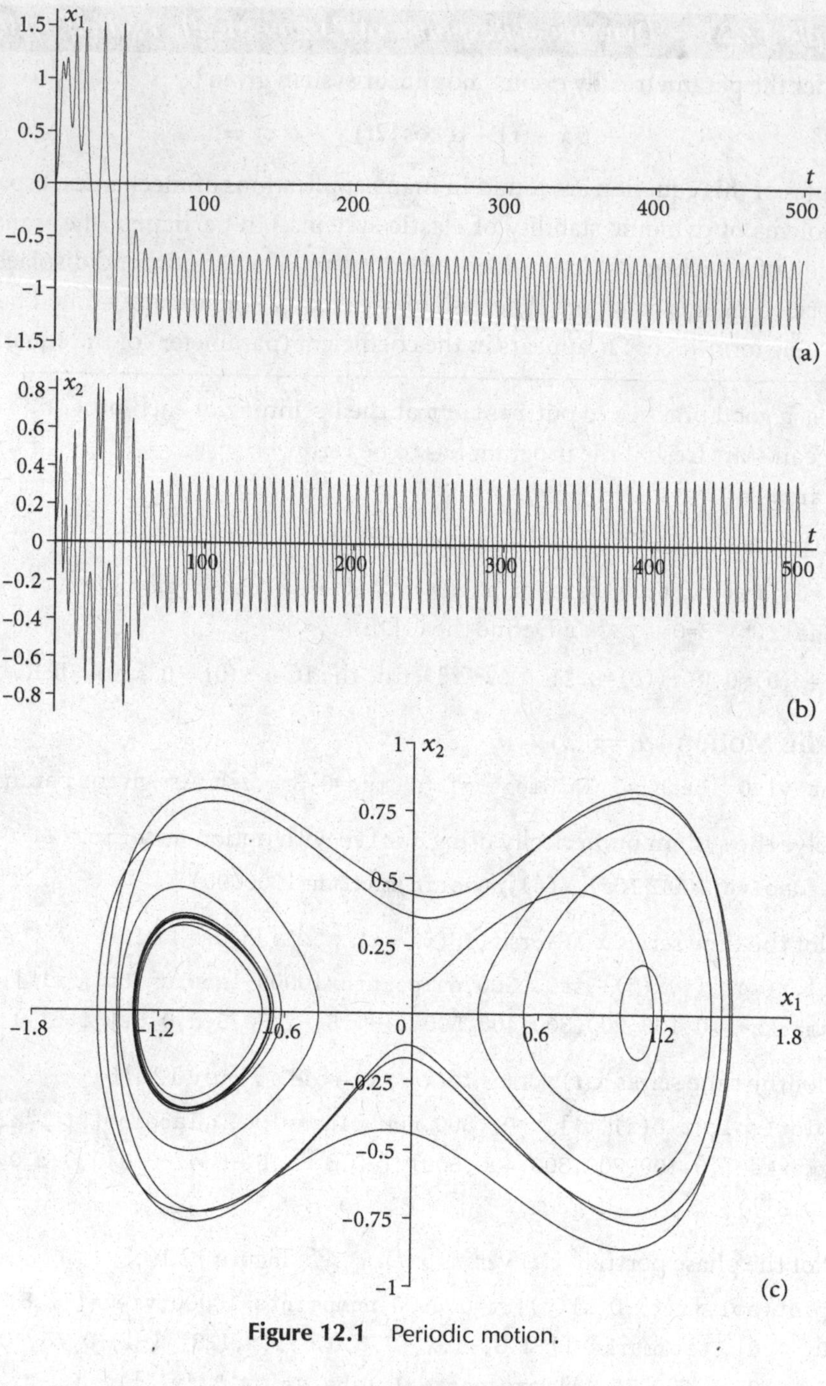

Figure 12.1 Periodic motion.

✍ Solve the system numerically using `dsolve` with option `numeric`.

```
>sol:=dsolve({ODE,ICs},x(t),numeric,maxfun=1000000):
```

✍ Plot the time series $x(t)$ versus t. ✍ Figure 12.2(a)

```
>odeplot(sol,[t,x(t)],t=0..500,numpoints=10000,labels=["t","x1"],
 tickmarks=[[0,100,200,300,400,500],[-1.5,-1,-0.5,0,0.5,1,1.5]]);
```

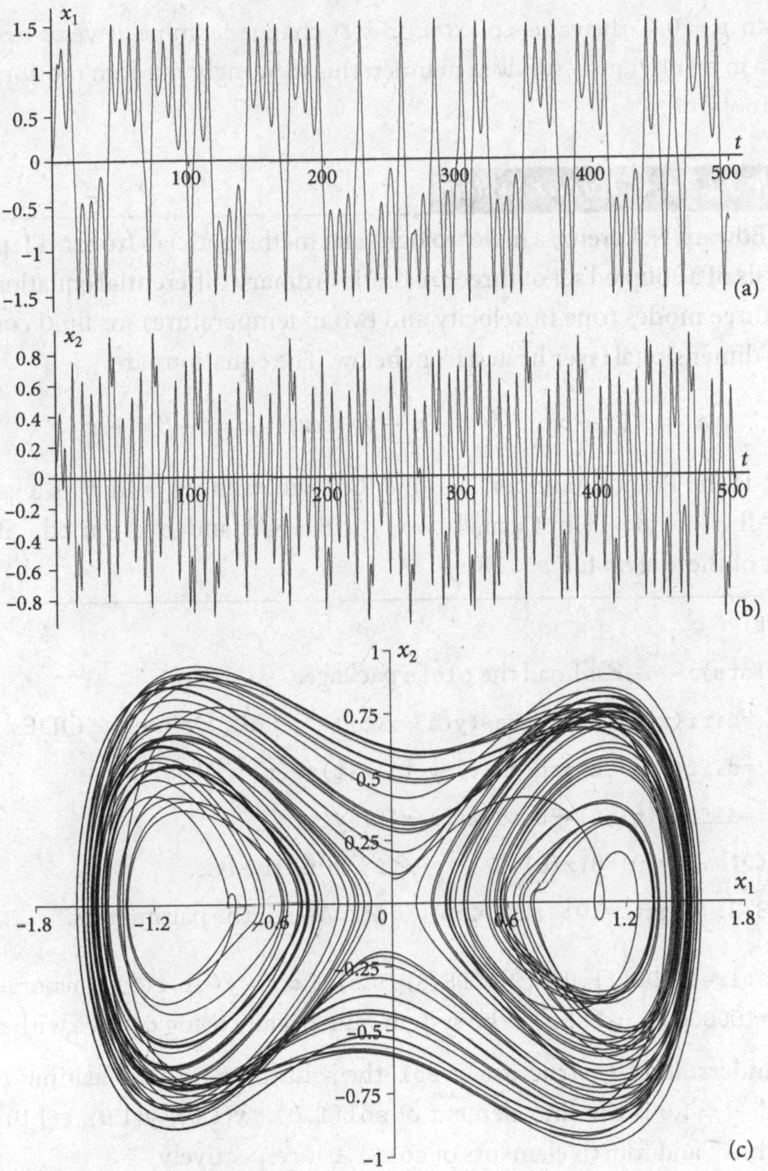

Figure 12.2 Chaotic motion.

✍ Plot the time series $\dot{x}(t)$ versus t. ✍ Figure 12.2(b)

```
>odeplot(sol,[t,D(x)(t)],t=0..500,numpoints=10000,labels=["t","x2"],
 tickmarks=[[0,100,200,300,400,500],[-0.8,-0.6,-0.4,-0.2,0,0.2,0.4,
 0.6,0.8]]);
```

✍ Plot the phase portrait $\dot{x}(t)$ versus $x(t)$. ✍ Figure 12.2(c)

```
>odeplot(sol,[x(t),D(x)(t)],t=0..500,numpoints=10000,view=[-1.8..1.8,
 -1.0..1.0],tickmarks=[[-1.8,-1.2,-0.6,0.6,1.2,1.8],[-1,-0.75,-0.5,
 -0.25,0.25,0.5,0.75,1]],axes=normal,labels=["x1","x2"]);
```

✍ When $\mu = 0.4$, the values of $x(t)$ or $\dot{x}(t)$ change from positive to negative or vice versa in an obviously random manner; this seemingly random motion is called chaotic motion.

Example 12.24 — Lorenz System

In 1963, Edward N. Lorenz, a meteorologist and mathematician from MIT, presented an analysis of a coupled set of three quadratic ordinary differential equations representing three modes (one in velocity and two in temperature) for fluid convection in a two-dimensional layer heated from below. The equations are

$$\dot{x} = \sigma(y - x), \quad \dot{y} = \rho x - y - xz, \quad \dot{z} = -\beta x + xy,$$

where σ is the Prandtl number, ρ is the Rayleigh number, and β is a geometric factor. All σ, ρ, $\beta > 0$, but usually $\sigma = 10$, $\beta = 8/3$ and ρ is varied. Study the behavior of the system for $\rho = 28$.

```
>restart:
>with(plots):
```
✍ Load the `plots` package.
```
>ODE[1]:=diff(x(t),t)=sigma*(y(t)-x(t)):
```
✍ Define the ODEs.
```
>ODE[2]:=diff(y(t),t)=rho*x(t)-y(t)-x(t)*z(t):
>ODE[3]:=diff(z(t),t)=-beta*z(t)+x(t)*y(t):
>ICs:=x(0)=-8,y(0)=8,z(0)=27:
```
✍ Define the ICs.
```
>beta:=8/3: sigma:=10: rho:=28:
```
✍ Assign the parameters.
```
>sol:=dsolve({ODE[1],ODE[2],ODE[3],ICs},{x(t),y(t),z(t)},numeric,
  maxfun=1000000):
```
✍ Solve the system numerically using `dsolve` with `numeric`.

✍ To understand the structure of `sol`, the solution `sol(1.0)` at time $t = 1.0$ is displayed. $t = 1.0$ is the first element of `sol(1.0)`. $x(1.0)$, $y(1.0)$, $z(1.0)$ are the second, third, and fourth elements of `sol(1.0)`, respectively.
```
>sol(1.0);
```
$$[t = 1.0,\ x(t) = 9.057106766983,\ y(t) = 14.55890078678,\ z(t) = 18.41514672820]$$

✍ Plot the time series $x(t)$, $y(t)$, and $z(t)$ versus t. ✍ Figure 12.3
```
>odeplot(sol,[t,x(t)],t=0..100,numpoints=10000,labels=["t","x"],
  tickmarks=[[0,20,40,60,80,100],[-15,-10,-5,0,5,10,15]]);
>odeplot(sol,[t,y(t)],t=0..100,numpoints=10000,labels=["t","y"],
  tickmarks=[[0,20,40,60,80,100],[-20,-10,0,10,20]]);
>odeplot(sol,[t,z(t)],t=0..100,numpoints=10000,labels=["t","z"],
  tickmarks=[[0,20,40,60,80,100],[0,10,20,30,40]]);
```

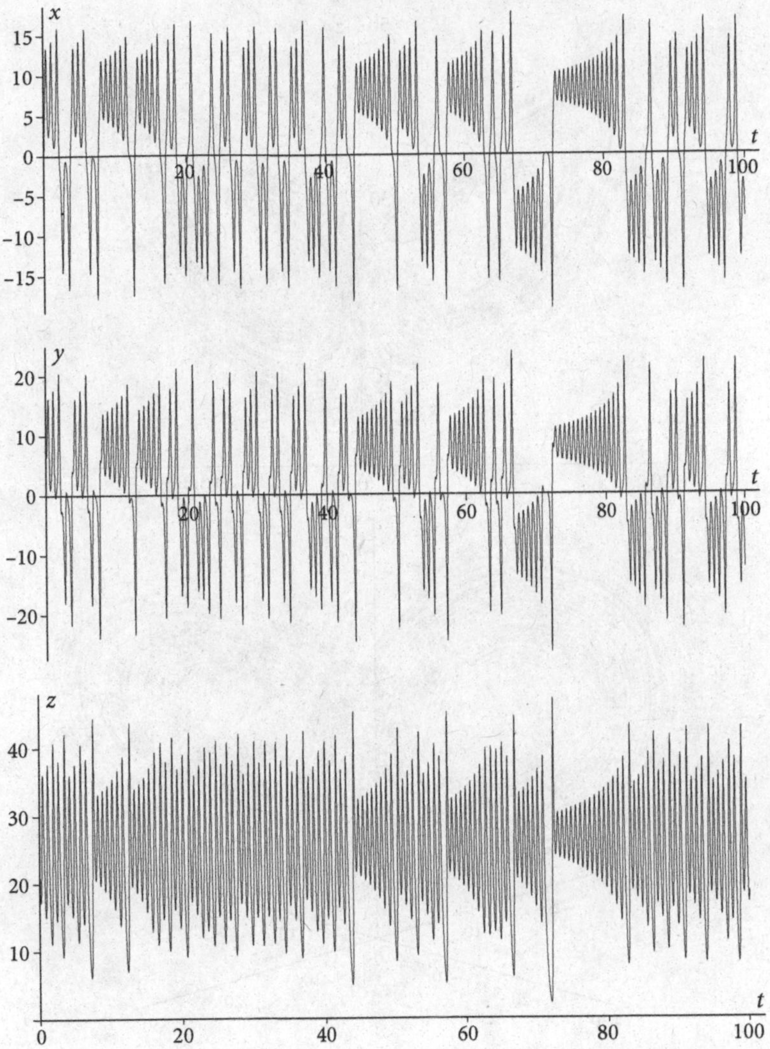

Figure 12.3 Lorenz system.

✍ Plot $z(t)$ versus $x(t)$. ✍ Figure 12.4(a)

```
>odeplot(sol,[x(t),z(t)],t=0..150,numpoints=100000,labels=["x","z"],
 view=[-20..20,0..50],tickmarks=[[-20,-10,0,10,20],[0,10,20,30,40,50]]);
```

✍ 3D plot of the variations of $x(t)$, $y(t)$, $z(t)$ with time t. ✍ Figure 12.4(b)

```
>odeplot(sol,[x(t),y(t),z(t)],t=0..150,numpoints=100000,
 orientation=[-45,65],view=[-30..30,-30..30,0..50],thickness=0,
 tickmarks=[[-30,-20,-10,0,10,20,30],[-30,-20,-10,0,10,20,30],
 [0,10,20,30,40,50]],axes=normal,labels=["x","y","z"]);
```

Remarks: When the solution is plotted in the xz-plane, the butterfly pattern is traced in "real time": the moving solution point $P(x(t), y(t), z(t))$ appears to undergo a random number of oscillations on the right followed by a random

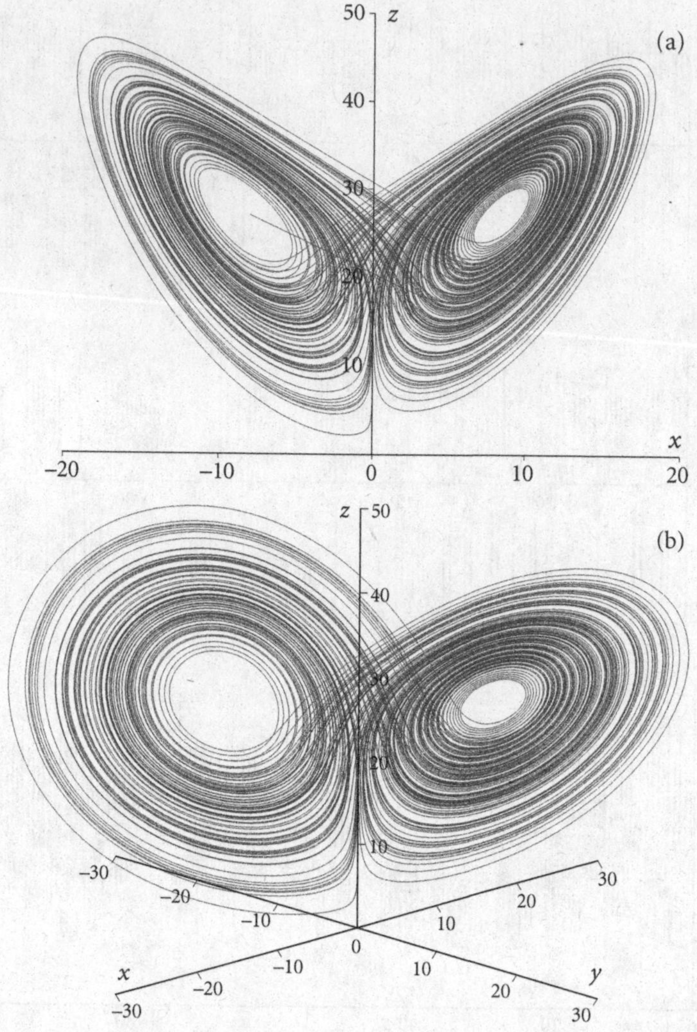

Figure 12.4 Lorenz system.

number of oscillations on the left, then a random number of oscillations on the right followed by random number of oscillations on the left, and so on. Given the meteorological origin of the Lorenz system, one naturally thinks of a random number of clear days followed by a random number of rainy days, then a random number of clear days followed by a random number of rainy days, and so on.

Solutions of Lorenz system are sensitive to the initial conditions—small variations of the initial condition may produce large variations in the long-term behavior of the system. This notion of sensitive dependence on initial conditions is the so-called butterfly effect in chaos theory—the flap of a butterfly's wings in Brazil may set off a tornado in Texas.

The difference between the solutions with only small difference in the initial conditions can be plotted to demonstrate sensitive dependence of the solutions of

Lorenz system to the initial conditions. Initially, the two solutions seem coincident. After a period of time, the difference between the two solutions is as large as the value of the solution.

✍ Define a new set of ICs. $x(0)$ is changed from -8 to -8.00001.

```
>ICs2:=x(0)=-8.00001,y(0)=8,z(0)=27:
>sol2:=dsolve({ODE[1],ODE[2],ODE[3],ICs2},{x(t),y(t),z(t)},numeric,
 maxfun=1000000):
```
✍ Solve the system again with new initial conditions ICs2.

```
>N:=3000:
```
✍ Number of points in the plot.

```
>for n from 0 by 1 to N do
>    T[n]:=n/100.:
```
✍ $t=0.00,\ 0.01,\ 0.02,\ldots,29.99,\ 30.00$.

```
>    delta[n]:=rhs(sol(T[n])[2])-rhs(sol2(T[n])[2]):
>end do:
```

✍ delta[n] is the difference between the solutions of $x(t)$ solved with two different sets of ICs at time T[n].

```
>T_list:=[seq(T[n],n=0..N)];
```
✍ Create a list of times.

T_list := [0.00, 0.01, ..., 29.99, 30.00]

```
>X_list:=[seq(delta[n],n=0..N)];
```
✍ Create a list of the differences of $x(t)$.

X_list := $[0.00001,\ 0.9056\times10^{-5},\ \ldots,\ 16.81918330,\ 16.71847501]$

```
>P:=zip('[ ]',T_list,X_list);
```
✍ Merge the lists T_list and X_list using zip.

P := $[[0.,0.00001],[0.01,0.9056\times10^{-5}],\ldots,[29.99,16.81918330],[30.,16.71847501]]$

✍ Plot data points using pointplot in the plots package with the data in a list of lists of the form $[[t_0,x_0],[t_1,x_1],\ldots,[t_n,x_n]]$.

```
>pointplot(P,style=line,color=black,tickmarks=[[0,5,10,15,20,25,30],
 [-30,-20,-10,0,10,20,30]],labels=["t","delta"],
 labelfont=[TIMES,ITALIC,14],axesfont=[TIMES,ROMAN,12]);
```

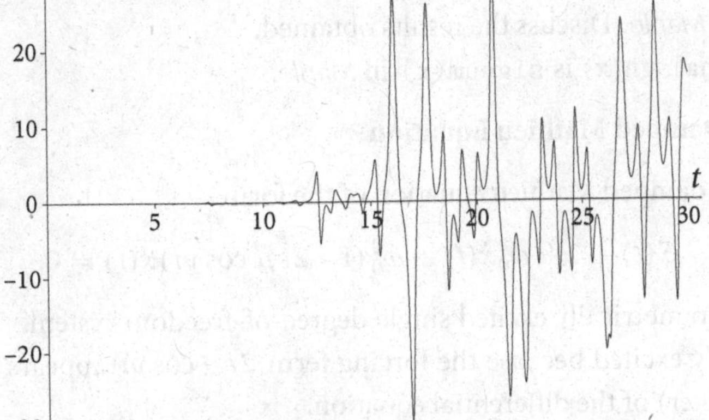

Problems

Most of the exercise and example problems presented in this book can serve as practicing problems using *Maple*.

12.1 For the example of *dynamical response of parametrically excited system*, study the dynamics of the system for $\mu = 0.34$, 0.35, and 0.57 by plotting the time series $x(t)$, $\dot{x}(t)$, and phase portraits $\dot{x}$-x using *Maple*. Discuss the results obtained.

12.2 Coulomb Dry Friction

Consider the mass-spring system moving on a rough surface as shown.

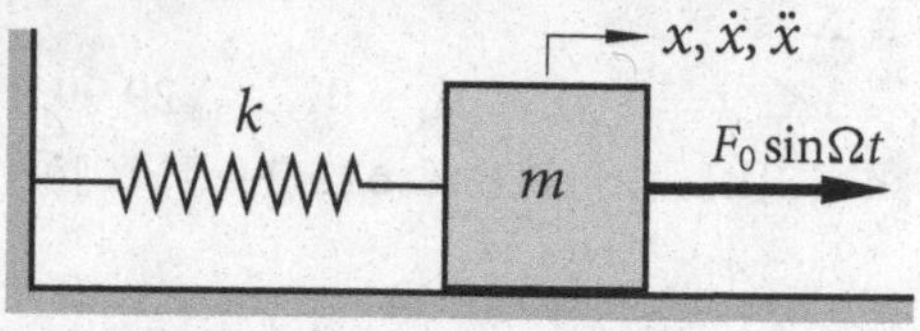

The equation of motion is given by

$$m\ddot{x} + \mu F_n \,\mathrm{sgn}(\dot{x}) + kx = F_0 \sin \Omega t,$$

where sgn(·) is the signum function defined as

$$\mathrm{sgn}(\dot{x}) = \begin{cases} +1, & \dot{x} > 0; \\ 0, & \dot{x} = 0; \\ -1, & \dot{x} < 0. \end{cases}$$

The term $\mu F_n \,\mathrm{sgn}(\dot{x})$ is the dry friction between the mass and the rough surface, and is called Coulomb damping. However, it is not proportional to the velocity, but rather depends only on the algebraic sign (direction) of the velocity. The equation of motion is consequently nonlinear.

Rewrite the equation of motion as

$$\ddot{x} + \xi \,\mathrm{sgn}(\dot{x}) + \omega_0^2 x = f_0 \sin \Omega t, \qquad \xi = \frac{\mu F_n}{m}, \quad \omega_0^2 = \frac{k}{m}, \quad f_0 = \frac{F_0}{m}.$$

Suppose the parameters $\xi = 0.1$, $\omega_0 = 2$, $f_0 = 1$, and the system is at rest at time $t = 0$. Plot the response $x(t)$ of the system, for $\Omega = 1.5$, 1.9, 1.97, 2, 2.1, 2.3, 2.5, and 4, using *Maple*. Discuss the results obtained.

✍ Note that sgn(x) is `signum(x)` in *Maple*.

12.3 The Damped Mathieu Equation

Consider the damped Mathieu equation of the form

$$\ddot{x}(t) + 2\varepsilon\zeta\omega_0\dot{x}(t) + \omega_0^2(1 - 2\varepsilon\mu \cos \nu t)x(t) = 0,$$

which is a parametrically excited single degree-of-freedom system. The system is parametrically excited because the forcing term $2\varepsilon\mu \cos \nu t$ appears in the coefficient (parameter) of the differential equation.

The Mathieu equation finds many applications in engineering applications. In particular, the transverse vibration of an elastic column subjected to harmonic axial load $P(t) = p_0 \cos \nu t$ is governed by the Mathieu equation.

It can be shown that, approximately, when the excitation frequency ν is in the following region

$$1 - \varepsilon\left(\frac{\mu^2}{4} - \zeta^2\right)^{1/2} < \frac{\nu}{2\omega_0} < 1 + \varepsilon\left(\frac{\mu^2}{4} - \zeta^2\right)^{1/2},$$

where $\varepsilon > 0$ is a small parameter, parametric resonance occurs, in which the response $x(t)$ grows exponentially.

For the parameters $\varepsilon = 0.1$, $\zeta = 0.1$, $\mu = 1$, $\omega_0 = 1$, the instability region is shown in the following figure

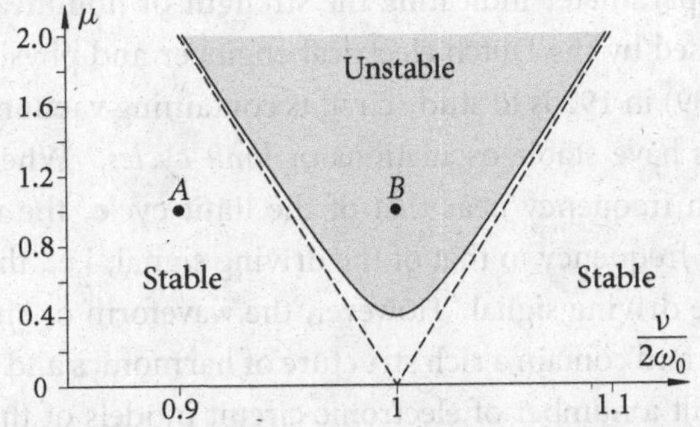

- For Case A, $\nu = 1.8$ and the system is in the *stable* region, in which the amplitude of the response *decays* exponentially.
- For Case B, $\nu = 2.0$ and the system is in the *unstable* region, in which the amplitude of the response *grows* exponentially.

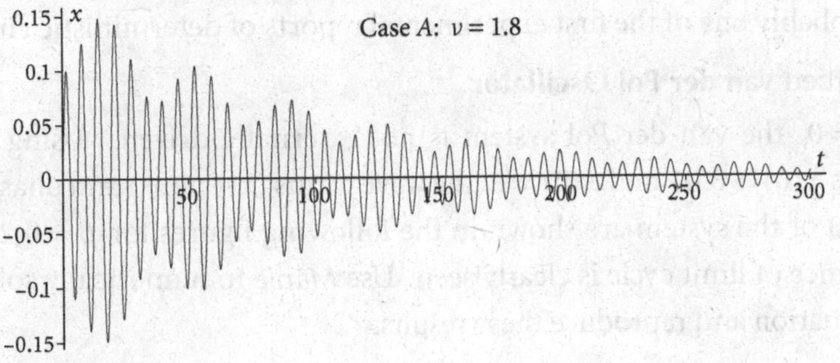

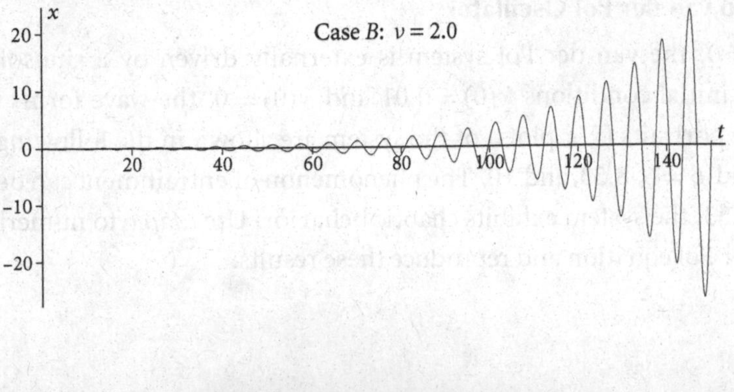

1. For these two cases, verify the stability of the response by numerically solving the equation of motion and plotting the responses using *Maple*. Use the initial conditions $x(0) = 0$ and $\dot{x}(0) = 0.1$.
2. Numerically determine by trial-and-error (accurate to the thousandth decimal place) the instability region, i.e., the value of ν_L and ν_U such that the system is in parametric resonance when $\nu_L \leqslant \nu \leqslant \nu_U$.

12.4 The van der Pol Oscillator

The van der Pol oscillator is the second-order ordinary equation with nonlinear damping given by

$$\ddot{x} + \beta(x^2 - 1)\dot{x} + x = a\cos\Omega t,$$

where $\beta > 0$ is a parameter indicating the strength of nonlinear damping. This model was proposed by the Dutch electrical engineer and physicist Balthasar van der Pol (1889–1959) in 1920s to study circuits containing vacuum tubes. He found that these circuits have stable oscillations or *limit cycles*. When they are driven with a signal with frequency near that of the limit cycle, the resulting periodic response shifts its frequency to that of the driving signal, i.e., the circuit becomes "entrained" to the driving signal. However, the waveform or signal shape can be quite complicated and contain a rich structure of harmonics and subharmonics.

Van der Pol built a number of electronic circuit models of the human heart to study the stability of heart dynamics. A real heart driven by a pacemaker is modeled by the circuit driven by an external signal. He was interested in finding out, using his entrainment work, how to stabilize a heart's irregular beating.

Van der Pol and his colleague van der Mark reported that an "irregular noise" was heard at certain driving frequencies between the natural entrainment frequencies. This is probably one of the first experimental reports of deterministic chaos.

The Unforced van der Pol Oscillator

When $a = 0$, the van der Pol system is not externally driven. Using the initial conditions $x(0) = 0.01$ and $\dot{x}(0) = 0$, the wave forms (x-t plots) and phase portraits ($\dot{x}$-x plots) of the system are shown in the following figures for $\beta = 0.2$, 1, and 5. The existence of limit cycle is clearly seen. Use *Maple* to numerically solve the van der Pol equation and reproduce these results.

The Forced van der Pol Oscillator

When $a \neq 0$, the van der Pol system is externally driven by a sinusoidal signal. Using the initial conditions $x(0) = 0.01$ and $\dot{x}(0) = 0$, the wave forms (x-t plots) and phase portraits ($\dot{x}$-x plots) of the system are shown in the following figure for $a = 1.2$ and $\beta = 6$, 8.53, and 10. The phenomenon of entrainment can be observed. For $\beta = 8.53$, the system exhibits chaotic behavior. Use *Maple* to numerically solve the van der Pol equation and reproduce these results.

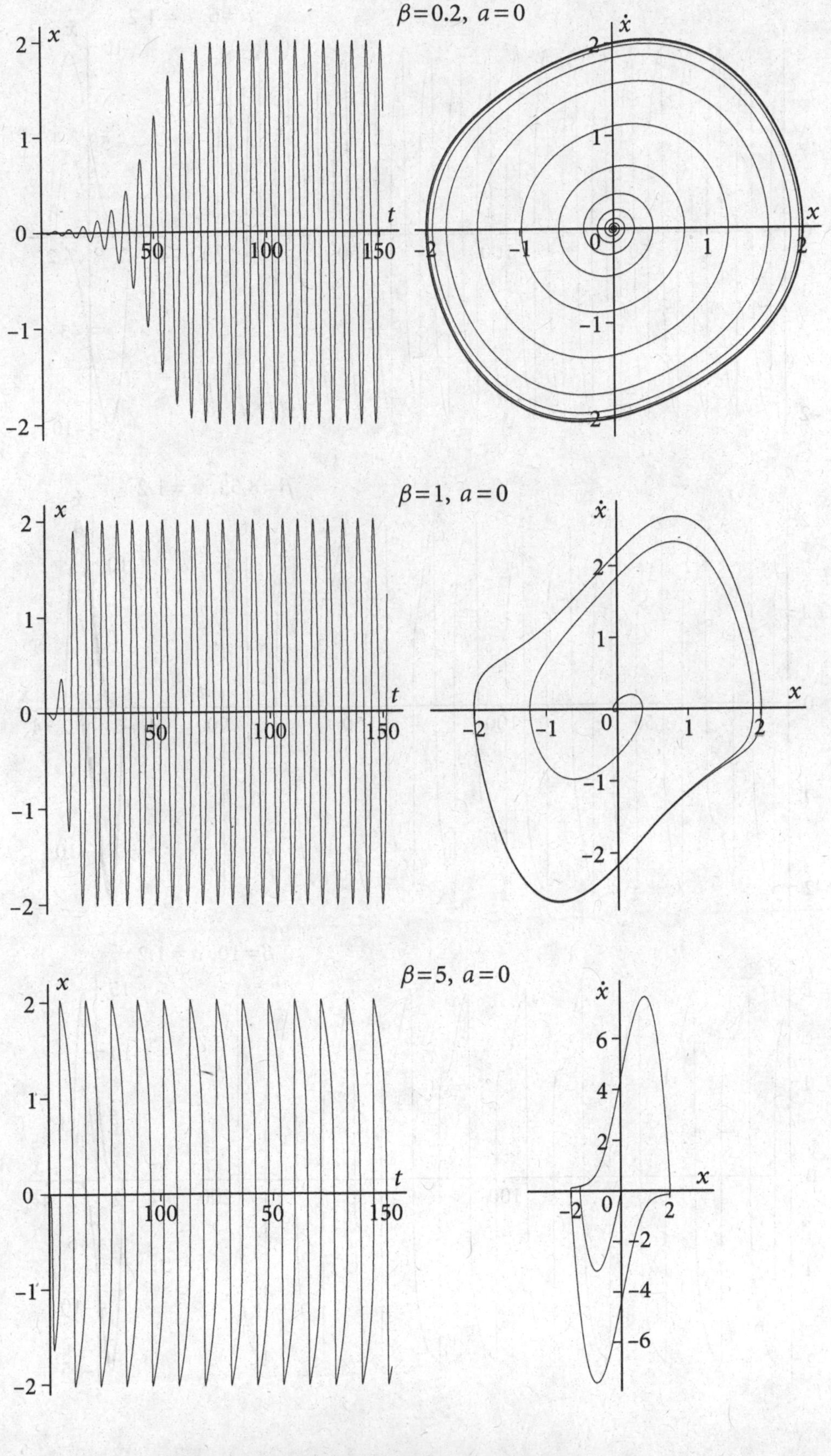
β=0.2, a=0
x
t
ẋ
x
β=1, a=0
x
t
ẋ
x
β=5, a=0
x
t
ẋ
x

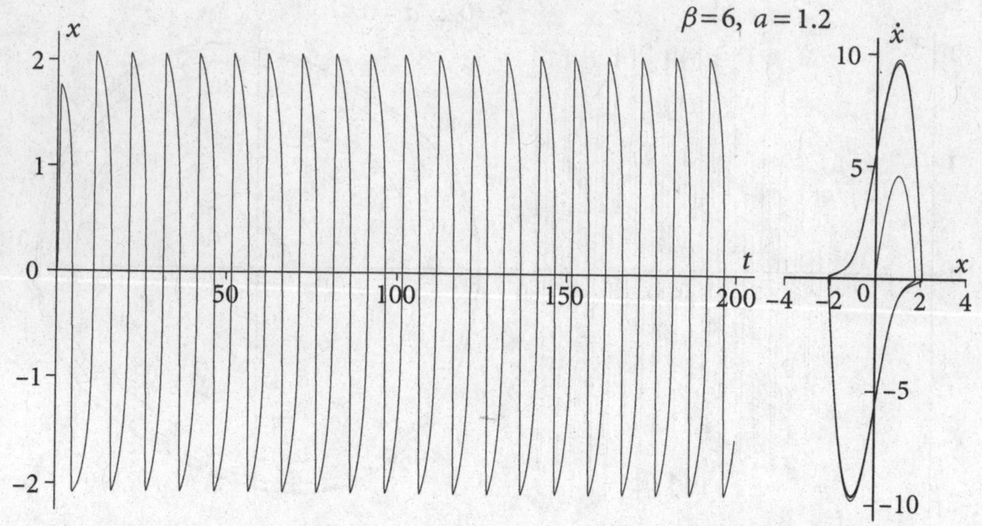
$\beta = 6,\ a = 1.2$
x
t
$\dot{x}$
x

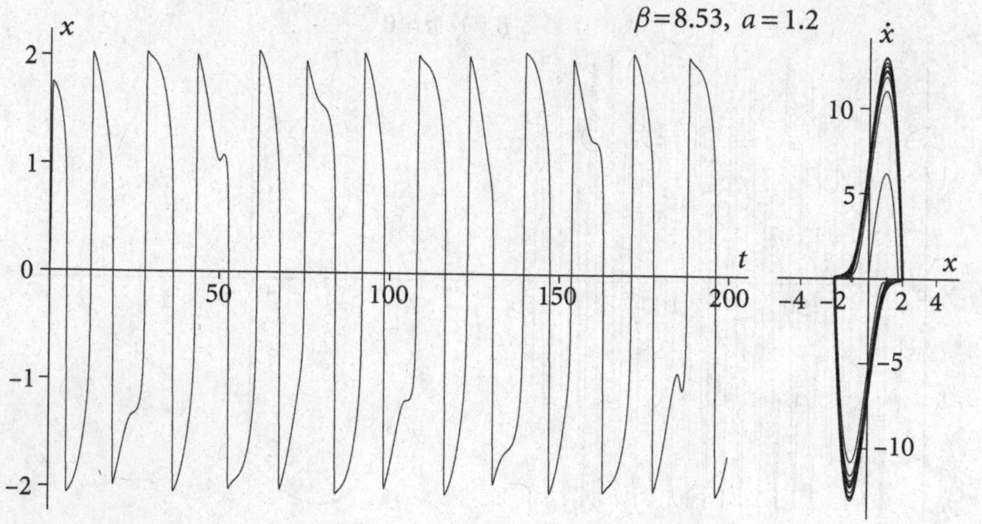
$\beta = 8.53,\ a = 1.2$
x
t
$\dot{x}$
x

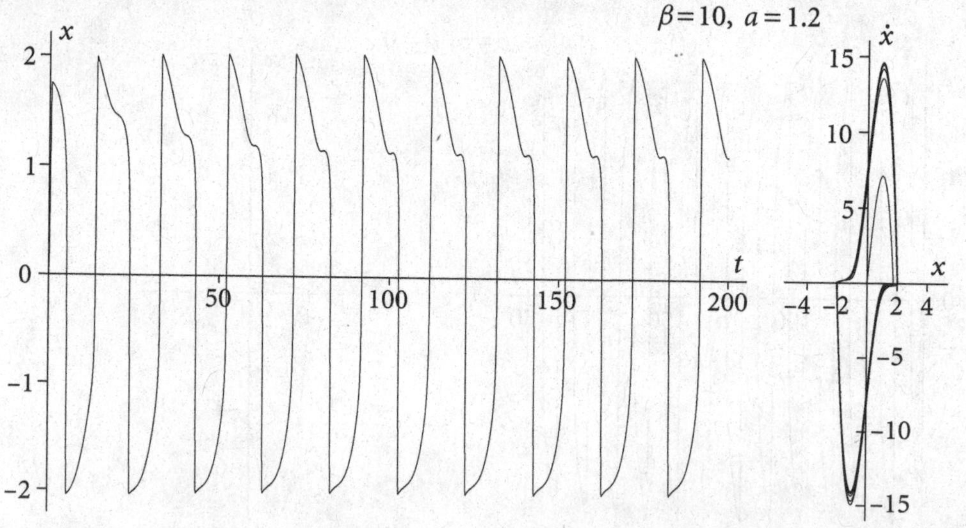
$\beta = 10,\ a = 1.2$
x
t
$\dot{x}$
x

APPENDIX A

Tables of Mathematical Formulas

A.1 Table of Trigonometric Identities

Trigonometric Functions	Hyperbolic Functions
1. $\sin^2\theta + \cos^2\theta = 1$	$\cosh^2 x - \sinh^2 x = 1$
$\Longrightarrow \quad 1 + \tan^2\theta = \sec^2\theta$	$\Longrightarrow \quad 1 - \tanh^2 x = \operatorname{sech}^2 x$
$\Longrightarrow \quad 1 + \cot^2\theta = \csc^2\theta$	$\Longrightarrow \quad \coth^2 x - 1 = \operatorname{csch}^2 x$
2. $\sin^2\theta = \dfrac{1 - \cos 2\theta}{2}$	$\sinh^2 x = \dfrac{\cosh 2x - 1}{2}$
3. $\cos^2\theta = \dfrac{1 + \cos 2\theta}{2}$	$\cosh^2 x = \dfrac{\cosh 2x + 1}{2}$
4. $\sin 2\theta = 2\sin\theta\cos\theta$	$\sinh 2x = 2\sinh x\cosh x$
5. $\cos 2\theta = \cos^2\theta - \sin^2\theta$	$\cosh 2x = \cosh^2 x + \sinh^2 x$
$= 2\cos^2\theta - 1$	$= 2\cosh^2 x - 1$
$= 1 - 2\sin^2\theta$	$= 1 + 2\sinh^2 x$

6. $\sin(A \pm B) = \sin A\cos B \pm \cos A\sin B$

7. $\cos(A \pm B) = \cos A\cos B \mp \sin A\sin B$

8. $\sin A \sin B = -\frac{1}{2}\left[\cos(A+B) - \cos(A-B)\right]$

9. $\cos A \cos B = \frac{1}{2}\left[\cos(A+B) + \cos(A-B)\right]$

10. $\sin A \cos B = \frac{1}{2}\left[\sin(A+B) + \sin(A-B)\right]$

11. $\cos A \sin B = \frac{1}{2}\left[\sin(A+B) - \sin(A-B)\right]$

12. $\sin A + \sin B = 2\sin\frac{A+B}{2}\cos\frac{A-B}{2}$

13. $\sin A - \sin B = 2\cos\frac{A+B}{2}\sin\frac{A-B}{2}$

14. $\cos A + \cos B = 2\cos\frac{A+B}{2}\cos\frac{A-B}{2}$

15. $\cos A - \cos B = -2\sin\frac{A+B}{2}\sin\frac{A-B}{2}$

16. Euler's Formula: $e^{i\theta} = \cos\theta + i\sin\theta$

17. $z^n = a(\cos\theta + i\sin\theta)$

$$\Longrightarrow z = \sqrt[n]{a}\left(\cos\frac{\theta+2k\pi}{n} + i\sin\frac{\theta+2k\pi}{n}\right), \quad k = 0, 1, \ldots, n-1$$

A.2 Table of Derivatives

1. $\dfrac{d}{dx}x^n = nx^{n-1}$

2. $\dfrac{d}{dx}e^x = e^x \implies \dfrac{d}{dx}a^x = a^x \ln a, \qquad \because a^x = e^{x \ln a}$

3. $\dfrac{d}{dx}\ln x = \dfrac{1}{x} \implies \dfrac{d}{dx}\log_a x = \dfrac{1}{x \ln a}, \qquad \because \log_a x = \dfrac{\ln x}{\ln a}$

Trigonometric Functions

4. $\dfrac{d}{dx}\sin x = \cos x$

5. $\dfrac{d}{dx}\cos x = -\sin x$

6. $\dfrac{d}{dx}\tan x = \dfrac{1}{\cos^2 x} = \sec^2 x$

7. $\dfrac{d}{dx}\cot x = -\dfrac{1}{\sin^2 x} = -\csc^2 x$

8. $\dfrac{d}{dx}\sec x = \dfrac{\sin x}{\cos^2 x} = \tan x \sec x$

9. $\dfrac{d}{dx}\cot x = -\dfrac{\cos x}{\sin^2 x} = -\cot x \csc x$

Hyperbolic Functions

$\dfrac{d}{dx}\sinh x = \cosh x$

$\dfrac{d}{dx}\cosh x = \sinh x$

$\dfrac{d}{dx}\tanh x = \dfrac{1}{\cosh^2 x} = \text{sech}^2 x$

$\dfrac{d}{dx}\coth x = -\dfrac{1}{\sinh^2 x} = -\text{csch}^2 x$

$\dfrac{d}{dx}\text{sech}\, x = -\dfrac{\sinh x}{\cosh^2 x} = -\tanh x\, \text{sech}\, x$

$\dfrac{d}{dx}\text{csch}\, x = -\dfrac{\cosh x}{\sinh^2 x} = -\coth x\, \text{csch}\, x$

Inverse Trigonometric Functions

10. $\dfrac{d}{dx}\sin^{-1} x = \dfrac{1}{\sqrt{1-x^2}} \qquad \dfrac{d}{dx}\cos^{-1} x = -\dfrac{1}{\sqrt{1-x^2}}$

11. $\dfrac{d}{dx}\tan^{-1} x = \dfrac{1}{1+x^2} \qquad \dfrac{d}{dx}\cot^{-1} x = -\dfrac{1}{1+x^2}$

12. $\dfrac{d}{dx}\sec^{-1} x = \dfrac{1}{x\sqrt{x^2-1}} \qquad \dfrac{d}{dx}\csc^{-1} x = -\dfrac{1}{x\sqrt{x^2-1}}$

A.3 Table of Integrals

1. $\int x^n \, dx = \dfrac{x^{n+1}}{n+1}, \quad n \neq -1$

2. $\int \dfrac{1}{x} \, dx = \ln|x|, \quad x \neq 0$ $\qquad \int \ln x \, dx = x(\ln x - 1), \quad x > 0$

3. $\int e^{ax} \, dx = \dfrac{1}{a} e^{ax}$ $\qquad \int b^{ax} \, dx = \dfrac{b^{ax}}{a \ln b}, \quad b > 0$

Trigonometric Functions | **Hyperbolic Functions**

4. $\int \sin x \, dx = -\cos x$ $\qquad \int \sinh x \, dx = \cosh x$

5. $\int \cos x \, dx = \sin x$ $\qquad \int \cosh x \, dx = \sinh x$

6. $\int \tan x \, dx = -\ln|\cos x|$ $\qquad \int \tanh x \, dx = \ln \cosh x$

7. $\int \cot x \, dx = \ln|\sin x|$ $\qquad \int \coth x \, dx = \ln|\sinh x|$

8. $\int \sec x \, dx = \ln|\sec x + \tan x| = \ln\left|\tan\left(\dfrac{\pi}{4} + \dfrac{x}{2}\right)\right| = \ln\left|\cot\left(\dfrac{\pi}{4} - \dfrac{x}{2}\right)\right|$ $\qquad \int \operatorname{sech} x \, dx = \tan^{-1}(\sinh x)$

9. $\int \csc x \, dx = \ln|\csc x - \cot x| = \ln\left|\tan \dfrac{x}{2}\right|$ $\qquad \int \operatorname{csch} x \, dx = \ln\left|\tanh \dfrac{x}{2}\right|$

10. $\int \sin^2 x \, dx = \dfrac{x}{2} - \dfrac{1}{4} \sin 2x$ $\qquad \int \sinh^2 x \, dx = \dfrac{1}{4} \sinh 2x - \dfrac{x}{2}$

11. $\int \cos^2 x \, dx = \dfrac{x}{2} + \dfrac{1}{4} \sin 2x$ $\qquad \int \cosh^2 x \, dx = \dfrac{1}{4} \sinh 2x + \dfrac{x}{2}$

12. $\int \tan^2 x \, dx = \tan x - x$ $\qquad \int \tanh^2 x \, dx = -\tanh x + \dfrac{1}{2} \ln\left|\dfrac{\tanh x + 1}{\tanh x - 1}\right|$

13. $\int \cot^2 x \, dx = -\cot x - x$ $\qquad \int \coth^2 x \, dx = -\coth x + \dfrac{1}{2} \ln\left|\dfrac{\coth x + 1}{\coth x - 1}\right|$

14. $\int \sec^2 x \, dx = \tan x$ $\qquad \int \operatorname{sech}^2 x \, dx = \tanh x$

15. $\int \csc^2 x \, dx = -\cot x$ $\qquad \int \operatorname{csch}^2 x \, dx = -\coth x$

16. $$\int x \sin ax \, dx = \frac{1}{a^2} \sin ax - \frac{x}{a} \cos ax$$

17. $$\int x^2 \sin ax \, dx = \frac{2x}{a^2} \sin ax - \frac{a^2x^2 - 2}{a^3} \cos ax$$

18. $$\int x^3 \sin ax \, dx = \frac{3a^2x^2 - 6}{a^4} \sin ax - \frac{a^2x^3 - 6x}{a^3} \cos ax$$

19. $$\int x^4 \sin ax \, dx = \frac{4a^2x^3 - 24x}{a^4} \sin ax - \frac{a^4x^4 - 12a^2x^2 + 24}{a^5} \cos ax$$

20. $$\int x \cos ax \, dx = \frac{x}{a} \sin ax + \frac{1}{a^2} \cos ax$$

21. $$\int x^2 \cos ax \, dx = \frac{a^2x^2 - 2}{a^3} \sin ax + \frac{2x}{a^2} \cos ax$$

22. $$\int x^3 \cos ax \, dx = \frac{a^2x^3 - 6x}{a^3} \sin ax + \frac{3a^2x^2 - 6}{a^4} \cos ax$$

23. $$\int x^4 \cos ax \, dx = \frac{a^4x^4 - 12a^2x^2 + 24}{a^5} \sin ax + \frac{4a^2x^3 - 24x}{a^4} \cos ax$$

24. $$\int e^{bx} \sin ax \, dx = e^{bx} \frac{b \sin ax - a \cos ax}{a^2 + b^2}$$

25. $$\int e^{bx} \cos ax \, dx = e^{bx} \frac{a \sin ax + b \cos ax}{a^2 + b^2}$$

26. $$\int x e^{ax} \, dx = \frac{ax - 1}{a^2} e^{ax}$$

27. $$\int x^2 e^{ax} \, dx = \frac{a^2x^2 - 2ax + 2}{a^3} e^{ax}$$

28. $$\int x^3 e^{ax} \, dx = \frac{a^3x^3 - 3a^2x^2 + 6ax - 6}{a^4} e^{ax}$$

29. $$\int x^4 e^{ax} \, dx = \frac{a^4x^4 - 4a^3x^3 + 12a^2x^2 - 24ax + 24}{a^5} e^{ax}$$

30. $$\int x^5 e^{ax} \, dx = \frac{a^5x^5 - 5a^4x^4 + 20a^3x^3 - 60a^2x^2 + 120ax - 120}{a^6} e^{ax}$$

31. $$\int \frac{1}{a^2 + x^2} \, dx = \frac{1}{a} \tan^{-1} \frac{x}{a}, \quad a > 0$$

32. $$\int \frac{1}{a^2 - x^2} \, dx = \frac{1}{2a} \ln \frac{a + x}{a - x} \quad \text{or} \quad \frac{1}{a} \tanh^{-1} \frac{x}{a}, \quad |x| < |a|$$

33. $\displaystyle\int \frac{1}{x^2-a^2}\,dx = \frac{1}{2a}\ln\frac{x-a}{x+a} \quad\text{or}\quad -\frac{1}{a}\coth^{-1}\frac{x}{a}, \quad |x|>|a|$

34. $\displaystyle\int \frac{1}{\sqrt{a^2-x^2}}\,dx = \sin^{-1}\frac{x}{a} \quad\text{or}\quad -\cos^{-1}\frac{x}{a}$

35. $\displaystyle\int \frac{1}{\sqrt{x^2-a^2}}\,dx = \cosh^{-1}\frac{x}{a} \quad\text{or}\quad \ln\left|x+\sqrt{x^2-a^2}\right|$

36. $\displaystyle\int \frac{1}{\sqrt{x^2+a^2}}\,dx = \sinh^{-1}\frac{x}{a} \quad\text{or}\quad \ln\left(x+\sqrt{x^2+a^2}\right)$

37. $\displaystyle\int \sqrt{a^2-x^2}\,dx = \frac{x}{2}\sqrt{a^2-x^2}+\frac{a^2}{2}\sin^{-1}\frac{x}{a}$

38. $\displaystyle\int \sqrt{x^2+a^2}\,dx = \frac{x}{2}\sqrt{x^2+a^2}+\frac{a^2}{2}\sinh^{-1}\frac{x}{a}$

or $\displaystyle\frac{x}{2}\sqrt{x^2+a^2}+\frac{a^2}{2}\ln\left(x+\sqrt{x^2+a^2}\right)$

39. $\displaystyle\int \sqrt{x^2-a^2}\,dx = \frac{x}{2}\sqrt{x^2-a^2}-\frac{a^2}{2}\cosh^{-1}\frac{x}{a}$

or $\displaystyle\frac{x}{2}\sqrt{x^2-a^2}-\frac{a^2}{2}\ln\left|x+\sqrt{x^2-a^2}\right|$

40. $\displaystyle\int \frac{1}{x\sqrt{a^2-x^2}}\,dx = -\frac{1}{a}\cosh^{-1}\frac{a}{x} \quad\text{or}\quad -\frac{1}{a}\ln\left|\frac{a+\sqrt{a^2-x^2}}{x}\right|$

41. $\displaystyle\int \frac{1}{x\sqrt{x^2+a^2}}\,dx = -\frac{1}{a}\sinh^{-1}\frac{a}{x} \quad\text{or}\quad -\frac{1}{a}\ln\left|\frac{a+\sqrt{x^2+a^2}}{x}\right|$

42. $\displaystyle\int \frac{1}{x\sqrt{x^2-a^2}}\,dx = \frac{1}{a}\cos^{-1}\frac{a}{x} \quad\text{or}\quad \frac{1}{a}\sec^{-1}\frac{x}{a}$

43. $\displaystyle\int \frac{1}{\sqrt{2ax-x^2}}\,dx = \cos^{-1}\left(1-\frac{x}{a}\right) \quad\text{or}\quad \sin^{-1}\left(\frac{x}{a}-1\right)$

44. $\displaystyle\int \sqrt{2ax-x^2}\,dx = \frac{x-a}{2}\sqrt{2ax-x^2}+\frac{a^2}{2}\sin^{-1}\left(\frac{x}{a}-1\right)$

45. $\displaystyle\int_0^{\frac{\pi}{2}} \begin{Bmatrix}\sin^n\theta\\ \cos^n\theta\end{Bmatrix} d\theta = \frac{(n-1)!!}{n!!}\times\begin{cases}\frac{1}{2}\pi, & \text{if } n \text{ is an even integer}\\ 1, & \text{if } n \text{ is an odd integer}\end{cases}$

46. $\displaystyle\int_0^{\frac{\pi}{2}} \sin^n\theta\cos^m\theta\,d\theta = \frac{(n-1)!!\,(m-1)!!}{(n+m)!!}\times\begin{cases}\frac{1}{2}\pi, & n, m \text{ even integers}\\ 1, & \text{otherwise}\end{cases}$

✍ $\displaystyle n!! = \begin{cases} n\cdot(n-2)\cdots5\cdot3\cdot1, & n>0 \text{ odd integer}\\ n\cdot(n-2)\cdots6\cdot4\cdot2, & n>0 \text{ even integer}\\ 1, & n=0\end{cases}$

A.4 Table of Laplace Transforms

Properties of Laplace Transform $\mathcal{L}\{f(t)\} = F(s)$

1. Laplace Transform of Derivatives

$$\mathcal{L}\{f'(t)\} = sF(s) - f(0)$$

$$\mathcal{L}\{f''(t)\} = s^2F(s) - sf(0) - f'(0)$$

$$\cdots \quad \cdots$$

$$\mathcal{L}\{f^{(n)}(t)\} = s^nF(s) - s^{n-1}f(0) - s^{n-2}f'(0) - \cdots - sf^{(n-2)}(0) - f^{(n-1)}(0), \quad n = 1, 2, \ldots$$

2. Laplace Transform of Integrals

$$\mathcal{L}\left\{\int_0^t \cdots \int_0^t f(u)\,(\mathrm{d}u)^n\right\} = \frac{F(s)}{s^n}$$

3. Property of Shifting

$$\mathcal{L}\{\mathrm{e}^{at}f(t)\} = F(s-a)$$

4. Property of Differentiation

$$\mathcal{L}\{t^nf(t)\} = (-1)^n\frac{\mathrm{d}^nF(s)}{\mathrm{d}s^n}, \quad n = 1, 2, \ldots$$

5. Property of Integration

$$\mathcal{L}\left\{\frac{f(t)}{t^n}\right\} = \int_s^\infty \cdots \int_s^\infty F(s)\,(\mathrm{d}s)^n, \quad n = 1, 2, \ldots$$

6. Convolution Integral

$$\mathcal{L}\left\{\int_0^t f(u)\,g(t-u)\,\mathrm{d}u\right\} = F(s)\,G(s)$$

7. Heaviside Function

$$\mathcal{L}\{H(t-a)\} = \frac{1}{s}\mathrm{e}^{-as}, \qquad \mathcal{L}\{f(t-a)\,H(t-a)\} = \mathrm{e}^{-as}F(s)$$

8. Dirac Delta Function

$$\mathcal{L}\{\delta(t-a)\} = \mathrm{e}^{-as}, \qquad \mathcal{L}\{f(t)\,\delta(t-a)\} = \mathrm{e}^{-as}f(a)$$

	$f(t)$	$\mathscr{L}\{f(t)\} = F(s)$
1.	1	$\frac{1}{s}, \quad s>0$
2.	$t^n, \quad n=1,2,\ldots$	$\frac{n!}{s^{n+1}}, \quad s>0$
3.	$t^\nu, \quad \nu>-1$	$\frac{\Gamma(\nu+1)}{s^{\nu+1}}, \quad s>0$
4.	$\sin\omega t$	$\frac{\omega}{s^2+\omega^2}, \quad s>0$
5.	$\cos\omega t$	$\frac{s}{s^2+\omega^2}, \quad s>0$
6.	$t\sin\omega t$	$\frac{2\omega s}{(s^2+\omega^2)^2}, \quad s>0$
7.	$t\cos\omega t$	$\frac{s^2-\omega^2}{(s^2+\omega^2)^2}, \quad s>0$
8.	e^{at}	$\frac{1}{s-a}, \quad s>a$
9.	$\sinh\omega t$	$\frac{\omega}{s^2-\omega^2}, \quad s>\lvert\omega\rvert$
10.	$\cosh\omega t$	$\frac{s}{s^2-\omega^2}, \quad s>\lvert\omega\rvert$
11.	$\sinh^2\omega t$	$\frac{2\omega^2}{s(s^2-4\omega^2)}, \quad s>2\lvert\omega\rvert$
12.	$\cosh^2\omega t$	$\frac{s^2-2\omega^2}{s(s^2-4\omega^2)}, \quad s>2\lvert\omega\rvert$
13.	$t^{\nu-1}\sinh\omega t, \ \mathcal{R}e(\nu)>-1, \ \nu\neq 0$	$\frac{\Gamma(\nu)}{2}\left[\frac{1}{(s-\omega)^\nu}-\frac{1}{(s+\omega)^\nu}\right], \quad s>\lvert\omega\rvert$
14.	$t^{\nu-1}\cosh\omega t, \ \mathcal{R}e(\nu)>0$	$\frac{\Gamma(\nu)}{2}\left[\frac{1}{(s-\omega)^\nu}+\frac{1}{(s+\omega)^\nu}\right], \quad s>\lvert\omega\rvert$
15.	$f(a^t), \quad a>0$	$\frac{1}{a}F\left(\frac{s}{a}\right)$
16.	$\int_t^\infty \frac{f(u)}{u}\,du$	$\frac{1}{s}\int_0^s F(u)\,du$

A.5 Table of Inverse Laplace Transforms

Properties of Inverse Laplace Transform $\mathscr{L}^{-1}\{F(s)\} = f(t)$

1. Property of Shifting

$$\mathscr{L}^{-1}\{F(s-a)\} = e^{at} f(t)$$

2. Property of Differentiation

$$\mathscr{L}^{-1}\left\{\frac{d^n F(s)}{ds^n}\right\} = (-1)^n t^n f(t), \quad n = 1, 2, \ldots$$

3. Property of Integration

$$\mathscr{L}^{-1}\left\{\int_s^\infty \cdots \int_s^\infty F(s)\,(ds)^n\right\} = \frac{f(t)}{t^n}, \quad n = 1, 2, \ldots$$

4. Convolution Integral

$$\mathscr{L}^{-1}\{F(s)\,G(s)\} = \int_0^t f(u)\,g(t-u)\,du = \int_0^t g(u)\,f(t-u)\,du$$

5. Heaviside Function

$$\mathscr{L}^{-1}\{e^{-as} F(s)\} = f(t-a)\,H(t-a)$$

6. Dirac Delta Function

$$\mathscr{L}^{-1}\{1\} = \delta(t), \qquad \mathscr{L}^{-1}\{e^{-as}\} = \delta(t-a)$$

	$F(s)$	$\mathscr{L}^{-1}\{F(s)\} = f(t)$
1.	$\dfrac{\omega}{s^2+\omega^2}$	$\sin \omega t$
2.	$\dfrac{s}{s^2+\omega^2}$	$\cos \omega t$
3.	$\dfrac{1}{s^n}, \quad n = 1, 2, \ldots$	$\dfrac{1}{(n-1)!}\, t^{n-1}$
4.	$\dfrac{1}{s^{n+\frac{1}{2}}}, \quad n = 0, 1, \ldots$	$\dfrac{2^{2^n} n!}{\sqrt{\pi}\,(2n)!}\, t^{n-\frac{1}{2}}$
5.	$\dfrac{1}{(s-a)(s-b)}, \quad a \neq b$ $\dfrac{1}{s^2-a^2}$	$\dfrac{1}{a-b}\left(e^{at} - e^{bt}\right)$ $\dfrac{1}{a} \sinh at$
6.	$\dfrac{s}{(s-a)(s-b)}, \quad a \neq b$ $\dfrac{s}{s^2-a^2}$	$\dfrac{1}{a-b}\left(a\,e^{at} - b\,e^{bt}\right)$ $\cosh at$
7.	$\dfrac{1}{(s-a)(s-b)(s-c)}, \quad a \neq b \neq c$	$-\dfrac{(b-c)\,e^{at} + (c-a)\,e^{bt} + (a-b)\,e^{ct}}{(a-b)(b-c)(c-a)}$
8.	$\dfrac{1}{s^4+4a^4}$	$\dfrac{1}{4a^3}(\sin at \cosh at - \cos at \sinh at)$
9.	$\dfrac{s}{s^4+4a^4}$	$\dfrac{1}{2a^2} \sin at \sinh at$
10.	$\dfrac{s^2}{s^4+4a^4}$	$\dfrac{1}{2a}(\sin at \cosh at + \cos at \sinh at)$
11.	$\dfrac{s^3}{s^4+4a^4}$	$\cos at \cosh at$
12.	$\dfrac{1}{s\,(s^4+4a^4)}$	$\dfrac{1 - \cos at \cosh at}{4a^4}$
13.	$\dfrac{1}{s^2\,(s^4+4a^4)}$	$\dfrac{2at - (\sin at \cosh at + \cos at \sinh at)}{8a^5}$
14.	$\dfrac{1}{s^3\,(s^4+4a^4)}$	$\dfrac{a^2t^2 - \sin at \sinh at}{8a^6}$
15.	$\dfrac{1}{s^4\,(s^4+4a^4)}$	$\dfrac{2a^3t^3 - 3(\sin at \cosh at - \cos at \sinh at)}{48a^7}$

	$F(s)$	$\mathscr{L}^{-1}\{F(s)\} = f(t)$
16.	$\dfrac{1}{(s^2+a^2)^2}$	$\dfrac{1}{2a^3}(\sin at - at\cos at)$
17.	$\dfrac{s}{(s^2+a^2)^2}$	$\dfrac{1}{2a}t\sin at$
18.	$\dfrac{1}{(s^2+a^2)^3}$	$\dfrac{1}{8a^5}\left[(3-a^2t^2)\sin at - 3at\cos at\right]$
19.	$\dfrac{s}{(s^2+a^2)^3}$	$\dfrac{1}{8a^3}t(\sin at - at\cos at)$
20.	$\dfrac{1}{(s^2+a^2)(s^2+b^2)}$	$\dfrac{a\sin bt - b\sin at}{ab(a^2-b^2)}$
21.	$\dfrac{s}{(s^2+a^2)(s^2+b^2)}$	$\dfrac{\cos bt - \cos at}{a^2-b^2}$
22.	$\dfrac{s^2}{(s^2+a^2)(s^2+b^2)}$	$\dfrac{a\sin at - b\sin bt}{a^2-b^2}$
23.	$\dfrac{s^3}{(s^2+a^2)(s^2+b^2)}$	$\dfrac{a^2\cos at - b^2\cos bt}{a^2-b^2}$
24.	$\dfrac{1}{s^4-a^4}$	$\dfrac{1}{2a^3}(\sinh at - \sin at)$
25.	$\dfrac{s}{s^4-a^4}$	$\dfrac{1}{2a^2}(\cosh at - \cos at)$
26.	$\dfrac{s^2}{s^4-a^4}$	$\dfrac{1}{2a}(\sinh at + \sin at)$
27.	$\dfrac{s^3}{s^4-a^4}$	$\dfrac{1}{2}(\cosh at + \cos at)$
28.	$\dfrac{1}{s^3+8a^3}$	$\dfrac{e^{at}(\sqrt{3}\sin\sqrt{3}at - \cos\sqrt{3}at) + e^{-2at}}{12a^2}$
29.	$\dfrac{s}{s^3+8a^3}$	$\dfrac{e^{at}(\sqrt{3}\sin\sqrt{3}at + \cos\sqrt{3}at) - e^{-2at}}{6a}$
30.	$\dfrac{s^2}{s^3+8a^3}$	$\dfrac{2e^{at}\cos\sqrt{3}at + e^{-2at}}{3}$

Index

A

Amplitude, 195, 201, 208
  modulated, 208
Analytic function, 393
Ascending motion of a rocket problem, 421
Automobile ignition circuit problem, 209
Auxiliary equation, 144
Average slope method, 437, 446, 453

B

Backward Euler method, 436, 446, 449, 453
Bar with variable cross-section problem, 121
Base excitation, 190–191
Beam-column, 218
  boundary conditions, 219
  equation of equilibrium, 220
  Laplace transform, 280
Beam
  flexural motion, 465
    equation of motion, 465
    forced vibration, 471
    free vibration, 466
Beams on elastic foundation, 283
  boundary conditions, 284
  equation of equilibrium, 284
  problem, 284, 288
Beat, 208
Bernoulli differential equation, 58, 75
Bessel's differential equation, 390, 408, 420, 424–426
  application, 418
  series solution, 408, 418
  solution, 418
Bessel function, 418, 425
  Maple, 512
  of the first kind, 410–411, 415, 418, 425
  of the second kind, 412–413, 415, 418, 425
Body cooling in air problem, 87
Boundary value problem, 10
Buckling of a tapered column problem, 418
  Maple, 513
Buckling
  column, 221
  tapered column, 418
    Maple, 513
Bullet through a plate problem, 94

C

Cable of a suspension bridge problem, 100
Cable under self-weight, 102
Capacitance, 108
Capacitor, 108
Chain moving problem, 123, 125
Characteristic equation, 143–144, 147, 151, 180, 305, 344, 460
  complex roots, 147
  linear differential equation, 144, 180
  multiple degrees-of-freedom system, 344
  real distinct roots, 143–144
  repeated roots, 151
Characteristic number, 144, 460
Circuit, 108, 209, 275, 278, 372–373, 375
  first-order, 113
  parallel RC, 110
  parallel RL, 111
  parallel RLC, 209
  RC, 109
  RL, 109
  second-order, 213
    Laplace transform, 275, 278
  series RC, 110
  series RL, 110
  series RLC, 209
  system of linear differential equations
    Laplace transform, 373
    matrix method, 375
Clairaut equation, 67
Complementary solution
  linear differential equation, 142–144, 147, 152, 180, 460
  matrix method, 326, 350
    complex eigenvalues, 328–329, 349
    distinct eigenvalues, 326–327, 349
    multiple eigenvalues, 330–331, 349
  method of operator

system of linear differential equations, 304–305, 348
Complex numbers, 148
Constant slope method, 432–433, 445
Convolution integral, 258
Cooling, 87
Cover-up method, 260
Cramer's Rule, 307
Critically damped system, 199
Laplace transform, 279
Cumulative error, 435, 452–453

D

D'Alembert's Principle, 91, 223, 229, 358–359, 364–365
***D*-operator, 140, 162**
inverse, 162
properties, 141
Damped natural circular frequency, 194, 196
Damped natural frequency, 196
Damper, 189, 302
Damping, 189, 302
Damping coefficient, 193–194, 198–199
modal, 347
Damping force, 189–190, 302
Dashpot damper, 189, 302
Defective (deficient) matrix, 331
Degree-of-freedom system
four, 362
infinitely many, 468
multiple, 301, 303, 344
damped forced vibration, 346
orthogonality of mode shapes, 345
undamped forced vibration, 346
undamped free vibration, 344
single, 188, 191
Laplace transform, 268
response, 193
two, 357, 377
Derivatives table, 533
Differential equation
boundary value problem, 10
definition, 6
existence and uniqueness, 11–12
general solution, 8
homogeneous, 7, 20
initial value problem, 10
numerical solution, 431
linear, 6–7, 140
constant coefficients, 7, 140
variable coefficients, 7, 178, 183
Maple, 498–499
Mathematica, 498
nonhomogeneous, 7
nonlinear, 6
numerical solution, 431
average slope method, 437, 453
backward Euler method, 436, 449, 453
constant slope method, 432–433
cumulative error, 435, 452–453
error analysis, 434
Euler method, 432–433, 449, 452
explicit method, 452
forward Euler method, 432–433, 452
GNU Scientific Library, 453
implicit method, 437–438, 453
improved Euler method, 437, 449, 453
improved Euler predictor-corrector method, 439, 452
IMSL Library, 453
local error, 434–435, 453
Maple, 453
Mathematica, 453
Matlab, 453
midpoint method, 441
NAG Library, 453
Numerical Recipes, 453
predictor-corrector technique, 438
roundoff error, 434
Runge-Kutta-Fehlberg method, 517–518
Runge-Kutta method, 440–444, 452
trapezoidal rule method, 438
truncation error, 434–435, 453
order, 6
ordinary, 6
partial, 8
particular solution, 8
series solution, 390
Frobenius series, 403, 405, 425
Fuchs' Theorem, 405
indicial equation, 404, 425
Maple, 512
ordinary point, 394, 397
regular singular point, 403
singular solution, 19
stiff, 449–450, 453
Maple, 517
Dirac delta function, 254–256
concentrated load, 471
Laplace transform, 256
Maple, 507
properties, 255

Direction field, 11
Displacement meter problem, 229
Dynamic magnification factor (DMF), 202, 370
Dynamical response of parametrically excited system problem, 518

E

Earthquake, 190
Eigenvalue, 326, 328–331, 349
 Maple, 510
Eigenvector, 326, 328–331, 349
 generalized, 331, 349
 Maple, 510
Error analysis, 434
Euler's formula, 147, 149
Euler constant, 413
Euler differential equation, 178, 183
Euler method, 432–433, 445, 449, 452
Exact differential equation, 31–33, 76
Example
 ascending motion of a rocket, 421
 automobile ignition circuit, 209
 bar with variable cross-section, 121
 beam-column
  Laplace transform, 280
 beams on elastic foundation, 284, 288
 body cooling in air, 87
 buckling of a tapered column, 418
  Maple, 513
 bullet through a plate, 94
 cable of a suspension bridge, 100
 chain moving, 123, 125
 displacement meter, 229
 dynamical response of parametrically excited system, 518
 ferry boat, 120
 float and cable, 107
 flywheel vibration, 227
 free flexural vibration of a simply supported beam, 468
 heating in a building, 88
 jet engine vibration, 223
 Lorenz system, 522
 object falling in air, 95
 particular moving in a plane, 300
 piston vibration, 224
 reservoir pollution, 127
 second-order circuit
  Laplace transform, 275
 single degree-of-freedom system under blast force, 273
 single degree-of-freedom system under sinusoidal excitation, 270
 two degrees-of-freedom system, 357
 vehicle passing a speed bump, 213
  Laplace transform, 272
 vibration of an automobile, 362
 water leaking, 126
 water tower, 220
Excitation frequency, 202, 204, 208
Existence and uniqueness theorem, 12
Explicit method, 452
Externally applied force, 189, 191

F

Ferry boat problem, 120
Finned surface, 476–477, 488
First-order circuit, 113
 problem, 111–112
First-order differential equation
 Bernoulli, 58, 75
 Clairaut, 67
 exact, 31–33, 76
 homogeneous, 20, 75
 inspection, 45, 76
 integrating factor, 31, 39–40, 76
 integrating factor by groups, 48, 77
 linear, 55, 75
 Maple, 499
 separation of variables, 16, 20, 75
 solvable for dependent variable, 61, 77
 solvable for independent variable, 61–62, 77
 special transformation, 25, 77
Flexural motion of beam, 465
 equation of motion, 465
 forced vibration, 471
  separation of variables, 471
 free vibration, 466
  infinitely many degrees-of-freedom system, 468
  separation of variables, 466
Float and cable problem, 107
Flywheel vibration problem, 227
Forced vibration, 193
 multiple degrees-of-freedom system, 346
 single degree-of-freedom system, 200
  Laplace transform, 270, 278
 two-story shear building, 380
Forward Euler method, 432–433, 445, 452
Fourier's equation in one-dimension, 474, 480, 483
Fourier's Law of Heat Conduction, 473

Fourier integral, 485
cosine integral, 485
sine integral, 485–486
Fourier series, 470, 485
cosine series, 471
sine series, 469, 471, 479, 482, 491–492
Free flexural vibration of a simply supported beam problem, 468
Free vibration, 193
multiple degrees-of-freedom system, 344
single degree-of-freedom system
critically damped, 199
Laplace transform, 269, 278
overdamped, 199
undamped, 194–195
underdamped, 196
two-story shear building, 378
Frequency equation, 344, 378
multiple degrees-of-freedom system, 344
Frobenius series, 403, 405, 425
Fuchs' Theorem, 405
Fundamental Theorem, 141

G

Gamma function, 410
Maple, 512
Gauss-Jordan method, 335
General solution, 8, 380
linear differential equation, 142
matrix method, 335, 350
method of operator, 312, 348
Generalized eigenvector, 331, 349
GNU Scientific Library, 453
Grouping terms method, 34

H

Hanging cable, 97
cable under self-weight, 102
problem, 106
suspension bridge, 97
Heat conduction, 473–474, 476
boundary conditions, 475
homogeneous, 476
insulated, 475
of the first kind, 475
of the second kind, 475
of the third kind, 475
coefficient of thermal conductivity, 473
equation, 473
one-dimensional transient, 474
three-dimensional steady-state, 475
two-dimensional steady-state, 474
finned surface, 476–477, 488
Fourier's equation in one-dimension, 474, 480, 483
Fourier's Law, 473
heat transfer coefficient, 476
Laplace's equation
in three-dimensions, 475, 488
in two-dimensions, 474, 477
one-dimensional transient, 480
on a semi-infinite interval, 483
thermal diffusivity, 474
three-dimensional steady-state, 488
two-dimensional steady-state, 476
Heating, 87
Heating in a building problem, 88
Heaviside step function, 249–250, 255
inverse Laplace transform, 257
Laplace transform, 249, 252
Maple, 507
Higher-order differential equation
dependent variable absent, 70, 77
immediately integrable, 68, 77
independent variable absent, 72, 78
Maple, 499
Homogeneous
differential equation, 7, 20
first-order differential equation, 20, 75
Hyperbolic functions, 145

I

Implicit method, 438, 453
Improved Euler method, 437, 446, 449, 453
Improved Euler predictor-corrector method, 439, 446, 452
Impulse-Momentum Principle, 91, 125, 254, 422
Impulse function, 254–255
IMSL Library, 453
Indicial equation, 404, 425
Inductance, 109
Inductor, 109
Inertia force, 91, 230–231, 359, 364–365
Inertia moment, 223, 229, 359, 364
Infinitely many degrees-of-freedom system, 468
Initial value problem, 10
numerical solution, 431
Inspection method, 45
Integral transform, 244
Integrals table, 534
Integrating factor, 31, 39–40, 76
Integrating factor by groups, 48, 77

Inverse Laplace transform, 257
convolution integral, 258
Heaviside step function, 257
Maple, 507
properties, 257, 539
table, 257, 539–540

J

Jet engine, 192
Jet engine vibration problem, 223

K

Kirchhoff's Current Law (KCL), 108, 110–113, 209, 212–213, 277, 373
Kirchhoff's Voltage Law (KVL), 108, 110, 209, 211, 276–277, 373

L

Landing gear, 192
Laplace's equation
in three-dimensions, 475, 488
in two-dimensions, 474, 477
Laplace transform, 244
beam-column, 280
beams on elastic foundation, 283
convolution integral, 258
definition, 244
Dirac delta function, 256
Heaviside step function, 249, 252
inverse, 257
properties, 257, 539
table, 539
linear differential equation, 263
Maple, 507
properties, 245, 537
second-order circuit, 275
single degree-of-freedom system, 268, 278
blast force, 273
forced vibration, 270, 278
free vibration, 269, 278
system of linear differential equations, 318, 348
table, 245, 537–538
vehicle passing a speed bump problem, 272
Legendre equation, 397
Linear differential equation, 140
auxiliary equation, 144
characteristic equation, 143–144, 147, 151, 180, 460
complex roots, 147
real distinct roots, 143–144
repeated roots, 151
characteristic number, 144, 460
complementary solution, 142–144, 147, 152, 180, 460
constant coefficients, 7, 140
general solution, 142
Laplace transform, 263
Maple, 506, 512
particular solution, 142, 153, 181
method of operator, 162–164, 166, 169, 171, 181–182
method of undetermined coefficients, 153, 181,
method of variation of parameters, 173, 181–182
variable coefficients, 7, 178, 183
second-order homogeneous, 403
Linear first-order differential equation, 55, 75
Lipschitz condition, 12
Local error, 434–435, 453
Lorenz system problem, 522

M

***Maple*, 453, 498**
`algsubs`, 502–503
`assuming`, 502, 504
`BesselJ`, 512–513
`BesselY`, 512–513
`collect`, 502, 504, 506, 508–510, 512, 514–515, 517
`cos`, 506, 508, 510
`exp`, 506, 508–510
`Heaviside`, 508–509
`ln`, 504, 517
`sin`, 506, 508, 510
`combine`, 502, 504, 508
`convert`, 507, 512, 514–517
`factorial`, 512, 517
`parfrac`, 507
`polynom`, 512, 514–517
`cos`, 499, 502, 509, 519
`D`, 506
`(D@@n)(y)(a)`, 506
`D(y)(a)`, 506
`diff`, 499–504, 506, 508–509, 514–516, 518–519, 522
`Dirac`, 507
do loop, 525
`dsolve`, 498–510, 514–520, 522, 525
`'formal_solution'`, 515,–516
`implicit`, 499–503, 505
`method=laplace`, 507–508

numeric, 517–520, 522, 525,–520, 522, 525,–518
series, 514,–517
Eigenvectors, 510–511, 517
eval, 505
evalf, 512, 514, 518
exp, 500, 506–509
fsolve, 514
GAMMA, 512
gamma, 512
Heaviside, 507–509
I, 510
implicitplot, 504
labels, 504
numpoints, 504
thickness, 504
tickmarks, 504
view, 504
invlaplace, 507
JordanForm, 511
laplace, 507
lhs, 501–503
ln, 501, 506, 512–513
map, 500
Matrix, 510–511, 517
numer, 501–503
odeplot, 519–523
axes, 519, 521, 523
labels, 519–523
numpoints, 519–523
orientation, 523
thickness, 523
tickmarks, 519–523
view, 519, 521, 523
odetest, 500, 504, 510
Order, 515
plot, 513
numpoints, 513
pointplot, 525
axesfont, 525
color, 525
labelfont, 525
labels, 525
style, 525
tickmarks, 525
positive, 502, 504
restart, 519, 522
rhs, 505, 525
RootOf, 505
seq, 525
series, 512, 514
simplify, 502, 505, 512, 517
sqrt, 502
sin, 499, 502, 506–507, 509
solve, 504–505
sqrt, 501
subs, 504–505, 512–513, 517
unapply, 505, 518
with, 504, 507, 510–511, 517, 519, 522
inttrans, 507
LinearAlgebra, 510–511, 517
plots, 504, 519, 522
zip, 525
Mass-spring-damper system, 191
***Mathematica*, 453, 498**
***Matlab*, 453**
Matrix method, 325, 349
complementary solution, 326, 350
complex eigenvalues, 328–329, 349
distinct eigenvalues, 326–327, 349
multiple eigenvalues, 330–331, 349
general solution, 335, 350
particular solution, 334, 350
method of variation of parameters, 334
system of linear differential equations, 325, 349
Matrix
Cramer's Rule, 307
damping, 303, 346–347
defective (deficient), 331
eigenvalue, 326, 328–331, 349
eigenvector, 326, 328–331, 349
Gauss-Jordan method, 335
generalized eigenvector, 331, 349
inverse, 335–336
Gauss-Jordan method, 335
Maple, 510
mass, 303, 344–346
modal, 345, 380
stiffness, 303, 344–346
Mechanical vibration, 357
Method of grouping terms, 34
Method of inspection, 45, 76
Method of operator
linear differential equations, 162, 181
polynomial, 166, 181
Shift Theorem, 164, 182
system of linear differential equations, 304–305, 307–308, 347–348
Theorem 1, 163, 182
Theorem 2, 164, 182
Theorem 3, 169, 182
Theorem 4, 171, 182

Method of separation of variables
first-order differential equation, 16, 20, 75
partial differential equation, 458, 492
Method of undetermined coefficients, 153, 181
exception, 159
Method of variation of parameters
linear differential equations, 173, 181–182
system of linear differential equations, 314, 334
Midpoint method, 441
Modal damping coefficient, 347
Modal frequency, 344, 378
Modal matrix, 345, 380
Mode shape, 344, 379–380, 468
orthogonality, 345, 380
Moment-curvature relationship, 219, 419, 465
Moment of inertia, 223, 228, 358, 362, 419, 465
Parallel Axis Theorem, 223, 228
Motion, 91
Multiple degrees-of-freedom system, 301, 344
damped forced vibration, 346
equations of motion, 303
orthogonality of mode shapes, 345
undamped forced vibration, 346
undamped free vibration, 344

N

NAG Library, 453
Natural circular frequency, 193, 195
damped, 194
undamped, 193
Natural frequency, 195, 204, 208
Natural purification in a stream, 114
Newton's Law of Cooling, 87
Newton's Second Law, 2, 91, 93, 95–96, 123, 190, 214, 224, 226, 301–302, 366–367, 465
Numerical Recipes, 453
Numerical solution, 431
average slope method, 437, 446, 453
backward Euler method, 436, 446, 449, 453
conditionally stable, 449
constant slope method, 432–433, 445
cumulative error, 435, 452–453
error analysis, 434
Euler method, 432–433, 445, 449, 452
explicit method, 452
forward Euler method, 432–433, 445, 452
GNU Scientific Library, 453
implicit method, 437–438, 453
improved Euler method, 437, 446, 449, 453
improved Euler predictor-corrector method, 439, 446, 452
IMSL Library, 453
local error, 434–435, 453
Maple, 453, 517
Mathematica, 453
Matlab, 453
midpoint method, 441
NAG Library, 453
Numerical Recipes, 453
predictor-corrector technique, 438
roundoff error, 434
Runge-Kutta-Fehlberg method, 517–518
Runge-Kutta method, 440
fourth-order, 443–444, 446, 452
second-order, 441–442, 446, 452
stability, 449, 451, 453
stepsize, 431, 453
system of differential equations, 445
trapezoidal rule method, 438
truncation error, 434–435, 453
unconditionally stable, 449
unstable, 449

O

Object falling in air problem, 95
Ohm's Law, 108
Operator
D, 140, 162
inverse, 162
properties, 141
method, 162, 304
linear differential equations, 162, 181
polynomial, 166, 181
Shift Theorem, 164, 182
system of linear differential equations, 304–305, 307–308, 347–348
Theorem 1, 163, 182
Theorem 2, 164, 182
Theorem 3, 169, 182
Theorem 4, 171, 182
Ordinary differential equation, 6
Ordinary point, 396, 424
Orthogonality
mode shape, 380
sine and cosine functions, 470, 472, 482
Overdamped system, 199
Laplace transform, 279

P

Parallel Axis Theorem, 223, 228
Parallel circuit
RC, 110

RL, 111
RLC, 209
Partial differential equation, 8, 457
separation of variables, 458, 492
simple, 457
Partial fractions, 259
cover-up method, 260
Maple, 507
Particular moving in a plane problem, 300
polar coordinates, 301
rectangular coordinates, 300
Particular solution, 8, 193, 200
linear differential equation, 142, 153, 181
matrix method, 334, 350
method of variation of parameters, 334
method of operator
linear differential equations, 162, 181
polynomial, 166, 181
Shift Theorem, 164, 182
system of linear differential equations, 307–308, 348
Theorem 1, 163, 182
Theorem 2, 164, 182
Theorem 3, 169, 182
Theorem 4, 171, 182
method of undetermined coefficients, 153, 181
exception, 159
method of variation of parameters
linear differential equations, 173, 181–182
system of linear differential equations, 314, 334
Period, 195
Phase angle, 195, 201
Piston vibration problem, 224
Power series, 391
convergence, 391
Maple, 512
operation, 392
radius of convergence, 391
Predictor-corrector technique, 438

R

RC circuit, 109
Reservoir pollution problem, 127
Resisting medium, 91
Resonance, 203–204, 208, 382–383
RL circuit, 109
RLC circuit, 209
Roundoff error, 434
Runge-Kutta-Fehlberg method, 517–518
Runge-Kutta method, 440
fourth-order, 443–444, 446, 452
second-order, 441–442, 446, 452

S

Second-order circuit, 213
Laplace transform, 275, 278
problem, 211, 275
Separation of variables method
first-order differential equation, 16, 20, 75
partial differential equation, 458, 492
Series circuit
RL, 110
RLC, 209
RC, 110
Series solution, 390
Bessel's differential equation, 408, 418
Frobenius series, 403, 405, 425
Fuchs' Theorem, 405
indicial equation, 404, 425
Legendre equation, 397
linear differential equation, 403
Maple, 512, 514
ordinary point, 394, 397
regular singular point, 403
Shear building
multiple story, 301
single story, 188–191
two-story, 377
forced vibration, 380
free vibration, 378
Shear force, 189–190, 302
Single degree-of-freedom system, 188, 191, 193
blast force
Laplace transform, 273
problem, 273
critically damped, 199
Laplace transform, 279
Laplace transform, 268, 278
blast force, 273
forced vibration, 270, 278
free vibration, 269, 278
sinusoidal excitation, 270
overdamped, 199
Laplace transform, 279
problem, 226
response, 193
sinusoidal excitation
Laplace transform, 270
problem, 270
undamped, 194, 204, 208
underdamped, 194, 200

Laplace transform, 278
Singular point, 403
irregular, 403
regular, 403, 425
Singular solution, 19
Sinusoidal excitation, 200, 204
Special function
Maple, 512
Special transformation
first-order differential equation, 25, 77
Steady-state solution, 200
Stepsize, 431, 453
Stiff differential equation, 449–450, 453
Maple, 517
Stiffness, 189
Stream, 114
Suspension bridge, 97
System of differential equations
Maple, 509
numerical solution, 445
average slope method, 446
backward Euler method, 446
constant slope method, 445
Euler method, 445
forward Euler method, 445
fourth-order Runge-Kutta method, 446
improved Euler method, 446
improved Euler predictor-corrector method, 446
second-order Runge-Kutta method, 446
System of linear differential equations
complementary solution
complex eigenvalues, 328–329, 349
distinct eigenvalues, 326–327, 349
matrix method, 326, 350
method of operator, 304–305, 348
multiple eigenvalues, 330–331, 349
general solution
matrix method, 335, 350
method of operator, 312, 348
Laplace transform, 318, 348
matrix method, 325, 349
method of operator, 304, 347
characteristic equation, 305
particular solution
matrix method, 334, 350
method of operator, 307–308, 348
method of variation of parameters, 314, 334

T

Table
derivatives, 533
integrals, 534
inverse Laplace transform, 539
Laplace transform, 537
trigonometric identities, 531
Taipei 101, 371–372
Taylor series, 432, 440
Transient solution, 200
Trapezoidal rule method, 438
Trigonometric identities table, 531
Truncation error, 434–435, 453
Tuned mass damper (TMD), 366–367, 370–372
Two degrees-of-freedom system
problem, 357
shear building, 377
forced vibration, 380
free vibration, 378

U

Undamped system, 194, 204, 208
Underdamped system, 194, 200
Laplace transform, 278
Undetermined coefficients method, 153, 181

V

Variation of parameters method
linear differential equations, 173, 181–182
system of linear differential equations, 314, 334
Vehicle passing a speed bump problem, 213
Laplace transform, 272
Vibration, 188–191, 193, 213, 272, 301, 344, 357, 377, 468
Vibration absorber, 366–367, 370–372
Vibration of an automobile problem, 362
Viscous dashpot damper, 189

W

Water leaking problem, 126
Water tower problem, 220

Printed in the United States
By Bookmasters